A Hierarchy of Particle Physics

" *TheStructure of Particles* "

Edited by

Paul F. Kisak

Virginia, USA

Visit our website at: https://www.createspace.com/5789454

Printed in The United States of America

First Trade Edition: 2015
10 9 8 7 6 5 4 3 2 1

Black & White on White paper
612 pages

ISBN-13: 978-1518684906
ISBN-10: 1518684904

Virginia, USA

Chapter 1

Elementary particle

This article is about the physics concept. For the novel, see The Elementary Particles.

In particle physics, an **elementary particle** or **fundamental particle** is a particle whose substructure is unknown, thus

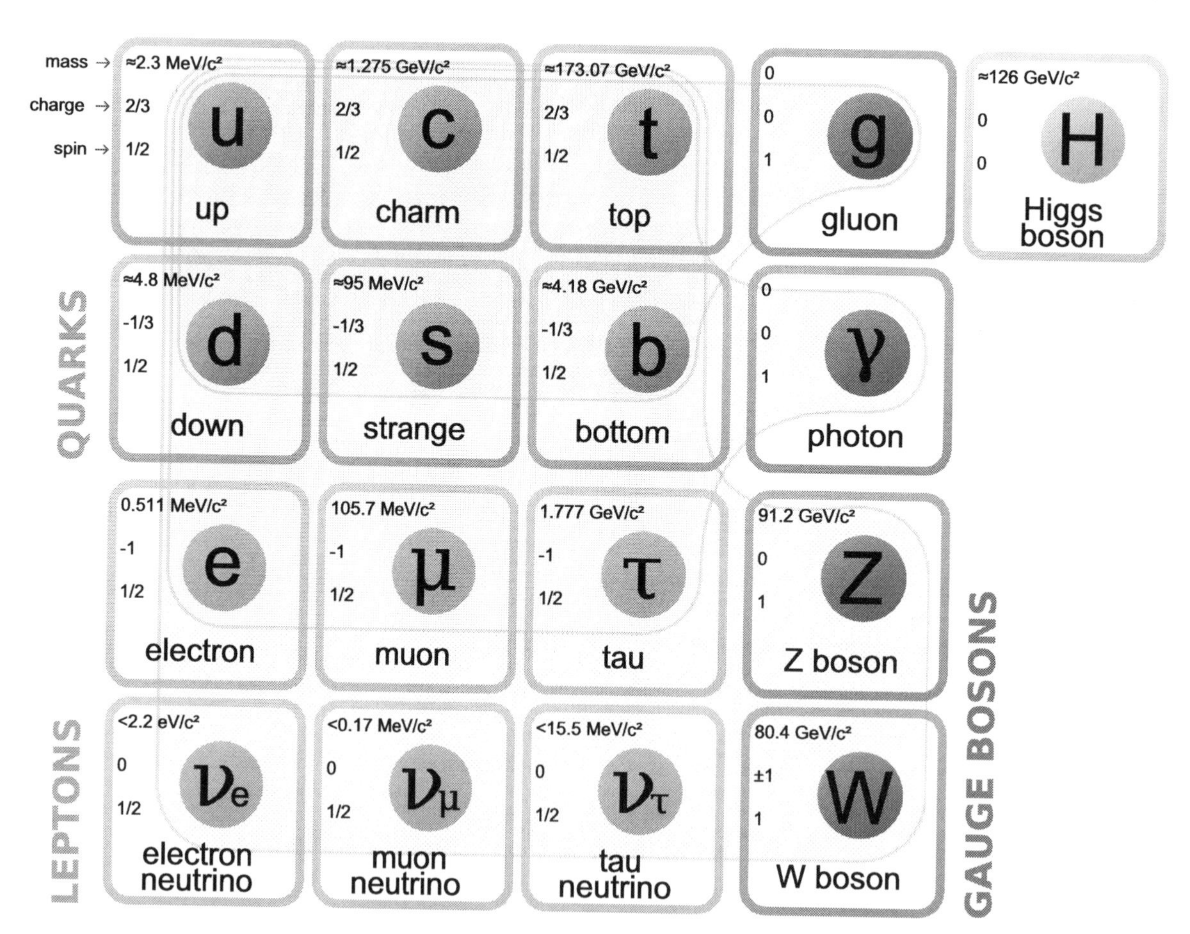

Elementary particles included in the Standard Model

it is unknown whether it is composed of other particles.[1] Known elementary particles include the fundamental fermions (quarks, leptons, antiquarks, and antileptons), which generally are "matter particles" and "antimatter particles", as well as the fundamental bosons (gauge bosons and Higgs boson), which generally are "force particles" that mediate interactions among fermions.[1] A particle containing two or more elementary particles is a *composite particle*.

Everyday matter is composed of atoms, once presumed to be matter's elementary particles—*atom* meaning "indivisible" in Greek—although the atom's existence remained controversial until about 1910, as some leading physicists regarded molecules as mathematical illusions, and matter as ultimately composed of energy.[1][2] Soon, subatomic constituents of the atom were identified. As the 1930s opened, the electron and the proton had been observed, along with the photon, the particle of electromagnetic radiation.[1] At that time, the recent advent of quantum mechanics was radically altering the conception of particles, as a single particle could seemingly span a field as would a wave, a paradox still eluding satisfactory explanation.[3][4][5]

Via quantum theory, protons and neutrons were found to contain quarks—up quarks and down quarks—now considered elementary particles.[1] And within a molecule, the electron's three degrees of freedom (charge, spin, orbital) can separate via wavefunction into three quasiparticles (holon, spinon, orbiton).[6] Yet a free electron—which, not orbiting an atomic nucleus, lacks orbital motion—appears unsplittable and remains regarded as an elementary particle.[6]

Around 1980, an elementary particle's status as indeed elementary—an *ultimate constituent* of substance—was mostly discarded for a more practical outlook,[1] embodied in particle physics' Standard Model, science's most experimentally successful theory.[5][7] Many elaborations upon and theories beyond the Standard Model, including the extremely popular supersymmetry, double the number of elementary particles by hypothesizing that each known particle associates with a "shadow" partner far more massive,[8][9] although all such superpartners remain undiscovered.[7][10] Meanwhile, an elementary boson mediating gravitation—the graviton—remains hypothetical.[1]

1.1 Overview

Main article: Standard Model
See also: Physics beyond the Standard Model
All elementary particles are—depending on their *spin*—either bosons or fermions. These are differentiated via the spin–

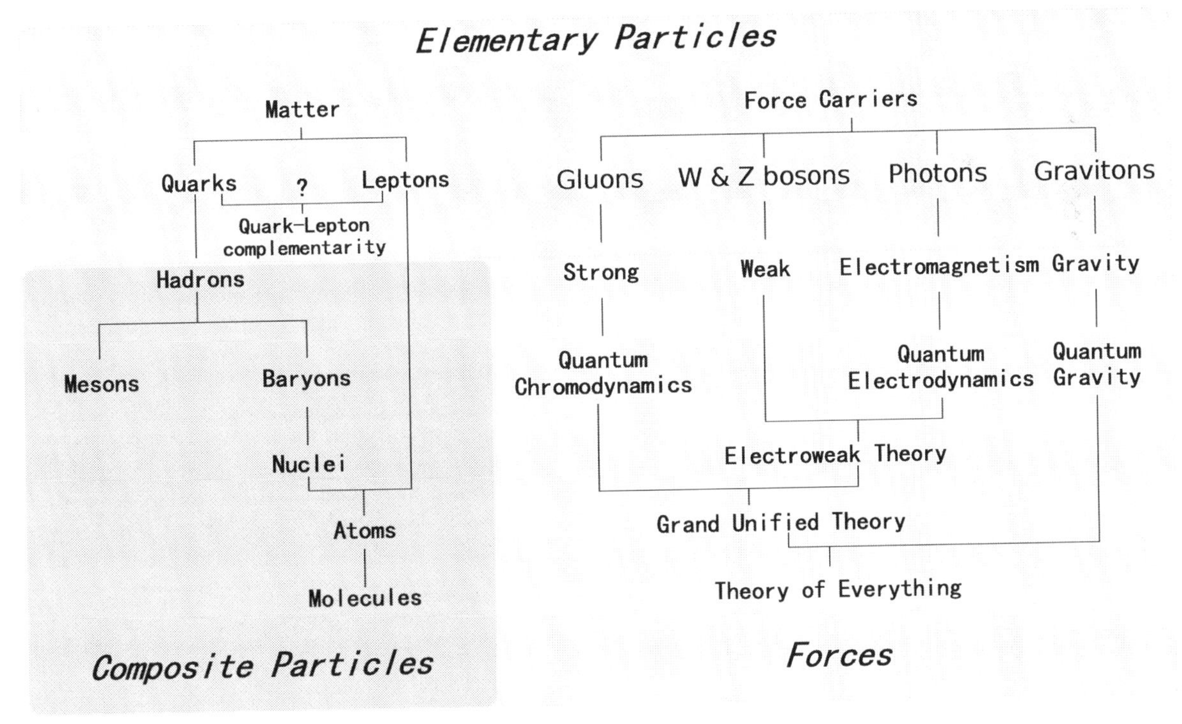

An overview of the various families of elementary and composite particles, and the theories describing their interactions

statistics theorem of quantum statistics. Particles of *half-integer* spin exhibit Fermi–Dirac statistics and are fermions.[1] Particles of *integer* spin, in other words full-integer, exhibit Bose–Einstein statistics and are bosons.[1]

Elementary fermions:

- Matter particles
  - Quarks:
    - up, down
    - charm, strange
    - top, bottom
  - Leptons:
    - electron, electron neutrino (a.k.a., "neutrino")
    - muon, muon neutrino
    - tau, tau neutrino
- Antimatter particles
  - Antiquarks
  - Antileptons

Elementary bosons:

- Force particles (gauge bosons):
  - photon
  - gluon (numbering eight)[1]
  - W^{+}, W^{-}, and Z^{0} bosons
  - graviton (hypothetical)[1]
- Scalar boson
  - Higgs boson

A particle's mass is quantified in units of energy versus the electron's (electronvolts). Through conversion of energy into mass, any particle can be produced through collision of other particles at high energy,[1][11] although the output particle might not contain the input particles, for instance matter creation from colliding photons. Likewise, the composite fermions protons were collided at nearly light speed to produce a Higgs boson, which elementary boson is far more massive.[11] The most massive elementary particle, the top quark, rapidly decays, but apparently does not contain, lighter particles.

When probed at energies available in experiments, particles exhibit spherical sizes. In operating particle physics' Standard Model, elementary particles are usually represented for predictive utility as point particles, which, as zero-dimensional, lack spatial extension. Though extremely successful, the Standard Model is limited to the microcosm by its omission of gravitation, and has some parameters arbitrarily added but unexplained.[12] Seeking to resolve those shortcomings, string theory posits that elementary particles are ultimately composed of one-dimensional energy strings whose absolute minimum size is the Planck length.

1.2 Common elementary particles

Main article: cosmic abundance of elements

According to the current models of big bang nucleosynthesis, the primordial composition of visible matter of the universe should be about 75% hydrogen and 25% helium-4 (in mass). Neutrons are made up of one up and two down quark, while protons are made of two up and one down quark. Since the other common elementary particles (such as electrons, neutrinos, or weak bosons) are so light or so rare when compared to atomic nuclei, we can neglect their mass contribution to the observable universe's total mass. Therefore, one can conclude that most of the visible mass of the universe consists of protons and neutrons, which, like all baryons, in turn consist of up quarks and down quarks.

Some estimates imply that there are roughly 10^{80} baryons (almost entirely protons and neutrons) in the observable universe.[13][14][15]

The number of protons in the observable universe is called the Eddington number.

In terms of number of particles, some estimates imply that nearly all the matter, excluding dark matter, occurs in neutrinos, and that roughly 10^{86} elementary particles of matter exist in the visible universe, mostly neutrinos.[15] Other estimates imply that roughly 10^{97} elementary particles exist in the visible universe (not including dark matter), mostly photons, gravitons, and other massless force carriers.[15]

1.3 Standard Model

Main article: Standard Model

The Standard Model of particle physics contains 12 flavors of elementary fermions, plus their corresponding antiparticles, as well as elementary bosons that mediate the forces and the Higgs boson, which was reported on July 4, 2012, as having been likely detected by the two main experiments at the LHC (ATLAS and CMS). However, the Standard Model is widely considered to be a provisional theory rather than a truly fundamental one, since it is not known if it is compatible with Einstein's general relativity. There may be hypothetical elementary particles not described by the Standard Model, such as the graviton, the particle that would carry the gravitational force, and sparticles, supersymmetric partners of the ordinary particles.

1.3.1 Fundamental fermions

Main article: Fermion

The 12 fundamental fermionic flavours are divided into three generations of four particles each. Six of the particles are quarks. The remaining six are leptons, three of which are neutrinos, and the remaining three of which have an electric charge of −1: the electron and its two cousins, the muon and the tau.

Antiparticles

Main article: Antimatter

There are also 12 fundamental fermionic antiparticles that correspond to these 12 particles. For example, the antielectron (positron) $e+$ is the electron's antiparticle and has an electric charge of +1.

Quarks

Main article: Quark

Isolated quarks and antiquarks have never been detected, a fact explained by confinement. Every quark carries one of three color charges of the strong interaction; antiquarks similarly carry anticolor. Color-charged particles interact via gluon exchange in the same way that charged particles interact via photon exchange. However, gluons are themselves color-charged, resulting in an amplification of the strong force as color-charged particles are separated. Unlike the electromagnetic force, which diminishes as charged particles separate, color-charged particles feel increasing force.

However, color-charged particles may combine to form color neutral composite particles called hadrons. A quark may pair up with an antiquark: the quark has a color and the antiquark has the corresponding anticolor. The color and anticolor cancel out, forming a color neutral meson. Alternatively, three quarks can exist together, one quark being "red",

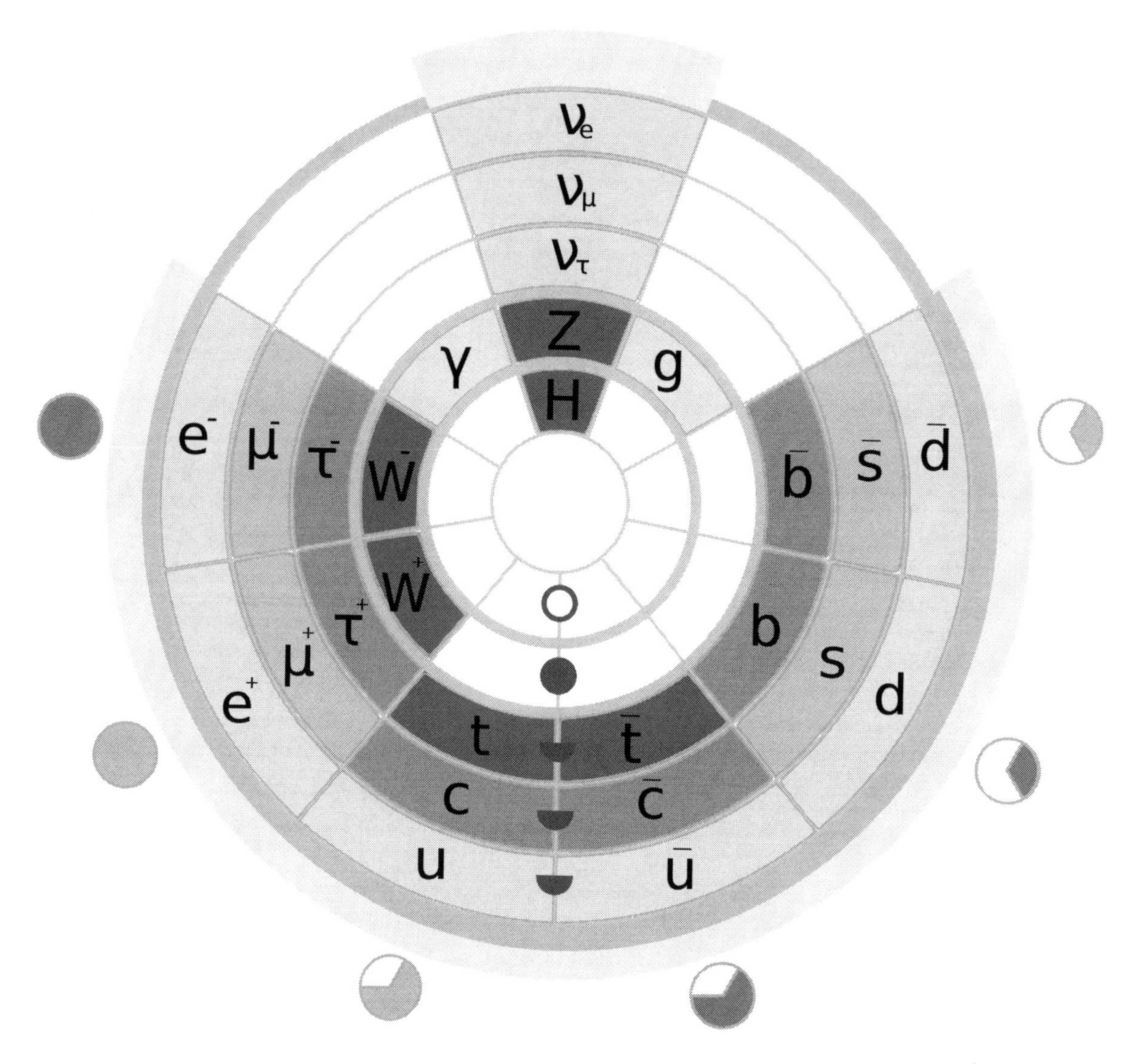

Graphic representation of the standard model. Spin, charge, mass and participation in different force interactions are shown. Click on the image to see the full description

another "blue", another "green". These three colored quarks together form a color-neutral baryon. Symmetrically, three antiquarks with the colors "antired", "antiblue" and "antigreen" can form a color-neutral antibaryon.

Quarks also carry fractional electric charges, but, since they are confined within hadrons whose charges are all integral, fractional charges have never been isolated. Note that quarks have electric charges of either +2/3 or −1/3, whereas antiquarks have corresponding electric charges of either −2/3 or +1/3.

Evidence for the existence of quarks comes from deep inelastic scattering: firing electrons at nuclei to determine the distribution of charge within nucleons (which are baryons). If the charge is uniform, the electric field around the proton should be uniform and the electron should scatter elastically. Low-energy electrons do scatter in this way, but, above a particular energy, the protons deflect some electrons through large angles. The recoiling electron has much less energy and a jet of particles is emitted. This inelastic scattering suggests that the charge in the proton is not uniform but split among smaller charged particles: quarks.

1.3.2 Fundamental bosons

Main article: Boson

In the Standard Model, vector (spin−1) bosons (gluons, photons, and the W and Z bosons) mediate forces, whereas the Higgs boson (spin-0) is responsible for the intrinsic mass of particles. Bosons differ from fermions in the fact that multiple bosons can occupy the same quantum state (Pauli exclusion principle). Also, bosons can be either elementary, like photons, or a combination, like mesons. The spin of bosons are integers instead of half integers.

Gluons

Main article: Gluon

Gluons mediate the strong interaction, which join quarks and thereby form hadrons, which are either baryons (three quarks) or mesons (one quark and one antiquark). Protons and neutrons are baryons, joined by gluons to form the atomic nucleus. Like quarks, gluons exhibit colour and anticolour—unrelated to the concept of visual color—sometimes in combinations, altogether eight variations of gluons.

Electroweak bosons

Main articles: W and Z bosons and Photon

There are three weak gauge bosons: W^+, W^-, and Z^0; these mediate the weak interaction. The W bosons are known for their mediation in nuclear decay. The W^- converts a neutron into a proton then decay into an electron and electron antineutrino pair. The Z^0 does not convert charge but rather changes momentum and is the only mechanism for elastically scattering neutrinos. The weak gauge bosons were discovered due to momentum change in electrons from neutrino-Z exchange. The massless photon mediates the electromagnetic interaction. These four gauge bosons form the electroweak interaction among elementary particles.

Higgs boson

Main article: Higgs boson

Although the weak and electromagnetic forces appear quite different to us at everyday energies, the two forces are theorized to unify as a single electroweak force at high energies. This prediction was clearly confirmed by measurements of cross-sections for high-energy electron-proton scattering at the HERA collider at DESY. The differences at low energies is a consequence of the high masses of the W and Z bosons, which in turn are a consequence of the Higgs mechanism. Through the process of spontaneous symmetry breaking, the Higgs selects a special direction in electroweak space that causes three electroweak particles to become very heavy (the weak bosons) and one to remain massless (the photon). On 4 July 2012, after many years of experimentally searching for evidence of its existence, the Higgs boson was announced to have been observed at CERN's Large Hadron Collider. Peter Higgs who first posited the existence of the Higgs boson was present at the announcement.[16] The Higgs boson is believed to have a mass of approximately 125 GeV.[17] The statistical significance of this discovery was reported as 5-sigma, which implies a certainty of roughly 99.99994%. In particle physics, this is the level of significance required to officially label experimental observations as a discovery. Research into the properties of the newly discovered particle continues.

Graviton

Main article: Graviton

The graviton is hypothesized to mediate gravitation, but remains undiscovered and yet is sometimes included in tables of elementary particles.[1] Its spin would be two—thus a boson—and it would lack charge or mass. Besides mediating an extremely feeble force, the graviton would have its own antiparticle and rapidly annihilate, rendering its detection extremely difficult even if it exists.

1.4 Beyond the Standard Model

Although experimental evidence overwhelmingly confirms the predictions derived from the Standard Model, some of its parameters were added arbitrarily, not determined by a particular explanation, which remain mysteries, for instance the hierarchy problem. Theories beyond the Standard Model attempt to resolve these shortcomings.

1.4.1 Grand unification

Main article: Grand Unified Theory

One extension of the Standard Model attempts to combine the electroweak interaction with the strong interaction into a single 'grand unified theory' (GUT). Such a force would be spontaneously broken into the three forces by a Higgs-like mechanism. The most dramatic prediction of grand unification is the existence of X and Y bosons, which cause proton decay. However, the non-observation of proton decay at the Super-Kamiokande neutrino observatory rules out the simplest GUTs, including SU(5) and SO(10).

1.4.2 Supersymmetry

Main article: Supersymmetry

Supersymmetry extends the Standard Model by adding another class of symmetries to the Lagrangian. These symmetries exchange fermionic particles with bosonic ones. Such a symmetry predicts the existence of supersymmetric particles, abbreviated as *sparticles*, which include the sleptons, squarks, neutralinos, and charginos. Each particle in the Standard Model would have a superpartner whose spin differs by 1/2 from the ordinary particle. Due to the breaking of supersymmetry, the sparticles are much heavier than their ordinary counterparts; they are so heavy that existing particle colliders would not be powerful enough to produce them. However, some physicists believe that sparticles will be detected by the Large Hadron Collider at CERN.

1.4.3 String theory

Main article: String theory

String theory is a model of physics where all “particles” that make up matter are composed of strings (measuring at the Planck length) that exist in an 11-dimensional (according to M-theory, the leading version) universe. These strings vibrate at different frequencies that determine mass, electric charge, color charge, and spin. A string can be open (a line) or closed in a loop (a one-dimensional sphere, like a circle). As a string moves through space it sweeps out something called a *world sheet*. String theory predicts 1- to 10-branes (a 1-brane being a string and a 10-brane being a 10-dimensional object) that prevent tears in the “fabric” of space using the uncertainty principle (E.g., the electron orbiting a hydrogen atom has the probability, albeit small, that it could be anywhere else in the universe at any given moment).

String theory proposes that our universe is merely a 4-brane, inside which exist the 3 space dimensions and the 1 time dimension that we observe. The remaining 6 theoretical dimensions either are very tiny and curled up (and too small to be macroscopically accessible) or simply do not/cannot exist in our universe (because they exist in a grander scheme called the "multiverse" outside our known universe).

Some predictions of the string theory include existence of extremely massive counterparts of ordinary particles due to vibrational excitations of the fundamental string and existence of a massless spin-2 particle behaving like the graviton.

1.4.4 Technicolor

Main article: Technicolor (physics)

Technicolor theories try to modify the Standard Model in a minimal way by introducing a new QCD-like interaction. This means one adds a new theory of so-called Techniquarks, interacting via so called Technigluons. The main idea is that the Higgs-Boson is not an elementary particle but a bound state of these objects.

1.4.5 Preon theory

Main article: Preon

According to preon theory there are one or more orders of particles more fundamental than those (or most of those) found in the Standard Model. The most fundamental of these are normally called preons, which is derived from "pre-quarks". In essence, preon theory tries to do for the Standard Model what the Standard Model did for the particle zoo that came before it. Most models assume that almost everything in the Standard Model can be explained in terms of three to half a dozen more fundamental particles and the rules that govern their interactions. Interest in preons has waned since the simplest models were experimentally ruled out in the 1980s.

1.4.6 Acceleron theory

Accelerons are the hypothetical subatomic particles that integrally link the newfound mass of the neutrino and to the dark energy conjectured to be accelerating the expansion of the universe.[18]

In theory, neutrinos are influenced by a new force resulting from their interactions with accelerons. Dark energy results as the universe tries to pull neutrinos apart.[18]

1.5 See also

- Asymptotic freedom
- List of particles
- Physical ontology
- Quantum field theory
- Quantum gravity
- Quantum triviality
- UV fixed point

1.6 Notes

[1] Sylvie Braibant; Giorgio Giacomelli; Maurizio Spurio (2012). *Particles and Fundamental Interactions: An Introduction to Particle Physics* (2nd ed.). Springer. pp. 1–3. ISBN 978-94-007-2463-1.

[2] Ronald Newburgh; Joseph Peidle; Wolfgang Rueckner (2006). "Einstein, Perrin, and the reality of atoms: 1905 revisited" (PDF). *American Journal of Physics.* **74** (6): 478–481. Bibcode:2006AmJPh..74..478N. doi:10.1119/1.2188962.

[3] Friedel Weinert (2004). *The Scientist as Philosopher: Philosophical Consequences of Great Scientific Discoveries*. Springer. p. 43. ISBN 978-3-540-20580-7.

[4] Friedel Weinert (2004). *The Scientist as Philosopher: Philosophical Consequences of Great Scientific Discoveries*. Springer. pp. 57–59. ISBN 978-3-540-20580-7.

[5] Meinard Kuhlmann (24 Jul 2013). "Physicists debate whether the world is made of particles or fields—or something else entirely". *Scientific American*.

[6] Zeeya Merali (18 Apr 2012). "Not-quite-so elementary, my dear electron: Fundamental particle 'splits' into quasiparticles, including the new 'orbiton'". *Nature*. doi:10.1038/nature.2012.10471.

[7] Ian O'Neill (24 Jul 2013). "LHC discovery maims supersymmetry, again". *Discovery News*. Retrieved 2013-08-28.

[8] Particle Data Group. "Unsolved mysteries—supersymmetry". *The Particle Adventure*. Berkeley Lab. Retrieved 2013-08-28.

[9] National Research Council (2006). *Revealing the Hidden Nature of Space and Time: Charting the Course for Elementary Particle Physics*. National Academies Press. p. 68. ISBN 978-0-309-66039-6.

[10] "CERN latest data shows no sign of supersymmetry—yet". *Phys.Org*. 25 Jul 2013. Retrieved 2013-08-28.

[11] Ryan Avent (19 Jul 2012). "The Q&A: Brian Greene—Life after the Higgs". *The Economist*. Retrieved 2013-08-28.

[12] Sylvie Braibant; Giorgio Giacomelli; Maurizio Spurio (2012). *Particles and Fundamental Interactions: An Introduction to Particle Physics* (2nd ed.). Springer. p. 384. ISBN 978-94-007-2463-1.

[13] Frank Heile. "Is the Total Number of Particles in the Universe Stable Over Long Periods of Time?". 2014.

[14] Jared Brooks. "Galaxies and Cosmology". 2014. p. 4, equation 16.

[15] Robert Munafo (24 Jul 2013). "Notable Properties of Specific Numbers". Retrieved 2013-08-28.

[16] Lizzy Davies (4 July 2014). "Higgs boson announcement live: CERN scientists discover subatomic particle". *The Guardian*. Retrieved 2012-07-06.

[17] Lucas Taylor (4 Jul 2014). "Observation of a new particle with a mass of 125 GeV". CMS. Retrieved 2012-07-06.

[18] "New theory links neutrino's slight mass to accelerating Universe expansion". *ScienceDaily*. 28 Jul 2004. Retrieved 2008-06-05.

1.7 Further reading

1.7.1 General readers

- Feynman, R.P. & Weinberg, S. (1987) *Elementary Particles and the Laws of Physics: The 1986 Dirac Memorial Lectures.* Cambridge Univ. Press.
- Ford, Kenneth W. (2005) *The Quantum World.* Harvard Univ. Press.
- Brian Greene (1999). *The Elegant Universe*. W.W.Norton & Company. ISBN 0-393-05858-1.
- John Gribbin (2000) *Q is for Quantum – An Encyclopedia of Particle Physics*. Simon & Schuster. ISBN 0-684-85578-X.
- Oerter, Robert (2006) *The Theory of Almost Everything: The Standard Model, the Unsung Triumph of Modern Physics.* Plume.
- Schumm, Bruce A. (2004) *Deep Down Things: The Breathtaking Beauty of Particle Physics.* Johns Hopkins University Press. ISBN 0-8018-7971-X.

- Martinus Veltman (2003). *Facts and Mysteries in Elementary Particle Physics*. World Scientific. ISBN 981-238-149-X.
- Frank Close (2004). *Particle Physics: A Very Short Introduction*. Oxford: Oxford University Press. ISBN 0-19-280434-0.
- Seiden, Abraham (2005). *Particle Physics – A Comprehensive Introduction*. Addison Wesley. ISBN 0-8053-8736-6.

1.7.2 Textbooks

- Bettini, Alessandro (2008) *Introduction to Elementary Particle Physics*. Cambridge Univ. Press. ISBN 978-0-521-88021-3
- Coughlan, G. D., J. E. Dodd, and B. M. Gripaios (2006) *The Ideas of Particle Physics: An Introduction for Scientists*, 3rd ed. Cambridge Univ. Press. An undergraduate text for those not majoring in physics.
- Griffiths, David J. (1987) *Introduction to Elementary Particles*. John Wiley & Sons. ISBN 0-471-60386-4.
- Kane, Gordon L. (1987). *Modern Elementary Particle Physics*. Perseus Books. ISBN 0-201-11749-5.
- Perkins, Donald H. (2000) *Introduction to High Energy Physics*, 4th ed. Cambridge Univ. Press.

1.8 External links

The most important address about the current experimental and theoretical knowledge about elementary particle physics is the Particle Data Group, where different international institutions collect all experimental data and give short reviews over the contemporary theoretical understanding.

- Particle Data Group

other pages are:

- Greene, Brian, "*Elementary particles*", The Elegant Universe, NOVA (PBS)
- particleadventure.org, a well-made introduction also for non physicists
- CERNCourier: Season of Higgs and melodrama
- Pentaquark information page
- Interactions.org, particle physics news
- Symmetry Magazine, a joint Fermilab/SLAC publication
- "Sized Matter: perception of the extreme unseen", Michigan University project for artistic visualisation of subatomic particles
- Elementary Particles made thinkable, an interactive visualisation allowing physical properties to be compared

Chapter 2

Timeline of particle discoveries

This is a **timeline of subatomic particle discoveries**, including all particles thus far discovered which appear to be elementary (that is, indivisible) given the best available evidence. It also includes the discovery of composite particles and antiparticles that were of particular historical importance.

More specifically, the inclusion criteria are:

- Elementary particles from the Standard Model of particle physics that have so far been observed. The Standard Model is the most comprehensive existing model of particle behavior. All Standard Model particles including the Higgs boson have been verified, and all other observed particles are combinations of two or more Standard Model particles.
- Antiparticles which were historically important to the development of particle physics, specifically the positron and antiproton. The discovery of these particles required very different experimental methods from that of their ordinary matter counterparts, and provided evidence that *all* particles had antiparticles—an idea that is fundamental to quantum field theory, the modern mathematical framework for particle physics. In the case of most subsequent particle discoveries, the particle and its anti-particle were discovered essentially simultaneously.
- Composite particles which were the first particle discovered containing a particular elementary constituent, or whose discovery was critical to the understanding of particle physics.

2.1 See also

- List of mesons
- List of baryons
- List of particles
- physics

2.2 References

[1] Hockberger, P. E. (2002). "A history of ultraviolet photobiology for humans, animals and microorganisms". *Photochem. Photobiol.* **76** (6): 561–579. doi:10.1562/0031-8655(2002)0760561AHOUPF2.0.CO2. ISSN 0031-8655. PMID 12511035.

[2] The ozone layer protects humans from this. Lyman, T. (1914). "Victor Schumann". *Astrophysical Journal* **38**: 1–4. Bibcode: doi:10.1086/142050.

[3] W.C. Röntgen (1895). "Über ein neue Art von Strahlen. Vorlaufige Mitteilung". *Sitzber. Physik. Med. Ges.* **137**: 1. as translated in A. Stanton (1896). "On a New Kind of Rays". *Nature* **53** (1369): 274–276. Bibcode:1896Natur..53R.274.. doi:10.1038/053274b0.

[4] J.J. Thomson (1897). "Cathode Rays". *Philosophical Magazine* **44** (269): 293–316. doi:10.1080/14786449708621070.

[5] E. Rutherford (1899). "Uranium Radiation and the Electrical Conduction Produced by it". *Philosophical Magazine* **47** (284): 109–163. doi:10.1080/14786449908621245.

[6] P. Villard (1900). "Sur la Réflexion et la Réfraction des Rayons Cathodiques et des Rayons Déviables du Radium". *Comptes Rendus de l'Académie des Sciences* **130**: 1010.

[7] E. Rutherford (1911). "The Scattering of α- and β- Particles by Matter and the Structure of the Atom". *Philosophical Magazine* **21** (125): 669–688. doi:10.1080/14786440508637080.

[8] E. Rutherford (1919). "Collision of α Particles with Light Atoms IV. An Anomalous Effect in Nitrogen". *Philosophical Magazine* **37**: 581.

[9] J. Chadwick (1932). "Possible Existence of a Neutron". *Nature* **129** (3252): 312. Bibcode:1932Natur.129Q.312C.doi:100.

[10]E. Rutherford (1920). "Nuclear Constitution of Atoms".*Proceedings of the Royal Society* A**97**(686): 374–400.. doi:10.1098/rspa.1920.0040.

[11]C.D. Anderson (1932). "The Apparent Existence of Easily Deflectable Positives".*Science***76**(1967): 238–9.Bibcode:1932Sci. doi:10.1126/science.76.1967.238. PMID 17731542.

[12] S.H. Neddermeyer, C.D. Anderson (1937). "Note on the nature of Cosmic-Ray Particles". *Physical Review* **51** (10): 884–886. Bibcode:1937PhRv...51..884N. doi:10.1103/PhysRev.51.884.

[13] M. Conversi, E. Pancini, O. Piccioni (1947). "On the Disintegration of Negative Muons". *Physical Review* **71** (3): 209–210. Bibcode:1947PhRv...71..209C. doi:10.1103/PhysRev.71.209.

[14] C.D. Anderson (1935). "On the Interaction of Elementary Particles". *Proceedings of the Physico-Mathematical Society of Japan* **17**: 48.

[15] G.D. Rochester, C.C. Butler (1947). "Evidence for the Existence of New Unstable Elementary Particles". *Nature* **160** (4077): 855–857. Bibcode:1947Natur.160..855R. doi:10.1038/160855a0.

[16] The Strange Quark

[17] O. Chamberlain, E. Segrè, C. Wiegand, T. Ypsilantis (1955). "Observation of Antiprotons". *Physical Review* **100** (3): 947–950. Bibcode:1955PhRv..100..947C. doi:10.1103/PhysRev.100.947.

[18]F. Reines, C.L. Cowan (1956). "The Neutrino".*Nature***178**(4531): 446–449.Bibcode:1956Natur.178..446R.doi:10.1038/1.

[19] G. Danby et al. (1962). "Observation of High-Energy Neutrino Reactions and the Existence of Two Kinds of Neutrinos". *Physical Review Letters* **9** (1): 36–44. Bibcode:1962PhRvL...9...36D. doi:10.1103/PhysRevLett.9.36.

[20] R. Nave. "The Xi Baryon". Hyperphysics. Retrieved 20 June 2009.

[21] E.D. Bloom et al. (1969). "High-Energy Inelastic *e–p* Scattering at 6° and 10°". *Physical Review Letters* **23** (16): 930–934. Bibcode:1969PhRvL..23..930B. doi:10.1103/PhysRevLett.23.930.

[22] M. Breidenbach et al. (1969). "Observed Behavior of Highly Inelastic Electron-Proton Scattering". *Physical Review Letters* **23** (16): 935–939. Bibcode:1969PhRvL..23..935B. doi:10.1103/PhysRevLett.23.935.

[23] J.J. Aubert et al. (1974). "Experimental Observation of a Heavy Particle *J*". *Physical Review Letters* **33** (23): 1404–1406. Bibcode:1974PhRvL..33.1404A. doi:10.1103/PhysRevLett.33.1404.

[24] J.-E. Augustin et al. (1974). "Discovery of a Narrow Resonance in e^+e^- Annihilation". *Physical Review Letters* **33** (23): 1406–1408. Bibcode:1974PhRvL..33.1406A. doi:10.1103/PhysRevLett.33.1406.

[25]B.J. Bjørken, S.L. Glashow (1964). "Elementary Particles and SU(4)".*Physics Letters***11**(3): 255–257.Bibcode:1964PhL..... doi:10.1016/0031-9163(64)90433-0.

[26] M.L. Perl et al. (1975). “Evidence for Anomalous Lepton Production in e^+–e^- Annihilation”. *Physical Review Letters* **35** (22): 1489–1492. Bibcode:1975PhRvL..35.1489P. doi:10.1103/PhysRevLett.35.1489.

[27] S.W. Herb et al. (1977). “Observation of a Dimuon Resonance at 9.5 GeV in 400-GeV Proton-Nucleus Collisions”. *Physical Review Letters* **39** (5): 252–255. Bibcode:1977PhRvL..39..252H. doi:10.1103/PhysRevLett.39.252.

[28] D.P. Barber et al. (1979). “Discovery of Three-Jet Events and a Test of Quantum Chromodynamics at PETRA”. *Physical Review Letters* **43** (12): 830–833. Bibcode:1979PhRvL..43..830B. doi:10.1103/PhysRevLett.43.830.

[29] J.J. Aubert *et al.* (European Muon Collaboration) (1983). “The ratio of the nucleon structure functions F_2^N for iron and deuterium”. *Physics Letters B* **123** (3–4): 275–278. Bibcode:1983PhLB..123..275A. doi:10.1016/0370-2693(83)90437-9.

[30] G. Arnison *et al.* (UA1 collaboration) (1983). “Experimental observation of lepton pairs of invariant mass around 95 GeV/c^2 at the CERN SPS collider”. *Physics Letters B* **126** (5): 398–410. Bibcode:1983PhLB..126..398A. doi:
0.

[31] F. Abe *et al.* (CDF collaboration) (1995). “Observation of Top quark production in p–p Collisions with the Collider Detector at Fermilab”. *Physical Review Letters* **74** (14): 2626–2631. arXiv:hep-ex/9503002. Bibcode:1995PhRvL..74.2626A. doi:10.1103/PhysRevLett.74.2626. PMID 10057978.

[32] S. Arabuchi *et al.* (D0 collaboration) (1995). “Observation of the Top Quark”. *Physical Review Letters* **74** (14): 2632–2637. arXiv:hep-ex/9503003. Bibcode:1995PhRvL..74.2632A. doi:10.1103/PhysRevLett.74.2632. PMID 10057979.

[33] G. Baur et al. (1996). “Production of Antihydrogen”. *Physics Letters B* **368** (3): 251–258. Bibcode:1996PhLB..368..251B. doi:10.1016/0370-2693(96)00005-6.

[34] “Physicists Find First Direct Evidence for Tau Neutrino at Fermilab” (Press release). Fermilab. 20 July 2000. Retrieved 20 March 2010.

[35] Boyle, Alan (4 July 2012). “Milestone in Higgs quest: Scientists find new particle”. *MSNBC* (MSNBC). Retrieved 5 July 2012.

- V.V. Ezhela et al. (1996). *Particle Physics: One Hundred Years of Discoveries: An Annotated Chronological Bibliography*. Springer–Verlag. ISBN 1-56396-642-5.

Chapter 3

List of particles

This is a list of the different types of particles found or believed to exist in the whole of the universe. For individual lists of the different particles, see the list below.

3.1 Elementary particles

Main article: Elementary particle

Elementary particles are particles with no measurable internal structure; that is, they are not composed of other particles. They are the fundamental objects of quantum field theory. Many families and sub-families of elementary particles exist. Elementary particles are classified according to their spin. Fermions have half-integer spin while bosons have integer spin. All the particles of the Standard Model have been experimentally observed, recently including the Higgs boson.[1][2]

3.1.1 Fermions

Main article: Fermion

Fermions are one of the two fundamental classes of particles, the other being bosons. Fermion particles are described by Fermi–Dirac statistics and have quantum numbers described by the Pauli exclusion principle. They include the quarks and leptons, as well as any composite particles consisting of an odd number of these, such as all baryons and many atoms and nuclei.

Fermions have half-integer spin; for all known elementary fermions this is $\frac{1}{2}$. All known fermions, except neutrinos, are also Dirac fermions; that is, each known fermion has its own distinct antiparticle. It is not known whether the neutrino is a Dirac fermion or a Majorana fermion.[3] Fermions are the basic building blocks of all matter. They are classified according to whether they interact via the color force or not. In the Standard Model, there are 12 types of elementary fermions: six quarks and six leptons.

Quarks

Main article: Quark

Quarks are the fundamental constituents of hadrons and interact via the strong interaction. Quarks are the only known carriers of fractional charge, but because they combine in groups of three (baryons) or in groups of two with antiquarks (mesons), only integer charge is observed in nature. Their respective antiparticles are the antiquarks, which are identical except for the fact that they carry the opposite electric charge (for example the up quark carries charge $+2/3$, while the up antiquark carries charge $-2/3$), color charge, and baryon number. There are six flavors of quarks; the three positively charged quarks are called “up-type quarks” and the three negatively charged quarks are called “down-type quarks”.

Leptons

Main article: Leptons

Leptons do not interact via the strong interaction. Their respective antiparticles are the antileptons which are identical, except for the fact that they carry the opposite electric charge and lepton number. The antiparticle of an electron is an antielectron, which is nearly always called a "positron" for historical reasons. There are six leptons in total; the three charged leptons are called “electron-like leptons”, while the neutral leptons are called "neutrinos". Neutrinos are known to oscillate, so that neutrinos of definite flavor do not have definite mass, rather they exist in a superposition of mass eigenstates. The hypothetical heavy right-handed neutrino, called a "sterile neutrino", has been left off the list.

3.1.2 Bosons

Main article: Boson

Bosons are one of the two fundamental classes of particles, the other being fermions. Bosons are characterized by Bose–Einstein statistics and all have integer spins. Bosons may be either elementary, like photons and gluons, or composite, like mesons.

The fundamental forces of nature are mediated by gauge bosons, and mass is believed to be created by the Higgs field. According to the Standard Model the elementary bosons are:

The graviton is added to the list although it is not predicted by the Standard Model, but by other theories in the framework of quantum field theory. Furthermore, gravity is non-renormalizable. There are a total of eight independent gluons. The Higgs boson is postulated by the electroweak theory primarily to explain the origin of particle masses. In a process known as the "Higgs mechanism", the Higgs boson and the other gauge bosons in the Standard Model acquire mass via spontaneous symmetry breaking of the SU(2) gauge symmetry. The Minimal Supersymmetric Standard Model (MSSM) predicts several Higgs bosons. A new particle expected to be the Higgs boson was observed at the CERN/LHC on March 14, 2013, around the energy of 126.5GeV with an accuracy of close to five sigma (99.9999%, which is accepted as definitive). The Higgs mechanism giving mass to other particles has not been observed yet.

3.1.3 Hypothetical particles

Supersymmetric theories predict the existence of more particles, none of which have been confirmed experimentally as of 2014:

Note: just as the photon, Z boson and $W^{\pm}$ bosons are superpositions of the B^0, W^0, W^1, and W^2 fields – the photino, zino, and wino$^{\pm}$ are superpositions of the bino^0, wino^0, wino^1, and wino^2 by definition.

No matter if one uses the original gauginos or this superpositions as a basis, the only predicted physical particles are neutralinos and charginos as a superposition of them together with the Higgsinos.

Other theories predict the existence of additional bosons:

Mirror particles are predicted by theories that restore parity symmetry.

"Magnetic monopole" is a generic name for particles with non-zero magnetic charge. They are predicted by some GUTs.

"Tachyon" is a generic name for hypothetical particles that travel faster than the speed of light and have an imaginary rest mass.

Preons were suggested as subparticles of quarks and leptons, but modern collider experiments have all but ruled out their existence.

Kaluza–Klein towers of particles are predicted by some models of extra dimensions. The extra-dimensional momentum is manifested as extra mass in four-dimensional spacetime.

3.2 Composite particles

3.2.1 Hadrons

Main article: Hadron

Hadrons are defined as strongly interacting composite particles. Hadrons are either:

- Composite fermions, in which case they are called baryons.
- Composite bosons, in which case they are called mesons.

Quark models, first proposed in 1964 independently by Murray Gell-Mann and George Zweig (who called quarks “aces”), describe the known hadrons as composed of valence quarks and/or antiquarks, tightly bound by the color force, which is mediated by gluons. A “sea” of virtual quark-antiquark pairs is also present in each hadron.

Baryons

See also: List of baryons

Ordinary baryons (composite fermions) contain three valence quarks or three valence antiquarks each.

- Nucleons are the fermionic constituents of normal atomic nuclei:
  - Protons, composed of two up and one down quark (uud)
  - Neutrons, composed of two down and one up quark (ddu)
- Hyperons, such as the Λ, Σ, Ξ, and Ω particles, which contain one or more strange quarks, are short-lived and heavier than nucleons. Although not normally present in atomic nuclei, they can appear in short-lived hypernuclei.
- A number of charmed and bottom baryons have also been observed.

Some hints at the existence of exotic baryons have been found recently; however, negative results have also been reported. Their existence is uncertain.

- Pentaquarks consist of four valence quarks and one valence antiquark.

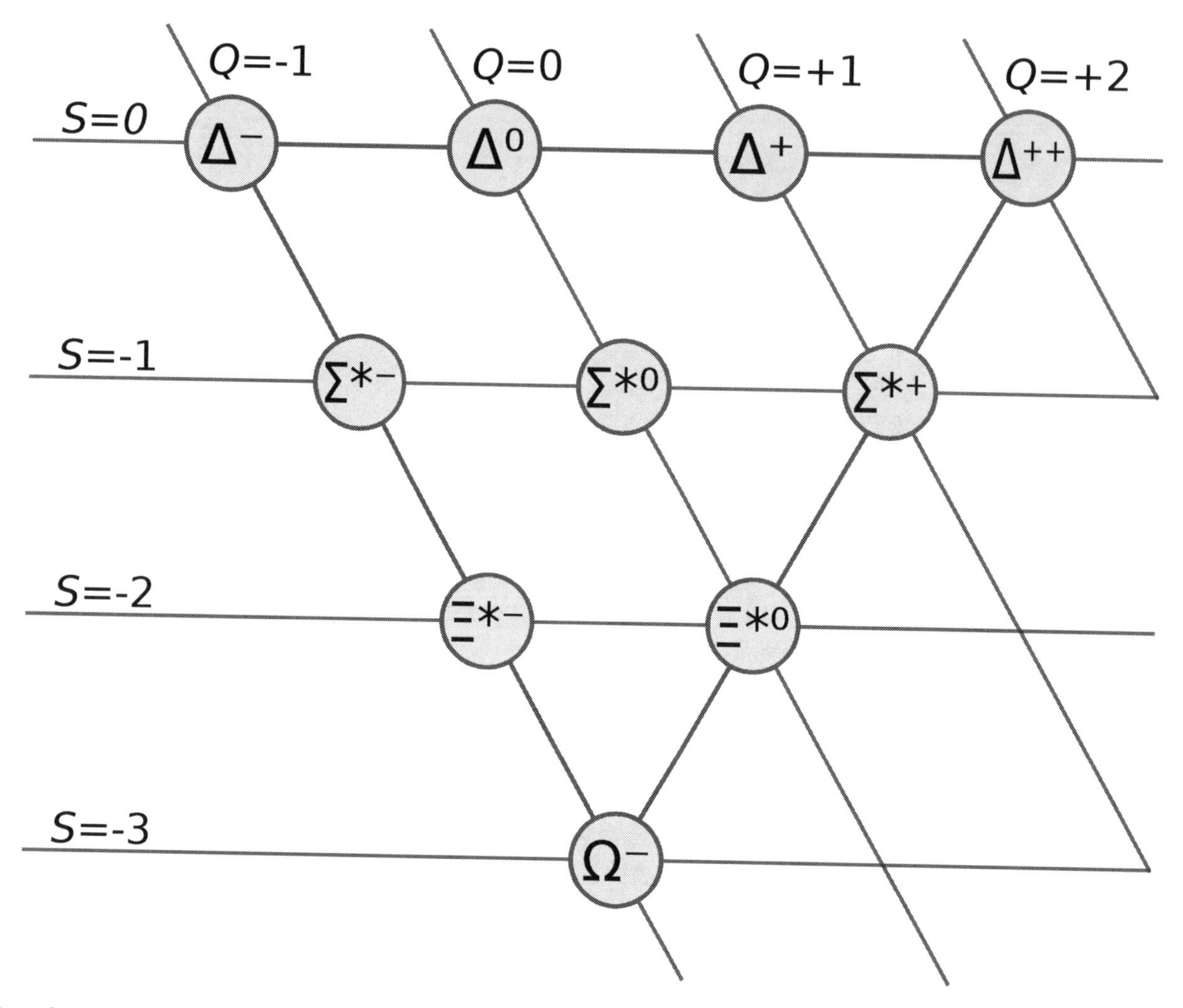

A combination of three u, d or s-quarks with a total spin of $^3/_2$ form the so-called "baryon decuplet".

Mesons

See also: List of mesons

Ordinary mesons are made up of a valence quark and a valence antiquark. Because mesons have spin of 0 or 1 and are not themselves elementary particles, they are "composite" bosons. Examples of mesons include the pion, kaon, and the J/ψ. In quantum hydrodynamic models, mesons mediate the residual strong force between nucleons.

At one time or another, positive signatures have been reported for all of the following exotic mesons but their existences have yet to be confirmed.

- A tetraquark consists of two valence quarks and two valence antiquarks;
- A glueball is a bound state of gluons with no valence quarks;
- Hybrid mesons consist of one or more valence quark-antiquark pairs and one or more real gluons.

3.2.2 Atomic nuclei

Atomic nuclei consist of protons and neutrons. Each type of nucleus contains a specific number of protons and a specific

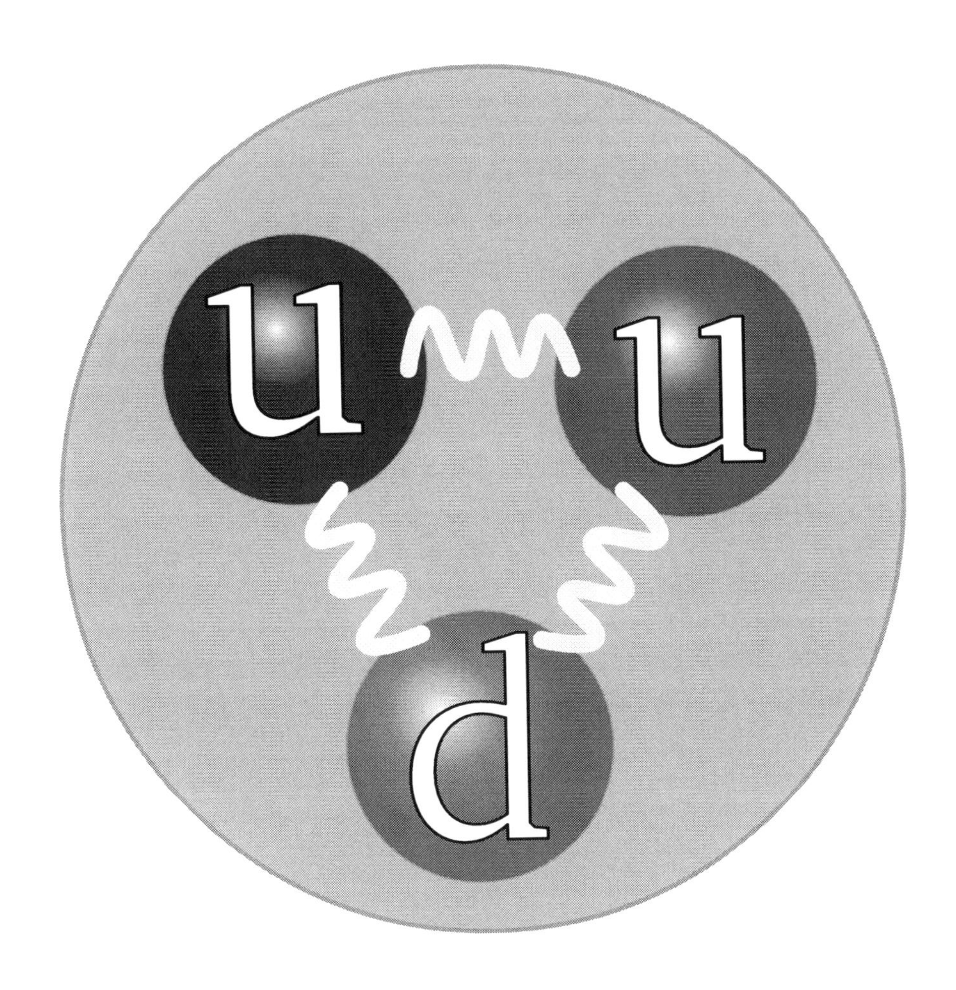

Proton quark structure: 2 up quarks and 1 down quark. The gluon tubes or flux tubes are now known to be Y shaped.

number of neutrons, and is called a "nuclide" or "isotope". Nuclear reactions can change one nuclide into another. See table of nuclides for a complete list of isotopes.

3.2.3 Atoms

Atoms are the smallest neutral particles into which matter can be divided by chemical reactions. An atom consists of a small, heavy nucleus surrounded by a relatively large, light cloud of electrons. Each type of atom corresponds to a specific chemical element. To date, 118 elements have been discovered, while only the elements 1-112,114, and 116 have received official names.

The atomic nucleus consists of protons and neutrons. Protons and neutrons are, in turn, made of quarks.

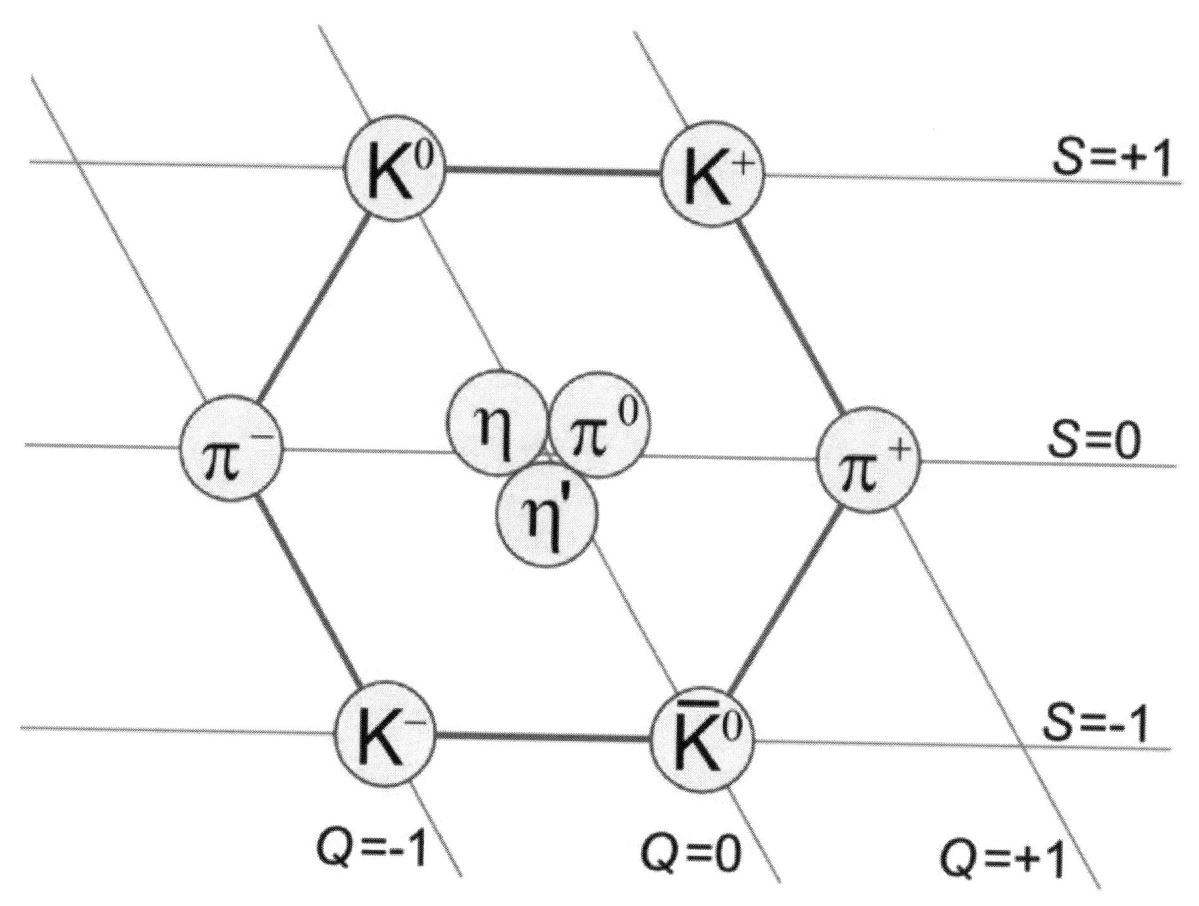

Mesons of spin 0 form a nonet

3.2.4 Molecules

Molecules are the smallest particles into which a non-elemental substance can be divided while maintaining the physical properties of the substance. Each type of molecule corresponds to a specific chemical compound. Molecules are a composite of two or more atoms. See list of compounds for a list of molecules.

3.3 Condensed matter

The field equations of condensed matter physics are remarkably similar to those of high energy particle physics. As a result, much of the theory of particle physics applies to condensed matter physics as well; in particular, there are a selection of field excitations, called quasi-particles, that can be created and explored. These include:

- Phonons are vibrational modes in a crystal lattice.
- Excitons are bound states of an electron and a hole.
- Plasmons are coherent excitations of a plasma.
- Polaritons are mixtures of photons with other quasi-particles.
- Polarons are moving, charged (quasi-) particles that are surrounded by ions in a material.
- Magnons are coherent excitations of electron spins in a material.

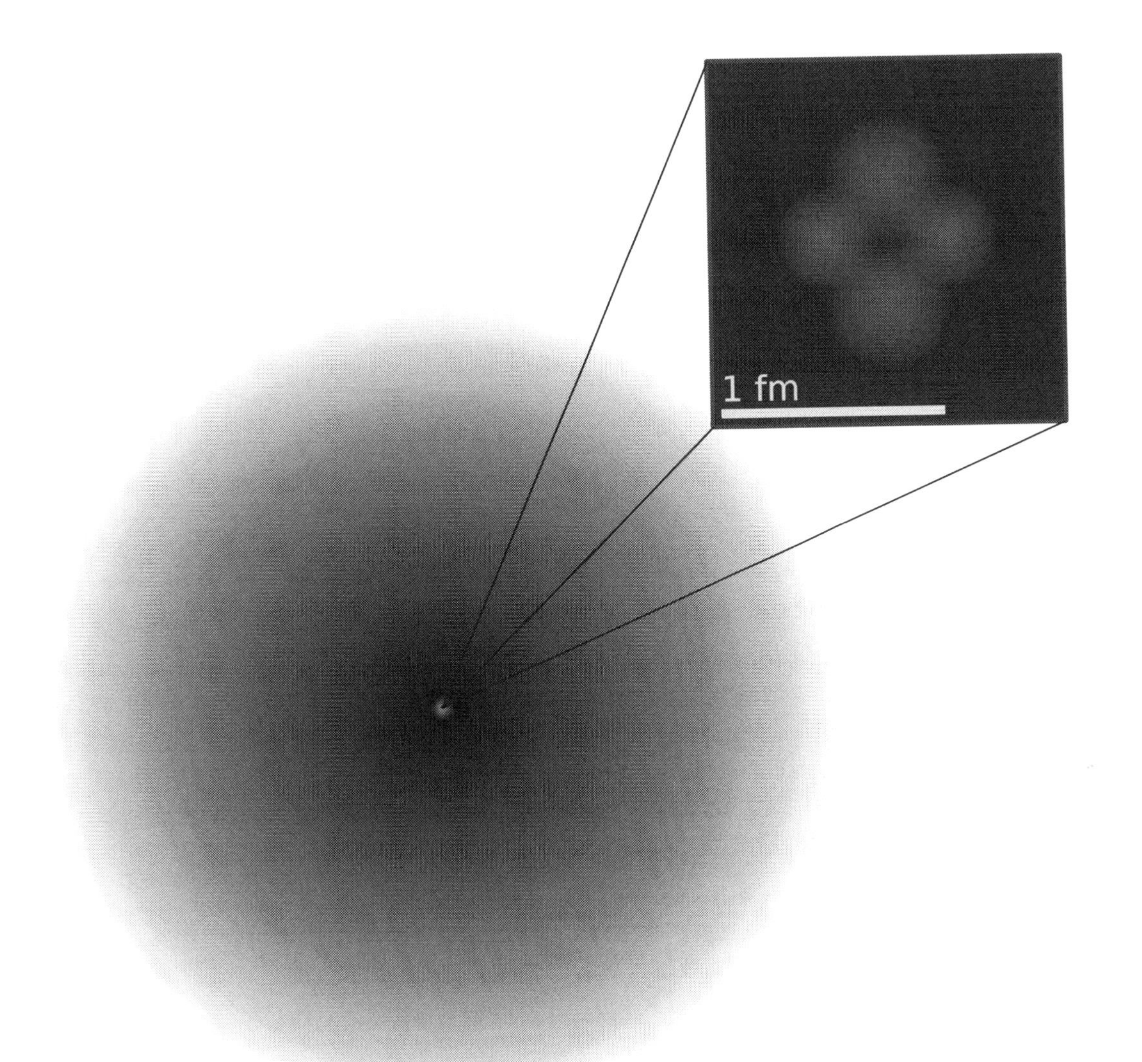

A semi-accurate depiction of the helium atom. In the nucleus, the protons are in red and neutrons are in purple. In reality, the nucleus is also spherically symmetrical.

3.4 Other

- An anyon is a generalization of fermion and boson in two-dimensional systems like sheets of graphene that obeys braid statistics.
- A plekton is a theoretical kind of particle discussed as a generalization of the braid statistics of the anyon to dimension > 2.
- A WIMP (weakly interacting massive particle) is any one of a number of particles that might explain dark matter (such as the neutralino or the axion).
- The pomeron, used to explain the elastic scattering of hadrons and the location of Regge poles in Regge theory.

- The skyrmion, a topological solution of the pion field, used to model the low-energy properties of the nucleon, such as the axial vector current coupling and the mass.
- A genon is a particle existing in a closed timelike world line where spacetime is curled as in a Frank Tipler or Ronald Mallett time machine.
- A goldstone boson is a massless excitation of a field that has been spontaneously broken. The pions are quasi-goldstone bosons (quasi- because they are not exactly massless) of the broken chiral isospin symmetry of quantum chromodynamics.
- A goldstino is a goldstone fermion produced by the spontaneous breaking of supersymmetry.
- An instanton is a field configuration which is a local minimum of the Euclidean action. Instantons are used in nonperturbative calculations of tunneling rates.
- A dyon is a hypothetical particle with both electric and magnetic charges.
- A geon is an electromagnetic or gravitational wave which is held together in a confined region by the gravitational attraction of its own field energy.
- An inflaton is the generic name for an unidentified scalar particle responsible for the cosmic inflation.
- A spurion is the name given to a "particle" inserted mathematically into an isospin-violating decay in order to analyze it as though it conserved isospin.
- What is called "true muonium", a bound state of a muon and an antimuon, is a theoretical exotic atom which has never been observed.

3.5 Classification by speed

- A tardyon or bradyon travels slower than light and has a non-zero rest mass.
- A luxon travels at the speed of light and has no rest mass.
- A tachyon (mentioned above) is a hypothetical particle that travels faster than the speed of light and has an imaginary rest mass.

3.6 See also

- Acceleron
- List of baryons
- List of compounds for a list of molecules.
- List of fictional elements, materials, isotopes and atomic particles
- List of mesons
- Periodic table for an overview of atoms.
- Standard Model for the current theory of these particles.
- Table of nuclides
- Timeline of particle discoveries

3.7 References

[1] Observation of a new boson at a mass of 125 GeV with the CMS experiment at the LHC (2013). *arXiv:1207.7235*.

[2] Observation of a new particle in the search for the Standard Model Higgs boson with the ATLAS detector at the LHC (2012). *arXiv:1207.7214*.

[3] B. Kayser, *Two Questions About Neutrinos*, arXiv:1012.4469v1 [hep-ph] (2010).

[4] R. Maartens (2004). *Brane-World Gravity* (PDF). *Living Reviews in Relativity* **7**. p. 7. Also available in web format at http://www.livingreviews.org/lrr-2004-7.

- C. Amsler *et al.* (Particle Data Group) (2008). "Review of Particle Physics". *Physics Letters B* **667** (1–5): 1. Bibcode:2008PhLB..667....1P. doi:10.1016/j.physletb.2008.07.018. *(All information on this list, and more, can be found in the extensive, biannually-updated review by the Particle Data Group)*

Chapter 4

Fermion

In particle physics, a **fermion** (a name coined by Paul Dirac[1] from the surname of Enrico Fermi) is any particle characterized by Fermi–Dirac statistics. These particles obey the Pauli exclusion principle. Fermions include all quarks and leptons, as well as any composite particle made of an odd number of these, such as all baryons and many atoms and nuclei. Fermions differ from bosons, which obey Bose–Einstein statistics.

A fermion can be an elementary particle, such as the electron, or it can be a composite particle, such as the proton. According to the spin-statistics theorem in any reasonable relativistic quantum field theory, particles with integer spin are bosons, while particles with half-integer spin are neutrons fermions.

Besides this spin characteristic, fermions have another specific property: they possess conserved baryon or lepton quantum numbers. Therefore what is usually referred as the spin statistics relation is in fact a spin statistics-quantum number relation.[2]

As a consequence of the Pauli exclusion principle, only one fermion can occupy a particular quantum state at any given time. If multiple fermions have the same spatial probability distribution, then at least one property of each fermion, such as its spin, must be different. Fermions are usually associated with matter, whereas bosons are generally force carrier particles, although in the current state of particle physics the distinction between the two concepts is unclear. At low temperature fermions show superfluidity for uncharged particles and superconductivity for charged particles. Composite fermions, such as protons and neutrons, are the key building blocks of everyday matter. Weakly interacting fermions can also display bosonic behavior under extreme conditions, such as superconductivity.

4.1 Elementary fermions

The Standard Model recognizes two types of elementary fermions, quarks and leptons. In all, the model distinguishes 24 different fermions. There are six quarks (up, down, strange, charm, bottom and top quarks), and six leptons (electron, electron neutrino, muon, muon neutrino, tau particle and tau neutrino), along with the corresponding antiparticle of each of these.

Mathematically, fermions come in three types - Weyl fermions (massless), Dirac fermions (massive), and Majorana fermions (each its own antiparticle). Most Standard Model fermions are believed to be Dirac fermions, although it is unknown at this time whether the neutrinos are Dirac or Majorana fermions. Dirac fermions can be treated as a combination of two Weyl fermions.[3]:106 So far there is no known example of Weyl fermion in particle physics. In July 2015, Weyl fermions have been experimentally realized in Weyl semimetals.

Enrico Fermi

4.2 Composite fermions

See also: List of particles § Composite particles

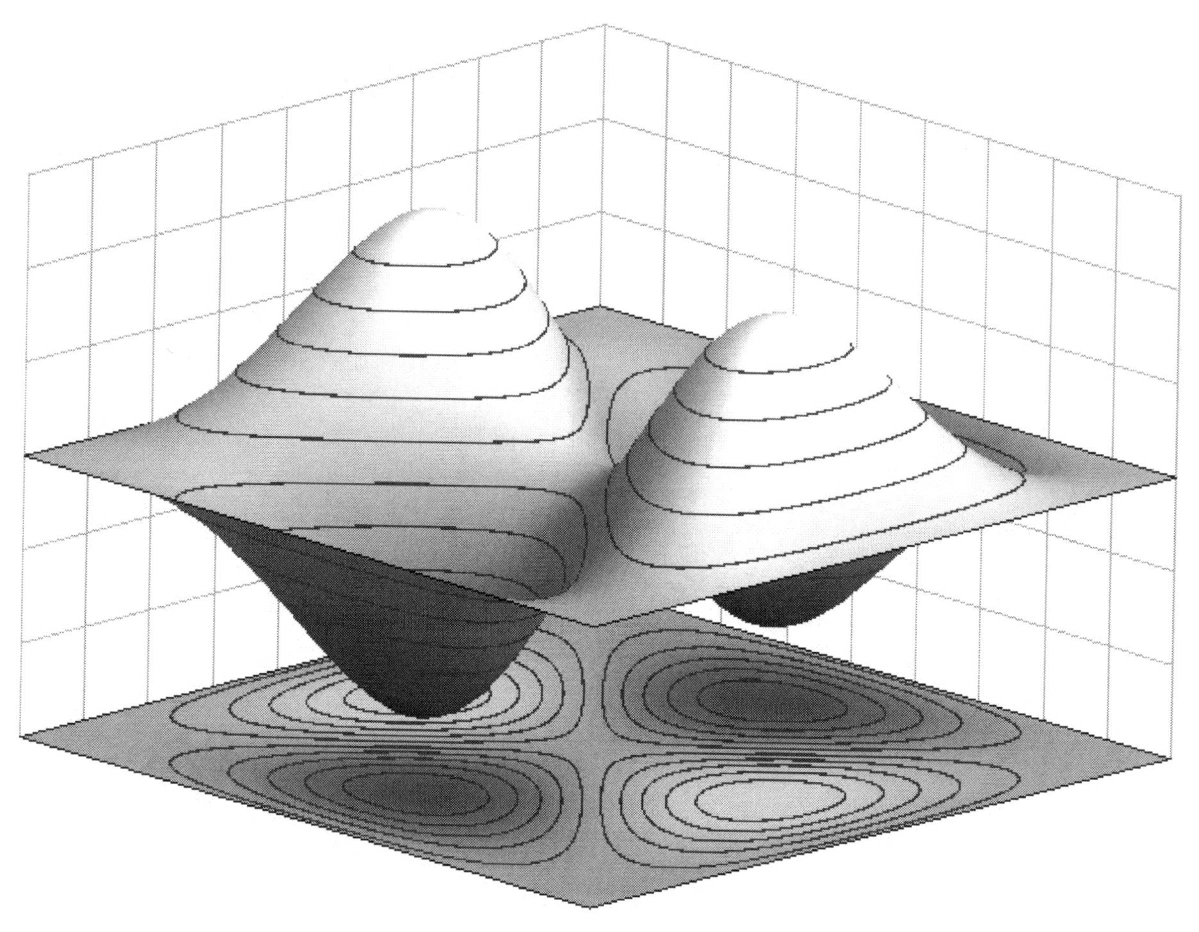

Antisymmetric wavefunction for a (fermionic) 2-particle state in an infinite square well potential.

Composite particles (such as hadrons, nuclei, and atoms) can be bosons or fermions depending on their constituents. More precisely, because of the relation between spin and statistics, a particle containing an odd number of fermions is itself a fermion. It will have half-integer spin.

Examples include the following:

- A baryon, such as the proton or neutron, contains three fermionic quarks and thus it is a fermion.
- The nucleus of a carbon-13 atom contains six protons and seven neutrons and is therefore a fermion.
- The atom helium-3 (^{3}He) is made of two protons, one neutron, and two electrons, and therefore it is a fermion.

The number of bosons within a composite particle made up of simple particles bound with a potential has no effect on whether it is a boson or a fermion.

Fermionic or bosonic behavior of a composite particle (or system) is only seen at large (compared to size of the system) distances. At proximity, where spatial structure begins to be important, a composite particle (or system) behaves according to its constituent makeup.

Fermions can exhibit bosonic behavior when they become loosely bound in pairs. This is the origin of superconductivity and the superfluidity of helium-3: in superconducting materials, electrons interact through the exchange of phonons, forming Cooper pairs, while in helium-3, Cooper pairs are formed via spin fluctuations.

The quasiparticles of the fractional quantum Hall effect are also known as composite fermions, which are electrons with an even number of quantized vortices attached to them.

4.2.1 Skyrmions

Main article: Skyrmion

In a quantum field theory, there can be field configurations of bosons which are topologically twisted. These are coherent states (or solitons) which behave like a particle, and they can be fermionic even if all the constituent particles are bosons. This was discovered by Tony Skyrme in the early 1960s, so fermions made of bosons are named skyrmions after him.

Skyrme's original example involved fields which take values on a three-dimensional sphere, the original nonlinear sigma model which describes the large distance behavior of pions. In Skyrme's model, reproduced in the large N or string approximation to quantum chromodynamics (QCD), the proton and neutron are fermionic topological solitons of the pion field.

Whereas Skyrme's example involved pion physics, there is a much more familiar example in quantum electrodynamics with a magnetic monopole. A bosonic monopole with the smallest possible magnetic charge and a bosonic version of the electron will form a fermionic dyon.

The analogy between the Skyrme field and the Higgs field of the electroweak sector has been used[4] to postulate that all fermions are skyrmions. This could explain why all known fermions have baryon or lepton quantum numbers and provide a physical mechanism for the Pauli exclusion principle.

4.3 See also

4.4 Notes

[1] Notes on Dirac's lecture *Developments in Atomic Theory* at Le Palais de la Découverte, 6 December 1945, UKNATARCHI Dirac Papers BW83/2/257889. See note 64 on page 331 in "The Strangest Man: The Hidden Life of Paul Dirac, Mystic of the Atom" by Graham Farmelo

[2] Physical Review D volume 87, page 0550003, year 2013, author Weiner, Richard M., title "Spin-statistics-quantum number connection and supersymmetry" arxiv:1302.0969

[3] T. Morii; C. S. Lim; S. N. Mukherjee (1 January 2004). *The Physics of the Standard Model and Beyond*. World Scientific. ISBN 978-981-279-560-1.

[4] Weiner, Richard M. (2010). "The Mysteries of Fermions". *International Journal of Theoretical Physics* **49** (5): 1174–1180. arXiv:0901.3816. Bibcode:2010IJTP...49.1174W. doi:10.1007/s10773-010-0292-7.

Chapter 5

Quark

This article is about the particle. For other uses, see Quark (disambiguation).

A **quark** (/ˈkwɔrk/ or /ˈkwɑrk/) is an elementary particle and a fundamental constituent of matter. Quarks combine to form composite particles called hadrons, the most stable of which are protons and neutrons, the components of atomic nuclei.[1] Due to a phenomenon known as *color confinement*, quarks are never directly observed or found in isolation; they can be found only within hadrons, such as baryons (of which protons and neutrons are examples), and mesons.[2][3] For this reason, much of what is known about quarks has been drawn from observations of the hadrons themselves.

Quarks have various intrinsic properties, including electric charge, mass, color charge and spin. Quarks are the only elementary particles in the Standard Model of particle physics to experience all four fundamental interactions, also known as *fundamental forces* (electromagnetism, gravitation, strong interaction, and weak interaction), as well as the only known particles whose electric charges are not integer multiples of the elementary charge.

There are six types of quarks, known as *flavors*: up, down, strange, charm, top, and bottom.[4] Up and down quarks have the lowest masses of all quarks. The heavier quarks rapidly change into up and down quarks through a process of particle decay: the transformation from a higher mass state to a lower mass state. Because of this, up and down quarks are generally stable and the most common in the universe, whereas strange, charm, bottom, and top quarks can only be produced in high energy collisions (such as those involving cosmic rays and in particle accelerators). For every quark flavor there is a corresponding type of antiparticle, known as an *antiquark*, that differs from the quark only in that some of its properties have equal magnitude but opposite sign.

The quark model was independently proposed by physicists Murray Gell-Mann and George Zweig in 1964.[5] Quarks were introduced as parts of an ordering scheme for hadrons, and there was little evidence for their physical existence until deep inelastic scattering experiments at the Stanford Linear Accelerator Center in 1968.[6][7] Accelerator experiments have provided evidence for all six flavors. The top quark was the last to be discovered at Fermilab in 1995.[5]

5.1 Classification

See also: Standard Model

The Standard Model is the theoretical framework describing all the currently known elementary particles. This model contains six flavors of quarks (q), named up (u), down (d), strange (s), charm (c), bottom (b), and top (t).[4] Antiparticles of quarks are called *antiquarks*, and are denoted by a bar over the symbol for the corresponding quark, such as u for an up antiquark. As with antimatter in general, antiquarks have the same mass, mean lifetime, and spin as their respective quarks, but the electric charge and other charges have the opposite sign.[8]

Quarks are spin-$\frac{1}{2}$ particles, implying that they are fermions according to the spin-statistics theorem. They are subject to the Pauli exclusion principle, which states that no two identical fermions can simultaneously occupy the same quantum state. This is in contrast to bosons (particles with integer spin), any number of which can be in the same state.[9] Unlike

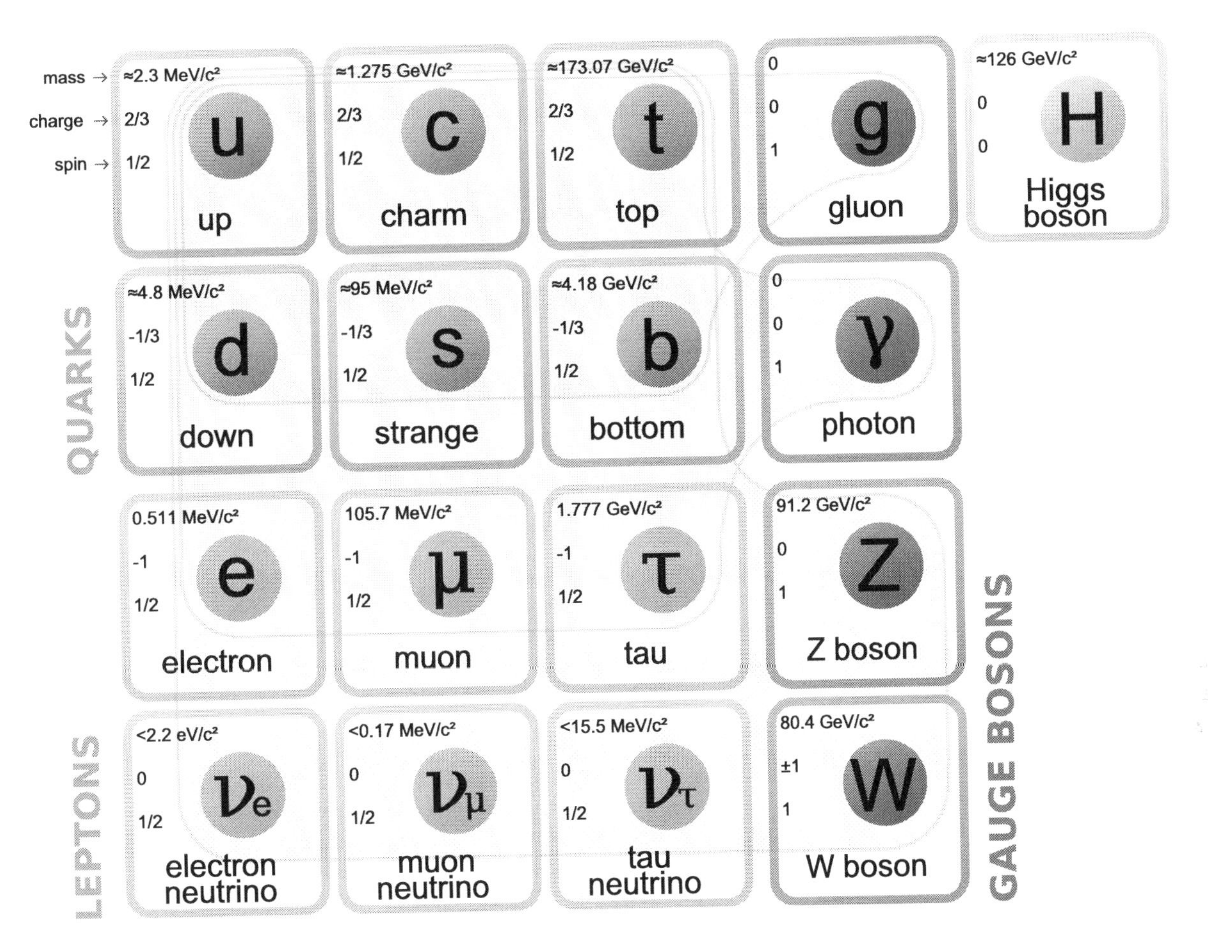

Six of the particles in the Standard Model are quarks (shown in purple). Each of the first three columns forms a generation of matter.

leptons, quarks possess color charge, which causes them to engage in the strong interaction. The resulting attraction between different quarks causes the formation of composite particles known as *hadrons* (see "Strong interaction and color charge" below).

The quarks which determine the quantum numbers of hadrons are called *valence quarks*; apart from these, any hadron may contain an indefinite number of virtual (or *sea*) quarks, antiquarks, and gluons which do not influence its quantum numbers.[10] There are two families of hadrons: baryons, with three valence quarks, and mesons, with a valence quark and an antiquark.[11] The most common baryons are the proton and the neutron, the building blocks of the atomic nucleus.[12] A great number of hadrons are known (see list of baryons and list of mesons), most of them differentiated by their quark content and the properties these constituent quarks confer. The existence of “exotic” hadrons with more valence quarks, such as tetraquarks (qqqq) and pentaquarks (qqqqq), has been conjectured[13] but not proven.[nb 1][13][14] However, on 13 July 2015, the LHCb collaboration at CERN reported results consistent with pentaquark states.[15]

Elementary fermions are grouped into three generations, each comprising two leptons and two quarks. The first generation includes up and down quarks, the second strange and charm quarks, and the third bottom and top quarks. All searches for a fourth generation of quarks and other elementary fermions have failed,[16] and there is strong indirect evidence that no more than three generations exist.[nb 2][17] Particles in higher generations generally have greater mass and less stability, causing them to decay into lower-generation particles by means of weak interactions. Only first-generation (up and down) quarks occur commonly in nature. Heavier quarks can only be created in high-energy collisions (such as in those involving cosmic rays), and decay quickly; however, they are thought to have been present during the first fractions of a second after the Big Bang, when the universe was in an extremely hot and dense phase (the quark epoch). Studies of heavier quarks are conducted in artificially created conditions, such as in particle accelerators.[18]

Having electric charge, mass, color charge, and flavor, quarks are the only known elementary particles that engage in all four fundamental interactions of contemporary physics: electromagnetism, gravitation, strong interaction, and weak interaction.[12] Gravitation is too weak to be relevant to individual particle interactions except at extremes of energy (Planck energy) and distance scales (Planck distance). However, since no successful quantum theory of gravity exists, gravitation is not described by the Standard Model.

See the table of properties below for a more complete overview of the six quark flavors' properties.

5.2 History

The quark model was independently proposed by physicists Murray Gell-Mann[19] (pictured) and George Zweig[20][21] in 1964.[5] The proposal came shortly after Gell-Mann's 1961 formulation of a particle classification system known as the *Eightfold Way*—or, in more technical terms, SU(3) flavor symmetry.[22] Physicist Yuval Ne'eman had independently developed a scheme similar to the Eightfold Way in the same year.[23][24]

At the time of the quark theory's inception, the "particle zoo" included, amongst other particles, a multitude of hadrons. Gell-Mann and Zweig posited that they were not elementary particles, but were instead composed of combinations of quarks and antiquarks. Their model involved three flavors of quarks, up, down, and strange, to which they ascribed properties such as spin and electric charge.[19][20][21] The initial reaction of the physics community to the proposal was mixed. There was particular contention about whether the quark was a physical entity or a mere abstraction used to explain concepts that were not fully understood at the time.[25]

In less than a year, extensions to the Gell-Mann–Zweig model were proposed. Sheldon Lee Glashow and James Bjorken predicted the existence of a fourth flavor of quark, which they called *charm*. The addition was proposed because it allowed for a better description of the weak interaction (the mechanism that allows quarks to decay), equalized the number of known quarks with the number of known leptons, and implied a mass formula that correctly reproduced the masses of the known mesons.[26]

In 1968, deep inelastic scattering experiments at the Stanford Linear Accelerator Center (SLAC) showed that the proton contained much smaller, point-like objects and was therefore not an elementary particle.[6][7][27] Physicists were reluctant to firmly identify these objects with quarks at the time, instead calling them "partons"—a term coined by Richard Feynman.[28][29][30] The objects that were observed at SLAC would later be identified as up and down quarks as the other flavors were discovered.[31] Nevertheless, "parton" remains in use as a collective term for the constituents of hadrons (quarks, antiquarks, and gluons).

The strange quark's existence was indirectly validated by SLAC's scattering experiments: not only was it a necessary component of Gell-Mann and Zweig's three-quark model, but it provided an explanation for the kaon (K) and pion (π) hadrons discovered in cosmic rays in 1947.[32]

In a 1970 paper, Glashow, John Iliopoulos and Luciano Maiani presented further reasoning for the existence of the as-yet undiscovered charm quark.[33][34] The number of supposed quark flavors grew to the current six in 1973, when Makoto Kobayashi and Toshihide Maskawa noted that the experimental observation of CP violation[nb 3][35] could be explained if there were another pair of quarks.

Charm quarks were produced almost simultaneously by two teams in November 1974 (see November Revolution)—one at SLAC under Burton Richter, and one at Brookhaven National Laboratory under Samuel Ting. The charm quarks were observed bound with charm antiquarks in mesons. The two parties had assigned the discovered meson two different symbols, J and ψ; thus, it became formally known as the J/ψ meson. The discovery finally convinced the physics community of the quark model's validity.[30]

In the following years a number of suggestions appeared for extending the quark model to six quarks. Of these, the 1975 paper by Haim Harari[36] was the first to coin the terms *top* and *bottom* for the additional quarks.[37]

In 1977, the bottom quark was observed by a team at Fermilab led by Leon Lederman.[38][39] This was a strong indicator of the top quark's existence: without the top quark, the bottom quark would have been without a partner. However, it was not until 1995 that the top quark was finally observed, also by the CDF[40] and DØ[41] teams at Fermilab.[5] It had a mass much larger than had been previously expected,[42] almost as large as that of a gold atom.[43]

Murray Gell-Mann at TED in 2007. Gell-Mann and George Zweig proposed the quark model in 1964.

5.3 Etymology

For some time, Gell-Mann was undecided on an actual spelling for the term he intended to coin, until he found the word *quark* in James Joyce's book *Finnegans Wake*:

> Three quarks for Muster Mark!
> Sure he has not got much of a bark

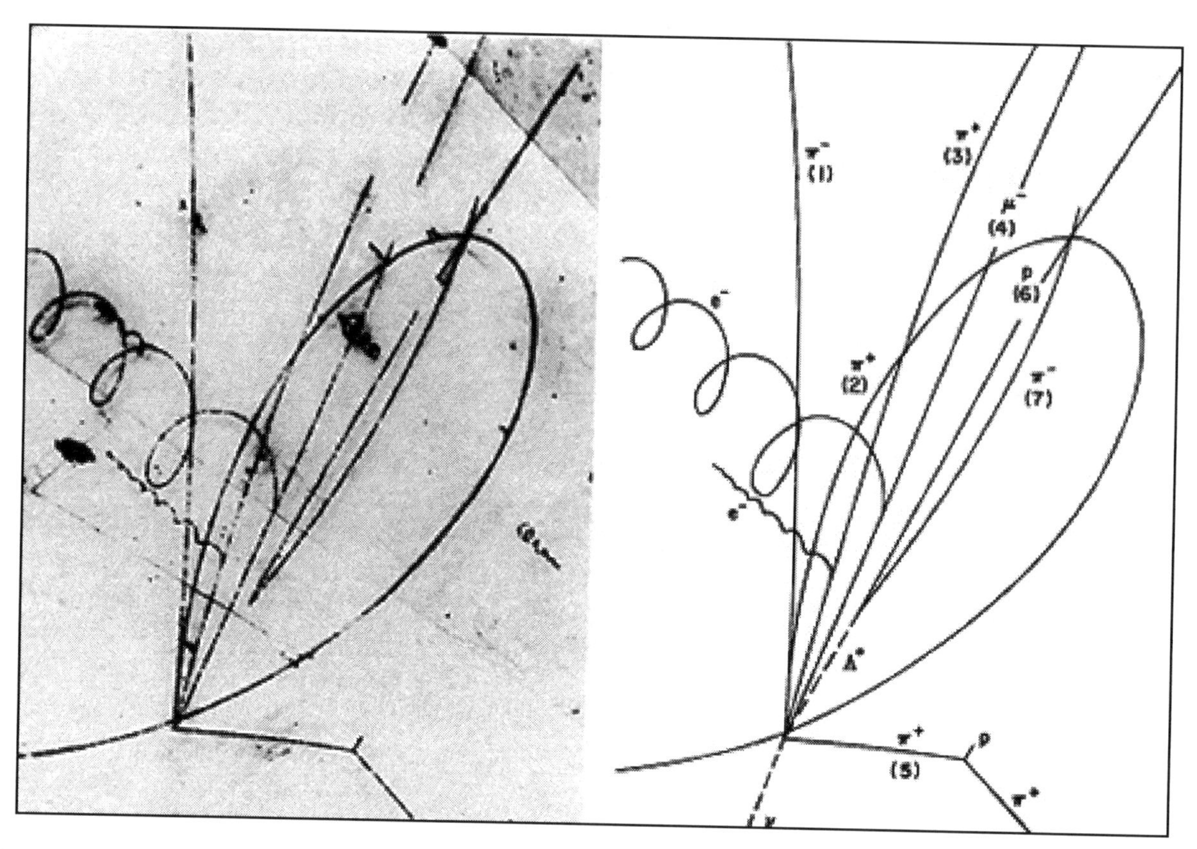

Photograph of the event that led to the discovery of the Σ++ c baryon, at the Brookhaven National Laboratory in 1974

> And sure any he has it's all beside the mark.
> — James Joyce, *Finnegans Wake*[44]

Gell-Mann went into further detail regarding the name of the quark in his book *The Quark and the Jaguar*:[45]

> In 1963, when I assigned the name "quark" to the fundamental constituents of the nucleon, I had the sound first, without the spelling, which could have been "kwork". Then, in one of my occasional perusals of *Finnegans Wake*, by James Joyce, I came across the word "quark" in the phrase "Three quarks for Muster Mark". Since "quark" (meaning, for one thing, the cry of the gull) was clearly intended to rhyme with "Mark", as well as "bark" and other such words, I had to find an excuse to pronounce it as "kwork". But the book represents the dream of a publican named Humphrey Chimpden Earwicker. Words in the text are typically drawn from several sources at once, like the "portmanteau" words in "Through the Looking-Glass". From time to time, phrases occur in the book that are partially determined by calls for drinks at the bar. I argued, therefore, that perhaps one of the multiple sources of the cry "Three quarks for Muster Mark" might be "Three quarts for Mister Mark", in which case the pronunciation "kwork" would not be totally unjustified. In any case, the number three fitted perfectly the way quarks occur in nature.

Zweig preferred the name *ace* for the particle he had theorized, but Gell-Mann's terminology came to prominence once the quark model had been commonly accepted.[46]

The quark flavors were given their names for several reasons. The up and down quarks are named after the up and down components of isospin, which they carry.[47] Strange quarks were given their name because they were discovered to be components of the strange particles discovered in cosmic rays years before the quark model was proposed; these particles were deemed "strange" because they had unusually long lifetimes.[48] Glashow, who coproposed charm quark with Bjorken, is quoted as saying, “We called our construct the 'charmed quark', for we were fascinated and pleased by the symmetry it brought to the subnuclear world.”[49] The names “bottom” and “top”, coined by Harari, were chosen because they are “logical partners for up and down quarks”.[36][37][48] In the past, bottom and top quarks were sometimes referred to as “beauty” and “truth” respectively, but these names have somewhat fallen out of use.[50] While “truth” never did catch on, accelerator complexes devoted to massive production of bottom quarks are sometimes called "beauty factories".[51]

5.4 Properties

5.4.1 Electric charge

See also: Electric charge

Quarks have fractional electric charge values – either $1/3$ or $2/3$ times the elementary charge (e), depending on flavor. Up, charm, and top quarks (collectively referred to as *up-type quarks*) have a charge of $+2/3$ e, while down, strange, and bottom quarks (*down-type quarks*) have $-1/3$ e. Antiquarks have the opposite charge to their corresponding quarks; up-type antiquarks have charges of $-2/3$ e and down-type antiquarks have charges of $+1/3$ e. Since the electric charge of a hadron is the sum of the charges of the constituent quarks, all hadrons have integer charges: the combination of three quarks (baryons), three antiquarks (antibaryons), or a quark and an antiquark (mesons) always results in integer charges.[52] For example, the hadron constituents of atomic nuclei, neutrons and protons, have charges of 0 e and +1 e respectively; the neutron is composed of two down quarks and one up quark, and the proton of two up quarks and one down quark.[12]

5.4.2 Spin

See also: Spin (physics)

Spin is an intrinsic property of elementary particles, and its direction is an important degree of freedom. It is sometimes visualized as the rotation of an object around its own axis (hence the name "spin"), though this notion is somewhat misguided at subatomic scales because elementary particles are believed to be point-like.[53]

Spin can be represented by a vector whose length is measured in units of the reduced Planck constant $\hbar$ (pronounced “h bar”). For quarks, a measurement of the spin vector component along any axis can only yield the values $+\hbar/2$ or $-\hbar/2$; for this reason quarks are classified as spin-$1/2$ particles.[54] The component of spin along a given axis – by convention the z axis – is often denoted by an up arrow ↑ for the value $+1/2$ and down arrow ↓ for the value $-1/2$, placed after the symbol for flavor. For example, an up quark with a spin of $+1/2$ along the z axis is denoted by u↑.[55]

5.4.3 Weak interaction

Main article: Weak interaction

A quark of one flavor can transform into a quark of another flavor only through the weak interaction, one of the four fundamental interactions in particle physics. By absorbing or emitting a W boson, any up-type quark (up, charm, and top quarks) can change into any down-type quark (down, strange, and bottom quarks) and vice versa. This flavor transformation mechanism causes the radioactive process of beta decay, in which a neutron (n) “splits” into a proton (p), an electron (e−) and an electron antineutrino (ν
e) (see picture). This occurs when one of the down quarks in the neutron (udd) decays into an up quark by emitting a virtual W− boson, transforming the neutron into a proton (uud). The W− boson then decays into an electron and an electron antineutrino.[56]

Both beta decay and the inverse process of *inverse beta decay* are routinely used in medical applications such as positron emission tomography (PET) and in experiments involving neutrino detection.

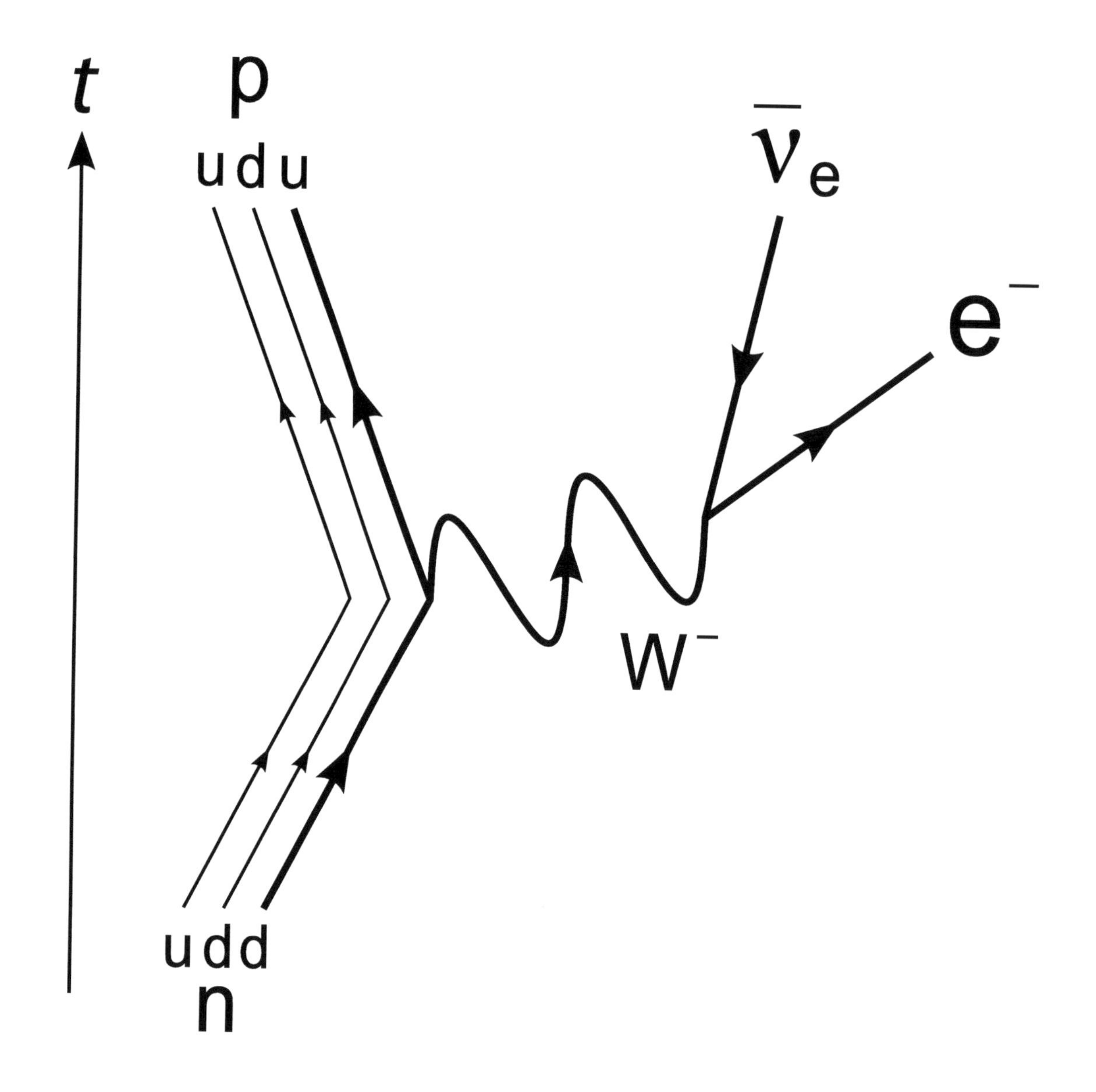

Feynman diagram of beta decay with time flowing upwards. The CKM matrix (discussed below) encodes the probability of this and other quark decays.

While the process of flavor transformation is the same for all quarks, each quark has a preference to transform into the quark of its own generation. The relative tendencies of all flavor transformations are described by a mathematical table, called the Cabibbo–Kobayashi–Maskawa matrix (CKM matrix). Enforcing unitarity, the approximate magnitudes of the entries of the CKM matrix are:[57]

$$\begin{bmatrix} |V_{\mathrm{ud}}| & |V_{\mathrm{us}}| & |V_{\mathrm{ub}}| \\ |V_{\mathrm{cd}}| & |V_{\mathrm{cs}}| & |V_{\mathrm{cb}}| \\ |V_{\mathrm{td}}| & |V_{\mathrm{ts}}| & |V_{\mathrm{tb}}| \end{bmatrix} \approx \begin{bmatrix} 0.974 & 0.225 & 0.003 \\ 0.225 & 0.973 & 0.041 \\ 0.009 & 0.040 & 0.999 \end{bmatrix},$$

where V_{ij} represents the tendency of a quark of flavor i to change into a quark of flavor j (or vice versa).[nb 4]

There exists an equivalent weak interaction matrix for leptons (right side of the W boson on the above beta decay diagram), called the Pontecorvo–Maki–Nakagawa–Sakata matrix (PMNS matrix).[58] Together, the CKM and PMNS matrices

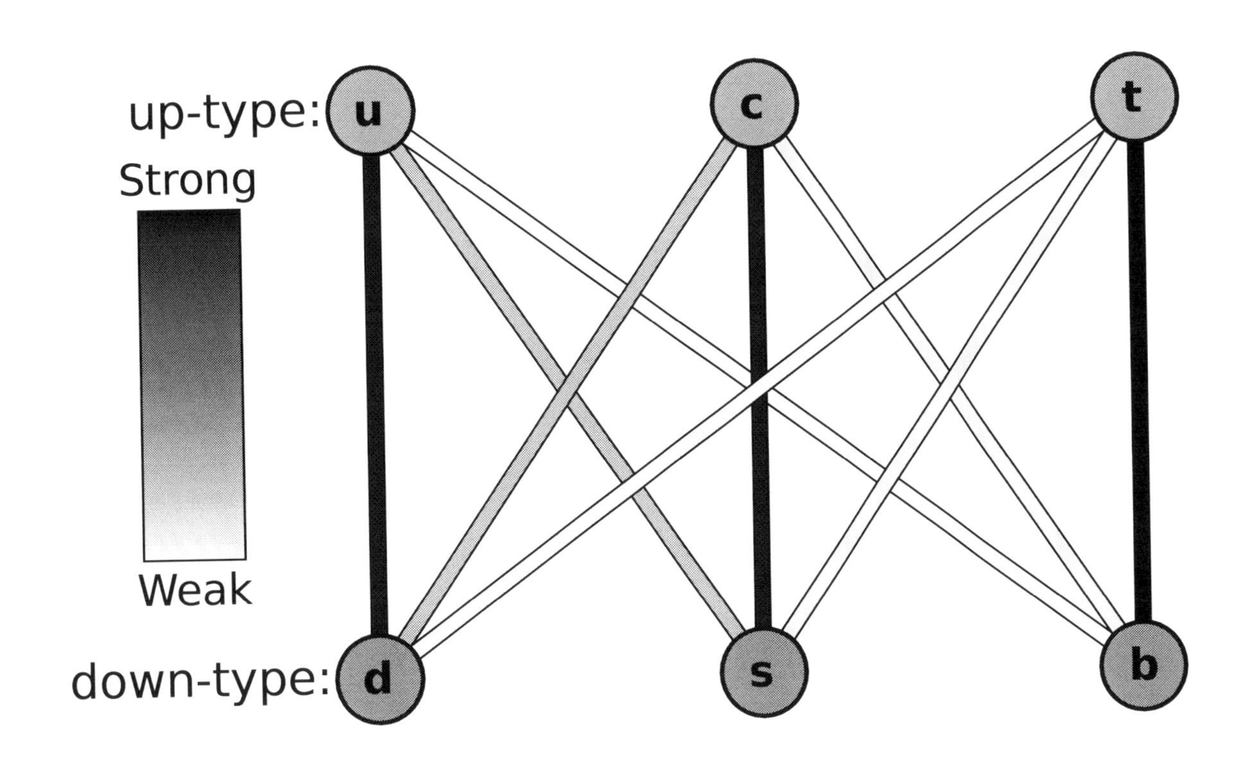

The strengths of the weak interactions between the six quarks. The "intensities" of the lines are determined by the elements of the CKM matrix.

describe all flavor transformations, but the links between the two are not yet clear.[59]

5.4.4 Strong interaction and color charge

See also: Color charge and Strong interaction

According to quantum chromodynamics (QCD), quarks possess a property called *color charge*. There are three types of color charge, arbitrarily labeled *blue*, *green*, and *red*.[nb 5] Each of them is complemented by an anticolor – *antiblue*, *antigreen*, and *antired*. Every quark carries a color, while every antiquark carries an anticolor.[60]

The system of attraction and repulsion between quarks charged with different combinations of the three colors is called strong interaction, which is mediated by force carrying particles known as *gluons*; this is discussed at length below. The theory that describes strong interactions is called quantum chromodynamics (QCD). A quark, which will have a single color value, can form a bound system with an antiquark carrying the corresponding anticolor. The result of two attracting quarks will be color neutrality: a quark with color charge ξ plus an antiquark with color charge $-\xi$ will result in a color charge of 0 (or "white" color) and the formation of a meson. This is analogous to the additive color model in basic optics. Similarly, the combination of three quarks, each with different color charges, or three antiquarks, each with anticolor charges, will result in the same "white" color charge and the formation of a baryon or antibaryon.[61]

In modern particle physics, gauge symmetries – a kind of symmetry group – relate interactions between particles (see gauge theories). Color SU(3) (commonly abbreviated to $SU(3)_c$) is the gauge symmetry that relates the color charge in quarks and is the defining symmetry for quantum chromodynamics.[62] Just as the laws of physics are independent of which directions in space are designated x, y, and z, and remain unchanged if the coordinate axes are rotated to a new orientation, the physics of quantum chromodynamics is independent of which directions in three-dimensional color space are identified as blue, red, and green. $SU(3)_c$ color transformations correspond to "rotations" in color space (which, mathematically speaking, is a complex space). Every quark flavor f, each with subtypes fB, fG, fR corresponding to the quark colors,[63] forms a triplet: a three-component quantum field which transforms under the fundamental representation of $SU(3)_c$.[64] The requirement that $SU(3)_c$ should be local – that is, that its transformations be allowed to vary with space and time – determines the properties of the strong interaction, in particular the existence of eight gluon types to act as its force carriers.[62][65]

5.4.5 Mass

See also: Invariant mass

Two terms are used in referring to a quark's mass: *current quark mass* refers to the mass of a quark by itself, while *constituent quark mass* refers to the current quark mass plus the mass of the gluon particle field surrounding the quark.[66] These masses typically have very different values. Most of a hadron's mass comes from the gluons that bind the constituent quarks together, rather than from the quarks themselves. While gluons are inherently massless, they possess energy – more specifically, quantum chromodynamics binding energy (QCBE) – and it is this that contributes so greatly to the overall mass of the hadron (see mass in special relativity). For example, a proton has a mass of approximately 938 MeV/c^2, of which the rest mass of its three valence quarks only contributes about 11 MeV/c^2; much of the remainder can be attributed to the gluons' QCBE.[67][68]

The Standard Model posits that elementary particles derive their masses from the Higgs mechanism, which is related to the Higgs boson. Physicists hope that further research into the reasons for the top quark's large mass of ~173 GeV/c^2, almost the mass of a gold atom,[67][69] might reveal more about the origin of the mass of quarks and other elementary particles.[70]

5.4.6 Table of properties

See also: Flavor (particle physics)

The following table summarizes the key properties of the six quarks. Flavor quantum numbers (isospin (I_3), charm (C), strangeness (S, not to be confused with spin), topness (T), and bottomness (B')) are assigned to certain quark flavors, and denote qualities of quark-based systems and hadrons. The baryon number (B) is $+1/3$ for all quarks, as baryons are made of three quarks. For antiquarks, the electric charge (Q) and all flavor quantum numbers (B, I_3, C, S, T, and B') are of opposite sign. Mass and total angular momentum (J; equal to spin for point particles) do not change sign for the antiquarks.

J = total angular momentum, B = baryon number, Q = electric charge, I_3 = isospin, C = charm, S = strangeness, T = topness, B' = bottomness.

* Notation such as 4190+180
−60 denotes measurement uncertainty. In the case of the top quark, the first uncertainty is statistical in nature, and the second is systematic.

5.5 Interacting quarks

See also: Color confinement and Gluon

As described by quantum chromodynamics, the strong interaction between quarks is mediated by gluons, massless vector gauge bosons. Each gluon carries one color charge and one anticolor charge. In the standard framework of particle interactions (part of a more general formulation known as perturbation theory), gluons are constantly exchanged between quarks through a virtual emission and absorption process. When a gluon is transferred between quarks, a color change occurs in both; for example, if a red quark emits a red–antigreen gluon, it becomes green, and if a green quark absorbs a red–antigreen gluon, it becomes red. Therefore, while each quark's color constantly changes, their strong interaction is preserved.[71][72][73]

Since gluons carry color charge, they themselves are able to emit and absorb other gluons. This causes *asymptotic freedom*: as quarks come closer to each other, the chromodynamic binding force between them weakens.[74] Conversely, as the distance between quarks increases, the binding force strengthens. The color field becomes stressed, much as an elastic band is stressed when stretched, and more gluons of appropriate color are spontaneously created to strengthen the field. Above a certain energy threshold, pairs of quarks and antiquarks are created. These pairs bind with the quarks being separated, causing new hadrons to form. This phenomenon is known as *color confinement*: quarks never appear in isolation.[72][75] This process of hadronization occurs before quarks, formed in a high energy collision, are able to interact in any other way. The only exception is the top quark, which may decay before it hadronizes.[76]

5.5.1 Sea quarks

Hadrons, along with the *valence quarks* (q
v) that contribute to their quantum numbers, contain virtual quark–antiquark (qq) pairs known as *sea quarks* (q
s). Sea quarks form when a gluon of the hadron's color field splits; this process also works in reverse in that the annihilation of two sea quarks produces a gluon. The result is a constant flux of gluon splits and creations colloquially known as "the sea".[77] Sea quarks are much less stable than their valence counterparts, and they typically annihilate each other within the interior of the hadron. Despite this, sea quarks can hadronize into baryonic or mesonic particles under certain circumstances.[78]

5.5.2 Other phases of quark matter

Main article: QCD matter

Under sufficiently extreme conditions, quarks may become deconfined and exist as free particles. In the course of asymptotic freedom, the strong interaction becomes weaker at higher temperatures. Eventually, color confinement would be lost and an extremely hot plasma of freely moving quarks and gluons would be formed. This theoretical phase of matter is called quark–gluon plasma.[81] The exact conditions needed to give rise to this state are unknown and have been the subject of a great deal of speculation and experimentation. A recent estimate puts the needed temperature at $(1.90 \pm 0.02) \times 10^{12}$ kelvin.[82] While a state of entirely free quarks and gluons has never been achieved (despite numerous attempts by CERN in the 1980s and 1990s),[83] recent experiments at the Relativistic Heavy Ion Collider have yielded evidence for liquid-like quark matter exhibiting "nearly perfect" fluid motion.[84]

The quark–gluon plasma would be characterized by a great increase in the number of heavier quark pairs in relation to the number of up and down quark pairs. It is believed that in the period prior to 10^{-6} seconds after the Big Bang (the quark epoch), the universe was filled with quark–gluon plasma, as the temperature was too high for hadrons to be stable.[85]

Given sufficiently high baryon densities and relatively low temperatures – possibly comparable to those found in neutron stars – quark matter is expected to degenerate into a Fermi liquid of weakly interacting quarks. This liquid would be characterized by a condensation of colored quark Cooper pairs, thereby breaking the local $SU(3)_c$ symmetry. Because quark Cooper pairs harbor color charge, such a phase of quark matter would be color superconductive; that is, color charge would be able to pass through it with no resistance.[86]

5.6 See also

- Color–flavor locking
- Neutron magnetic moment
- Leptons
- Preons – Hypothetical particles which were once postulated to be subcomponents of quarks and leptons

- Quarkonium – Mesons made of a quark and antiquark of the same flavor
- Quark star – A hypothetical degenerate neutron star with extreme density
- Quark–lepton complementarity – Possible fundamental relation between quarks and leptons

5.7 Notes

[1] Several research groups claimed to have proven the existence of tetraquarks and pentaquarks in the early 2000s. While the status of tetraquarks is still under debate, all known pentaquark candidates have previously been established as non-existent.

[2] The main evidence is based on the resonance width of the Z0 boson, which constrains the 4th generation neutrino to have a mass greater than ~45 GeV/c^2. This would be highly contrasting with the other three generations' neutrinos, whose masses cannot exceed 2 MeV/c^2.

[3] CP violation is a phenomenon which causes weak interactions to behave differently when left and right are swapped (P symmetry) and particles are replaced with their corresponding antiparticles (C symmetry).

[4] The actual probability of decay of one quark to another is a complicated function of (amongst other variables) the decaying quark's mass, the masses of the decay products, and the corresponding element of the CKM matrix. This probability is directly proportional (but not equal) to the magnitude squared ($|Vij|^2$) of the corresponding CKM entry.

[5] Despite its name, color charge is not related to the color spectrum of visible light.

5.8 References

[1] "Quark (subatomic particle)". *Encyclopædia Britannica*. Retrieved 2008-06-29.

[2] R. Nave. "Confinement of Quarks". *HyperPhysics*. Georgia State University, Department of Physics and Astronomy. Retrieved 2008-06-29.

[3] R. Nave. "Bag Model of Quark Confinement". *HyperPhysics*. Georgia State University, Department of Physics and Astronomy. Retrieved 2008-06-29.

[4] R. Nave. "Quarks". *HyperPhysics*. Georgia State University, Department of Physics and Astronomy. Retrieved 2008-06-29.

[5] B. Carithers, P. Grannis (1995). "Discovery of the Top Quark" (PDF). *Beam Line* (SLAC) **25** (3): 4–16. Retrieved 2008-09-23.

[6] E.D. Bloom et al. (1969). "High-Energy Inelastic *e–p* Scattering at 6° and 10°". *Physical Review Letters* **23** (16): 930–934. Bibcode:1969PhRvL..23..930B. doi:10.1103/PhysRevLett.23.930.

[7] M. Breidenbach et al. (1969). "Observed Behavior of Highly Inelastic Electron–Proton Scattering". *Physical Review Letters* **23** (16): 935–939. Bibcode:1969PhRvL..23..935B. doi:10.1103/PhysRevLett.23.935.

[8] S.S.M. Wong (1998). *Introductory Nuclear Physics* (2nd ed.). Wiley Interscience. p. 30. ISBN 0-471-23973-9.

[9] K.A. Peacock (2008). *The Quantum Revolution*. Greenwood Publishing Group. p. 125. ISBN 0-313-33448-X.

[10] B. Povh, C. Scholz, K. Rith, F. Zetsche (2008). *Particles and Nuclei*. Springer. p. 98. ISBN 3-540-79367-4.

[11] Section 6.1. in P.C.W. Davies (1979). *The Forces of Nature*. Cambridge University Press. ISBN 0-521-22523-X.

[12] M. Munowitz (2005). *Knowing*. Oxford University Press. p. 35. ISBN 0-19-516737-6.

[13] W.-M. Yao (Particle Data Group) et al. (2006). "Review of Particle Physics: Pentaquark Update" (PDF). *Journal of Physics G* **33** (1): 1–1232. arXiv:astro-ph/0601168. Bibcode:2006JPhG...33....1Y. doi:10.1088/0954-3899/33/1/001.

[14] C. Amsler (Particle Data Group) et al. (2008). "Review of Particle Physics: Pentaquarks" (PDF). *Physics Letters B* **667** (1): 1–1340. Bibcode:2008PhLB..667....1P. doi:10.1016/j.physletb.2008.07.018.
C. Amsler (Particle Data Group) et al. (2008). "Review of Particle Physics: New Charmonium-Like States" (PDF). *Physics Letters B* **667** (1): 1–1340. Bibcode:2008PhLB..667....1P. doi:10.1016/j.physletb.2008.07.018.
E.V. Shuryak (2004). *The QCD Vacuum, Hadrons and Superdense Matter*. World Scientific. p. 59. ISBN 981-238-574-6.

[15] R. Aaij et al. (LHCb collaboration) (2015). "Observation of J/ψp resonances consistent with pentaquark states in Λ0 b→J/ψK−
p decays". *Physical Review Letters* **115** (7). doi:10.1103/PhysRevLett.115.072001.

[16] C. Amsler (Particle Data Group) et al. (2008). "Review of Particle Physics: b′ (4th Generation) Quarks, Searches for" (PDF). *Physics Letters B* **667** (1): 1–1340. Bibcode:2008PhLB..667....1P. doi:10.1016/j.physletb.2008.07.018.
C. Amsler (Particle Data Group) et al. (2008). "Review of Particle Physics: t′ (4th Generation) Quarks, Searches for" (PDF). *Physics Letters B* **667** (1): 1–1340. Bibcode:2008PhLB..667....1P. doi:10.1016/j.physletb.2008.07.018.

[17] D. Decamp; Deschizeaux, B.; Lees, J.-P.; Minard, M.-N.; Crespo, J.M.; Delfino, M.; Fernandez, E.; Martinez, M. et al. (1989). "Determination of the number of light neutrino species". *Physics Letters B* **231** (4): 519. Bibcode:1989PhLB..231..519D. doi:10.1016/0370-2693(89)90704-1.
A. Fisher (1991). "Searching for the Beginning of Time: Cosmic Connection". *Popular Science* **238** (4): 70.
J.D. Barrow (1997) [1994]. "The Singularity and Other Problems". *The Origin of the Universe* (Reprint ed.). Basic Books. ISBN 978-0-465-05314-8.

[18] D.H. Perkins (2003). *Particle Astrophysics*. Oxford University Press. p. 4. ISBN 0-19-850952-9.

[19]M. Gell-Mann (1964). "A Schematic Model of Baryons and Mesons".*Physics Letters***8**(3): 214–215.Bibcode:1964PhL. doi:10.1016/S0031-9163(64)92001-3.

[20] G. Zweig (1964). "An SU(3) Model for Strong Interaction Symmetry and its Breaking" (PDF). *CERN Report No.8182/TH.401*.

[21] G. Zweig (1964). "An SU(3) Model for Strong Interaction Symmetry and its Breaking: II" (PDF). *CERN Report No.8419/TH.412*.

[22] M. Gell-Mann (2000) [1964]. "The Eightfold Way: A theory of strong interaction symmetry". In M. Gell-Mann, Y. Ne'eman. *The Eightfold Way*. Westview Press. p. 11. ISBN 0-7382-0299-1.
Original: M. Gell-Mann (1961). "The Eightfold Way: A theory of strong interaction symmetry". *Synchrotron Laboratory Report CTSL-20* (California Institute of Technology).

[23] Y. Ne'eman (2000) [1964]. "Derivation of strong interactions from gauge invariance". In M. Gell-Mann, Y. Ne'eman. *The Eightfold Way*. Westview Press. ISBN 0-7382-0299-1.
Original Y. Ne'eman (1961). "Derivation of strong interactions from gauge invariance".*Nuclear Physics***26**(2): 222.BibN.doi: 10.1016/0029-5582(61)90134-1.

[24] R.C. Olby, G.N. Cantor (1996). *Companion to the History of Modern Science*. Taylor & Francis. p. 673. ISBN 0-415-14578-3.

[25] A. Pickering (1984). *Constructing Quarks*. University of Chicago Press. pp. 114–125. ISBN 0-226-66799-5.

[26]B.J. Bjorken, S.L. Glashow; Glashow (1964). "Elementary Particles and SU(4)".*Physics Letters***11**(3): 255–257.Bibcode:1964P. doi:10.1016/0031-9163(64)90433-0.

[27] J.I. Friedman. "The Road to the Nobel Prize". Hue University. Retrieved 2008-09-29.

[28]R.P. Feynman (1969). "Very High-Energy Collisions of Hadrons".*Physical Review Letters***23**(24): 1415–1417.Bibcode:19. doi:10.1103/PhysRevLett.23.1415.

[29] S. Kretzer et al. (2004). "CTEQ6 Parton Distributions with Heavy Quark Mass Effects". *Physical Review D* **69** (11): 114005. arXiv:hep-ph/0307022. Bibcode:2004PhRvD..69k4005K. doi:10.1103/PhysRevD.69.114005.

[30] D.J. Griffiths (1987). *Introduction to Elementary Particles*. John Wiley & Sons. p. 42. ISBN 0-471-60386-4.

[31] M.E. Peskin, D.V. Schroeder (1995). *An introduction to quantum field theory*. Addison–Wesley. p. 556. ISBN 0-201-50397-2.

[32] V.V. Ezhela (1996). *Particle physics*. Springer. p. 2. ISBN 1-56396-642-5.

[33] S.L. Glashow, J. Iliopoulos, L. Maiani; Iliopoulos; Maiani (1970). "Weak Interactions with Lepton–Hadron Symmetry". *Physical Review D* **2** (7): 1285–1292. Bibcode:1970PhRvD...2.1285G. doi:10.1103/PhysRevD.2.1285.

[34] D.J. Griffiths (1987). *Introduction to Elementary Particles*. John Wiley & Sons. p. 44. ISBN 0-471-60386-4.

[35] M. Kobayashi, T. Maskawa; Maskawa (1973). "CP-Violation in the Renormalizable Theory of Weak Interaction". *Progress of Theoretical Physics* **49** (2): 652–657. Bibcode:1973PThPh..49..652K. doi:10.1143/PTP.49.652.

[36] H. Harari (1975). "A new quark model for hadrons". *Physics Letters B* **57B** (3): 265. Bibcode:1975PhLB...57..265H. doi:10.1016/0370-2693(75)90072-6.

[37] K.W. Staley (2004). *The Evidence for the Top Quark*. Cambridge University Press. pp. 31–33. ISBN 978-0-521-82710-2.

[38] S.W. Herb et al. (1977). "Observation of a Dimuon Resonance at 9.5 GeV in 400-GeV Proton-Nucleus Collisions". *Physical Review Letters* **39** (5): 252. Bibcode:1977PhRvL..39..252H. doi:10.1103/PhysRevLett.39.252.

[39] M. Bartusiak (1994). *A Positron named Priscilla*. National Academies Press. p. 245. ISBN 0-309-04893-1.

[40] F. Abe (CDF Collaboration) et al. (1995). "Observation of Top Quark Production in pp Collisions with the Collider Detector at Fermilab". *Physical Review Letters* **74** (14): 2626–2631. Bibcode:1995PhRvL..74.2626A. doi:10.1103/PhysRevLett.74.2626. PMID 10057978.

[41] S. Abachi (DØ Collaboration) et al. (1995). "Search for High Mass Top Quark Production in pp Collisions at $\sqrt{s}$ = 1.8 TeV". *Physical Review Letters* **74** (13): 2422–2426. Bibcode:1995PhRvL..74.2422A. doi:10.1103/PhysRevLett.74.2422.

[42] K.W. Staley (2004). *The Evidence for the Top Quark*. Cambridge University Press. p. 144. ISBN 0-521-82710-8.

[43] "New Precision Measurement of Top Quark Mass". Brookhaven National Laboratory News. 2004. Retrieved 2013-11-03.

[44] J. Joyce (1982) [1939]. *Finnegans Wake*. Penguin Books. p. 383. ISBN 0-14-006286-6.

[45] M. Gell-Mann (1995). *The Quark and the Jaguar: Adventures in the Simple and the Complex*. Henry Holt and Co. p. 180. ISBN 978-0-8050-7253-2.

[46] J. Gleick (1992). *Genius: Richard Feynman and modern physics*. Little Brown and Company. p. 390. ISBN 0-316-90316-7.

[47] J.J. Sakurai (1994). S.F Tuan, ed. *Modern Quantum Mechanics* (Revised ed.). Addison–Wesley. p. 376. ISBN 0-201-53929-2.

[48] D.H. Perkins (2000). *Introduction to high energy physics*. Cambridge University Press. p. 8. ISBN 0-521-62196-8.

[49] M. Riordan (1987). *The Hunting of the Quark: A True Story of Modern Physics*. Simon & Schuster. p. 210. ISBN 978-0-671-50466-3.

[50] F. Close (2006). *The New Cosmic Onion*. CRC Press. p. 133. ISBN 1-58488-798-2.

[51] J.T. Volk et al. (1987). "Letter of Intent for a Tevatron Beauty Factory" (PDF). Fermilab Proposal #783.

[52] G. Fraser (2006). *The New Physics for the Twenty-First Century*. Cambridge University Press. p. 91. ISBN 0-521-81600-9.

[53] "The Standard Model of Particle Physics". BBC. 2002. Retrieved 2009-04-19.

[54] F. Close (2006). *The New Cosmic Onion*. CRC Press. pp. 80–90. ISBN 1-58488-798-2.

[55] D. Lincoln (2004). *Understanding the Universe*. World Scientific. p. 116. ISBN 981-238-705-6.

[56] "Weak Interactions". *Virtual Visitor Center*. Stanford Linear Accelerator Center. 2008. Retrieved 2008-09-28.

[57] K. Nakamura et al. (2010). "Review of Particles Physics: The CKM Quark-Mixing Matrix" (PDF). *J. Phys. G* **37** (75021): 150.

[58] Z. Maki, M. Nakagawa, S. Sakata (1962). "Remarks on the Unified Model of Elementary Particles". *Progress of Theoretical Physics* **28** (5): 870. Bibcode:1962PThPh..28..870M. doi:10.1143/PTP.28.870.

[59] B.C. Chauhan, M. Picariello, J. Pulido, E. Torrente-Lujan (2007). "Quark–lepton complementarity, neutrino and standard model data predict θPMNS
13 = 9°+1°
−2°". *European Physical Journal* **C50** (3): 573–578. arXiv:hep-ph/0605032. Bibcode:2007EPJC...50..573C. doi:
007-0212-z.

[60] R. Nave. "The Color Force". *HyperPhysics*. Georgia State University, Department of Physics and Astronomy. Retrieved 2009-04-26.

[61] B.A. Schumm (2004). *Deep Down Things*. Johns Hopkins University Press. pp. 131–132. ISBN 0-8018-7971-X. OCLC 55229065.

[62] Part III of M.E. Peskin, D.V. Schroeder (1995). *An Introduction to Quantum Field Theory*. Addison–Wesley. ISBN 0-201-50397-2.

[63] V. Icke (1995). *The force of symmetry*. Cambridge University Press. p. 216. ISBN 0-521-45591-X.

[64] M.Y. Han (2004). *A story of light*. World Scientific. p. 78. ISBN 981-256-034-3.

[65] C. Sutton. "Quantum chromodynamics (physics)". *Encyclopædia Britannica Online*. Retrieved 2009-05-12.

[66] A. Watson (2004). *The Quantum Quark*. Cambridge University Press. pp. 285–286. ISBN 0-521-82907-0.

[67] K.A. Olive *et al.* (Particle Data Group), Chin. Phys. **C38**, 090001 (2014) (URL: http://pdg.lbl.gov)

[68] W. Weise, A.M. Green (1984). *Quarks and Nuclei*. World Scientific. pp. 65–66. ISBN 9971-966-61-1.

[69] D. McMahon (2008). *Quantum Field Theory Demystified*. McGraw–Hill. p. 17. ISBN 0-07-154382-1.

[70] S.G. Roth (2007). *Precision electroweak physics at electron–positron colliders*. Springer. p. VI. ISBN 3-540-35164-7.

[71] R.P. Feynman (1985). *QED: The Strange Theory of Light and Matter* (1st ed.). Princeton University Press. pp. 136–137. ISBN 0-691-08388-6.

[72] M. Veltman (2003). *Facts and Mysteries in Elementary Particle Physics*. World Scientific. pp. 45–47. ISBN 981-238-149-X.

[73] F. Wilczek, B. Devine (2006). *Fantastic Realities*. World Scientific. p. 85. ISBN 981-256-649-X.

[74] F. Wilczek, B. Devine (2006). *Fantastic Realities*. World Scientific. pp. 400ff. ISBN 981-256-649-X.

[75] T. Yulsman (2002). *Origin*. CRC Press. p. 55. ISBN 0-7503-0765-X.

[76] F. Garberson (2008). "Top Quark Mass and Cross Section Results from the Tevatron". arXiv:0808.0273 [hep-ex].

[77] J. Steinberger (2005). *Learning about Particles*. Springer. p. 130. ISBN 3-540-21329-5.

[78] C.-Y. Wong (1994). *Introduction to High-energy Heavy-ion Collisions*. World Scientific. p. 149. ISBN 981-02-0263-6.

[79] S.B. Rüester, V. Werth, M. Buballa, I.A. Shovkovy, D.H. Rischke; Werth; Buballa; Shovkovy; Rischke (2005). "The phase diagram of neutral quark matter: Self-consistent treatment of quark masses". *Physical Review D* **72** (3): 034003. arXiv:hep-ph/0503184. Bibcode:2005PhRvD..72c4004R. doi:10.1103/PhysRevD.72.034004.

[80] M.G. Alford, K. Rajagopal, T. Schaefer, A. Schmitt; Schmitt; Rajagopal; Schäfer (2008). "Color superconductivity in dense quark matter". *Reviews of Modern Physics* **80** (4): 1455–1515. arXiv:0709.4635. Bibcode:2008RvMP...80.1455A. doi:

[81]S. Mrowczynski (1998). "Quark–Gluon Plasma".*Acta Physica Polonica B***29**: 3711.arXiv:nucl-th/9905005.Bibcode:.

[82] Z. Fodor, S.D. Katz; Katz (2004). "Critical point of QCD at finite T and μ, lattice results for physical quark masses". *Journal of High Energy Physics* **2004** (4): 50. arXiv:hep-lat/0402006. Bibcode:2004JHEP...04..050F. doi:10.1088/1126-6708/2004/04/050.

[83] U. Heinz, M. Jacob (2000). "Evidence for a New State of Matter: An Assessment of the Results from the CERN Lead Beam Programme". arXiv:nucl-th/0002042.

[84] "RHIC Scientists Serve Up "Perfect" Liquid". Brookhaven National Laboratory News. 2005. Retrieved 2009-05-22.

[85] T. Yulsman (2002). *Origins: The Quest for Our Cosmic Roots*. CRC Press. p. 75. ISBN 0-7503-0765-X.

[86] A. Sedrakian, J.W. Clark, M.G. Alford (2007). *Pairing in fermionic systems*. World Scientific. pp. 2–3. ISBN 981-256-907-3.

5.9 Further reading

- A. Ali, G. Kramer; Kramer (2011). "JETS and QCD: A historical review of the discovery of the quark and gluon jets and its impact on QCD".*European Physical Journal H***36**(2): 245.arXiv:1012.2288.Bibcode:2011E.doi: 10.1140/epjh/e2011-10047-1.
- D.J. Griffiths (2008). *Introduction to Elementary Particles* (2nd ed.). Wiley–VCH. ISBN 3-527-40601-8.
- I.S. Hughes (1985). *Elementary particles* (2nd ed.). Cambridge University Press. ISBN 0-521-26092-2.
- R. Oerter (2005). *The Theory of Almost Everything: The Standard Model, the Unsung Triumph of Modern Physics.* Pi Press. ISBN 0-13-236678-9.
- A. Pickering (1984). *Constructing Quarks: A Sociological History of Particle Physics.* The University of Chicago Press. ISBN 0-226-66799-5.
- B. Povh (1995). *Particles and Nuclei: An Introduction to the Physical Concepts.* Springer–Verlag. ISBN 0-387-59439-6.
- M. Riordan (1987). *The Hunting of the Quark: A true story of modern physics.* Simon & Schuster. ISBN 0-671-64884-5.
- B.A. Schumm (2004). *Deep Down Things: The Breathtaking Beauty of Particle Physics.* Johns Hopkins University Press. ISBN 0-8018-7971-X.

5.10 External links

- 1969 Physics Nobel Prize lecture by Murray Gell-Mann
- 1976 Physics Nobel Prize lecture by Burton Richter
- 1976 Physics Nobel Prize lecture by Samuel C.C. Ting
- 2008 Physics Nobel Prize lecture by Makoto Kobayashi
- 2008 Physics Nobel Prize lecture by Toshihide Maskawa
- The Top Quark And The Higgs Particle by T.A. Heppenheimer – A description of CERN's experiment to count the families of quarks.
- Bowley, Roger; Copeland, Ed. "Quarks". *Sixty Symbols.* Brady Haran for the University of Nottingham.

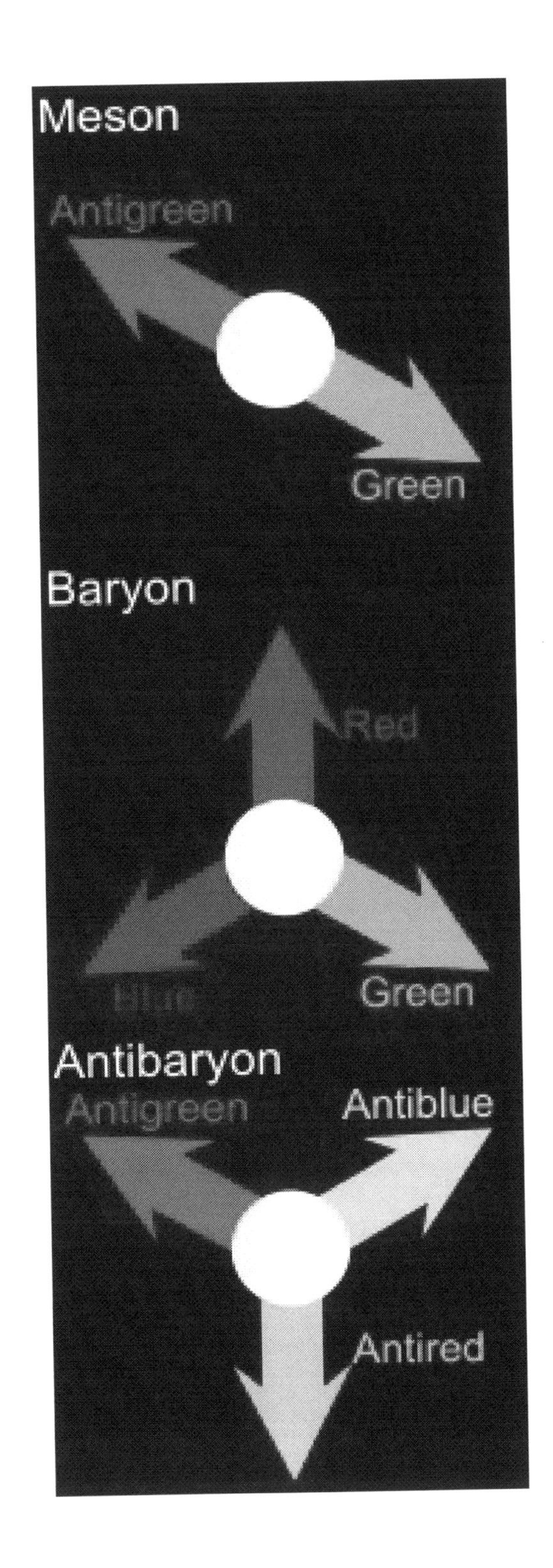

All types of hadrons have zero total color charge.

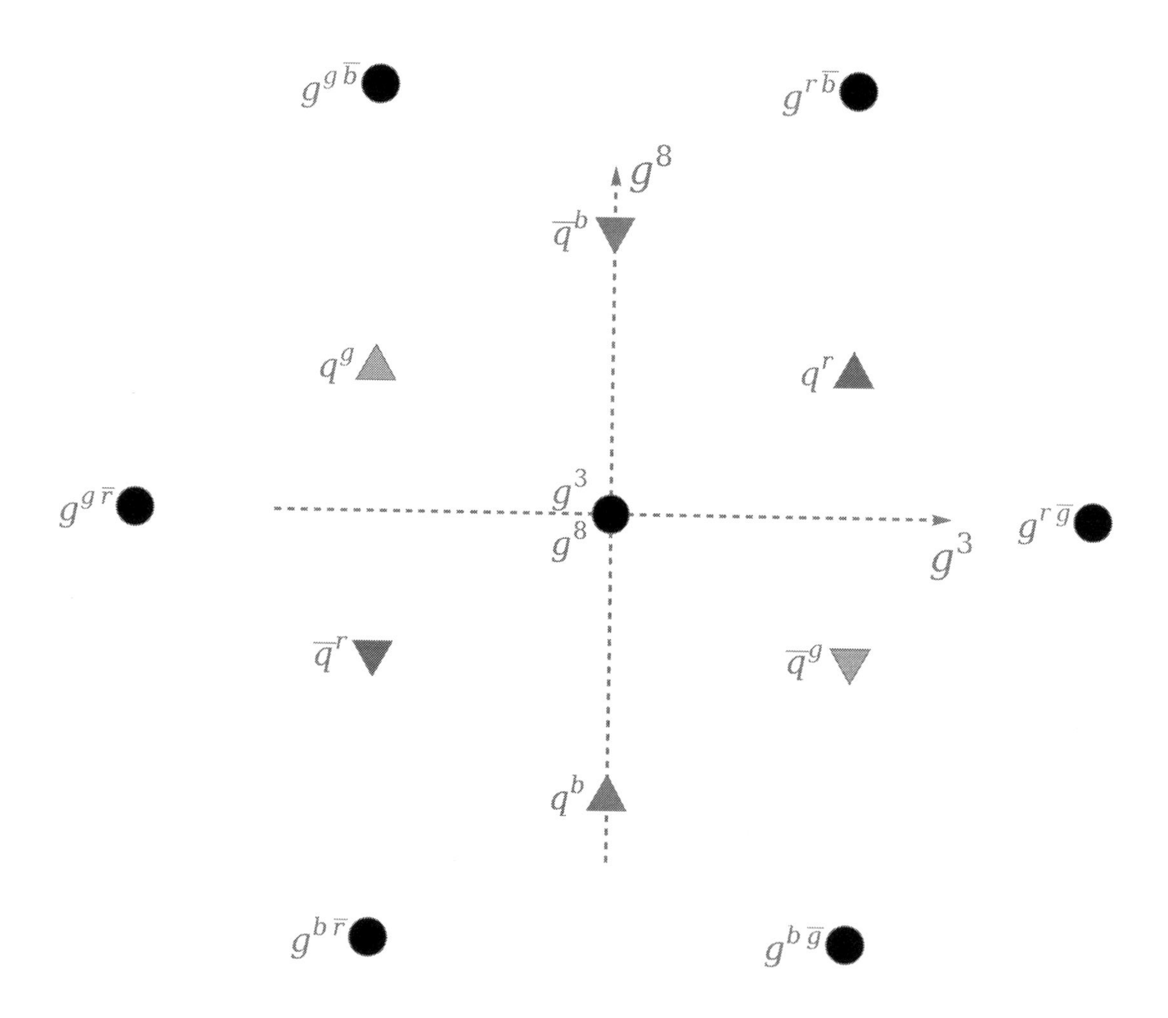

The pattern of strong charges for the three colors of quark, three antiquarks, and eight gluons (with two of zero charge overlapping).

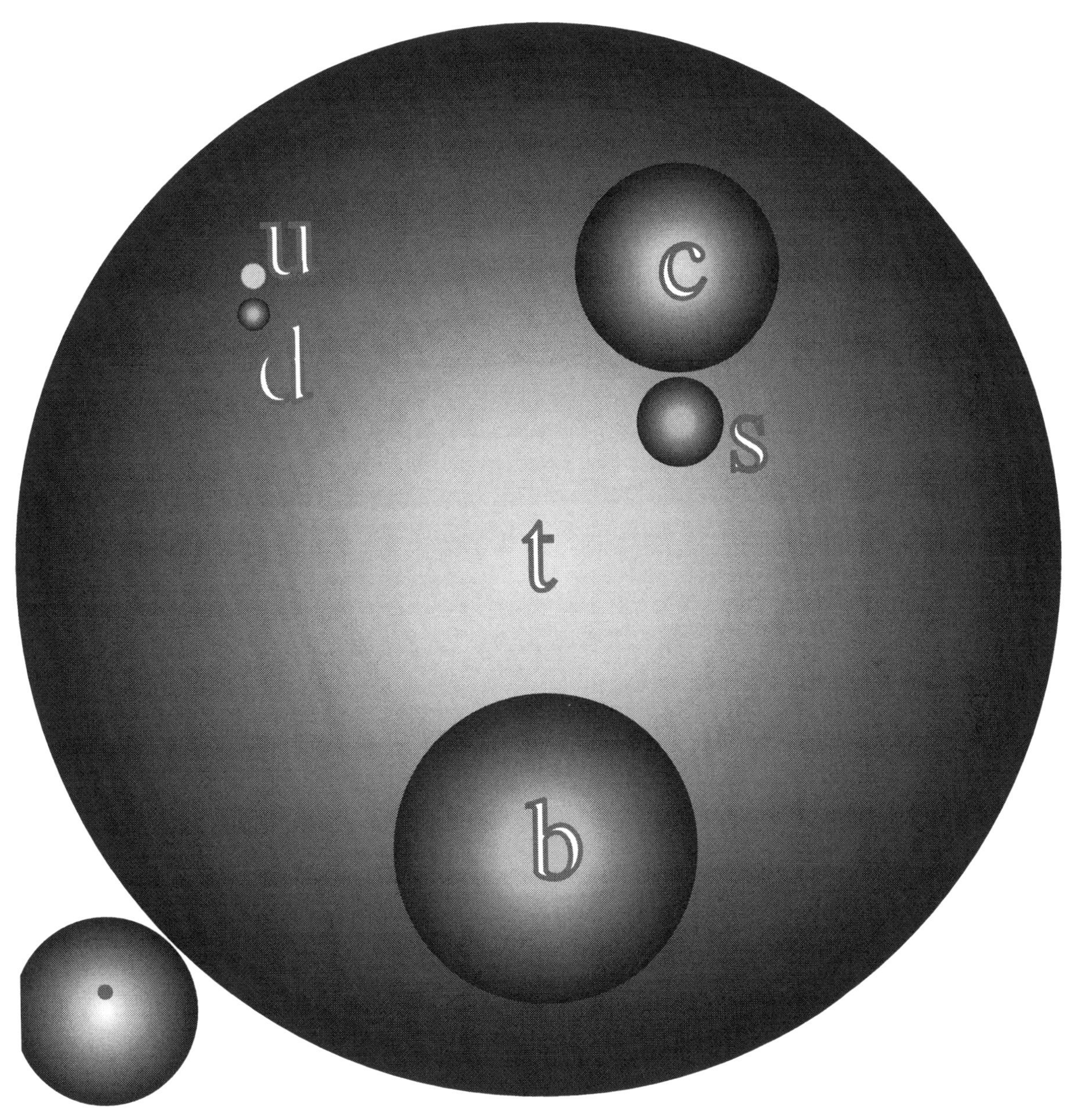

Current quark masses for all six flavors in comparison, as balls of proportional volumes. Proton and electron (red) are shown in bottom left corner for scale

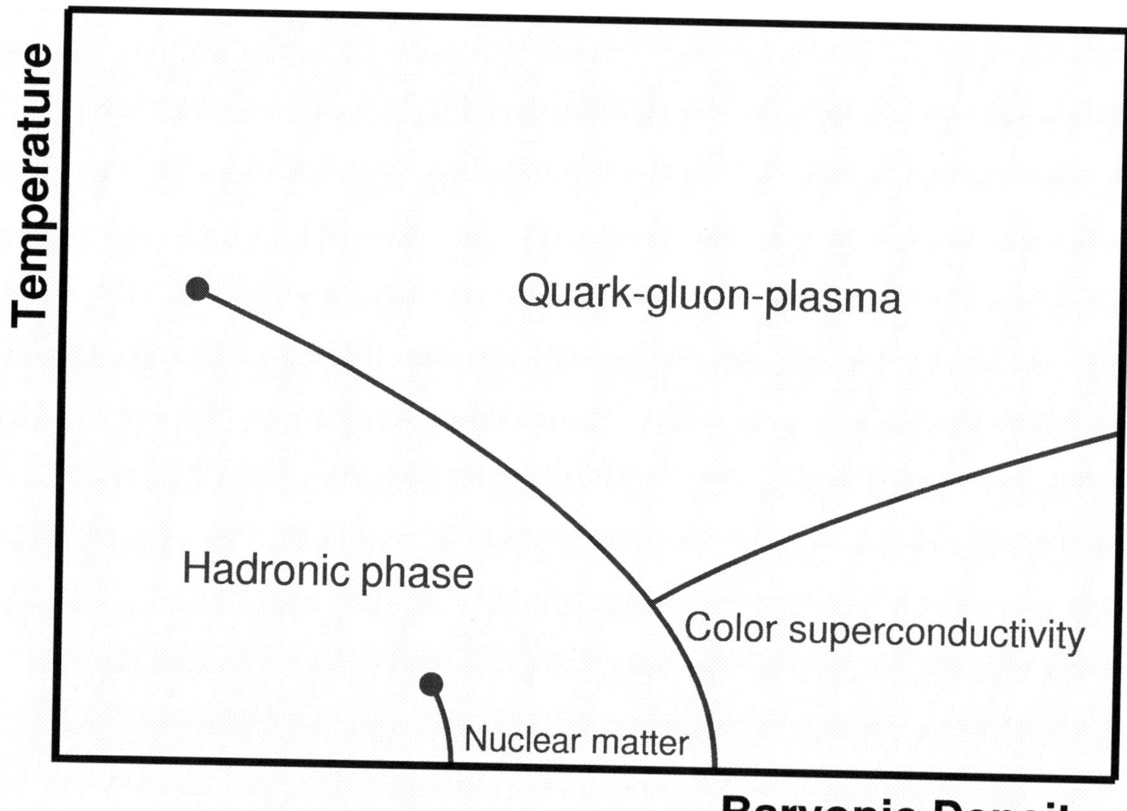

A qualitative rendering of the phase diagram of quark matter. The precise details of the diagram are the subject of ongoing research.[79][80]

Dedication

This book is dedicated to

The University of Michigan
- all of it -

Contents

Chapter 6

Up quark

The **up quark** or **u quark** (symbol: *u*) is the lightest of all quarks, a type of elementary particle, and a major constituent of matter. It, along with the down quark, forms the neutrons (one up quark, two down quarks) and protons (two up quarks, one down quark) of atomic nuclei. It is part of the first generation of matter, has an electric charge of $+\frac{2}{3}\,e$ and a bare mass of 1.8–3.0 MeV/c^2. Like all quarks, the up quark is an elementary fermion with spin-$\frac{1}{2}$, and experiences all four fundamental interactions: gravitation, electromagnetism, weak interactions, and strong interactions. The antiparticle of the up quark is the **up antiquark** (sometimes called *antiup quark* or simply *antiup*), which differs from it only in that some of its properties have equal magnitude but opposite sign.

Its existence (along with that of the down and strange quarks) was postulated in 1964 by Murray Gell-Mann and George Zweig to explain the *Eightfold Way* classification scheme of hadrons. The up quark was first observed by experiments at the Stanford Linear Accelerator Center in 1968.

6.1 History

In the beginnings of particle physics (first half of the 20th century), hadrons such as protons, neutrons and pions were thought to be elementary particles. However, as new hadrons were discovered, the 'particle zoo' grew from a few particles in the early 1930s and 1940s to several dozens of them in the 1950s. The relationships between each of them were unclear until 1961, when Murray Gell-Mann[2] and Yuval Ne'eman[3] (independently of each other) proposed a hadron classification scheme called the *Eightfold Way*, or in more technical terms, SU(3) flavor symmetry.

This classification scheme organized the hadrons into isospin multiplets, but the physical basis behind it was still unclear. In 1964, Gell-Mann[4] and George Zweig[5][6] (independently of each other) proposed the quark model, then consisting only of up, down, and strange quarks.[7] However, while the quark model explained the Eightfold Way, no direct evidence of the existence of quarks was found until 1968 at the Stanford Linear Accelerator Center.[8][9] Deep inelastic scattering experiments indicated that protons had substructure, and that protons made of three more-fundamental particles explained the data (thus confirming the quark model).[10]

At first people were reluctant to describe the three bodies as quarks, instead preferring Richard Feynman's parton description,[11][12][13] but over time the quark theory became accepted (see *November Revolution*).[14]

6.2 Mass

Despite being extremely common, the bare mass of the up quark is not well determined, but probably lies between 1.8 and 3.0 MeV/c^2.[1] Lattice QCD calculations give a more precise value: 2.01±0.14 MeV/c^2.[15]

When found in mesons (particles made of one quark and one antiquark) or baryons (particles made of three quarks), the 'effective mass' (or 'dressed' mass) of quarks becomes greater because of the binding energy caused by the gluon field between each quark (see mass–energy equivalence).The bare mass of up quarks is so light, it cannot be straightforwardly calculated because relativistic effects have to be taken into account.

6.3 See also

- Down quark
- Isospin
- Quark model
- Quantum Mechanics

6.4 References

[1] J. Beringer *et al.* (Particle Data Group) (2012). "PDGLive Particle Summary 'Quarks (u, d, s, c, b, t, b', t', Free)'" (PDF). Particle Data Group. Retrieved 2013-02-21.

[2] M. Gell-Mann (2000) [1964]. "The Eightfold Way: A theory of strong interaction symmetry". In M. Gell-Mann, Y. Ne'eman. *The Eightfold Way*. Westview Press. p. 11. ISBN 0-7382-0299-1.
Original: M. Gell-Mann (1961). "The Eightfold Way: A theory of strong interaction symmetry". *Synchrotron Laboratory Report CTSL-20* (California Institute of Technology)

[3] Y. Ne'eman (2000) [1964]. "Derivation of strong interactions from gauge invariance". In M. Gell-Mann, Y. Ne'eman. *The Eightfold Way*. Westview Press. ISBN 0-7382-0299-1.
Original Y.Ne'eman(1961). "Derivation of strong interactions from gauge invariance".*Nuclear Physics***26**(2):(61)90134-1.

[4] M.Gell-Mann(1964). "A Schematic Model of Baryons and Mesons".*Physics Letters***8**(3): 214–215. Bibcode:1964PhL..... doi:10.1016/S0031-9163(64)92001-3.

[5] G. Zweig (1964). "An SU(3) Model for Strong Interaction Symmetry and its Breaking". *CERN Report No.8181/Th 8419*.

[6] G. Zweig (1964). "An SU(3) Model for Strong Interaction Symmetry and its Breaking: II". *CERN Report No.8419/Th 8412*.

[7] B. Carithers, P. Grannis (1995). "Discovery of the Top Quark" (PDF). *Beam Line* (SLAC) **25** (3): 4–16. Retrieved 2008-09-23.

[8] E. D. Bloom; Coward, D.; Destaebler, H.; Drees, J.; Miller, G.; Mo, L.; Taylor, R.; Breidenbach, M. et al. (1969). "High-Energy Inelastic *e–p* Scattering at 6° and 10°". *Physical Review Letters* **23** (16): 930–934. Bibcode:1969PhRvL..23..930B. doi:10.1103/PhysRevLett.23.930.

[9] M. Breidenbach; Friedman, J.; Kendall, H.; Bloom, E.; Coward, D.; Destaebler, H.; Drees, J.; Mo, L.; Taylor, R. et al. (1969). "Observed Behavior of Highly Inelastic Electron–Proton Scattering". *Physical Review Letters* **23** (16): 935–939. Bibcode:1969PhRvL..23..935B. doi:10.1103/PhysRevLett.23.935.

[10] J. I. Friedman. "The Road to the Nobel Prize". Hue University. Retrieved 2008-09-29.

[11] R.P.Feynman(1969). "Very High-Energy Collisions of Hadrons".*Physical Review Letters***23**(24): 1415–1417. Bibcode:19F. doi:10.1103/PhysRevLett.23.1415.

[12] S. Kretzer; Lai, H.; Olness, Fredrick; Tung, W. et al. (2004). "CTEQ6 Parton Distributions with Heavy Quark Mass Effects". *Physical Review D***69**(11): 114005. arXiv:hep-ph/0307022. Bibcode:2004PhRvD..69k4005K.doi:10.1103/PhysRevD.69.5.

[13] D. J. Griffiths (1987). *Introduction to Elementary Particles*. John Wiley & Sons. p. 42. ISBN 0-471-60386-4.

[14] M. E. Peskin, D. V. Schroeder (1995). *An introduction to quantum field theory*. Addison–Wesley. p. 556. ISBN 0-201-50397-2.

[15] Cho, Adrian (April 2010). "Mass of the Common Quark Finally Nailed Down". Science Magazine.

6.5 Further reading

- A. Ali, G. Kramer; Kramer (2011). "JETS and QCD: A historical review of the discovery of the quark and gluon jets and its impact on QCD". *European Physical Journal H* **36** (2): 245. A. doi:10.1140/epjh/e2011-10047-1.
- R. Nave. "Quarks". *HyperPhysics*. Georgia State University, Department of Physics and Astronomy. Retrieved 2008-06-29.
- A. Pickering (1984). *Constructing Quarks*. University of Chicago Press. pp. 114–125. ISBN 0-226-66799-5.

Chapter 7

Down quark

The **down quark** or **d quark** (symbol: d) is the second-lightest of all quarks, a type of elementary particle, and a major constituent of matter. Together with the up quark, it forms the neutrons (one up quark, two down quarks) and protons (two up quarks, one down quark) of atomic nuclei. It is part of the first generation of matter, has an electric charge of $-\frac{1}{3}$ e and a bare mass of 4.8+0.5
−0.3 MeV/c^2.[1] Like all quarks, the down quark is an elementary fermion with spin-$\frac{1}{2}$, and experiences all four fundamental interactions: gravitation, electromagnetism, weak interactions, and strong interactions. The antiparticle of the down quark is the **down antiquark** (sometimes called *antidown quark* or simply *antidown*), which differs from it only in that some of its properties have equal magnitude but opposite sign.

Its existence (along with that of the up and strange quarks) was postulated in 1964 by Murray Gell-Mann and George Zweig to explain the *Eightfold Way* classification scheme of hadrons. The down quark was first observed by experiments at the Stanford Linear Accelerator Center in 1968.

7.1 History

In the beginnings of particle physics (first half of the 20th century), hadrons such as protons, neutrons, and pions were thought to be elementary particles. However, as new hadrons were discovered, the 'particle zoo' grew from a few particles in the early 1930s and 1940s to several dozens of them in the 1950s. The relationships between each of them was unclear until 1961, when Murray Gell-Mann[2] and Yuval Ne'eman[3] (independently of each other) proposed a hadron classification scheme called the *Eightfold Way*, or in more technical terms, SU(3) flavor symmetry.

This classification scheme organized the hadrons into isospin multiplets, but the physical basis behind it was still unclear. In 1964, Gell-Mann[4] and George Zweig[5][6] (independently of each other) proposed the quark model, then consisting only of up, down, and strange quarks.[7] However, while the quark model explained the Eightfold Way, no direct evidence of the existence of quarks was found until 1968 at the Stanford Linear Accelerator Center.[8][9] Deep inelastic scattering experiments indicated that protons had substructure, and that protons made of three more-fundamental particles explained the data (thus confirming the quark model).[10]

At first people were reluctant to identify the three-bodies as quarks, instead preferring Richard Feynman's parton description,[but over time the quark theory became accepted(see*November Revolution*).[14]

7.2 Mass

Despite being extremely common, the bare mass of the down quark is not well determined, but probably lies between 4.5 and 5.310^0 MeV/c^2.[1] Lattice QCD calculations give a more precise value: 4.79±0.16 MeV/c^2.[15]

When found in mesons (particles made of one quark and one antiquark) or baryons (particles made of three quarks), the 'effective mass' (or 'dressed' mass) of quarks becomes greater because of the binding energy caused by the gluon field between quarks (see mass–energy equivalence). For example, the effective mass of down quarks in a proton is around 330 MeV/c^2. Because the bare mass of down quarks is so small, it cannot be straightforwardly calculated because relativistic effects have to be taken into account.

7.3 See also

- Up quark
- Isospin
- Quark model

7.4 References

[1] J. Beringer (Particle Data Group) et al. (2013). "PDGLive Particle Summary 'Quarks (u, d, s, c, b, t, b′, t′, Free)'" (PDF). Particle Data Group. Retrieved 2013-07-23.

[2] M. Gell-Mann (2000) [1964]. "The Eightfold Way: A theory of strong interaction symmetry". In M. Gell-Mann, Y. Ne'eman. *The Eightfold Way*. Westview Press. p. 11. ISBN 0-7382-0299-1.
Original: M. Gell-Mann (1961). "The Eightfold Way: A theory of strong interaction symmetry". *Synchrotron Laboratory Report CTSL-20* (California Institute of Technology).

[3] Y. Ne'eman (2000) [1964]. "Derivation of strong interactions from gauge invariance". In M. Gell-Mann, Y. Ne'eman. *The Eightfold Way*. Westview Press. ISBN 0-7382-0299-1.
Original Y. Ne'eman (1961). "Derivation of strong interactions from gauge invariance". *Nuclear Physics* **26** (2): 222. B 2N.doi:10.1016/0029-5582(61)90134-1.

[4] M. Gell-Mann (1964). "A Schematic Model of Baryons and Mesons". *Physics Letters* **8** (3): 214–215. Bibcode:1964PhL.....8..214G. doi:10.1016/S0031-9163(64)92001-3.

[5] G. Zweig (1964). "An SU(3) Model for Strong Interaction Symmetry and its Breaking". *CERN Report No.8181/Th 8419*.

[6] G. Zweig (1964). "An SU(3) Model for Strong Interaction Symmetry and its Breaking: II". *CERN Report No.8419/Th 8412*.

[7] B. Carithers, P. Grannis (1995). "Discovery of the Top Quark" (PDF). *Beam Line* (SLAC) **25** (3): 4–16. Retrieved 2008-09-23.

[8] E. D. Bloom; Coward, D.; Destaebler, H.; Drees, J.; Miller, G.; Mo, L.; Taylor, R.; Breidenbach, M. et al. (1969). "High-Energy Inelastic *e–p* Scattering at 6° and 10°". *Physical Review Letters* **23** (16): 930–934. Bibcode:1969PhRvL..23..930B. doi:10.1103/PhysRevLett.23.930.

[9] M. Breidenbach; Friedman, J.; Kendall, H.; Bloom, E.; Coward, D.; Destaebler, H.; Drees, J.; Mo, L.; Taylor, R. et al. (1969). "Observed Behavior of Highly Inelastic Electron–Proton Scattering". *Physical Review Letters* **23** (16): 935–939. Bibcode:1969PhRvL..23..935B. doi:10.1103/PhysRevLett.23.935.

[10] J. I. Friedman. "The Road to the Nobel Prize". Hue University. Retrieved 2008-09-29.

[11] R.P.Feynman(1969). "Very High-Energy Collisions of Hadrons".*Physical Review Letters***23**(24): 1415–1417. B doi:10.1103/PhysRevLett.23.1415.

[12] S. Kretzer; Lai, H.; Olness, Fredrick; Tung, W. et al. (2004). "CTEQ6 Parton Distributions with Heavy Quark Mass Effects". *Physical Review D* **69** (11): 114005. arXiv:hep-ph/0307022. Bibcode:2004PhRvD..69k4005K. doi:10.1103/PhysRevD.69.114005.

[13] D. J. Griffiths (1987). *Introduction to Elementary Particles*. John Wiley & Sons. p. 42. ISBN 0-471-60386-4.

[14] M. E. Peskin, D. V. Schroeder (1995). *An introduction to quantum field theory*. Addison–Wesley. p. 556. ISBN 0-201-50397-2.

[15] Cho, Adrian (April 2010). "Mass of the Common Quark Finally Nailed Down". Science Magazine.

7.5 Further reading

- A. Ali, G. Kramer; Kramer (2011). "JETS and QCD: A historical review of the discovery of the quark and gluon jets and its impact on QCD". *European Physical Journal H* **36** (2): 245. arXiv:1012.2288. Bibcode:2011EPJH... .doi:10.1140/epjh/e2011-10047-1.
- R. Nave. "Quarks". *HyperPhysics*. Georgia State University, Department of Physics and Astronomy. Retrieved 2008-06-29.
- A. Pickering (1984). *Constructing Quarks*. University of Chicago Press. pp. 114–125. ISBN 0-226-66799-5.

Chapter 8

Charm quark

The **charm quark** or **c quark** (from its symbol, c) is the third most massive of all quarks, a type of elementary particle. Charm quarks are found in hadrons, which are subatomic particles made of quarks. Example of hadrons containing charm quarks include the J/ψ meson (J/ψ), D mesons (D), charmed Sigma baryons (Σ
c), and other charmed particles.

It, along with the strange quark is part of the second generation of matter, and has an electric charge of $+\frac{2}{3}\,e$ and a bare mass of 1.29+0.05
−0.11 GeV/c^2.[1] Like all quarks, the charm quark is an elementary fermion with spin-$\frac{1}{2}$, and experiences all four fundamental interactions: gravitation, electromagnetism, weak interactions, and strong interactions. The antiparticle of the charm quark is the **charm antiquark** (sometimes called *anticharm quark* or simply *anticharm*), which differs from it only in that some of its properties have equal magnitude but opposite sign.

The existence of a fourth quark had been speculated by a number of authors around 1964 (for instance by James Bjorken and Sheldon Glashow[4]), but its prediction is usually credited to Sheldon Glashow, John Iliopoulos and Luciano Maiani in 1970 (see GIM mechanism).[5] The first charmed particle (a particle containing a charm quark) to be discovered was the J/ψ meson. It was discovered by a team at the Stanford Linear Accelerator Center (SLAC), led by Burton Richter,[6] and one at the Brookhaven National Laboratory (BNL), led by Samuel Ting.[7]

The 1974 discovery of the J/ψ (and thus the charm quark) ushered in a series of breakthroughs which are collectively known as the *November Revolution*.

8.1 Hadrons containing charm quarks

Main articles: List of baryons and list of mesons

Some of the hadrons containing charm quarks include:

- D mesons contain a charm quark (or its antiparticle) and an up or down quark.
- D
  s mesons contain a charm quark and a strange quark.
- There are many charmonium states, for example the J/ψ particle. These consist of a charm quark and its antiparticle.
- Charmed baryons have been observed, and are named in analogy with strange baryons (e.g. Λ+
  c).

8.2 See also

- Quark model

8.3 Notes

[1] K. Nakamura *et al.* (Particle Data Group) (2011). "PDGLive Particle Summary 'Quarks (u, d, s, c, b, t, b′, t′, Free)'" (PDF). Particle Data Group. Retrieved 2011-08-08.

[2] Carl Rod Nave. "Transformation of Quark Flavors by the Weak Interaction". Retrieved 2010-12-06. The c quark has about 5% probability of decaying into a d quark instead of an s quark.

[3] K. Nakamura et al. (2010). "Review of Particles Physics: The CKM Quark-Mixing Matrix" (PDF). *J. Phys. G* **37** (75021): 150.

[4] B.J.Bjorken,S.L.Glashow;Glashow(1964). "Elementary particles and SU(4)".*Physics Letters***11**(3): 255–257. BibcB. doi:10.1016/0031-9163(64)90433-0.

[5] S.L. Glashow, J. Iliopoulos, L. Maiani; Iliopoulos; Maiani (1970). "Weak Interactions with Lepton–Hadron Symmetry". *Physical Review D* **2** (7): 1285–1292. Bibcode:1970PhRvD...2.1285G. doi:10.1103/PhysRevD.2.1285.

[6] J.-E. Augustin; Boyarski, A.; Breidenbach, M.; Bulos, F.; Dakin, J.; Feldman, G.; Fischer, G.; Fryberger, D.; Hanson, G.; Jean-Marie, B.; Larsen, R.; Lüth, V.; Lynch, H.; Lyon, D.; Morehouse, C.; Paterson, J.; Perl, M.; Richter, B.; Rapidis, P.; Schwitters, R.; Tanenbaum, W.; Vannucci, F.; Abrams, G.; Briggs, D.; Chinowsky, W.; Friedberg, C.; Goldhaber, G.; Hollebeek, R.; Kadyk, J.; Lulu, B. (1974). "Discovery of a Narrow Resonance in e^+e^- Annihilation". *Physical Review Letters* **33** (23): 1406. Bibcode:1974PhRvL..33.1406A. doi:10.1103/PhysRevLett.33.1406.

[7] J.J.Aubert et al. (1974). "Experimental Observation of a Heavy Particle*J*".*Physical Review Letters***33**(23): 1404. BibcodeA. doi:10.1103/PhysRevLett.33.1404.

8.4 Further reading

- R. Nave. "Quarks". *HyperPhysics*. Georgia State University, Department of Physics and Astronomy. Retrieved 2008-06-29.

- A. Pickering (1984). *Constructing Quarks*. University of Chicago Press. pp. 114–125. ISBN 0-226-66799-5.

Chapter 9

Strange quark

The **strange quark** or **s quark** (from its symbol, *s*) is the third-lightest of all quarks, a type of elementary particle. Strange quarks are found in subatomic particles called hadrons. Example of hadrons containing strange quarks include kaons (K), strange D mesons (D
s), Sigma baryons (Σ), and other strange particles.

Along with the charm quark, it is part of the second generation of matter, and has an electric charge of $-\frac{1}{3}$ *e* and a bare mass of 95+5
−5 MeV/c^2.[1] Like all quarks, the strange quark is an elementary fermion with spin-½,and experiences all 4 fundamental interactions: gravitation, electromagnetism, weak interactions, and strong interactions. The antiparticle of the strange quark is the **strange antiquark** (sometimes called *antistrange quark* or simply *antistrange*), which differs from it only in that some of its properties have equal magnitude but opposite sign.

The first strange particle (a particle containing a strange quark) was discovered in 1947 (kaons), but the existence of the strange quark itself (and that of the up and down quarks) was only postulated in 1964 by Murray Gell-Mann and George Zweig to explain the *Eightfold Way* classification scheme of hadrons. The first evidence for the existence of quarks came in 1968, in deep inelastic scattering experiments at the Stanford Linear Accelerator Center. These experiments confirmed the existence of up and down quarks, and by extension, strange quarks, as they were required to explain the Eightfold Way.

9.1 History

In the beginnings of particle physics (first half of the 20th century), hadrons such as protons, neutron and pions were thought to be elementary particles. However, new hadrons were discovered, the 'particle zoo' grew from a few particles in the early 1930s and 1940s to several dozens of them in the 1950s. However some particles were much longer lived than others; most particles decayed through the strong interaction and had lifetimes of around 10^{-23} seconds. But when they decayed through the weak interactions, they had lifetimes of around 10^{-10} seconds to decay. While studying these decays Murray Gell-Mann (in 1953)[2][3] and Kazuhiko Nishijima (in 1955)[4] developed the concept of *strangeness* (which Nishijima called *eta-charge*, after the eta meson (η)) which explained the 'strangeness' of the longer-lived particles. The Gell-Mann–Nishijima formula is the result of these efforts to understand strange decays.

However, the relationships between each particles and the physical basis behind the strangeness property was still unclear. In 1961, Gell-Mann[5] and Yuval Ne'eman[6] (independently of each other) proposed a hadron classification scheme called the *Eightfold Way*, or in more technical terms, SU(3) flavor symmetry. This ordered hadrons into isospin multiplets. The physical basis behind both isospin and strangeness was only explained in 1964, when Gell-Mann[7] and George Zweig[8][9] (independently of each other) proposed the quark model, then consisting only of up, down, and strange quarks.[10] Up and down quarks were the carriers of isospin, while the strange quark carried strangeness. While the quark model explained the Eightfold Way, no direct evidence of the existence of quarks was found until 1968 at the Stanford Linear Accelerator Center.[11][12] Deep inelastic scattering experiments indicated that protons had substructure, and that protons made of three more-fundamental particles explained the data (thus confirming the quark model).[13]

At first people were reluctant to identify the three-bodies as quarks, instead preferring Richard Feynman's parton description, but over time the quark theory became accepted(see*November Revolution*).[17]

9.2 See also

- Quark model
- Strange matter
- Strangeness production
- Strangelet
- Strange star

9.3 References

[1] J. Beringer *et al.* (Particle Data Group) (2012). "PDGLive Particle Summary 'Quarks (u, d, s, c, b, t, b′, t′, Free)'" (PDF). Particle Data Group. Retrieved 2012-11-30.

[2] M. Gell-Mann (1953). "Isotopic Spin and New Unstable Particles". *Physical Review* **92** (3): 833. Bibcode:1953PhRv...92..833G. doi:10.1103/PhysRev.92.833.

[3] G. Johnson (2000). *Strange Beauty: Murray Gell-Mann and the Revolution in Twentieth-Century Physics*. Random House. p. 119. ISBN 0-679-43764-9. By the end of the summer... [Gell-Mann] completed his first paper, "Isotopic Spin and Curious Particles" and send it of to *Physical Review*. The editors hated the title, so he amended it to "Strange Particles". They wouldn't go for that either—never mind that almost everybody used the term—suggesting insteand "Isotopic Spin and New Unstable Particles".

[4] K. Nishijima, Kazuhiko (1955). "Charge Independence Theory of V Particles". *Progress of Theoretical Physics* **13** (3): 285. Bibcode:1955PThPh..13..285N. doi:10.1143/PTP.13.285.

[5] M. Gell-Mann (2000) [1964]. "The Eightfold Way: A theory of strong interaction symmetry". In M. Gell-Mann, Y. Ne'eman. *The Eightfold Way*. Westview Press. p. 11. ISBN 0-7382-0299-1.
Original: M. Gell-Mann (1961). "The Eightfold Way: A theory of strong interaction symmetry". *Synchrotron Laboratory Report CTSL-20* (California Institute of Technology)

[6] Y. Ne'eman (2000) [1964]. "Derivation of strong interactions from gauge invariance". In M. Gell-Mann, Y. Ne'eman. *The Eightfold Way*. Westview Press. ISBN 0-7382-0299-1.
Original Y. Ne'eman (1961). "Derivation of strong interactions from gauge invariance". *Nuclear Physics* **26** (2): 222. Bi N.doi:10.1016/0029-5582(61)90134-1.

[7] M. Gell-Mann (1964). "A Schematic Model of Baryons and Mesons". *Physics Letters* **8** (3): 214–215. Bibcode:1964PhL.....8..214G. doi:10.1016/S0031-9163(64)92001-3.

[8] G. Zweig (1964). "An SU(3) Model for Strong Interaction Symmetry and its Breaking". *CERN Report No.8181/Th 8419*.

[9] G. Zweig (1964). "An SU(3) Model for Strong Interaction Symmetry and its Breaking: II". *CERN Report No.8419/Th 8412*.

[10] B. Carithers, P. Grannis (1995). "Discovery of the Top Quark" (PDF). *Beam Line* (SLAC) **25** (3): 4–16. Retrieved 2008-09-23.

[11] E. D. Bloom; Coward, D.; Destaebler, H.; Drees, J.; Miller, G.; Mo, L.; Taylor, R.; Breidenbach, M. et al. (1969). "High-Energy Inelastic *e–p* Scattering at 6° and 10°". *Physical Review Letters* **23** (16): 930–934. Bibcode:1969PhRvL..23..930B. doi:10.1103/PhysRevLett.23.930.

[12] M. Breidenbach; Friedman, J.; Kendall, H.; Bloom, E.; Coward, D.; Destaebler, H.; Drees, J.; Mo, L.; Taylor, R. et al. (1969). "Observed Behavior of Highly Inelastic Electron–Proton Scattering". *Physical Review Letters* **23** (16): 935–939. Bibcode:1969PhRvL..23..935B. doi:10.1103/PhysRevLett.23.935.

[13] J. I. Friedman. "The Road to the Nobel Prize". Hue University. Retrieved 2008-09-29.

[14] R.P.Feynman(1969). "Very High-Energy Collisions of Hadrons".*Physical Review Letters***23**(24): 1415–1417. Bibcode:1969 doi:10.1103/PhysRevLett.23.1415.

[15] S. Kretzer; Lai, H.; Olness, Fredrick; Tung, W. et al. (2004). "CTEQ6 Parton Distributions with Heavy Quark Mass Effects". *Physical Review D***69**(11): 114005. arXiv:hep-th/0307022. Bibcode:2004PhRvD..69k4005K.doi:10.1103/PhysRevD.69.15.

[16] D. J. Griffiths (1987). *Introduction to Elementary Particles*. John Wiley & Sons. p. 42. ISBN 0-471-60386-4.

[17] M. E. Peskin, D. V. Schroeder (1995). *An introduction to quantum field theory*. Addison–Wesley. p. 556. ISBN 0-201-50397-2.

9.4 Further reading

- R. Nave. "Quarks". *HyperPhysics*. Georgia State University, Department of Physics and Astronomy. Retrieved 2008-06-29.
- A. Pickering (1984). *Constructing Quarks*. University of Chicago Press. pp. 114–125. ISBN 0-226-66799-5.

Chapter 10

Top quark

The **top quark**, also known as the **t quark** (symbol: t) or **truth quark**, is an elementary particle and a fundamental constituent of matter. Like all quarks, the top quark is an elementary fermion with spin-$\frac{1}{2}$, and experiences all four fundamental interactions: gravitation, electromagnetism, weak interactions, and strong interactions. It has an electric charge of $+\frac{2}{3}\,e$,[2] and is the most massive of all observed elementary particles. It has a mass of 173.34 ± 0.27 (stat) ± 0.71 (syst)10^0 GeV/c^2,[1] which is about the same mass as an atom of tungsten. The antiparticle of the top quark is the **top antiquark** (symbol: t, sometimes called *antitop quark* or simply *antitop*), which differs from it only in that some of its properties have equal magnitude but opposite sign.

The top quark interacts primarily by the strong interaction, but can only decay through the weak force. It decays to a W boson and either a bottom quark (most frequently), a strange quark, or, on the rarest of occasions, a down quark. The Standard Model predicts its mean lifetime to be roughly 5×10^{-25} s.[3] This is about a twentieth of the timescale for strong interactions, and therefore it does not form hadrons, giving physicists a unique opportunity to study a "bare" quark (all other quarks hadronize, meaning that they combine with other quarks to form hadrons, and can only be observed as such). Because it is so massive, the properties of the top quark allow predictions to be made of the mass of the Higgs boson under certain extensions of the Standard Model (see Mass and coupling to the Higgs boson below). As such, it is extensively studied as a means to discriminate between competing theories.

Its existence (and that of the bottom quark) was postulated in 1973 by Makoto Kobayashi and Toshihide Maskawa to explain the observed CP violations in kaon decay,[4] and was discovered in 1995 by the CDF[5] and DØ[6] experiments at Fermilab. Kobayashi and Maskawa won the 2008 Nobel Prize in Physics for the prediction of the top and bottom quark, which together form the third generation of quarks.[7]

10.1 History

In 1973, Makoto Kobayashi and Toshihide Maskawa predicted the existence of a third generation of quarks to explain observed CP violations in kaon decay.[4] The names top and bottom were introduced by Haim Harari in 1975,[8][9] to match the names of the first generation of quarks (up and down) reflecting the fact that the two were the 'up' and 'down' component of a weak isospin doublet.[10] The top quark was sometimes called *truth quark* in the past, but over time *top quark* became the predominant use.[11]

The proposal of Kobayashi and Maskawa heavily relied on the GIM mechanism put forward by Sheldon Lee Glashow, John Iliopoulos and Luciano Maiani,[12] which predicted the existence of the then still unobserved charm quark. When in November 1974 teams at Brookhaven National Laboratory (BNL) and the Stanford Linear Accelerator Center (SLAC) simultaneously announced the discovery of the J/ψ meson, it was soon after identified as a bound state of the missing charm quark with its antiquark. This discovery allowed the GIM mechanism to become part of the Standard Model.[13] With the acceptance of the GIM mechanism, Kobayashi and Maskawa's prediction also gained in credibility. Their case was further strengthened by the discovery of the tau by Martin Lewis Perl's team at SLAC between 1974 and 1978.[14] This announced a third generation of leptons, breaking the new symmetry between leptons and quarks introduced by the

GIM mechanism. Restoration of the symmetry implied the existence of a fifth and sixth quark.

It was in fact not long until a fifth quark, the bottom, was discovered by the E288 experiment team, led by Leon Lederman at Fermilab in 1977.[15][16][17] This strongly suggested that there must also be a sixth quark, the top, to complete the pair. It was known that this quark would be heavier than the bottom, requiring more energy to create in particle collisions, but the general expectation was that the sixth quark would soon be found. However, it took another 18 years before the existence of the top was confirmed.[18]

Early searches for the top quark at SLAC and DESY (in Hamburg) came up empty-handed. When, in the early eighties, the Super Proton Synchrotron (SPS) at CERN discovered the W boson and the Z boson, it was again felt that the discovery of the top was imminent. As the SPS gained competition from the Tevatron at Fermilab there was still no sign of the missing particle, and it was announced by the group at CERN that the top mass must be at least 41 GeV/c^2. After a race between CERN and Fermilab to discover the top, the accelerator at CERN reached its limits without creating a single top, pushing the lower bound on its mass up to 77 GeV/c^2.[18]

The Tevatron was (until the start of LHC operation at CERN in 2009) the only hadron collider powerful enough to produce top quarks. In order to be able to confirm a future discovery, a second detector, the DØ detector, was added to the complex (in addition to the Collider Detector at Fermilab (CDF) already present). In October 1992, the two groups found their first hint of the top, with a single creation event that appeared to contain the top. In the following years, more evidence was collected and on April 22, 1994, the CDF group submitted their paper presenting tentative evidence for the existence of a top quark with a mass of about 175 GeV/c^2. In the meantime, DØ had found no more evidence than the suggestive event in 1992. A year later, on March 2, 1995, after having gathered more evidence and a reanalysis of the DØ data (who had been searching for a much lighter top), the two groups jointly reported the discovery of the top with a certainty of 99.9998% at a mass of 176±18 GeV/c^2.[5][6][18]

In the years leading up to the top quark discovery, it was realized that certain precision measurements of the electroweak vector boson masses and couplings are very sensitive to the value of the top quark mass. These effects become much larger for higher values of the top mass and therefore could indirectly see the top quark even if it could not be directly produced in any experiment at the time. The largest effect from the top quark mass was on the T parameter and by 1994 the precision of these indirect measurements had led to a prediction of the top quark mass to be between 145 GeV/c^2 and 185 GeV/c^2. It is the development of techniques that ultimately allowed such precision calculations that led to Gerardus 't Hooft and Martinus Veltman winning the Nobel Prize in physics in 1999.[19][20]

10.2 Properties

- At the final Tevatron energy of 1.96 TeV, top–antitop pairs were produced with a cross section of about 7 picobarns (pb).[21] The Standard Model prediction (at next-to-leading order with m_t = 175 GeV/c^2) is 6.7–7.5 pb.
- The W bosons from top quark decays carry polarization from the parent particle, hence pose themselves as a unique probe to top polarization.
- In the Standard Model, the top quark is predicted to have a spin quantum number of ½ and electric charge +⅔. A first measurement of the top quark charge has been published, resulting in approximately 90% confidence limit that the top quark charge is indeed +⅔.[22]

10.3 Production

Because top quarks are very massive, large amounts of energy are needed to create one. The only way to achieve such high energies is through high energy collisions. These occur naturally in the Earth's upper atmosphere as cosmic rays collide with particles in the air, or can be created in a particle accelerator. In 2011, after the Tevatron ceased operations, the Large Hadron Collider at CERN became the only accelerator that generates a beam of sufficient energy to produce top quarks, with a center-of-mass energy of 7 TeV.

There are multiple processes that can lead to the production of a top quark. The most common is production of a top–antitop pair via strong interactions. In a collision, a highly energetic gluon is created, which subsequently decays into a top and antitop. This process was responsible for the majority of the top events at Tevatron and was the process observed when the top was first discovered in 1995.[23] It is also possible to produce pairs of top–antitop through the decay of an intermediate photon or Z-boson. However, these processes are predicted to be much rarer and have a virtually identical experimental signature in a hadron collider like Tevatron.

A distinctly different process is the production of single tops via weak interaction. This can happen in two ways (called channels): either an intermediate W-boson decays into a top and antibottom quark ("s-channel") or a bottom quark (probably created in a pair through the decay of a gluon) transforms to top quark by exchanging a W-boson with an up or down quark ("t-channel"). The first evidence for these processes was published by the DØ collaboration in December 2006,[24] and in March 2009 the CDF[25] and DØ[23] collaborations released twin papers with the definitive observation of these processes. The main significance of measuring these production processes is that their frequency is directly proportional to the $| V_{tb} |^2$ component of the CKM matrix.

10.4 Decay

The only known way that a top quark can decay is through the weak interaction producing a W-boson and a down-type quark (down, strange, or bottom). Because of its enormous mass, the top quark is extremely short-lived with a predicted lifetime of only 5×10^{-25} s.[3] As a result top quarks do not have time to form hadrons before they decay, as other quarks do. This provides physicists with the unique opportunity to study the behavior of a "bare" quark.

In particular, it is possible to directly determine the branching ratio $\Gamma(W^+b) / \Gamma(W^+q\ (q = b,s,d))$. The best current determination of this ratio is 0.91±0.04.[26] Since this ratio is equal to $| V_{tb} |^2$ according to the Standard Model, this gives another way of determining the CKM element $| V_{tb} |$, or in combination with the determination of $| V_{tb} |$ from single top production provides tests for the assumption that the CKM matrix is unitary.[27]

The Standard Model also allows more exotic decays, but only at one loop level, meaning that they are extremely suppressed. In particular, it is possible for a top quark to decay into another up-type quark (an up or a charm) by emitting a photon or a Z-boson.[28] Searches for these exotic decay modes have provided no evidence for their existence in accordance with expectations from the Standard Model. The branching ratios for these decays have been determined to be less than 5.9 in 1,000 for photonic decay and less than 2.1 in 1,000 for Z-boson decay at 95% confidence.[26]

10.5 Mass and coupling to the Higgs boson

The Standard Model describes fermion masses through the Higgs mechanism. The Higgs boson has a Yukawa coupling to the left- and right-handed top quarks. After electroweak symmetry breaking (when the Higgs acquires a vacuum expectation value), the left- and right-handed components mix, becoming a mass term.

$$\mathcal{L} = y_t h q u^c \rightarrow \frac{y_t v}{\sqrt{2}}(1 + h^0/v) u u^c$$

The top quark Yukawa coupling has a value of

$$y_t = \sqrt{2} m_t / v \simeq 1$$

where $v = 246$ GeV is the value of the Higgs vacuum expectation value.

10.5.1 Yukawa couplings

See also: Beta function (physics)

In the Standard Model, all of the quark and lepton Yukawa couplings are small compared to the top quark Yukawa coupling. Understanding this hierarchy in the fermion masses is an open problem in theoretical physics. Yukawa couplings are not constants and their values change depending on the energy scale (distance scale) at which they are measured. The dynamics of Yukawa couplings are determined by the renormalization group equation.

One of the prevailing views in particle physics is that the size of the top quark Yukawa coupling is determined by the renormalization group, leading to the "quasi-infrared fixed point."

The Yukawa couplings of the up, down, charm, strange and bottom quarks, are hypothesized to have small values at the extremely high energy scale of grand unification, 10^{15} GeV. They increase in value at lower energy scales, at which the quark masses are generated by the Higgs. The slight growth is due to corrections from the QCD coupling. The corrections from the Yukawa couplings are negligible for the lower mass quarks.

If, however, a quark Yukawa coupling has a large value at very high energies, its Yukawa corrections will evolve and cancel against the QCD corrections. This is known as a (quasi-) infrared fixed point. No matter what the initial starting value of the coupling is, if it is sufficiently large it will reach this fixed point value. The corresponding quark mass is then predicted.

The top quark Yukawa coupling lies very near the infrared fixed point of the Standard Model. The renormalization group equation is:

$$\mu\frac{\partial}{\partial\mu}y_t \approx \frac{y_t}{16\pi^2}\left(\frac{9}{2}y_t^2 - 8g_3^2 - \frac{9}{4}g_2^2 - \frac{17}{20}g_1^2\right),$$

where g_3 is the color gauge coupling, g_2 is the weak isospin gauge coupling, and g_1 is the weak hypercharge gauge coupling. This equation describes how the Yukawa coupling changes with energy scale μ. Solutions to this equation for large initial values y_t cause the right-hand side of the equation to quickly approach zero, locking y_t to the QCD coupling g_3. The value of the fixed point is fairly precisely determined in the Standard Model, leading to a top quark mass of 230 GeV. However, if there is more than one Higgs doublet, the mass value will be reduced by Higgs mixing angle effects in an unpredicted way.

In the minimal supersymmetric extension of the Standard Model (MSSM), there are two Higgs doublets and the renormalization group equation for the top quark Yukawa coupling is slightly modified:

$$\mu\frac{\partial}{\partial\mu}y_t \approx \frac{y_t}{16\pi^2}\left(6y_t^2 + y_b^2 - \frac{16}{3}g_3^2 - 3g_2^2 - \frac{13}{15}g_1^2\right),$$

where y_b is the bottom quark Yukawa coupling. This leads to a fixed point where the top mass is smaller, 170–200 GeV. The uncertainty in this prediction arises because the bottom quark Yukawa coupling can be amplified in the MSSM. Some theorists believe this is supporting evidence for the MSSM.

The quasi-infrared fixed point has subsequently formed the basis of top quark condensation theories of electroweak symmetry breaking in which the Higgs boson is composite at *extremely* short distance scales, composed of a pair of top and antitop quarks.

10.6 See also

- CDF experiment
- Topness
- Top quark condensate
- Topcolor
- Quark model

10.7 References

[1] The ATLAS, CDF, CMS, D0 Collaborations (2014). "First combination of Tevatron and LHC measurements of the top-quark mass". Retrieved 2014-03-19.

[2] S. Willenbrock (2003). "The Standard Model and the Top Quark". In H.B Prosper and B. Danilov (eds.). *Techniques and Concepts of High-Energy Physics XII*. NATO Science Series **123**. Kluwer Academic. pp. 1–41. arXiv:hep-ph/0211067v3. ISBN 1-4020-1590-9.

[3] A.Quadt(2006). "Top quark physics at hadron colliders".*European Physical Journal C***48**(3): 835–1000. Bibcode:2006EPJC...4. doi:10.1140/epjc/s2006-02631-6.

[4] M. Kobayashi, T. Maskawa (1973). "*CP*-Violation in the Renormalizable Theory of Weak Interaction". *Progress of Theoretical Physics* **49** (2): 652. Bibcode:1973PThPh..49..652K. doi:10.1143/PTP.49.652.

[5] F. Abe *et al.* (CDF Collaboration) (1995). "Observation of Top Quark Production in pp Collisions with the Collider Detector at Fermilab". *Physical Review Letters* **74** (14): 2626–2631. Bibcode:1995PhRvL..74.2626A. doi:10.1103/PhysRevLett.74.2626. PMID 10057978.

[6] S. Abachi *et al.* (DØ Collaboration) (1995). "Search for High Mass Top Quark Production in pp Collisions at $\sqrt{s}$ = 1.8 TeV". *Physical Review Letters* **74** (13): 2422–2426. Bibcode:1995PhRvL..74.2422A. doi:10.1103/PhysRevLett.74.2422.

[7] "2008 Nobel Prize in Physics". The Nobel Foundation. 2008. Retrieved 2009-09-11.

[8] H.Harari(1975). "A new quark model for hadrons".*Physics Letters B***57**(3): 265. Bibcode:1975PhLB...57..265H.doi:10.1016-2693(75)90072-6.

[9] K.W. Staley (2004). *The Evidence for the Top Quark*. Cambridge University Press. pp. 31–33. ISBN 978-0-521-82710-2.

[10] D.H. Perkins (2000). *Introduction to high energy physics*. Cambridge University Press. p. 8. ISBN 0-521-62196-8.

[11] F. Close (2006). *The New Cosmic Onion*. CRC Press. p. 133. ISBN 1-58488-798-2.

[12] S.L. Glashow, J. Iliopoulous, L. Maiani (1970). "Weak Interactions with Lepton–Hadron Symmetry". *Physical Review D* **2** (7): 1285–1292. Bibcode:1970PhRvD...2.1285G. doi:10.1103/PhysRevD.2.1285.

[13] A. Pickering (1999). *Constructing Quarks: A Sociological History of Particle Physics*. University of Chicago Press. pp. 253–254. ISBN 978-0-226-66799-7.

[14] M.L. Perl et al. (1975). "Evidence for Anomalous Lepton Production in e+e– Annihilation". *Physical Review Letters* **35** (22): 1489. Bibcode:1975PhRvL..35.1489P. doi:10.1103/PhysRevLett.35.1489.

[15] "Discoveries at Fermilab – Discovery of the Bottom Quark" (Press release). Fermilab. 7 August 1977. Retrieved 2009-07-24.

[16] L.M. Lederman (2005). "Logbook: Bottom Quark". *Symmetry Magazine* **2** (8).

[17] S.W. Herb et al. (1977). "Observation of a Dimuon Resonance at 9.5 GeV in 400-GeV Proton-Nucleus Collisions". *Physical Review Letters* **39** (5): 252. Bibcode:1977PhRvL..39..252H. doi:10.1103/PhysRevLett.39.252.

[18] T.M. Liss, P.L. Tipton (1997). "The Discovery of the Top Quark" (PDF). *Scientific American*: 54–59.

[19] "The Nobel Prize in Physics 1999". The Nobel Foundation. Retrieved 2009-09-10.

[20] "The Nobel Prize in Physics 1999, Press Release" (Press release). The Nobel Foundation. 12 October 1999. Retrieved 2009-09-10.

[21] D. Chakraborty (DØ and CDF collaborations) (2002). *Top quark and W/Z results from the Tevatron* (PDF). Rencontres de Moriond. p. 26.

[22] V.M. Abazov *et al.* (DØ Collaboration) (2007). "Experimental discrimination between charge 2*e*/3 top quark and charge 4*e*/3 exotic quark production scenarios". *Physical Review Letters* **98** (4): 041801. arXiv:hep-ex/0608044. Bibcode:2007Ph A.doi:10.1103/PhysRevLett.98.041801. PMID17358756.

[23] V.M. Abazov *et al.* (DØ Collaboration) (2009). "Observation of Single Top Quark Production". *Physical Review Letters* **103** (9). arXiv:0903.0850. Bibcode:2009PhRvL.103i2001A. doi:10.1103/PhysRevLett.103.092001.

[24] V.M. Abazov *et al.* (DØ Collaboration) (2007). "Evidence for production of single top quarks and first direct measurement of $|V_{tb}|$". *Physical Review Letters* **98**(18): 181802. arXiv:hep-ex/0612052. Bibcode:2007PhRvL..98r1802A.doi:10.1103/P 2.PMID17501561.

[25] T. Aaltonen *et al.* (CDF Collaboration) (2009). "First Observation of Electroweak Single Top Quark Production". *Physical Review Letters* **103** (9). arXiv:0903.0885. Bibcode:2009PhRvL.103i2002A. doi:10.1103/PhysRevLett.103.092002.

[26] J. Beringer *et al.* (Particle Data Group) (2012). "PDGLive Particle Summary 'Quarks (u, d, s, c, b, t, b', t', Free)'" (PDF). Particle Data Group. Retrieved 2013-07-23.

[27] V.M. Abazov *et al.* (DØ Collaboration) (2008). "Simultaneous measurement of the ratio B(t→Wb)/B(t→Wq) and the top-quark pair production cross section with the DØ detector at $\sqrt{s}$ = 1.96 TeV". *Physical Review Letters* **100** (19): 192003. arXiv:0801.1326. Bibcode:2008PhRvL.100s2003A. doi:10.1103/PhysRevLett.100.192003.

[28] S. Chekanov *et al.* (ZEUS Collaboration) (2003). "Search for single-top production in ep collisions at HERA". *Physics Letters B* **559** (3–4): 153. arXiv:hep-ex/0302010. Bibcode:2003PhLB..559..153Z. doi:10.1016/S0370-2693(03)00333-2.

10.8 Further reading

- Frank Fiedler; for the D0; CDF Collaborations (June 2005). "Top Quark Production and Properties at the Tevatron". arXiv:hep-ex/0506005 [hep-ex].
- R. Nave. "Quarks". *HyperPhysics*. Georgia State University, Department of Physics and Astronomy. Retrieved 2008-06-29.
- A. Pickering (1984). *Constructing Quarks*. University of Chicago Press. pp. 114–125. ISBN 0-226-66799-5.

10.9 External links

- Top quark on arxiv.org
- Tevatron Electroweak Working Group
- Top quark information on Fermilab website
- Logbook pages from CDF and DZero collaborations' top quark discovery
- Scientific American article on the discovery of the top quark
- Public Homepage of Top Quark Analysis Results from DØ Collaboration at Fermilab
- Public Homepage of Top Quark Analysis Results from CDF Collaboration at Fermilab
- Harvard Magazine article about the 1994 top quark discovery
- 1999 Nobel Prize in Physics

Chapter 11

Bottom quark

The **bottom quark** or **b quark**, also known as the **beauty quark**, is a third-generation quark with a charge of $-\frac{1}{3}\,e$. Although all quarks are described in a similar way by quantum chromodynamics, the bottom quark's large bare mass (around 4.2 GeV/c^2,[3] a bit more than four times the mass of a proton), combined with low values of the CKM matrix elements V_{ub} and V_{cb}, gives it a distinctive signature that makes it relatively easy to identify experimentally (using a technique called B-tagging). Because three generations of quark are required for CP violation (see CKM matrix), mesons containing the bottom quark are the easiest particles to use to investigate the phenomenon; such experiments are being performed at the BaBar, Belle and LHCb experiments. The bottom quark is also notable because it is a product in almost all top quark decays, and is a frequent decay product for the Higgs boson.

The bottom quark was theorized in 1973 by physicists Makoto Kobayashi and Toshihide Maskawa to explain CP violation.[1] The name "bottom" was introduced in 1975 by Haim Harari.[4][5] The bottom quark was discovered in 1977 by the Fermilab E288 experiment team led by Leon M. Lederman, when collisions produced bottomonium.[2][6][7] Kobayashi and Maskawa won the 2008 Nobel Prize in Physics for their explanation of CP-violation.[8][9] On its discovery, there were efforts to name the bottom quark "beauty", but "bottom" became the predominant usage.

The bottom quark can decay into either an up quark or charm quark via the weak interaction. Both these decays are suppressed by the CKM matrix, making lifetimes of most bottom particles (~10^{-12} s) somewhat higher than those of charmed particles (~10^{-13} s), but lower than those of strange particles (from ~10^{-10} to ~10^{-8} s).

11.1 Hadrons containing bottom quarks

Main articles: list of baryons and list of mesons

Some of the hadrons containing bottom quarks include:

- B mesons contain a bottom quark (or its antiparticle) and an up or down quark.
- B
  c and B
  s mesons contain a bottom quark along with a charm quark or strange quark respectively.
- There are many bottomonium states, for example the ϒ meson and $\chi_b(3P)$, the first particle discovered in LHC. These consist of a bottom quark and its antiparticle.
- Bottom baryons have been observed, and are named in analogy with strange baryons (e.g. Λ0
  b).

11.2 See also

- Quark model

11.3 References

[1] M. Kobayashi; T. Maskawa (1973). "CP-Violation in the Renormalizable Theory of Weak Interaction". *Progress of Theoretical Physics* **49** (2): 652–657. Bibcode:1973PThPh..49..652K. doi:10.1143/PTP.49.652.

[2] "Discoveries at Fermilab – Discovery of the Bottom Quark" (Press release). Fermilab. 7 August 1977. Retrieved 2009-07-24.

[3] J. Beringer (Particle Data Group) et al. (2012). "PDGLive Particle Summary 'Quarks (u, d, s, c, b, t, b′, t′, Free)'" (PDF). Particle Data Group. Retrieved 2012-12-18.

[4] H.Harari(1975). "A new quark model for hadrons".*Physics Letters B*57(3): 265. Bibcode:1975PhLB...57..265H.doi:10.1-2693(75)90072-6.

[5] K.W. Staley (2004). *The Evidence for the Top Quark*. Cambridge University Press. pp. 31–33. ISBN 978-0-521-82710-2.

[6] L.M. Lederman (2005). "Logbook: Bottom Quark". *Symmetry Magazine* **2** (8).

[7] S.W. Herb; Hom, D.; Lederman, L.; Sens, J.; Snyder, H.; Yoh, J.; Appel, J.; Brown, B.; Brown, C.; Innes, W.; Ueno, K.; Yamanouchi, T.; Ito, A.; Jöstlein, H.; Kaplan, D.; Kephart, R. et al. (1977). "Observation of a Dimuon Resonance at 9.5 GeV in 400-GeV Proton-Nucleus Collisions". *Physical Review Letters* **39** (5): 252. Bibcode:1977PhRvL..39..252H. doi:10.1103/PhysRevLett.39.252.

[8] 2008 Physics Nobel Prize lecture by Makoto Kobayashi

[9] 2008 Physics Nobel Prize lecture by Toshihide Maskawa

11.4 Further reading

- L. Lederman (1978). "The Upsilon Particle". *Scientific American* **239** (4): 72. doi:10.1038/scientificamerican1078-72.
- R. Nave. "Quarks". *HyperPhysics*. Georgia State University, Department of Physics and Astronomy. Retrieved 2008-06-29.
- A. Pickering (1984). *Constructing Quarks*. University of Chicago Press. pp. 114–125. ISBN 0-226-66799-5.
- J. Yoh (1997). "The Discovery of the b Quark at Fermilab in 1977: The Experiment Coordinator's Story" (PDF). *Proceedings of Twenty Beautiful Years of Bottom Physics*. Fermilab. Retrieved 2009-07-24.

11.5 External links

- History of the discovery of the bottom quark / Upsilon meson

Chapter 12

Lepton

For other uses, see Lepton (disambiguation).

A **lepton** is an elementary, half-integer spin (spin ½) particle that does not undergo strong interactions, but is subject to the Pauli exclusion principle.[1] The best known of all leptons is the electron, which is directly tied to all chemical properties. Two main classes of leptons exist: charged leptons (also known as the *electron-like* leptons), and neutral leptons (better known as neutrinos). Charged leptons can combine with other particles to form various composite particles such as atoms and positronium, while neutrinos rarely interact with anything, and are consequently rarely observed.

There are six types of leptons, known as *flavours*, forming three *generations*.[2] The first generation is the *electronic leptons*, comprising the electron (e−) and electron neutrino (ν
e); the second is the *muonic leptons*, comprising the muon (μ−) and muon neutrino (ν
μ); and the third is the *tauonic leptons*, comprising the tau (τ−) and the tau neutrino (ν
τ). Electrons have the least mass of all the charged leptons. The heavier muons and taus will rapidly change into electrons through a process of particle decay: the transformation from a higher mass state to a lower mass state. Thus electrons are stable and the most common charged lepton in the universe, whereas muons and taus can only be produced in high energy collisions (such as those involving cosmic rays and those carried out in particle accelerators).

Leptons have various intrinsic properties, including electric charge, spin, and mass. Unlike quarks however, leptons are not subject to the strong interaction, but they are subject to the other three fundamental interactions: gravitation, electromagnetism (excluding neutrinos, which are electrically neutral), and the weak interaction. For every lepton flavor there is a corresponding type of antiparticle, known as antilepton, that differs from the lepton only in that some of its properties have equal magnitude but opposite sign. However, according to certain theories, neutrinos may be their own antiparticle, but it is not currently known whether this is the case or not.

The first charged lepton, the electron, was theorized in the mid-19th century by several scientists[3][4][5] and was discovered in 1897 by J. J. Thomson.[6] The next lepton to be observed was the muon, discovered by Carl D. Anderson in 1936, which was classified as a meson at the time.[7] After investigation, it was realized that the muon did not have the expected properties of a meson, but rather behaved like an electron, only with higher mass. It took until 1947 for the concept of "leptons" as a family of particle to be proposed.[8] The first neutrino, the electron neutrino, was proposed by Wolfgang Pauli in 1930 to explain certain characteristics of beta decay.[8] It was first observed in the Cowan–Reines neutrino experiment conducted by Clyde Cowan and Frederick Reines in 1956.[8][9] The muon neutrino was discovered in 1962 by Leon M. Lederman, Melvin Schwartz and Jack Steinberger,[10] and the tau discovered between 1974 and 1977 by Martin Lewis Perl and his colleagues from the Stanford Linear Accelerator Center and Lawrence Berkeley National Laboratory.[11] The tau neutrino remained elusive until July 2000, when the DONUT collaboration from Fermilab announced its discovery.[12][13]

Leptons are an important part of the Standard Model. Electrons are one of the components of atoms, alongside protons and neutrons. Exotic atoms with muons and taus instead of electrons can also be synthesized, as well as lepton–antilepton particles such as positronium.

12.1 Etymology

The name *lepton* comes from the Greek λεπτός *leptós*, "fine, small, thin" (neuter form: λεπτόν *leptón*);[14][15] the earliest attested form of the word is the Mycenaean Greek ?????, *re-po-to*, written in Linear B syllabic script.[16] *Lepton* was first used by physicist Léon Rosenfeld in 1948:[17]

> Following a suggestion of Prof. C. Møller, I adopt — as a pendant to "nucleon" — the denomination "lepton" (from λεπτός, small, thin, delicate) to denote a particle of small mass.

The etymology incorrectly implies that all the leptons are of small mass. When Rosenfeld named them, the only known leptons were electrons and muons, which are in fact of small mass — the mass of an electron (0.511 MeV/c^2)[18] and the mass of a muon (with a value of 105.7 MeV/c^2)[19] are fractions of the mass of the "heavy" proton (938.3 MeV/c^2).[20] However, the mass of the tau (discovered in the mid 1970s) (1777 MeV/c^2)[21] is nearly twice that of the proton, and about 3,500 times that of the electron.

12.2 History

See also: Electron § Discovery, Muon § History and Tau (particle) § History

The first lepton identified was the electron, discovered by J.J. Thomson and his team of British physicists in 1897.[22][23]

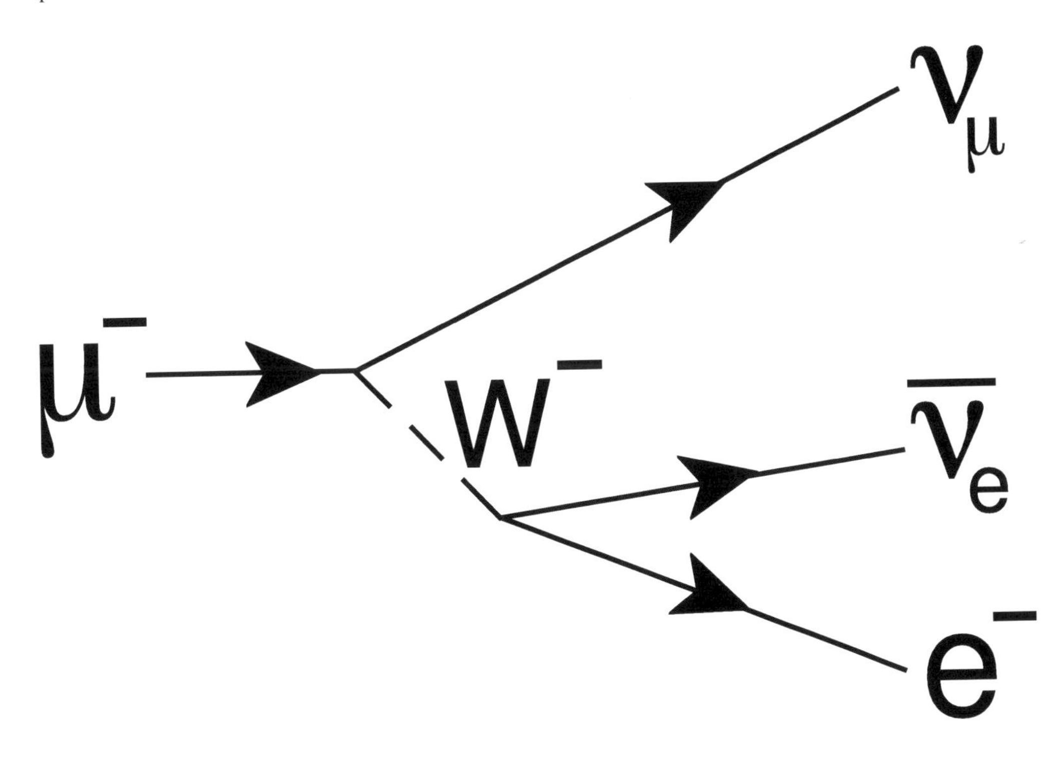

A muon transmutes into a muon neutrino by emitting a W− boson. The W− boson subsequently decays into an electron and an electron antineutrino.

Then in 1930 Wolfgang Pauli postulated the electron neutrino to preserve conservation of energy, conservation of momentum, and conservation of angular momentum in beta decay.[24] Pauli theorized that an undetected particle was carrying away the difference between the energy, momentum, and angular momentum of the initial and observed final particles. The electron neutrino was simply called the neutrino, as it was not yet known that neutrinos came in different flavours (or different "generations").

Nearly 40 years after the discovery of the electron, the muon was discovered by Carl D. Anderson in 1936. Due to its mass, it was initially categorized as a meson rather than a lepton.[25] It later became clear that the muon was much more similar to the electron than to mesons, as muons do not undergo the strong interaction, and thus the muon was reclassified: electrons, muons, and the (electron) neutrino were grouped into a new group of particles – the leptons. In 1962 Leon M. Lederman, Melvin Schwartz and Jack Steinberger showed that more than one type of neutrino exists by first detecting interactions of the muon neutrino, which earned them the 1988 Nobel Prize, although by then the different flavours of neutrino had already been theorized.[26]

The tau was first detected in a series of experiments between 1974 and 1977 by Martin Lewis Perl with his colleagues at the SLAC LBL group.[27] Like the electron and the muon, it too was expected to have an associated neutrino. The first evidence for tau neutrinos came from the observation of "missing" energy and momentum in tau decay, analogous to the "missing" energy and momentum in beta decay leading to the discovery of the electron neutrino. The first detection of tau neutrino interactions was announced in 2000 by the DONUT collaboration at Fermilab, making it the latest particle of the Standard Model to have been directly observed,[28] apart from the Higgs boson, which probably has been discovered in 2012.

Although all present data is consistent with three generations of leptons, some particle physicists are searching for a fourth generation. The current lower limit on the mass of such a fourth charged lepton is 100.8 GeV/c^2,[29] while its associated neutrino would have a mass of at least 45.0 GeV/c^2.[30]

12.3 Properties

12.3.1 Spin and chirality

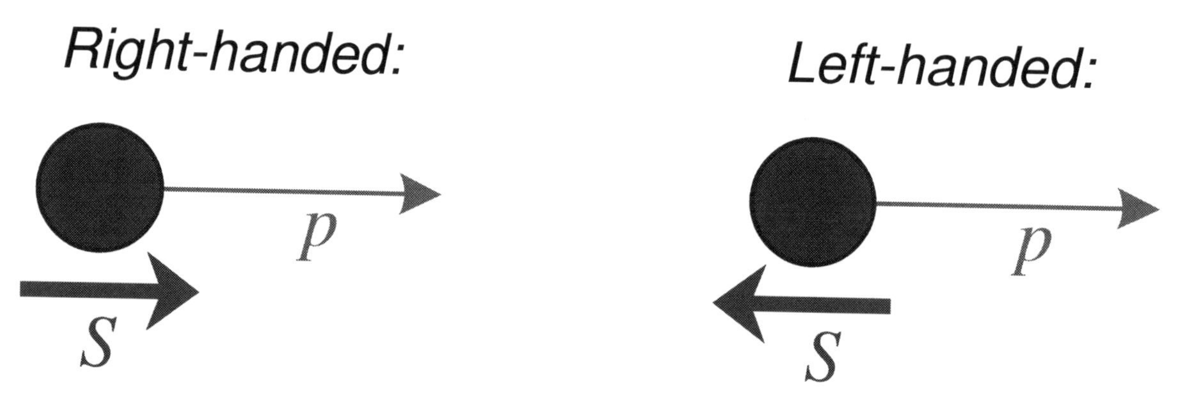

Left-handed and right-handed helicities

Leptons are spin-½ particles. The spin-statistics theorem thus implies that they are fermions and thus that they are subject to the Pauli exclusion principle; no two leptons of the same species can be in exactly the same state at the same time. Furthermore, it means that a lepton can have only two possible spin states, namely up or down.

A closely related property is chirality, which in turn is closely related to a more easily visualized property called helicity. The helicity of a particle is the direction of its spin relative to its momentum; particles with spin in the same direction as their momentum are called *right-handed* and otherwise they are called *left-handed*. When a particle is mass-less, the direction of its momentum relative to its spin is frame independent, while for massive particles it is possible to 'overtake' the particle by a Lorentz transformation flipping the helicity. Chirality is a technical property (defined through the transformation behaviour under the Poincaré group) that agrees with helicity for (approximately) massless particles and is still well defined for massive particles.

In many quantum field theories—such as quantum electrodynamics and quantum chromodynamics—left and right-handed fermions are identical. However in the Standard Model left-handed and right-handed fermions are treated asymmetrically. Only left-handed fermions participate in the weak interaction, while there are no right-handed neutrinos. This is an example of parity violation. In the literature left-handed fields are often denoted by a capital L subscript (e.g. e–L) and right-handed fields are denoted by a capital R subscript.

12.3.2 Electromagnetic interaction

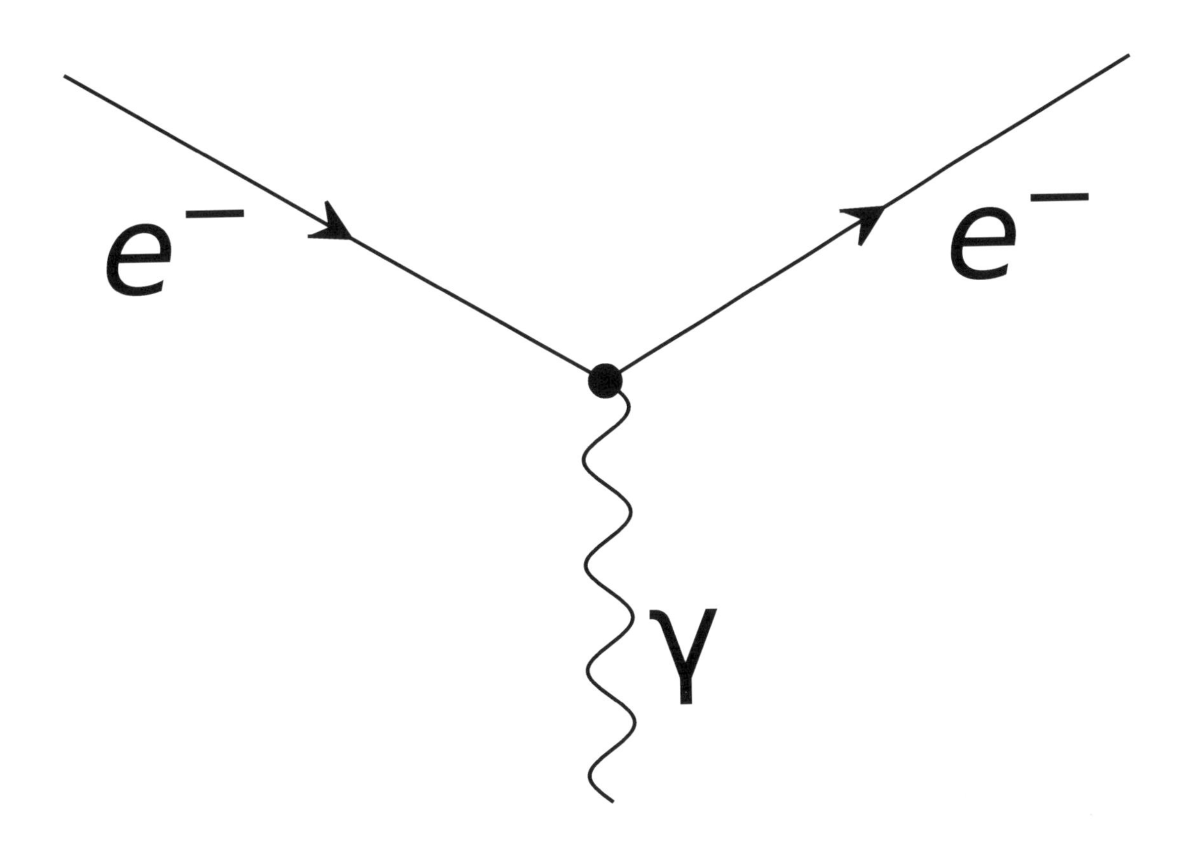

Lepton–photon interaction

One of the most prominent properties of leptons is their electric charge, Q. The electric charge determines the strength of their electromagnetic interactions. It determines the strength of the electric field generated by the particle (see Coulomb's law) and how strongly the particle reacts to an external electric or magnetic field (see Lorentz force). Each generation contains one lepton with $Q = -e$ (conventionally the charge of a particle is expressed in units of the elementary charge) and one lepton with zero electric charge. The lepton with electric charge is commonly simply referred to as a 'charged lepton' while the neutral lepton is called a neutrino. For example the first generation consists of the electron e− with a negative electric charge and the electrically neutral electron neutrino ν
e.

In the language of quantum field theory the electromagnetic interaction of the charged leptons is expressed by the fact that the particles interact with the quantum of the electromagnetic field, the photon. The Feynman diagram of the electron-photon interaction is shown on the right.

Because leptons possess an intrinsic rotation in the form of their spin, charged leptons generate a magnetic field. The size of their magnetic dipole moment μ is given by,

$$\mu = g\frac{Q\hbar}{4m},$$

where m is the mass of the lepton and g is the so-called g-factor for the lepton. First order approximation quantum mechanics predicts that the g-factor is 2 for all leptons. However, higher order quantum effects caused by loops in Feynman diagrams introduce corrections to this value. These corrections, referred to as the anomalous magnetic dipole moment, are very sensitive to the details of a quantum field theory model and thus provide the opportunity for precision tests of the standard model. The theoretical and measured values for the electron anomalous magnetic dipole moment are within agreement within eight significant figures.[31]

12.3.3 Weak Interaction

In the Standard Model the left-handed charged lepton and the left-handed neutrino are arranged in doublet (ν eL, e−L) that transforms in the spinor representation ($T = ½$) of the weak isospin SU(2) gauge symmetry. This means that these particles are eigenstates of the isospin projection T_3 with eigenvalues $½$ and $-½$ respectively. In the meantime, the right-handed charged lepton transforms as a weak isospin scalar ($T = 0$) and thus does not participate in the weak interaction, while there is no right-handed neutrino at all.

The Higgs mechanism recombines the gauge fields of the weak isospin SU(2) and the weak hypercharge U(1) symmetries to three massive vector bosons (W+, W−, Z0) mediating the weak interaction, and one massless vector boson, the photon, responsible for the electromagnetic interaction. The electric charge Q can be calculated from the isospin projection T_3 and weak hypercharge YW through the Gell-Mann–Nishijima formula,

$$Q = T_3 + YW/2$$

To recover the observed electric charges for all particles the left-handed weak isospin doublet (ν eL, e−L) must thus have YW = −1, while the right-handed isospin scalar e−
R must have YW = −2. The interaction of the leptons with the massive weak interaction vector bosons is shown in the figure on the left.

12.3.4 Mass

In the Standard Model each lepton starts out with no intrinsic mass. The charged leptons (i.e. the electron, muon, and tau) obtain an effective mass through interaction with the Higgs field, but the neutrinos remain massless. For technical reasons the masslessness of the neutrinos implies that there is no mixing of the different generations of charged leptons as there is for quarks. This is in close agreement with current experimental observations.[32]

However, it is known from experiments – most prominently from observed neutrino oscillations[33] – that neutrinos do in fact have some very small mass, probably less than 2 eV/c^2.[34] This implies the existence of physics beyond the Standard Model. The currently most favoured extension is the so-called seesaw mechanism, which would explain both why the left-handed neutrinos are so light compared to the corresponding charged leptons, and why we have not yet seen any right-handed neutrinos.

12.3.5 Leptonic numbers

Main article: Lepton number

The members of each generation's weak isospin doublet are assigned leptonic numbers that are conserved under the Standard Model.[35] Electrons and electron neutrinos have an *electronic number* of $L_e = 1$, while muons and muon neutrinos have a *muonic number* of $L\mu = 1$, while tau particles and tau neutrinos have a *tauonic number* of $L\tau = 1$. The antileptons have their respective generation's leptonic numbers of −1.

Conservation of the leptonic numbers means that the number of leptons of the same type remains the same, when particles interact. This implies that leptons and antileptons must be created in pairs of a single generation. For example, the following processes are allowed under conservation of leptonic numbers:

$$e- + e+ \rightarrow \gamma + \gamma,$$

$$\begin{pmatrix} \nu_e \\ e^- \end{pmatrix}, \begin{pmatrix} \nu_\mu \\ \mu^- \end{pmatrix}, \begin{pmatrix} \nu_\tau \\ \tau^- \end{pmatrix}$$

Each generation forms a weak isospin doublet.

τ− + τ+ → Z0 + Z0,

but not these:

γ → e− + μ+,

W− → e− + ν

τ,

Z0 → μ− + τ+.

However, neutrino oscillations are known to violate the conservation of the individual leptonic numbers. Such a violation is considered to be smoking gun evidence for physics beyond the Standard Model. A much stronger conservation law is the conservation of the total number of leptons (L), conserved even in the case of neutrino oscillations, but even it is still violated by a tiny amount by the chiral anomaly.

12.4 Universality

The coupling of the leptons to gauge bosons are flavour-independent (i.e., the interactions between leptons and gauge bosons are the same for all leptons).[35] This property is called *lepton universality* and has been tested in measurements of the tau and muon lifetimes and of Z boson partial decay widths, particularly at the Stanford Linear Collider (SLC) and Large Electron-Positron Collider (LEP) experiments.[36]:241–243[37]:138

The decay rate (Γ) of muons through the process μ− → e− + ν
e + ν
μ is approximately given by an expression of the form (see muon decay for more details)[35]

$$\Gamma\left(\mu^- \to e^- + \bar{\nu_e} + \nu_\mu\right) = K_1 G_F^2 m_\mu^5,$$

where K_1 is some constant, and GF is the Fermi coupling constant. The decay rate of tau particles through the process
τ− → e− + ν
e + ν
τ is given by an expression of the same form[35]

$$\Gamma\left(\tau^- \to e^- + \bar{\nu_e} + \nu_\tau\right) = K_2 G_F^2 m_\tau^5,$$

where K_2 is some constant. Muon–Tauon universality implies that $K_1 = K_2$. On the other hand, electron–muon universality implies[35]

$$\Gamma\left(\tau^- \to e^- + \bar{\nu_e} + \nu_\tau\right) = \Gamma\left(\tau^- \to \mu^- + \bar{\nu_\mu} + \nu_\tau\right).$$

This explains why the branching ratios for the electronic mode (17.85%) and muonic (17.36%) mode of tau decay are equal (within error).[21]

Universality also accounts for the ratio of muon and tau lifetimes. The lifetime of a lepton (τ_l) is related to the decay rate by[35]

$$\tau_l = \frac{B\left(l^- \to e^- + \bar{\nu_e} + \nu_l\right)}{\Gamma\left(l^- \to e^- + \bar{\nu_e} + \nu_l\right)},$$

where $B(x \to y)$ and $\Gamma(x \to y)$ denotes the branching ratios and the resonance width of the process $x \to y$.

The ratio of tau and muon lifetime is thus given by[35]

$$\frac{\tau_\tau}{\tau_\mu} = \frac{B\left(\tau^- \to e^- + \bar{\nu_e} + \nu_\tau\right)}{B\left(\mu^- \to e^- + \bar{\nu_e} + \nu_\mu\right)}\left(\frac{m_\mu}{m_\tau}\right)^5.$$

Using the values of the 2008 *Review of Particle Physics* for the branching ratios of muons[19] and tau[21] yields a lifetime ratio of ~1.29×10^{-7}, comparable to the measured lifetime ratio of ~1.32×10^{-7}. The difference is due to K_1 and K_2 not actually being constants; they depend on the mass of leptons.

12.5 Table of leptons

12.6 See also

- Koide formula
- List of particles
- Preons – hypothetical particles which were once postulated to be subcomponents of quarks and leptons

12.7 Notes

[1] "Lepton (physics)". *Encyclopædia Britannica*. Retrieved 2010-09-29.

[2] R. Nave. "Leptons". *HyperPhysics*. Georgia State University, Department of Physics and Astronomy. Retrieved 2010-09-29.

[3] W.V. Farrar (1969). "Richard Laming and the Coal-Gas Industry, with His Views on the Structure of Matter". *Annals of Science* **25** (3): 243–254. doi:10.1080/00033796900200141.

[4] T. Arabatzis (2006). *Representing Electrons: A Biographical Approach to Theoretical Entities*. University of Chicago Press. pp. 70–74. ISBN 0-226-02421-0.

[5] J.Z. Buchwald, A. Warwick (2001). *Histories of the Electron: The Birth of Microphysics*. MIT Press. pp. 195–203. ISBN 0-262-52424-4.

[6] J.J. Thomson (1897). "Cathode Rays". *Philosophical Magazine* **44** (269): 293. doi:10.1080/14786449708621070.

[7] S.H. Neddermeyer, C.D. Anderson; Anderson (1937). "Note on the Nature of Cosmic-Ray Particles". *Physical Review* **51** (10): 884–886. Bibcode:1937PhRv...51..884N. doi:10.1103/PhysRev.51.884.

[8] "The Reines-Cowan Experiments: Detecting the Poltergeist" (PDF). *Los Alamos Science* **25**: 3. 1997. Retrieved 2010-02-10.

[9] F. Reines, C.L. Cowan, Jr.; Cowan (1956). "The Neutrino". *Nature* **178** (4531): 446. Bibcode:1956Natur.178..446R. doi:10.1038/178446a0.

[10] G. Danby; Gaillard, J-M.; Goulianos, K.; Lederman, L.; Mistry, N.; Schwartz, M.; Steinberger, J. et al. (1962). "Observation of high-energy neutrino reactions and the existence of two kinds of neutrinos". *Physical Review Letters* **9**: 36. Bibcode:1962PhRv .doi:10.1103/PhysRevLett.9.36.

[11] M.L. Perl; Abrams, G.; Boyarski, A.; Breidenbach, M.; Briggs, D.; Bulos, F.; Chinowsky, W.; Dakin, J.; Feldman, G.; Friedberg, C.; Fryberger, D.; Goldhaber, G.; Hanson, G.; Heile, F.; Jean-Marie, B.; Kadyk, J.; Larsen, R.; Litke, A.; Lüke, D.; Lulu, B.; Lüth, V.; Lyon, D.; Morehouse, C.; Paterson, J.; Pierre, F.; Pun, T.; Rapidis, P.; Richter, B.; Sadoulet, B. et al. (1975). "Evidence for Anomalous Lepton Production in e+e− Annihilation". *Physical Review Letters* **35** (22): 1489. Bibcode:1975PhRvL..35.1489P. doi:10.1103/PhysRevLett.35.1489.

[12] "Physicists Find First Direct Evidence for Tau Neutrino at Fermilab" (Press release). Fermilab. 20 July 2000.

[13] K. Kodama *et al.* (DONUT Collaboration); Kodama; Ushida; Andreopoulos; Saoulidou; Tzanakos; Yager; Baller; Boehnlein; Freeman; Lundberg; Morfin; Rameika; Yun; Song; Yoon; Chung; Berghaus; Kubantsev; Reay; Sidwell; Stanton; Yoshida; Aoki; Hara; Rhee; Ciampa; Erickson; Graham et al. (2001). "Observation of tau neutrino interactions". *Physics Letters B* **504** (3): 218. arXiv:hep-ex/0012035. Bibcode:2001PhLB..504..218D. doi:10.1016/S0370-2693(01)00307-0.

[14] "lepton". *Online Etymology Dictionary.*

[15] λεπτός. Liddell, Henry George; Scott, Robert; *A Greek–English Lexicon* at the Perseus Project.

[16] Found on the KN L 693 and PY Un 1322 tablets. "The Linear B word re-po-to". Palaeolexicon. Word study tool of ancient languages. Raymoure, K.A. "re-po-to". *Minoan Linear A & Mycenaean Linear B.* Deaditerranean. "KN 693 L (103)". "PY 1322 Un + fr. (Cii)". *DĀMOS: Database of Mycenaean at Oslo.* University of Oslo.

[17] L. Rosenfeld (1948)

[18] C. Amsler *et al.* (2008): Particle listings – e−

[19] C. Amsler *et al.* (2008): Particle listings – μ−

[20] C. Amsler *et al.* (2008): Particle listings – p+

[21] C. Amsler *et al.* (2008): Particle listings – τ−

[22] S. Weinberg (2003)

[23] R. Wilson (1997)

[24] K. Riesselmann (2007)

[25] S.H. Neddermeyer, C.D. Anderson (1937)

[26] I.V. Anicin (2005)

[27] M.L. Perl et al. (1975)

[28] K. Kodama (2001)

[29] C. Amsler *et al.* (2008) Heavy Charged Leptons Searches

[30] C. Amsler *et al.* (2008) Searches for Heavy Neutral Leptons

[31] M.E. Peskin, D.V. Schroeder (1995), p. 197

[32] M.E. Peskin, D.V. Schroeder (1995), p. 27

[33] Y. Fukuda *et al.* (1998)

[34] C.Amsler et al. (2008): Particle listings – Neutrino properties

[35] B.R. Martin, G. Shaw (1992)

[36] J. P. Cumalat (1993). *Physics in Collision 12.* Atlantica Séguier Frontières. ISBN 978-2-86332-129-4.

[37] G Fraser (1 January 1998). *The Particle Century.* CRC Press. ISBN 978-1-4200-5033-2.

[38] J. Peltoniemi, J. Sarkamo (2005)

12.8 References

- C. Amsler *et al.* (Particle Data Group); Amsler; Doser; Antonelli; Asner; Babu; Baer; Band; Barnett; Bergren; Beringer; Bernardi; Bertl; Bichsel; Biebel; Bloch; Blucher; Blusk; Cahn; Carena; Caso; Ceccucci; Chakraborty; Chen; Chivukula; Cowan; Dahl; d'Ambrosio; Damour et al. (2008). "Review of Particle Physics". *Physics Letters B* **667**: 1. Bibcode:2008PhLB..667....1P. doi:10.1016/j.physletb.2008.07.018.
- I.V. Anicin (2005). "The Neutrino – Its Past, Present and Future". *SFIN (Institute of Physics, Belgrade) year XV, Series A: Conferences, No. A2 (2002) 3–59*: 3172. arXiv:physics/0503172. Bibcode:2005physics...3172A.
- Y.Fukuda; Hayakawa, T.; Ichihara, E.; Inoue, K.; Ishihara, K.; Ishino, H.; Itow, Y.; Kajita, T. et al. (1998). "Evidence for Oscillation of Atmospheric Neutrinos". *Physical Review Letters* **81** (8): 1562–1567. arXiv:hep-ex/9807003. Bibcode:1998PhRvL..81.1562F. doi:10.1103/PhysRevLett.81.1562.
- K. Kodama; Ushida, N.; Andreopoulos, C.; Saoulidou, N.; Tzanakos, G.; Yager, P.; Baller, B.; Boehnlein, D.; Freeman, W.; Lundberg, B.; Morfin, J.; Rameika, R.; Yun, J.C.; Song, J.S.; Yoon, C.S.; Chung, S.H.; Berghaus, P.; Kubantsev, M.; Reay, N.W.; Sidwell, R.; Stanton, N.; Yoshida, S.; Aoki, S.; Hara, T.; Rhee, J.T.; Ciampa, D.; Erickson, C.; Graham, M.; Heller, K. et al. (2001). "Observation of tau neutrino interactions". *Physics Letters B* **504** (3): 218. arXiv:hep-ex/0012035. Bibcode:2001PhLB..504..218D. doi:10.1016/S0370-2693(01)00307-0.
- B.R. Martin, G. Shaw (1992). "Chapter 2 – Leptons, quarks and hadrons". *Particle Physics*. John Wiley & Sons. pp. 23–47. ISBN 0-471-92358-3.
- S.H. Neddermeyer, C.D. Anderson; Anderson (1937). "Note on the Nature of Cosmic-Ray Particles". *Physical Review* **51** (10): 884–886. Bibcode:1937PhRv...51..884N. doi:10.1103/PhysRev.51.884.
- J. Peltoniemi, J. Sarkamo (2005). "Laboratory measurements and limits for neutrino properties". *The Ultimate Neutrino Page*. Retrieved 2008-11-07.
- M.L. Perl; Abrams, G.; Boyarski, A.; Breidenbach, M.; Briggs, D.; Bulos, F.; Chinowsky, W.; Dakin, J. et al. (1975). "Evidence for Anomalous Lepton Production in e^+–e^- Annihilation". *Physical Review Letters* **35** (22): 1489–1492. Bibcode:1975PhRvL..35.1489P. doi:10.1103/PhysRevLett.35.1489.
- M.E. Peskin, D.V. Schroeder (1995). *Introduction to Quantum Field Theory*. Westview Press. ISBN 0-201-50397-2.
- K. Riesselmann (2007). "Logbook: Neutrino Invention". *Symmetry Magazine* **4** (2).
- L. Rosenfeld (1948). *Nuclear Forces*. Interscience Publishers. p. xvii.
- R. Shankar (1994). "Chapter 2 – Rotational Invariance and Angular Momentum". *Principles of Quantum Mechanics* (2nd ed.). Springer. pp. 305–352. ISBN 978-0-306-44790-7.
- S. Weinberg (2003). *The Discovery of Subatomic Particles*. Cambridge University Press. ISBN 0-521-82351-X.
- R. Wilson (1997). *Astronomy Through the Ages: The Story of the Human Attempt to Understand the Universe*. CRC Press. p. 138. ISBN 0-7484-0748-0.

12.9 External links

- Particle Data Group homepage. The PDG compiles authoritative information on particle properties.
- Leptons, a summary of leptons from *Hyperphysics*.

[14] "The Nobel Prize in Physics 1936". Retrieved 2010-01-21.

[15] GILMER, PENNY J. (19 July 2011). "IRÈNE JOLIOT-CURIE, A NOBEL LAUREATE IN ARTIFICIAL RADIOACTIVITY" (PDF). p. 8. Retrieved 13 July 2013.

[16] "Antimatter caught streaming from thunderstorms on Earth". BBC. 11 January 2011. Archived from the original on 12 January 2011. Retrieved 11 January 2011.

[17] Adriani, O.; Barbarino, G. C.; Bazilevskaya, G. A.; Bellotti, R. et al. (2011). "The Discovery of Geomagnetically Trapped Cosmic-Ray Antiprotons". *The Astrophysical Journal Letters* **737** (2): L29. arXiv:1107.4882v1. Bibcode:2011ApJ...737L..29A. doi:10.1088/2041-8205/737/2/L29.

[18] Than, Ker (10 August 2011). "Antimatter Found Orbiting Earth—A First". National Geographic Society. Retrieved 12 August 2011.

[19] "What's the Matter with Antimatter?". NASA. 29 May 2000. Archived from the original on 4 June 2008. Retrieved 24 May 2008.

[20] "Radiation and Radioactive Decay. Radioactive Human Body". Harvard Natural Sciences Lecture Demonstrations. Retrieved 2011-05-18.

[21] Winteringham, F. P. W; Effects, F.A.O. Standing Committee on Radiation, Land And Water Development Division, Food and Agriculture Organization of the United Nations (1989). *Radioactive fallout in soils, crops and food: a background review*. Food & Agriculture Org. p. 32. ISBN 978-92-5-102877-3.

[22] Engelkemeir, DW; KF Flynn; LE Glendenin (1962). "Positron Emission in the Decay of K^{40}". *Physical Review* **126** (5): 1818. Bibcode:1962PhRv..126.1818E. doi:10.1103/PhysRev.126.1818.

[23] L. Accardo et al. (AMS Collaboration) (18 September 2014). "High Statistics Measurement of the Positron Fraction in Primary Cosmic Rays of 0.5–500 GeV with the Alpha Magnetic Spectrometer on the International Space Station" (PDF). *Physical Review Letters* **113**: 121101. Bibcode:2014PhRvL.113l1101A. doi:10.1103/PhysRevLett.113.121101.

[24] Schirber, Michael. "Synopsis: More Dark Matter Hints from Cosmic Rays?". American Physical Society. Retrieved 21 September 2014.

[25] "New results from the Alpha Magnetic$Spectrometer on the International Space Station" (PDF). *AMS-02 at NASA*. Retrieved 21 September 2014.

[26] Aguilar, M.; Alberti, G.; Alpat, B.; Alvino, A.; Ambrosi, G.; Andeen, K.; Anderhub, H.; Arruda, L.; Azzarello, P.; Bachlechner, A.; Barao, F.; Baret, B.; Barrau, A.; Barrin, L.; Bartoloni, A.; Basara, L.; Basili, A.; Batalha, L.; Bates, J.; Battiston, R.; Bazo, J.; Becker, R.; Becker, U.; Behlmann, M.; Beischer, B.; Berdugo, J.; Berges, P.; Bertucci, B.; Bigongiari, G. et al. (2013). "First Result from the Alpha Magnetic Spectrometer on the International Space Station: Precision Measurement of the Positron Fraction in Primary Cosmic Rays of 0.5–350 GeV". *Physical Review Letters* **110** (14): 141102. Bibcode:2013PhRvL.110n1102A. doi:10.1103/PhysRevLett.110.141102.

[27] AMS Collaboration; Aguilar, M.; Alcaraz, J.; Allaby, J.; Alpat, B.; Ambrosi, G.; Anderhub, H.; Ao, L. et al. (August 2002). "The Alpha Magnetic Spectrometer (AMS) on the International Space Station: Part I – results from the test flight on the space shuttle". *Physics Reports* **366** (6): 331–405. Bibcode:2002PhR...366..331A. doi:10.1016/S0370-1573(02)00013-3.

[28] Bland, E. (1 December 2008). "Laser technique produces bevy of antimatter". MSNBC. Retrieved 2009-07-16. The LLNL scientists created the positrons by shooting the lab's high-powered Titan laser onto a one-millimeter-thick piece of gold.

[29] "Laser creates billions of antimatter particles". *Cosmos Online*.

[30] Phelps, Michael E. (2006). *PET: physics, instrumentation, and scanners*. Springer. pp. 2–3. ISBN 0-387-32302-3.

[31] "Introduction to Positron Research". *St. Olaf College*.

13.7 External links

- What is a Positron? (from the Frequently Asked Questions :: Center for Antimatter-Matter Studies)
- Website about positrons and antimatter
- Positron information search at SLAC
- Positron Annihilation as a method of experimental physics used in materials research.
- New production method to produce large quantities of positrons
- Website about antimatter (positrons, positronium and antihydrogen). Positron Laboratory, Como, Italy
- Website of the AEgIS: Antimatter Experiment: Gravity, Interferometry, Spectroscopy, CERN
- Synopsis: Tabletop Particle Accelerator ... new tabletop method for generating electron-positron streams.

In August 2014, Decision Sciences International Corporation announced it had been awarded a contract by Toshiba for use of its muon tracking detectors in reclaiming the Fukushima nuclear complex.[21] The Fukushima Daiichi Tracker (FDT) was proposed to make a few months of muon measurements to show the distribution of the reactor cores.

In December 2014, Tepco reported that they would be using two different muon imaging techniques at Fukushima, "Muon Scanning Method" on Unit 1 (the most badly damaged, where the fuel may have left the reactor vessel) and "Muon Scattering Method" on Unit 2.[22]

The International Research Institute for Nuclear Decommissioning IRID in Japan and the High Energy Accelerator Research Organization KEK call the method they developed for Unit 1 the **muon permeation method**; 1,200 optical fibers for wavelength conversion light up when muons come into contact with them.[23] After a month of data collection, it is hoped to reveal the location and amount of fuel debris still inside the reactor. The measurements began in February 2015.[24]

14.8 See also

- Muonic atoms
- Muon spin spectroscopy
- Muon-catalyzed fusion
- Muon Tomography
- Mu2e, an experiment to detect neutrinoless conversion of muons to electrons
- List of particles

14.9 References

[1] J. Beringer et al. (Particle Data Group) (2012). "PDGLive Particle Summary 'Leptons (e, mu, tau, ... neutrinos ...)'" (PDF). Particle Data Group. Retrieved 2013-01-12.

[2] New Evidence for the Existence of a Particle Intermediate Between the Proton and Electron", Phys. Rev. 52, 1003 (1937).

[3] Yukaya Hideka, On the Interaction of Elementary Particles 1, Proceedings of the Physico-Mathematical Society of Japan (3) 17, 48, pp 139–148 (1935). (Read 17 November 1934)

[4] S. Carroll (2004). *Spacetime and Geometry: An Introduction to General Relativity*. Addison Wesley. p. 204

[5] Mark Wolverton (September 2007). "Muons for Peace: New Way to Spot Hidden Nukes Gets Ready to Debut". *Scientific American* **297** (3): 26–28. doi:10.1038/scientificamerican0907-26.

[6] "Physicists Announce Latest Muon g-2 Measurement" (Press release). Brookhaven National Laboratory. 30 July 2002. Retrieved 2009-11-14.

[7] J. Adam (MEG Collaboration) et al. (2013). "New Constraint on the Existence of the mu+ -> e+ gamma Decay". *Physical Review Letters* **110** (20): 201801. arXiv:1303.0754. Bibcode:2013PhRvL.110t1801A. doi:10.1103/PhysRevLett.110.201801.

[8] Fleming, D. G.; Arseneau, D. J.; Sukhorukov, O.; Brewer, J. H.; Mielke, S. L.; Schatz, G. C.; Garrett, B. C.; Peterson, K. A.; Truhlar, D. G. (28 Jan 2011). "Kinetic Isotope Effects for the Reactions of Muonic Helium and Muonium with H2". *Science* **331** (6016): 448–450. Bibcode:2011Sci...331..448F. doi:10.1126/science.1199421. PMID 21273484.

[9] Moncada, F.; Cruz, D.; Reyes, A. "Muonic alchemy: Transmuting elements with the inclusion of negative muons". *Chemical Physics Letters* **539**: 209–213. Bibcode:2012CPL...539..209M. doi:10.1016/j.cplett.2012.04.062.

[10] Moncada, F.; Cruz, D.; Reyes, A (10 May 2013). "Electronic properties of atoms and molecules containing one and two negative muons". *Chemical Physics Letters* **570**: 16–21. Bibcode:2013CPL...570...16M. doi:10.1016/j.cplett.2013.03.004.

[11] TRIUMF Muonic Hydrogen collaboration. "A brief description of Muonic Hydrogen research". Retrieved 2010-11-7

[12] Pohl, Randolf et al. "*The Size of the Proton*" *Nature* 466, 213–216 (8 July 2010)

[13] "The Muon g-2 Experiment Home Page". G-2.bnl.gov. 2004-01-08. Retrieved 2012-01-06.

[14] "(from the July 2007 review by Particle Data Group)" (PDF). Retrieved 2012-01-06.

[15] Hagiwara, K; Martin, A; Nomura, D; Teubner, T (2007). "Improved predictions for g–2g–2 of the muon and αQED(MZ2)". *Physics Letters B* **649** (2–3): 173. arXiv:hep-ph/0611102. Bibcode:2007PhLB..649..173H. doi:10.1016/j.physletb.2007.04.012.

[16] "Revolutionary muon experiment to begin with 3,200-mile move of 50-foot-wide particle storage ring". May 8, 2013. Retrieved Mar 16, 2015.

[17] "Decision Sciences Corp".

[18] George, E.P. (July 1, 1955). "Cosmic rays measure overburden of tunnel". *Commonwealth Engineer*: 455.

[19] Alvarez,L.W. (1970). "Search for hidden chambers in the pyramids using cosmic rays".*Science***167**: 832. Bibcode:1970Sci...1. doi:10.1126/science.167.3919.832.

[20] Konstantin N. Borozdin, Gary E. Hogan, Christopher Morris, William C. Priedhorsky, Alexander Saunders, Larry J. Schultz & Margaret E. Teasdale. "Radiographic imaging with cosmic-ray muons". Nature.

[21] http://www.decisionsciencescorp.com/ds-awarded-toshiba-contract-fukushima-daiichi-nuclear-project/

[22] Tepco to start "scanning" inside of Reactor 1 in early February by using muon Fukushima Diary

[23] "Muon measuring instrument production for "muon permeation method" and its review by international experts". IRID.or.jp.

[24] Muon Scans Begin At Fukushima Daiichi - SimplyInfo

- S.H. Neddermeyer, C.D. Anderson; Anderson (1937). "Note on the Nature of Cosmic-Ray Particles". *Physical Review* **51** (10): 884–886. Bibcode:1937PhRv...51..884N. doi:10.1103/PhysRev.51.884.
- J.C. Street, E.C. Stevenson; Stevenson (1937). "New Evidence for the Existence of a Particle of Mass Intermediate Between the Proton and Electron". *Physical Review* **52** (9): 1003–1004. Bibcode:1937PhRv...52.1003S. doi:10.1103/PhysRev.52.1003.
- G. Feinberg, S. Weinberg; Weinberg (1961). "Law of Conservation of Muons". *Physical Review Letters* **6** (7): 381–383. Bibcode:1961PhRvL...6..381F. doi:10.1103/PhysRevLett.6.381.
- Serway & Faughn (1995). *College Physics* (4th ed.). Saunders. p. 841.
- M. Knecht (2003). "The Anomalous Magnetic Moments of the Electron and the Muon". In B. Duplantier, V. Rivasseau. *Poincaré Seminar 2002: Vacuum Energy – Renormalization*. Progress in Mathematical Physics **30**. Birkhäuser Verlag. p. 265. ISBN 3-7643-0579-7.
- E. Derman (2004). *My Life As A Quant*. Wiley. pp. 58–62.

14.10 External links

- Muon anomalous magnetic moment and supersymmetry
- g-2 (muon anomalous magnetic moment) experiment
- muLan (Measurement of the Positive Muon Lifetime) experiment
- The Review of Particle Physics
- The TRIUMF Weak Interaction Symmetry Test
- The MEG Experiment (Search for the decay Muon → Positron + Gamma)
- King, Philip. "Making Muons". *Backstage Science*. Brady Haran.

Chapter 15

Tau (particle)

Not to be confused with the τ^+ of the τ–θ puzzle, which is now identified as a kaon.

The **tau** (τ), also called the **tau lepton**, **tau particle** or **tauon**, is an elementary particle similar to the electron, with negative electric charge and a spin of $\frac{1}{2}$. Together with the electron, the muon, and the three neutrinos, it is a lepton. Like all elementary particles with half-integral spin, the tau has a corresponding antiparticle of opposite charge but equal mass and spin, which in the tau's case is the **antitau** (also called the *positive tau*). Tau particles are denoted by $\tau-$ and the antitau by $\tau+$.

Tau leptons have a lifetime of 2.9×10^{-13} s and a mass of 1776.82 MeV/c^2 (compared to 105.7 MeV/c^2 for muons and 0.511 MeV/c^2 for electrons). Since their interactions are very similar to those of the electron, a tau can be thought of as a much heavier version of the electron. Because of their greater mass, tau particles do not emit as much bremsstrahlung radiation as electrons; consequently they are potentially highly penetrating, much more so than electrons. However, because of their short lifetime, the range of the tau is mainly set by their decay length, which is too small for bremsstrahlung to be noticeable: their penetrating power appears only at ultra high energy (above PeV energies).[4]

As with the case of the other charged leptons, the tau has an associated tau neutrino, denoted by ν
τ.

15.1 History

The tau was detected in a series of experiments between 1974 and 1977 by Martin Lewis Perl with his colleagues at the SLAC-LBL group.[2] Their equipment consisted of SLAC's then-new e+–e− colliding ring, called SPEAR, and the LBL magnetic detector. They could detect and distinguish between leptons, hadrons and photons. They did not detect the tau directly, but rather discovered anomalous events:

"*We have discovered 64 events of the form*

$$e+ + e- \rightarrow e\pm + \mu\mp + \text{at least two undetected particles}$$

for which we have no conventional explanation."

The need for at least two undetected particles was shown by the inability to conserve energy and momentum with only one. However, no other muons, electrons, photons, or hadrons were detected. It was proposed that this event was the production and subsequent decay of a new particle pair:

$$e+ + e- \rightarrow \tau+ + \tau- \rightarrow e\pm + \mu\mp + 4\nu$$

This was difficult to verify, because the energy to produce the $\tau+\tau-$ pair is similar to the threshold for D meson production.

Work done at DESY-Hamburg, and with the Direct Electron Counter (DELCO) at SPEAR, subsequently established the mass and spin of the tau.

The symbol τ was derived from the Greek *τρίτον* (*triton*, meaning "third" in English), since it was the third charged lepton discovered.[5]

Martin Perl shared the 1995 Nobel Prize in Physics with Frederick Reines. The latter was awarded his share of the prize for experimental discovery of the neutrino.

15.2 Tau decay

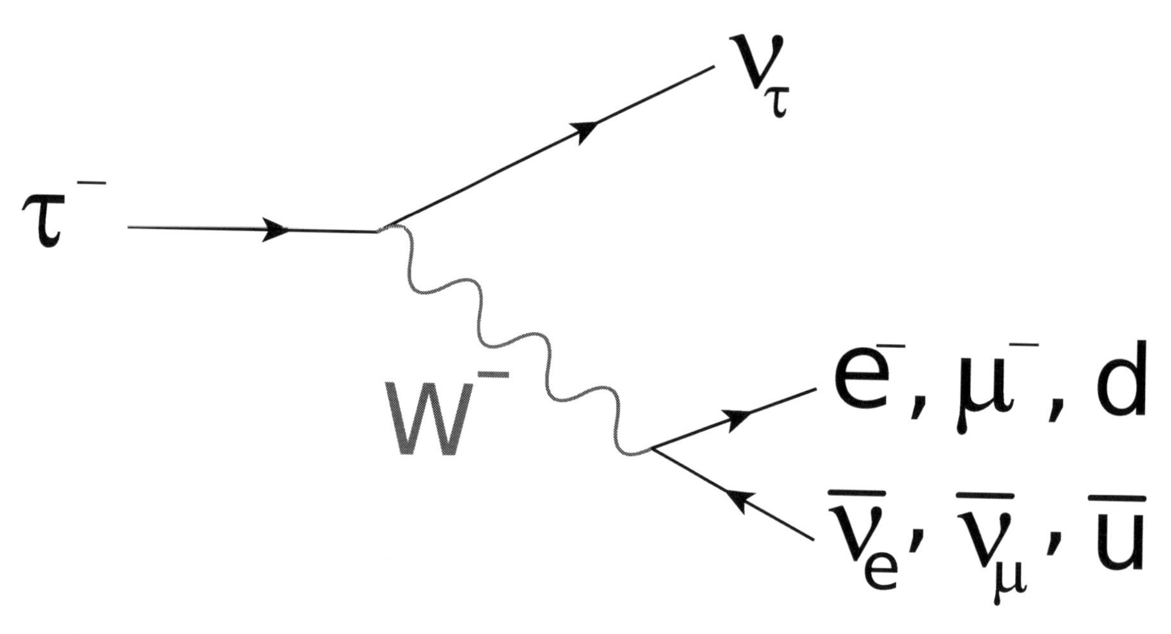

Feynman diagram of the common decays of the tau by emission of a W boson.

The tau is the only lepton that can decay into hadrons – the other leptons do not have the necessary mass. Like the other decay modes of the tau, the hadronic decay is through the weak interaction.[6]

The branching ratio of the dominant hadronic tau decays are:[3]

- 25.52% for decay into a charged pion, a neutral pion, and a tau neutrino;
- 10.83% for decay into a charged pion and a tau neutrino;
- 9.30% for decay into a charged pion, two neutral pions, and a tau neutrino;
- 8.99% for decay into three charged pions (of which two have the same electrical charge) and a tau neutrino;
- 2.70% for decay into three charged pions (of which two have the same electrical charge), a neutral pion, and a tau neutrino;
- 1.05% for decay into three neutral pions, a charged pion, and a tau neutrino.

In total, the tau lepton will decay hadronically approximately 64.79% of the time.

Since the tauonic lepton number is conserved in weak decays, a tau neutrino is always created when a tau decays.[6]

The branching ratio of the common purely leptonic tau decays are:[3]

- 17.82% for decay into a tau neutrino, electron and electron antineutrino;
- 17.39% for decay into a tau neutrino, muon and muon antineutrino.

The similarity of values of the two branching ratios is a consequence of lepton universality.

15.3 Exotic atoms

The tau lepton is predicted to form exotic atoms like other charged subatomic particles. One of such, called **tauonium** by the analogy to muonium, consists in antitauon and an electron: τ+e–.[7]

Another one is an onium atom τ+τ– called *true tauonium* and is difficult to detect due to tau's extremely short lifetime at low (non-relativistic) energies needed to form this atom. Its detection is important for quantum electrodynamics.[7]

15.4 See also

- Koide formula

15.5 References

[1] L. B. Okun (1980). *Leptons and Quarks*. V.I. Kisin (trans.). North-Holland Publishing. p. 103. ISBN 978-0444869241.

[2] Perl, M. L.; Abrams, G.; Boyarski, A.; Breidenbach, M.; Briggs, D.; Bulos, F.; Chinowsky, W.; Dakin, J. et al. (1975). "Evidence for Anomalous Lepton Production in e+e– Annihilation". *Physical Review Letters* **35** (22): 1489. Bibcode:1975PhRvL..3 .doi:10.1103/PhysRevLett.35.1489.

[3] J. Beringer *et al.* (Particle Data Group) (2012). "Review of Particle Physics". *Journal of Physics G* **86** (1): 581–651. Bibcode:2012PhRvD..86a0001B. doi:10.1103/PhysRevD.86.010001. lchapter= ignored (help)

[4] D. Fargion, P.G. De Sanctis Lucentini, M. De Santis, M. Grossi (2004). "Tau Air Showers from Earth". *The Astrophysical Journal* **613** (2): 1285. arXiv:hep-ph/0305128. Bibcode:2004ApJ...613.1285F. doi:10.1086/423124.

[5] M.L. Perl (1977). "Evidence for, and properties of, the new charged heavy lepton" (PDF). In T. Thanh Van (ed.). *Proceedings of the XII Rencontre de Moriond*. SLAC-PUB-1923.

[6] Riazuddin (2009). "Non-standard interactions" (PDF). *NCP 5th Particle Physics Sypnoisis* (Islamabad,: Riazuddin, Head of High-Energy Theory Group at National Center for Physics) **1** (1): 1–25.

[7] Brodsky, Stanley J.; Lebed, Richard F. (2009). "Production of the Smallest QED Atom: True Muonium ($\mu^+\mu^-$)". *Physical Review Letters* **102** (21): 213401. arXiv:0904.2225. Bibcode:2009PhRvL.102u3401B. doi:10.1103/PhysRevLett.102.213401.

15.6 External links

- Nobel Prize in Physics 1995
- Perl's logbook showing tau discovery
- A Tale of Three Papers gives the covers of the three original papers announcing the discovery.

Chapter 16

Neutrino

For other uses, see Neutrino (disambiguation).

A **neutrino** (/nuːˈtriːnoʊ/ or /njuːˈtriːnoʊ/, in Italian [neuˈtrino]) is an electrically neutral elementary particle[4] with half-integer spin. The neutrino (meaning "little neutral one" in Italian) is denoted by the Greek letter ν (*nu*). All evidence suggests that neutrinos have mass but that their masses are tiny, even compared to other subatomic particles. They are the only identified candidate for dark matter, specifically hot dark matter.[5]

Neutrinos are leptons, along with the charged electrons, muons, and taus, and come in three flavors: electron neutrinos (ν
e), muon neutrinos (ν
μ), and tau neutrinos (ν
τ). Each flavor is also associated with an antiparticle, called an "antineutrino", which also has no electric charge and half-integer spin. Neutrinos are produced in a way that conserves lepton number; i.e., for every electron neutrino produced, a positron (anti-electron) is produced, and for every electron antineutrino produced, an electron is produced as well.

Neutrinos do not carry any electric charge, which means that they are not affected by the electromagnetic force that acts on charged particles, and are leptons, so they are not affected by the strong force that acts on particles inside atomic nuclei. Neutrinos are therefore affected only by the weak subatomic force and by gravity. The weak force is a very short-range interaction, and gravity is extremely weak on the subatomic scale. Thus, neutrinos typically pass through normal matter unimpeded and undetected.

Neutrinos can be created in several ways, including in certain types of radioactive decay, in nuclear reactions such as those that take place in the Sun, in nuclear reactors, when cosmic rays hit atoms and in supernovas. The majority of neutrinos in the vicinity of the earth are from nuclear reactions in the Sun. In fact, about 65 billion (6.5×10^{10}) solar neutrinos per second pass through every square centimeter perpendicular to the direction of the Sun in the region of the Earth.[6]

Neutrinos are now understood to oscillate between different flavors in flight. That is, an electron neutrino produced in a beta decay reaction may arrive in a detector as a muon or tau neutrino. This oscillation requires that the different neutrino flavors have different masses, although these masses have been shown to be tiny. From cosmological measurements, we know that the sum of the three neutrino masses must be less than one millionth that of the electron.[7]

16.1 History

16.1.1 Pauli's proposal

The neutrino[nb 1] was postulated first by Wolfgang Pauli in 1930 to explain how beta decay could conserve energy, momentum, and angular momentum (spin). In contrast to Niels Bohr, who proposed a statistical version of the conservation laws to explain the event, Pauli hypothesized an undetected particle that he called a "neutron" in keeping with convention employed for naming both the proton and the electron, which in 1930 were known to be respective products though a small percent of potassium, (0.0117%) is the single most abundant radioisotope in the human body. In a human body of 70 kg mass, about 4,400 nuclei of ^{40}K decay per second.[20] The activity of natural potassium is 31 Bq/g.[21] About 0.001% of these ^{40}K decays produce about 4000 natural positrons per day in the human body.[22] These positrons soon find an electron, undergo annihilation, and produce pairs of 511 keV gamma rays, in a process similar (but much lower intensity) to that which happens during a PET scan nuclear medicine procedure.

13.2.1 Observation in cosmic rays

Main article: Cosmic ray

Satellite experiments have found evidence of positrons (as well as a few antiprotons) in primary cosmic rays, amounting to less than 1% of the particles in primary cosmic rays. These do not appear to be the products of large amounts of antimatter from the Big Bang, or indeed complex antimatter in the universe (evidence for which is lacking, see below). Rather, the antimatter in cosmic rays appear to consist of only these two elementary particles, probably made in energetic processes long after the Big Bang.

Preliminary results from the presently operating Alpha Magnetic Spectrometer (*AMS-02*) on board the International Space Station show that positrons in the cosmic rays arrive with no directionality, and with energies that range from 10 GeV to 250 GeV. In September, 2014, new results with almost twice as much data were presented in a talk at CERN and published in Physical Review Letters.[23][24] A new measurement of positron fraction up to 500 GeV was reported, showing that positron fraction peaks at a maximum of about 16% of total electron+positron events, around an energy of 275 ± 32 GeV. At higher energies, up to 500 GeV, the ratio of positrons to electrons begins to fall again. The absolute flux of positrons also begins to fall before 500 GeV, but peaks at energies far higher than electron energies, which peak about 10 GeV.[25] These results on interpretation have been suggested to be due to positron production in annihilation events of massive dark matter particles.[26]

Positrons, like anti-protons, do not appear to originate from any hypothetical "antimatter" regions of the universe. On the contrary, there is no evidence of complex antimatter atomic nuclei, such as antihelium nuclei (i.e., anti-alpha particles), in cosmic rays. These are actively being searched for. A prototype of the *AMS-02* designated *AMS-01*, was flown into space aboard the Space Shuttle *Discovery* on STS-91 in June 1998. By not detecting any antihelium at all, the *AMS-01* established an upper limit of 1.1×10^{-6} for the antihelium to helium flux ratio.[27]

13.3 Artificial production

New research has dramatically increased the quantity of positrons that experimentalists can produce. Physicists at the Lawrence Livermore National Laboratory in California have used a short, ultra-intense laser to irradiate a millimetre-thick gold target and produce more than 100 billion positrons.[28][29]

13.4 Applications

Certain kinds of particle accelerator experiments involve colliding positrons and electrons at relativistic speeds. The high impact energy and the mutual annihilation of these matter/antimatter opposites create a fountain of diverse subatomic particles. Physicists study the results of these collisions to test theoretical predictions and to search for new kinds of particles.

Gamma rays, emitted indirectly by a positron-emitting radionuclide (tracer), are detected in positron emission tomography (PET) scanners used in hospitals. PET scanners create detailed three-dimensional images of metabolic activity within the human body.[30]

An experimental tool called positron annihilation spectroscopy (PAS) is used in materials research to detect variations in density, defects, displacements, or even voids, within a solid material.[31]

13.5 See also

- Beta particle
- Radioactive decay
- List of particles
- Positron emission tomography
- Positronium
- Proton
- Positronic brain

13.6 References

13.6.1 Notes

[1] The fractional version's denominator is the inverse of the decimal value (along with its relative standard uncertainty of 4.2×10^{-10}).

13.6.2 Citations

[1] The original source for CODATA is:

Mohr, P.J.; Taylor, B.N.; Newell, D.B. (2006). "CODATA recommended values of the fundamental physical constants". *Reviews of Modern Physics* **80** (2): 633–730. arXiv:0801.0028. Bibcode:2008RvMP...80..633M. doi:10.1103/RevModPhys.80.633.
Individual physical constants from the CODATA are available at:

"The NIST Reference on Constants, Units and Uncertainty". National Institute of Standards and Technology. Retrieved 2013-10-24.

[2] P. A. M. Dirac. "The quantum theory of the electron" (PDF).

[3] P. A. M. Dirac. "A Theory of Electrons and Protons" (PDF).

[4] Frank Close (2009). *Antimatter*. Oxford University Press. p. 46. ISBN 978-0-19-955016-6.

[5] P.A.M.Dirac(1931). "Quantised Singularities in the Quantum Field".*Proc.R.Soc.Lond.A***133**(821): 2–3. Bibcode:1931RSP. doi:10.1098/rspa.1931.0130.

[6] Feynman,Richard(1949). "The Theory of Positrons".*Physical Review***76**(76): 749. Bibcode:1949PhRv...76..749F.doi:10.9.

[7] Feynman, Richard (1965-12-11). *The Development of the Space-Time View of Quantum Electrodynamics* (Speech). Nobel Lecture. Retrieved 2007-01-02.

[8] Nambu, Yoichiro (1950). "The Use of the Proper Time in Quantum Electrodynamics I". *Progress in Theoretical Physics* **5** (5): 82. Bibcode:1950PThPh...5...82N. doi:10.1143/PTP.5.82.

[9] Frank Close. *Antimatter*. Oxford University Press. pp. 50–52. ISBN 978-0-19-955016-6.

[10] *general chemistry*. Taylor & Francis. 1943. p. 660. GGKEY:0PYLHBL5D4L. Retrieved 15 June 2011.

[11] Cowan, Eugene (1982). "The Picture That Was Not Reversed". *Engineering & Science* **46** (2): 6–28.

[12] Jagdish Mehra; Helmut Rechenberg (2000). *The Historical Development of Quantum Theory, Volume 6: The Completion of Quantum Mechanics 1926–1941*. Springer. p. 804. ISBN 978-0-387-95175-1.

[13] Anderson, Carl D. (1933). "The Positive Electron". *Physical Review* **43** (6): 491–494. Bibcode:1933PhRv...43..491A. doi:10.1103/PhysRev.43.491.

Chapter 13

Positron

For other uses, see Positron (disambiguation).

The **positron** or **antielectron** is the antiparticle or the antimatter counterpart of the electron. The positron has an electric charge of +1 *e*, a spin of ½, and has the same mass as an electron. When a low-energy positron collides with a low-energy electron, annihilation occurs, resulting in the production of two or more gamma ray photons (see electron–positron annihilation).

Positrons may be generated by positron emission radioactive decay (through weak interactions), or by pair production from a sufficiently energetic photon which is interacting with an atom in a material.

13.1 History

13.1.1 Theory

In 1928, Paul Dirac published a paper[2] proposing that electrons can have both a positive charge and negative energy. This paper introduced the Dirac equation, a unification of quantum mechanics, special relativity, and the then-new concept of electron spin to explain the Zeeman effect. The paper did not explicitly predict a new particle, but did allow for electrons having either positive or negative energy as solutions. Hermann Weyl then published "Gravitation and the Electron" (Proceedings of the National Academy of Sciences of the United States of America, Vol. 15, No. 4-Apr. 15, 1929, pp. 323–334) discussing the mathematical implications of the negative energy solution. The positive-energy solution explained experimental results, but Dirac was puzzled by the equally valid negative-energy solution that the mathematical model allowed. Quantum mechanics did not allow the negative energy solution to simply be ignored, as classical mechanics often did in such equations; the dual solution implied the possibility of an electron spontaneously jumping between positive and negative energy states. However, no such transition had yet been observed experimentally. He referred to the issues raised by this conflict between theory and observation as "difficulties" that were "unresolved".

Dirac wrote a follow-up paper in December 1929[3] that attempted to explain the unavoidable negative-energy solution for the relativistic electron. He argued that "... an electron with negative energy moves in an external [electromagnetic] field as though it carries a positive charge." He further asserted that all of space could be regarded as a "sea" of negative energy states that were filled, so as to prevent electrons jumping between positive energy states (negative electric charge) and negative energy states (positive charge). The paper also explored the possibility of the proton being an island in this sea, and that it might actually be a negative-energy electron. Dirac acknowledged that the proton having a much greater mass than the electron was a problem, but expressed "hope" that a future theory would resolve the issue.

Robert Oppenheimer argued strongly against the proton being the negative-energy electron solution to Dirac's equation. He asserted that if it were, the hydrogen atom would rapidly self-destruct.[4] Persuaded by Oppenheimer's argument, Dirac published a paper in 1931 that predicted the existence of an as-yet unobserved particle that he called an "anti-electron" that would have the same mass as an electron and that would mutually annihilate upon contact with an electron.[5]

Feynman, and earlier Stueckelberg, proposed an interpretation of the positron as an electron moving backward in time,[6] reinterpreting the negative-energy solutions of the Dirac equation. Electrons moving backward in time would have a positive electric charge. Wheeler invoked this concept to explain the identical properties shared by all electrons, suggesting that "they are all the same electron" with a complex, self-intersecting worldline.[7] Yoichiro Nambu later applied it to all production and annihilation of particle-antiparticle pairs, stating that "the eventual creation and annihilation of pairs that may occur now and then is no creation or annihilation, but only a change of direction of moving particles, from past to future, or from future to past."[8] The backwards in time point of view is nowadays accepted as completely equivalent to other pictures, but it does not have anything to do with the macroscopic terms "cause" and "effect", which do not appear in a microscopic physical description.

13.1.2 Experimental clues and discovery

Dmitri Skobeltsyn first observed the positron in 1929.[9][10] While using a Wilson cloud chamber[11] to try to detect gamma radiation in cosmic rays, Skobeltsyn detected particles that acted like electrons but curved in the opposite direction in an applied magnetic field.[10]

Likewise, in 1929 Chung-Yao Chao, a graduate student at Caltech, noticed some anomalous results that indicated particles behaving like electrons, but with a positive charge, though the results were inconclusive and the phenomenon was not pursued.[12]

Carl David Anderson discovered the positron on August 2, 1932,[13] for which he won the Nobel Prize for Physics in 1936.[14] Anderson did not coin the term *positron*, but allowed it at the suggestion of the Physical Review journal editor to which he submitted his discovery paper in late 1932. The positron was the first evidence of antimatter and was discovered when Anderson allowed cosmic rays to pass through a cloud chamber and a lead plate. A magnet surrounded this apparatus, causing particles to bend in different directions based on their electric charge. The ion trail left by each positron appeared on the photographic plate with a curvature matching the mass-to-charge ratio of an electron, but in a direction that showed its charge was positive.[15]

Anderson wrote in retrospect that the positron could have been discovered earlier based on Chung-Yao Chao's work, if only it had been followed up.[12] Frédéric and Irène Joliot-Curie in Paris had evidence of positrons in old photographs when Anderson's results came out, but they had dismissed them as protons.[15]

13.2 Natural production

Main article: Positron emission

Positrons are produced naturally in β^+ decays of naturally occurring radioactive isotopes (for example, potassium-40) and in interactions of gamma quanta (emitted by radioactive nuclei) with matter. Antineutrinos are another kind of antiparticle created by natural radioactivity (β^- decay). Many different kinds of antiparticles are also produced by (and contained in) cosmic rays. Recent (as of January 2011) research by the American Astronomical Society has discovered antimatter (positrons) originating above thunderstorm clouds; positrons are produced in gamma-ray flashes created by electrons accelerated by strong electric fields in the clouds.[16] Antiprotons have also been found to exist in the Van Allen Belts around the Earth by the PAMELA module.[17][18]

Antiparticles, of which the most common are positrons due to their low mass, are also produced in any environment with a sufficiently high temperature (mean particle energy greater than the pair production threshold). During the period of baryogenesis, when the universe was extremely hot and dense, matter and antimatter were continually produced and annihilated. The presence of remaining matter, and absence of detectable remaining antimatter,[19] also called baryon asymmetry, is attributed to CP-violation: a violation of the CP-symmetry relating matter to antimatter. The exact mechanism of this violation during baryogenesis remains a mystery.

Positrons production from radioactive $\beta+$ decay, can be considered both artificial and natural production, as the generation of the radioisotope can be natural or artificial. Perhaps the best known naturally-occurring radioisotope which produces positrons is potassium-40, a long-lived isotope of potassium which occurs as a primordial isotope of potassium, and even

Chapter 14

Muon

The **muon** (/ˈmjuːɒn/; from the Greek letter mu (μ) used to represent it) is an elementary particle similar to the electron, with electric charge of −1 *e* and a spin of ½, but with a much greater mass (105.7 MeV/c^2). It is classified as a lepton, together with the electron (mass 0.511 MeV/c^2), the tau (mass 1776.82 MeV/c^2), and the three neutrinos (electron neutrino ν
e, muon neutrino ν
μ and tau neutrino ν
τ). As is the case with other leptons, the muon is not believed to have any sub-structure—that is, it is not thought to be composed of any simpler particles.

The muon is an unstable subatomic particle with a mean lifetime of 2.2 μs. Among all known unstable subatomic particles, only the neutron (lasting around 15 minutes) and some atomic nuclei have a longer decay lifetime; others decay significantly faster. The decay of the muon (as well as of the neutron, the longest-lived unstable baryon), is mediated by the weak interaction exclusively. Muon decay always produces at least three particles, which must include an electron of the same charge as the muon and two neutrinos of different types.

Like all elementary particles, the muon has a corresponding antiparticle of opposite charge (+1 *e*) but equal mass and spin: the **antimuon** (also called a *positive muon*). Muons are denoted by μ− and antimuons by μ+. Muons were previously called **mu mesons**, but are not classified as mesons by modern particle physicists (see § History), and that name is no longer used by the physics community.

Muons have a mass of 105.7 MeV/c^2, which is about 207 times that of the electron. Due to their greater mass, muons are not as sharply accelerated when they encounter electromagnetic fields, and do not emit as much bremsstrahlung (deceleration radiation). This allows muons of a given energy to penetrate far more deeply into matter than electrons, since the deceleration of electrons and muons is primarily due to energy loss by the bremsstrahlung mechanism. As an example, so-called "secondary muons", generated by cosmic rays hitting the atmosphere, can penetrate to the Earth's surface, and even into deep mines.

Because muons have a very large mass and energy compared with the decay energy of radioactivity, they are never produced by radioactive decay. They are, however, produced in copious amounts in high-energy interactions in normal matter, in certain particle accelerator experiments with hadrons, or naturally in cosmic ray interactions with matter. These interactions usually produce pi mesons initially, which most often decay to muons.

As with the case of the other charged leptons, the muon has an associated muon neutrino, denoted by ν
μ, which is not the same particle as the electron neutrino, and does not participate in the same nuclear reactions.

14.1 History

Muons were discovered by Carl D. Anderson and Seth Neddermeyer at Caltech in 1936, while studying cosmic radiation. Anderson had noticed particles that curved differently from electrons and other known particles when passed through a magnetic field. They were negatively charged but curved less sharply than electrons, but more sharply than protons, for particles of the same velocity. It was assumed that the magnitude of their negative electric charge was equal to that of the electron, and so to account for the difference in curvature, it was supposed that their mass was greater than an electron but smaller than a proton. Thus Anderson initially called the new particle a *mesotron*, adopting the prefix *meso-* from the Greek word for "mid-". The existence of the muon was confirmed in 1937 by J. C. Street and E. C. Stevenson's cloud chamber experiment.[2]

A particle with a mass in the meson range had been predicted before the discovery of any mesons, by theorist Hideki Yukawa:[3]

> "It seems natural to modify the theory of Heisenberg and Fermi in the following way. The transition of a heavy particle from neutron state to proton state is not always accompanied by the emission of light particles. The transition is sometimes taken up by another heavy particle."

Because of its mass, the mu meson was initially thought to be Yukawa's particle, but it later proved to have the wrong properties. Yukawa's predicted particle, the pi meson, was finally identified in 1947 (again from cosmic ray interactions), and shown to differ from the earlier-discovered mu meson by having the correct properties to be a particle which mediated the nuclear force.

With two particles now known with the intermediate mass, the more general term *meson* was adopted to refer to any such particle within the correct mass range between electrons and nucleons. Further, in order to differentiate between the two different types of mesons after the second meson was discovered, the initial mesotron particle was renamed the *mu meson* (the Greek letter μ (*mu*) corresponds to *m*), and the new 1947 meson (Yukawa's particle) was named the pi meson.

As more types of mesons were discovered in accelerator experiments later, it was eventually found that the mu meson significantly differed not only from the pi meson (of about the same mass), but also from all other types of mesons. The difference, in part, was that mu mesons did not interact with the nuclear force, as pi mesons did (and were required to do, in Yukawa's theory). Newer mesons also showed evidence of behaving like the pi meson in nuclear interactions, but not like the mu meson. Also, the mu meson's decay products included both a neutrino and an antineutrino, rather than just one or the other, as was observed in the decay of other charged mesons.

In the eventual Standard Model of particle physics codified in the 1970s, all mesons other than the mu meson were understood to be hadrons—that is, particles made of quarks—and thus subject to the nuclear force. In the quark model, a *meson* was no longer defined by mass (for some had been discovered that were very massive—more than nucleons), but instead were particles composed of exactly two quarks (a quark and antiquark), unlike the baryons, which are defined as particles composed of three quarks (protons and neutrons were the lightest baryons). Mu mesons, however, had shown themselves to be fundamental particles (leptons) like electrons, with no quark structure. Thus, mu mesons were not mesons at all, in the new sense and use of the term *meson* used with the quark model of particle structure.

With this change in definition, the term *mu meson* was abandoned, and replaced whenever possible with the modern term *muon*, making the term mu meson only historical. In the new quark model, other types of mesons sometimes continued to be referred to in shorter terminology (e.g., *pion* for pi meson), but in the case of the muon, it retained the shorter name and was never again properly referred to by older "mu meson" terminology.

The eventual recognition of the "mu meson" muon as a simple "heavy electron" with no role at all in the nuclear interaction, seemed so incongruous and surprising at the time, that Nobel laureate I. I. Rabi famously quipped, "Who ordered that?"

In the Rossi–Hall experiment (1941), muons were used to observe the time dilation (or alternately, length contraction) predicted by special relativity, for the first time.

14.2 Muon sources

On Earth, most naturally occurring muons are created by quasars and supernovas, which consist mostly of protons, many arriving from deep space at very high energy[4]

> About 10,000 muons reach every square meter of the earth's surface a minute; these charged particles form as by-products of cosmic rays colliding with molecules in the upper atmosphere. Traveling at relativistic

speeds, muons can penetrate tens of meters into rocks and other matter before attenuating as a result of absorption or deflection by other atoms.[5]

When a cosmic ray proton impacts atomic nuclei in the upper atmosphere, pions are created. These decay within a relatively short distance (meters) into muons (their preferred decay product), and muon neutrinos. The muons from these high energy cosmic rays generally continue in about the same direction as the original proton, at a velocity near the speed of light. Although their lifetime *without* relativistic effects would allow a half-survival distance of only about 456 m (2,197 μs×ln(2) × 0,9997×c) at most (as seen from Earth) the time dilation effect of special relativity (from the viewpoint of the Earth) allows cosmic ray secondary muons to survive the flight to the Earth's surface, since in the Earth frame, the muons have a longer half life due to their velocity. From the viewpoint (inertial frame) of the muon, on the other hand, it is the length contraction effect of special relativity which allows this penetration, since in the muon frame, its lifetime is unaffected, but the length contraction causes distances through the atmosphere and Earth to be far shorter than these distances in the Earth rest-frame. Both effects are equally valid ways of explaining the fast muon's unusual survival over distances.

Since muons are unusually penetrative of ordinary matter, like neutrinos, they are also detectable deep underground (700 meters at the Soudan 2 detector) and underwater, where they form a major part of the natural background ionizing radiation. Like cosmic rays, as noted, this secondary muon radiation is also directional.

The same nuclear reaction described above (i.e. hadron-hadron impacts to produce pion beams, which then quickly decay to muon beams over short distances) is used by particle physicists to produce muon beams, such as the beam used for the muon $g - 2$ experiment.[6]

14.3 Muon decay

See also: Michel parameters

Muons are unstable elementary particles and are heavier than electrons and neutrinos but lighter than all other matter particles. They decay via the weak interaction. Because lepton numbers must be conserved, one of the product neutrinos of muon decay must be a muon-type neutrino and the other an electron-type antineutrino (antimuon decay produces the corresponding antiparticles, as detailed below). Because charge must be conserved, one of the products of muon decay is always an electron of the same charge as the muon (a positron if it is a positive muon). Thus all muons decay to at least an electron, and two neutrinos. Sometimes, besides these necessary products, additional other particles that have no net charge and spin of zero (e.g., a pair of photons, or an electron-positron pair), are produced.

The dominant muon decay mode (sometimes called the Michel decay after Louis Michel) is the simplest possible: the muon decays to an electron, an electron antineutrino, and a muon neutrino. Antimuons, in mirror fashion, most often decay to the corresponding antiparticles: a positron, an electron neutrino, and a muon antineutrino. In formulaic terms, these two decays are:

μ− → e− + ν
e + ν
μ
μ+ → e+ + ν
e + ν
μ

The mean lifetime, τ = 1/Γ, of the (positive) muon is (2.1969811±0.0000022) μs.[1] The equality of the muon and antimuon lifetimes has been established to better than one part in 10^4.

The muon decay width which follows from Fermi's golden rule follows Sargent's law of fifth-power dependence on $m\mu$,

$$\Gamma = \frac{G_F^2 m_\mu^5}{192\pi^3} I\left(\frac{m_e^2}{m_\mu^2}\right),$$

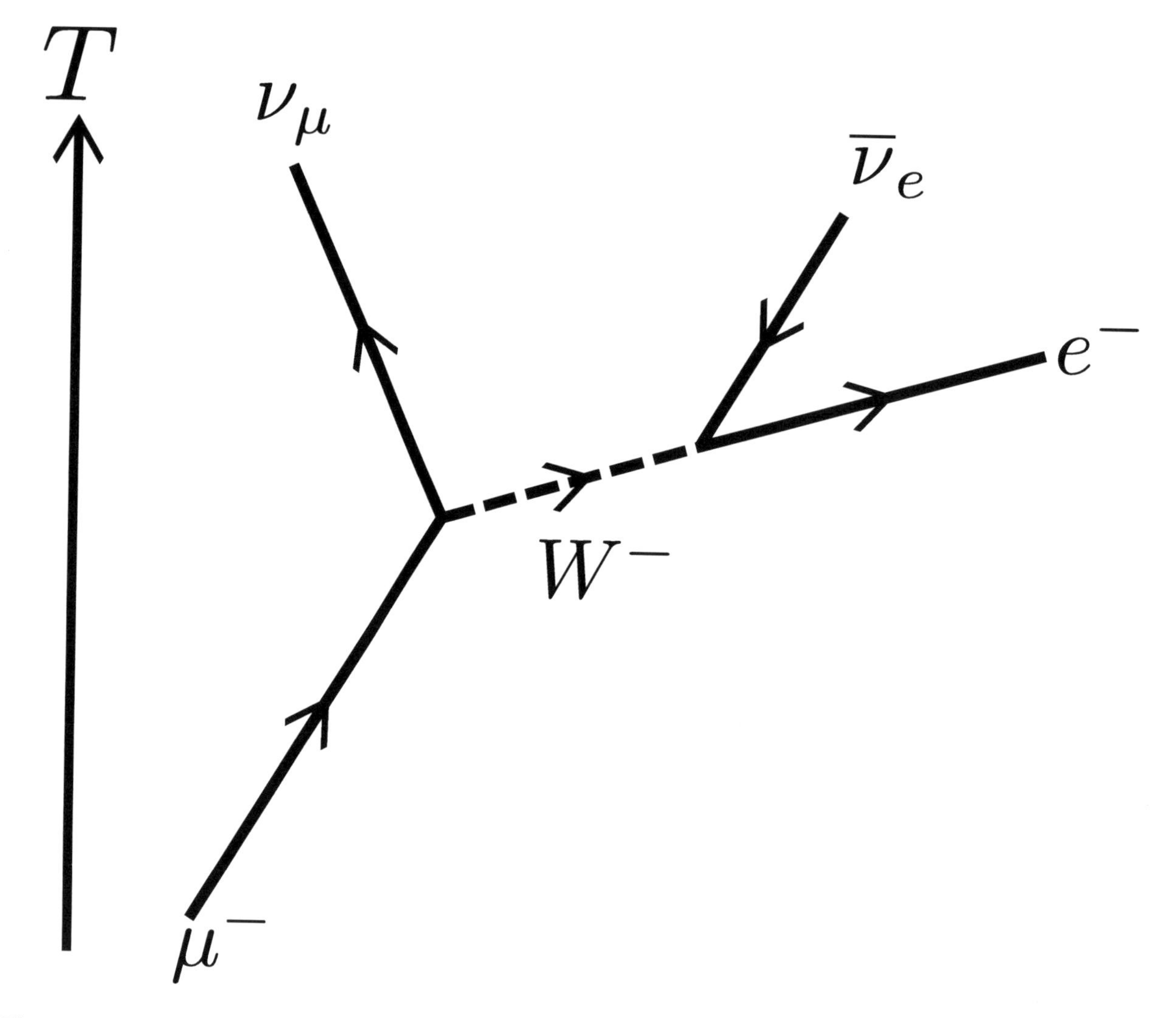

The most common decay of the muon

where $I(x) = 1 - 8x - 12x^2 \ln x + 8x^3 - x^4$, G_F is the Fermi coupling constant and $x = 2E_e/m_\mu c^2$ is the fraction of the maximum energy transmitted to the electron.

The decay distributions of the electron in muon decays have been parameterised using the so-called Michel parameters. The values of these four parameters are predicted unambiguously in the Standard Model of particle physics, thus muon decays represent a good test of the space-time structure of the weak interaction. No deviation from the Standard Model predictions has yet been found.

For the decay of the muon, the expected decay distribution for the Standard Model values of Michel parameters is

$$\frac{d^2\Gamma}{dx\, d\cos\theta} \sim x^2[(3 - 2x) + P_\mu \cos\theta(1 - 2x)]$$

where θ is the angle between the muon's polarization vector $\mathbf{P}_\mu$ and the decay-electron momentum vector, and $P_\mu = |\mathbf{P}_\mu|$ is the fraction of muons that are forward-polarized. Integrating this expression over electron energy gives the angular distribution of the daughter electrons:

$$\frac{d\Gamma}{d\cos\theta} \sim 1 - \frac{1}{3}P_\mu \cos\theta.$$

The electron energy distribution integrated over the polar angle (valid for $x < 1$) is

$$\frac{d\Gamma}{dx} \sim (3x^2 - 2x^3).$$

Due to the muons decaying by the weak interaction, parity conservation is violated. Replacing the $\cos\theta$ term in the expected decay values of the Michel Parameters with a $\cos\omega t$ term, where ω is the Larmor frequency from Larmor precession of the muon in a uniform magnetic field, given by:

$\omega = \frac{egB}{2m}$

where m is mass of the muon, e is charge, g is the muon g-factor and B is applied field.

A change in the electron distribution computed using the standard, unprecessional, Michel Parameters can be seen displaying a periodicity of π radians. This can be shown to physically correspond to a phase change of π, introduced in the electron distribution as the angular momentum is changed by the action of the charge conjugation operator, which is conserved by the weak interaction.

The observation of Parity violation in muon decay can be compared to the concept of violation of parity in weak interactions in general as an extension of The Wu Experiment, as well as the change of angular momentum introduced by a phase change of π corresponding to the charge-parity operator being invariant in this interaction. This fact is true for all lepton interactions in The Standard Model.

Certain neutrino-less decay modes are kinematically allowed but forbidden in the Standard Model. Examples forbidden by lepton flavour conservation are:

μ− → e− + γ and

μ− → e− + e+ + e− .

Observation of such decay modes would constitute clear evidence for theories beyond the Standard Model. Upper limits for the branching fractions of such decay modes were measured in many experiments starting more than 50 years ago. The current upper limit for the μ+ → e+ + γ branching fraction was measured 2013 in the MEG experiment and is 5.7×10^{-13}.[7]

14.4 Muonic atoms

The muon was the first elementary particle discovered that does not appear in ordinary atoms. Negative muons can, however, form muonic atoms (also called mu-mesic atoms), by replacing an electron in ordinary atoms. Muonic hydrogen atoms are much smaller than typical hydrogen atoms because the much larger mass of the muon gives it a much more localized ground-state wavefunction than is observed for the electron. In multi-electron atoms, when only one of the electrons is replaced by a muon, the size of the atom continues to be determined by the other electrons, and the atomic size is nearly unchanged. However, in such cases the orbital of the muon continues to be smaller and far closer to the nucleus than the atomic orbitals of the electrons.

Muonic helium is created by substituting a muon for one of the electrons in helium-4. The muon orbits much closer to the nucleus, so muonic helium can therefore be regarded like an isotope of helium whose nucleus consists of two neutrons, two protons and a muon, with a single electron outside. Colloquially, it could be called "helium 4.1", since the mass of the muon is roughly 0.1 amu. Chemically, muonic helium, possessing an unpaired valence electron, can bond with other atoms, and behaves more like a hydrogen atom than an inert helium atom.[8][9][10]

A positive muon, when stopped in ordinary matter, can also bind an electron and form an exotic atom known as muonium (Mu) atom, in which the muon acts as the nucleus. The positive muon, in this context, can be considered a pseudo-isotope of hydrogen with one ninth of the mass of the proton. Because the reduced mass of muonium, and hence its Bohr radius, is very close to that of hydrogen, this short-lived "atom" behaves chemically — to a first approximation — like hydrogen, deuterium and tritium.

14.5 Use in measurement of the proton charge radius

The recent culmination of a twelve year experiment at investigating the proton's charge radius involved the use of muonic hydrogen. This form of hydrogen is composed of a muon orbiting a proton.[11] The Lamb shift in muonic hydrogen was measured by driving the muon from its 2s state up to an excited 2p state using a laser. The frequency of the photon required to induce this transition was revealed to be 50 terahertz which, according to present theories of quantum electrodynamics, yields a value of 0.84184 ± 0.00067 femtometres for the charge radius of the proton.[12]

14.6 Anomalous magnetic dipole moment

The anomalous magnetic dipole moment is the difference between the experimentally observed value of the magnetic dipole moment and the theoretical value predicted by the Dirac equation. The measurement and prediction of this value is very important in the precision tests of QED (quantum electrodynamics). The E821 experiment[13] at Brookhaven National Laboratory (BNL) studied the precession of muon and anti-muon in a constant external magnetic field as they circulated in a confining storage ring. E821 reported the following average value[14] in 2006:

$$a = \frac{g-2}{2} = 0.00116592080(54)(33)$$

where the first errors are statistical and the second systematic.

The prediction for the value of the muon anomalous magnetic moment includes three parts:

$$\alpha\mu^{SM} = \alpha\mu^{QED} + \alpha\mu^{EW} + \alpha\mu^{had}.$$

The difference between the g-factors of the muon and the electron is due to their difference in mass. Because of the muon's larger mass, contributions to the theoretical calculation of its anomalous magnetic dipole moment from Standard Model weak interactions and from contributions involving hadrons are important at the current level of precision, whereas these effects are not important for the electron. The muon's anomalous magnetic dipole moment is also sensitive to contributions from new physics beyond the Standard Model, such as supersymmetry. For this reason, the muon's anomalous magnetic moment is normally used as a probe for new physics beyond the Standard Model rather than as a test of QED.[15] A new experiment at Fermilab using the E821 magnet will improve the precision of this measurement.[16]

14.7 Muon radiography and tomography

Main article: Muon tomography

Since muons are much more deeply penetrating than X-rays or gamma rays, muon imaging can be used with much thicker material or, with cosmic ray sources, larger objects. An important advantage of muon non-ionizing radiation is that it is safe for humans, plants, and animals. One example is commercial muon tomography used to image entire cargo containers to detect shielded nuclear material, as well as explosives or other contraband.[17]

The technique of muon transmission radiography based on cosmic ray sources was first used in the 1950s to measure the depth of the overburden of a tunnel in Australia[18] and in the 1960s to search for possible hidden chambers in the Pyramid of Chephren in Giza.[19]

In 2003, the scientists at Los Alamos National Laboratory developed a new imaging technique: **muon scattering tomography**. With muon scattering tomography, both incoming and outgoing trajectories for each particle are reconstructed, such as with sealed aluminum drift tubes.[20] Since the development of this technique, several companies have started to use it.

for alpha and beta decay. He considered that the new particle was emitted from the nucleus together with the electron or beta particle in the process of beta decay.[8][nb 2]

James Chadwick discovered a much more massive nuclear particle in 1932 and also named it a neutron, leaving two kinds of particles with the same name. Pauli earlier had used the term "neutron" for both the particle that conserved energy in beta decay, and a presumed neutral particle in the nucleus.[nb 3] The word "neutrino" entered the international vocabulary through Enrico Fermi, who used it during a conference in Paris in July 1932 and at the Solvay Conference in October 1933, where also Pauli employed it. The name (the Italian equivalent of "little neutral one") was jokingly coined by Edoardo Amaldi during a conversation with Fermi at the Institute of physics of via Panisperna in Rome, in order to distinguish this light neutral particle from Chadwick's neutron. [9]

In Fermi's theory of beta decay, Chadwick's large neutral particle could decay to a proton, electron, and the smaller neutral particle (flavored as an electron antineutrino):

n0 → p+ + e− + ν
e

Fermi's paper, written in 1934, unified Pauli's neutrino with Paul Dirac's positron and Werner Heisenberg's neutron–proton model and gave a solid theoretical basis for future experimental work. However, the journal Nature rejected Fermi's paper, saying that the theory was "too remote from reality". He submitted the paper to an Italian journal, which accepted it, but the general lack of interest in his theory at that early date caused him to switch to experimental physics.[10][11]

Nevertheless, even in 1934 there were hints that Bohr's idea that the energy conservation laws were not followed, was incorrect. At the Solvay conference of 1934, the first measurements of the energy spectra of beta decay were reported, and these spectra were found to impose a strict limit on the energy of electrons from each type of beta decay. Such a limit was not expected if the conservation of energy was not upheld, in which case any amount of energy would be expected to be statistically available in at least a few decays. The natural explanation of the beta decay spectrum as first measured in 1934 was that only a limited (and conserved) amount of energy was available, and a new particle was sometimes taking a varying fraction of this limited energy, leaving the rest for the beta particle. Pauli made use of the occasion to publicly emphasize that the still-undetected "neutrino" must be an actual particle.

16.1.2 Direct detection

In 1942 Wang Ganchang first proposed the use of beta capture to experimentally detect neutrinos.[12] In the 20 July 1956 issue of *Science*, Clyde Cowan, Frederick Reines, F. B. Harrison, H. W. Kruse, and A. D. McGuire published confirmation that they had detected the neutrino,[13][14] a result that was rewarded almost forty years later with the 1995 Nobel Prize.[15]

In this experiment, now known as the Cowan–Reines neutrino experiment, antineutrinos created in a nuclear reactor by beta decay reacted with protons to produce neutrons and positrons:

ν
e + p+ → n0 + e+

The positron quickly finds an electron, and they annihilate each other. The two resulting gamma rays (γ) are detectable. The neutron can be detected by its capture on an appropriate nucleus, releasing a gamma ray. The coincidence of both events – positron annihilation and neutron capture – gives a unique signature of an antineutrino interaction.

16.1.3 Neutrino flavor

The antineutrino discovered by Cowan and Reines is the antiparticle of the electron neutrino. In 1962, Leon M. Lederman, Melvin Schwartz and Jack Steinberger showed that more than one type of neutrino exists by first detecting interactions of the muon neutrino (already hypothesised with the name *neutretto*),[16] which earned them the 1988 Nobel Prize in Physics. When the third type of lepton, the tau, was discovered in 1975 at the Stanford Linear Accelerator Center, it too was expected to have an associated neutrino (the tau neutrino). First evidence for this third neutrino type came from

Clyde Cowan conducting the neutrino experiment c. 1956

the observation of missing energy and momentum in tau decays analogous to the beta decay leading to the discovery of the electron neutrino. The first detection of tau neutrino interactions was announced in summer of 2000 by the DONUT collaboration at Fermilab; its existence had already been inferred by both theoretical consistency and experimental data from the Large Electron–Positron Collider.

16.1.4 Solar neutrino problem

Main article: Solar neutrino problem

Starting in the late 1960s, several experiments found that the number of electron neutrinos arriving from the Sun was between one third and one half the number predicted by the Standard Solar Model. This discrepancy, which became known as the solar neutrino problem, remained unresolved for some thirty years. It was resolved by discovery of neutrino oscillation and mass. (The Standard Model of particle physics had assumed that neutrinos are massless and cannot change flavor. However, if neutrinos had mass, they could change flavor, or *oscillate* between flavors).

16.1.5 Oscillation

A practical method for investigating neutrino oscillations was first suggested by Bruno Pontecorvo in 1957 using an analogy with kaon oscillations; over the subsequent 10 years he developed the mathematical formalism and the modern formulation of vacuum oscillations. In 1985 Stanislav Mikheyev and Alexei Smirnov (expanding on 1978 work by Lincoln Wolfenstein) noted that flavor oscillations can be modified when neutrinos propagate through matter. This so-called Mikheyev–Smirnov–Wolfenstein effect (MSW effect) is important to understand because many neutrinos emitted by fusion in the Sun pass through the dense matter in the solar core (where essentially all solar fusion takes place) on their way to detectors on Earth.

Starting in 1998, experiments began to show that solar and atmospheric neutrinos change flavors (see Super-Kamiokande and Sudbury Neutrino Observatory). This resolved the solar neutrino problem: the electron neutrinos produced in the Sun had partly changed into other flavors which the experiments could not detect.

Although individual experiments, such as the set of solar neutrino experiments, are consistent with non-oscillatory mechanisms of neutrino flavor conversion, taken altogether, neutrino experiments imply the existence of neutrino oscillations. Especially relevant in this context are the reactor experiment KamLAND and the accelerator experiments such as MINOS. The KamLAND experiment has indeed identified oscillations as the neutrino flavor conversion mechanism involved in the solar electron neutrinos. Similarly MINOS confirms the oscillation of atmospheric neutrinos and gives a better determination of the mass squared splitting.[17]

16.1.6 Supernova neutrinos

See also: Supernova Early Warning System

Raymond Davis, Jr. and Masatoshi Koshiba were jointly awarded the 2002 Nobel Prize in Physics; Davis for his pioneer work on cosmic neutrinos and Koshiba for the first real time observation of supernova neutrinos. The detection of solar neutrinos, and of neutrinos of the SN 1987A supernova in 1987 marked the beginning of neutrino astronomy. In an average supernova, approximately 10^{57} (an Octodecillion) neutrinos are released.

16.2 Properties and reactions

The neutrino has half-integer spin ($\hbar/2$) and is therefore a fermion. Neutrinos interact primarily through the weak force. The discovery of neutrino flavor oscillations implies that neutrinos have mass. The existence of a neutrino mass strongly suggests the existence of a tiny neutrino magnetic moment[18] of the order of 10^{-19} μB, allowing the possibility that neutrinos may interact electromagnetically as well. An experiment done by C. S. Wu at Columbia University showed that neutrinos always have left-handed chirality.[19] It is very hard to uniquely identify neutrino interactions among the natural background of radioactivity. For this reason, in early experiments a special reaction channel was chosen to facilitate the identification: the interaction of an antineutrino with one of the hydrogen nuclei in the water molecules. A hydrogen nucleus is a single proton, so simultaneous nuclear interactions, which would occur within a heavier nucleus, don't need to be considered for the detection experiment. Within a cubic metre of water placed right outside a nuclear reactor, only relatively few such interactions can be recorded, but the setup is now used for measuring the reactor's plutonium production rate.

16.2.1 Mikheyev–Smirnov–Wolfenstein effect

Main article: Mikheyev–Smirnov–Wolfenstein effect

Neutrinos traveling through matter, in general, undergo a process analogous to light traveling through a transparent material. This process is not directly observable because it does not produce ionizing radiation, but gives rise to the MSW effect. Only a small fraction of the neutrino's energy is transferred to the material.

16.2.2 Nuclear reactions

Neutrinos can interact with a nucleus, changing it to another nucleus. This process is used in radiochemical neutrino detectors. In this case, the energy levels and spin states within the target nucleus have to be taken into account to estimate the probability for an interaction. In general the interaction probability increases with the number of neutrons and protons within a nucleus.

16.2.3 Induced fission

Very much like neutrons do in nuclear reactors, neutrinos can induce fission reactions within heavy nuclei.[20] So far, this reaction has not been measured in a laboratory, but is predicted to happen within stars and supernovae. The process affects the abundance of isotopes seen in the universe.[21] Neutrino fission of deuterium nuclei has been observed in the Sudbury Neutrino Observatory, which uses a heavy water detector.

16.2.4 Types

There are three known types (*flavors*) of neutrinos: electron neutrino ν
e, muon neutrino ν
μ and tau neutrino ν
τ, named after their partner leptons in the Standard Model (see table at right). The current best measurement of the number of neutrino types comes from observing the decay of the Z boson. This particle can decay into any light neutrino and its antineutrino, and the more types of light neutrinos[nb 4] available, the shorter the lifetime of the Z boson. Measurements of the Z lifetime have shown that the number of light neutrino types is 3.[18] The correspondence between the six quarks in the Standard Model and the six leptons, among them the three neutrinos, suggests to physicists' intuition that there should be exactly three types of neutrino. However, actual proof that there are only three kinds of neutrinos remains an elusive goal of particle physics.

The possibility of *sterile* neutrinos—relatively light neutrinos which do not participate in the weak interaction but which could be created through flavor oscillation (see below)—is unaffected by these Z-boson-based measurements, and the existence of such particles is in fact hinted by experimental data from the LSND experiment. However, the currently running MiniBooNE experiment suggested, until recently, that sterile neutrinos are not required to explain the experimental data,[22] although the latest research into this area is on-going and anomalies in the MiniBooNE data may allow for exotic neutrino types, including sterile neutrinos.[23] A recent re-analysis of reference electron spectra data from the Institut Laue-Langevin[24] has also hinted at a fourth, sterile neutrino.[25]

Recently analyzed data from the Wilkinson Microwave Anisotropy Probe of the cosmic background radiation is compatible with either three or four types of neutrinos. It is hoped that the addition of two more years of data from the probe will resolve this uncertainty.[26]

16.2.5 Antineutrinos

Antineutrinos, the antiparticles of neutrinos, are neutral particles produced in nuclear beta decay. These are emitted during beta particle emissions, in which a neutron decays into a proton, electron, and antineutrino. They have a spin of ½, and are part of the lepton family of particles. All antineutrinos observed thus far possess right-handed helicity (i.e. only one of the two possible spin states has ever been seen), while neutrinos are left-handed. Antineutrinos, like neutrinos, interact with other matter only through the gravitational and weak forces, making them very difficult to detect experimentally. Neutrino oscillation experiments indicate that antineutrinos have mass, but beta decay experiments constrain that mass to this value. Thus, there exists at least one neutrino mass eigenstate with a mass of at least 0.04 eV.[45]

In 2009, lensing data of a galaxy cluster were analyzed to predict a neutrino mass of about 1.5 eV.[46] This surprisingly high value requires that the three neutrino masses be nearly equal, with neutrino oscillations of order meV. The masses lie below the Mainz-Troitsk upper bound of 2.2 eV for the electron antineutrino.[47] The latter will be tested in 2015 in the KATRIN experiment, that searches for a mass between 0.2 eV and 2 eV.

A number of efforts are under way to directly determine the absolute neutrino mass scale in laboratory experiments. The methods applied involve nuclear beta decay (KATRIN and MARE).

On 31 May 2010, OPERA researchers observed the first tau neutrino candidate event in a muon neutrino beam, the first time this transformation in neutrinos had been observed, providing further evidence that they have mass.[48]

In July 2010 the 3-D MegaZ DR7 galaxy survey reported that they had measured a limit of the combined mass of the three neutrino varieties to be less than 0.28 eV.[49] A tighter upper bound yet for this sum of masses, 0.23 eV, was reported in March 2013 by the Planck collaboration,[50] whereas a February 2014 result estimates the sum as 0.320 ± 0.081 eV based on discrepancies between the cosmological consequences implied by Planck's detailed measurements of the Cosmic Microwave Background and predictions arising from observing other phenomena, combined with the assumption that neutrinos are responsible for the observed weaker gravitational lensing than would be expected from massless neutrinos.[51]

If the neutrino is a Majorana particle, the mass may be calculated by finding the half life of neutrinoless double-beta decay of certain nuclei. As of 2015, the lowest upper limit on the Majorana mass of the neutrino has been set by KamLAND-Zen: 0.12–0.25 eV.[52]

16.2.9 Size

Standard Model neutrinos are fundamental point-like particles. An effective size can be defined using their electroweak cross section (apparent size in electroweak interaction). The average electroweak characteristic size is $r^2 = n \times 10^{-33}$ cm^2 ($n \times 1$ nanobarn), where $n = 3.2$ for electron neutrino, $n = 1.7$ for muon neutrino and $n = 1.0$ for tau neutrino; it depends on no other properties than mass.[53] However, this is best understood as being relevant only to probability of scattering. Since the neutrino does not interact electromagnetically, and is defined quantum mechanically by a wavefunction, it does not have a size in the same sense as everyday objects.[54] Furthermore, processes that produce neutrinos impart such high energies to them that they travel at almost the speed of light. Nevertheless, neutrinos are fermions, and thus obey the Pauli exclusion principle, i.e. that increasing their density forces them into progressively higher momentum states.

16.2.10 Chirality

Experimental results show that (nearly) all produced and observed neutrinos have left-handed helicities (spins antiparallel to momenta), and all antineutrinos have right-handed helicities, within the margin of error. In the massless limit, it means that only one of two possible chiralities is observed for either particle. These are the only chiralities included in the Standard Model of particle interactions.

It is possible that their counterparts (right-handed neutrinos and left-handed antineutrinos) simply do not exist. If they do, their properties are substantially different from observable neutrinos and antineutrinos. It is theorized that they are either very heavy (on the order of GUT scale—see *Seesaw mechanism*), do not participate in weak interaction (so-called sterile neutrinos), or both.

The existence of nonzero neutrino masses somewhat complicates the situation. Neutrinos are produced in weak interactions as chirality eigenstates. However, chirality of a massive particle is not a constant of motion; helicity is, but the chirality operator does not share eigenstates with the helicity operator. Free neutrinos propagate as mixtures of left- and right-handed helicity states, with mixing amplitudes on the order of $m\nu/E$. This does not significantly affect the experiments, because neutrinos involved are nearly always ultrarelativistic, and thus mixing amplitudes are vanishingly small. For example, most solar neutrinos have energies on the order of 100 keV–1 MeV, so the fraction of neutrinos with "wrong" helicity among them cannot exceed 10^{-10}.[55][56]

16.3 Sources

16.3.1 Artificial

Reactor neutrinos

Nuclear reactors are the major source of human-generated neutrinos. Antineutrinos are made in the beta-decay of neutron-rich daughter fragments in the fission process. Generally, the four main isotopes contributing to the antineutrino flux are 235U, 238U, 239Pu and 241Pu (i.e. via the antineutrinos emitted during beta-minus decay of their respective fission fragments). The average nuclear fission releases about 200 MeV of energy, of which roughly 4.5% (or about 9 MeV)[57] is radiated away as antineutrinos. For a typical nuclear reactor with a thermal power of 4000 MW, meaning that the core produces this much heat, and an electrical power generation of 1300 MW, the total power production from fissioning atoms is actually 4185 MW, of which 185 MW is radiated away as antineutrino radiation and never appears in the engineering. This is to say, 185 MW of fission energy is *lost* from this reactor and does not appear as heat available to run turbines, since antineutrinos penetrate all building materials practically without interaction.[nb 5]

The antineutrino energy spectrum depends on the degree to which the fuel is burned (plutonium-239 fission antineutrinos on average have slightly more energy than those from uranium-235 fission), but in general, the *detectable* antineutrinos from fission have a peak energy between about 3.5 and 4 MeV, with a maximum energy of about 10 MeV.[58] There is no established experimental method to measure the flux of low-energy antineutrinos. Only antineutrinos with an energy above threshold of 1.8 MeV can be uniquely identified (see *neutrino detection* below). An estimated 3% of all antineutrinos from a nuclear reactor carry an energy above this threshold. Thus, an average nuclear power plant may generate over 10^{20} antineutrinos per second above this threshold, but also a much larger number (97%/3% = ~30 times this number) below the energy threshold, which cannot be seen with present detector technology.

Accelerator neutrinos

Some particle accelerators have been used to make neutrino beams. The technique is to collide protons with a fixed target, producing charged pions or kaons. These unstable particles are then magnetically focused into a long tunnel where they decay while in flight. Because of the relativistic boost of the decaying particle, the neutrinos are produced as a beam rather than isotropically. Efforts to construct an accelerator facility where neutrinos are produced through muon decays are ongoing.[59] Such a setup is generally known as a neutrino factory.

Nuclear bombs

Nuclear bombs also produce very large quantities of neutrinos. Fred Reines and Clyde Cowan considered the detection of neutrinos from a bomb prior to their search for reactor neutrinos; a fission reactor was recommended as a better alternative by Los Alamos physics division leader J.M.B. Kellogg.[60] Fission bombs produce antineutrinos (from the fission process), and fusion bombs produce both neutrinos (from the fusion process) and antineutrinos (from the initiating fission explosion).

16.3.2 Geologic

Main article: Geoneutrino

Neutrinos are part of the natural background radiation. In particular, the decay chains of 238U and 232Th isotopes, as well as40K, include beta decays which emit antineutrinos. These so-called geoneutrinos can provide valuable information on the Earth's interior. A first indication for geoneutrinos was found by the KamLAND experiment in 2005. KamLAND's main background in the geoneutrino measurement are the antineutrinos coming from reactors. Several future experiments aim at improving the geoneutrino measurement and these will necessarily have to be far away from reactors.

be very small. A neutrino–antineutrino interaction has been suggested in attempts to form a composite photon with the neutrino theory of light.

Because antineutrinos and neutrinos are neutral particles, it is possible that they are actually the same particle. Particles that have this property are known as Majorana particles. Majorana neutrinos have the property that the neutrino and antineutrino could be distinguished only by chirality; what experiments observe as a difference between the neutrino and antineutrino could simply be due to one particle with two possible chiralities. If neutrinos are indeed Majorana particles, neutrinoless double beta decay, as well as a range of other lepton number violating phenomena, would be allowed. Several experiments have been and are being conducted to search for this process.

Researchers around the world have begun to investigate the possibility of using antineutrinos for reactor monitoring in the context of preventing the proliferation of nuclear weapons.[27][28][29]

Antineutrinos were first detected as a result of their interaction with protons in a large tank of water. This was installed next to a nuclear reactor as a controllable source of the antineutrinos. (See: Cowan–Reines neutrino experiment)

Only antineutrinos, not neutrinos, take part in the Glashow resonance.

16.2.6 Flavor oscillations

Main article: Neutrino oscillation

Neutrinos are most often created or detected with a well defined flavor (electron, muon, tau). However, in a phenomenon known as neutrino flavor oscillation, neutrinos are able to oscillate among the three available flavors while they propagate through space. Specifically, this occurs because the neutrino flavor eigenstates are not the same as the neutrino mass eigenstates (simply called 1, 2, 3). This allows for a neutrino that was produced as an electron neutrino at a given location to have a calculable probability to be detected as either a muon or tau neutrino after it has traveled to another location. This quantum mechanical effect was first hinted by the discrepancy between the number of electron neutrinos detected from the Sun's core failing to match the expected numbers, dubbed as the "solar neutrino problem". In the Standard Model the existence of flavor oscillations implies nonzero differences between the neutrino masses, because the amount of mixing between neutrino flavors at a given time depends on the differences between their squared masses. There are other possibilities in which neutrino can oscillate even if they are massless. If Lorentz invariance is not an exact symmetry, neutrinos can experience Lorentz-violating oscillations.[30]

It is possible that the neutrino and antineutrino are in fact the same particle, a hypothesis first proposed by the Italian physicist Ettore Majorana. The neutrino could transform into an antineutrino (and vice versa) by flipping the orientation of its spin state.[31]

This change in spin would require the neutrino and antineutrino to have nonzero mass, and therefore travel slower than light, because such a spin flip, caused only by a change in point of view, can take place only if inertial frames of reference exist that move faster than the particle: such a particle has a spin of one orientation when seen from a frame which moves slower than the particle, but the opposite spin when observed from a frame that moves faster than the particle.

On July 19, 2013 the results from the T2K experiment presented at the European Physical Society Conference on High Energy Physics in Stockholm, Sweden, confirmed neutrino oscillation theory.[32][33]

16.2.7 Speed

Main article: Measurements of neutrino speed

Before neutrinos were found to oscillate, they were generally assumed to be massless, propagating at the speed of light. According to the theory of special relativity, the question of neutrino velocity is closely related to their mass. If neutrinos are massless, they must travel at the speed of light. However, if they have mass, they cannot reach the speed of light.

Also some Lorentz-violating variants of quantum gravity might allow faster-than-light neutrinos. A comprehensive framework for Lorentz violations is the Standard-Model Extension (SME).

In the early 1980s, first measurements of neutrino speed were done using pulsed pion beams (produced by pulsed proton beams hitting a target). The pions decayed producing neutrinos, and the neutrino interactions observed within a time window in a detector at a distance were consistent with the speed of light. This measurement was repeated in 2007 using the MINOS detectors, which found the speed of 3 GeV neutrinos to be, at the 99% confidence level, in the range between 0.999976 c and 1.000126 c. The central value of 1.000051c is higher than the speed of light but is also consistent with a velocity of exactly c or even slightly less. This measurement set an upper bound on the mass of the muon neutrino of 50 MeV at 99% confidence.[34][35] After the detectors for the project were upgraded in 2012, MINOS refined their initial result and found agreement with the speed of light, with the difference in the arrival time of neutrinos and light of −0.0006% (±0.0012%).[36]

A similar observation was made, on a much larger scale, with supernova 1987A (SN 1987A). 10-MeV antineutrinos from the supernova were detected within a time window that was consistent with the speed of light for the neutrinos. Currently, the question of whether or not neutrinos have mass cannot be decided; their speed is (as yet) indistinguishable from the speed of light.

In September 2011, the OPERA collaboration released calculations showing velocities of 17-GeV and 28-GeV neutrinos exceeding the speed of light in their experiments (see Faster-than-light neutrino anomaly). In November 2011, OPERA repeated its experiment with changes so that the speed could be determined individually for each detected neutrino. The results showed the same faster-than-light speed. However, in February 2012 reports came out that the results may have been caused by a loose fiber optic cable attached to one of the atomic clocks which measured the departure and arrival times of the neutrinos. An independent recreation of the experiment in the same laboratory by ICARUS found no discernible difference between the speed of a neutrino and the speed of light.[37] In June 2012, CERN announced that new measurements conducted by four Gran Sasso experiments (OPERA, ICARUS, Borexino and LVD) found agreement between the speed of light and the speed of neutrinos, finally refuting the initial OPERA result.[38]

16.2.8 Mass

The Standard Model of particle physics assumed that neutrinos are massless. However the experimentally established phenomenon of neutrino oscillation, which mixes neutrino flavour states with neutrino mass states (analogously to CKM mixing), requires neutrinos to have nonzero masses.[22] Massive neutrinos were originally conceived by Bruno Pontecorvo in the 1950s. Enhancing the basic framework to accommodate their mass is straightforward by adding a right-handed Lagrangian. This can be done in two ways. If, like other fundamental Standard Model particles, mass is generated by the Dirac mechanism, then the framework would require an SU(2) singlet. This particle would have no other Standard Model interactions (apart from the Yukawa interactions with the neutral component of the Higgs doublet), so is called a sterile neutrino. Or, mass can be generated by the Majorana mechanism, which would require the neutrino and antineutrino to be the same particle.

The strongest upper limit on the masses of neutrinos comes from cosmology: the Big Bang model predicts that there is a fixed ratio between the number of neutrinos and the number of photons in the cosmic microwave background. If the total energy of all three types of neutrinos exceeded an average of 50 eV per neutrino, there would be so much mass in the universe that it would collapse.[39] This limit can be circumvented by assuming that the neutrino is unstable; however, there are limits within the Standard Model that make this difficult. A much more stringent constraint comes from a careful analysis of cosmological data, such as the cosmic microwave background radiation, galaxy surveys, and the Lyman-alpha forest. These indicate that the summed masses of the three neutrinos must be less than 0.3 eV.[40]

In 1998, research results at the Super-Kamiokande neutrino detector determined that neutrinos can oscillate from one flavor to another, which requires that they must have a nonzero mass.[41] While this shows that neutrinos have mass, the absolute neutrino mass scale is still not known. This is because neutrino oscillations are sensitive only to the difference in the squares of the masses.[42] The best estimate of the difference in the squares of the masses of mass eigenstates 1 and 2 was published by KamLAND in 2005: Δm2
21 = 0.000079 eV^2.[43] In 2006, the MINOS experiment measured oscillations from an intense muon neutrino beam, determining the difference in the squares of the masses between neutrino mass eigenstates 2 and 3. The initial results indicate |Δm2
32| = 0.0027 eV^2, consistent with previous results from Super-Kamiokande.[44] Since |Δm2
32| is the difference of two squared masses, at least one of them has to have a value which is at least the square root of

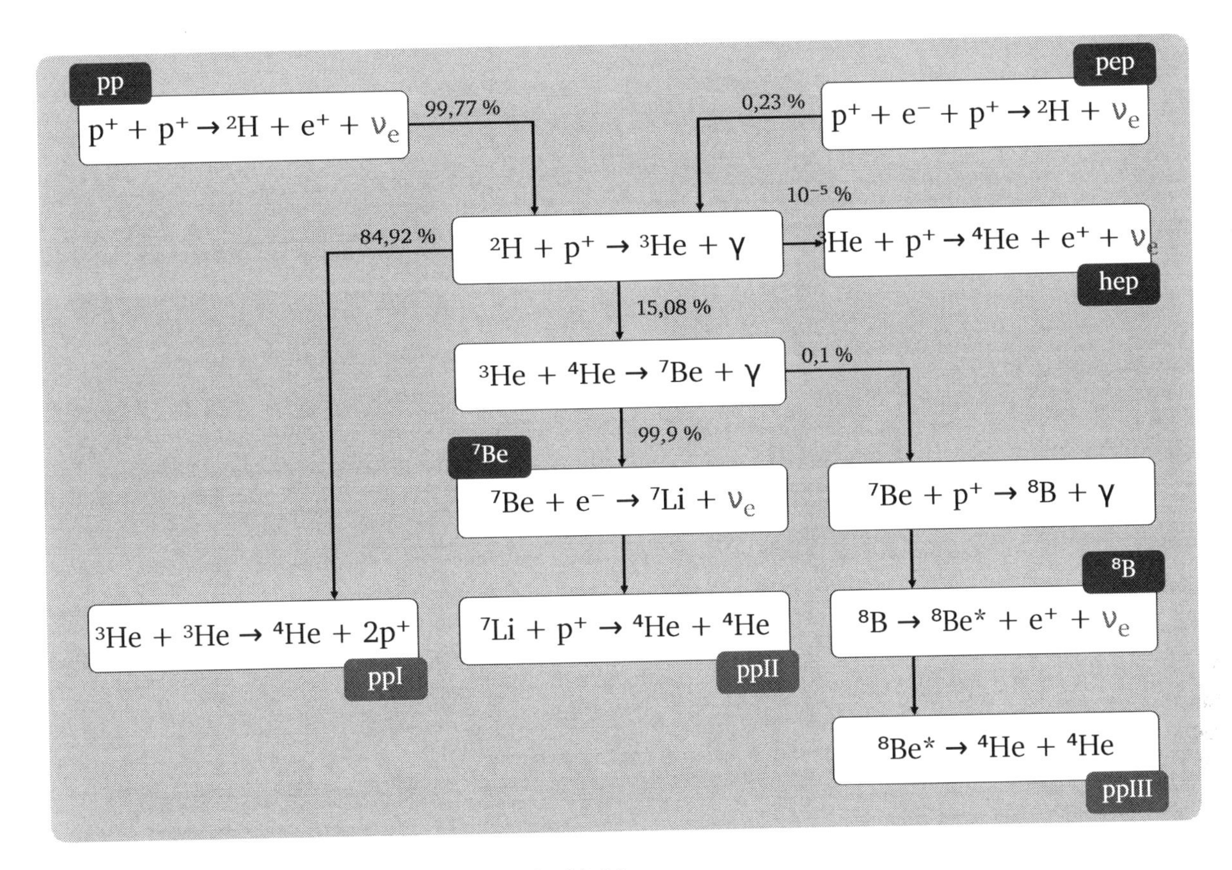

Solar neutrinos (proton–proton chain) in the Standard Solar Model

16.3.3 Atmospheric

Atmospheric neutrinos result from the interaction of cosmic rays with atomic nuclei in the Earth's atmosphere, creating showers of particles, many of which are unstable and produce neutrinos when they decay. A collaboration of particle physicists from Tata Institute of Fundamental Research (India), Osaka City University (Japan) and Durham University (UK) recorded the first cosmic ray neutrino interaction in an underground laboratory in Kolar Gold Fields in India in 1965.

16.3.4 Solar

Solar neutrinos originate from the nuclear fusion powering the Sun and other stars. The details of the operation of the Sun are explained by the Standard Solar Model. In short: when four protons fuse to become one helium nucleus, two of them have to convert into neutrons, and each such conversion releases one electron neutrino.

The Sun sends enormous numbers of neutrinos in all directions. Each second, about 65 billion (6.5×10^{10}) solar neutrinos pass through every square centimeter on the part of the Earth that faces the Sun.[6] Since neutrinos are insignificantly absorbed by the mass of the Earth, the surface area on the side of the Earth opposite the Sun receives about the same number of neutrinos as the side facing the Sun.

16.3.5 Supernovae

In 1966 Colgate and White[61] calculated that neutrinos carry away most of the gravitational energy released by the collapse of massive stars, events now categorized as Type Ib and Ic and Type II supernovae. When such stars collapse, matter densities at the core becomes so high (10^{17} kg/m^3) that the degeneracy of electrons is not enough to prevent protons and electrons from combining to form a neutron and an electron neutrino. A second and more important neutrino source

SN 1987A

is the thermal energy (100 billion kelvins) of the newly formed neutron core, which is dissipated via the formation of neutrino–antineutrino pairs of all flavors.[62]

Colgate and White's theory of supernova neutrino production was confirmed in 1987, when neutrinos from supernova 1987A were detected. The water-based detectors Kamiokande II and IMB detected 11 and 8 antineutrinos of thermal origin,[62] respectively, while the scintillator-based Baksan detector found 5 neutrinos (lepton number = 1) of either thermal or electron-capture origin, in a burst lasting less than 13 seconds. The neutrino signal from the supernova arrived at earth several hours before the arrival of the first electromagnetic radiation, as expected from the evident fact that the latter emerges along with the shock wave. The exceptionally feeble interaction with normal matter allowed the neutrinos to pass through the churning mass of the exploding star, while the electromagnetic photons were slowed.

Because neutrinos interact so little with matter, it is thought that a supernova's neutrino emissions carry information about the innermost regions of the explosion. Much of the *visible* light comes from the decay of radioactive elements produced by the supernova shock wave, and even light from the explosion itself is scattered by dense and turbulent gases, and thus delayed. The neutrino burst is expected to reach Earth before any electromagnetic waves, including visible light, gamma rays or radio waves. The exact time delay depends on the velocity of the shock wave and on the thickness of the outer layer of the star. For a Type II supernova, astronomers expect the neutrino flood to be released seconds after the stellar core collapse, while the first electromagnetic signal may emerge hours later, after the explosion shock wave has had time to reach the surface of the star. The SNEWS project uses a network of neutrino detectors to monitor the sky for candidate supernova events; the neutrino signal will provide a useful advance warning of a star exploding in the Milky Way.

Although neutrinos pass through the outer gases of a supernova without scattering, they provide information about the deeper supernova core with evidence that here, even neutrinos scatter to a significant extent. In a supernova core the densities are those of a neutron star (which is expected to be formed in this type of supernova),[63] becoming large enough to influence the duration of the neutrino signal by delaying some neutrinos. The length of the neutrino signal from SN 1987A, some 13 seconds, was far longer than it would take in theory for neutrinos to pass directly through the neutrino-generating core of a supernova, expected to be only 32 kilometers in diameter SN 1987A. The number of neutrinos counted was also consistent with a total neutrino energy of 2.2×10^{46} joules, which was estimated to be nearly all of the total energy of the supernova.[64]

16.3.6 Supernova remnants

The energy of supernova neutrinos ranges from a few to several tens of MeV. However, the sites where cosmic rays are accelerated are expected to produce neutrinos that are at least one million times more energetic, produced from turbulent gaseous environments left over by supernova explosions: the supernova remnants. The origin of the cosmic rays was attributed to supernovas by Walter Baade and Fritz Zwicky; this hypothesis was refined by Vitaly L. Ginzburg and Sergei I. Syrovatsky who attributed the origin to supernova remnants, and supported their claim by the crucial remark, that the cosmic ray losses of the Milky Way is compensated, if the efficiency of acceleration in supernova remnants is about 10 percent. Ginzburg and Syrovatskii's hypothesis is supported by the specific mechanism of "shock wave acceleration" happening in supernova remnants, which is consistent with the original theoretical picture drawn by Enrico Fermi, and is receiving support from observational data. The very-high-energy neutrinos are still to be seen, but this branch of neutrino astronomy is just in its infancy. The main existing or forthcoming experiments that aim at observing very-high-energy neutrinos from our galaxy are Baikal, AMANDA, IceCube, ANTARES, NEMO and Nestor. Related information is provided by very-high-energy gamma ray observatories, such as VERITAS, HESS and MAGIC. Indeed, the collisions of cosmic rays are supposed to produce charged pions, whose decay give the neutrinos, and also neutral pions, whose decay give gamma rays: the environment of a supernova remnant is transparent to both types of radiation.

Still-higher-energy neutrinos, resulting from the interactions of extragalactic cosmic rays, could be observed with the Pierre Auger Observatory or with the dedicated experiment named ANITA.

16.3.7 Big Bang

Main article: Cosmic neutrino background

It is thought that, just like the cosmic microwave background radiation left over from the Big Bang, there is a background of low-energy neutrinos in our Universe. In the 1980s it was proposed that these may be the explanation for the dark matter thought to exist in the universe. Neutrinos have one important advantage over most other dark matter candidates: it is known that they exist. However, this idea also has serious problems.

From particle experiments, it is known that neutrinos are very light. This means that they easily move at speeds close to the speed of light. For this reason, dark matter made from neutrinos is termed "hot dark matter". The problem is that being fast moving, the neutrinos would tend to have spread out evenly in the universe before cosmological expansion made them cold enough to congregate in clumps. This would cause the part of dark matter made of neutrinos to be smeared out and unable to cause the large galactic structures that we see.

Further, these same galaxies and groups of galaxies appear to be surrounded by dark matter that is not fast enough to escape from those galaxies. Presumably this matter provided the gravitational nucleus for formation. This implies that neutrinos cannot make up a significant part of the total amount of dark matter.

From cosmological arguments, relic background neutrinos are estimated to have density of 56 of each type per cubic centimeter and temperature 1.9 K (1.7×10^{-4} eV) if they are massless, much colder if their mass exceeds 0.001 eV. Although their density is quite high, they have not yet been observed in the laboratory, as their energy is below thresholds of most detection methods, and due to extremely low neutrino interaction cross-sections at sub-eV energies. In contrast, boron-8 solar neutrinos—which are emitted with a higher energy—have been detected definitively despite having a space density that is lower than that of relic neutrinos by some 6 orders of magnitude.

16.4 Detection

Main article: Neutrino detector

Neutrinos cannot be detected directly, because they do not ionize the materials they are passing through (they do not carry electric charge and other proposed effects, like the MSW effect, do not produce traceable radiation). A unique reaction to identify antineutrinos, sometimes referred to as inverse beta decay, as applied by Reines and Cowan (see below), requires a very large detector in order to detect a significant number of neutrinos. All detection methods require the neutrinos to carry a minimum threshold energy. So far, there is no detection method for low-energy neutrinos, in the sense that potential neutrino interactions (for example by the MSW effect) cannot be uniquely distinguished from other causes. Neutrino detectors are often built underground in order to isolate the detector from cosmic rays and other background radiation.

Antineutrinos were first detected in the 1950s near a nuclear reactor. Reines and Cowan used two targets containing a solution of cadmium chloride in water. Two scintillation detectors were placed next to the cadmium targets. Antineutrinos with an energy above the threshold of 1.8 MeV caused charged current interactions with the protons in the water, producing positrons and neutrons. This is very much like β+ decay, where energy is used to convert a proton into a neutron, a positron (e+) and an electron neutrino (ν
e) is emitted:

From known β+ decay:

Energy + p → n + e+ + ν
e

In the Cowan and Reines experiment, instead of an outgoing neutrino, you have an incoming antineutrino (ν
e) from a nuclear reactor:

Energy (>1.8 MeV) + p + ν
e → n + e+

The resulting positron annihilation with electrons in the detector material created photons with an energy of about 0.5 MeV. Pairs of photons in coincidence could be detected by the two scintillation detectors above and below the target. The neutrons were captured by cadmium nuclei resulting in gamma rays of about 8 MeV that were detected a few microseconds after the photons from a positron annihilation event.

Since then, various detection methods have been used. Super Kamiokande is a large volume of water surrounded by photomultiplier tubes that watch for the Cherenkov radiation emitted when an incoming neutrino creates an electron or muon in the water. The Sudbury Neutrino Observatory is similar, but uses heavy water as the detecting medium, which uses the same effects, but also allows the additional reaction any-flavor neutrino photo-dissociation of deuterium, resulting in a free neutron which is then detected from gamma radiation after chlorine-capture. Other detectors have consisted of large volumes of chlorine or gallium which are periodically checked for excesses of argon or germanium, respectively, which are created by electron-neutrinos interacting with the original substance. MINOS uses a solid plastic scintillator coupled to photomultiplier tubes, while Borexino uses a liquid pseudocumene scintillator also watched by photomultiplier tubes and the proposed NOvA detector will use liquid scintillator watched by avalanche photodiodes. The IceCube Neutrino Observatory uses 1 km^3 of the Antarctic ice sheet near the south pole with photomultiplier tubes distributed throughout the volume.

16.5 Motivation for scientific interest

Neutrinos' low mass and neutral charge mean they interact exceedingly weakly with other particles and fields. This feature of weak interaction interests scientists because it means neutrinos can be used to probe environments that other radiation (such as light or radio waves) cannot penetrate.

Using neutrinos as a probe was first proposed in the mid-20th century as a way to detect conditions at the core of the Sun. The solar core cannot be imaged directly because electromagnetic radiation (such as light) is diffused by the great amount and density of matter surrounding the core. On the other hand, neutrinos pass through the Sun with few interactions. Whereas photons emitted from the solar core may require 40,000 years to diffuse to the outer layers of the Sun, neutrinos generated in stellar fusion reactions at the core cross this distance practically unimpeded at nearly the speed of light.[65][66]

Neutrinos are also useful for probing astrophysical sources beyond the Solar System because they are the only known particles that are not significantly attenuated by their travel through the interstellar medium. Optical photons can be obscured or diffused by dust, gas, and background radiation. High-energy cosmic rays, in the form of swift protons and atomic nuclei, are unable to travel more than about 100 megaparsecs due to the Greisen–Zatsepin–Kuzmin limit (GZK cutoff). Neutrinos, in contrast, can travel even greater distances barely attenuated.

The galactic core of the Milky Way is fully obscured by dense gas and numerous bright objects. Neutrinos produced in the galactic core might be measurable by Earth-based neutrino telescopes.

Another important use of the neutrino is in the observation of supernovae, the explosions that end the lives of highly massive stars. The core collapse phase of a supernova is an extremely dense and energetic event. It is so dense that no known particles are able to escape the advancing core front except for neutrinos. Consequently, supernovae are known to release approximately 99% of their radiant energy in a short (10-second) burst of neutrinos.[67] These neutrinos are a very useful probe for core collapse studies.

The rest mass of the neutrino (see above) is an important test of cosmological and astrophysical theories (see *Dark matter*). The neutrino's significance in probing cosmological phenomena is as great as any other method, and is thus a major focus of study in astrophysical communities.[68]

The study of neutrinos is important in particle physics because neutrinos typically have the lowest mass, and hence are examples of the lowest-energy particles theorized in extensions of the Standard Model of particle physics.

In November 2012 American scientists used a particle accelerator to send a coherent neutrino message through 780 feet of rock. This marks the first use of neutrinos for communication, and future research may permit binary neutrino messages to be sent immense distances through even the densest materials, such as the Earth's core.[69]

16.6 See also

- List of neutrino experiments

16.7 Notes

[1] More specifically, the electron neutrino.

[2] Niels Bohr was notably opposed to this interpretation of beta decay and was ready to accept that energy, momentum and angular momentum were not conserved quantities.

[3] These events necessitated renaming Pauli's less massive, momentum-conserving particle.

[4] In this context, "light neutrino" means neutrinos with less than half the mass of the Z boson.

[5] Typically about one third of the heat which is deposited in a reactor core is available to be converted to electricity, and a 4000 MW reactor would produce only 2700 MW of actual heat, with the rest being converted to its 1300 MW of electric power production.

16.8 References

[1] "Astronomers Accurately Measure the Mass of Neutrinos for the First Time". *scitechdaily.com*. Image credit:NASA, ESA, and J. Lotz, M. Mountain, A. Koekemoer, and the HFF Team (STScI). February 10, 2014. Archived from the original on May 7, 2014. Retrieved May 7, 2014.

[2] Foley, James A. (February 10, 2014). "Mass of Neutrinos Accurately Calculated for First Time, Physicists Report". *natureworldnews.com*. Image credit: . via Wikimedia Commons. Archived from the original on May 7, 2014. Retrieved May 7, 2014.

[3] Battye, Richard A.; Moss, Adam (2014). "Evidence for Massive Neutrinos from Cosmic Microwave Background and Lensing Observations".*Physical Review Letters***112**(5): 051303. arXiv:1308.5870v2. Bibcode:2014PhRvL.112e1303B.doi:10.110 1303.PMID24580586.

[4] "Neutrino". *Glossary for the Research Perspectives of the Max Planck Society*. Max Planck Gesellschaft. Retrieved 2012-03-27.

[5] Dodelson, Scott; Widrow, Lawrence M. (1994). "Sterile neutrinos as dark matter" **72** (17).

[6] Bahcall, John N.; Serenelli, Aldo M.; Basu, Sarbani (2005). "New Solar Opacities, Abundances, Helioseismology, and Neutrino Fluxes". *The Astrophysical Journal* **621** (1): L85–8. arXiv:astro-ph/0412440. Bibcode:2005ApJ...621L..85B. doi:10.1086/428929.

[7] Olive, K. A. "Sum of Neutrino Masses" (PDF). *Chinese Physics C*.

[8] Brown,Laurie M. (1978). "The idea of the neutrino".*Physics Today***31**(9): 23–8. Bibcode:1978PhT....31i..23B.doi:10.1063/1..

[9] E. Amaldi (1984). "From the discovery of the neutron to the discovery of nuclear fission". *Phys. Rep.* **111** (1–4): 306.

[10] F. Close (2010). *Neutrino*. Oxford University Press. ISBN 978-0-19-957459-9.

[11] E. Fermi (1934). "Versuch einer Theorie der β-Strahlen. I". *Zeitschrift für Physik A* **88** (3–4): 161. Bibcode:1934ZPhy...88..161F. doi:10.1007/BF01351864. Translated in F. L. Wilson (1968). "Fermi's Theory of Beta Decay" (PDF). *American Journal of Physics* **36** (12): 1150. Bibcode:1968AmJPh..36.1150W. doi:10.1119/1.1974382.

[12] K.-C.Wang(1942). "A Suggestion on the Detection of the Neutrino".*Physical Review***61**(1–2): 97. Bibcode:1942PhRv...61W. doi:10.1103/PhysRev.61.97.

[13] C. L. Cowan Jr.; F. Reines; F. B. Harrison; H. W. Kruse et al. (1956). "Detection of the Free Neutrino: a Confirmation". *Science* **124** (3212): 103–4. Bibcode:1956Sci...124..103C. doi:10.1126/science.124.3212.103. PMID 17796274.

[14] K. Winter (2000). *Neutrino physics*. Cambridge University Press. p. 38ff. ISBN 978-0-521-65003-8. This source reproduces the 1956 paper.

[15] "The Nobel Prize in Physics 1995". The Nobel Foundation. Retrieved 29 June 2010.

[16] I. V. Anicin (2005). "The Neutrino – Its Past, Present and Future". arXiv:physics/0503172.

[17] M. Maltoni; T. Schwetz; M. Tórtola; J. W. F. Valle (2004). "Status of global fits to neutrino oscillations". *New Journal of Physics* **6** (1): 122. arXiv:hep-ph/0405172. Bibcode:2004NJPh....6..122M. doi:10.1088/1367-2630/6/1/122.

[18] Particle Data Group; Eidelman, S.; Hayes, K. G.; Olive, K. A.; Aguilar-Benitez, M.; Amsler, C.; Asner, D.; Babu, K. S.; Barnett, R. M.; Beringer, J.; Burchat, P. R.; Carone, C. D.; Caso, S.; Conforto, G.; Dahl, O.; d'Ambrosio, G.; Doser, M.; Feng, J. L.; Gherghetta, T.; Gibbons, L.; Goodman, M.; Grab, C.; Groom, D. E.; Gurtu, A.; Hagiwara, K.; Hernández-Rey, J. J.; Hikasa, K.; Honscheid, K.; Jawahery, H. et al. (2004). "Review of Particle Physics". *Physics Letters B* **592**: 1–5. arXiv:astro-ph/0406663. Bibcode:2004PhLB..592....1P. doi:10.1016/j.physletb.2004.06.001.

[19] S.M. Caroll (25 March 2009). "Ada Lovelace Day: Chien-Shiung Wu". *Discover Magazine*. Retrieved 2011-09-23.

[20] Kolbe, E.; Langanke, K.; Fuller, G. M. (2004). "Neutrino-Induced Fission of Neutron-Rich Nuclei". *Physical Review Letters* **92** (11): 111101. arXiv:astro-ph/0308350. Bibcode:2004PhRvL..92k1101K. doi:10.1103/PhysRevLett.92.111101. PMID 15089120.

[21] Kelić, A.; Zinner, N.; Kolbe, E.; Langanke, K.; Schmidt, K.-H. (2005). "Cross sections and fragment distributions from neutrino-induced fission on r-process nuclei". *Physics Letters B* **616** (1–2): 48–58. arXiv:hep-ex/0312045. Bibcode:2005P 8K.doi:10.1016/j.physletb.2005.04.074.

[22] Karagiorgi, G.; Aguilar-Arevalo, A.; Conrad, J. M.; Shaevitz, M. H.; Whisnant, K.; Sorel, M.; Barger, V. (2007). "LeptonicCPviolation studies at MiniBooNE in the (3+2) sterile neutrino oscillation hypothesis". *Physical Review D* **75**: 013011. arXiv:hep-ph/0609177. Bibcode:2007PhRvD..75a3011K. doi:10.1103/PhysRevD.75.013011.

[23] M. Alpert (2007). "Dimensional Shortcuts". *Scientific American*. Retrieved 2009-10-31.

[24] Mueller, Th. A.; Lhuillier, D.; Fallot, M.; Letourneau, A.; Cormon, S.; Fechner, M.; Giot, L.; Lasserre, T.; Martino, J.; Mention, G.; Porta, A.; Yermia, F. (2011). "Improved predictions of reactor antineutrino spectra". *Physical Review C* **83** (5): 054615. arXiv:1101.2663. Bibcode:2011PhRvC..83e4615M. doi:10.1103/PhysRevC.83.054615.

[25] Mention, G.; Fechner, M.; Lasserre, Th.; Mueller, Th. A.; Lhuillier, D.; Cribier, M.; Letourneau, A. (2011). "Reactor antineutrino anomaly".*Physical Review D***83**(7): 073006. arXiv:1101.2755. Bibcode:2011PhRvD..83g3006M.doi:10.11033006.

[26] R. Cowen (2 February 2010). "Ancient Dawn's Early Light Refines the Age of the Universe". *Science News*. Retrieved 2010-02-03.

[27] neutrinos.llnl.gov "LLNL/SNL Applied Antineutrino Physics Project. LLNL-WEB-204112". 2006.

[28] apc.univ-paris7.fr "Applied Antineutrino Physics 2007 workshop". 2007.

[29] "New Tool To Monitor Nuclear Reactors Developed". ScienceDaily. 13 March 2008. Retrieved 2008-03-16.

[30] Alan Kostelecký, V.; Mewes, Matthew (2004). "Lorentz andCPTviolation in neutrinos". *Physical Review D* **69**: 016005. arXiv:hep-ph/0309025. Bibcode:2004PhRvD..69a6005A. doi:10.1103/PhysRevD.69.016005.

[31] C. Giunti; C.W. Kim (2007). *Fundamentals of neutrino physics and astrophysics*. Oxford University Press. p. 255. ISBN 0-19-850871-9.

[32] "Neutrino shape-shift points to new physics" *Physics News*, 19 July 2013.

[33] "Neutrino 'flavour' flip confirmed" *BBC News*, 19 July 2013.

[34] Adamson, P.; Andreopoulos, C.; Arms, K. E.; Armstrong, R.; Auty, D. J.; Avvakumov, S.; Ayres, D. S.; Baller, B.; Barish, B.; Barnes, P. D.; Barr, G.; Barrett, W. L.; Beall, E.; Becker, B. R.; Belias, A.; Bergfeld, T.; Bernstein, R. H.; Bhattacharya, D.; Bishai, M.; Blake, A.; Bock, B.; Bock, G. J.; Boehm, J.; Boehnlein, D. J.; Bogert, D.; Border, P. M.; Bower, C.; Buckley-Geer, E.; Cabrera, A. et al. (2007). "Measurement of neutrino velocity with the MINOS detectors and NuMI neutrino beam". *Physical Review D* **76** (7): 072005. arXiv:0706.0437. Bibcode:2007PhRvD..76g2005A. doi:10.1103/PhysRevD.76.072005.

[35] D. Overbye (22 September 2011). "Tiny neutrinos may have broken cosmic speed limit". *New York Times*. That group found, although with less precision, that the neutrino speeds were consistent with the speed of light.

[36] Hesla, Leah (June 8, 2012). "MINOS reports new measurement of neutrino velocity". Fermilab today. Retrieved April 2, 2015.

[37] Antonello, M.; Aprili, P.; Baiboussinov, B.; Baldo Ceolin, M.; Benetti, P.; Calligarich, E.; Canci, N.; Centro, S.; Cesana, A.; Cieślik, K.; Cline, D.B.; Cocco, A.G.; Dabrowska, A.; Dequal, D.; Dermenev, A.; Dolfini, R.; Farnese, C.; Fava, A.; Ferrari, A.; Fiorillo, G.; Gibin, D.; Gigli Berzolari, A.; Gninenko, S.; Guglielmi, A.; Haranczyk, M.; Holeczek, J.; Ivashkin, A.; Kisiel, J.; Kochanek, I. et al. (2012). "Measurement of the neutrino velocity with the ICARUS detector at the CNGS beam". *Physics Letters B* **713**: 17–22. arXiv:1203.3433. Bibcode:2012PhLB..713...17I. doi:10.1016/j.physletb.2012.05.033.

[38] "Neutrinos sent from CERN to Gran Sasso respect the cosmic speed limit, experiments confirm" (Press release). CERN. June 8, 2012. Retrieved April 2, 2015.

[39] Hut, P.; Olive, K.A. (1979). "A cosmological upper limit on the mass of heavy neutrinos". *Physics Letters B* **87** (1–2): 144–6. Bibcode:1979PhLB...87..144H. doi:10.1016/0370-2693(79)90039-X.

[40] Goobar, Ariel; Hannestad, Steen; Mörtsell, Edvard; Tu, Huitzu (2006). "The neutrino mass bound from WMAP 3 year data, the baryon acoustic peak, the SNLS supernovae and the Lyman-α forest". *Journal of Cosmology and Astroparticle Physics* **2006** (6): 019. arXiv:astro-ph/0602155. Bibcode:2006JCAP...06..019G. doi:10.1088/1475-7516/2006/06/019.

[41] Fukuda, Y.; Hayakawa, ; Ichihara, E.; Inoue, K.; Ishihara, K.; Ishino, H.; Itow, Y.; Kajita, T.; Kameda, J.; Kasuga, S.; Kobayashi, K.; Kobayashi, Y.; Koshio, Y.; Martens, K.; Miura, M.; Nakahata, M.; Nakayama, S.; Okada, A.; Oketa, M.; Okumura, K.; Ota, M.; Sakurai, N.; Shiozawa, M.; Suzuki, Y.; Takeuchi, Y.; Totsuka, Y.; Yamada, S.; Earl, M.; Habig, A. et al. (1998). "Measurements of the Solar Neutrino Flux from Super-Kamiokande's First 300 Days". *Physical Review Letters* **81** (6): 1158. arXiv:hep-ex/9805021. Bibcode:1998PhRvL..81.1158F. doi:10.1103/PhysRevLett.81.1158.

[42] Mohapatra, R N; Antusch, S; Babu, K S; Barenboim, G; Chen, M-C; De Gouvêa, A; De Holanda, P; Dutta, B; Grossman, Y; Joshipura, A; Kayser, B; Kersten, J; Keum, Y Y; King, S F; Langacker, P; Lindner, M; Loinaz, W; Masina, I; Mocioiu, I; Mohanty, S; Murayama, H; Pascoli, S; Petcov, S T; Pilaftsis, A; Ramond, P; Ratz, M; Rodejohann, W; Shrock, R; Takeuchi, T et al. (2007). "Theory of neutrinos: A white paper". *Reports on Progress in Physics* **70** (11): 1757. arXiv:hep-ph/0510213. Bibcode:2007RPPh...70.1757M. doi:10.1088/0034-4885/70/11/R02.

[43] Araki, T.; Eguchi, K.; Enomoto, S.; Furuno, K.; Ichimura, K.; Ikeda, H.; Inoue, K.; Ishihara, K.; Iwamoto, T.; Kawashima, T.; Kishimoto, Y.; Koga, M.; Koseki, Y.; Maeda, T.; Mitsui, T.; Motoki, M.; Nakajima, K.; Ogawa, H.; Owada, K.; Ricol, J.-S.; Shimizu, I.; Shirai, J.; Suekane, F.; Suzuki, A.; Tada, K.; Tajima, O.; Tamae, K.; Tsuda, Y.; Watanabe, H. et al. (2005). "Measurement of Neutrino Oscillation with KamLAND: Evidence of Spectral Distortion". *Physical Review Letters* **94** (8): 081801. arXiv:hep-ex/0406035. Bibcode:2005PhRvL..94h1801A. doi:10.1103/PhysRevLett.94.081801. PMID 15783875.

[44] "MINOS experiment sheds light on mystery of neutrino disappearance" (Press release). Fermilab. 30 March 2006. Retrieved 2007-11-25.

[45] Amsler, C.; Doser, M.; Antonelli, M.; Asner, D.M.; Babu, K.S.; Baer, H.; Band, H.R.; Barnett, R.M.; Bergren, E.; Beringer, J.; Bernardi, G.; Bertl, W.; Bichsel, H.; Biebel, O.; Bloch, P.; Blucher, E.; Blusk, S.; Cahn, R.N.; Carena, M.; Caso, C.; Ceccucci, A.; Chakraborty, D.; Chen, M.-C.; Chivukula, R.S.; Cowan, G.; Dahl, O.; d'Ambrosio, G.; Damour, T.; De Gouvêa, A. et al. (2008). "Review of Particle Physics".*Physics Letters B***667**: 1. Bibcode:2008PhLB..667....1P.doi:10.1016/j.physletb.2008.07.08.

[46] Nieuwenhuizen, Th. M. (2009). "Do non-relativistic neutrinos constitute the dark matter?". *EPL* **86** (5): 59001. arXiv:0812.4552. Bibcode:2009EL.....8659001N. doi:10.1209/0295-5075/86/59001.

[47] "The most sensitive analysis on the neutrino mass [...] is compatible with a neutrino mass of zero. Considering its uncertainties this value corresponds to an upper limit on the electron neutrino mass of $m < 2.2$ eV/c^2 (95% Confidence Level)" The Mainz Neutrino Mass Experiment

[48] Agafonova, N.; Aleksandrov, A.; Altinok, O.; Ambrosio, M.; Anokhina, A.; Aoki, S.; Ariga, A.; Ariga, T.; Autiero, D.; Badertscher, A.; Bagulya, A.; Bendhabi, A.; Bertolin, A.; Besnier, M.; Bick, D.; Boyarkin, V.; Bozza, C.; Brugière, T.; Brugnera, R.; Brunet, F.; Brunetti, G.; Buontempo, S.; Cazes, A.; Chaussard, L.; Chernyavsky, M.; Chiarella, V.; Chon-Sen, N.; Chukanov, A.; Ciesielski, R. et al. (2010). "Observation of a first ντ candidate event in the OPERA experiment in the CNGS beam".*Physics Letters B***691**(3): 138–45. arXiv:1006.1623. Bibcode:2010PhLB..691..138A.doi:10.1016/j.physletb.2010.06.02.

[49] Thomas, Shaun A.; Abdalla, Filipe B.; Lahav, Ofer (2010). "Upper Bound of 0.28 eV on Neutrino Masses from the Largest Photometric Redshift Survey". *Physical Review Letters* **105** (3): 031301. arXiv:0911.5291. Bibcode:2010PhRvL.105c1301T. doi:10.1103/PhysRevLett.105.031301. PMID 20867754.

[50] Planck Collaboration, P. A. R.; Ade, P. A. R.; Aghanim, N.; Armitage-Caplan, C.; Arnaud, M.; Ashdown, M.; Atrio-Barandela, F.; Aumont, J.; Baccigalupi, C.; Banday, A. J.; Barreiro, R. B.; Bartlett, J. G.; Battaner, E.; Benabed, K.; Benoît, A.; Benoit-Lévy, A.; Bernard, J.-P.; Bersanelli, M.; Bielewicz, P.; Bobin, J.; Bock, J. J.; Bonaldi, A.; Bond, J. R.; Borrill, J.; Bouchet, F. R.; Bridges, M.; Bucher, M.; Burigana, C.; Butler, R. C. et al. (2013). "Planck 2013 results. XVI. Cosmological parameters". *Astronomy & Astrophysics* **1303**: 5076. arXiv:1303.5076. Bibcode:2013arXiv1303.5076P. doi:10.1051/0004-6361/201321591.

[51] Battye, Richard A.; Moss, Adam (2014). "Evidence for Massive Neutrinos from Cosmic Microwave Background and Lensing Observations".*Physical Review Letters***112**(5): 051303. arXiv:1308.5870. Bibcode:2014PhRvL.112e1303B.doi:10.1103 51303.PMID24580586.

[52] A. Gando et al. (KamLAND-Zen Collaboration) (Feb 7, 2013). "Limit on Neutrinoless ββ Decay of Xe136 from the First Phase of KamLAND-Zen and Comparison with the Positive Claim in Ge76". *Phys. Rev. Lett. 110, 062502.* Bibcode:2013PhRvL. 110f2502G.doi:10.1103/PhysRevLett.11

[53] Lucio, J. L.; Rosado, A.; Zepeda, A. (1985). "Characteristic size for the neutrino".*Physical Review D***31** (5): 1091– . Bibcode:1985PhRvD..31.1091L. doi:10.1103/PhysRevD.31.1091. PMID 9955801.

[54] Choi, Charles Q. (2 June 2009). "Particles Larger Than Galaxies Fill the Universe?". *National Geographic News.*

[55] B. Kayser (2005). "Neutrino mass, mixing, and flavor change" (PDF). Particle Data Group. Retrieved 2007-11-25.

[56] S.M. Bilenky; C. Giunti (2001). "Lepton Numbers in the framework of Neutrino Mixing". *International Journal of Modern Physics A* **16** (24): 3931–3949. arXiv:hep-ph/0102320. Bibcode:2001IJMPA..16.3931B. doi:10.1142/S0217751X01004967.

[57] "Nuclear Fission and Fusion, and Nuclear Interactions". NLP National Physical Laboratory. 2008. Retrieved 2009-06-25.

[58] A. Bernstein; Wang, Y.; Gratta, G.; West, T. (2002). "Nuclear reactor safeguards and monitoring with antineutrino detectors". *Journal of Applied Physics* **91** (7): 4672. arXiv:nucl-ex/0108001. Bibcode:2002JAP....91.4672B. doi:10.1063/1.1452775.

[59] A. Bandyopadhyay et al. (ISS Physics Working Group) et al. (2007). "Physics at a future Neutrino Factory and super-beam facility". *Reports on Progress in Physics* **72** (10): 6201. arXiv:0710.4947. Bibcode:2009RPPh...72j6201B. doi:10.1088/0034-4885/72/10/106201.

[60] F. Reines; C. Cowan, Jr. (1997). "The Reines-Cowan Experiments: Detecting the Poltergeist" (PDF). *Los Alamos Science* **25**: 3.

[61] S. A. Colgate & R. H. White (1966). "The Hydrodynamic Behavior of Supernova Explosions". *The Astrophysical Journal* **143**: 626. Bibcode:1966ApJ...143..626C. doi:10.1086/148549.

[62] A.K. Mann (1997). *Shadow of a star: The neutrino story of Supernova 1987A*. W. H. Freeman. p. 122. ISBN 0-7167-3097-9.

[63] Products of the 1987A supernova

[64] Diameter of neutrino-generating core, and total neutrino power of SN 1987A

[65] J.N. Bahcall (1989). *Neutrino Astrophysics*. Cambridge University Press. ISBN 0-521-37975-X.

[66] D.R. David Jr. (2003). "Nobel Lecture: A half-century with solar neutrinos". *Reviews of Modern Physics* **75** (3): 10. Bibcode:2003RvMP...75..985D. doi:10.1103/RevModPhys.75.985.

[67] "Physics – Supernova Starting Gun: Neutrinos". Focus.aps.org. 2009-07-17. Retrieved 2012-04-05.

[68] G.B.Gelmini;A.Kusenko;T.J.Weiler(May2010). "Through Neutrino Eyes".*Scientific American***302**(5): 38–45. Bibcode:2010G. doi:10.1038/scientificamerican0510-38.

[69] Stancil, D. D.; Adamson, P.; Alania, M.; Aliaga, L. et al. (2012). "Demonstration of Communication Using Neutrinos" (PDF). *Modern Physics Letters A* **27** (12): 1250077. arXiv:1203.2847. Bibcode:2012MPLA...2750077S. doi:10.1142/S02177370. Lay summary–*Popular Science* (March15, 2012).

16.9 Bibliography

- Adam, T.; *et al.* (OPERA collaboration) (2011). "Measurement of the neutrino velocity with the OPERA detector in the CNGS beam". arXiv:1109.4897 [hep-ex].
- Alberico, W. M.; Bilenky, S. M. (2004). "Neutrino Oscillations, Masses And Mixing". *Physics of Particles and Nuclei* **35**: 297–323. arXiv:hep-ph/0306239. Bibcode:2003hep.ph....6239A.
- Bahcall, J. N. (1989). *Neutrino Astrophysics*. Cambridge University Press. ISBN 0-521-35113-8.
- Bumfiel, G. (1 October 2001). "The Milky Way's Hidden Black Hole". *Scientific American*. Retrieved 2010-04-23.
- Close, F. (2010). *Neutrino*. Oxford University Press. ISBN 978-0-19-957459-9.
- Griffiths, D. J. (1987). *Introduction to Elementary Particles*. John Wiley & Sons. ISBN 0-471-60386-4.
- Perkins, D. H. (1999). *Introduction to High Energy Physics*. Cambridge University Press. ISBN 0-521-62196-8.
- Povh, B. (1995). *Particles and Nuclei: An Introduction to the Physical Concepts*. Springer-Verlag. ISBN 0-387-59439-6.
- Riazuddin (2005). "Neutrinos" (PDF). National Center for Physics.
- Schopper, H. F. (1966). *Weak interactions and nuclear beta decay*. North-Holland.
- Tammann, G. A.; Thielemann, F. K.; Trautmann, D. (2003). "Opening new windows in observing the Universe". Europhysics News. Retrieved 2006-06-08.
- Tipler, P.; Llewellyn, R. (2002). *Modern Physics* (4th ed.). W. H. Freeman. ISBN 0-7167-4345-0.
- Tomonaga, S.-I. (1997). *The Story of Spin*. University of Chicago Press.
- Zuber, K. (2003). *Neutrino Physics*. IOP Publishing. ISBN 978-0-7503-0750-5.

16.10 External links

- “What’s a Neutrino?", Dave Casper (University of California, Irvine)
- Neutrino unbound: On-line review and e-archive on Neutrino Physics and Astrophysics
- Nova: The Ghost Particle: Documentary on US public television from WGBH
- Measuring the density of the earth’s core with neutrinos
- Universe submerged in a sea of chilled neutrinos, *New Scientist*, 5 March 2008
- What’s a neutrino?
- Search for neutrinoless double beta decay with enriched 76Ge in Gran Sasso 1990–2003
- Neutrino caught in the act of changing from muon-type to tau-type, CERN press release
- Cosmic Weight Gain: A Wispy Particle Bulks Up by George Johnson
- Neutrino 'ghost particle' sized up by astronomers BBC News 22 June 2010
- Pillar of physics challenged
- Merrifield, Michael; Copeland, Ed; Bowley, Roger (2010). “Neutrinos”. *Sixty Symbols*. Brady Haran for the University of Nottingham.
- The Neutrino with Dr. Clyde L. Cowan (Lecture on Project Poltergeist by Clyde Cowan)
- Nuclear Reactor as the Source of Antineutrinos

Chapter 17

Boson

For other uses, see Boson (disambiguation).

In quantum mechanics, a **boson** (/ˈboʊsɒn/,[1] /ˈboʊzɒn/[2]) is a particle that follows Bose–Einstein statistics. Bosons make up one of the two classes of particles, the other being fermions.[3] The name boson was coined by Paul Dirac[4] to commemorate the contribution of the Indian physicist Satyendra Nath Bose[5][6] in developing, with Einstein, Bose–Einstein statistics—which theorizes the characteristics of elementary particles.[7] Examples of bosons include fundamental particles such as photons, gluons, and W and Z bosons (the four force-carrying gauge bosons of the Standard Model), the Higgs boson, and the still-theoretical graviton of quantum gravity; composite particles (e.g. mesons and stable nuclei of even mass number such as deuterium (with one proton and one neutron, mass number = 2), helium-4, or lead-208[Note 1]); and some quasiparticles (e.g. Cooper pairs, plasmons, and phonons).[8]:130

An important characteristic of bosons is that their statistics do not restrict the number of them that occupy the same quantum state. This property is exemplified by helium-4 when it is cooled to become a superfluid.[9] Unlike bosons, two identical fermions cannot occupy the same quantum space. Whereas the elementary particles that make up matter (i.e. leptons and quarks) are fermions, the elementary bosons are force carriers that function as the 'glue' holding matter together.[10] This property holds for all particles with integer spin (s = 0, 1, 2 etc.) as a consequence of the spin–statistics theorem. When a gas of Bose particles is cooled down to absolute zero then the kinetic energy of the particles decreases to a negligible amount and they condense into a lowest energy level state. This state is called Bose-Einstein condensation. It is believed that this phenomenon is the secret behind superfluidity of liquids.

17.1 Types

Bosons may be either elementary, like photons, or composite, like mesons.

While most bosons are composite particles, in the Standard Model there are five bosons which are elementary:

- the four gauge bosons (γ · g · Z · W±)
- the only scalar boson (the Higgs boson (H0))

Additionally, the graviton (G) is a hypothetical elementary particle not incorporated in the Standard Model. If it exists, a graviton must be a boson, and could conceivably be a gauge boson.

Composite bosons are important in superfluidity and other applications of Bose–Einstein condensates. When a gas of Bose particles is cooled to absolute zero its kinetic energy decreases up to a negligible amount then the particles would condense into the lowest energy state. This phenomenon is known as Bose-Einstein condensation and it is believed that this phenomenon is the secret behind superfluidity of liquids.

Satyendra Nath Bose

17.2 Properties

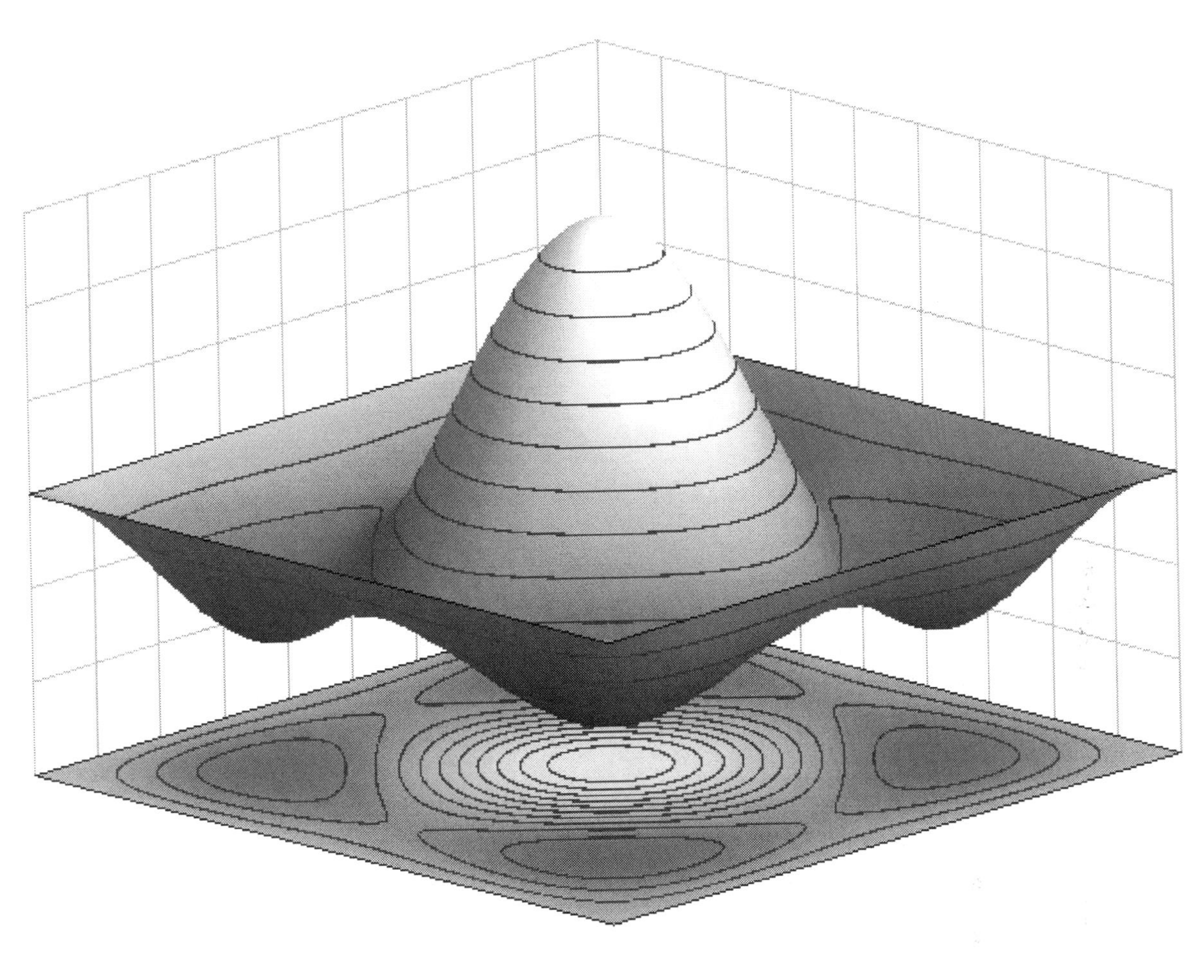

Symmetric wavefunction for a (bosonic) 2-particle state in an infinite square well potential.

Bosons differ from fermions, which obey Fermi–Dirac statistics. Two or more identical fermions cannot occupy the same quantum state (see Pauli exclusion principle).

Since bosons with the same energy can occupy the same place in space, bosons are often force carrier particles. Fermions are usually associated with matter (although in quantum physics the distinction between the two concepts is not clear cut).

Bosons are particles which obey Bose–Einstein statistics: when one swaps two bosons (of the same species), the wavefunction of the system is unchanged.[11] Fermions, on the other hand, obey Fermi–Dirac statistics and the Pauli exclusion principle: two fermions cannot occupy the same quantum state, resulting in a "rigidity" or "stiffness" of matter which includes fermions. Thus fermions are sometimes said to be the constituents of matter, while bosons are said to be the particles that transmit interactions (force carriers), or the constituents of radiation. The quantum fields of bosons are bosonic fields, obeying canonical commutation relations.

The properties of lasers and masers, superfluid helium-4 and Bose–Einstein condensates are all consequences of statistics of bosons. Another result is that the spectrum of a photon gas in thermal equilibrium is a Planck spectrum, one example of which is black-body radiation; another is the thermal radiation of the opaque early Universe seen today as microwave background radiation. Interactions between elementary particles are called fundamental interactions. The fundamental interactions of virtual bosons with real particles result in all forces we know.

All known elementary and composite particles are bosons or fermions, depending on their spin: particles with half-integer spin are fermions; particles with integer spin are bosons. In the framework of nonrelativistic quantum mechanics, this is a purely empirical observation. However, in relativistic quantum field theory, the spin–statistics theorem shows that half-integer spin particles cannot be bosons and integer spin particles cannot be fermions.[12]

In large systems, the difference between bosonic and fermionic statistics is only apparent at large densities—when their wave functions overlap. At low densities, both types of statistics are well approximated by Maxwell–Boltzmann statistics, which is described by classical mechanics.

17.3 Elementary bosons

See also: List of particles: Bosons

All observed elementary particles are either fermions or bosons. The observed elementary bosons are all gauge bosons: photons, W and Z bosons, gluons, and the Higgs boson.

- Photons are the force carriers of the electromagnetic field.
- W and Z bosons are the force carriers which mediate the weak force.
- Gluons are the fundamental force carriers underlying the strong force.
- Higgs Bosons give W and Z bosons mass via the Higgs mechanism. Their existence was confirmed by CERN on 14 March 2013.

Finally, many approaches to quantum gravity postulate a force carrier for gravity, the graviton, which is a boson of spin plus or minus two.

17.4 Composite bosons

See also: List of particles: Composite particles

Composite particles (such as hadrons, nuclei, and atoms) can be bosons or fermions depending on their constituents. More precisely, because of the relation between spin and statistics, a particle containing an even number of fermions is a boson, since it has integer spin.

Examples include the following:

- Any meson, since mesons contain one quark and one antiquark.
- The nucleus of a carbon-12 atom, which contains 6 protons and 6 neutrons.
- The helium-4 atom, consisting of 2 protons, 2 neutrons and 2 electrons.

The number of bosons within a composite particle made up of simple particles bound with a potential has no effect on whether it is a boson or a fermion.

17.5 To which states can bosons crowd?

Bose–Einstein statistics encourages identical bosons to crowd into one quantum state, but not any state is necessarily convenient for it. Aside of statistics, bosons can interact – for example, helium-4 atoms are repulsed by intermolecular force on a very close approach, and if one hypothesizes their condensation in a spatially-localized state, then gains from the statistics cannot overcome a prohibitive force potential. A spatially-delocalized state (i.e. with low $|\psi(x)|$) is preferable: if the number density of the condensate is about the same as in ordinary liquid or solid state, then the repulsive potential for the N-particle condensate in such state can be not higher than for a liquid or a crystalline lattice of the same N particles described without quantum statistics. Thus, Bose–Einstein statistics for a material particle is not a mechanism to bypass physical restrictions on the density of the corresponding substance, and superfluid liquid helium has the density comparable to the density of ordinary liquid matter. Spatially-delocalized states also permit for a low momentum according to uncertainty principle, hence for low kinetic energy; that's why superfluidity and superconductivity are usually observed in low temperatures.

Photons do not interact with themselves and hence do not experience this difference in states where to crowd (see squeezed coherent state).

17.6 See also

- Anyon
- Bose gas
- Identical particles
- Parastatistics
- Fermion

17.7 Notes

[1] Even-mass-number nuclides, which comprise 152/255 = ~ 60% of all stable nuclides, are bosons, i.e. they have integer spin. Almost all (148 of the 152) are even-proton, even-neutron (EE) nuclides, which necessarily have spin 0 because of pairing. The remainder of the stable bosonic nuclides are 5 odd-proton, odd-neutron stable nuclides (see even and odd atomic nuclei#Odd proton, odd neutron); these odd–odd bosons are: 2
1H, 6
3Li,10
5B, 14
7N and 180m
73Ta). All have nonzero integer spin.

17.8 References

[1] Wells, John C. (1990). *Longman pronunciation dictionary*. Harlow, England: Longman. ISBN 0582053838. entry "Boson"

[2] "boson". *Collins Dictionary*.

[3] Carroll, Sean (2007) *Dark Matter, Dark Energy: The Dark Side of the Universe*, Guidebook Part 2 p. 43, The Teaching Company, ISBN 1598033506 "...boson: A force-carrying particle, as opposed to a matter particle (fermion). Bosons can be piled on top of each other without limit. Examples include photons, gluons, gravitons, weak bosons, and the Higgs boson. The spin of a boson is always an integer, such as 0, 1, 2, and so on..."

[4] Notes on Dirac's lecture *Developments in Atomic Theory* at Le Palais de la Découverte, 6 December 1945, UKNATARCHI Dirac Papers BW83/2/257889. See note 64 to p. 331 in "The Strangest Man" by Graham Farmelo

[5] Daigle, Katy (10 July 2012). "India: Enough about Higgs, let's discuss the boson". *AP News*. Retrieved 10 July 2012.

[6] Bal, Hartosh Singh (19 September 2012). "The Bose in the Boson". *New York Times blog*. Retrieved 21 September 2012.

[7] "Higgs boson: The poetry of subatomic particles". *BBC News*. 4 July 2012. Retrieved 6 July 2012.

[8] Charles P. Poole, Jr. (11 March 2004). *Encyclopedic Dictionary of Condensed Matter Physics*. Academic Press. ISBN 978-0-08-054523-3.

[9] "boson". *Merriam-Webster Online Dictionary*. Retrieved 21 March 2010.

[10] Carroll, Sean. "Explain it in 60 seconds: Bosons". *Symmetry Magazine*. Fermilab/SLAC. Retrieved 15 February 2013.

[11] Srednicki, Mark (2007). *Quantum Field Theory*, Cambridge University Press, pp. 28–29, ISBN 978-0-521-86449-7.

[12] Sakurai, J.J. (1994). *Modern Quantum Mechanics* (Revised Edition), p. 362. Addison-Wesley, ISBN 0-201-53929-2.

Chapter 18

Gauge boson

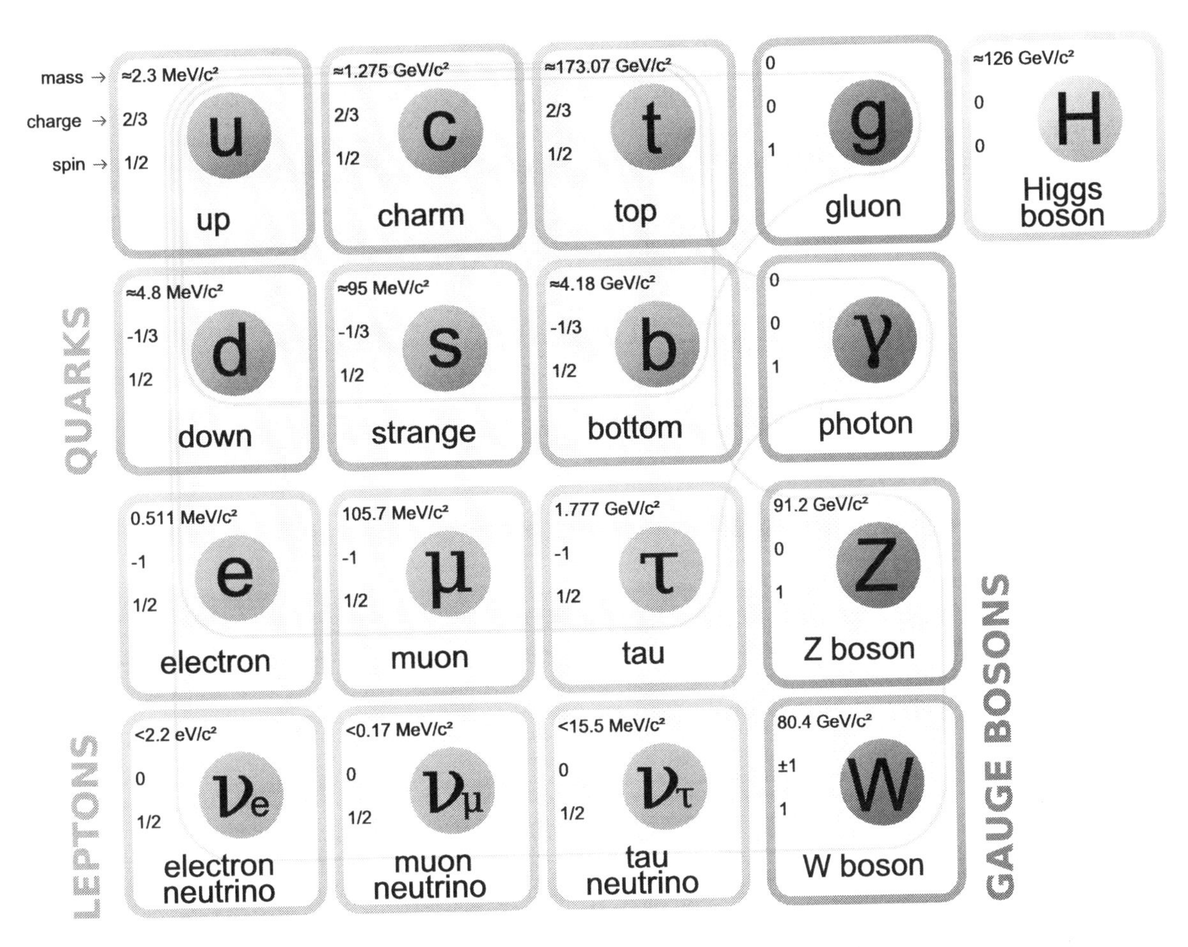

The Standard Model of elementary particles, with the gauge bosons in the fourth column in red

In particle physics, a **gauge boson** is a force carrier, a bosonic particle that carries any of the fundamental interactions of nature.[1][2] Elementary particles, whose interactions are described by a gauge theory, interact with each other by the exchange of gauge bosons—usually as virtual particles.

18.1 Gauge bosons in the Standard Model

The Standard Model of particle physics recognizes four kinds of gauge bosons: photons, which carry the electromagnetic interaction; W and Z bosons, which carry the weak interaction; and gluons, which carry the strong interaction.[3]

Isolated gluons do not occur at low energies because they are color-charged, and subject to color confinement.

18.1.1 Multiplicity of gauge bosons

In a quantized gauge theory, gauge bosons are quanta of the gauge fields. Consequently, there are as many gauge bosons as there are generators of the gauge field. In quantum electrodynamics, the gauge group is $U(1)$; in this simple case, there is only one gauge boson. In quantum chromodynamics, the more complicated group $SU(3)$ has eight generators, corresponding to the eight gluons. The three W and Z bosons correspond (roughly) to the three generators of $SU(2)$ in GWS theory.

18.1.2 Massive gauge bosons

For technical reasons involving gauge invariance, gauge bosons are described mathematically by field equations for massless particles. Therefore, at a naïve theoretical level all gauge bosons are required to be massless, and the forces that they describe are required to be long-ranged. The conflict between this idea and experimental evidence that the weak interaction has a very short range requires further theoretical insight.

According to the Standard Model, the W and Z bosons gain mass via the Higgs mechanism. In the Higgs mechanism, the four gauge bosons (of $SU(2)\times U(1)$ symmetry) of the unified electroweak interaction couple to a Higgs field. This field undergoes spontaneous symmetry breaking due to the shape of its interaction potential. As a result, the universe is permeated by a nonzero Higgs vacuum expectation value (VEV). This VEV couples to three of the electroweak gauge bosons (the Ws and Z), giving them mass; the remaining gauge boson remains massless (the photon). This theory also predicts the existence of a scalar Higgs boson, which has been observed in experiments that were reported on 4 July 2012.[4]

18.2 Beyond the Standard Model

18.2.1 Grand unification theories

A grand unified theory predicts additional gauge bosons named X and Y bosons. The hypothetical X and Y bosons direct interactions between quarks and leptons, hence violating conservation of baryon number and causing proton decay. Such bosons would be even more massive than W and Z bosons due to symmetry breaking. Analysis of data collected from such sources as the Super-Kamiokande neutrino detector has yielded no evidence of X and Y bosons.

18.2.2 Gravitons

The fourth fundamental interaction, gravity, may also be carried by a boson, called the graviton. In the absence of experimental evidence and a mathematically coherent theory of quantum gravity, it is unknown whether this would be a gauge boson or not. The role of gauge invariance in general relativity is played by a similar symmetry: diffeomorphism invariance.

18.2.3 W' and Z' bosons

Main article: W' and Z' bosons

W' and Z' bosons refer to hypothetical new gauge bosons (named in analogy with the Standard Model W and Z bosons).

18.3 See also

- 1964 PRL symmetry breaking papers
- Boson
- Glueball
- Quantum chromodynamics
- Quantum electrodynamics

18.4 References

[1] Gribbin, John (2000). *Q is for Quantum – An Encyclopedia of Particle Physics*. Simon & Schuster. ISBN 0-684-85578-X.

[2] Clark, John, E.O. (2004). *The Essential Dictionary of Science*. Barnes & Noble. ISBN 0-7607-4616-8.

[3] Veltman, Martinus (2003). *Facts and Mysteries in Elementary Particle Physics*. World Scientific. ISBN 981-238-149-X.

[4] "CERN experiments observe particle consistent with long-sought Higgs boson". CERN. Retrieved 4 July 2012.

18.5 External links

- Explanation of gauge boson and gauge fields by Christopher T. Hill

Chapter 19

Photon

This article is about the elementary particle of light. For other uses, see Photon (disambiguation).

A **photon** is an elementary particle, the quantum of light and all other forms of electromagnetic radiation. It is the force carrier for the electromagnetic force, even when static via virtual photons. The effects of this force are easily observable at the microscopic and at the macroscopic level, because the photon has zero rest mass; this allows long distance interactions. Like all elementary particles, photons are currently best explained by quantum mechanics and exhibit wave–particle duality, exhibiting properties of waves and of particles. For example, a single photon may be refracted by a lens or exhibit wave interference with itself, but also act as a particle giving a definite result when its position is measured. Waves and quanta, being two observable aspects of a single phenomenon cannot have their true nature described in terms of any mechanical model. [2] A representation of this dual property of light, which assumes certain points on the wave front to be the seat of the energy is also impossible. Thus, the quanta in a light wave cannot be spatially localized. Some defined physical parameters of a photon are listed.

The modern photon concept was developed gradually by Albert Einstein in the first years of the 20th century to explain experimental observations that did not fit the classical wave model of light. In particular, the photon model accounted for the frequency dependence of light's energy, and explained the ability of matter and radiation to be in thermal equilibrium. It also accounted for anomalous observations, including the properties of black-body radiation, that other physicists, most notably Max Planck, had sought to explain using *semiclassical models*, in which light is still described by Maxwell's equations, but the material objects that emit and absorb light do so in amounts of energy that are *quantized* (i.e., they change energy only by certain particular discrete amounts and cannot change energy in any arbitrary way). Although these semiclassical models contributed to the development of quantum mechanics, many further experiments[3][4] starting with Compton scattering of single photons by electrons, first observed in 1923, validated Einstein's hypothesis that *light itself* is quantized. In 1926 the optical physicist Frithiof Wolfers and the chemist Gilbert N. Lewis coined the name *photon* for these particles, and after 1927, when Arthur H. Compton won the Nobel Prize for his scattering studies, most scientists accepted the validity that quanta of light have an independent existence, and the term *photon* for light quanta was accepted.

In the Standard Model of particle physics, photons and other elementary particles are described as a necessary consequence of physical laws having a certain symmetry at every point in spacetime. The intrinsic properties of particles, such as charge, mass and spin, are determined by the properties of this gauge symmetry. The photon concept has led to momentous advances in experimental and theoretical physics, such as lasers, Bose–Einstein condensation, quantum field theory, and the probabilistic interpretation of quantum mechanics. It has been applied to photochemistry, high-resolution microscopy, and measurements of molecular distances. Recently, photons have been studied as elements of quantum computers and for applications in optical imaging and optical communication such as quantum cryptography.

19.1 Nomenclature

In 1900, the German physicist Max Planck was working on black-body radiation and suggested that the energy in electromagnetic waves could only be released in "packets" of energy. In his 1901 article [5] in Annalen der Physik he called these packets "energy elements". The word *quanta* (singular *quantum*) was used even before 1900 to mean particles or amounts of different quantities, including electricity. Later, in 1905, Albert Einstein went further by suggesting that electromagnetic waves could only exist in these discrete wave-packets.[6] He called such a wave-packet *the light quantum* (German: *das Lichtquant*).[Note 1] The name *photon* derives from the Greek word for light, φῶς (transliterated *phôs*). Arthur Compton used *photon* in 1928, referring to Gilbert N. Lewis.[7] The same name was used earlier, by the American physicist and psychologist Leonard T. Troland, who coined the word in 1916, in 1921 by the Irish physicist John Joly, in 1924 by the French physiologist René Wurmser (1890-1993) and in 1926 by the French physicist Frithiof Wolfers (1891-1971).[8] The name was suggested initially as a unit related to the illumination of the eye and the resulting sensation of light and was used later on in a physiological context. Although Wolfers's and Lewis's theories were never accepted, as they were contradicted by many experiments, the new name was adopted very soon by most physicists after Compton used it.[8][Note 2]

In physics, a photon is usually denoted by the symbol γ (the Greek letter gamma). This symbol for the photon probably derives from gamma rays, which were discovered in 1900 by Paul Villard,[9][10] named by Ernest Rutherford in 1903, and shown to be a form of electromagnetic radiation in 1914 by Rutherford and Edward Andrade.[11] In chemistry and optical engineering, photons are usually symbolized by $h\nu$, the energy of a photon, where h is Planck's constant and the Greek letter ν (nu) is the photon's frequency. Much less commonly, the photon can be symbolized by hf, where its frequency is denoted by f.

19.2 Physical properties

See also: Special relativity and Photonic molecule

A photon is massless,[Note 3] has no electric charge,[12] and is stable. A photon has two possible polarization states. In the momentum representation, which is preferred in quantum field theory, a photon is described by its wave vector, which determines its wavelength λ and its direction of propagation. A photon's wave vector may not be zero and can be represented either as a spatial 3-vector or as a (relativistic) four-vector; in the latter case it belongs to the light cone (pictured). Different signs of the four-vector denote different circular polarizations, but in the 3-vector representation one should account for the polarization state separately; it actually is a spin quantum number. In both cases the space of possible wave vectors is three-dimensional.

The photon is the gauge boson for electromagnetism,[13]:29-30 and therefore all other quantum numbers of the photon (such as lepton number, baryon number, and flavour quantum numbers) are zero.[14] Also, the photon does not obey the Pauli exclusion principle.[15]:1221

Photons are emitted in many natural processes. For example, when a charge is accelerated it emits synchrotron radiation. During a molecular, atomic or nuclear transition to a lower energy level, photons of various energy will be emitted, from radio waves to gamma rays. A photon can also be emitted when a particle and its corresponding antiparticle are annihilated (for example, electron–positron annihilation).[15]:572, 1114, 1172

In empty space, the photon moves at c (the speed of light) and its energy and momentum are related by $E = pc$, where p is the magnitude of the momentum vector $\mathbf{p}$. This derives from the following relativistic relation, with $m = 0$:[16]

$$E^2 = p^2c^2 + m^2c^4.$$

The energy and momentum of a photon depend only on its frequency (ν) or inversely, its wavelength (λ):

$$E = \hbar\omega = h\nu = \frac{hc}{\lambda}$$

$$\boldsymbol{p} = \hbar\boldsymbol{k},$$

where $\boldsymbol{k}$ is the wave vector (where the wave number $k = |\boldsymbol{k}| = 2\pi/\lambda$), $\omega = 2\pi\nu$ is the angular frequency, and $\hbar = h/2\pi$ is the reduced Planck constant.[17]

Since $\boldsymbol{p}$ points in the direction of the photon's propagation, the magnitude of the momentum is

$$p = \hbar k = \frac{h\nu}{c} = \frac{h}{\lambda}.$$

The photon also carries spin angular momentum that does not depend on its frequency.[18] The magnitude of its spin is $\sqrt{2}\hbar$ and the component measured along its direction of motion, its helicity, must be $\pm\hbar$. These two possible helicities, called right-handed and left-handed, correspond to the two possible circular polarization states of the photon.[19]

To illustrate the significance of these formulae, the annihilation of a particle with its antiparticle in free space must result in the creation of at least *two* photons for the following reason. In the center of momentum frame, the colliding antiparticles have no net momentum, whereas a single photon always has momentum (since it is determined, as we have seen, only by the photon's frequency or wavelength—which cannot be zero). Hence, conservation of momentum (or equivalently, translational invariance) requires that at least two photons are created, with zero net momentum. (However, it is possible if the system interacts with another particle or field for annihilation to produce one photon, as when a positron annihilates with a bound atomic electron, it is possible for only one photon to be emitted, as the nuclear Coulomb field breaks translational symmetry.)[20]:64-65 The energy of the two photons, or, equivalently, their frequency, may be determined from conservation of four-momentum. Seen another way, the photon can be considered as its own antiparticle. The reverse process, pair production, is the dominant mechanism by which high-energy photons such as gamma rays lose energy while passing through matter.[21] That process is the reverse of "annihilation to one photon" allowed in the electric field of an atomic nucleus.

The classical formulae for the energy and momentum of electromagnetic radiation can be re-expressed in terms of photon events. For example, the pressure of electromagnetic radiation on an object derives from the transfer of photon momentum per unit time and unit area to that object, since pressure is force per unit area and force is the change in momentum per unit time.[22]

19.2.1 Experimental checks on photon mass

Current commonly accepted physical theories imply or assume the photon to be strictly massless. If the photon is not a strictly massless particle, it would not move at the exact speed of light in vacuum, *c*. Its speed would be lower and depend on its frequency. Relativity would be unaffected by this; the so-called speed of light, *c*, would then not be the actual speed at which light moves, but a constant of nature which is the maximum speed that any object could theoretically attain in space-time.[23] Thus, it would still be the speed of space-time ripples (gravitational waves and gravitons), but it would not be the speed of photons.

If a photon did have non-zero mass, there would be other effects as well. Coulomb's law would be modified and the electromagnetic field would have an extra physical degree of freedom. These effects yield more sensitive experimental probes of the photon mass than the frequency dependence of the speed of light. If Coulomb's law is not exactly valid, then that would cause the presence of an electric field inside a hollow conductor when it is subjected to an external electric field. This thus allows one to test Coulomb's law to very high precision.[24] A null result of such an experiment has set a limit of $m \lesssim 10^{-14}$ eV/c^2.[25]

Sharper upper limits have been obtained in experiments designed to detect effects caused by the galactic vector potential. Although the galactic vector potential is very large because the galactic magnetic field exists on very long length scales, only the magnetic field is observable if the photon is massless. In case of a massive photon, the mass term $\frac{1}{2}m^2 A_\mu A^\mu$ would affect the galactic plasma. The fact that no such effects are seen implies an upper bound on the photon mass of $m < 3\times10^{-27}$ eV/c^2.[26] The galactic vector potential can also be probed directly by measuring the torque exerted on a magnetized ring.[27] Such methods were used to obtain the sharper upper limit of 10^{-18}eV/c^2 (the equivalent of 1.07×10^{-27} atomic mass units) given by the Particle Data Group.[28]

These sharp limits from the non-observation of the effects caused by the galactic vector potential have been shown to be model dependent.[29] If the photon mass is generated via the Higgs mechanism then the upper limit of $m \lesssim 10^{-14}$ eV/c^2 from the test of Coulomb's law is valid.

Photons inside superconductors do develop a nonzero effective rest mass; as a result, electromagnetic forces become short-range inside superconductors.[30]

See also: Supernova/Acceleration Probe

19.3 Historical development

Main article: Light

In most theories up to the eighteenth century, light was pictured as being made up of particles. Since particle models cannot easily account for the refraction, diffraction and birefringence of light, wave theories of light were proposed by René Descartes (1637),[31] Robert Hooke (1665),[32] and Christiaan Huygens (1678);[33] however, particle models remained dominant, chiefly due to the influence of Isaac Newton.[34] In the early nineteenth century, Thomas Young and August Fresnel clearly demonstrated the interference and diffraction of light and by 1850 wave models were generally accepted.[35] In 1865, James Clerk Maxwell's prediction[36] that light was an electromagnetic wave—which was confirmed experimentally in 1888 by Heinrich Hertz's detection of radio waves[37]—seemed to be the final blow to particle models of light.

The Maxwell wave theory, however, does not account for *all* properties of light. The Maxwell theory predicts that the energy of a light wave depends only on its intensity, not on its frequency; nevertheless, several independent types of experiments show that the energy imparted by light to atoms depends only on the light's frequency, not on its intensity. For example, some chemical reactions are provoked only by light of frequency higher than a certain threshold; light of frequency lower than the threshold, no matter how intense, does not initiate the reaction. Similarly, electrons can be ejected from a metal plate by shining light of sufficiently high frequency on it (the photoelectric effect); the energy of the ejected electron is related only to the light's frequency, not to its intensity.[38][Note 4]

At the same time, investigations of blackbody radiation carried out over four decades (1860–1900) by various researchers[culminated inMax Planck'shypothesis[5][40]that the energy of*any*system that absorbs or emits electromagnetic radiation of frequency$\nu$is an integer multiple of an energy quantum$E = h\nu$. As shown by Albert Einstein,[6][41] some form of energy quantization *must* be assumed to account for the thermal equilibrium observed between matter and electromagnetic radiation; for this explanation of the photoelectric effect, Einstein received the 1921 Nobel Prize in physics.[42]

Since the Maxwell theory of light allows for all possible energies of electromagnetic radiation, most physicists assumed initially that the energy quantization resulted from some unknown constraint on the matter that absorbs or emits the radiation. In 1905, Einstein was the first to propose that energy quantization was a property of electromagnetic radiation itself.[6] Although he accepted the validity of Maxwell's theory, Einstein pointed out that many anomalous experiments could be explained if the *energy* of a Maxwellian light wave were localized into point-like quanta that move independently of one another, even if the wave itself is spread continuously over space.[6] In 1909[41] and 1916,[43] Einstein showed that, if Planck's law of black-body radiation is accepted, the energy quanta must also carry momentum $p = h/\lambda$, making them full-fledged particles. This photon momentum was observed experimentally[44] by Arthur Compton, for which he received the Nobel Prize in 1927. The pivotal question was then: how to unify Maxwell's wave theory of light with its experimentally observed particle nature? The answer to this question occupied Albert Einstein for the rest of his life,[45] and was solved in quantum electrodynamics and its successor, the Standard Model (see Second quantization and The photon as a gauge boson, below).

19.4 Einstein's light quantum

Unlike Planck, Einstein entertained the possibility that there might be actual physical quanta of light—what we now call photons. He noticed that a light quantum with energy proportional to its frequency would explain a number of troubling puzzles and paradoxes, including an unpublished law by Stokes, the ultraviolet catastrophe, and of course the photoelectric effect. Stokes's law said simply that the frequency of fluorescent light cannot be greater than the frequency of the light (usually ultraviolet) inducing it. Einstein eliminated the ultraviolet catastrophe by imagining a gas of photons behaving like a gas of electrons that he had previously considered. He was advised by a colleague to be careful how he wrote up this paper, in order to not challenge Planck too directly, as he was a powerful figure, and indeed the warning was justified, as Planck never forgave him for writing it.[46]

19.5 Early objections

Einstein's 1905 predictions were verified experimentally in several ways in the first two decades of the 20th century, as recounted in Robert Millikan's Nobel lecture.[47] However, before Compton's experiment[44] showing that photons carried momentum proportional to their wave number (or frequency) (1922), most physicists were reluctant to believe that electromagnetic radiation itself might be particulate. (See, for example, the Nobel lectures of Wien,[39] Planck[40] and Millikan.[47]) Instead, there was a widespread belief that energy quantization resulted from some unknown constraint on the matter that absorbs or emits radiation. Attitudes changed over time. In part, the change can be traced to experiments such as Compton scattering, where it was much more difficult not to ascribe quantization to light itself to explain the observed results.[48]

Even after Compton's experiment, Niels Bohr, Hendrik Kramers and John Slater made one last attempt to preserve the Maxwellian continuous electromagnetic field model of light, the so-called BKS model.[49] To account for the data then available, two drastic hypotheses had to be made:

1. **Energy and momentum are conserved only on the average in interactions between matter and radiation, not in elementary processes such as absorption and emission.** This allows one to reconcile the discontinuously changing energy of the atom (jump between energy states) with the continuous release of energy into radiation.
2. **Causality is abandoned**. For example, spontaneous emissions are merely emissions induced by a "virtual" electromagnetic field.

However, refined Compton experiments showed that energy–momentum is conserved extraordinarily well in elementary processes; and also that the jolting of the electron and the generation of a new photon in Compton scattering obey causality to within 10 ps. Accordingly, Bohr and his co-workers gave their model "as honorable a funeral as possible".[45] Nevertheless, the failures of the BKS model inspired Werner Heisenberg in his development of matrix mechanics.[50]

A few physicists persisted[51] in developing semiclassical models in which electromagnetic radiation is not quantized, but matter appears to obey the laws of quantum mechanics. Although the evidence for photons from chemical and physical experiments was overwhelming by the 1970s, this evidence could not be considered as *absolutely* definitive; since it relied on the interaction of light with matter, a sufficiently complicated theory of matter could in principle account for the evidence. Nevertheless, *all* semiclassical theories were refuted definitively in the 1970s and 1980s by photon-correlation experiments.[Note 5] Hence, Einstein's hypothesis that quantization is a property of light itself is considered to be proven.

19.6 Wave–particle duality and uncertainty principles

See also: Wave–particle duality, Squeezed coherent state, Uncertainty principle and De Broglie–Bohm theory
Photons, like all quantum objects, exhibit wave-like and particle-like properties. Their dual wave–particle nature can be difficult to visualize. The photon displays clearly wave-like phenomena such as diffraction and interference on the length scale of its wavelength. For example, a single photon passing through a double-slit experiment lands on the screen exhibiting interference phenomena but only if no measure was made on the actual slit being run across. To account for the particle interpretation that phenomenon is called probability distribution but behaves according to Maxwell's equations.[52] However, experiments confirm that the photon is *not* a short pulse of electromagnetic radiation; it does not spread out as it propagates, nor does it divide when it encounters a beam splitter.[53] Rather, the photon seems to be a point-like particle since it is absorbed or emitted *as a whole* by arbitrarily small systems, systems much smaller than its wavelength, such as an atomic nucleus ($\approx 10^{-15}$ m across) or even the point-like electron. Nevertheless, the photon is *not* a point-like particle whose trajectory is shaped probabilistically by the electromagnetic field, as conceived by Einstein and others; that hypothesis was also refuted by the photon-correlation experiments cited above. According to our present understanding, the electromagnetic field itself is produced by photons, which in turn result from a local gauge symmetry and the laws of quantum field theory (see the Second quantization and Gauge boson sections below).

A key element of quantum mechanics is Heisenberg's uncertainty principle, which forbids the simultaneous measurement of the position and momentum of a particle along the same direction. Remarkably, the uncertainty principle for charged, material particles *requires* the quantization of light into photons, and even the frequency dependence of the photon's energy and momentum. An elegant illustration is Heisenberg's thought experiment for locating an electron with an ideal microscope.[54] The position of the electron can be determined to within the resolving power of the microscope, which is given by a formula from classical optics

$$\Delta x \sim \frac{\lambda}{\sin\theta}$$

where θ is the aperture angle of the microscope. Thus, the position uncertainty Δx can be made arbitrarily small by reducing the wavelength λ. The momentum of the electron is uncertain, since it received a "kick" Δp from the light scattering from it into the microscope. If light were *not* quantized into photons, the uncertainty Δp could be made arbitrarily small by reducing the light's intensity. In that case, since the wavelength and intensity of light can be varied independently, one could simultaneously determine the position and momentum to arbitrarily high accuracy, violating the uncertainty principle. By contrast, Einstein's formula for photon momentum preserves the uncertainty principle; since the photon is scattered anywhere within the aperture, the uncertainty of momentum transferred equals

$$\Delta p \sim p_{\text{photon}} \sin\theta = \frac{h}{\lambda}\sin\theta$$

giving the product $\Delta x \Delta p \sim h$, which is Heisenberg's uncertainty principle. Thus, the entire world is quantized; both matter and fields must obey a consistent set of quantum laws, if either one is to be quantized.[55]

The analogous uncertainty principle for photons forbids the simultaneous measurement of the number n of photons (see Fock state and the Second quantization section below) in an electromagnetic wave and the phase ϕ of that wave

$$\Delta n \Delta\phi > 1$$

See coherent state and squeezed coherent state for more details.

Both (photons and material) particles such as electrons create analogous interference patterns when passing through a double-slit experiment. For photons, this corresponds to the interference of a Maxwell light wave whereas, for material particles, this corresponds to the interference of the Schrödinger wave equation. Although this similarity might suggest that Maxwell's equations are simply Schrödinger's equation for photons, most physicists do not agree.[56][57] For one thing, they are mathematically different; most obviously, Schrödinger's one equation solves for a complex field, whereas Maxwell's four equations solve for real fields. More generally, the normal concept of a Schrödinger probability wave function cannot be applied to photons.[58] Being massless, they cannot be localized without being destroyed; technically, photons cannot have a position eigenstate $|\mathbf{r}\rangle$, and, thus, the normal Heisenberg uncertainty principle $\Delta x \Delta p > h/2$ does not pertain to photons. A few substitute wave functions have been suggested for the photon,[59][60][61][62] but they have not come into general use. Instead, physicists generally accept the second-quantized theory of photons described below, quantum electrodynamics, in which photons are quantized excitations of electromagnetic modes.

Another interpretation, that avoids duality, is the De Broglie–Bohm theory: known also as the *pilot-wave model*, the photon in this theory is both, wave and particle.[63] *"This idea seems to me so natural and simple, to resolve the wave-particle dilemma in such a clear and ordinary way, that it is a great mystery to me that it was so generally ignored"*,[64] J.S.Bell.

19.7 Bose–Einstein model of a photon gas

Main articles: Bose gas, Bose–Einstein statistics, Spin-statistics theorem and Gas in a box

In 1924, Satyendra Nath Bose derived Planck's law of black-body radiation without using any electromagnetism, but rather a modification of coarse-grained counting of phase space.[65] Einstein showed that this modification is equivalent to assuming that photons are rigorously identical and that it implied a "mysterious non-local interaction",[66][67] now understood as the requirement for a symmetric quantum mechanical state. This work led to the concept of coherent states and the development of the laser. In the same papers, Einstein extended Bose's formalism to material particles (bosons) and predicted that they would condense into their lowest quantum state at low enough temperatures; this Bose–Einstein condensation was observed experimentally in 1995.[68] It was later used by Lene Hau to slow, and then completely stop, light in 1999[69] and 2001.[70]

The modern view on this is that photons are, by virtue of their integer spin, bosons (as opposed to fermions with half-integer spin). By the spin-statistics theorem, all bosons obey Bose–Einstein statistics (whereas all fermions obey Fermi–Dirac statistics).[71]

19.8 Stimulated and spontaneous emission

Main articles: Stimulated emission and Laser

In 1916, Einstein showed that Planck's radiation law could be derived from a semi-classical, statistical treatment of photons and atoms, which implies a relation between the rates at which atoms emit and absorb photons. The condition follows from the assumption that light is emitted and absorbed by atoms independently, and that the thermal equilibrium is preserved by interaction with atoms. Consider a cavity in thermal equilibrium and filled with electromagnetic radiation and atoms that can emit and absorb that radiation. Thermal equilibrium requires that the energy density $\rho(\nu)$ of photons with frequency ν (which is proportional to their number density) is, on average, constant in time; hence, the rate at which photons of any particular frequency are *emitted* must equal the rate of *absorbing* them.[72]

Einstein began by postulating simple proportionality relations for the different reaction rates involved. In his model, the rate R_{ji} for a system to *absorb* a photon of frequency ν and transition from a lower energy E_j to a higher energy E_i is proportional to the number N_j of atoms with energy E_j and to the energy density $\rho(\nu)$ of ambient photons with that frequency,

$$R_{ji} = N_j B_{ji} \rho(\nu)$$

where B_{ji} is the rate constant for absorption. For the reverse process, there are two possibilities: spontaneous emission of a photon, and a return to the lower-energy state that is initiated by the interaction with a passing photon. Following Einstein's approach, the corresponding rate R_{ij} for the emission of photons of frequency ν and transition from a higher energy E_i to a lower energy E_j is

$$R_{ij} = N_i A_{ij} + N_i B_{ij} \rho(\nu)$$

where A_{ij} is the rate constant for emitting a photon spontaneously, and B_{ij} is the rate constant for emitting it in response to ambient photons (induced or stimulated emission). In thermodynamic equilibrium, the number of atoms in state i and that of atoms in state j must, on average, be constant; hence, the rates R_{ji} and R_{ij} must be equal. Also, by arguments analogous to the derivation of Boltzmann statistics, the ratio of N_i and N_j is $g_i/g_j \exp{(E_j - E_i)/kT)}$, where $g_{i,j}$ are the degeneracy of the state i and that of j, respectively, $E_{i,j}$ their energies, k the Boltzmann constant and T the system's temperature. From this, it is readily derived that $g_i B_{ij} = g_j B_{ji}$ and

$$A_{ij} = \frac{8\pi h\nu^3}{c^3} B_{ij}.$$

The A and Bs are collectively known as the *Einstein coefficients*.[73]

Einstein could not fully justify his rate equations, but claimed that it should be possible to calculate the coefficients A_{ij} , B_{ji} and B_{ij} once physicists had obtained "mechanics and electrodynamics modified to accommodate the quantum hypothesis".[74] In fact, in 1926, Paul Dirac derived the B_{ij} rate constants in using a semiclassical approach,[75] and, in 1927, succeeded in deriving *all* the rate constants from first principles within the framework of quantum theory.[76][77] Dirac's work was the foundation of quantum electrodynamics, i.e., the quantization of the electromagnetic field itself. Dirac's approach is also called *second quantization* or quantum field theory;[78][79][80] earlier quantum mechanical treatments only treat material particles as quantum mechanical, not the electromagnetic field.

Einstein was troubled by the fact that his theory seemed incomplete, since it did not determine the *direction* of a spontaneously emitted photon. A probabilistic nature of light-particle motion was first considered by Newton in his treatment of birefringence and, more generally, of the splitting of light beams at interfaces into a transmitted beam and a reflected beam. Newton hypothesized that hidden variables in the light particle determined which path it would follow.[34] Similarly, Einstein hoped for a more complete theory that would leave nothing to chance, beginning his separation[45] from quantum mechanics. Ironically, Max Born's probabilistic interpretation of the wave function[81][82] was inspired by Einstein's later work searching for a more complete theory.[83]

19.9 Second quantization and high energy photon interactions

Main article: Quantum field theory

In 1910, Peter Debye derived Planck's law of black-body radiation from a relatively simple assumption.[84] He correctly decomposed the electromagnetic field in a cavity into its Fourier modes, and assumed that the energy in any mode was an integer multiple of $h\nu$, where ν is the frequency of the electromagnetic mode. Planck's law of black-body radiation follows immediately as a geometric sum. However, Debye's approach failed to give the correct formula for the energy fluctuations of blackbody radiation, which were derived by Einstein in 1909.[41]

In 1925, Born, Heisenberg and Jordan reinterpreted Debye's concept in a key way.[85] As may be shown classically, the Fourier modes of the electromagnetic field—a complete set of electromagnetic plane waves indexed by their wave vector **k** and polarization state—are equivalent to a set of uncoupled simple harmonic oscillators. Treated quantum mechanically, the energy levels of such oscillators are known to be $E = nh\nu$, where ν is the oscillator frequency. The key new step was to identify an electromagnetic mode with energy $E = nh\nu$ as a state with n photons, each of energy $h\nu$. This approach gives the correct energy fluctuation formula.

Dirac took this one step further.[76][77] He treated the interaction between a charge and an electromagnetic field as a small perturbation that induces transitions in the photon states, changing the numbers of photons in the modes, while conserving energy and momentum overall. Dirac was able to derive Einstein's A_{ij} and B_{ij} coefficients from first principles, and showed that the Bose–Einstein statistics of photons is a natural consequence of quantizing the electromagnetic field correctly (Bose's reasoning went in the opposite direction; he derived Planck's law of black-body radiation by *assuming* B–E statistics). In Dirac's time, it was not yet known that all bosons, including photons, must obey Bose–Einstein statistics.

Dirac's second-order perturbation theory can involve virtual photons, transient intermediate states of the electromagnetic field; the static electric and magnetic interactions are mediated by such virtual photons. In such quantum field theories, the probability amplitude of observable events is calculated by summing over *all* possible intermediate steps, even ones that are unphysical; hence, virtual photons are not constrained to satisfy $E = pc$, and may have extra polarization states; depending on the gauge used, virtual photons may have three or four polarization states, instead of the two states of real photons. Although these transient virtual photons can never be observed, they contribute measurably to the probabilities of observable events. Indeed, such second-order and higher-order perturbation calculations can give apparently infinite contributions to the sum. Such unphysical results are corrected for using the technique of renormalization.

Other virtual particles may contribute to the summation as well; for example, two photons may interact indirectly through virtual electron–positron pairs.[86] In fact, such photon-photon scattering (see two-photon physics), as well as electron-photon scattering, is meant to be one of the modes of operations of the planned particle accelerator, the International Linear Collider.[87]

In modern physics notation, the quantum state of the electromagnetic field is written as a Fock state, a tensor product of the states for each electromagnetic mode

$$|n_{k_0}\rangle \otimes |n_{k_1}\rangle \otimes \cdots \otimes |n_{k_n}\rangle \ldots$$

where $|n_{k_i}\rangle$ represents the state in which n_{k_i} photons are in the mode k_i . In this notation, the creation of a new photon in mode k_i (e.g., emitted from an atomic transition) is written as $|n_{k_i}\rangle \rightarrow |n_{k_i}+1\rangle$. This notation merely expresses the concept of Born, Heisenberg and Jordan described above, and does not add any physics.

19.10 The hadronic properties of the photon

Measurements of the interaction between energetic photons and hadrons show that the interaction is much more intense than expected by the interaction of merely photons with the hadron's electric charge. Furthermore, the interaction of energetic photons with protons is similar to the interaction of photons with neutrons[88] in spite of the fact that the electric charge structures of protons and neutrons are substantially different.

A theory called Vector Meson Dominance (VMD) was developed to explain this effect. According to VMD, the photon is a superposition of the pure electromagnetic photon (which interacts only with electric charges) and vector meson.[89]

However, if experimentally probed at very short distances, the intrinsic structure of the photon is recognized as a flux of quark and gluon components, quasi-free according to asymptotic freedom in QCD and described by the photon structure function.[90][91] A comprehensive comparison of data with theoretical predictions is presented in a recent review.[92]

19.11 The photon as a gauge boson

Main article: Gauge theory

The electromagnetic field can be understood as a gauge field, i.e., as a field that results from requiring that a gauge symmetry holds independently at every position in spacetime.[93] For the electromagnetic field, this gauge symmetry is the Abelian U(1) symmetry of complex numbers of absolute value 1, which reflects the ability to vary the phase of a complex number without affecting observables or real valued functions made from it, such as the energy or the Lagrangian.

The quanta of an Abelian gauge field must be massless, uncharged bosons, as long as the symmetry is not broken; hence, the photon is predicted to be massless, and to have zero electric charge and integer spin. The particular form of the electromagnetic interaction specifies that the photon must have spin ±1; thus, its helicity must be $\pm\hbar$. These two spin components correspond to the classical concepts of right-handed and left-handed circularly polarized light. However, the transient virtual photons of quantum electrodynamics may also adopt unphysical polarization states.[93]

In the prevailing Standard Model of physics, the photon is one of four gauge bosons in the electroweak interaction; the other three are denoted W^+, W^- and Z^0 and are responsible for the weak interaction. Unlike the photon, these gauge bosons have mass, owing to a mechanism that breaks their SU(2) gauge symmetry. The unification of the photon with W and Z gauge bosons in the electroweak interaction was accomplished by Sheldon Glashow, Abdus Salam and Steven Weinberg, for which they were awarded the 1979 Nobel Prize in physics.[94][95][96] Physicists continue to hypothesize grand unified theories that connect these four gauge bosons with the eight gluon gauge bosons of quantum chromodynamics; however, key predictions of these theories, such as proton decay, have not been observed experimentally.[97]

19.12 Contributions to the mass of a system

See also: Mass in special relativity and General relativity

The energy of a system that emits a photon is *decreased* by the energy E of the photon as measured in the rest frame of the emitting system, which may result in a reduction in mass in the amount E/c^2 . Similarly, the mass of a system that absorbs a photon is *increased* by a corresponding amount. As an application, the energy balance of nuclear reactions involving photons is commonly written in terms of the masses of the nuclei involved, and terms of the form E/c^2 for the gamma photons (and for other relevant energies, such as the recoil energy of nuclei).[98]

This concept is applied in key predictions of quantum electrodynamics (QED, see above). In that theory, the mass of electrons (or, more generally, leptons) is modified by including the mass contributions of virtual photons, in a technique known as renormalization. Such "radiative corrections" contribute to a number of predictions of QED, such as the magnetic dipole moment of leptons, the Lamb shift, and the hyperfine structure of bound lepton pairs, such as muonium and positronium.[99]

Since photons contribute to the stress–energy tensor, they exert a gravitational attraction on other objects, according to the theory of general relativity. Conversely, photons are themselves affected by gravity; their normally straight trajectories may be bent by warped spacetime, as in gravitational lensing, and their frequencies may be lowered by moving to a higher gravitational potential, as in the Pound–Rebka experiment. However, these effects are not specific to photons; exactly the same effects would be predicted for classical electromagnetic waves.[100]

19.13 Photons in matter

See also: Group velocity and Photochemistry

Any 'explanation' of how photons travel through matter has to explain why different arrangements of matter are transparent or opaque at different wavelengths (light through carbon as diamond or not, as graphite) and why individual photons behave in the same way as large groups. Explanations that invoke 'absorption' and 're-emission' have to provide an explanation for the directionality of the photons (diffraction, reflection) and further explain how entangled photon pairs can travel through matter without their quantum state collapsing.

The simplest explanation is that light that travels through transparent matter does so at a lower speed than *c*, the speed of light in a vacuum. In addition, light can also undergo scattering and absorption. There are circumstances in which heat transfer through a material is mostly radiative, involving emission and absorption of photons within it. An example would be in the core of the Sun. Energy can take about a million years to reach the surface.[101] However, this phenomenon is distinct from scattered radiation passing diffusely through matter, as it involves local equilibrium between the radiation and the temperature. Thus, the time is how long it takes the *energy* to be transferred, not the *photons* themselves. Once in open space, a photon from the Sun takes only 8.3 minutes to reach Earth. The factor by which the speed of light is decreased in a material is called the refractive index of the material. In a classical wave picture, the slowing can be explained by the light inducing electric polarization in the matter, the polarized matter radiating new light, and the new light interfering with the original light wave to form a delayed wave. In a particle picture, the slowing can instead be described as a blending of the photon with quantum excitation of the matter (quasi-particles such as phonons and excitons) to form a polariton; this polariton has a nonzero effective mass, which means that it cannot travel at *c*.

Alternatively, photons may be viewed as *always* traveling at *c*, even in matter, but they have their phase shifted (delayed or advanced) upon interaction with atomic scatters: this modifies their wavelength and momentum, but not speed.[102] A light wave made up of these photons does travel slower than the speed of light. In this view the photons are "bare", and are scattered and phase shifted, while in the view of the preceding paragraph the photons are "dressed" by their interaction with matter, and move without scattering or phase shifting, but at a lower speed.

Light of different frequencies may travel through matter at different speeds; this is called dispersion. In some cases, it can result in extremely slow speeds of light in matter. The effects of photon interactions with other quasi-particles may be observed directly in Raman scattering and Brillouin scattering.[103]

Photons can also be absorbed by nuclei, atoms or molecules, provoking transitions between their energy levels. A classic example is the molecular transition of retinal $C_{20}H_{28}O$, which is responsible for vision, as discovered in 1958 by Nobel laureate biochemist George Wald and co-workers. The absorption provokes a cis-trans isomerization that, in combination with other such transitions, is transduced into nerve impulses. The absorption of photons can even break chemical bonds, as in the photodissociation of chlorine; this is the subject of photochemistry.[104][105] Analogously, gamma rays can in some circumstances dissociate atomic nuclei in a process called photodisintegration.

19.14 Technological applications

Photons have many applications in technology. These examples are chosen to illustrate applications of photons *per se*, rather than general optical devices such as lenses, etc. that could operate under a classical theory of light. The laser is an extremely important application and is discussed above under stimulated emission.

Individual photons can be detected by several methods. The classic photomultiplier tube exploits the photoelectric effect: a photon landing on a metal plate ejects an electron, initiating an ever-amplifying avalanche of electrons. Charge-coupled device chips use a similar effect in semiconductors: an incident photon generates a charge on a microscopic capacitor that can be detected. Other detectors such as Geiger counters use the ability of photons to ionize gas molecules, causing a detectable change in conductivity.[106]

Planck's energy formula $E = h\nu$ is often used by engineers and chemists in design, both to compute the change in energy resulting from a photon absorption and to predict the frequency of the light emitted for a given energy transition. For example, the emission spectrum of a fluorescent light bulb can be designed using gas molecules with different electronic energy levels and adjusting the typical energy with which an electron hits the gas molecules within the bulb.[Note 6]

Under some conditions, an energy transition can be excited by "two" photons that individually would be insufficient. This allows for higher resolution microscopy, because the sample absorbs energy only in the region where two beams of different colors overlap significantly, which can be made much smaller than the excitation volume of a single beam (see two-photon excitation microscopy). Moreover, these photons cause less damage to the sample, since they are of lower energy.[107]

In some cases, two energy transitions can be coupled so that, as one system absorbs a photon, another nearby system "steals" its energy and re-emits a photon of a different frequency. This is the basis of fluorescence resonance energy transfer, a technique that is used in molecular biology to study the interaction of suitable proteins.[108]

Several different kinds of hardware random number generator involve the detection of single photons. In one example, for each bit in the random sequence that is to be produced, a photon is sent to a beam-splitter. In such a situation, there are two possible outcomes of equal probability. The actual outcome is used to determine whether the next bit in the sequence is "0" or "1".[109][110]

19.15 Recent research

See also: Quantum optics

Much research has been devoted to applications of photons in the field of quantum optics. Photons seem well-suited to be elements of an extremely fast quantum computer, and the quantum entanglement of photons is a focus of research. Nonlinear optical processes are another active research area, with topics such as two-photon absorption, self-phase modulation, modulational instability and optical parametric oscillators. However, such processes generally do not require the assumption of photons *per se*; they may often be modeled by treating atoms as nonlinear oscillators. The nonlinear process of spontaneous parametric down conversion is often used to produce single-photon states. Finally, photons are essential in some aspects of optical communication, especially for quantum cryptography.[Note 7]

19.16 See also

- Advanced Photon Source at Argonne National Laboratory
- Ballistic photon
- Doppler shift
- Electromagnetic radiation
- HEXITEC

- Laser
- Light
- Luminiferous aether
- Medipix
- Phonons
- Photon counting
- Photon energy
- Photon polarization
- Photonic molecule
- Photography
- Photonics
- Quantum optics
- Single photon sources
- Static forces and virtual-particle exchange
- Two-photon physics
- EPR paradox
- Dirac equation

19.17 Notes

[1] Although the 1967 Elsevier translation of Planck's Nobel Lecture interprets Planck's *Lichtquant* as "photon", the more literal 1922 translation by Hans Thacher Clarke and Ludwik Silberstein *The origin and development of the quantum theory*, The Clarendon Press, 1922 (here) uses "light-quantum". No evidence is known that Planck himself used the term "photon" by 1926 (see also this note).

[2] Isaac Asimov credits Arthur Compton with defining quanta of energy as photons in 1923. Asimov, I. (1966). *The Neutrino, Ghost Particle of the Atom*. Garden City (NY): Doubleday. ISBN 0-380-00483-6. LCCN 66017073. and Asimov, I. (1966). *The Universe From Flat Earth To Quasar*. New York (NY): Walker. ISBN 0-8027-0316-X. LCCN 66022515.

[3] The mass of the photon is believed to be exactly zero, based on experiment and theoretical considerations described in the article. Some sources also refer to the *relativistic mass* concept, which is just the energy scaled to units of mass. For a photon with wavelength λ or energy E, this is $h/\lambda c$ or E/c^2. This usage for the term "mass" is no longer common in scientific literature. Further info: What is the mass of a photon? http://math.ucr.edu/home/baez/physics/ParticleAndNuclear/photon_mass.html

[4] The phrase "no matter how intense" refers to intensities below approximately 10^{13} W/cm^2 at which point perturbation theory begins to break down. In contrast, in the intense regime, which for visible light is above approximately 10^{14} W/cm^2, the classical wave description correctly predicts the energy acquired by electrons, called ponderomotive energy. (See also: Boreham *et al.* (1996). "Photon density and the correspondence principle of electromagnetic interaction".) By comparison, sunlight is only about 0.1 W/cm^2.

[5] These experiments produce results that cannot be explained by any classical theory of light, since they involve anticorrelations that result from the quantum measurement process. In 1974, the first such experiment was carried out by Clauser, who reported a violation of a classical Cauchy–Schwarz inequality. In 1977, Kimble *et al.* demonstrated an analogous anti-bunching effect of photons interacting with a beam splitter; this approach was simplified and sources of error eliminated in the photon-anticorrelation experiment of Grangier *et al.* (1986). This work is reviewed and simplified further in Thorn *et al.* (2004). (These references are listed below under #Additional references.)

[6] An example is US Patent Nr. 5212709.

[7] Introductory-level material on the various sub-fields of quantum optics can be found in Fox, M. (2006). *Quantum Optics: An Introduction*. Oxford University Press. ISBN 0-19-856673-5.

19.18 References

[1] Amsler, C. (Particle Data Group); Amsler; Doser; Antonelli; Asner; Babu; Baer; Band; Barnett; Bergren; Beringer; Bernardi; Bertl; Bichsel; Biebel; Bloch; Blucher; Blusk; Cahn; Carena; Caso; Ceccucci; Chakraborty; Chen; Chivukula; Cowan; Dahl; d'Ambrosio; Damour et al. (2008). "Review of Particle Physics: Gauge and Higgs bosons" (PDF). *Physics Letters B* **667**: 1. Bibcode:2008PhLB..667....1P. doi:10.1016/j.physletb.2008.07.018.

[2] Joos, George (1951). *Theoretical Physics*. London and Glasgow: Blackie and Son Limited. p. 679.

[3] Kimble, H.J.; Dagenais, M.; Mandel, L.; Dagenais; Mandel (1977). "Photon Anti-bunching in Resonance Fluorescence". *Physical Review Letters* **39** (11): 691–695. Bibcode:1977PhRvL..39..691K. doi:10.1103/PhysRevLett.39.691.

[4] Grangier, P.; Roger, G.; Aspect, A.; Roger; Aspect (1986). "Experimental Evidence for a Photon Anticorrelation Effect on a Beam Splitter: A New Light on Single-Photon Interferences". *Europhysics Letters* **1** (4): 173–179. Bibcode:1986EL......1..173G. doi:10.1209/0295-5075/1/4/004.

[5] Planck, M. (1901). "On the Law of Distribution of Energy in the Normal Spectrum". *Annalen der Physik* **4** (3): 553–563. Bibcode:1901AnP...309..553P. doi:10.1002/andp.19013090310. Archived from the original on 2008-04-18.

[6] Einstein, A. (1905). "Über einen die Erzeugung und Verwandlung des Lichtes betreffenden heuristischen Gesichtspunkt" (PDF). *Annalen der Physik* (in German) **17** (6): 132–148. Bibcode:1905AnP...322..132E. doi:10.1002/andp.19053220607.. An English translation is available from Wikisource.

[7] "Discordances entre l'expérience et la théorie électromagnétique du rayonnement." In Électrons et Photons. Rapports et Discussions de Cinquième Conseil de Physique, edited by Institut International de Physique Solvay. Paris: Gauthier-Villars, pp. 55-85.

[8] Helge Kragh: *Photon: New light on an old name*. Arxiv, 2014-2-28

[9] Villard, P. (1900). "Sur la réflexion et la réfraction des rayons cathodiques et des rayons déviables du radium". *Comptes Rendus des Séances de l'Académie des Sciences* (in French) **130**: 1010–1012.

[10] Villard, P. (1900). "Sur le rayonnement du radium". *Comptes Rendus des Séances de l'Académie des Sciences* (in French) **130**: 1178–1179.

[11] Rutherford, E.; Andrade, E.N.C. (1914). "The Wavelength of the Soft Gamma Rays from Radium B". *Philosophical Magazine* **27** (161): 854–868. doi:10.1080/14786440508635156.

[12] Kobychev, V.V.; Popov, S.B. (2005). "Constraints on the photon charge from observations of extragalactic sources". *Astronomy Letters* **31** (3): 147–151. arXiv:hep-ph/0411398. Bibcode:2005AstL...31..147K. doi:10.1134/1.1883345.

[13] Role as gauge boson and polarization section 5.1 inAitchison, I.J.R.; Hey, A.J.G. (1993). *Gauge Theories in Particle Physics*. IOP Publishing. ISBN 0-85274-328-9.

[14] See p.31 inAmsler, C. et al. (2008). "Review of Particle Physics". *Physics Letters B* **667**: 1–1340. Bibcode:2008PhLB..667....1P. doi:10.1016/j.physletb.2008.07.018.

[15] Halliday, David; Resnick, Robert; Walker, Jerl (2005), *Fundamental of Physics* (7th ed.), USA: John Wiley and Sons, Inc., ISBN 0-471-23231-9

[16] See section 1.6 in Alonso, M.; Finn, E.J. (1968). *Fundamental University Physics Volume III: Quantum and Statistical Physics*. Addison-Wesley. ISBN 0-201-00262-0.

[17] Davison E. Soper, Electromagnetic radiation is made of photons, Institute of Theoretical Science, University of Oregon

[18] This property was experimentally verified by Raman and Bhagavantam in 1931: Raman, C.V.; Bhagavantam, S. (1931). "Experimental proof of the spin of the photon" (PDF). *Indian Journal of Physics* **6**: 353.

[19] Burgess, C.; Moore, G. (2007). "1.3.3.2". *The Standard Model. A Primer*. Cambridge University Press. ISBN 0-521-86036-9.

[20] Griffiths, David J. (2008), *Introduction to Elementary Particles* (2nd revised ed.), WILEY-VCH, ISBN 978-3-527-40601-2

[21] E.g., section 9.3 in Alonso, M.; Finn, E.J. (1968). *Fundamental University Physics Volume III: Quantum and Statistical Physics*. Addison-Wesley.

[22] E.g., Appendix XXXII in Born, M. (1962). *Atomic Physics*. Blackie & Son. ISBN 0-486-65984-4.

[23] Mermin,David(February1984). "Relativity without light".*American Journal of Physics***52**(2): 119–124. Bibcode:1984AmJPh. doi:10.1119/1.13917.

[24] Plimpton, S.; Lawton, W. (1936). "A Very Accurate Test of Coulomb's Law of Force Between Charges". *Physical Review* **50** (11): 1066. Bibcode:1936PhRv...50.1066P. doi:10.1103/PhysRev.50.1066.

[25] Williams, E.; Faller, J.; Hill, H. (1971). "New Experimental Test of Coulomb's Law: A Laboratory Upper Limit on the Photon Rest Mass". *Physical Review Letters* **26** (12): 721. Bibcode:1971PhRvL..26..721W. doi:10.1103/PhysRevLett.26.721.

[26] Chibisov,G V(1976). "Astrophysical upper limits on the photon rest mass".*Soviet Physics Uspekhi***19**(7): 624. Bibcode:197C. doi:10.1070/PU1976v019n07ABEH005277.

[27] Lakes, Roderic (1998). "Experimental Limits on the Photon Mass and Cosmic Magnetic Vector Potential". *Physical Review Letters* **80** (9): 1826. Bibcode:1998PhRvL..80.1826L. doi:10.1103/PhysRevLett.80.1826.

[28] Amsler, C; Doser, M; Antonelli, M; Asner, D; Babu, K; Baer, H; Band, H; Barnett, R et al. (2008). "Review of Particle Physics*". *Physics Letters B* **667**: 1. Bibcode:2008PhLB..667....1P. doi:10.1016/j.physletb.2008.07.018. Summary Table

[29] Adelberger, Eric; Dvali, Gia; Gruzinov, Andrei (2007). "Photon-Mass Bound Destroyed by Vortices". *Physical Review Letters* **98** (1): 010402. arXiv:hep-ph/0306245. Bibcode:2007PhRvL..98a0402A. doi:10.1103/PhysRevLett.98.010402. PMID 17358459. preprint

[30] Wilczek, Frank (2010). *The Lightness of Being: Mass, Ether, and the Unification of Forces*. Basic Books. p. 212. ISBN 978-0-465-01895-6.

[31] Descartes, R. (1637). *Discours de la méthode (Discourse on Method)* (in French). Imprimerie de Ian Maire. ISBN 0-268-00870-1.

[32] Hooke, R. (1667). *Micrographia: or some physiological descriptions of minute bodies made by magnifying glasses with observations and inquiries thereupon ...* London (UK): Royal Society of London. ISBN 0-486-49564-7.

[33] Huygens, C. (1678). *Traité de la lumière* (in French).. An English translation is available from Project Gutenberg

[34] Newton, I. (1952) [1730]. *Opticks* (4th ed.). Dover (NY): Dover Publications. Book II, Part III, Propositions XII–XX; Queries 25–29. ISBN 0-486-60205-2.

[35] Buchwald, J.Z. (1989). *The Rise of the Wave Theory of Light: Optical Theory and Experiment in the Early Nineteenth Century*. University of Chicago Press. ISBN 0-226-07886-8. OCLC 18069573.

[36] Maxwell, J.C. (1865). "A Dynamical Theory of the Electromagnetic Field". *Philosophical Transactions of the Royal Society* **155**: 459–512. Bibcode:1865RSPT..155..459C. doi:10.1098/rstl.1865.0008. This article followed a presentation by Maxwell on 8 December 1864 to the Royal Society.

[37] Hertz, H. (1888). "Über Strahlen elektrischer Kraft". *Sitzungsberichte der Preussischen Akademie der Wissenschaften (Berlin)* (in German) **1888**: 1297–1307.

[38] Frequency-dependence of luminiscence p. 276f., photoelectric effect section 1.4 in Alonso, M.; Finn, E.J. (1968). *Fundamental University Physics Volume III: Quantum and Statistical Physics*. Addison-Wesley. ISBN 0-201-00262-0.

[39] Wien, W. (1911). "Wilhelm Wien Nobel Lecture".

[40] Planck, M. (1920). "Max Planck's Nobel Lecture".

[41] Einstein, A. (1909). "Über die Entwicklung unserer Anschauungen über das Wesen und die Konstitution der Strahlung" (PDF). *Physikalische Zeitschrift* (in German) **10**: 817–825.. An English translation is available from Wikisource.

[42] Presentation speech by Svante Arrhenius for the 1921 Nobel Prize in Physics, December 10, 1922. Online text from [nobelprize.org], The Nobel Foundation 2008. Access date 2008-12-05.

[43] Einstein, A. (1916). "Zur Quantentheorie der Strahlung". *Mitteilungen der Physikalischen Gesellschaft zu Zürich* **16**: 47. Also *Physikalische Zeitschrift*, **18**, 121–128 (1917). (German)

[44] Compton, A. (1923). "A Quantum Theory of the Scattering of X-rays by Light Elements". *Physical Review* **21** (5): 483–502. Bibcode:1923PhRv...21..483C. doi:10.1103/PhysRev.21.483.

[45] Pais, A. (1982). *Subtle is the Lord: The Science and the Life of Albert Einstein*. Oxford University Press. ISBN 0-19-853907-X.

[46] *Einstein and the Quantum: The Quest of the Valiant Swabian*, A. Douglas Stone, Princeton University Press, 2013.

[47] Millikan, R.A (1924). "Robert A. Millikan's Nobel Lecture".

[48] Hendry, J. (1980). "The development of attitudes to the wave-particle duality of light and quantum theory, 1900–1920". *Annals of Science* **37** (1): 59–79. doi:10.1080/00033798000200121.

[49] Bohr, N.; Kramers, H.A.; Slater, J.C. (1924). "The Quantum Theory of Radiation". *Philosophical Magazine* **47**: 785–802. doi:10.1080/14786442408565262. Also *Zeitschrift für Physik*, **24**, 69 (1924).

[50] Heisenberg, W. (1933). "Heisenberg Nobel lecture".

[51] Mandel, L. (1976). E. Wolf, ed. "The case for and against semiclassical radiation theory". *Progress in Optics*. Progress in Optics (North-Holland) **13**: 27–69. doi:10.1016/S0079-6638(08)70018-0. ISBN 978-0-444-10806-7.

[52] Taylor, G.I. (1909). *Interference fringes with feeble light. Proceedings of the Cambridge Philosophical Society* **15**: 114–115.

[53] Saleh, B. E. A. and Teich, M. C. (2007). *Fundamentals of Photonics*. Wiley. ISBN 0-471-35832-0.

[54] Heisenberg, W. (1927). "Über den anschaulichen Inhalt der quantentheoretischen Kinematik und Mechanik". *Zeitschrift für Physik* (in German) **43** (3–4): 172–198. Bibcode:1927ZPhy...43..172H. doi:10.1007/BF01397280.

[55] E.g., p. 10f. in Schiff, L.I. (1968). *Quantum Mechanics* (3rd ed.). McGraw-Hill. ASIN B001B3MINM. ISBN 0-07-055287-8.

[56] Kramers, H.A. (1958). *Quantum Mechanics*. Amsterdam: North-Holland. ASIN B0006AUW5C. ISBN 0-486-49533-7.

[57] Bohm, D. (1989) [1954]. *Quantum Theory*. Dover Publications. ISBN 0-486-65969-0.

[58] Newton, T.D.; Wigner, E.P. (1949). "Localized states for elementary particles". *Reviews of Modern Physics* **21** (3): 400–406. Bibcode:1949RvMP...21..400N. doi:10.1103/RevModPhys.21.400.

[59] Bialynicki-Birula, I. (1994). "On the wave function of the photon" (PDF). *Acta Physica Polonica A* **86**: 97–116.

[60] Sipe,J.E. (1995). "Photon wave functions".*Physical ReviewA***52**(3): 1875–1883. Bibcode:1995PhRvA..52.1875S.doi:1875.

[61] Bialynicki-Birula, I. (1996). "Photon wave function". *Progress in Optics*. Progress in Optics **36**: 245–294. doi:10.1016/S0079-6638(08)70316-0. ISBN 978-0-444-82530-8.

[62] Scully, M.O.; Zubairy, M.S. (1997). *Quantum Optics*. Cambridge (UK): Cambridge University Press. ISBN 0-521-43595-1.

[63] The best illustration is the Couder experiment, demonstrating the behaviour of a mechanical analog, see https://www.youtube.com/watch?v=W9yWv5dqSKk

[64] Bell, J. S., "Speakable and Unspeakable in Quantum Mechanics", Cambridge: Cambridge University Press, 1987.

[65] Bose,S.N.(1924). "Plancks Gesetz und Lichtquantenhypothese".*Zeitschrift für Physik*(in German)**26**: 178–181. Bibcode78B. doi:10.1007/BF01327326.

[66] Einstein, A. (1924). "Quantentheorie des einatomigen idealen Gases". *Sitzungsberichte der Preussischen Akademie der Wissenschaften (Berlin), Physikalisch-mathematische Klasse* (in German) **1924**: 261–267.

[67] Einstein, A. (1925). "Quantentheorie des einatomigen idealen Gases, Zweite Abhandlung". *Sitzungsberichte der Preussischen Akademie der Wissenschaften (Berlin), Physikalisch-mathematische Klasse* (in German) **1925**: 3–14. doi:10.1002/3527608958.ch28. ISBN 978-3-527-60895-9.

[68] Anderson, M.H.; Ensher, J.R.; Matthews, M.R.; Wieman, C.E.; Cornell, E.A. (1995). "Observation of Bose–Einstein Condensation in a Dilute Atomic Vapor". *Science* **269** (5221): 198–201. Bibcode:1995Sci...269..198A. doi:10.1126/science.269 8.JSTOR2888436. PMID 17789847.

[69] "Physicists Slow Speed of Light". News.harvard.edu (1999-02-18). Retrieved on 2015-05-11.

[70] "Light Changed to Matter, Then Stopped and Moved". photonics.com (February 2007). Retrieved on 2015-05-11.

[71] Streater, R.F.; Wightman, A.S. (1989). *PCT, Spin and Statistics, and All That*. Addison-Wesley. ISBN 0-201-09410-X.

[72] Einstein, A. (1916). "Strahlungs-emission und -absorption nach der Quantentheorie". *Verhandlungen der Deutschen Physikalischen Gesellschaft* (in German) **18**: 318–323. Bibcode:1916DPhyG..18..318E.

[73] Section 1.4 in Wilson, J.; Hawkes, F.J.B. (1987). *Lasers: Principles and Applications*. New York: Prentice Hall. ISBN 0-13-523705-X.

[74] P. 322 in Einstein, A. (1916). "Strahlungs-emission und -absorption nach der Quantentheorie". *Verhandlungen der Deutschen Physikalischen Gesellschaft* (in German) **18**: 318–323. Bibcode:1916DPhyG..18..318E.:

> Die Konstanten A_m^n and B_m^n würden sich direkt berechnen lassen, wenn wir im Besitz einer im Sinne der Quantenhypothese modifizierten Elektrodynamik und Mechanik wären."

[75] Dirac, P.A.M. (1926). "On the Theory of Quantum Mechanics". *Proceedings of the Royal Society A* **112** (762): 661–677. Bibcode:1926RSPSA.112..661D. doi:10.1098/rspa.1926.0133.

[76] Dirac, P.A.M. (1927). "The Quantum Theory of the Emission and Absorption of Radiation" (PDF). *Proceedings of the Royal Society A* **114** (767): 243–265. Bibcode:1927RSPSA.114..243D. doi:10.1098/rspa.1927.0039.

[77] Dirac,P.A.M.(1927b).*The Quantum Theory of Dispersion*.*Proceedings of the Royal SocietyA***114**: 710–728. doi:10.1098/rspa1.

[78] Heisenberg,W.;Pauli,W.(1929). "Zur Quantentheorie der Wellenfelder".*Zeitschrift für Physik*(in German)**56**: 1. BibcodeH. doi:10.1007/BF01340129.

[79] Heisenberg, W.; Pauli, W. (1930). "Zur Quantentheorie der Wellenfelder". *Zeitschrift für Physik* (in German) **59** (3–4): 139. Bibcode:1930ZPhy...59..168H. doi:10.1007/BF01341423.

[80] Fermi, E. (1932). "Quantum Theory of Radiation" (PDF). *Reviews of Modern Physics* **4**: 87. Bibcode:1932RvMP....4...87F. doi:10.1103/RevModPhys.4.87.

[81] Born, M. (1926). "Zur Quantenmechanik der Stossvorgänge" (PDF). *Zeitschrift für Physik* (in German) **37** (12): 863–867. Bibcode:1926ZPhy...37..863B. doi:10.1007/BF01397477.

[82] Born,M.(1926). "Quantenmechanik der Stossvorgänge".*Zeitschrift für Physik*(in German)**38**(11–12): 803. Bibcode:1926ZP. doi:10.1007/BF01397184.

[83] Pais, A. (1986). *Inward Bound: Of Matter and Forces in the Physical World*. Oxford University Press. p. 260. ISBN 0-19-851997-4. Specifically, Born claimed to have been inspired by Einstein's never-published attempts to develop a "ghost-field" theory, in which point-like photons are guided probabilistically by ghost fields that follow Maxwell's equations.

[84] Debye, P. (1910). "Der Wahrscheinlichkeitsbegriff in der Theorie der Strahlung". *Annalen der Physik* (in German) **33** (16): 1427–1434. Bibcode:1910AnP...338.1427D. doi:10.1002/andp.19103381617.

[85] Born, M.; Heisenberg, W.; Jordan, P. (1925). "Quantenmechanik II". *Zeitschrift für Physik* (in German) **35** (8–9): 557–615. Bibcode:1926ZPhy...35..557B. doi:10.1007/BF01379806.

[86] Photon-photon-scattering section 7-3-1, renormalization chapter 8-2 in Itzykson, C.; Zuber, J.-B. (1980). *Quantum Field Theory*. McGraw-Hill. ISBN 0-07-032071-3.

[87] Weiglein, G. (2008). "Electroweak Physics at the ILC". *Journal of Physics: Conference Series* **110** (4): 042033. arXiv:0711.3003. Bibcode:2008JPhCS.110d2033W. doi:10.1088/1742-6596/110/4/042033.

[88] Bauer, T. H.; Spital, R. D.; Yennie, D. R.; Pipkin, F. M. (1978). "The hadronic properties of the photon in high-energy interactions". *Reviews of Modern Physics* **50** (2): 261. Bibcode:1978RvMP...50..261B. doi:10.1103/RevModPhys.50.261.

[89] Sakurai, J. J. (1960). "Theory of strong interactions". *Annals of Physics* **11**: 1. Bibcode:1960AnPhy..11....1S. doi:10.1016/0003-4916(60)90126-3.

[90] Walsh,T.F.;Zerwas,P. (1973). "Two-photon processes in the parton model".*Physics Letters B***44**(2): 195. Bibcode:1973P. doi:10.1016/0370-2693(73)90520-0.

[91] Witten, E. (1977). "Anomalous cross section for photon-photon scattering in gauge theories". *Nuclear Physics B* **120** (2): 189. Bibcode:1977NuPhB.120..189W. doi:10.1016/0550-3213(77)90038-4.

[92] Nisius, R. (2000). "The photon structure from deep inelastic electron–photon scattering". *Physics Reports* **332** (4–6): 165. Bibcode:2000PhR...332..165N. doi:10.1016/S0370-1573(99)00115-5.

[93] Ryder, L.H. (1996). *Quantum field theory* (2nd ed.). Cambridge University Press. ISBN 0-521-47814-6.

[94] Sheldon Glashow Nobel lecture, delivered 8 December 1979.

[95] Abdus Salam Nobel lecture, delivered 8 December 1979.

[96] Steven Weinberg Nobel lecture, delivered 8 December 1979.

[97] E.g., chapter 14 in Hughes, I. S. (1985). *Elementary particles* (2nd ed.). Cambridge University Press. ISBN 0-521-26092-2.

[98] E.g., section 10.1 in Dunlap, R.A. (2004). *An Introduction to the Physics of Nuclei and Particles.* Brooks/Cole. ISBN 0-534-39294-6.

[99] Radiative correction to electron mass section 7-1-2, anomalous magnetic moments section 7-2-1, Lamb shift section 7-3-2 and hyperfine splitting in positronium section 10-3 in Itzykson, C.; Zuber, J.-B. (1980). *Quantum Field Theory.* McGraw-Hill. ISBN 0-07-032071-3.

[100] E. g. sections 9.1 (gravitational contribution of photons) and 10.5 (influence of gravity on light) in Stephani, H.; Stewart, J. (1990). *General Relativity: An Introduction to the Theory of Gravitational Field.* Cambridge University Press. pp. 86 ff, 108 ff. ISBN 0-521-37941-5.

[101] Naeye, R. (1998). *Through the Eyes of Hubble: Birth, Life and Violent Death of Stars.* CRC Press. ISBN 0-7503-0484-7. OCLC 40180195.

[102] Ch 4 in Hecht, Eugene (2001). *Optics.* Addison Wesley. ISBN 978-0-8053-8566-3.

[103] Polaritons section 10.10.1, Raman and Brillouin scattering section 10.11.3 in Patterson, J.D.; Bailey, B.C. (2007). *Solid-State Physics: Introduction to the Theory.* Springer. pp. 569 ff, 580 ff. ISBN 3-540-24115-9.

[104] E.g., section 11-5 C in Pine, S.H.; Hendrickson, J.B.; Cram, D.J.; Hammond, G.S. (1980). *Organic Chemistry* (4th ed.). McGraw-Hill. ISBN 0-07-050115-7.

[105] Nobel lecture given by G. Wald on December 12, 1967, online at nobelprize.org: The Molecular Basis of Visual Excitation.

[106] Photomultiplier section 1.1.10, CCDs section 1.1.8, Geiger counters section 1.3.2.1 in Kitchin, C.R. (2008). *Astrophysical Techniques.* Boca Raton (FL): CRC Press. ISBN 1-4200-8243-4.

[107] Denk, W.; Svoboda, K. (1997). "Photon upmanship: Why multiphoton imaging is more than a gimmick". *Neuron* **18** (3): 351–357. doi:10.1016/S0896-6273(00)81237-4. PMID 9115730.

[108] Lakowicz, J.R. (2006). *Principles of Fluorescence Spectroscopy.* Springer. pp. 529 ff. ISBN 0-387-31278-1.

[109] Jennewein, T.; Achleitner, U.; Weihs, G.; Weinfurter, H.; Zeilinger, A. (2000). "A fast and compact quantum random number generator". *Review of Scientific Instruments* **71** (4): 1675–1680. arXiv:quant-ph/9912118. Bibcode:2000RScI...71.1675J. doi:10.1063/1.1150518.

[110] Stefanov, A.; Gisin, N.; Guinnard, O.; Guinnard, L.; Zbiden, H. (2000). "Optical quantum random number generator". *Journal of Modern Optics* **47** (4): 595–598. doi:10.1080/095003400147908.

19.19 Additional references

By date of publication:

- Clauser, J.F. (1974). "Experimental distinction between the quantum and classical field-theoretic predictions for the photoelectric effect". *Physical Review D* **9** (4): 853–860. Bibcode:1974PhRvD...9..853C. doi:10.1103/PhysRevD.9.853.
- Kimble, H.J.; Dagenais, M.; Mandel, L. (1977). "Photon Anti-bunching in Resonance Fluorescence". *Physical Review Letters* **39** (11): 691–695. Bibcode:1977PhRvL..39..691K. doi:10.1103/PhysRevLett.39.691.
- Pais, A. (1982). *Subtle is the Lord: The Science and the Life of Albert Einstein*. Oxford University Press.
- Feynman, Richard (1985). *QED: The Strange Theory of Light and Matter*. Princeton University Press. ISBN 978-0-691-12575-6.
- Grangier, P.; Roger, G.; Aspect, A. (1986). "Experimental Evidence for a Photon Anticorrelation Effect on a Beam Splitter: A New Light on Single-Photon Interferences". *Europhysics Letters* **1** (4): 173–179. Bibcode:1986EL......1..173G. doi:10.1209/0295-5075/1/4/004.
- Lamb, W.E. (1995). "Anti-photon". *Applied Physics B* **60** (2–3): 77–84. Bibcode:1995ApPhB..60...77L. doi:10.1007/BF01135846.
- Special supplemental issue of *Optics and Photonics News* (vol. 14, October 2003) article web link
  - Roychoudhuri, C.; Rajarshi, R. (2003). "The nature of light: what is a photon?". *Optics and Photonics News* **14**: S1 (Supplement).
  - Zajonc, A. "Light reconsidered". *Optics and Photonics News* **14**: S2–S5 (Supplement).
  - Loudon, R. "What is a photon?". *Optics and Photonics News* **14**: S6–S11 (Supplement).
  - Finkelstein, D. "What is a photon?". *Optics and Photonics News* **14**: S12–S17 (Supplement).
  - Muthukrishnan, A.; Scully, M.O.; Zubairy, M.S. "The concept of the photon—revisited". *Optics and Photonics News* **14**: S18–S27 (Supplement).
  - Mack, H.; Schleich, W.P.. "A photon viewed from Wigner phase space". *Optics and Photonics News* **14**: S28–S35 (Supplement).
- Glauber, R. (2005). "One Hundred Years of Light Quanta" (PDF). *2005 Physics Nobel Prize Lecture*.
- Hentschel, K. (2007). "Light quanta: The maturing of a concept by the stepwise accretion of meaning". *Physics and Philosophy* **1** (2): 1–20.

Education with single photons:

- Thorn, J.J.; Neel, M.S.; Donato, V.W.; Bergreen, G.S.; Davies, R.E.; Beck, M. (2004). "Observing the quantum behavior of light in an undergraduate laboratory" (PDF). *American Journal of Physics* **72** (9): 1210–1219. Bibcode:2004AmJPh..72.1210T. doi:10.1119/1.1737397.
- Bronner, P.; Strunz, Andreas; Silberhorn, Christine; Meyn, Jan-Peter (2009). "Interactive screen experiments with single photons". *European Journal of Physics* **30** (2): 345–353. Bibcode:2009EJPh...30..345B. doi:10.1088/0143-0807/30/2/014.

19.20 External links

- The dictionary definition of photon at Wiktionary
- Media related to Photon at Wikimedia Commons

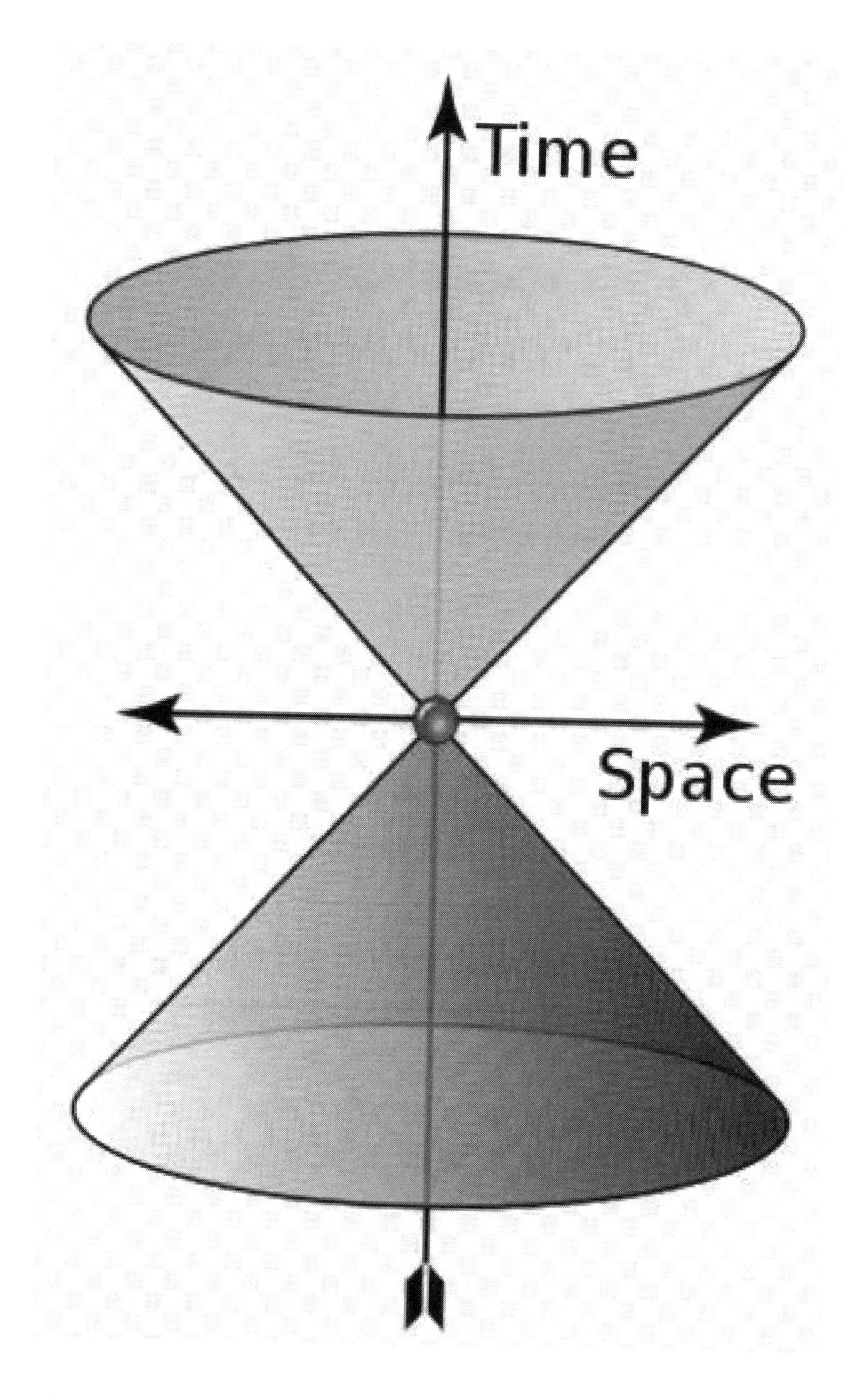

The cone shows possible values of wave 4-vector of a photon. The "time" axis gives the angular frequency (rad·s−1) and the "space" axes represent the angular wavenumber (rad·m−1). Green and indigo represent left and right polarization

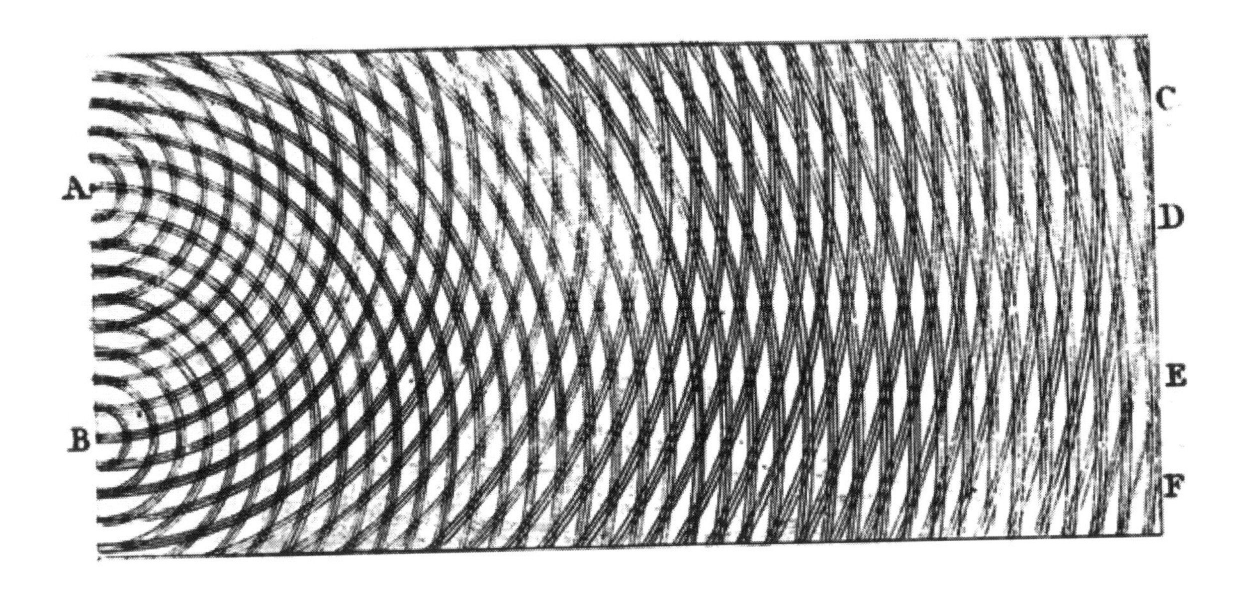

Thomas Young's double-slit experiment in 1801 showed that light can act as a wave, helping to invalidate early particle theories of light.[15]:964

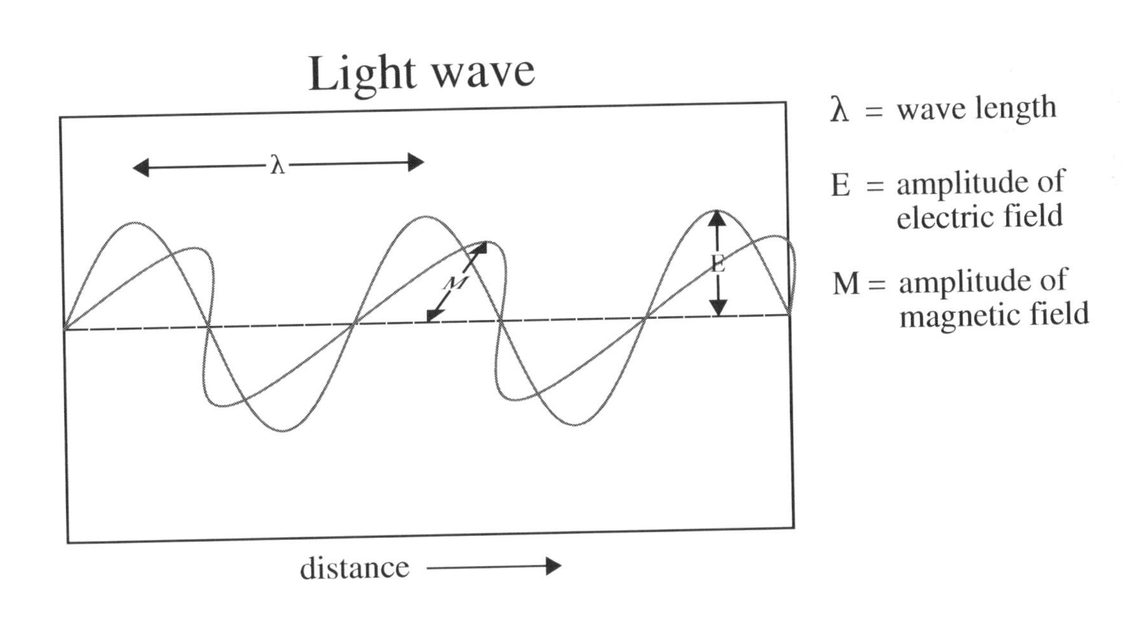

In 1900, Maxwell's theoretical model of light as oscillating electric and magnetic fields seemed complete. However, several observations could not be explained by any wave model of electromagnetic radiation, leading to the idea that light-energy was packaged into quanta *described by E=hν. Later experiments showed that these light-quanta also carry momentum and, thus, can be considered particles: the* photon *concept was born, leading to a deeper understanding of the electric and magnetic fields themselves.*

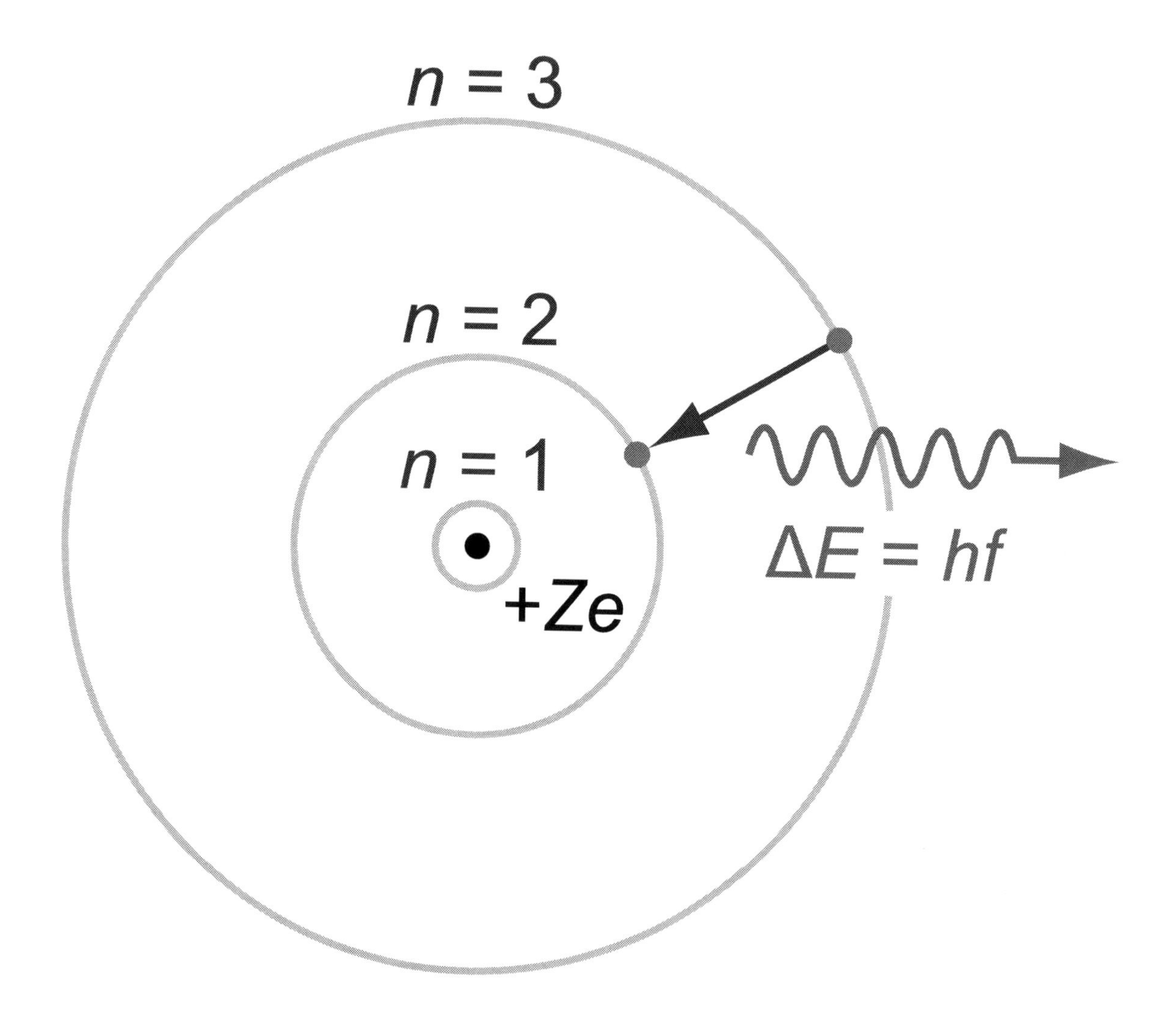

Up to 1923, most physicists were reluctant to accept that light itself was quantized. Instead, they tried to explain photon behavior by quantizing only matter, *as in the Bohr model of the hydrogen atom (shown here). Even though these semiclassical models were only a first approximation, they were accurate for simple systems and they led to quantum mechanics.*

Photons in a Mach–Zehnder interferometer exhibit wave-like interference and particle-like detection at single-photon detectors.

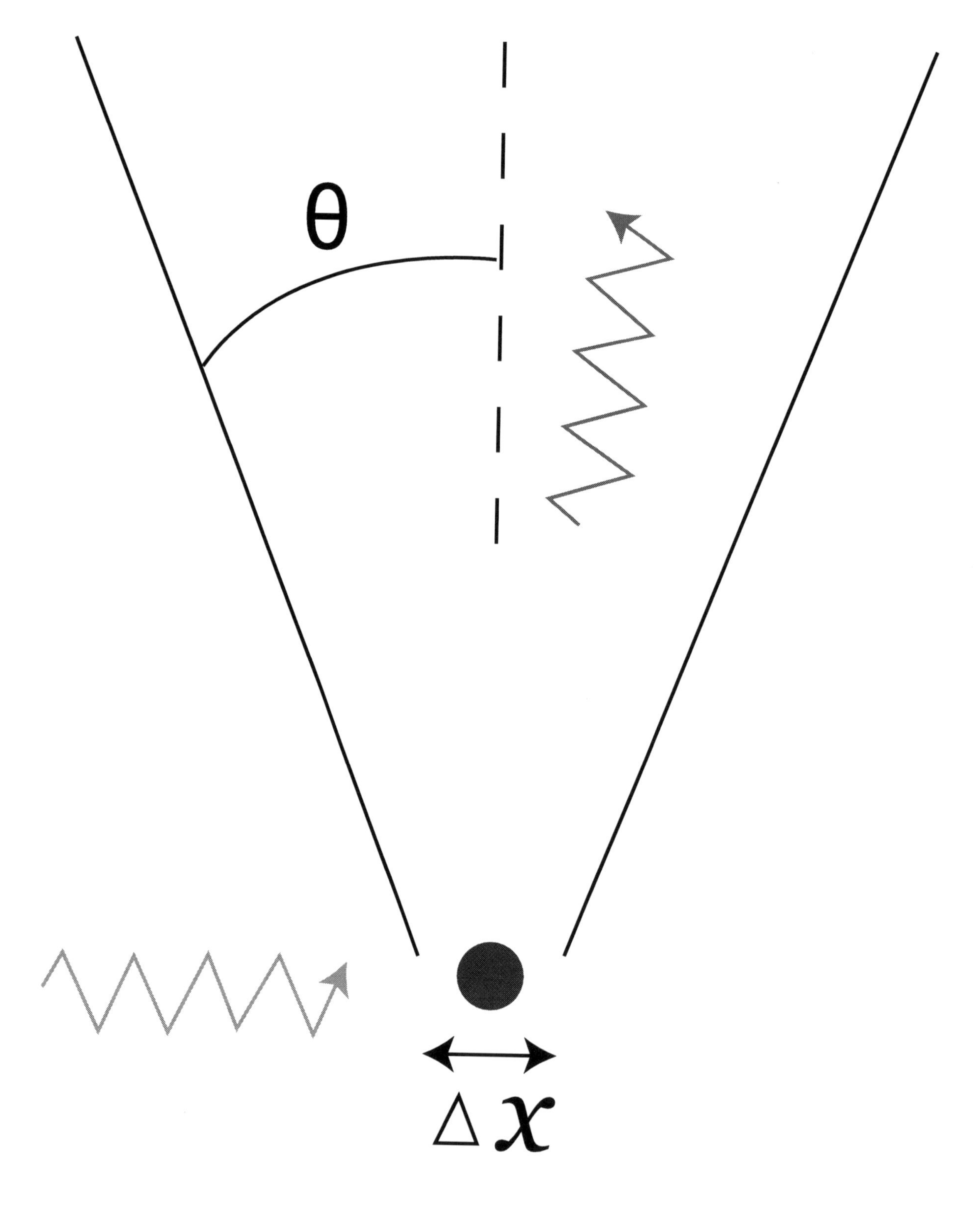

Heisenberg's thought experiment for locating an electron (shown in blue) with a high-resolution gamma-ray microscope. The incoming gamma ray (shown in green) is scattered by the electron up into the microscope's aperture angle θ. The scattered gamma ray is shown in red. Classical optics shows that the electron position can be resolved only up to an uncertainty Δx that depends on θ and the wavelength λ of the incoming light.

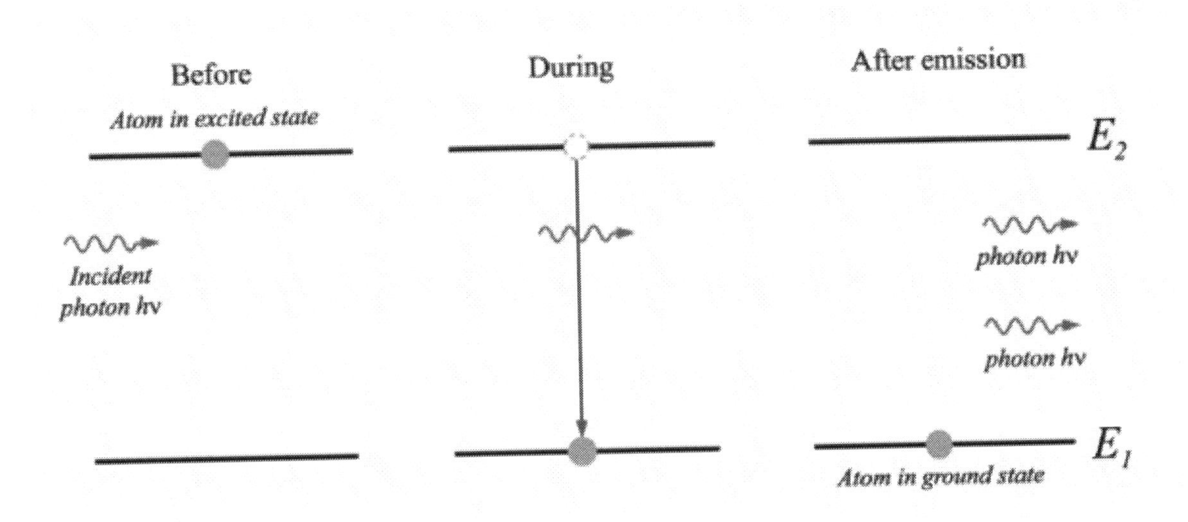

Stimulated emission (in which photons "clone" themselves) was predicted by Einstein in his kinetic analysis, and led to the development of the laser. Einstein's derivation inspired further developments in the quantum treatment of light, which led to the statistical interpretation of quantum mechanics.

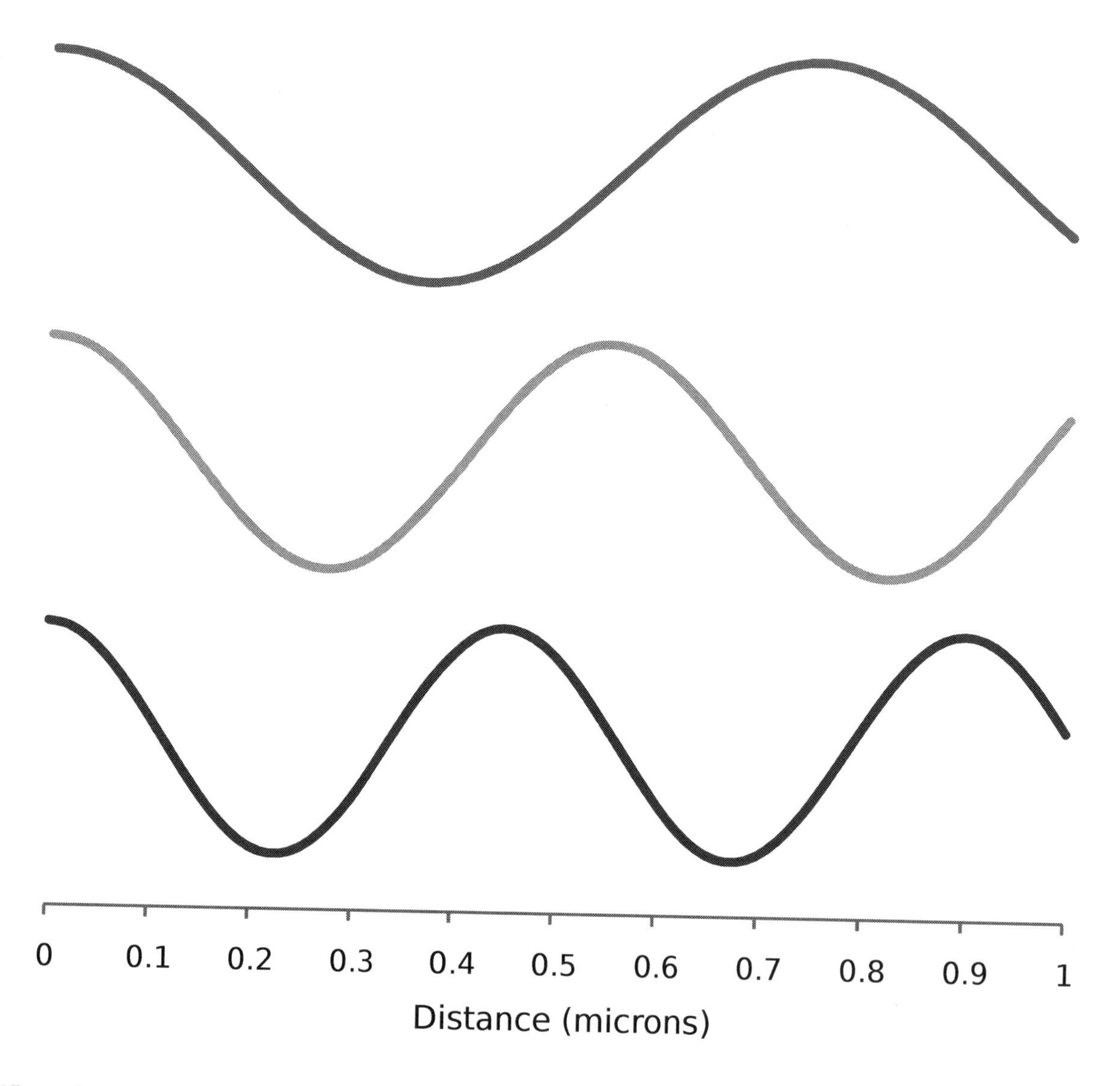

Different electromagnetic modes *(such as those depicted here) can be treated as independent simple harmonic oscillators. A photon corresponds to a unit of energy E=hν in its electromagnetic mode.*

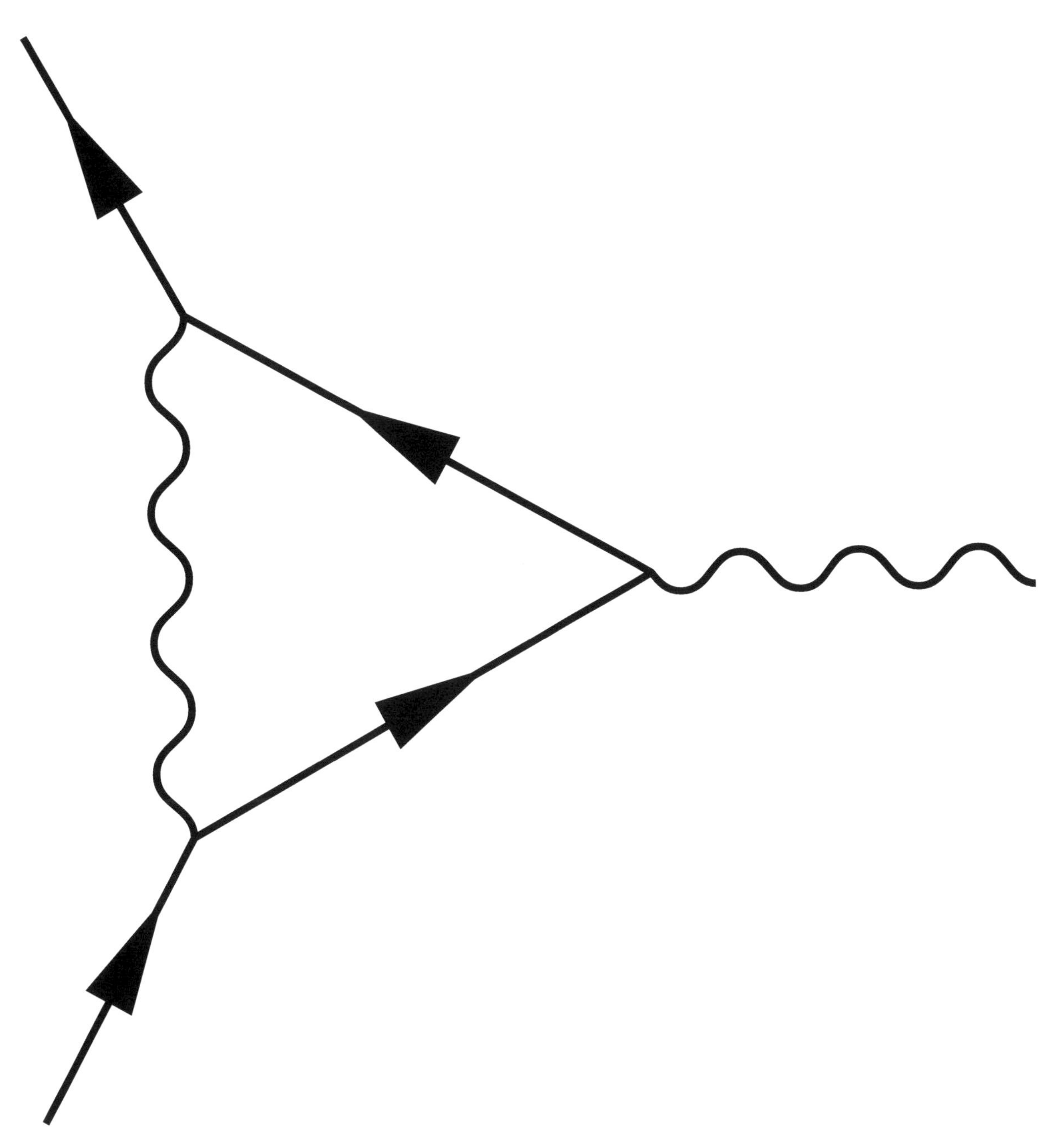

In quantum field theory, the probability of an event is computed by summing the probability amplitude (a complex number) for all possible ways in which the event can occur, as in the Feynman diagram shown here; the probability equals the square of the modulus of the total amplitude.

Chapter 20

Gluon

Gluons /ˈɡluːɒnz/ are elementary particles that act as the exchange particles (or gauge bosons) for the strong force between quarks, analogous to the exchange of photons in the electromagnetic force between two charged particles.[6]

In technical terms, gluons are vector gauge bosons that mediate strong interactions of quarks in quantum chromodynamics (QCD). Gluons themselves carry the color charge of the strong interaction. This is unlike the photon, which mediates the electromagnetic interaction but lacks an electric charge. Gluons therefore participate in the strong interaction in addition to mediating it, making QCD significantly harder to analyze than QED (quantum electrodynamics).

20.1 Properties

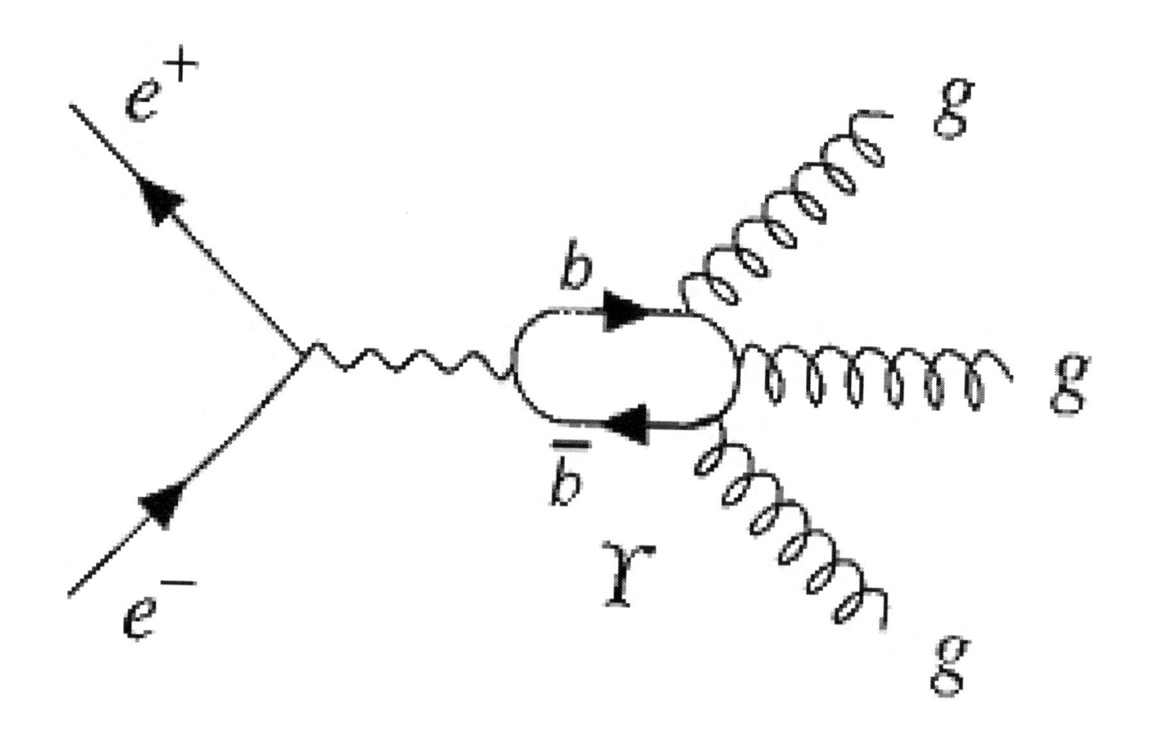

Diagram 2: $e^+e^- \rightarrow \Upsilon(9.46) \rightarrow 3g$

The gluon is a vector boson; like the photon, it has a spin of 1. While massive spin-1 particles have three polarization states, massless gauge bosons like the gluon have only two polarization states because gauge invariance requires the polarization to be transverse. In quantum field theory, unbroken gauge invariance requires that gauge bosons have zero mass (experiment limits the gluon's rest mass to less than a few meV/c^2). The gluon has negative intrinsic parity.

20.2 Numerology of gluons

Unlike the single photon of QED or the three W and Z bosons of the weak interaction, there are eight independent types of gluon in QCD.

This may be difficult to understand intuitively. Quarks carry three types of color charge; antiquarks carry three types of anticolor. Gluons may be thought of as carrying both color and anticolor, but to correctly understand how they are combined, it is necessary to consider the mathematics of color charge in more detail.

20.2.1 Color charge and superposition

In quantum mechanics, the states of particles may be added according to the principle of superposition; that is, they may be in a "combined state" with a *probability*, if some particular quantity is measured, of giving several different outcomes. A relevant illustration in the case at hand would be a gluon with a color state described by:

$(r\bar{b} + b\bar{r})/\sqrt{2}.$

This is read as "red–antiblue plus blue–antired". (The factor of the square root of two is required for normalization, a detail that is not crucial to understand in this discussion.) If one were somehow able to make a direct measurement of the color of a gluon in this state, there would be a 50% chance of it having red-antiblue color charge and a 50% chance of blue-antired color charge.

20.2.2 Color singlet states

It is often said that the stable strongly interacting particles (such as the proton and the neutron, i.e. hadrons) observed in nature are "colorless", but more precisely they are in a "color singlet" state, which is mathematically analogous to a *spin* singlet state.[7] Such states allow interaction with other color singlets, but not with other color states; because long-range gluon interactions do not exist, this illustrates that gluons in the singlet state do not exist either.[7]

The color singlet state is:[7]

$(r\bar{r} + b\bar{b} + g\bar{g})/\sqrt{3}.$

In words, if one could measure the color of the state, there would be equal probabilities of it being red-antired, blue-antiblue, or green-antigreen.

20.2.3 Eight gluon colors

There are eight remaining independent color states, which correspond to the "eight types" or "eight colors" of gluons. Because states can be mixed together as discussed above, there are many ways of presenting these states, which are known as the "color octet". One commonly used list is:[7]

These are equivalent to the Gell-Mann matrices; the translation between the two is that red-antired is the upper-left matrix entry, red-antiblue is the upper middle entry, blue-antigreen is the middle right entry, and so on. The critical feature of these particular eight states is that they are linearly independent, and also independent of the singlet state; there is no way to add any combination of states to produce any other. (It is also impossible to add them to make rr, gg, or bb[8] otherwise the forbidden singlet state could also be made.) There are many other possible choices, but all are mathematically equivalent, at least equally complex, and give the same physical results.

20.2.4 Group theory details

Technically, QCD is a gauge theory with SU(3) gauge symmetry. Quarks are introduced as spinor fields in N_f flavors, each in the fundamental representation (triplet, denoted **3**) of the color gauge group, SU(3). The gluons are vector fields in the adjoint representation (octets, denoted **8**) of color SU(3). For a general gauge group, the number of force-carriers (like photons or gluons) is always equal to the dimension of the adjoint representation. For the simple case of SU(N), the dimension of this representation is $N^2 - 1$.

In terms of group theory, the assertion that there are no color singlet gluons is simply the statement that quantum chromodynamics has an SU(3) rather than a U(3) symmetry. There is no known *a priori* reason for one group to be preferred over the other, but as discussed above, the experimental evidence supports SU(3).[7] The U(1) group for electromagnetic field combines with a slightly more complicated group known as SU(2),S stands for "special", which means the corresponding matrices have derterminant 1.

20.3 Confinement

Main article: Color confinement

Since gluons themselves carry color charge, they participate in strong interactions. These gluon-gluon interactions constrain color fields to string-like objects called "flux tubes", which exert constant force when stretched. Due to this force, quarks are confined within composite particles called hadrons. This effectively limits the range of the strong interaction to 1×10^{-15} meters, roughly the size of an atomic nucleus. Beyond a certain distance, the energy of the flux tube binding two quarks increases linearly. At a large enough distance, it becomes energetically more favorable to pull a quark-antiquark pair out of the vacuum rather than increase the length of the flux tube.

Gluons also share this property of being confined within hadrons. One consequence is that gluons are not directly involved in the nuclear forces between hadrons. The force mediators for these are other hadrons called mesons.

Although in the normal phase of QCD single gluons may not travel freely, it is predicted that there exist hadrons that are formed entirely of gluons — called glueballs. There are also conjectures about other exotic hadrons in which real gluons (as opposed to virtual ones found in ordinary hadrons) would be primary constituents. Beyond the normal phase of QCD (at extreme temperatures and pressures), quark–gluon plasma forms. In such a plasma there are no hadrons; quarks and gluons become free particles.

20.4 Experimental observations

Quarks and gluons (colored) manifest themselves by fragmenting into more quarks and gluons, which in turn hadronize into normal (colorless) particles, correlated in jets. As shown in 1978 summer conferences[2] the PLUTO detector at the electron-positron collider DORIS (DESY) produced the first evidence that the hadronic decays of the very narrow resonance Υ(9.46) could be interpreted as three-jet event topologies produced by three gluons. Later published analyses by the same experiment confirmed this interpretation and also the spin 1 nature of the gluon[9][10] (see also the recollection[2] and PLUTO experiments).

In summer 1979 at higher energies at the electron-positron collider PETRA (DESY) again three-jet topologies were observed, now interpreted as qq gluon bremsstrahlung, now clearly visible, by TASSO,[11] MARK-J[12] and PLUTO experiments[13] (later in 1980 also by JADE[14]). The spin 1 of the gluon was confirmed in 1980 by TASSO[15] and PLUTO experiments[16] (see also the review[3]). In 1991 a subsequent experiment at the LEP storage ring at CERN again confirmed this result.[17]

The gluons play an important role in the elementary strong interactions between quarks and gluons, described by QCD and studied particularly at the electron-proton collider HERA at DESY. The number and momentum distribution of the gluons in the proton (gluon density) have been measured by two experiments, H1 and ZEUS,[18] in the years 1996 till today (2012). The gluon contribution to the proton spin has been studied by the HERMES experiment at HERA.[19] The gluon density in the proton (when behaving hadronically) also has been measured.[20]

Color confinement is verified by the failure of free quark searches (searches of fractional charges). Quarks are normally produced in pairs (quark + antiquark) to compensate the quantum color and flavor numbers; however at Fermilab single production of top quarks has been shown (technically this still involves a pair production, but quark and antiquark are of different flavor).[21] No glueball has been demonstrated.

Deconfinement was claimed in 2000 at CERN SPS[22] in heavy-ion collisions, and it implies a new state of matter: quark–gluon plasma, less interacting than in the nucleus, almost as in a liquid. It was found at the Relativistic Heavy Ion Collider (RHIC) at Brookhaven in the years 2004–2010 by four contemporaneous experiments.[23] A quark–gluon plasma state has been confirmed at the CERN Large Hadron Collider (LHC) by the three experiments ALICE, ATLAS and CMS in 2010.[24]

20.5 See also

- Quark
- Hadron
- Meson
- Gauge boson
- Quark model
- Quantum chromodynamics
- Quark–gluon plasma
- Color confinement
- Glueball
- Gluon field
- Gluon field strength tensor
- Exotic hadrons
- Standard Model
- Three-jet events
- Deep inelastic scattering

20.6 References

[1] M.Gell-Mann(1962). “Symmetries of Baryons and Mesons”.*Physical Review***125**(3): 1067–1084. Bibcode:1962PhRv..1. doi:10.1103/PhysRev.125.1067.

[2] B.R. Stella and H.-J. Meyer (2011). "ϒ(9.46 GeV) and the gluon discovery (a critical recollection of PLUTO results)". *European Physical Journal H* **36** (2): 203–243. arXiv:1008.1869v3. Bibcode:2011EPJH...36..203S. doi:10.1140/epjh/e2011-10029-3.

[3] P. Söding (2010). “On the discovery of the gluon”. *European Physical Journal H* **35** (1): 3–28. Bibcode:2010EPJH...35....3S. doi:10.1140/epjh/e2010-00002-5.

[4] W.-M.Yao et al. (2006). “Review of Particle Physics”(PDF).*Journal of Physics G***33**: 1. arXiv:astro-ph/0601168. Bibc1Y. doi:10.1088/0954-3899/33/1/001.

[5] F. Yndurain (1995). “Limits on the mass of the gluon”. *Physics Letters B* **345** (4): 524. Bibcode:1995PhLB..345..524Y. doi:10.1016/0370-2693(94)01677-5.

[6] C.R. Nave. "The Color Force". *HyperPhysics*. Georgia State University, Department of Physics. Retrieved 2012-04-02.

[7] David Griffiths (1987). *Introduction to Elementary Particles*. John Wiley & Sons. pp. 280–281. ISBN 0-471-60386-4.

[8] J. Baez. "Why are there eight gluons and not nine?". Retrieved 2009-09-13.

[9] Ch. Berger *et al.* (PLUTO Collaboration) (1979). "Jet analysis of the ϒ(9.46) decay into charged hadrons". *Physics Letters B* **82** (3–4): 449. Bibcode:1979PhLB...82..449B. doi:10.1016/0370-2693(79)90265-X.

[10] Ch. Berger*et al.* (PLUTO Collaboration) (1981). "Topology of theϒdecay".*Zeitschrift für Physik C***8**(2): 101. Bibcode:191B. doi:10.1007/BF01547873.

[11] R. Brandelik *et al.* (TASSO collaboration) (1979). "Evidence for Planar Events in e^+e^- Annihilation at High Energies". *Physics Letters B* **86** (2): 243–249. Bibcode:1979PhLB...86..243B. doi:10.1016/0370-2693(79)90830-X.

[12] D.P. Barber *et al.* (MARK-J collaboration) (1979). "Discovery of Three-Jet Events and a Test of Quantum Chromodynamics at PETRA". *Physical Review Letters* **43** (12): 830. Bibcode:1979PhRvL..43..830B. doi:10.1103/PhysRevLett.43.830.

[13] Ch. Berger *et al.* (PLUTO Collaboration) (1979). "Evidence for Gluon Bremsstrahlung in e^+e^- Annihilations at High Energies". *Physics Letters B* **86** (3–4): 418. Bibcode:1979PhLB...86..418B. doi:10.1016/0370-2693(79)90869-4.

[14] W. Bartel *et al.* (JADE Collaboration) (1980). "Observation of planar three-jet events in e^+e^- annihilation and evidence for gluon bremsstrahlung". *Physics Letters B* **91**: 142. Bibcode:1980PhLB...91..142B. doi:10.1016/0370-2693(80)90680-2.

[15] R. Brandelik *et al.* (TASSO Collaboration) (1980). "Evidence for a spin-1 gluon in three-jet events". *Physics Letters B* **97** (3–4): 453. Bibcode:1980PhLB...97..453B. doi:10.1016/0370-2693(80)90639-5.

[16] Ch. Berger *et al.* (PLUTO Collaboration) (1980). "A study of multi-jet events in e^+e^- annihilation". *Physics Letters B* **97** (3–4): 459. Bibcode:1980PhLB...97..459B. doi:10.1016/0370-2693(80)90640-1.

[17] G. Alexander *et al.* (OPAL Collaboration) (1991). "Measurement of Three-Jet Distributions Sensitive to the Gluon Spin in e^+e^- Annihilations at $\sqrt{s} = 91$ GeV". *Zeitschrift für Physik C* **52** (4): 543. Bibcode:1991ZPhyC..52..543A. doi:10.1007/BF01562326.

[18] L. Lindeman (H1 and ZEUS collaborations) (1997). "Proton structure functions and gluon density at HERA". *Nuclear Physics B Proceedings Supplements* **64**: 179–183. Bibcode:1998NuPhS..64..179L. doi:10.1016/S0920-5632(97)01057-8.

[19] http://www-hermes.desy.de

[20] C. Adloff *et al.* (H1 collaboration) (1999). "Charged particle cross sections in the photoproduction and extraction of the gluon density in the photon". *European Physical Journal C* **10**: 363–372. arXiv:hep-ex/9810020. Bibcode:1999EPJC...10..363H. doi:10.1007/s100520050761.

[21] M. Chalmers (6 March 2009). "Top result for Tevatron". *Physics World*. Retrieved 2012-04-02.

[22] M.C. Abreu et al. (2000). "Evidence for deconfinement of quark and antiquark from the J/Ψ suppression pattern measured in Pb-Pb collisions at the CERN SpS". *Physics Letters B* **477**: 28–36. Bibcode:2000PhLB..477...28A. doi:10.1016/S0370-2693(00)00237-9.

[23] D. Overbye (15 February 2010). "In Brookhaven Collider, Scientists Briefly Break a Law of Nature". *New York Times*. Retrieved 2012-04-02.

[24] "LHC experiments bring new insight into primordial universe" (Press release). CERN. 26 November 2010. Retrieved 2012-04-02.

20.7 Further reading

- A. Ali and G. Kramer (2011). "JETS and QCD: A historical review of the discovery of the quark and gluon jets and its impact on QCD". *European Physical Journal H* **36** (2): 245–326. arXiv:1012.2288. Bibcode:2011EPJH...3 .doi:10.1140/epjh/e2011-10047-1.

Chapter 21

W and Z bosons

The **W and Z bosons** (together known as the **weak bosons** or, less specifically, the **intermediate vector bosons**) are the elementary particles that mediate the weak interaction; their symbols are W+, W−, and Z. The W bosons have a positive and negative electric charge of 1 elementary charge respectively and are each other's antiparticles. The Z boson is electrically neutral and is its own antiparticle. The three particles have a spin of 1, and the W bosons have a magnetic moment, while the Z has none. All three of these particles are very short-lived, with a half-life of about 3×10^{-25} s. Their discovery was a major success for what is now called the Standard Model of particle physics.

The W bosons are named after the ***w***eak force. The physicist Steven Weinberg named the additional particle the "Z particle",[3] later giving the explanation that it was the last additional particle needed by the model – the W bosons had already been named – and that it has *z*ero electric charge.[4]

The two **W bosons** are best known as mediators of neutrino absorption and emission, where their charge is associated with electron or positron emission or absorption, always causing nuclear transmutation. The Z boson is not involved in the absorption or emission of electrons and positrons.

The **Z boson** mediates the transfer of momentum, spin, and energy when neutrinos scatter *elastically* from matter, something that must happen without the production or absorption of new, charged particles. Such behaviour (which is almost as common as inelastic neutrino interactions) is seen in bubble chambers irradiated with neutrino beams. Whenever an electron simply "appears" in such a chamber as a new free particle suddenly moving with kinetic energy, and moves in the direction of the neutrinos as the apparent result of a new impulse, and this behavior happens more often when the neutrino beam is present, it is inferred to be a result of a neutrino interacting directly with the electron. Here, the neutrino simply strikes the electron and scatters away from it, transferring some of the neutrino's momentum to the electron. Since (i) neither neutrinos nor electrons are affected by the strong force, (ii) neutrinos are electrically neutral (therefore don't interact electromagnetically), and (iii) the incredibly small masses of these particles make any gravitational force between them negligible, such an interaction can only happen via the weak force. Since such an electron is not created from a nucleon, and is unchanged except for the new force impulse imparted by the neutrino, this weak force interaction between the neutrino and the electron must be mediated by a weak-force boson particle with no charge. Thus, this interaction requires a Z boson.

21.1 Basic properties

These bosons are among the heavyweights of the elementary particles. With masses of 80.4 GeV/c^2 and 91.2 GeV/c^2, respectively, the W and Z bosons are almost 100 times as large as the proton – heavier, even, than entire atoms of iron. The masses of these bosons are significant because they act as the force carriers of a quite short-range fundamental force: their high masses thus limit the range of the weak nuclear force. By way of contrast, the electromagnetic force has an infinite range, because its force carrier, the photon, has zero mass, and the same is supposed of the hypothetical graviton.

All three bosons have particle spin $s = 1$. The emission of a W+ or W− boson either raises or lowers the electric charge of the emitting particle by one unit, and also alters the spin by one unit. At the same time, the emission or absorption of a W boson can change the type of the particle – for example changing a strange quark into an up quark. The neutral Z boson cannot change the electric charge of any particle, nor can it change any other of the so-called "charges" (such as strangeness, baryon number, charm, etc.). The emission or absorption of a Z boson can only change the spin, momentum, and energy of the other particle. (See also *weak neutral current.*)

21.2 Weak nuclear force

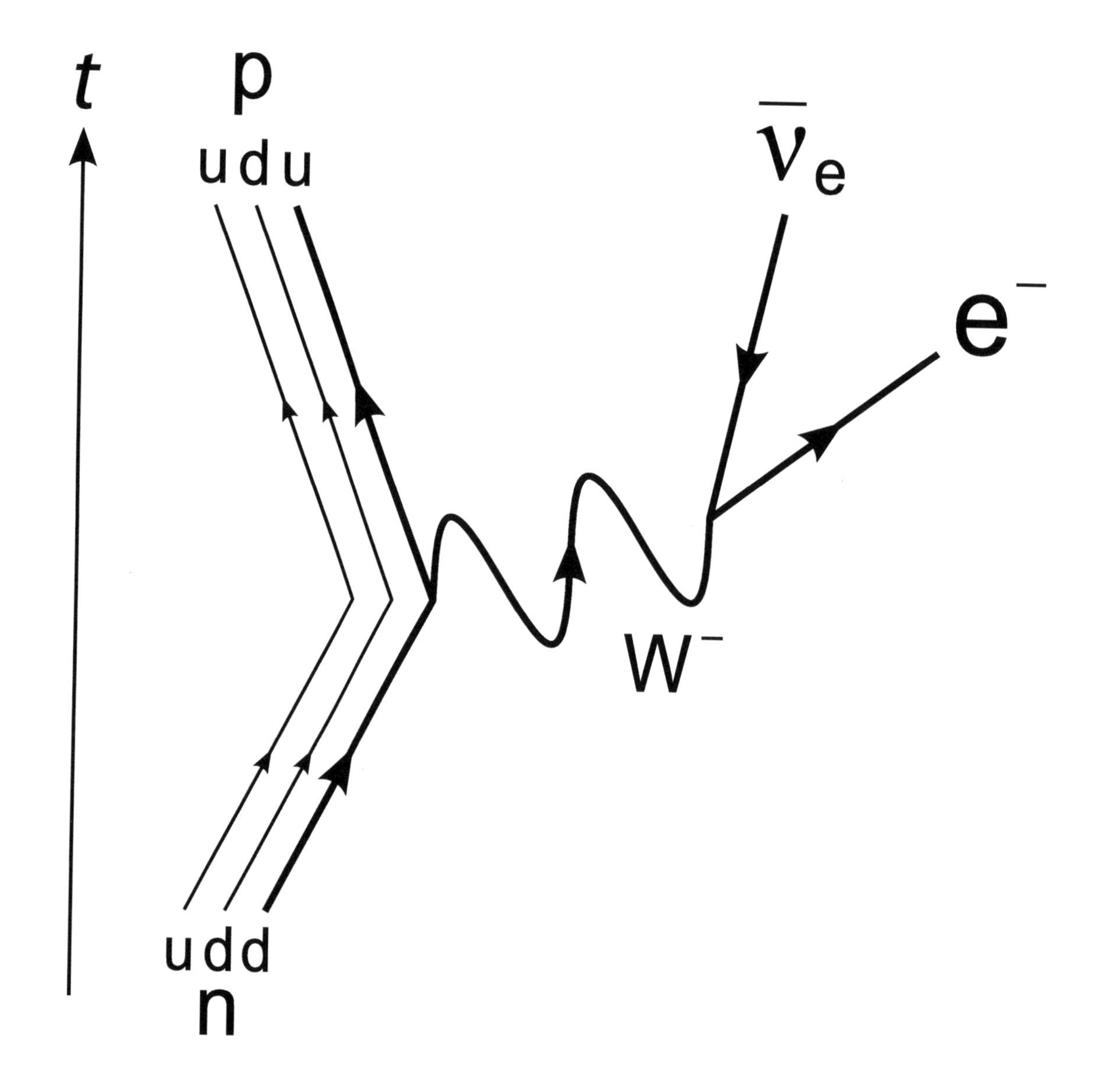

The Feynman diagram for beta decay of a neutron into a proton, electron, and electron antineutrino via an intermediate heavy W boson

The W and Z bosons are carrier particles that mediate the weak nuclear force, much as the photon is the carrier particle for the electromagnetic force.

21.2.1 W bosons

The W bosons are best known for their role in nuclear decay. Consider, for example, the beta decay of cobalt-60.

60
27Co $\rightarrow$ 60
28Ni^{+} + e− + ν
e

This reaction does not involve the whole cobalt-60 nucleus, but affects only one of its 33 neutrons. The neutron is converted into a proton while also emitting an electron (called a beta particle in this context) and an electron antineutrino:

n0 $\rightarrow$ p+ + e− + ν
e

Again, the neutron is not an elementary particle but a composite of an up quark and two down quarks (udd). It is in fact one of the down quarks that interacts in beta decay, turning into an up quark to form a proton (uud). At the most fundamental level, then, the weak force changes the flavour of a single quark:

d $\rightarrow$ u + W−

which is immediately followed by decay of the W− itself:

W− $\rightarrow$ e− + ν
e

21.2.2 Z boson

The Z boson is its own antiparticle. Thus, all of its flavour quantum numbers and charges are zero. The exchange of a Z boson between particles, called a neutral current interaction, therefore leaves the interacting particles unaffected, except for a transfer of momentum. Z boson interactions involving neutrinos have distinctive signatures: They provide the only known mechanism for elastic scattering of neutrinos in matter; neutrinos are almost as likely to scatter elastically (via Z boson exchange) as inelastically (via W boson exchange). The first prediction of Z bosons was made by Brazilian physicist José Leite Lopes in 1958,[5] by devising an equation which showed the analogy of the weak nuclear interactions with electromagnetism. Steve Weinberg, Sheldon Glashow and Abdus Salam used later these results to develop the electroweak unification,[6] in 1973. Weak neutral currents via Z boson exchange were confirmed shortly thereafter in 1974, in a neutrino experiment in the Gargamelle bubble chamber at CERN.

21.3 Predicting the W and Z

Following the spectacular success of quantum electrodynamics in the 1950s, attempts were undertaken to formulate a similar theory of the weak nuclear force. This culminated around 1968 in a unified theory of electromagnetism and weak interactions by Sheldon Glashow, Steven Weinberg, and Abdus Salam, for which they shared the 1979 Nobel Prize in Physics.[7] Their electroweak theory postulated not only the W bosons necessary to explain beta decay, but also a new Z boson that had never been observed.

The fact that the W and Z bosons have mass while photons are massless was a major obstacle in developing electroweak theory. These particles are accurately described by an SU(2) gauge theory, but the bosons in a gauge theory must be massless. As a case in point, the photon is massless because electromagnetism is described by a U(1) gauge theory. Some mechanism is required to break the SU(2) symmetry, giving mass to the W and Z in the process. One explanation, the Higgs mechanism, was forwarded by the 1964 PRL symmetry breaking papers. It predicts the existence of yet another

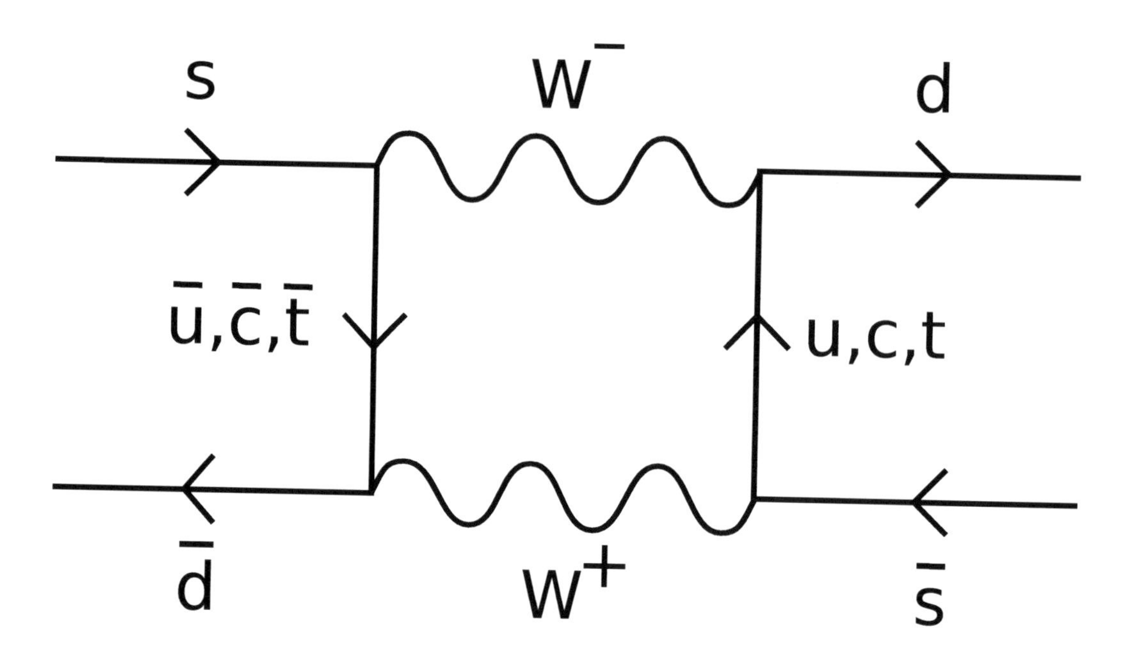

A Feynman diagram showing the exchange of a pair of W bosons. This is one of the leading terms contributing to neutral Kaon oscillation.

new particle; the Higgs boson. Of the four components of a Goldstone boson created by the Higgs field, three are "eaten" by the W^+, Z^0, and W^- bosons to form their longitudinal components and the remainder appears as the spin 0 Higgs boson.

The combination of the SU(2) gauge theory of the weak interaction, the electromagnetic interaction, and the Higgs mechanism is known as the Glashow-Weinberg-Salam model. These days it is widely accepted as one of the pillars of the Standard Model of particle physics. As of 13 December 2011, intensive search for the Higgs boson carried out at CERN has indicated that if the particle is to be found, it seems likely to be found around 125 GeV. On 4 July 2012, the CMS and the ATLAS experimental collaborations at CERN announced the discovery of a new particle with a mass of 125.3 ± 0.6 GeV that appears consistent with a Higgs boson.

21.4 Discovery

Unlike beta decay, the observation of neutral current interactions that involve particles *other than neutrinos* requires huge investments in particle accelerators and detectors, such as are available in only a few high-energy physics laboratories in the world (and then only after 1983). This is because Z-bosons behave in somewhat the same manner as photons, but do not become important until the energy of the interaction is comparable with the relatively huge mass of the Z boson.

The discovery of the W and Z bosons was considered a major success for CERN. First, in 1973, came the observation of neutral current interactions as predicted by electroweak theory. The huge Gargamelle bubble chamber photographed the tracks of a few electrons suddenly starting to move, seemingly of their own accord. This is interpreted as a neutrino interacting with the electron by the exchange of an unseen Z boson. The neutrino is otherwise undetectable, so the only observable effect is the momentum imparted to the electron by the interaction.

The discovery of the W and Z bosons themselves had to wait for the construction of a particle accelerator powerful enough to produce them. The first such machine that became available was the Super Proton Synchrotron, where unambiguous signals of W bosons were seen in January 1983 during a series of experiments made possible by Carlo Rubbia and Simon van der Meer. The actual experiments were called UA1 (led by Rubbia) and UA2 (led by Pierre Darriulat),[8] and were the collaborative effort of many people. Van der Meer was the driving force on the accelerator end (stochastic cooling). UA1 and UA2 found the Z boson a few months later, in May 1983. Rubbia and van der Meer were promptly awarded

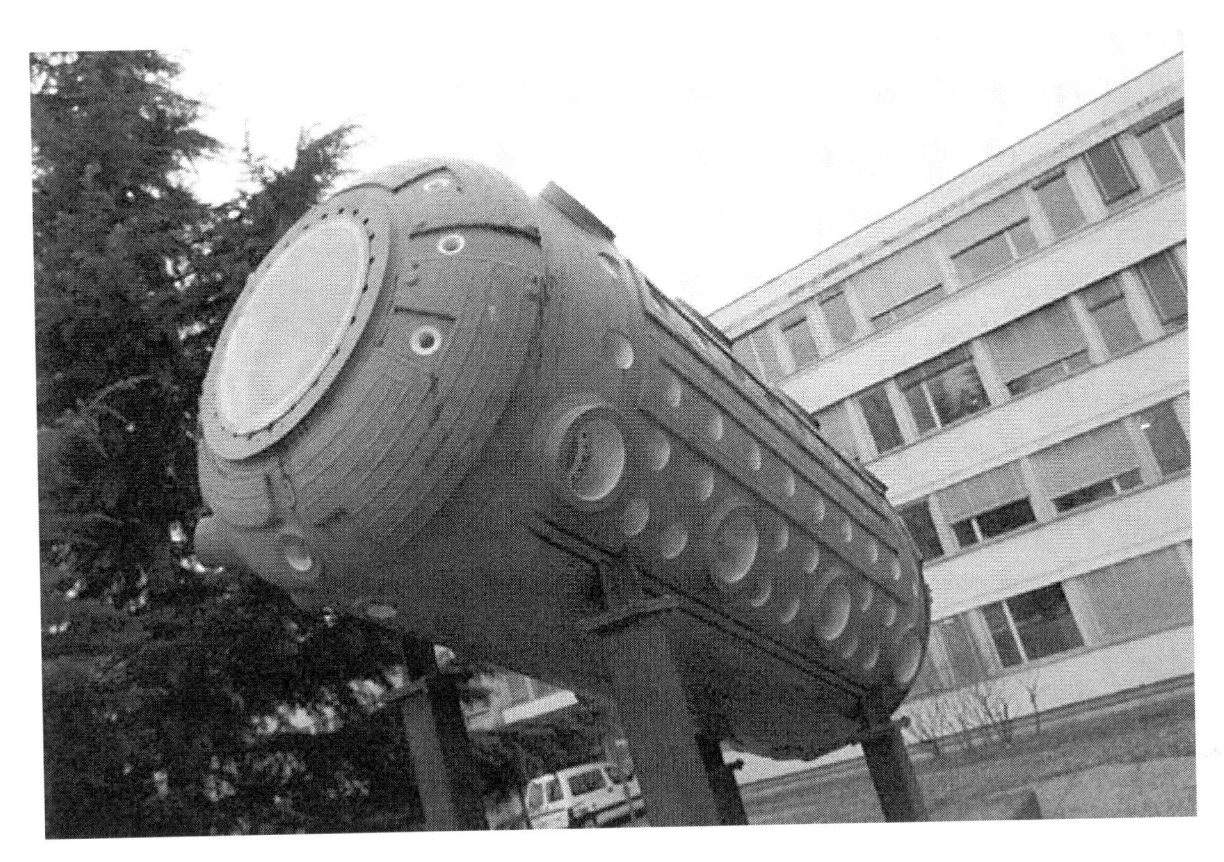

The Gargamelle bubble chamber, now exhibited at CERN

the 1984 Nobel Prize in Physics, a most unusual step for the conservative Nobel Foundation.[9]

The W+, W−, and Z0 bosons, together with the photon (γ), comprise the four gauge bosons of the electroweak interaction.

21.5 Decay

The W and Z bosons decay to fermion–antifermion pairs but neither the W nor the Z bosons can decay into the higher-mass top quark. Neglecting phase space effects and higher order corrections, simple estimates of their branching fractions can be calculated from the coupling constants.

21.5.1 W bosons

W bosons can decay to a lepton and neutrino or to an up-type quark and a down-type quark. The decay width of the W boson to a quark–antiquark pair is proportional to the corresponding squared CKM matrix element and the number of quark colours, *NC* = 3. The decay widths for the W bosons are then proportional to:

Here, e+, μ+, τ+ denote the three flavours of leptons (more exactly, the positive charged antileptons). ν
e, ν
μ, ν
τ denote the three flavours of neutrinos. The other particles, starting with u and d, all denote quarks and antiquarks (factor *NC* is applied). The various *Vij* denote the corresponding CKM matrix coefficients.

Unitarity of the CKM matrix implies that $|V_{ud}|^2 + |V_{us}|^2 + |V_{ub}|^2 = |V_{cd}|^2 + |V_{cs}|^2 + |V_{cb}|^2 = 1$. Therefore the leptonic branching ratios of the W boson are approximately *B*(e+ν
e) = *B*(μ+ν

μ) = B(τ+ν
τ) = $^1/_9$. The hadronic branching ratio is dominated by the CKM-favored ud and cs final states. The sum of the hadronic branching ratios has been measured experimentally to be 67.60±0.27%, with $B(l^+\nu_l)$ = 10.80±0.09%.[1]

21.5.2 Z bosons

Z bosons decay into a fermion and its antiparticle. As the Z-boson is a mixture of the pre-symmetry-breaking W^0 and B^0 bosons (see weak mixing angle), each vertex factor includes a factor $T_3 - Qsin^2\theta W$, where T_3 is the third component of the weak isospin of the fermion, Q is the electric charge of the fermion (in units of the elementary charge), and θW is the weak mixing angle. Because the weak isospin is different for fermions of different chirality, either left-handed or right-handed, the coupling is different as well.

The **relative** strengths of each coupling can be estimated by considering that the decay rates include the square of these factors, and all possible diagrams (e.g. sum over quark families, and left and right contributions). This is just an estimate, as we are considering only tree-level diagrams in the Fermi theory.

Here, L and R denote the left- and right-handed chiralities of the fermions respectively. (The right-handed neutrinos do not exist in the standard model. However, in some extensions beyond the standard model they do.) The notation $x = sin^2\theta W$ is used.

21.6 See also

- Bose–Einstein statistics
- Boson
- List of particles
- Standard Model (mathematical formulation)
- W' and Z' bosons
- X and Y bosons: analogous pair of bosons predicted by the Grand Unified Theory

21.7 References

[1] J. Beringer et al. (2012). "2012 Review of Particle Physics - Gauge and Higgs Bosons" (PDF). *Physical Review D* **86**: 1. Bibcode:2012PhRvD..86a0001B. doi:10.1103/PhysRevD.86.010001.

[2] (PDF) http://pdg.lbl.gov/2013/reviews/rpp2013-rev-w-mass.pdf. Missing or empty |title= (help)

[3] Steven Weinberg, A Model of Leptons, Phys. Rev. Lett. 19, 1264–1266 (1967) – the electroweak unification paper.

[4] Weinberg, Steven (1993). *Dreams of a Final Theory: the search for the fundamental laws of nature*. Vintage Press. p. 94. ISBN 0-09-922391-0.

[5] "Forty years of the first attempt at the electroweak unification and of the prediction of the weak neutral boson".

[6] "The Nobel Prize in Physics 1979". Nobel Foundation. Retrieved 2008-09-10.

[7] Nobel Prize in Physics for 1979 (see also Nobel Prize in Physics on Wikipedia)

[8] The UA2 Collaboration collection

[9] 1984 Nobel Prize in physics

[10] C. Amsler et al. (Particle Data Group), PL B667, 1 (2008) and 2009 partial update for the 2010 edition

21.8 External links

- The Review of Particle Physics, the ultimate source of information on particle properties.
- The W and Z particles: a personal recollection by Pierre Darriulat
- When CERN saw the end of the alphabet by Daniel Denegri
- W and Z particles at Hyperphysics

Chapter 22

Scalar boson

A **scalar boson** is a boson whose spin equals zero. *Boson* means that it has an integer-valued spin; the *scalar* fixes this value to 0.

The name "scalar boson" arises from quantum field theory. It refers to the particular transformation properties under Lorentz transformation.

22.1 Examples

- Various known composite particles are scalar bosons, e.g. the alpha particle and the pi meson. Among the scalar mesons, one distinguishes between the scalar and pseudoscalar mesons, which refers to their transformation property under parity.
- The only fundamental scalar boson in the standard model of elementary particle physics is the Higgs boson, whose existence was confirmed on 14 March 2013 at the Large Hadron Collider. As a result of this confirmation, the 2013 Nobel Prize in physics was awarded to Peter Higgs and François Englert.
- One very popular quantum field theory, which uses scalar bosonic fields and is introduced in many introductory books to quantum field theories[1] for pedagogical reasons, is the so-called φ^4-theory. It usually serves as a toy model to introduce the basic concepts of the field.

22.2 See also

- Scalar meson
- Scalar field theory
- Vector boson
- Higgs Boson

22.3 References

[1] Michael E. Peskin and Daniel V. Schroeder (1995). *An Introduction to Quantum Field Theory*. Westview Press. ISBN 0-201-50397-2.

Chapter 23

Higgs boson

The **Higgs boson** or **Higgs particle** is an elementary particle in the Standard Model of particle physics. It is the quantum excitation of the **Higgs field**[6][7]—a fundamental field of crucial importance to particle physics theory,[7] first suspected to exist in the 1960s, that unlike other known fields such as the electromagnetic field, takes a non-zero constant value almost everywhere. The question of the Higgs field's existence has been the last unverified part of the Standard Model of particle physics and, according to some, "the central problem in particle physics".[8][9] The presence of this field, now believed to be confirmed, explains why some fundamental particles have mass when, based on the symmetries controlling their interactions, they should be massless. The existence of the Higgs field would also resolve several other long-standing puzzles, such as the reason for the weak force's extremely short range.

Although it is hypothesized that the Higgs field permeates the entire Universe, evidence for its existence has been very difficult to obtain. In principle, the Higgs field can be detected through its excitations, manifest as Higgs particles, but these are extremely difficult to produce and detect. The importance of this fundamental question led to a 40 year search, and the construction of one of the world's most expensive and complex experimental facilities to date, CERN's Large Hadron Collider,[10] able to create Higgs bosons and other particles for observation and study. On 4 July 2012, the discovery of a new particle with a mass between 125 and 127 GeV/c^2 was announced; physicists suspected that it was the Higgs boson.[11][12][13] Since then, however, the particle had been shown to behave, interact, and decay in many of the ways predicted by the Standard Model. It was also tentatively confirmed to have even parity and zero spin,[1] two fundamental attributes of a Higgs boson. This appears to be the first elementary scalar particle discovered in nature.[14] More data are needed to verify that the discovered particle has properties matching those predicted for the Higgs boson by the Standard Model, or whether, as predicted by some theories, multiple Higgs bosons exist.[3]

The Higgs boson is named after Peter Higgs, one of six physicists who, in 1964, proposed the mechanism that suggested the existence of such a particle. On December 10, 2013, two of them, Peter Higgs and François Englert, were awarded the Nobel Prize in Physics for their work and prediction (Englert's co-researcher Robert Brout had died in 2011 and the Nobel Prize is not ordinarily given posthumously).[15] Although Higgs's name has come to be associated with this theory, several researchers between about 1960 and 1972 each independently developed different parts of it. In mainstream media the Higgs boson has often been called the "God particle", from a 1993 book on the topic; the nickname is strongly disliked by many physicists, including Higgs, who regard it as sensationalistic.[16][17][18]

In the Standard Model, the Higgs particle is a boson with no spin, electric charge, or colour charge. It is also very unstable, decaying into other particles almost immediately. It is a quantum excitation of one of the four components of the Higgs field. The latter constitutes a scalar field, with two neutral and two electrically charged components that form a complex doublet of the weak isospin SU(2) symmetry. The Higgs field is tachyonic (this does not refer to faster-than-light speeds, it means that symmetry-breaking through condensation of a particle must occur under certain conditions), and has a "Mexican hat" shaped potential with nonzero strength everywhere (including otherwise empty space), which in its vacuum state breaks the weak isospin symmetry of the electroweak interaction. When this happens, three components of the Higgs field are "absorbed" by the SU(2) and U(1) gauge bosons (the "Higgs mechanism") to become the longitudinal components of the now-massive W and Z bosons of the weak force. The remaining electrically neutral component separately couples to other particles known as fermions (via Yukawa couplings), causing these to acquire mass as well. Some versions of the theory predict more than one kind of Higgs fields and bosons. Alternative "Higgsless" models would have been considered if the Higgs boson was not discovered.

23.1 A non-technical summary

23.1.1 "Higgs" terminology

23.1.2 Overview

Physicists explain the properties and forces between elementary particles in terms of the Standard Model—a widely accepted and "remarkably" accurate[21] framework based on gauge invariance and symmetries, believed to explain almost everything in the known universe, other than gravity.[22] But by around 1960 all attempts to create a gauge invariant theory for two of the four fundamental forces had consistently failed at one crucial point: although gauge invariance seemed extremely important, it seemed to make any theory of electromagnetism and the weak force go haywire, by demanding that either many particles with mass were massless or that non-existent forces and massless particles had to exist. Scientists had no idea how to get past this point.

In 1962 physicist Philip Anderson wrote a paper that built upon work by Yoichiro Nambu concerning "broken symmetries" in superconductivity and particle physics. He suggested that "broken symmetries" might also be the missing piece needed to solve the problems of gauge invariance. In 1964 a theory was created almost simultaneously by 3 different groups of researchers, that showed Anderson's suggestion was possible - the gauge theory and "mass problems" could indeed be resolved if an unusual kind of field, now generally called the "Higgs field", existed throughout the universe; if the Higgs field did exist, it would apparently cause existing particles to acquire mass instead of new massless particles being formed. Although these ideas did not gain much initial support or attention, by 1972 they had been developed into a comprehensive theory and proved capable of giving "sensible" results that accurately described particles known at the time, and which accurately predicted of several other particles discovered during the following years.[Note 7] During the 1970s these theories rapidly became the "standard model". There was not yet any direct evidence that the Higgs field actually existed, but even without proof of the field, the accuracy of its predictions led scientists to believe the theory might be true. By the 1980s the question whether or not the Higgs field existed had come to be regarded as one of the most important unanswered questions in particle physics.

If Higgs field could be shown to exist, it would be a monumental discovery for science and human knowledge, and would open doorways to new knowledge in many disciplines. If not, then other more complicated theories would need to be considered. The simplest means to test the existence of the Higgs field would be a search for a new elementary particle that the field would have to give off, a particle known as "Higgs bosons" or the "Higgs particle". This particle would be extremely difficult to find. After significant technological advancements, by the 1990s two large experimental installations were being designed and constructed that allowed to search for the Higgs boson.

While several symmetries in nature are spontaneously broken through a form of the Higgs mechanism, in the context of the Standard Model the term "Higgs mechanism" almost always means symmetry breaking of the electroweak field. It is considered confirmed, but revealing the exact cause has been difficult. Various analogies have also been invented to describe the Higgs field and boson, including analogies with well-known symmetry breaking effects such as the rainbow and prism, electric fields, ripples, and resistance of macro objects moving through media, like people moving through crowds or some objects moving through syrup or molasses. However, analogies based on simple resistance to motion are inaccurate as the Higgs field does not work by resisting motion.

23.2 Significance

23.2.1 Scientific impact

Evidence of the Higgs field and its properties has been extremely significant scientifically, for many reasons. The Higgs boson's importance is largely that it is able to be examined using existing knowledge and experimental technology, as a way to confirm and study the entire Higgs field theory.[6][7] Conversely, proof that the Higgs field and boson do not exist would also have been significant. In discussion form, the relevance includes:

23.2.2 Practical and technological impact of discovery

As yet, there are no known immediate technological benefits of finding the Higgs particle. However, a common pattern for fundamental discoveries is for practical applications to follow later, once the discovery has been explored further, at which point they become the basis for new technologies of importance to society.[44][45][46]

The challenges in particle physics have furthered major technological of widespread importance. For example, the World Wide Web began as a project to improve CERN's communication system. CERN's requirement to process massive amounts of data produced by the Large Hadron Collider also led to contributions to the fields of distributed and cloud computing.

23.3 History

See also: 1964 PRL symmetry breaking papers, Higgs mechanism and History of quantum field theory
Particle physicists study matter made from fundamental particles whose interactions are mediated by exchange particles - gauge bosons - acting as force carriers. At the beginning of the 1960s a number of these particles had been discovered or proposed, along with theories suggesting how they relate to each other, some of which had already been reformulated as field theories in which the objects of study are not particles and forces, but quantum fields and their symmetries.[47]:150 However, attempts to unify known fundamental forces such as the electromagnetic force and the weak nuclear force were known to be incomplete. One known omission was that gauge invariant approaches, including non-abelian models such as Yang–Mills theory (1954), which held great promise for unified theories, also seemed to predict known massive particles as massless.[48] Goldstone's theorem, relating to continuous symmetries within some theories, also appeared to rule out many obvious solutions,[49] since it appeared to show that zero-mass particles would have to also exist that were "simply not seen".[50] According to Guralnik, physicists had "no understanding" how these problems could be overcome.[50]

Particle physicist and mathematician Peter Woit summarised the state of research at the time:

> "Yang and Mills work on non-abelian gauge theory had one huge problem: in perturbation theory it has massless particles which don't correspond to anything we see. One way of getting rid of this problem is now fairly well-understood, the phenomenon of confinement realized in QCD, where the strong interactions get rid of the massless "gluon" states at long distances. By the very early sixties, people had begun to understand another source of massless particles: spontaneous symmetry breaking of a continuous symmetry. What Philip Anderson realized and worked out in the summer of 1962 was that, when you have *both* gauge symmetry *and* spontaneous symmetry breaking, the Nambu–Goldstone massless mode can combine with the massless gauge field modes to produce a physical massive vector field. This is what happens in superconductivity, a subject about which Anderson was (and is) one of the leading experts." *[text condensed]* [48]

The Higgs mechanism is a process by which vector bosons can get rest mass *without* explicitly breaking gauge invariance, as a byproduct of spontaneous symmetry breaking.[51][52] The mathematical theory behind spontaneous symmetry breaking was initially conceived and published within particle physics by Yoichiro Nambu in 1960,[53] the concept that such a mechanism could offer a possible solution for the "mass problem" was originally suggested in 1962 by Philip Anderson (who had previously written papers on broken symmetry and its outcomes in superconductivity[54] and concluded in his 1963 paper on Yang-Mills theory that *"considering the superconducting analog... [t]hese two types of bosons seem capable of canceling each other out... leaving finite mass bosons"*),[55]:4–5[56] and Abraham Klein and Benjamin Lee showed in March 1964 that Goldstone's theorem could be avoided this way in at least some non-relativistic cases and speculated it might be possible in truly relativistic cases.[57]

Nobel Prize Laureate Peter Higgs in Stockholm, December 2013

These approaches were quickly developed into a full relativistic model, independently and almost simultaneously, by three groups of physicists: by François Englert and Robert Brout in August 1964;[58] by Peter Higgs in October 1964;[59] and by Gerald Guralnik, Carl Hagen, and Tom Kibble (GHK) in November 1964.[60] Higgs also wrote a short but important[51] response published in September 1964 to an objection by Gilbert,[61] which showed that if calculating within the radiation gauge, Goldstone's theorem and Gilbert's objection would become inapplicable.[Note 11] (Higgs later described Gilbert's objection as prompting his own paper.[62]) Properties of the model were further considered by Guralnik in 1965,[63] by Higgs in 1966,[64] by Kibble in 1967,[65] and further by GHK in 1967.[66] The original three 1964 papers showed that when a gauge theory is combined with an additional field that spontaneously breaks the symmetry, the gauge bosons can consistently acquire a finite mass.[51][52][67] In 1967, Steven Weinberg[68] and Abdus Salam[69] independently showed how a Higgs mechanism could be used to break the electroweak symmetry of Sheldon Glashow's unified model for the weak and electromagnetic interactions[70] (itself an extension of work by Schwinger), forming what became the Standard Model of particle physics. Weinberg was the first to observe that this would also provide mass terms for the fermions.[71] [Note 12]

However, the seminal papers on spontaneous breaking of gauge symmetries were at first largely ignored, because it was widely believed that the (non-Abelian gauge) theories in question were a dead-end, and in particular that they could not be renormalised. In 1971–72, Martinus Veltman and Gerard 't Hooft proved renormalisation of Yang–Mills was possible in two papers covering massless, and then massive, fields.[71] Their contribution, and others' work on the renormalization group - including "substantial" theoretical work by Russian physicists Ludvig Faddeev, Andrei Slavnov, Efim Fradkin and Igor Tyutin[72] - was eventually "enormously profound and influential",[73] but even with all key elements of the eventual theory published there was still almost no wider interest. For example, Coleman found in a study that "essentially no-one paid any attention" to Weinberg's paper prior to 1971[74] and discussed by David Politzer in his 2004 Nobel speech.[73] – now the most cited in particle physics[75] – and even in 1970 according to Politzer, Glashow's teaching of the weak interaction contained no mention of Weinberg's, Salam's, or Glashow's own work.[73] In practice, Politzer states, almost everyone learned of the theory due to physicist Benjamin Lee, who combined the work of Veltman and 't Hooft with insights by others, and popularised the completed theory.[73] In this way, from 1971, interest and acceptance "exploded" [73] and the ideas were quickly absorbed in the mainstream.[71][73]

The resulting electroweak theory and Standard Model have accurately predicted (among other things) weak neutral currents, three bosons, the top and charm quarks, and with great precision, the mass and other properties of some of these.[Note 7] Many of those involved eventually won Nobel Prizes or other renowned awards. A 1974 paper and comprehensive review in *Reviews of Modern Physics* commented that "while no one doubted the [mathematical] correctness of these arguments, no one quite believed that nature was diabolically clever enough to take advantage of them",[76]:9 adding that the theory had so far produced accurate answers that accorded with experiment, but it was unknown whether the theory was fundamentally correct.[76]:9,36(footnote),43–44,47 By 1986 and again in the 1990s it became possible to write that understanding and proving the Higgs sector of the Standard Model was "the central problem today in particle physics".[8][9]

23.3.1 Summary and impact of the PRL papers

The three papers written in 1964 were each recognised as milestone papers during *Physical Review Letters* 's 50th anniversary celebration.[67] Their six authors were also awarded the 2010 J. J. Sakurai Prize for Theoretical Particle Physics for this work.[77] (A controversy also arose the same year, because in the event of a Nobel Prize only up to three scientists could be recognised, with six being credited for the papers.[78]) Two of the three PRL papers (by Higgs and by GHK) contained equations for the hypothetical field that eventually would become known as the Higgs field and its hypothetical quantum, the Higgs boson.[59][60] Higgs' subsequent 1966 paper showed the decay mechanism of the boson; only a massive boson can decay and the decays can prove the mechanism.

In the paper by Higgs the boson is massive, and in a closing sentence Higgs writes that "an essential feature" of the theory "is the prediction of incomplete multiplets of scalar and vector bosons".[59] (Frank Close comments that 1960s gauge theorists were focused on the problem of massless *vector* bosons, and the implied existence of a massive *scalar* boson was not seen as important; only Higgs directly addressed it.[79]:154, 166, 175) In the paper by GHK the boson is massless and decoupled from the massive states.[60] In reviews dated 2009 and 2011, Guralnik states that in the GHK model the boson is massless only in a lowest-order approximation, but it is not subject to any constraint and acquires mass at higher orders, and adds that the GHK paper was the only one to show that there are no massless Goldstone bosons in the model and to give a complete analysis of the general Higgs mechanism.[50][80] All three reached similar conclusions, despite their very different approaches: Higgs' paper essentially used classical techniques, Englert and Brout's involved calculating vacuum polarization in perturbation theory around an assumed symmetry-breaking vacuum state, and GHK used operator formalism and conservation laws to explore in depth the ways in which Goldstone's theorem may be worked around.[51]

23.4 Theoretical properties

Main article: Higgs mechanism

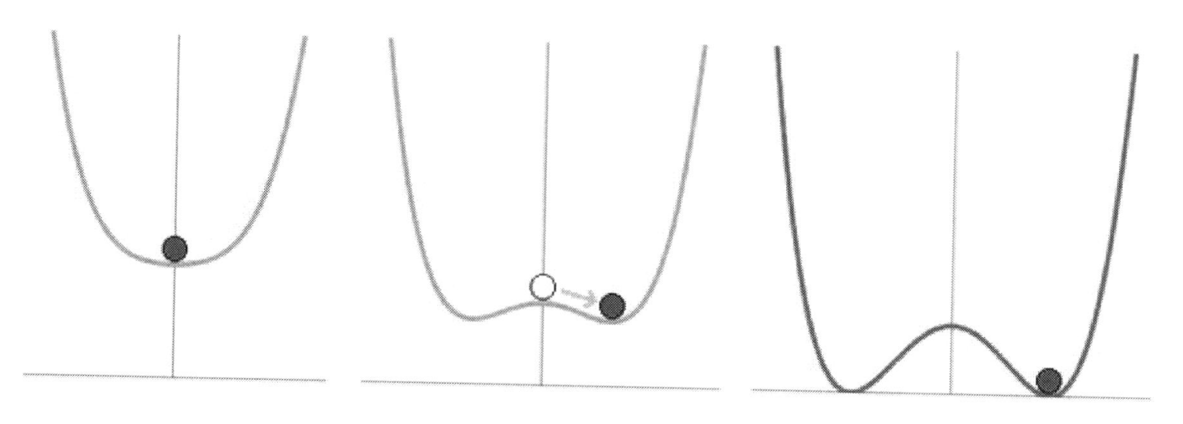

"Symmetry breaking illustrated": – At high energy levels (left) *the ball settles in the center, and the result is symmetrical. At lower energy levels* (right), *the overall "rules" remain symmetrical, but the "Mexican hat" potential comes into effect: "local" symmetry inevitably becomes broken since eventually the ball must at random roll one way or another.*

23.4.1 Theoretical need for the Higgs

Gauge invariance is an important property of modern particle theories such as the Standard Model, partly due to its success in other areas of fundamental physics such as electromagnetism and the strong interaction (quantum chromodynamics). However, there were great difficulties in developing gauge theories for the weak nuclear force or a possible unified electroweak interaction. Fermions with a mass term would violate gauge symmetry and therefore cannot be gauge invariant. (This can be seen by examining the Dirac Lagrangian for a fermion in terms of left and right handed components; we find none of the spin-half particles could ever flip helicity as required for mass, so they must be massless.[Note 13]) W and Z bosons are observed to have mass, but a boson mass term contains terms, which clearly depend on the choice of gauge and therefore these masses too cannot be gauge invariant. Therefore, it seems that *none* of the standard model fermions *or* bosons could "begin" with mass as an inbuilt property except by abandoning gauge invariance. If gauge invariance were to be retained, then these particles had to be acquiring their mass by some other mechanism or interaction. Additionally, whatever was giving these particles their mass, had to not "break" gauge invariance as the basis for other parts of the theories where it worked well, *and* had to not require or predict unexpected massless particles and long-range forces (seemingly an inevitable consequence of Goldstone's theorem) which did not actually seem to exist in nature.

A solution to all of these overlapping problems came from the discovery of a previously unnoticed borderline case hidden in the mathematics of Goldstone's theorem,[Note 11] that under certain conditions it *might* theoretically be possible for a symmetry to be broken *without* disrupting gauge invariance and *without* any new massless particles or forces, and having "sensible" (renormalisable) results mathematically: this became known as the Higgs mechanism.

The Standard Model hypothesizes a field which is responsible for this effect, called the Higgs field (symbol: ϕ), which has the unusual property of a non-zero amplitude in its ground state; i.e., a non-zero vacuum expectation value. It can have this effect because of its unusual "Mexican hat" shaped potential whose lowest "point" is not at its "centre". Below a certain extremely high energy level the existence of this non-zero vacuum expectation spontaneously breaks electroweak gauge symmetry which in turn gives rise to the Higgs mechanism and triggers the acquisition of mass by those particles interacting with the field. This effect occurs because scalar field components of the Higgs field are "absorbed" by the massive bosons as degrees of freedom, and couple to the fermions via Yukawa coupling, thereby producing the expected mass terms. In effect when symmetry breaks under these conditions, the Goldstone bosons that arise *interact* with the Higgs field (and with other particles capable of interacting with the Higgs field) instead of becoming new massless particles, the intractable problems of both underlying theories "neutralise" each other, and the residual outcome is that elementary particles acquire a consistent mass based on how strongly they interact with the Higgs field. It is the simplest known process capable of giving mass to the gauge bosons while remaining compatible with gauge theories.[81] Its quantum would be a scalar boson, known as the Higgs boson.[82]

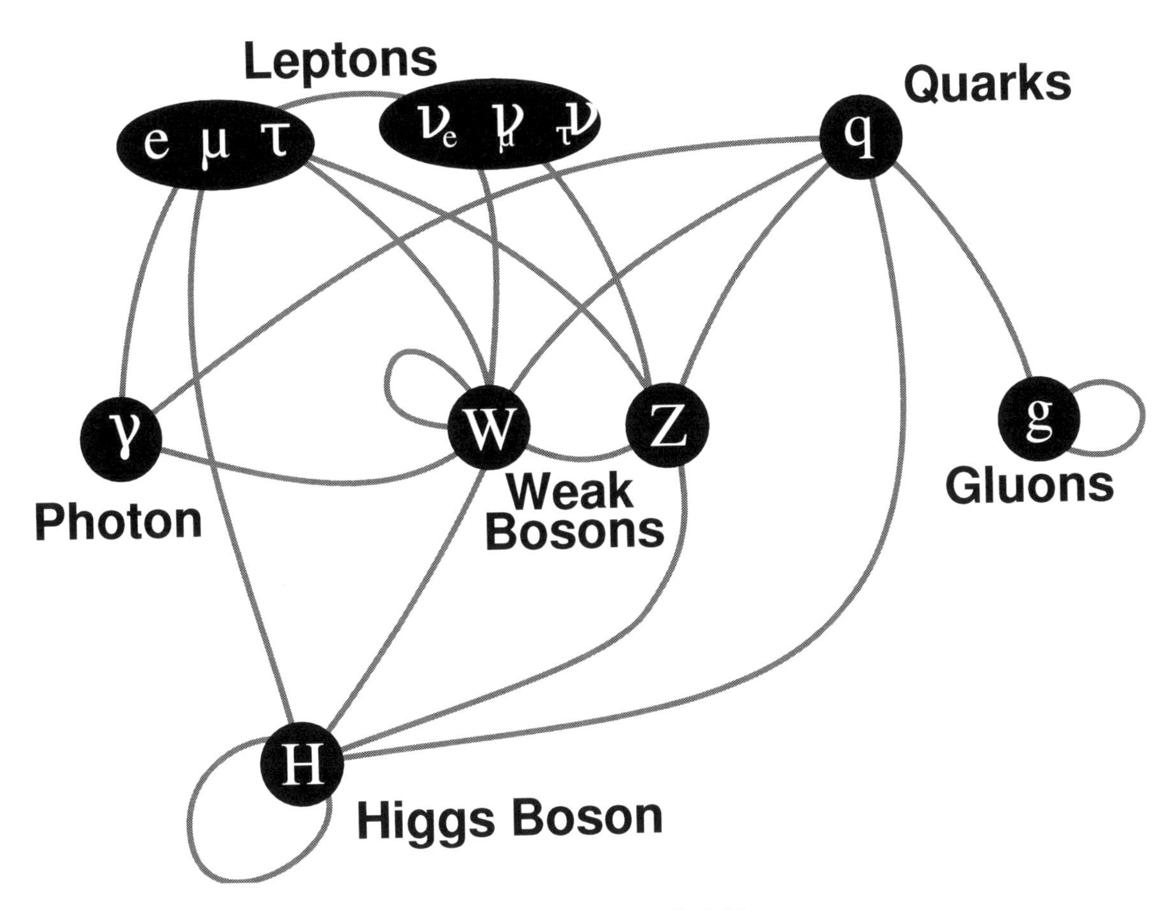

Summary of interactions between certain particles described by the Standard Model.

23.4.2 Properties of the Higgs field

In the Standard Model, the Higgs field is a scalar tachyonic field – 'scalar' meaning it does not transform under Lorentz transformations, and 'tachyonic' meaning the field (but not the particle) has imaginary mass and in certain configurations must undergo symmetry breaking. It consists of four components, two neutral ones and two charged component fields. Both of the charged components and one of the neutral fields are Goldstone bosons, which act as the longitudinal third-polarization components of the massive W^+, W^-, and Z bosons. The quantum of the remaining neutral component corresponds to (and is theoretically realised as) the massive Higgs boson,[83] this component can interact with fermions via Yukawa coupling to give them mass, as well.

Mathematically, the Higgs field has imaginary mass and is therefore a tachyonic field.[84] While tachyons (particles that move faster than light) are a purely hypothetical concept, fields with imaginary mass have come to play an important role in modern physics.[85][86] Under no circumstances do any excitations ever propagate faster than light in such theories — the presence or absence of a tachyonic mass has no effect whatsoever on the maximum velocity of signals (there is no violation of causality).[87] Instead of faster-than-light particles, the imaginary mass creates an instability:- any configuration in which one or more field excitations are tachyonic must spontaneously decay, and the resulting configuration contains no physical tachyons. This process is known as tachyon condensation, and is now believed to be the explanation for how the Higgs mechanism itself arises in nature, and therefore the reason behind electroweak symmetry breaking.

Although the notion of imaginary mass might seem troubling, it is only the field, and not the mass itself, that is quantized. Therefore, the field operators at spacelike separated points still commute (or anticommute), and information and particles still do not propagate faster than light.[88] Tachyon condensation drives a physical system that has reached a local limit and might naively be expected to produce physical tachyons, to an alternate stable state where no physical tachyons exist. Once a tachyonic field such as the Higgs field reaches the minimum of the potential, its quanta are not tachyons any more but rather are ordinary particles such as the Higgs boson.[89]

23.4.3 Properties of the Higgs boson

Since the Higgs field is scalar, the Higgs boson has no spin. The Higgs boson is also its own antiparticle and is CP-even, and has zero electric and colour charge.[90]

The Minimal Standard Model does not predict the mass of the Higgs boson.[91] If that mass is between 115 and 180 GeV/c^2, then the Standard Model can be valid at energy scales all the way up to the Planck scale (10^{19} GeV).[92] Many theorists expect new physics beyond the Standard Model to emerge at the TeV-scale, based on unsatisfactory properties of the Standard Model.[93] The highest possible mass scale allowed for the Higgs boson (or some other electroweak symmetry breaking mechanism) is 1.4 TeV; beyond this point, the Standard Model becomes inconsistent without such a mechanism, because unitarity is violated in certain scattering processes.[94]

It is also possible, although experimentally difficult, to estimate the mass of the Higgs boson indirectly. In the Standard Model, the Higgs boson has a number of indirect effects; most notably, Higgs loops result in tiny corrections to masses of W and Z bosons. Precision measurements of electroweak parameters, such as the Fermi constant and masses of W/Z bosons, can be used to calculate constraints on the mass of the Higgs. As of July 2011, the precision electroweak measurements tell us that the mass of the Higgs boson is likely to be less than about 161 GeV/c^2 at 95% confidence level (this upper limit would increase to 185 GeV/c^2 if the lower bound of 114.4 GeV/c^2 from the LEP-2 direct search is allowed for[95]). These indirect constraints rely on the assumption that the Standard Model is correct. It may still be possible to discover a Higgs boson above these masses if it is accompanied by other particles beyond those predicted by the Standard Model.[96]

23.4.4 Production

If Higgs particle theories are valid, then a Higgs particle can be produced much like other particles that are studied, in a particle collider. This involves accelerating a large number of particles to extremely high energies and extremely close to the speed of light, then allowing them to smash together. Protons and lead ions (the bare nuclei of lead atoms) are used at the LHC. In the extreme energies of these collisions, the desired esoteric particles will occasionally be produced and this can be detected and studied; any absence or difference from theoretical expectations can also be used to improve the theory. The relevant particle theory (in this case the Standard Model) will determine the necessary kinds of collisions and detectors. The Standard Model predicts that Higgs bosons could be formed in a number of ways,[97][98][99] although the probability of producing a Higgs boson in any collision is always expected to be very small—for example, only 1 Higgs boson per 10 billion collisions in the Large Hadron Collider.[Note 14] The most common expected processes for Higgs boson production are:

- *Gluon fusion.* If the collided particles are hadrons such as the proton or antiproton—as is the case in the LHC and Tevatron—then it is most likely that two of the gluons binding the hadron together collide. The easiest way to produce a Higgs particle is if the two gluons combine to form a loop of virtual quarks. Since the coupling of particles to the Higgs boson is proportional to their mass, this process is more likely for heavy particles. In practice it is enough to consider the contributions of virtual top and bottom quarks (the heaviest quarks). This process is the dominant contribution at the LHC and Tevatron being about ten times more likely than any of the other processes.[97][98]

- *Higgs Strahlung.* If an elementary fermion collides with an anti-fermion—e.g., a quark with an anti-quark or an electron with a positron—the two can merge to form a virtual W or Z boson which, if it carries sufficient energy, can then emit a Higgs boson. This process was the dominant production mode at the LEP, where an electron and a positron collided to form a virtual Z boson, and it was the second largest contribution for Higgs production at the Tevatron. At the LHC this process is only the third largest, because the LHC collides protons with protons, making a quark-antiquark collision less likely than at the Tevatron. Higgs Strahlung is also known as *associated production*.[97][98][99]

- *Weak boson fusion.* Another possibility when two (anti-)fermions collide is that the two exchange a virtual W or Z boson, which emits a Higgs boson. The colliding fermions do not need to be the same type. So, for example, an up quark may exchange a Z boson with an anti-down quark. This process is the second most important for the production of Higgs particle at the LHC and LEP.[97][99]

- *Top fusion*. The final process that is commonly considered is by far the least likely (by two orders of magnitude). This process involves two colliding gluons, which each decay into a heavy quark–antiquark pair. A quark and antiquark from each pair can then combine to form a Higgs particle.[97][98]

23.4.5 Decay

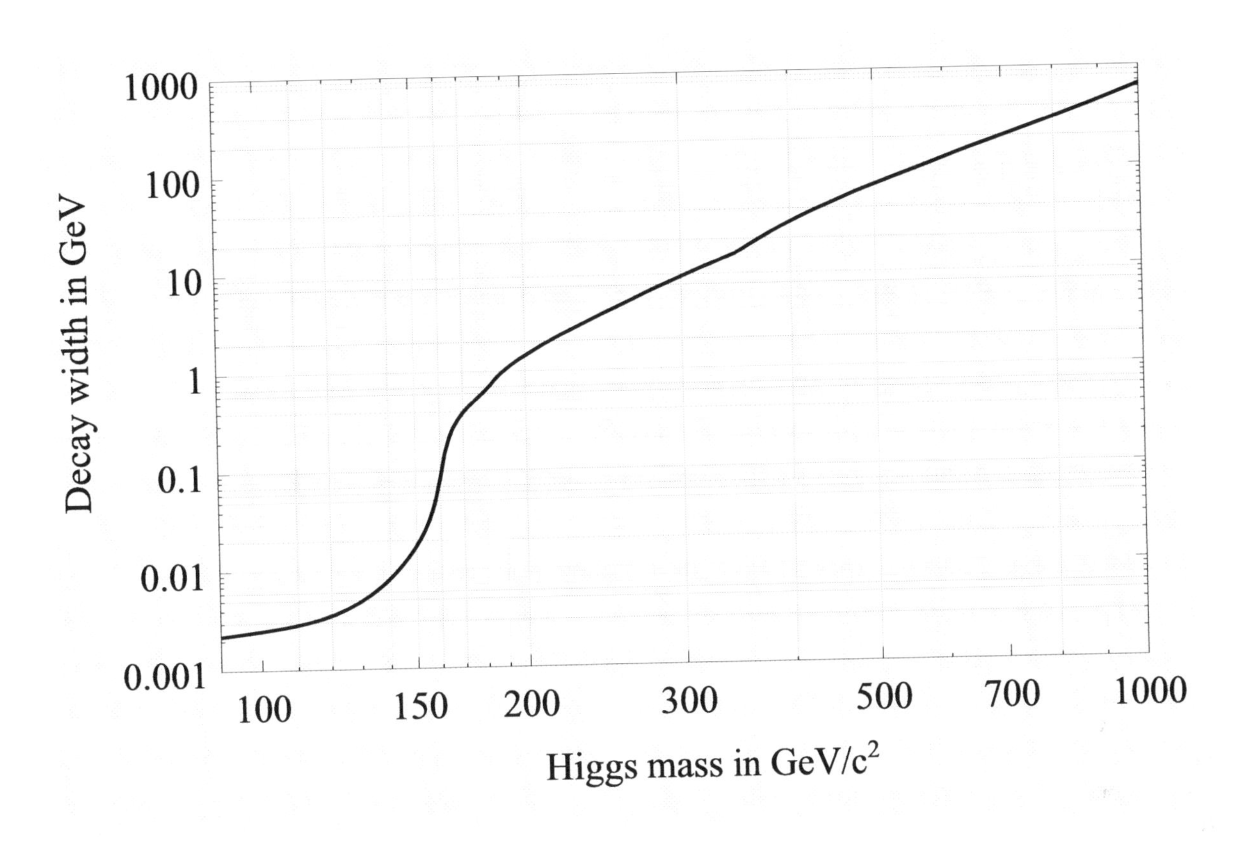

The Standard Model prediction for the decay width of the Higgs particle depends on the value of its mass.

Quantum mechanics predicts that if it is possible for a particle to decay into a set of lighter particles, then it will eventually do so.[101] This is also true for the Higgs boson. The likelihood with which this happens depends on a variety of factors including: the difference in mass, the strength of the interactions, etc. Most of these factors are fixed by the Standard Model, except for the mass of the Higgs boson itself. For a Higgs boson with a mass of 126 GeV/c^2 the SM predicts a mean life time of about 1.6×10^{-22} s.[Note 2]

Since it interacts with all the massive elementary particles of the SM, the Higgs boson has many different processes through which it can decay. Each of these possible processes has its own probability, expressed as the *branching ratio*; the fraction of the total number decays that follows that process. The SM predicts these branching ratios as a function of the Higgs mass (see plot).

One way that the Higgs can decay is by splitting into a fermion–antifermion pair. As general rule, the Higgs is more likely to decay into heavy fermions than light fermions, because the mass of a fermion is proportional to the strength of its interaction with the Higgs.[102] By this logic the most common decay should be into a top–antitop quark pair. However, such a decay is only possible if the Higgs is heavier than ~346 GeV/c^2, twice the mass of the top quark. For a Higgs mass of 126 GeV/c^2 the SM predicts that the most common decay is into a bottom–antibottom quark pair, which happens 56.1% of the time.[5] The second most common fermion decay at that mass is a tau–antitau pair, which happens only about 6% of the time.[5]

Another possibility is for the Higgs to split into a pair of massive gauge bosons. The most likely possibility is for the Higgs to decay into a pair of W bosons (the light blue line in the plot), which happens about 23.1% of the time for a

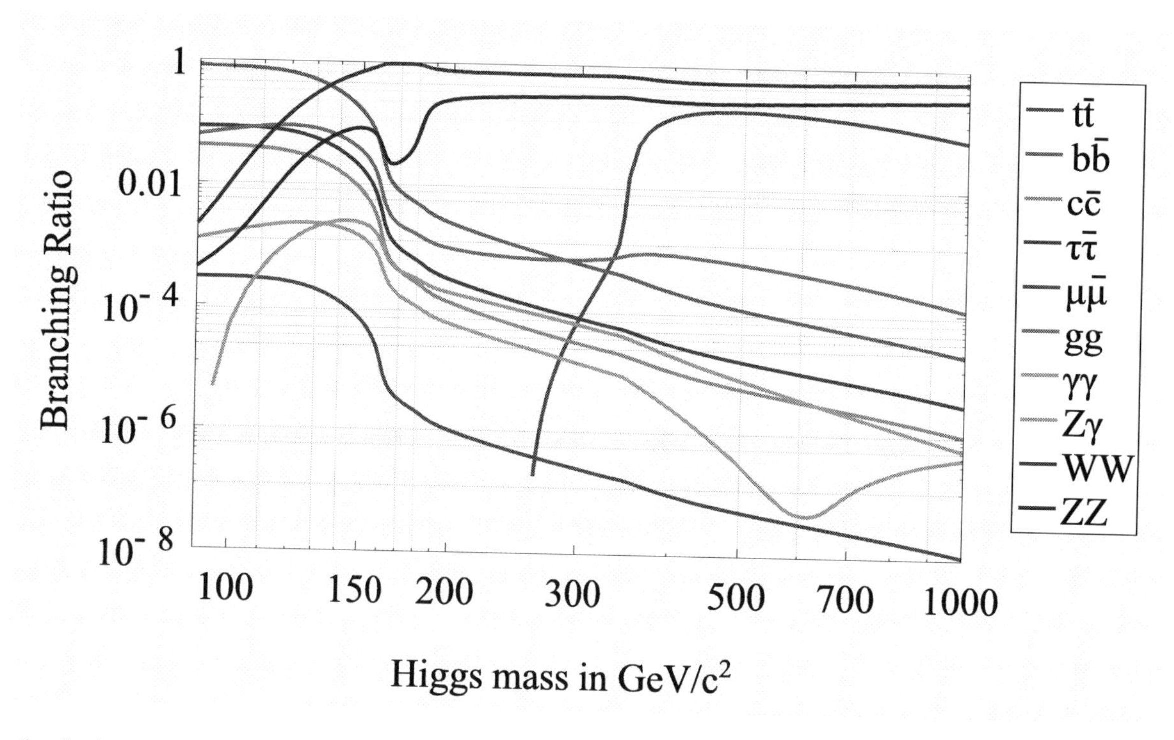

The Standard Model prediction for the branching ratios of the different decay modes of the Higgs particle depends on the value of its mass.

Higgs boson with a mass of 126 GeV/c^2.[5] The W bosons can subsequently decay either into a quark and an antiquark or into a charged lepton and a neutrino. However, the decays of W bosons into quarks are difficult to distinguish from the background, and the decays into leptons cannot be fully reconstructed (because neutrinos are impossible to detect in particle collision experiments). A cleaner signal is given by decay into a pair of Z-bosons (which happens about 2.9% of the time for a Higgs with a mass of 126 GeV/c^2),[5] if each of the bosons subsequently decays into a pair of easy-to-detect charged leptons (electrons or muons).

Decay into massless gauge bosons (i.e., gluons or photons) is also possible, but requires intermediate loop of virtual heavy quarks (top or bottom) or massive gauge bosons.[102] The most common such process is the decay into a pair of gluons through a loop of virtual heavy quarks. This process, which is the reverse of the gluon fusion process mentioned above, happens approximately 8.5% of the time for a Higgs boson with a mass of 126 GeV/c^2.[5] Much rarer is the decay into a pair of photons mediated by a loop of W bosons or heavy quarks, which happens only twice for every thousand decays.[5] However, this process is very relevant for experimental searches for the Higgs boson, because the energy and momentum of the photons can be measured very precisely, giving an accurate reconstruction of the mass of the decaying particle.[102]

23.4.6 Alternative models

Main article: Alternatives to the Standard Model Higgs

The Minimal Standard Model as described above is the simplest known model for the Higgs mechanism with just one Higgs field. However, an extended Higgs sector with additional Higgs particle doublets or triplets is also possible, and many extensions of the Standard Model have this feature. The non-minimal Higgs sector favoured by theory are the two-Higgs-doublet models (2HDM), which predict the existence of a quintet of scalar particles: two CP-even neutral Higgs bosons h^0 and H^0, a CP-odd neutral Higgs boson A^0, and two charged Higgs particles $H^{\pm}$. Supersymmetry ("SUSY") also predicts relations between the Higgs-boson masses and the masses of the gauge bosons, and could accommodate a 125 GeV/c^2 neutral Higgs boson.

The key method to distinguish between these different models involves study of the particles' interactions ("coupling") and exact decay processes ("branching ratios"), which can be measured and tested experimentally in particle collisions. In the Type-I 2HDM model one Higgs doublet couples to up and down quarks, while the second doublet does not couple to quarks. This model has two interesting limits, in which the lightest Higgs couples to just fermions ("gauge-phobic") or just gauge bosons ("fermiophobic"), but not both. In the Type-II 2HDM model, one Higgs doublet only couples to up-type quarks, the other only couples to down-type quarks.[103] The heavily researched Minimal Supersymmetric Standard Model (MSSM) includes a Type-II 2HDM Higgs sector, so it could be disproven by evidence of a Type-I 2HDM Higgs.

In other models the Higgs scalar is a composite particle. For example, in technicolor the role of the Higgs field is played by strongly bound pairs of fermions called techniquarks. Other models, feature pairs of top quarks (see top quark condensate). In yet other models, there is no Higgs field at all and the electroweak symmetry is broken using extra dimensions.[104][105]

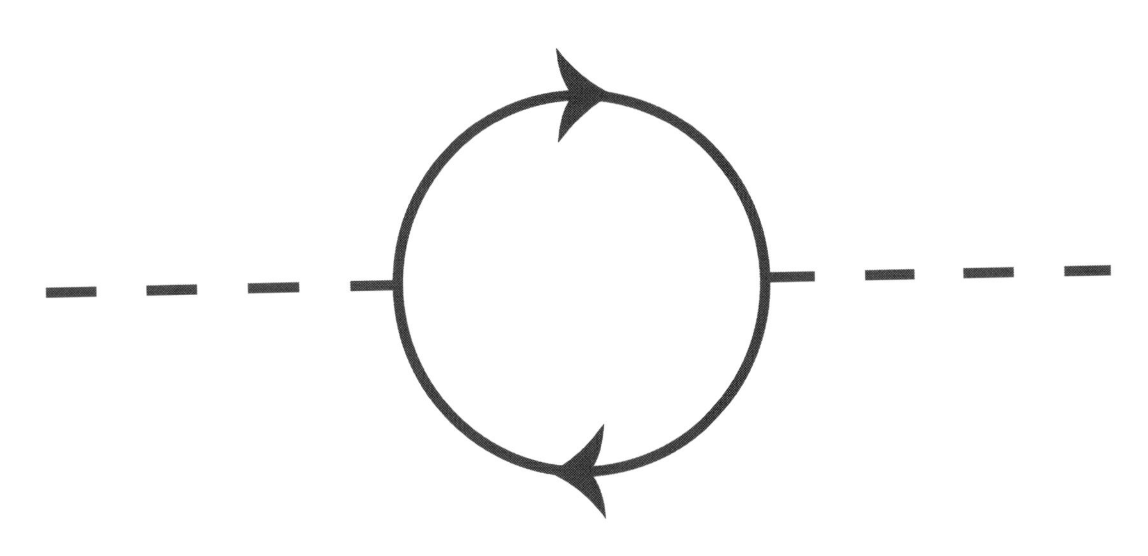

A one-loop Feynman diagram of the first-order correction to the Higgs mass. In the Standard Model the effects of these corrections are potentially enormous, giving rise to the so-called hierarchy problem.

23.4.7 Further theoretical issues and hierarchy problem

Main articles: Hierarchy problem and Hierarchy problem § The Higgs mass

The Standard Model leaves the mass of the Higgs boson as a parameter to be measured, rather than a value to be calculated. This is seen as theoretically unsatisfactory, particularly as quantum corrections (related to interactions with virtual particles) should apparently cause the Higgs particle to have a mass immensely higher than that observed, but at the same time the Standard Model requires a mass of the order of 100 to 1000 GeV to ensure unitarity (in this case, to unitarise longitudinal vector boson scattering).[106] Reconciling these points appears to require explaining why there is an almost-perfect cancellation resulting in the visible mass of ~ 125 GeV, and it is not clear how to do this. Because the weak force is about 10^{32} times stronger than gravity, and (linked to this) the Higgs boson's mass is so much less than the Planck mass or the grand unification energy, it appears that either there is some underlying connection or reason for these observations which is unknown and not described by the Standard Model, or some unexplained and extremely precise fine-tuning of parameters – however at present neither of these explanations is proven. This is known as a hierarchy problem.[107] More broadly, the hierarchy problem amounts to the worry that a future theory of fundamental particles and interactions should not have excessive fine-tunings or unduly delicate cancellations, and should allow masses of particles such as the Higgs boson to be calculable. The problem is in some ways unique to spin-0 particles (such as the Higgs boson), which can give rise to issues related to quantum corrections that do not affect particles with spin.[106] A number of solutions have been proposed, including supersymmetry, conformal solutions and solutions via extra dimensions such as braneworld models.

There are also issues of quantum triviality, which suggests that it may not be possible to create a consistent quantum field theory involving elementary scalar particles.

23.5 Experimental search

Main article: Search for the Higgs boson

To produce Higgs bosons, two beams of particles are accelerated to very high energies and allowed to collide within a particle detector. Occasionally, although rarely, a Higgs boson will be created fleetingly as part of the collision byproducts. Because the Higgs boson decays very quickly, particle detectors cannot detect it directly. Instead the detectors register all the decay products (the *decay signature*) and from the data the decay process is reconstructed. If the observed decay products match a possible decay process (known as a *decay channel*) of a Higgs boson, this indicates that a Higgs boson may have been created. In practice, many processes may produce similar decay signatures. Fortunately, the Standard Model precisely predicts the likelihood of each of these, and each known process, occurring. So, if the detector detects more decay signatures consistently matching a Higgs boson than would otherwise be expected if Higgs bosons did not exist, then this would be strong evidence that the Higgs boson exists.

Because Higgs boson production in a particle collision is likely to be very rare (1 in 10 billion at the LHC),[Note 14] and many other possible collision events can have similar decay signatures, the data of hundreds of trillions of collisions needs to be analysed and must "show the same picture" before a conclusion about the existence of the Higgs boson can be reached. To conclude that a new particle has been found, particle physicists require that the statistical analysis of two independent particle detectors each indicate that there is lesser than a one-in-a-million chance that the observed decay signatures are due to just background random Standard Model events—i.e., that the observed number of events is more than 5 standard deviations (sigma) different from that expected if there was no new particle. More collision data allows better confirmation of the physical properties of any new particle observed, and allows physicists to decide whether it is indeed a Higgs boson as described by the Standard Model or some other hypothetical new particle.

To find the Higgs boson, a powerful particle accelerator was needed, because Higgs bosons might not be seen in lower-energy experiments. The collider needed to have a high luminosity in order to ensure enough collisions were seen for conclusions to be drawn. Finally, advanced computing facilities were needed to process the vast amount of data (25 petabytes per year as at 2012) produced by the collisions.[109] For the announcement of 4 July 2012, a new collider known as the Large Hadron Collider was constructed at CERN with a planned eventual collision energy of 14 TeV—over seven times any previous collider—and over 300 trillion (3×10^{14}) LHC proton–proton collisions were analysed by the LHC Computing Grid, the world's largest computing grid (as of 2012), comprising over 170 computing facilities in a worldwide network across 36 countries.[109][110][111]

23.5.1 Search prior to 4 July 2012

The first extensive search for the Higgs boson was conducted at the Large Electron–Positron Collider (LEP) at CERN in the 1990s. At the end of its service in 2000, LEP had found no conclusive evidence for the Higgs.[Note 15] This implied that if the Higgs boson were to exist it would have to be heavier than 114.4 GeV/c^2.[112]

The search continued at Fermilab in the United States, where the Tevatron—the collider that discovered the top quark in 1995—had been upgraded for this purpose. There was no guarantee that the Tevatron would be able to find the Higgs, but it was the only supercollider that was operational since the Large Hadron Collider (LHC) was still under construction and the planned Superconducting Super Collider had been cancelled in 1993 and never completed. The Tevatron was only able to exclude further ranges for the Higgs mass, and was shut down on 30 September 2011 because it no longer could keep up with the LHC. The final analysis of the data excluded the possibility of a Higgs boson with a mass between 147 GeV/c^2 and 180 GeV/c^2. In addition, there was a small (but not significant) excess of events possibly indicating a Higgs boson with a mass between 115 GeV/c^2 and 140 GeV/c^2.[113]

The Large Hadron Collider at CERN in Switzerland, was designed specifically to be able to either confirm or exclude the existence of the Higgs boson. Built in a 27 km tunnel under the ground near Geneva originally inhabited by LEP, it was designed to collide two beams of protons, initially at energies of 3.5 TeV per beam (7 TeV total), or almost 3.6 times that of the Tevatron, and upgradeable to 2 × 7 TeV (14 TeV total) in future. Theory suggested if the Higgs boson existed, collisions at these energy levels should be able to reveal it. As one of the most complicated scientific instruments ever built, its operational readiness was delayed for 14 months by a magnet quench event nine days after its inaugural tests, caused by a faulty electrical connection that damaged over 50 superconducting magnets and contaminated the vacuum system.[114][115][116]

Data collection at the LHC finally commenced in March 2010.[117] By December 2011 the two main particle detectors at the LHC, ATLAS and CMS, had narrowed down the mass range where the Higgs could exist to around 116-130 GeV (ATLAS) and 115-127 GeV (CMS).[118][119] There had also already been a number of promising event excesses that had "evaporated" and proven to be nothing but random fluctuations. However, from around May 2011,[120] both experiments had seen among their results, the slow emergence of a small yet consistent excess of gamma and 4-lepton decay signatures and several other particle decays, all hinting at a new particle at a mass around 125 GeV.[120] By around November 2011, the anomalous data at 125 GeV was becoming "too large to ignore" (although still far from conclusive), and the team leaders at both ATLAS and CMS each privately suspected they might have found the Higgs.[120] On November 28, 2011, at an internal meeting of the two team leaders and the director general of CERN, the latest analyses were discussed outside their teams for the first time, suggesting both ATLAS and CMS might be converging on a possible shared result at 125 GeV, and initial preparations commenced in case of a successful finding.[120] While this information was not known publicly at the time, the narrowing of the possible Higgs range to around 115–130 GeV and the repeated observation of small but consistent event excesses across multiple channels at both ATLAS and CMS in the 124-126 GeV region (described as "tantalising hints" of around 2-3 sigma) were public knowledge with "a lot of interest".[121] It was therefore widely anticipated around the end of 2011, that the LHC would provide sufficient data to either exclude or confirm the finding of a Higgs boson by the end of 2012, when their 2012 collision data (with slightly higher 8 TeV collision energy) had been examined.[121][122]

23.5.2 Discovery of candidate boson at CERN

On 22 June 2012 CERN announced an upcoming seminar covering tentative findings for 2012,[126][127] and shortly afterwards (from around 1 July 2012 according to an analysis of the spreading rumour in social media[128]) rumours began to spread in the media that this would include a major announcement, but it was unclear whether this would be a stronger signal or a formal discovery.[129][130] Speculation escalated to a "fevered" pitch when reports emerged that Peter Higgs, who proposed the particle, was to be attending the seminar,[131][132] and that "five leading physicists" had been invited – generally believed to signify the five living 1964 authors – with Higgs, Englert, Guralnik, Hagen attending and Kibble confirming his invitation (Brout having died in 2011).[133][134]

On 4 July 2012 both of the CERN experiments announced they had independently made the same discovery:[135] CMS of a previously unknown boson with mass 125.3 ± 0.6 GeV/c^2[136][137] and ATLAS of a boson with mass 126.0 ± 0.6 GeV/c^2.[138][139] Using the combined analysis of two interaction types (known as 'channels'), both experiments independently reached a local significance of 5 sigma — implying that the probability of getting at least as strong a result by chance alone is less than 1 in 3 million. When additional channels were taken into account, the CMS significance was reduced to 4.9 sigma.[137]

The two teams had been working 'blinded' from each other from around late 2011 or early 2012,[120] meaning they did not discuss their results with each other, providing additional certainty that any common finding was genuine validation of a particle.[109] This level of evidence, confirmed independently by two separate teams and experiments, meets the formal level of proof required to announce a confirmed discovery.

On 31 July 2012, the ATLAS collaboration presented additional data analysis on the "observation of a new particle", including data from a third channel, which improved the significance to 5.9 sigma (1 in 588 million chance of obtaining at least as strong evidence by random background effects alone) and mass 126.0 ± 0.4 (stat) ± 0.4 (sys) GeV/c^2, [139] and CMS improved the significance to 5-sigma and mass 125.3 ± 0.4 (stat) ± 0.5 (sys) GeV/c^2.[136]

23.5.3 The new particle tested as a possible Higgs boson

Following the 2012 discovery, it was still unconfirmed whether or not the 125 GeV/c^2 particle was a Higgs boson. On one hand, observations remained consistent with the observed particle being the Standard Model Higgs boson, and the particle decayed into at least some of the predicted channels. Moreover, the production rates and branching ratios for the observed channels broadly matched the predictions by the Standard Model within the experimental uncertainties. However, the experimental uncertainties currently still left room for alternative explanations, meaning an announcement of the discovery of a Higgs boson would have been premature.[102] To allow more opportunity for data collection, the LHC's proposed 2012 shutdown and 2013–14 upgrade were postponed by 7 weeks into 2013.[140]

In November 2012, in a conference in Kyoto researchers said evidence gathered since July was falling into line with the basic Standard Model more than its alternatives, with a range of results for several interactions matching that theory's predictions.[141] Physicist Matt Strassler highlighted "considerable" evidence that the new particle is not a pseudoscalar negative parity particle (consistent with this required finding for a Higgs boson), "evaporation" or lack of increased significance for previous hints of non-Standard Model findings, expected Standard Model interactions with W and Z bosons, absence of "significant new implications" for or against supersymmetry, and in general no significant deviations to date from the results expected of a Standard Model Higgs boson.[142] However some kinds of extensions to the Standard Model would also show very similar results;[143] so commentators noted that based on other particles that are still being understood long after their discovery, it may take years to be sure, and decades to fully understand the particle that has been found.[141][142]

These findings meant that as of January 2013, scientists were very sure they had found an unknown particle of mass ~ 125 GeV/c^2, and had not been misled by experimental error or a chance result. They were also sure, from initial observations, that the new particle was some kind of boson. The behaviours and properties of the particle, so far as examined since July 2012, also seemed quite close to the behaviours expected of a Higgs boson. Even so, it could still have been a Higgs boson or some other unknown boson, since future tests could show behaviours that do not match a Higgs boson, so as of December 2012 CERN still only stated that the new particle was "consistent with" the Higgs boson,[11][13] and scientists did not yet positively say it was the Higgs boson.[144] Despite this, in late 2012, widespread media reports announced (incorrectly) that a Higgs boson had been confirmed during the year.[Note 16]

In January 2013, CERN director-general Rolf-Dieter Heuer stated that based on data analysis to date, an answer could be possible 'towards' mid-2013,[150] and the deputy chair of physics at Brookhaven National Laboratory stated in February 2013 that a "definitive" answer might require "another few years" after the collider's 2015 restart.[151] In early March 2013, CERN Research Director Sergio Bertolucci stated that confirming spin-0 was the major remaining requirement to determine whether the particle is at least some kind of Higgs boson.[152]

23.5.4 Preliminary confirmation of existence and current status

On 14 March 2013 CERN confirmed that:

> "CMS and ATLAS have compared a number of options for the spin-parity of this particle, and these all prefer no spin and even parity [two fundamental criteria of a Higgs boson consistent with the Standard Model]. This, coupled with the measured interactions of the new particle with other particles, strongly indicates that it is a Higgs boson." [1]

This also makes the particle the first elementary scalar particle to be discovered in nature.[14]

Examples of tests used to validate whether the 125 GeV particle is a Higgs boson:[142][153]

23.6 Public discussion

23.6.1 Naming

Names used by physicists

The name most strongly associated with the particle and field is the Higgs boson[79]:168 and Higgs field. For some time the particle was known by a combination of its PRL author names (including at times Anderson), for example the Brout–Englert–Higgs particle, the Anderson-Higgs particle, or the Englert–Brout–Higgs–Guralnik–Hagen–Kibble mechanism,[Note 17] and these are still used at times.[51][160] Fueled in part by the issue of recognition and a potential shared Nobel Prize,[160][161] the most appropriate name is still occasionally a topic of debate as at 2012.[160] (Higgs himself prefers to call the particle either by an acronym of all those involved, or "the scalar boson", or "the so-called Higgs particle".[161])

A considerable amount has been written on how Higgs' name came to be exclusively used. Two main explanations are offered.

Nickname

The Higgs boson is often referred to as the "God particle" in popular media outside the scientific community. The nickname comes from the title of the 1993 book on the Higgs boson and particle physics - The God Particle: If the Universe Is the Answer,What Is the Question? byNobel Physics prizewinnerandFermilabdirectorLeon Lederman. [21] Lederman wrote it in the context of failing US government support for the Superconducting Super Collider,[175] a part-constructed titanic[176][177] competitor to the Large Hadron Collider with planned collision energies of 2×20 TeV that was championed by Lederman since its 1983 inception[175][178][179] and shut down in 1993. The book sought in part to promote awareness of the significance and need for such a project in the face of its possible loss of funding.[180] Lederman, a leading researcher in the field, wanted to title his book "The Goddamn Particle: If the Universe is the Answer, What is the Question?" But his editor decided that the title was too controversial and convinced Lederman to change the title to "The God Particle: If the Universe is the Answer, What is the Question?"[181]

And since the Higgs Boson deals with how matter was formed at the time of the big bang, and since newspapers loved the term, the term "God particle" was used.

While media use of this term may have contributed to wider awareness and interest,[182] many scientists feel the name is inappropriate[16][17][183] since it is sensational hyperbole and misleads readers;[184] the particle also has nothing to do with God, leaves open numerous questions in fundamental physics, and does not explain the ultimate origin of the universe. Higgs, an atheist, was reported to be displeased and stated in a 2008 interview that he found it "embarrassing" because it was "the kind of misuse... which I think might offend some people".[184][185][186] Science writer Ian Sample stated in his 2010 book on the search that the nickname is "universally hate[d]" by physicists and perhaps the "worst derided" in the history of physics, but that (according to Lederman) the publisher rejected all titles mentioning "Higgs" as unimaginative and too unknown.[187]

Lederman begins with a review of the long human search for knowledge, and explains that his tongue-in-cheek title draws an analogy between the impact of the Higgs field on the fundamental symmetries at the Big Bang, and the apparent chaos of structures, particles, forces and interactions that resulted and shaped our present universe, with the biblical story of Babel in which the primordial single language of early Genesis was fragmented into many disparate languages and cultures.[188]

> Today ... we have the standard model, which reduces all of reality to a dozen or so particles and four forces. ... It's a hard-won simplicity [...and...] remarkably accurate. But it is also incomplete and, in fact, internally inconsistent... This boson is so central to the state of physics today, so crucial to our final understanding of the structure of matter, yet so elusive, that I have given it a nickname: the God Particle. Why God Particle? Two reasons. One, the publisher wouldn't let us call it the Goddamn Particle, though that might be a more appropriate title, given its villainous nature and the expense it is causing. And two, there is a connection, of sorts, to another book, a *much* older one...
> — Leon M. Lederman and Dick Teresi, *The God Particle: If the Universe is the Answer, What is the Question*[21] p. 22

Lederman asks whether the Higgs boson was added just to perplex and confound those seeking knowledge of the universe, and whether physicists will be confounded by it as recounted in that story, or ultimately surmount the challenge and understand "how beautiful is the universe [God has] made".[189]

Other proposals

A renaming competition by British newspaper *The Guardian* in 2009 resulted in their science correspondent choosing the name "the champagne bottle boson" as the best submission: "The bottom of a champagne bottle is in the shape of the Higgs potential and is often used as an illustration in physics lectures. So it's not an embarrassingly grandiose name, it is memorable, and [it] has some physics connection too."[190] The name *Higgson* was suggested as well, in an opinion piece in the Institute of Physics' online publication *physicsworld.com*.[191]

23.6.2 Media explanations and analogies

There has been considerable public discussion of analogies and explanations for the Higgs particle and how the field creates mass,[192][193] including coverage of explanatory attempts in their own right and a competition in 1993 for the best popular explanation by then-UK Minister for Science Sir William Waldegrave[194] and articles in newspapers worldwide.

An educational collaboration involving an LHC physicist and a High School Teachers at CERN educator suggests that dispersion of light – responsible for the rainbow and dispersive prism – is a useful analogy for the Higgs field's symmetry breaking and mass-causing effect.[195]

Matt Strassler uses electric fields as an analogy:[196]

> Some particles interact with the Higgs field while others don't. Those particles that feel the Higgs field act as if they have mass. Something similar happens in an electric field – charged objects are pulled around and neutral objects can sail through unaffected. So you can think of the Higgs search as an attempt to make waves in the Higgs field *[create Higgs bosons]* to prove it's really there.

A similar explanation was offered by *The Guardian*:[197]

> The Higgs boson is essentially a ripple in a field said to have emerged at the birth of the universe and to span the cosmos to this day ... The particle is crucial however: it is the smoking gun, the evidence required to show the theory is right.

The Higgs field's effect on particles was famously described by physicist David Miller as akin to a room full of political party workers spread evenly throughout a room: the crowd gravitates to and slows down famous people but does not slow down others.[Note 18] He also drew attention to well-known effects in solid state physics where an electron's effective mass can be much greater than usual in the presence of a crystal lattice.[198]

Analogies based on drag effects, including analogies of "syrup" or "molasses" are also well known, but can be somewhat misleading since they may be understood (incorrectly) as saying that the Higgs field simply resists some particles' motion but not others' – a simple resistive effect could also conflict with Newton's third law.[200]

23.6.3 Recognition and awards

There has been considerable discussion of how to allocate the credit if the Higgs boson is proven, made more pointed as a Nobel prize had been expected, and the very wide basis of people entitled to consideration. These include a range of theoreticians who made the Higgs mechanism theory possible, the theoreticians of the 1964 PRL papers (including Higgs himself), the theoreticians who derived from these, a working electroweak theory and the Standard Model itself, and also the experimentalists at CERN and other institutions who made possible the proof of the Higgs field and boson in reality. The Nobel prize has a limit of 3 persons to share an award, and some possible winners are already prize holders for other work, or are deceased (the prize is only awarded to persons in their lifetime). Existing prizes for works relating to the Higgs field, boson, or mechanism include:

Photograph of light passing through a dispersive prism: the rainbow effect arises because photons are not all affected to the same degree by the dispersive material of the prism.

- Nobel Prize in Physics (1979) – Glashow, Salam, and Weinberg, *for contributions to the theory of the unified weak and electromagnetic interaction between elementary particles* [201]
- Nobel Prize in Physics (1999) – 't Hooft and Veltman, *for elucidating the quantum structure of electroweak inter-*

actions in physics [202]

- Nobel Prize in Physics (2008) – Nambu (shared), *for the discovery of the mechanism of spontaneous broken symmetry in subatomic physics* [53]
- J. J. Sakurai Prize for Theoretical Particle Physics (2010) – Hagen, Englert, Guralnik, Higgs, Brout, and Kibble, *for elucidation of the properties of spontaneous symmetry breaking in four-dimensional relativistic gauge theory and of the mechanism for the consistent generation of vector boson masses* [77] (for the 1964 papers described above)
- Wolf Prize (2004) – Englert, Brout, and Higgs
- Nobel Prize in Physics (2013) - Peter Higgs and François Englert, *for the theoretical discovery of a mechanism that contributes to our understanding of the origin of mass of subatomic particles, and which recently was confirmed through the discovery of the predicted fundamental particle, by the ATLAS and CMS experiments at CERN's Large Hadron Collider* [203]

Additionally Physical Review Letters' 50-year review (2008) recognized the 1964 PRL symmetry breaking papers and Weinberg's 1967 paper *A model of Leptons* (the most cited paper in particle physics, as of 2012) "milestone Letters".[75]

Following reported observation of the Higgs-like particle in July 2012, several Indian media outlets reported on the supposed neglect of credit to Indian physicist Satyendra Nath Bose after whose work in the 1920s the class of particles "bosons" is named[204][205] (although physicists have described Bose's connection to the discovery as tenuous).[206]

23.7 Technical aspects and mathematical formulation

See also: Standard Model (mathematical formulation)

In the Standard Model, the Higgs field is a four-component scalar field that forms a complex doublet of the weak isospin SU(2) symmetry:

while the field has charge +1/2 under the weak hypercharge U(1) symmetry (in the convention where the electric charge, Q, the weak isospin, I_3, and the weak hypercharge, Y, are related by $Q = I_3 + Y$).[207]

The Higgs part of the Lagrangian is[207]

where W_μ^a and B_μ are the gauge bosons of the SU(2) and U(1) symmetries, g and g' their respective coupling constants, $\tau^a = \sigma^a/2$ (where σ^a are the Pauli matrices) a complete set generators of the SU(2) symmetry, and $\lambda > 0$ and $\mu^2 > 0$, so that the ground state breaks the SU(2) symmetry (see figure). The ground state of the Higgs field (the bottom of the potential) is degenerate with different ground states related to each other by a SU(2) gauge transformation. It is always possible to pick a gauge such that in the ground state $\phi^1 = \phi^2 = \phi^3 = 0$. The expectation value of ϕ^0 in the ground state (the vacuum expectation value or vev) is then $\langle \phi^0 \rangle = v$, where $v = \frac{|\mu|}{\sqrt{\lambda}}$. The measured value of this parameter is ~246 GeV/c^2.[102] It has units of mass, and is the only free parameter of the Standard Model that is not a dimensionless number. Quadratic terms in W_μ and B_μ arise, which give masses to the *W* and *Z* bosons:[207]

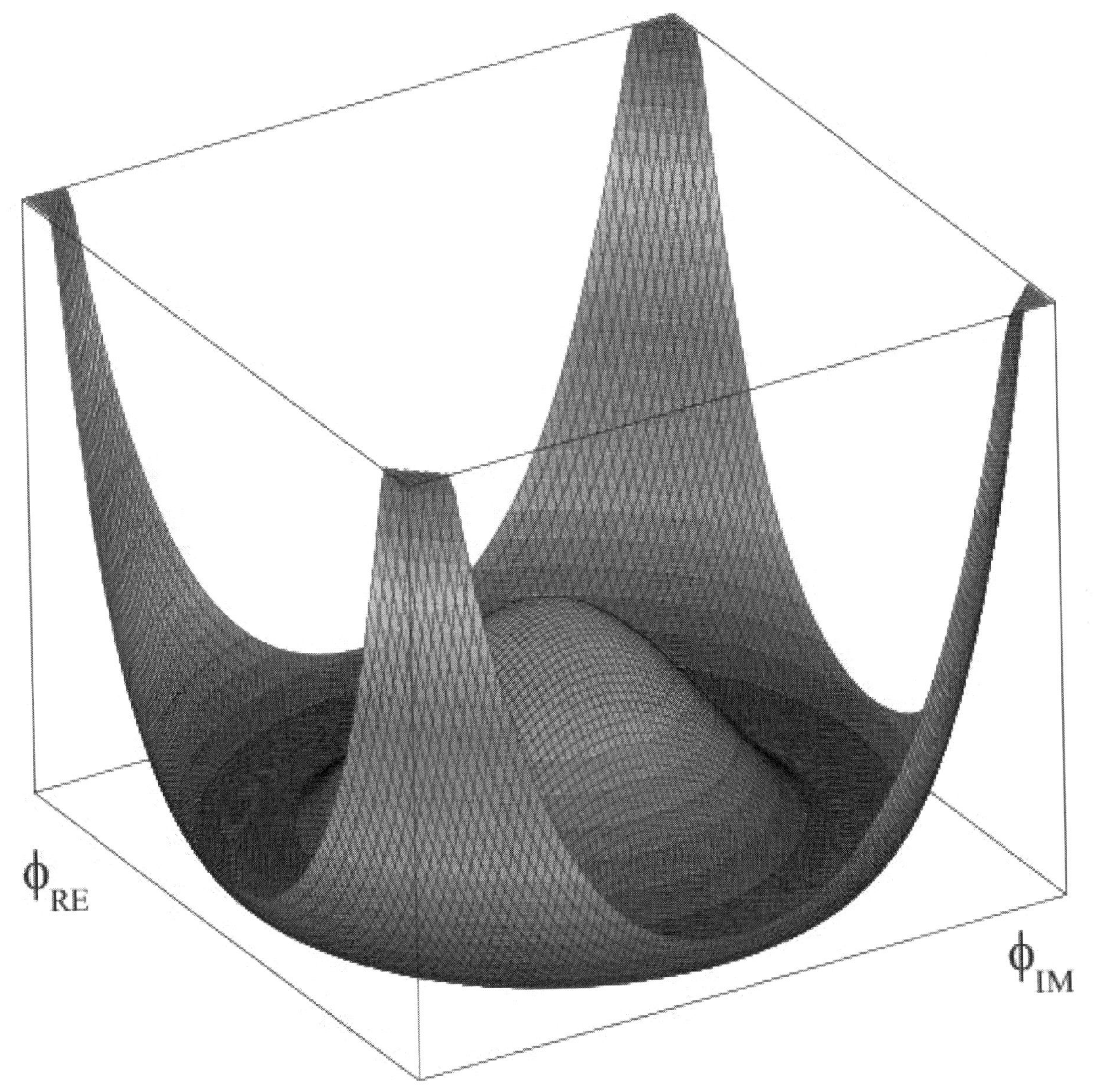

The potential for the Higgs field, plotted as function of ϕ^0 and ϕ^3 . It has a Mexican-hat *or* champagne-bottle profile *at the ground.*

with their ratio determining the Weinberg angle, $\cos\theta_W = \frac{M_W}{M_Z} = \frac{|g|}{\sqrt{g^2+g'^2}}$, and leave a massless U(1) photon, γ .

The quarks and the leptons interact with the Higgs field through Yukawa interaction terms:

where $(d, u, e, \nu)^i_{L,R}$ are left-handed and right-handed quarks and leptons of the *i*th generation, $\lambda^{ij}_{u,d,e}$ are matrices of Yukawa couplings where h.c. denotes the hermitian conjugate terms. In the symmetry breaking ground state, only the terms containing ϕ^0 remain, giving rise to mass terms for the fermions. Rotating the quark and lepton fields to the basis where the matrices of Yukawa couplings are diagonal, one gets

where the masses of the fermions are $m^i_{u,d,e} = \lambda^i_{u,d,e} v/\sqrt{2}$, and $\lambda^i_{u,d,e}$ denote the eigenvalues of the Yukawa matrices.[207]

23.8 See also

Standard Model

- Quantum gauge theory
- History of quantum field theory
- Introduction to quantum mechanics
- Noncommutative standard model and noncommutative geometry generally
- Standard Model (mathematical formulation) (and especially Standard Model fields overview and mass terms and the Higgs mechanism)

Other

- Bose–Einstein statistics
- Dalitz plot
- Higgs boson in fiction
- Quantum triviality
- ZZ diboson
- Scalar boson
- Stueckelberg action
- Tachyonic field

23.9 Notes

[1] Note that such events also occur due to other processes. Detection involves a statistically significant excess of such events at specific energies.

[2] In the Standard Model, the total decay width of a Higgs boson with a mass of 126 GeV/c^2 is predicted to be 4.21×10^{-3} GeV.[5] The mean lifetime is given by $\tau = \hbar/\Gamma$.

[3] The range of a force is inversely proportional to the mass of the particles transmitting it.[19] In the Standard Model, forces are carried by virtual particles. These particles' movement and interactions with each other are limited by the energy–time uncertainty principle. As a result, the more massive a single virtual particle is, the greater its energy, and therefore the shorter the distance it can travel. A particle's mass therefore determines the maximum distance at which it can interact with other particles and on any force it mediates. By the same token, the reverse is also true: massless and near-massless particles can carry long distance forces. *(See also: Compton wavelength and Static forces and virtual-particle exchange)* Since experiments have shown that the weak force acts over only a very short range, this implies that there must exist massive gauge bosons. And indeed, their masses have since been confirmed by measurement.

[4] It is quite common for a law of physics to hold true only if certain assumptions held true or only under certain conditions. For example, Newton's laws of motion apply only at speeds where relativistic effects are negligible; and laws related to conductivity, gases, and classical physics (as opposed to quantum mechanics) may apply only within certain ranges of size, temperature, pressure, or other conditions.

[5] Electroweak symmetry is broken by the Higgs field in its lowest energy state, called its "ground state". At high energy levels this does not happen, and the gauge bosons of the weak force would therefore be expected to be massless.

[6] By the 1960s, many had already started to see gauge theories as failing to explain particle physics because theorists had been unable to solve the mass problem or even explain how gauge theory could provide a solution. So the idea that the Standard Model – which relied on a Higgs field, not yet proved to exist – could be fundamentally incorrect. Against this, once the model was developed around 1972, no better theory existed, and its predictions and solutions were so accurate, that it became the preferred theory anyway. It then became crucial to science, to know whether it was *correct*.

[7] The success of the Higgs-based electroweak theory and Standard Model is illustrated by their predictions of the mass of two particles later detected: the W boson (predicted mass: 80.390 ± 0.018 GeV, experimental measurement: 80.387 ± 0.019 GeV), and the Z boson (predicted mass: 91.1874 ± 0.0021, experimental measurement: 91.1876 ± 0.0021 GeV). The existence of the Z boson was itself another prediction. Other accurate predictions included the weak neutral current, the gluon, and the top and charm quarks, all later proven to exist as the theory said.

[8] For example, Huffington Post/Reuters[35] and others[36][37]

[9] The bubble's effects would be expected to propagate across the universe at the speed of light from wherever it occurred. However space is vast – with even the nearest galaxy being over 2 million lightyears from us, and others being many billions of lightyears distant, so the effect of such an event would be unlikely to arise here for billions of years after first occurring.[39][40]

[10] If the Standard Model is valid, then the particles and forces we observe in our universe exist as they do, because of underlying quantum fields. Quantum fields can have states of differing stability, including 'stable', 'unstable' and 'metastable' states (the latter remain stable unless sufficiently perturbed). If a more stable vacuum state were able to arise, then existing particles and forces would no longer arise as they presently do. Different particles or forces would arise from (and be shaped by) whatever new quantum states arose. The world we know depends upon these particles and forces, so if this happened, everything around us, from subatomic particles to galaxies, and all fundamental forces, would be reconstituted into new fundamental particles and forces and structures. The universe would potentially lose all of its present structures and become inhabited by new ones (depending upon the exact states involved) based upon the same quantum fields.

[11] Goldstone's theorem only applies to gauges having manifest Lorentz covariance, a condition that took time to become questioned. But the process of quantisation requires a gauge to be fixed and at this point it becomes possible to choose a gauge such as the 'radiation' gauge which is not invariant over time, so that these problems can be avoided. According to Bernstein (1974, p.8):

> "the "radiation gauge" condition ∇·A(x) = 0 is clearly noncovariant, which means that if we wish to maintain transversality of the photon in all Lorentz frames, the photon field Aμ(x) cannot transform like a four-vector. This is no catastrophe, since the photon *field* is not an observable, and one can readily show that the S-matrix elements, which *are* observable have covariant structures .... in gauge theories one might arrange things so that one had a symmetry breakdown because of the noninvariance of the vacuum; but, because the Goldstone *et al.* proof breaks down, the zero mass Goldstone mesons need not appear." [Emphasis in original]

Bernstein (1974) contains an accessible and comprehensive background and review of this area, see external links

[12] A field with the "Mexican hat" potential $V(\phi) = \mu^2\phi^2 + \lambda\phi^4$ and $\mu^2 < 0$ has a minimum not at zero but at some non-zero value ϕ_0 . By expressing the action in terms of the field $\tilde{\phi} = \phi - \phi_0$ (where ϕ_0 is a constant independent of position), we find the Yukawa term has a component $g\phi_0\bar{\psi}\psi$. Since both g and ϕ_0 are constants, this looks exactly like the mass term for a fermion of mass $g\phi_0$. The field $\tilde{\phi}$ is then the Higgs field.

[13] In the Standard Model, the mass term arising from the Dirac Lagrangian for any fermion ψ is $-m\bar{\psi}\psi$. This is *not* invariant under the electroweak symmetry, as can be seen by writing ψ in terms of left and right handed components:

$$-m\bar{\psi}\psi = -m(\bar{\psi}_L\psi_R + \bar{\psi}_R\psi_L)$$

i.e., contributions from $\bar{\psi}_L\psi_L$ and $\bar{\psi}_R\psi_R$ terms do not appear. We see that the mass-generating interaction is achieved by constant flipping of particle chirality. Since the spin-half particles have no right/left helicity pair with the same SU(2) and SU(3) representation and the same weak hypercharge, then assuming these gauge charges are conserved in the vacuum, none of the spin-half particles could ever swap helicity. Therefore, in the absence of some other cause, all fermions must be massless.

[14] The example is based on the production rate at the LHC operating at 7 TeV. The total cross-section for producing a Higgs boson at the LHC is about 10 picobarn,[97] while the total cross-section for a proton–proton collision is 110 millibarn.[100]

[15] Just before LEP's shut down, some events that hinted at a Higgs were observed, but it was not judged significant enough to extend its run and delay construction of the LHC.

[16] Announced in articles in *Time*,[145] Forbes,[146] *Slate*,[147] *NPR*,[148] and others.[149]

[17] Other names have included: the "Anderson–Higgs" mechanism,[159] "Higgs–Kibble" mechanism (by Abdus Salam)[79] and "ABEGHHK'tH" mechanism [for Anderson, Brout, Englert, Guralnik, Hagen, Higgs, Kibble and 't Hooft] (by Peter Higgs).[79]

[18] In Miller's analogy, the Higgs field is compared to political party workers spread evenly throughout a room. There will be some people (in Miller's example an anonymous person) who pass through the crowd with ease, paralleling the interaction between the field and particles that do not interact with it, such as massless photons. There will be other people (in Miller's example the British prime minister) who would find their progress being continually slowed by the swarm of admirers crowding around, paralleling the interaction for particles that do interact with the field and by doing so, acquire a finite mass.[198][199]

23.10 References

[1] O'Luanaigh, C. (14 March 2013). "New results indicate that new particle is a Higgs boson". CERN. Retrieved 2013-10-09.

[2] Bryner, J. (14 March 2013). "Particle confirmed as Higgs boson". *NBC News*. Retrieved 2013-03-14.

[3] Heilprin, J. (14 March 2013). "Higgs Boson Discovery Confirmed After Physicists Review Large Hadron Collider Data at CERN". *The Huffington Post*. Retrieved 2013-03-14.

[4] ATLAS; CMS (26 March 2015). "Combined Measurement of the Higgs Boson Mass in pp Collisions at √s=7 and 8 TeV with the ATLAS and CMS Experiments". arXiv:1503.07589.

[5] LHC Higgs Cross Section Working Group; Dittmaier; Mariotti; Passarino; Tanaka; Alekhin; Alwall; Bagnaschi; Banfi (2012). "Handbook of LHC Higgs Cross Sections: 2. Differential Distributions". *CERN Report 2 (Tables A.1 – A.20)* **1201**: 3084. arXiv:1201.3084. Bibcode:2012arXiv1201.3084L.

[6] Onyisi, P. (23 October 2012). "Higgs boson FAQ". University of Texas ATLAS group. Retrieved 2013-01-08.

[7] Strassler, M. (12 October 2012). "The Higgs FAQ 2.0". *ProfMattStrassler.com*. Retrieved 2013-01-08. [Q] Why do particle physicists care so much about the Higgs particle?
[A] Well, actually, they don't. What they really care about is the Higgs *field*, because it is *so* important. [emphasis in original]

[8] José Luis Lucio and Arnulfo Zepeda (1987). *Proceedings of the II Mexican School of Particles and Fields, Cuernavaca-Morelos, 1986*. World Scientific. p. 29. ISBN 9971504340.

[9] Gunion, Dawson, Kane, and Haber (199). *The Higgs Hunter's Guide (1st ed.)*. pp. 11 (?). ISBN 9780786743186. – quoted as being in the first (1990) edition of the book by Peter Higgs in his talk "My Life as a Boson", 2001, ref#25.

[10] Strassler, M. (8 October 2011). "The Known Particles – If The Higgs Field Were Zero". *ProfMattStrassler.com*. Retrieved 13 November 2012. The Higgs field: so important it merited an entire experimental facility, the Large Hadron Collider, dedicated to understanding it.

[11] Biever, C. (6 July 2012). "It's a boson! But we need to know if it's the Higgs". *New Scientist*. Retrieved 2013-01-09. 'As a layman, I would say, I think we have it,' said Rolf-Dieter Heuer, director general of CERN at Wednesday's seminar announcing the results of the search for the Higgs boson. But when pressed by journalists afterwards on what exactly 'it' was, things got more complicated. 'We have discovered a boson – now we have to find out what boson it is'
Q: 'If we don't know the new particle is a Higgs, what do we know about it?' We know it is some kind of boson, says Vivek Sharma of CMS [...]
Q: 'are the CERN scientists just being too cautious? What would be enough evidence to call it a Higgs boson?' As there could be many different kinds of Higgs bosons, there's no straight answer.
[emphasis in original]

[12] Siegfried, T. (20 July 2012). "Higgs Hysteria". *Science News*. Retrieved 2012-12-09. In terms usually reserved for athletic achievements, news reports described the finding as a monumental milestone in the history of science.

[13] Del Rosso, A. (19 November 2012). "Higgs: The beginning of the exploration". *CERN Bulletin* (47–48). Retrieved 2013-01-09. Even in the most specialized circles, the new particle discovered in July is not yet being called the "Higgs boson". Physicists still hesitate to call it that before they have determined that its properties fit with those the Higgs theory predicts the Higgs boson has.

[14] Naik, G. (14 March 2013). "New Data Boosts Case for Higgs Boson Find". *The Wall Street Journal*. Retrieved 2013-03-15. 'We've never seen an elementary particle with spin zero,' said Tony Weidberg, a particle physicist at the University of Oxford who is also involved in the CERN experiments.

[15] Overbye, D. (8 October 2013). "For Nobel, They Can Thank the 'God Particle'". *The New York Times*. Retrieved 2013-11-03.

[16] Sample, I. (29 May 2009). "Anything but the God particle". *The Guardian*. Retrieved 2009-06-24.

[17] Evans, R. (14 December 2011). "The Higgs boson: Why scientists hate that you call it the 'God particle'". *National Post*. Retrieved 2013-11-03.

[18] The nickname occasionally has been satirised in mainstream media as well. Borowitz, Andy (July 13, 2012). "5 questions for the Higgs boson". *The New Yorker*.

[19] Shu, F. H. (1982). *The Physical Universe: An Introduction to Astronomy*. University Science Books. pp. 107–108. ISBN 978-0-935702-05-7.

[20] Shu, F. H. (1982). *The Physical Universe: An Introduction to Astronomy*. University Science Books. pp. 107–108. ISBN 978-0-935702-05-7.

[21] Leon M. Lederman and Dick Teresi (1993). *The God Particle: If the Universe is the Answer, What is the Question*. Houghton Mifflin Company.

[22] Heath, Nick, *The Cern tech that helped track down the God particle*, TechRepublic, 4 July 2012

[23] Rao, Achintya (2 July 2012). "Why would I care about the Higgs boson?". *CMS Public Website*. CERN. Retrieved 18 July 2012.

[24] Max Jammer, *Concepts of Mass in Contemporary Physics and Philosophy* (Princeton, NJ: Princeton University Press, 2000) pp.162–163, who provides many references in support of this statement.

[25] The Large Hadron Collider: Shedding Light on the Early Universe – lecture by R.-D. Heuer, CERN, Chios, Greece, 28 September 2011

[26] Alekhin, Djouadi and Moch, S.; Djouadi, A.; Moch, S. (2012-08-13). "The top quark and Higgs boson masses and the stability of the electroweak vacuum". *Physics Letters B* **716**: 214. arXiv:1207.0980. Bibcode:2012PhLB..716..214A. doi:10.10 .Retrieved20February2013.

[27] M.S. Turner, F. Wilczek (1982). "Is our vacuum metastable?". *Nature* **298** (5875): 633–634. Bibcode:1982Natur.298..633T. doi:10.1038/298633a0.

[28] S. Coleman and F. De Luccia (1980). "Gravitational effects on and of vacuum decay". *Physical Review* **D21** (12): 3305. Bibcode:1980PhRvD..21.3305C. doi:10.1103/PhysRevD.21.3305.

[29] M.Stone(1976). "Lifetime and decay of excited vacuum states".*Phys.Rev.D***14**(12): 3568–3573. Bibcode:1976PhRv8S. doi:10.1103/PhysRevD.14.3568.

[30] P.H.Frampton(1976). "Vacuum Instability and Higgs Scalar Mass".*Phys.Rev.Lett.***37**(21): 1378–1380. Bibcode:19778F. doi:10.1103/PhysRevLett.37.1378.

[31] P.H. Frampton (1977). "Consequences of Vacuum Instability in Quantum Field Theory". *Phys. Rev.* **D15** (10): 2922–28. Bibcode:1977PhRvD..15.2922F. doi:10.1103/PhysRevD.15.2922.

[32] Ellis, Espinosa, Giudice, Hoecker, & Riotto, J.; Espinosa, J.R.; Giudice, G.F.; Hoecker, A.; Riotto, A. (2009). "The Probable Fate of the Standard Model". *Phys. Lett. B* **679** (4): 369–375. arXiv:0906.0954. Bibcode:2009PhLB..679..369E. doi:10.1016/j.physletb.2009.07.054.

[33] Masina, Isabella (2013-02-12). "Higgs boson and top quark masses as tests of electroweak vacuum stability". *Phys. Rev. D* **87** (5): 53001. arXiv:1209.0393. Bibcode:2013PhRvD..87e3001M. doi:10.1103/PhysRevD.87.053001.

[34] Buttazzo, Degrassi, Giardino, Giudice, Sala, Salvio, Strumia (2013-07-12). "Investigating the near-criticality of the Higgs boson". *JHEP 1312 (2013) 089*. arXiv:1307.3536. Bibcode:2013JHEP...12..089B. doi:10.1007/JHEP12(2013)089.

[35] Irene Klotz (editing by David Adams and Todd Eastham) (2013-02-18). "Universe Has Finite Lifespan, Higgs Boson Calculations Suggest". *Huffington Post*. Reuters. Retrieved 21 February 2013. Earth will likely be long gone before any Higgs boson particles set off an apocalyptic assault on the universe

[36] Hoffman, Mark (2013-02-19). "Higgs Boson Will Destroy The Universe Eventually". *ScienceWorldReport*. Retrieved 21 February 2013.

[37] "Higgs boson will aid in creation of the universe – and how it will end". *Catholic Online/NEWS CONSORTIUM*. 2013-02-20. Retrieved 21 February 2013. [T]he Earth will likely be long gone before any Higgs boson particles set off an apocalyptic assault on the universe

[38] Salvio, Alberto (2015-04-09). "A Simple Motivated Completion of the Standard Model below the Planck Scale: Axions and Right-Handed Neutrinos".*Physics Letters B***743**: 428. arXiv:1501.03781. Bibcode:2015PhLB..743..428S.doi:10.1016/015.

[39] Boyle, Alan (2013-02-19). "Will our universe end in a 'big slurp'? Higgs-like particle suggests it might". *NBC News' Cosmic log*. Retrieved 21 February 2013. [T]he bad news is that its mass suggests the universe will end in a fast-spreading bubble of doom. The good news? It'll probably be tens of billions of years. The article quotes Fermilab's Joseph Lykken: "[T]he parameters for our universe, including the Higgs [and top quark's masses] suggest that we're just at the edge of stability, in a "metastable" state. Physicists have been contemplating such a possibility for more than 30 years. Back in 1982, physicists Michael Turner and Frank Wilczek wrote in Nature that "without warning, a bubble of true vacuum could nucleate somewhere in the universe and move outwards..."

[40] Peralta, Eyder (2013-02-19). "If Higgs Boson Calculations Are Right, A Catastrophic 'Bubble' Could End Universe". *npr – two way*. Retrieved 21 February 2013. Article cites Fermilab's Joseph Lykken: "The bubble forms through an unlikely quantum fluctuation, at a random time and place," Lykken tells us. "So in principle it could happen tomorrow, but then most likely in a very distant galaxy, so we are still safe for billions of years before it gets to us."

[41] Bezrukov; Shaposhnikov (2007-10-19). "The Standard Model Higgs boson as the inflaton". *Phys.Lett. B659 (2008) 703-706*. arXiv:0710.3755. Bibcode:2008PhLB..659..703B. doi:10.1016/j.physletb.2007.11.072.

[42] Salvio, Alberto (2013-08-09). "Higgs Inflation at NNLO after the Boson Discovery". *Phys.Lett. B727 (2013) 234-239*. arXiv:1308.2244. Bibcode:2013PhLB..727..234S. doi:10.1016/j.physletb.2013.10.042.

[43] Cole, K. (2000-12-14). "One Thing Is Perfectly Clear: Nothingness Is Perfect". *Los Angeles Times*. p. 'Science File'. Retrieved 17 January 2013. [T]he Higgs' influence (or the influence of something like it) could reach much further. For example, something like the Higgs—if not exactly the Higgs itself—may be behind many other unexplained "broken symmetries" in the universe as well ... In fact, something very much like the Higgs may have been behind the collapse of the symmetry that led to the Big Bang, which created the universe. When the forces first began to separate from their primordial sameness—taking on the distinct characters they have today—they released energy in the same way as water releases energy when it turns to ice. Except in this case, the freezing packed enough energy to blow up the universe. ... However it happened, the moral is clear: Only when the perfection shatters can everything else be born.

[44] Higgs Matters – Kathy Sykes, 30 Nove 2012

[45] Why the public should care about the Higgs Boson – Jodi Lieberman, American Physical Society (APS)

[46] Matt Strassler's blog – Why the Higgs particle matters 2 July 2012

[47] Sean Carroll (13 November 2012). *The Particle at the End of the Universe: How the Hunt for the Higgs Boson Leads Us to the Edge of a New World*. Penguin Group US. ISBN 978-1-101-60970-5.

[48] Woit, Peter (13 November 2010). "The Anderson–Higgs Mechanism". Dr. Peter Woit (Senior Lecturer in Mathematics Columbia University and Ph.D. particle physics). Retrieved 12 November 2012.

[49] Goldstone,J;Salam,Abdus;Weinberg,Steven(1962). "Broken Symmetries".*Physical Review***127**(3): 965–970. Bibc965G. doi:10.1103/PhysRev.127.965.

[50] Guralnik, G. S. (2011). "The Beginnings of Spontaneous Symmetry Breaking in Particle Physics". arXiv:1110.2253 [physics.hist-ph].

[51] Kibble,T.W.B. (2009). "Englert–Brout–Higgs–Guralnik–Hagen–Kibble Mechanism".*Scholarpedia***4**(1): 6441. B441K. doi:10.4249/scholarpedia.6441. Retrieved 2012-11-23.

[52] Kibble, T. W. B. "History of Englert–Brout–Higgs–Guralnik–Hagen–Kibble Mechanism (history)". *Scholarpedia* **4** (1): 8741. Bibcode:2009SchpJ...4.8741K. doi:10.4249/scholarpedia.8741. Retrieved 2012-11-23.

[53] The Nobel Prize in Physics 2008 – official Nobel Prize website.

[54] List of Anderson 1958–1959 papers referencing 'symmetry', at APS Journals

[55] Higgs, Peter (2010-11-24). "My Life as a Boson" (PDF). Talk given by Peter Higgs at Kings College, London, Nov 24 2010, expanding on a paper originally presented in 2001. Retrieved 17 January 2013. – the original 2001 paper can be found at: Duff and Liu, ed. (2003) [year of publication]. *2001 A Spacetime Odyssey: Proceedings of the Inaugural Conference of the Michigan Center for Theoretical Physics, Michigan, USA, 21–25 May 2001*. World Scientific. pp. 86–88. ISBN 9812382313. Retrieved 17 January 2013.

[56] Anderson, P. (1963). "Plasmons, gauge invariance and mass". *Physical Review* **130**: 439. Bibcode:1963PhRv..130..439A. doi:10.1103/PhysRev.130.439.

[57] Klein, A.; Lee, B. (1964). "Does Spontaneous Breakdown of Symmetry Imply Zero-Mass Particles?". *Physical Review Letters* **12** (10): 266. Bibcode:1964PhRvL..12..266K. doi:10.1103/PhysRevLett.12.266.

[58] Englert, François; Brout, Robert (1964). "Broken Symmetry and the Mass of Gauge Vector Mesons". *Physical Review Letters* **13** (9): 321–23. Bibcode:1964PhRvL..13..321E. doi:10.1103/PhysRevLett.13.321.

[59] Higgs, Peter (1964). "Broken Symmetries and the Masses of Gauge Bosons". *Physical Review Letters* **13** (16): 508–509. Bibcode:1964PhRvL..13..508H. doi:10.1103/PhysRevLett.13.508.

[60] Guralnik, Gerald; Hagen, C. R.; Kibble, T. W. B. (1964). "Global Conservation Laws and Massless Particles". *Physical Review Letters* **13** (20): 585–587. Bibcode:1964PhRvL..13..585G. doi:10.1103/PhysRevLett.13.585.

[61] Higgs,Peter(1964). "Broken symmetries,massless particles and gaugefields".*Physics Letters***12**(2): 132–133. Bibcode:192H. doi:10.1016/0031-9163(64)91136-9.

[62] Higgs, Peter (2010-11-24). "My Life as a Boson" (PDF). Talk given by Peter Higgs at Kings College, London, Nov 24 2010. Retrieved 17 January 2013. Gilbert ... wrote a response to [Klein and Lee's paper] saying 'No, you cannot do that in a relativistic theory. You cannot have a preferred unit time-like vector like that.' This is where I came in, because the next month was when I responded to Gilbert's paper by saying 'Yes, you can have such a thing' but only in a gauge theory with a gauge field coupled to the current.

[63] G.S. Guralnik (2011). "Gauge invariance and the Goldstone theorem – 1965 Feldafing talk". *Modern Physics Letters A* **26** (19): 1381–1392. arXiv:1107.4592. Bibcode:2011MPLA...26.1381G. doi:10.1142/S0217732311036188.

[64] Higgs, Peter (1966). "Spontaneous Symmetry Breakdown without Massless Bosons". *Physical Review* **145** (4): 1156–1163. Bibcode:1966PhRv..145.1156H. doi:10.1103/PhysRev.145.1156.

[65] Kibble,Tom(1967). "Symmetry Breaking in Non-Abelian Gauge Theories".*Physical Review***155**(5): 1554–1561. Bi54K. doi:10.1103/PhysRev.155.1554.

[66] "Guralnik, G S; Hagen, C R and Kibble, T W B (1967). Broken Symmetries and the Goldstone Theorem. Advances in Physics, vol. 2" (PDF).

[67] "Physical Review Letters – 50th Anniversary Milestone Papers". Physical Review Letters.

[68] S. Weinberg (1967). "A Model of Leptons". *Physical Review Letters* **19** (21): 1264–1266. Bibcode:1967PhRvL..19.1264W. doi:10.1103/PhysRevLett.19.1264.

[69] A. Salam (1968). N. Svartholm, ed. *Elementary Particle Physics: Relativistic Groups and Analyticity*. Eighth Nobel Symposium. Stockholm: Almquvist and Wiksell. p. 367.

[70] S.L.Glashow(1961). "Partial-symmetries of weak interactions".*Nuclear Physics***22**(4): 579–588. Bibcode:1961NucPh9G. doi:10.1016/0029-5582(61)90469-2.

[71] Ellis, John; Gaillard, Mary K.; Nanopoulos, Dimitri V. (2012). "A Historical Profile of the Higgs Boson". arXiv:1201.6045 [hep-ph].

[72] "Martin Veltman Nobel Lecture, December 12, 1999, p.391" (PDF). Retrieved 2013-10-09.

[73] Politzer, David. "The Dilemma of Attribution". *Nobel Prize lecture, 2004*. Nobel Prize. Retrieved 22 January 2013. Sidney Coleman published in Science magazine in 1979 a citation search he did documenting that essentially no one paid any attention to Weinberg's Nobel Prize winning paper until the work of 't Hooft (as explicated by Ben Lee). In 1971 interest in Weinberg's paper exploded. I had a parallel personal experience: I took a one-year course on weak interactions from Shelly Glashow in 1970, and he never even mentioned the Weinberg–Salam model or his own contributions.

[74] Coleman,Sidney(1979-12-14). "The1979Nobel Prize in Physics".*Science***206**(4424): 1290–1292. Bibcode:106.1290C. doi:10.1126/science.206.4424.1290.

[75] Letters from the Past – A PRL Retrospective (50 year celebration, 2008)

[76] Jeremy Bernstein (January 1974). "Spontaneous symmetry breaking, gauge theories, the Higgs mechanism and all that" (PDF). *Reviews of Modern Physics* **46** (1): 7. Bibcode:1974RvMP...46....7B. doi:10.1103/RevModPhys.46.7. Retrieved 2012-12-10.

[77] American Physical Society – "J. J. Sakurai Prize for Theoretical Particle Physics".

[78] Merali, Zeeya (4 August 2010). "Physicists get political over Higgs". *Nature Magazine*. Retrieved 28 December 2011.

[79] Close, Frank (2011). *The Infinity Puzzle: Quantum Field Theory and the Hunt for an Orderly Universe*. Oxford: Oxford University Press. ISBN 978-0-19-959350-7.

[80] G.S. Guralnik (2009). "The History of the Guralnik, Hagen and Kibble development of the Theory of Spontaneous Symmetry Breaking and Gauge Particles". *International Journal of Modern Physics A* **24** (14): 2601–2627. arXiv:0907.3466. Bibcode:2009IJMPA..24.2601G. doi:10.1142/S0217751X09045431.

[81] Peskin, Michael E.; Schroeder, Daniel V. (1995). *Introduction to Quantum Field Theory*. Reading, MA: Addison-Wesley Publishing Company. pp. 717–719 and 787–791. ISBN 0-201-50397-2.

[82] Peskin & Schroeder 1995, pp. 715–716

[83] Gunion, John (2000). *The Higgs Hunter's Guide* (illustrated, reprint ed.). Westview Press. pp. 1–3. ISBN 9780738203058.

[84] Lisa Randall, *Warped Passages: Unraveling the Mysteries of the Universe's Hidden Dimensions*, p.286: "People initially thought of tachyons as particles travelling faster than the speed of light...But we now know that a tachyon indicates an instability in a theory that contains it. Regrettably for science fiction fans, tachyons are not real physical particles that appear in nature."

[85] Sen,Ashoke(April2002). "Rolling Tachyon".*J.High Energy Phys.***2002**(0204): 048. arXiv:hep-th/0203211. Bibcod048S. doi:10.1088/1126-6708/2002/04/048.

[86] Kutasov, David; Marino, Marcos & Moore, Gregory W. (2000). "Some exact results on tachyon condensation in string field theory". *JHEP* **0010**: 045.

[87] Aharonov, Y.; Komar, A.; Susskind, L. (1969). "Superluminal Behavior, Causality, and Instability". *Phys. Rev.* (American Physical Society) **182** (5): 1400–1403. Bibcode:1969PhRv..182.1400A. doi:10.1103/PhysRev.182.1400.

[88] Feinberg,Gerald(1967). "Possibility of Faster-Than-Light Particles".*Physical Review***159**(5): 1089–1105. Bibcode:19689F. doi:10.1103/PhysRev.159.1089.

[89] Michael E. Peskin and Daniel V. Schroeder (1995). *An Introduction to Quantum Field Theory*, Perseus books publishing.

[90] Flatow, Ira (6 July 2012). "At Long Last, The Higgs Particle... Maybe". *NPR*. Retrieved 10 July 2012.

[91] "Explanatory Figures for the Higgs Boson Exclusion Plots". *ATLAS News*. CERN. Retrieved 6 July 2012.

[92] Bernardi, G.; Carena, M.; Junk, T. (2012). "Higgs Bosons: Theory and Searches" (PDF). p. 7.

[93] Lykken, Joseph D. (2009). "Beyond the Standard Model". *Proceedings of the 2009 European School of High-Energy Physics, Bautzen, Germany, 14 – 27 June 2009*. arXiv:1005.1676.

[94] Plehn, Tilman (2012). *Lectures on LHC Physics*. Lecture Notes is Physics **844**. Springer. Sec. 1.2.2. arXiv:0910.4122. ISBN 3642240399.

[95] "LEP Electroweak Working Group".

[96] Peskin, Michael E.; Wells, James D. (2001). "How Can a Heavy Higgs Boson be Consistent with the Precision Electroweak Measurements?".*Physical Review D***64**(9): 093003. arXiv:hep-ph/0101342. Bibcode:2001PhRvD..64i3003P.doi:10.11093003.

[97] Baglio, Julien; Djouadi, Abdelhak (2011). "Higgs production at the lHC". *Journal of High Energy Physics* **1103** (3): 055. arXiv:1012.0530. Bibcode:2011JHEP...03..055B. doi:10.1007/JHEP03(2011)055.

[98] Baglio, Julien; Djouadi, Abdelhak (2010). "Predictions for Higgs production at the Tevatron and the associated uncertainties". *Journal of High Energy Physics***1010**(10): 063. arXiv:1003.4266. Bibcode:2010JHEP...10..064B.doi:10.1007/JH)064.

[99] Teixeira-Dias (LEP Higgs working group), P. (2008). "Higgs boson searches at LEP". *Journal of.Physics: Conference Series* **110** (4): 042030. arXiv:0804.4146. Bibcode:2008JPhCS.110d2030T. doi:10.1088/1742-6596/110/4/042030.

[100] "Collisions". *LHC Machine Outreach*. CERN. Retrieved 26 July 2012.

[101] Asquith, Lily (22 June 2012). "Why does the Higgs decay?". *Life and Physics* (London: The Guardian). Retrieved 14 August 2012.

[102] "Higgs bosons: theory and searches" (PDF). *PDGLive*. Particle Data Group. 12 July 2012. Retrieved 15 August 2012.

[103] Branco, G. C.; Ferreira, P.M.; Lavoura, L.; Rebelo, M.N.; Sher, Marc; Silva, João P. (July 2012). "Theory and phenomenology of two-Higgs-doublet models". *Physics Reports* (Elsevier) **516** (1): 1–102. arXiv:1106.0034. Bibcode:2012PhR...516....1B. doi:10.1016/j.physrep.2012.02.002.

[104] Csaki, C.; Grojean, C.; Pilo, L.; Terning, J. (2004). "Towards a realistic model of Higgsless electroweak symmetry breaking". *Physical Review Letters***92**(10): 101802. arXiv:hep-ph/0308038. Bibcode:2004PhRvL..92j1802C.doi1802.PMID15089195.

[105] Csaki, C.; Grojean, C.; Pilo, L.; Terning, J.; Terning, John (2004). "Gauge theories on an interval: Unitarity without a Higgs". *Physical Review D***69**(5): 055006. arXiv:hep-ph/0305237. Bibcode:2004PhRvD..69e5006C.doi:10.1103/PhysRevD.66.

[106] "The Hierarchy Problem: why the Higgs has a snowball's chance in hell". Quantum Diaries. 2012-07-01. Retrieved 19 March 2013.

[107] "The Hierarchy Problem | Of Particular Significance". Profmattstrassler.com. Retrieved 2013-10-09.

[108] "Collisions". *LHC Machine Outreach*. CERN. Retrieved 26 July 2012.

[109] "Hunt for Higgs boson hits key decision point". MSNBC. 2012-12-06. Retrieved 2013-01-19.

[110] Worldwide LHC Computing Grid main page 14 November 2012: *"[A] global collaboration of more than 170 computing centres in 36 countries ... to store, distribute and analyse the ~25 Petabytes (25 million Gigabytes) of data annually generated by the Large Hadron Collider"*

[111] What is the Worldwide LHC Computing Grid? (Public 'About' page) 14 November 2012: *"Currently WLCG is made up of more than 170 computing centers in 36 countries...The WLCG is now the world's largest computing grid"*

[112] W.-M.Yao et al. (2006). "Review of Particle Physics"(PDF).*Journal of Physics G***33**: 1. arXiv:astro-ph/0601168. B.1Y. doi:10.1088/0954-3899/33/1/001.

[113] The CDF Collaboration, the D0 Collaboration, the Tevatron New Physics, Higgs Working Group (2012). "Updated Combination of CDF and D0 Searches for Standard Model Higgs Boson Production with up to 10.0 fb^{-1} of Data". arXiv:1207.0449 [hep-ex].

[114] "Interim Summary Report on the Analysis of the 19 September 2008 Incident at the LHC" (PDF). CERN. 15 October 2008. EDMS 973073. Retrieved 28 September 2009.

[115] "CERN releases analysis of LHC incident" (Press release). CERN Press Office. 16 October 2008. Retrieved 28 September 2009.

[116] "LHC to restart in 2009" (Press release). CERN Press Office. 5 December 2008. Retrieved 8 December 2008.

[117] "LHC progress report". *The Bulletin*. CERN. 3 May 2010. Retrieved 7 December 2011.

[118] "ATLAS experiment presents latest Higgs search status". *ATLAS homepage*. CERN. 13 December 2011. Retrieved 13 December 2011.

[119] Taylor, Lucas (13 December 2011). "CMS search for the Standard Model Higgs Boson in LHC data from 2010 and 2011". *CMS public website*. CERN. Retrieved 13 December 2011.

[120] Overbye, D. (5 March 2013). "Chasing The Higgs Boson". *The New York Times*. Retrieved 2013-03-05.

[121] "ATLAS and CMS experiments present Higgs search status" (Press release). CERN Press Office. 13 December 2011. Retrieved 14 September 2012. the statistical significance is not large enough to say anything conclusive. As of today what we see is consistent either with a background fluctuation or with the presence of the boson. Refined analyses and additional data delivered in 2012 by this magnificent machine will definitely give an answer

[122] "WLCG Public Website". CERN. Retrieved 29 October 2012.

[123] CMS collaboration (2014). "Precise determination of the mass of the Higgs boson and tests of compatibility of its couplings with the standard model predictions using proton collisions at 7 and 8 TeV". arXiv:1412.8662.

[124] ATLAS collaboration (2014). "Measurements of Higgs boson production and couplings in the four-lepton channel in pp collisions at center-of-mass energies of 7 and 8 TeV with the ATLAS detector". arXiv:1408.5191.

[125] ATLAS collaboration (2014). "Measurement of Higgs boson production in the diphoton decay channel in pp collisions at center-of-mass energies of 7 and 8 TeV with the ATLAS detector". arXiv:1408.7084.

[126] "Press Conference: Update on the search for the Higgs boson at CERN on 4 July 2012". Indico.cern.ch. 22 June 2012. Retrieved 4 July 2012.

[127] "CERN to give update on Higgs search". CERN. 22 June 2012. Retrieved 2 July 2011.

[128] "Scientists analyse global Twitter gossip around Higgs boson discovery". *phys.org (from arXiv)*. 2013-01-23. Retrieved 6 February 2013. – stated to be *" the first time scientists have been able to analyse the dynamics of social media on a global scale before, during and after the announcement of a major scientific discovery."* For the paper itself see: De Domenico, M.; Lima, A.; Mougel, P.; Musolesi, M. (2013). "The Anatomy of a Scientific Gossip". arXiv:1301.2952. Bibcode:2013NatSR...3E2980D. doi:10.1038/srep02980.

[129] "Higgs boson particle results could be a quantum leap". Times LIVE. 28 June 2012. Retrieved 4 July 2012.

[130] CERN prepares to deliver Higgs particle findings, Australian Broadcasting Corporation. Retrieved 4 July 2012.

[131] "God Particle Finally Discovered? Higgs Boson News At Cern Will Even Feature Scientist It's Named After". Huffingtonpost.co.uk. Retrieved 2013-01-19.

[132] Our Bureau (2012-07-04). "Higgs on way, theories thicken". Calcutta, India: Telegraphindia.com. Retrieved 2013-01-19.

[133] Thornhill, Ted (2013-07-03). "God Particle Finally Discovered? Higgs Boson News At Cern Will Even Feature Scientist It's Named After". *Huffington Post*. Retrieved 23 July 2013.

[134] Cooper, Rob (2013-07-01) [updated subsequently]. "God particle is 'found': Scientists at Cern expected to announce on Wednesday Higgs boson particle has been discovered". *Daily Mail* (London). Retrieved 23 July 2013. - States that "*"Five leading theoretical physicists have been invited to the event on Wednesday - sparking speculation that the particle has been discovered."*, including Higgs and Englert, and that Kibble - who was invited but unable to attend - "told the Sunday Times: 'My guess is that is must be a pretty positive result for them to be asking us out there'."

[135] Adrian Cho (13 July 2012). "Higgs Boson Makes Its Debut After Decades-Long Search". *Science* **337** (6091): 141–143. doi:10.1126/science.337.6091.141. PMID 22798574.

[136] CMS collaboration (2012). "Observation of a new boson at a mass of 125 GeV with the CMS experiment at the LHC". *Physics Letters B* **716** (1): 30–61. arXiv:1207.7235. Bibcode:2012PhLB..716...30C. doi:10.1016/j.physletb.2012.08.021.

[137] Taylor, Lucas (4 July 2012). "Observation of a New Particle with a Mass of 125 GeV". *CMS Public Website*. CERN. Retrieved 4 July 2012.

[138] "Latest Results from ATLAS Higgs Search". *ATLAS News*. CERN. 4 July 2012. Retrieved 4 July 2012.

[139] ATLAS collaboration (2012). "Observation of a New Particle in the Search for the Standard Model Higgs Boson with the ATLAS Detector at the LHC".*Physics Letters B***716**(1): 1–29. arXiv:1207.7214. Bibcode:2012PhLB..716....1A.doi:10.8.020.

[140] Gillies, James (23 July 2012). "LHC 2012 proton run extended by seven weeks". *CERN bulletin*. Retrieved 29 August 2012.

[141] "Higgs boson behaving as expected". *3 News NZ*. 15 November 2012.

[142] Strassler, Matt (2012-11-14). "Higgs Results at Kyoto". *Of Particular Significance: Conversations About Science with Theoretical Physicist Matt Strassler*. Prof. Matt Strassler's personal particle physics website. Retrieved 10 January 2013. ATLAS and CMS only just co-discovered this particle in July ... We will not know after today whether it is a Higgs at all, whether it is a Standard Model Higgs or not, or whether any particular speculative idea...is now excluded. [...] Knowledge about nature does not come easy. We discovered the top quark in 1995, and we are still learning about its properties today... we will still be learning important things about the Higgs during the coming few decades. We've no choice but to be patient.

[143] Sample, Ian (14 November 2012). “Higgs particle looks like a bog Standard Model boson, say scientists”. *The Guardian* (London). Retrieved 15 November 2012.

[144] “CERN experiments observe particle consistent with long-sought Higgs boson”. CERN press release. 4 July 2012. Retrieved 4 July 2012.

[145] “Person Of The Year 2012”. *Time*. 19 December 2012.

[146] “Higgs Boson Discovery Has Been Confirmed”. Forbes. Retrieved 2013-10-09.

[147] Slate Video Staff (2012-09-11). “Higgs Boson Confirmed; CERN Discovery Passes Test”. Slate.com. Retrieved 2013-10-09.

[148] “The Year Of The Higgs, And Other Tiny Advances In Science”. NPR. 2013-01-01. Retrieved 2013-10-09.

[149] “Confirmed: the Higgs boson does exist”. *The Sydney Morning Herald*. 4 July 2012.

[150] “AP CERN chief: Higgs boson quest could wrap up by midyear”. *MSNBC*. Associated Press. 2013-01-27. Retrieved 20 February 2013. Rolf Heuer, director of [CERN], said he is confident that “towards the middle of the year, we will be there.” – Interview by AP, at the World Economic Forum, 26 Jan 2013.

[151] Boyle, Alan (2013-02-16). “Will our universe end in a 'big slurp'? Higgs-like particle suggests it might”. *NBCNews.com – cosmic log*. Retrieved 20 February 2013. 'it’s going to take another few years’ after the collider is restarted to confirm definitively that the newfound particle is the Higgs boson.

[152] Gillies, James (2013-03-06). “A question of spin for the new boson”. CERN. Retrieved 7 March 2013.

[153] Adam Falkowski (writing as 'Jester') (2013-02-27). “When shall we call it Higgs?". Résonaances particle physics blog. Retrieved 7 March 2013.

[154] CMS Collaboration (February 2013). “Study of the Mass and Spin-Parity of the Higgs Boson Candidate via Its Decays to Z Boson Pairs”. *Phys. Rev. Lett.* (American Physical Society) **110** (8): 081803. arXiv:1212.6639. Bibcode:2013PhRvL.110h1803C. doi:10.1103/PhysRevLett.110.081803. Retrieved 15 September 2014.

[155] ATLAS Collaboration (7 October 2013). “Evidence for the spin-0 nature of the Higgs boson using ATLAS data”. *Phys. Lett. B* (American Physical Society) **726** (1-3): 120–144. Bibcode:2013PhLB..726..120A. doi:10.1016/j.physletb.2013.08.026. Retrieved 15 September 2014.

[156] “Higgs-like Particle in a Mirror”. American Physical Society. Retrieved 26 February 2013.

[157] The CMS Collaboration (2014-06-22). “Evidence for the direct decay of the 125 GeV Higgs boson to fermions”. Nature Publishing Group doi= 10.1038/nphys3005.

[158] Adam Falkowski (writing as 'Jester') (2012-12-13). “Twin Peaks in ATLAS”. Résonaances particle physics blog. Retrieved 24 February 2013.

[159] Liu, G. Z.; Cheng, G. (2002). “Extension of the Anderson-Higgs mechanism”. *Physical Review B* **65** (13): 132513. arXiv:cond-mat/0106070. Bibcode:2002PhRvB..65m2513L. doi:10.1103/PhysRevB.65.132513.

[160] Editorial (2012-03-21). “Mass appeal: As physicists close in on the Higgs boson, they should resist calls to change its name”. *Nature*. 483, 374 (7390): 374. Bibcode:2012Natur.483..374.. doi:10.1038/483374a. Retrieved 21 January 2013.

[161] Becker, Kate (2012-03-29). “A Higgs by Any Other Name”. “NOVA” (PBS) physics. Retrieved 21 January 2013.

[162] “Frequently Asked Questions: The Higgs!". *The Bulletin*. CERN. Retrieved 18 July 2012.

[163] Woit’s physics blog *“Not Even Wrong”*: Anderson on Anderson-Higgs 2013-04-13

[164] Sample, Ian (2012-07-04). “Higgs boson’s many great minds cause a Nobel prize headache”. *The Guardian* (London). Retrieved 23 July 2013.

[165] “Rochester’s Hagen Sakurai Prize Announcement” (Press release). University of Rochester. 2010.

[166] *C.R. Hagen Sakurai Prize Talk* (YouTube). 2010.

[167] Cho, A (2012-09-14). "Particle physics. Why the 'Higgs'?" (PDF). *Science* **337** (6100): 1287. doi:10.1126/science.337. .PMID 22984044. Lee ... apparently used the term 'Higgs Boson' as early as 1966... but what may have made the term stick is a seminal paper Steven Weinberg...published in1967...Weinberg acknowledged the mix-up in an essay in the*New York Review of Books*in May2012. (See also the original article in*New York Review of Books* [168] and Frank Close's 2011 book *The Infinity Puzzle*[79]:372)

[168] Weinberg, Steven (2012-05-10). "The Crisis of Big Science". *The New York Review of Books* (footnote 1). Retrieved 12 February 2013.

[169] Examples of early papers using the term "Higgs boson" include 'A phenomenological profile of the Higgs boson' (Ellis, Gaillard and Nanopoulos, 1976), 'Weak interaction theory and neutral currents' (Bjorken, 1977), and 'Mass of the Higgs boson' (Wienberg, received 1975)

[170] Leon Lederman; Dick Teresi (2006). *The God Particle: If the Universe Is the Answer, What Is the Question?*. Houghton Mifflin Harcourt. ISBN 0-547-52462-5.

[171] Kelly Dickerson (September 8, 2014). "Stephen Hawking Says 'God Particle' Could Wipe Out the Universe". livescience.com.

[172] Jim Baggott (2012). *Higgs: The invention and discovery of the 'God Particle'*. Oxford University Press. ISBN 978-0-19-165003-1.

[173] Scientific American Editors (2012). *The Higgs Boson: Searching for the God Particle*. Macmillan. ISBN 978-1-4668-2413-3.

[174] Ted Jaeckel (2007). *The God Particle: The Discovery and Modeling of the Ultimate Prime Particle*. Universal-Publishers. ISBN 978-1-58112-959-5.

[175] Aschenbach, Joy (1993-12-05). "No Resurrection in Sight for Moribund Super Collider : Science: Global financial partnerships could be the only way to salvage such a project. Some feel that Congress delivered a fatal blow". *Los Angeles Times*. Retrieved 16 January 2013. 'We have to keep the momentum and optimism and start thinking about international collaboration,' said Leon M. Lederman, the Nobel Prize-winning physicist who was the architect of the super collider plan

[176] "A Supercompetition For Illinois". *Chicago Tribune*. 1986-10-31. Retrieved 16 January 2013. The SSC, proposed by the U.S. Department of Energy in 1983, is a mind-bending project ... this gigantic laboratory ... this titanic project

[177] Diaz, Jesus (2012-12-15). "This Is [The] World's Largest Super Collider That Never Was". *Gizmodo*. Retrieved 16 January 2013. ...this titanic complex...

[178] Abbott, Charles (June 1987). "Illinois Issues journal, June 1987". p. 18. Lederman, who considers himself an unofficial propagandist for the super collider, said the SSC could reverse the physics brain drain in which bright young physicists have left America to work in Europe and elsewhere.

[179] Kevles, Dan. "Good-bye to the SSC: On the Life and Death of the Superconducting Super Collider" (PDF). *California Institute of Technology: "Engineering & Science"*. 58 no. 2 (Winter 1995): 16–25. Retrieved 16 January 2013. Lederman, one of the principal spokesmen for the SSC, was an accomplished high-energy experimentalist who had made Nobel Prize-winning contributions to the development of the Standard Model during the 1960s (although the prize itself did not come until 1988). He was a fixture at congressional hearings on the collider, an unbridled advocate of its merits.

[180] Calder, Nigel (2005). *Magic Universe:A Grand Tour of Modern Science*. pp. 369–370. ISBN 9780191622359. The possibility that the next big machine would create the Higgs became a carrot to dangle in front of funding agencies and politicians. A prominent American physicist, Leon lederman *[sic]*, advertised the Higgs as The God Particle in the title of a book published in 1993 ...Lederman was involved in a campaign to persuade the US government to continue funding the Superconducting Super Collider... the ink was not dry on Lederman's book before the US Congress decided to write off the billions of dollars already spent

[181] Lederman, Leon (1993). *The God Particle If the Universe Is the Answer, What Is the Question?* (PDF). Dell Publishing. p. Chapter 2, Page 2. ISBN 0-385-31211-3. Retrieved 30 July 2015.

[182] Alister McGrath, Higgs boson: the particle of faith, *The Daily Telegraph*, Published 15 December 2011. Retrieved 15 December 2011.

[183] Sample, Ian (3 March 2009). "Father of the God particle: Portrait of Peter Higgs unveiled". London: The Guardian. Retrieved 24 June 2009.

[184] Chivers, Tom (2011-12-13). "How the 'God particle' got its name". *The Telegraph* (London). Retrieved 2012-12-03.

[185] Key scientist sure "God particle" will be found soon Reuters news story. 7 April 2008.

[186] "Interview: the man behind the 'God particle'", New Scientist 13 September 2008, pp. 44–5 (original interview in the Guardian: Father of the 'God Particle', June 30, 2008)

[187] Sample, Ian (2010). *Massive: The Hunt for the God Particle*. pp. 148–149 and 278–279. ISBN 9781905264957.

[188] Cole, K. (2000-12-14). "One Thing Is Perfectly Clear: Nothingness Is Perfect". *Los Angeles Times*. p. 'Science File'. Retrieved 17 January 2013. Consider the early universe–a state of pure, perfect nothingness; a formless fog of undifferentiated stuff ... 'perfect symmetry' ... What shattered this primordial perfection? One likely culprit is the so-called Higgs field ... Physicist Leon Lederman compares the way the Higgs operates to the biblical story of Babel [whose citizens] all spoke the same language ... Like God, says Lederman, the Higgs differentiated the perfect sameness, confusing everyone (physicists included) ... [Nobel Prizewinner Richard] Feynman wondered why the universe we live in was so obviously askew ... Perhaps, he speculated, total perfection would have been unacceptable to God. And so, just as God shattered the perfection of Babel, 'God made the laws only nearly symmetrical'

[189] Lederman, p. 22 *et seq*:

> "Something we cannot yet detect and which, one might say, has been put there to test and confuse us ... The issue is whether physicists will be confounded by this puzzle or whether, in contrast to the unhappy Babylonians, we will continue to build the tower and, as Einstein put it, 'know the mind of God'."
>
> "And the Lord said, Behold the people are un-confounding my confounding. And the Lord sighed and said, Go to, let us go down, and there give them the God Particle so that they may see how beautiful is the universe I have made".

[190] Sample, Ian (12 June 2009). "Higgs competition: Crack open the bubbly, the God particle is dead". *The Guardian* (London). Retrieved 4 May 2010.

[191] Gordon, Fraser (5 July 2012). "Introducing the higgson". *physicsworld.com*. Retrieved 25 August 2012.

[192] Wolchover, Natalie (2012-07-03). "Higgs Boson Explained: How 'God Particle' Gives Things Mass". *Huffington Post*. Retrieved 21 January 2013.

[193] Oliver, Laura (2012-07-04). "Higgs boson: how would you explain it to a seven-year-old?". *The Guardian* (London). Retrieved 21 January 2013.

[194] Zimmer, Ben (2012-07-15). "Higgs boson metaphors as clear as molasses". *The Boston Globe*. Retrieved 21 January 2013.

[195] "The Higgs particle: an analogy for Physics classroom (section)". www.lhc-closer.es (a collaboration website of LHCb physicist Xabier Vidal and High School Teachers at CERN educator Ramon Manzano). Retrieved 2013-01-09.

[196] Flam, Faye (2012-07-12). "Finally – A Higgs Boson Story Anyone Can Understand". *The Philadelphia Inquirer (philly.com)*. Retrieved 21 January 2013.

[197] Sample, Ian (2011-04-28). "How will we know when the Higgs particle has been detected?". *The Guardian* (London). Retrieved 21 January 2013.

[198] Miller, David. "A quasi-political Explanation of the Higgs Boson; for Mr Waldegrave, UK Science Minister 1993". Retrieved 10 July 2012.

[199] Kathryn Grim. "Ten things you may not know about the Higgs boson". Symmetry Magazine. Retrieved 10 July 2012.

[200] David Goldberg, Associate Professor of Physics, Drexel University (2010-10-17). "What's the Matter with the Higgs Boson?". io9.com "Ask a physicist". Retrieved 21 January 2013.

[201] The Nobel Prize in Physics 1979 – official Nobel Prize website.

[202] The Nobel Prize in Physics 1999 – official Nobel Prize website.

[203] – official Nobel Prize website.

[204] Daigle, Katy (10 July 2012). "India: Enough about Higgs, let's discuss the boson". *AP News*. Retrieved 10 July 2012.

[205] Bal, Hartosh Singh (19 September 2012). "The Bose in the Boson". New York Times. Retrieved 21 September 2012.

[206] Alikhan, Anvar (16 July 2012). "The Spark In A Crowded Field". *Outlook India*. Retrieved 10 July 2012.

[207] Peskin & Schroeder 1995, Chapter 20

23.11 Further reading

- Nambu, Yoichiro; Jona-Lasinio, Giovanni (1961). "Dynamical Model of Elementary Particles Based on an Analogy with Superconductivity".*Physical Review***122**: 345–358. Bibcode:1961PhRv..122..345N.doi:10.11.
- Klein, Abraham; Lee, Benjamin W. (1964). "Does Spontaneous Breakdown of Symmetry Imply Zero-Mass Particles?". *Physical Review Letters* **12** (10): 266. Bibcode:1964PhRvL..12..266K. doi:10.1103/PhysRevLett.12.266.
- Anderson,Philip W.(1963). "Plasmons,Gauge Invariance,and Mass".*Physical Review***130**: 439. Bibcode:. doi:10.1103/PhysRev.130.439.
- Gilbert, Walter (1964). "Broken Symmetries and Massless Particles". *Physical Review Letters* **12** (25): 713. Bibcode:1964PhRvL..12..713G. doi:10.1103/PhysRevLett.12.713.
- Higgs, Peter (1964). "Broken Symmetries, Massless Particles and Gauge Fields". *Physics Letters* **12** (2): 132–133. Bibcode:1964PhL....12..132H. doi:10.1016/0031-9163(64)91136-9.
- Guralnik, Gerald S.; Hagen, C.R.; Kibble, Tom W.B. (1968). "Broken Symmetries and the Goldstone Theorem". In R.L. Cool and R.E. Marshak. *Advances in Physics, Vol. 2*. Interscience Publishers. pp. 567–708. ISBN 978-0470170571.

23.12 External links

23.12.1 Popular science, mass media, and general coverage

- Hunting the Higgs Boson at C.M.S. Experiment, at CERN
- The Higgs Boson" by the CERN exploratorium.
- "Particle Fever", documentary film about the search for the Higgs Boson.
- "The Atom Smashers", documentary film about the search for the Higgs Boson at Fermilab.
- Collected Articles at the *Guardian*
- Video (04:38) – CERN Announcement on 4 July 2012, of the discovery of a particle which is suspected will be a Higgs Boson.
- Video1 (07:44) + Video2 (07:44) – Higgs Boson Explained by CERN Physicist, Dr. Daniel Whiteson (16 June 2011).
- HowStuffWorks: What exactly is the Higgs Boson?
- Carroll, Sean. "Higgs Boson with Sean Carroll". *Sixty Symbols*. University of Nottingham.
- Overbye, Dennis (2013-03-05). "Chasing the Higgs Boson: How 2 teams of rivals at CERN searched for physics' most elusive particle". *New York Times Science pages*. Retrieved 22 July 2013. - New York Times "behind the scenes" style article on the Higgs' search at ATLAS and CMS
- The story of the Higgs theory by the authors of the PRL papers and others closely associated:
  - Higgs, Peter (2010). "My Life as a Boson" (PDF). Talk given at Kings College, London, Nov 24 2010. Retrieved 17 January 2013. (also:)
  - Kibble, Tom (2009). "Englert–Brout–Higgs–Guralnik–Hagen–Kibble mechanism (history)". Scholarpedia. Retrieved 17 January 2013. (also:)

  - Guralnik, Gerald (2009). "The History of the Guralnik, Hagen and Kibble development of the Theory of Spontaneous Symmetry Breaking and Gauge Particles". *International Journal of Modern Physics A* **24** (14): 2601–2627. arXiv:0907.3466. Bibcode:2009IJMPA..24.2601G. doi:10.1142/S0217751X09045431., Guralnik, Gerald (2011). "The Beginnings of Spontaneous Symmetry Breaking in Particle Physics. Proceedings of the DPF-2011 Conference, Providence, RI, 8–13 August 2011". arXiv:1110.2253v1 [physics.hist-ph]., and Guralnik, Gerald (2013). "Heretical Ideas that Provided the Cornerstone for the Standard Model of Particle Physics". SPG MITTEILUNGEN March 2013, No. 39, (p. 14), and Talk at Brown University about the 1964 PRL papers
  - Philip Anderson (not one of the PRL authors) on symmetry breaking in superconductivity and its migration into particle physics and the PRL papers

- Cartoon about the search
- Cham, Jorge (2014-02-19). "True Tales from the Road: The Higgs Boson Re-Explained". *Piled Higher and Deeper*. Retrieved 2014-02-25.

23.12.2 Significant papers and other

- Observation of a new particle in the search for the Standard Model Higgs Boson with the ATLAS detector at the LHC
- Observation of a new Boson at a mass of 125 GeV with the CMS experiment at the LHC
- Particle Data Group: Review of searches for Higgs Bosons.
- 2001, a spacetime odyssey: proceedings of the Inaugural Conference of the Michigan Center for Theoretical Physics : Michigan, USA, 21–25 May 2001, (p.86 – 88), ed. Michael J. Duff, James T. Liu, ISBN 978-981-238-231-3, containing Higgs' story of the Higgs Boson.
- A.A. Migdal & A.M. Polyakov, *Spontaneous Breakdown of Strong Interaction Symmetry and the Absence of Massless Particles*, Sov.J.-JETP 24,91 (1966) - example of a 1966 Russian paper on the subject.

23.12.3 Introductions to the field

- Spontaneous symmetry breaking, gauge theories, the Higgs mechanism and all that (Bernstein, *Reviews of Modern Physics* Jan 1974) - an introduction of 47 pages covering the development, history and mathematics of Higgs theories from around 1950 to 1974.

Chapter 24

Faddeev–Popov ghost

In physics, **Faddeev–Popov ghosts** (also called **gauge ghosts** or **ghost fields**) are additional fields which are introduced into gauge quantum field theories to maintain the consistency of the path integral formulation. They are named after Ludvig Faddeev and Victor Popov.[1][2]

There is also a more general meaning of the word "**ghost**" in theoretical physics, which is discussed below (see *general ghosts in theoretical physics*).

24.1 Overcounting in Feynman path integrals

The necessity for Faddeev–Popov ghosts follows from the requirement that in the path integral formulation, quantum field theories should yield unambiguous, non-singular solutions. This is not possible when a gauge symmetry is present since there is no procedure for selecting any one solution from a range of physically equivalent solutions, all related by a gauge transformation. The problem stems from the path integrals overcounting field configurations related by gauge symmetries, since those correspond to the same physical state; the measure of the path integrals contains a factor which does not allow obtaining various results directly from the original action using the regular methods (e.g., Feynman diagrams). It is possible, however, to modify the action, such that the regular methods will be applicable by adding some additional fields, which break the gauge symmetry, which are called the *ghost fields*. This technique is called the "Faddeev–Popov procedure" (see also BRST quantization). The ghost fields are a computational tool in that they do not correspond to any real particles in external states: they *only* appear as virtual particles in Feynman diagrams – or as the *absence* of some gauge configurations. However they are necessary to preserve unitarity.

The exact form or formulation of ghosts is dependent on the particular gauge chosen, although the same physical results are obtained with all the gauges. The Feynman-'t Hooft gauge is usually the simplest gauge for this purpose, and is assumed for the rest of this article.

24.2 Spin-statistics relation violated

The Faddeev–Popov ghosts violate the spin-statistics relation, which is another reason why they are often regarded as "non-physical" particles.

For example, in Yang–Mills theories (such as quantum chromodynamics) the ghosts are complex scalar fields (spin 0), but they anti-commute (like fermions).

In general, anti-commuting ghosts are associated with fermionic symmetries, while commuting ghosts are associated with bosonic symmetries.

24.3 Gauge fields and associated ghost fields

Every gauge field has an associated ghost, and where the gauge field acquires a mass via the Higgs mechanism, the associated ghost field acquires the same mass (in the Feynman-'t Hooft gauge only, not true for other gauges).

24.4 Appearance in Feynman diagrams

In Feynman diagrams the ghosts appear as closed loops wholly composed of 3-vertices, attached to the rest of the diagram via a gauge particle at each 3-vertex. Their contribution to the S-matrix is exactly cancelled (in the Feynman-'t Hooft gauge) by a contribution from a similar loop of gauge particles with only 3-vertex couplings or gauge attachments to the rest of the diagram.[3] (A loop of gauge particles not wholly composed of 3-vertex couplings is not cancelled by ghosts.) The opposite sign of the contribution of the ghost and gauge loops is due to them having opposite fermionic/bosonic natures. (Closed fermion loops have an extra −1 associated with them; bosonic loops don't.)

24.5 Ghost field Lagrangian

The Lagrangian for the ghost fields $c^a(x)$ in Yang–Mills theories (where a is an index in the adjoint representation of the gauge group) is given by

$$\mathcal{L}_{\text{ghost}} = \partial_\mu \bar{c}^a \partial^\mu c^a + g f^{abc} \left(\partial^\mu \bar{c}^a\right) A^b_\mu c^c \ .$$

The first term is a kinetic term like for regular complex scalar fields, and the second term describes the interaction with the gauge fields. Note that in *abelian* gauge theories (such as quantum electrodynamics) the ghosts do not have any effect since $f^{abc} = 0$ and, consequently, the ghost particles do not interact with the gauge fields.

24.6 General ghosts in theoretical physics

The Faddeev–Popov ghosts are sometimes referred to as “good ghosts”. The “bad ghosts” represent another, more general meaning of the word “ghost” in theoretical physics: states of negative norm—or fields with the wrong sign of the kinetic term, such as Pauli–Villars ghosts—whose existence allows the probabilities to be negative thus violating unitarity.

24.7 Changing the symmetry

Ghost particles could obtain the symmetry or break it in gauge fields. The “good ghost” particles actually obtain the symmetry by unchanging the “gauge fixing lagrangian” in a gauge transformation, while bad ghost particles break the symmetry by bringing in the non-abelian G-matrix which does change the symmetry, and this was the main reason to introduce the Gauge covariant and contravariant derivatives.

24.8 References

[1] L. D. Faddeev and V. N. Popov, (1967). “Feynman Diagrams for the Yang-Mills Field”, *Phys. Lett.* **B25** 29.

[2] W. F. Chen. Quantum Field Theory and Differential Geometry

[3] R. P. Feynman, (1963). “Quantum Theory of Gravitation”, *Acta Physica Polonica* **24**, 697–722. Feynman discovered empirically that “boxing” and simply dismissing these diagrams restored unitarity. "*Because, unfortunately, I also discovered in the*

process that the trouble is present in the Yang–Mills theory; and, secondly, I have incidentally discovered a tree–ring connection which is of very great interest and importance in the meson theories and so on. And so I'm stuck to have to continue this investigation, and of course you appreciate that this is the secret reason for doing any work, no matter how absurd and irrational and academic it looks: we all realize that no matter how small a thing is, if it has physical interest and is thought about carefully enough, you're bound to think of something that's good for something else."

24.9 External links

- Scholarpedia
- Copeland, Ed; Padilla, Antonio. "Ghost Particles". *Sixty Symbols*. Brady Haran for the University of Nottingham.

Chapter 25

Superpartner

In particle physics, a **Superpartner** (also **Sparticle**) is a hypothetical elementary particle. Supersymmetry is one of the synergistic theories in current high-energy physics that predicts the existence of these "shadow" particles.[1][2]

The word *superpartner* is a portmanteau of *supersymmetry* and *partner*. The word *sparticle* is a portmanteau of *supersymmetry* and *particle*.

25.1 Theoretical predictions

According to the supersymmetry theory, each fermion should have a partner boson, the fermion's superpartner, and each boson should have a partner fermion. Exact *unbroken* supersymmetry would predict that a particle and its superpartners would have the same mass. No superpartners of the Standard Model particles have yet been found. This may indicate that supersymmetry is incorrect, or it may also be the result of the fact that supersymmetry is not an exact, *unbroken* symmetry of nature. If superpartners are found, their masses would indicate the scale at which supersymmetry is broken.[1][3]

For particles that are real scalars (such as an axion), there is a fermion superpartner as well as a second, real scalar field. For axions, these particles are often referred to as axinos and saxions.

In extended supersymmetry there may be more than one superparticle for a given particle. For instance, with two copies of supersymmetry in four dimensions, a photon would have two fermion superpartners and a scalar superpartner.

In zero dimensions it is possible to have supersymmetry, but no superpartners. However, this is the only situation where supersymmetry does not imply the existence of superpartners.

25.2 Recreating superpartners

If the supersymmetry theory is correct, it should be possible to recreate these particles in high-energy particle accelerators. Doing so will not be an easy task; these particles may have masses up to a thousand times greater than their corresponding "real" particles.[1]

Some researchers have hoped the Large Hadron Collider at CERN might produce evidence for the existence of superpartner particles.[1] However, as of 2013, no such evidence has been found.[4]

25.3 See also

- Chargino
- Gluino

- Gravitino as a superpartner of the hypothetical graviton
- Neutralino
- Sfermion
- Higgsino

25.4 References

[1] Langacker, Paul (November 22, 2010). Sprouse, Gene D., ed. "Meet a superpartner at the LHC". *Physics* (New York: American Physical Society) **3** (98). Bibcode:2010PhyOJ...3...98L. doi:10.1103/Physics.3.98. ISSN 1943-2879. OCLC 233971234. Archived from the original on 2011-02-22. Retrieved 21 February 2011.

[2] Overbye, Dennis (May 15, 2007). "A Giant Takes On Physics' Biggest Questions". *The New York Times* (Manhattan, New York: Arthur Ochs Sulzberger, Jr.). p. F1. ISSN 0362-4331. OCLC 1645522. Retrieved 21 February 2011.

[3] Quigg, Chris (January 17, 2008). "Sidebar: Solving the Higgs Puzzle". *Scientific American* (Nature Publishing Group). ISSN 0036-8733. OCLC 1775222. Archived from the original on 2011-02-22. Retrieved 21 February 2011.

[4] Jamieson, Valerie (13 December 2013). "Higgs Nobel bash: I was at the party of the universe". *New Scientist*. Retrieved 20 December 2013. So far the Higgs hasn't given many supersymmetric clues.

25.5 External links

- Argonne National Laboratory
- Large Hadron Collider
- CERN homepage

Chapter 26

Gaugino

In particle physics, a **gaugino** is the hypothetical superpartner of a gauge field, as predicted by gauge theory combined with supersymmetry. They are fermions.

In the minimal supersymmetric extension of the standard model the following gauginos exist:

- The gluino (symbol g̃) is the superpartner of the gluon, and hence carries color charge.
- The gravitino (symbol G̃) is the supersymmetric partner of the graviton.
- The **winos** (symbol W± and W^3) are the superpartners of the W bosons of the SU(2)L gauge fields.
- The **bino** is the superpartner of the U(1) gauge field corresponding to weak hypercharge.

Gauginos mix with higgsinos, the superpartners of the Higgs field's degrees of freedom, to form linear combinations ("mass eigenstates") called neutralinos (electrically neutral) and charginos (electrically charged). In many models the lightest supersymmetric particle (LSP), often a neutralino such as the photino, is stable. In that case it is a WIMP and a candidate for dark matter.

26.1 References

- G. Bertone, D. Hooper, J. Silk (2005). "Particle Dark Matter: Evidence, Candidates and Constraints". *Physics Reports* **405**: 279–390. arXiv:hep-ph/0404175. Bibcode:2005PhR...405..279B. doi:10.1016/j.physrep.2004.08.031.

Chapter 27

Gluino

In supersymmetry, a **gluino** (symbol $\tilde{g}$) is the hypothetical supersymmetric partner of a gluon. Should they exist, gluinos are expected by supersymmetry theorists to be pair produced in particle accelerators such as the Large Hadron Collider.

In supersymmetric theories, gluinos are Majorana fermions and interact via the strong force as a color octet.[1] Gluinos have a lepton number 0, baryon number 0, and spin 1/2.

In models of supersymmetry that conserve R-parity, gluinos decay via the strong interaction to a squark and a quark, provided that an appropriate mass relation is satisfied. The squark subsequently decays to another quark and the lightest supersymmetric particle, LSP (which leaves the detector unseen). This means that a typical signal for a gluino at a hadron collider would be four jets plus missing energy.

However if gluinos are lighter than squarks, 3-body decay of a gluino to a neutralino and a quark antiquark pair is kinematically accessible through an off-shell squark.

27.1 Footnotes

[1] As there are 8 gluons of different color combinations, there are 8 gluinos of different color combinations, too.

Chapter 28

Gravitino

In supergravity theories combining general relativity and supersymmetry, the **gravitino** (G̃) is the gauge fermion supersymmetric partner of the graviton. It has been suggested as a candidate for dark matter.

If it exists, it is a fermion of spin $^3/_2$ and therefore obeys the Rarita-Schwinger equation. The gravitino field is conventionally written as $\psi\mu\alpha$ with $\mu = 0,1,2,3$ a four-vector index and $\alpha = 1,2$ a spinor index. For $\mu = 0$ one would get negative norm modes, as with every massless particle of spin 1 or higher. These modes are unphysical, and for consistency there must be a gauge symmetry which cancels these modes: $\delta\psi\mu\alpha = \partial\mu\varepsilon\alpha$ where $\varepsilon\alpha(x)$ is a spinor function of spacetime. This gauge symmetry is a local supersymmetry transformation, and the resulting theory is supergravity.

Thus the gravitino is the fermion mediating supergravity interactions, just as the photon is mediating electromagnetism, and the graviton is presumably mediating gravitation. Whenever supersymmetry is broken in supergravity theories, it acquires a mass which is determined by the scale at which supersymmetry is broken. This varies greatly between different models of supersymmetry breaking, but if supersymmetry is to solve the hierarchy problem of the Standard Model, the gravitino cannot be more massive than about 1 TeV/c^2.

28.1 Gravitino cosmological problem

If the gravitino indeed has a mass of the order of TeV, then it creates a problem in the standard model of cosmology, at least naïvely.[1][2][3][4]

One option is that the gravitino is stable. This would be the case if the gravitino is the lightest supersymmetric particle and R-parity is conserved (or nearly so). In this case the gravitino is a candidate for dark matter; as such gravitinos will have been created in the very early universe. However, one may calculate the density of gravitinos and it turns out to be much higher than the observed dark matter density.

The other option is that the gravitino is unstable. Thus the gravitinos mentioned above would decay and will not contribute to the observed dark matter density. However, since they decay only through gravitational interactions, their lifetime would be very long, of the order of Mpl^2 / m^3 in natural units, where Mpl is the Planck mass and m is the mass of a gravitino. For a gravitino mass of the order of TeV this would be 10^5 s, much later than the era of nucleosynthesis. At least one possible channel of decay must include either a photon, a charged lepton or a meson, each of which would be energetic enough to destroy a nucleus if it strikes one. One can show that enough such energetic particles will be created in the decay as to destroy almost all the nuclei created in the era of nucleosynthesis, in contrast with observations. In fact, in such a case the universe would have been made of hydrogen alone, and star formation would probably be impossible.

One possible solution to the cosmological gravitino problem is the split supersymmetry model, where the gravitino mass is much higher than the TeV scale, but other fermionic supersymmetric partners of standard model particles already appear at this scale.

Another solution is that R-parity is slightly violated and the gravitino is the lightest supersymmetric particle. This causes almost all supersymmetric particles in the early Universe to decay into Standard Model particles via R-parity violating interactions well before the synthesis of primordial nuclei; a small fraction however decay into gravitinos, whose half-life is orders of magnitude greater than the age of the Universe due to the suppression of the decay rate by the Planck scale and the small R-parity violating couplings.[5]

28.2 See also

- Supersymmetry

28.3 References

[1] T. Moroi, H. Murayama Cosmological constraints on the light stable gravitino Phys.Lett.B303:289–294,1993

[2] N. Okada, O. Seto A brane world cosmological solution to the gravitino problem Phys.Rev.D71:023517,2005

[3] A.de Gouvea,T.Moroi,H.MurayamaCosmology of Supersymmetric Models with Low-energy Gauge MediatRev.D56: 1281–1299,1997

[4] M. Endo Moduli Stabilization and Moduli-Induced Gravitino Problem talk given at SUSY'06, 12 June 2006

[5] F. Takayama and M. Yamaguchi, Phys. Lett. B 485 (2000)

Chapter 29

Photino

A **photino** is a hypothetical subatomic particle, the fermion WIMP superpartner of the photon predicted by supersymmetry. It is an example of a gaugino. Even though no photino has ever been observed so far, it is expected to be the lightest stable particle in the universe.[2]

29.1 Photino numbers

Photinos have a lepton number 0, baryon number 0, and spin 1/2. With an R-parity of -1 it is a possible candidate for dark matter. It mixes with the superpartners of the Z boson (zino) and the neutral higgs (higgsino) to the neutralino.

29.2 References

[1] STENGER, V. J. "Photinos from cosmic sources". nature.com. Retrieved 2015-09-24.

[2] "Tracking down the missing mass". *New Scientist* (1490): 32. 9 January 1986. Retrieved 24 September 2015.

Chapter 30

Higgsino

In particle physics, a **Higgsino**, symbol $\tilde{H}$, is the theoretical superpartner of theHiggs boson, as predicted by supersymmetry. The Higgsino is a Dirac fermion and that is a weak isodoublet with hypercharge half under the Standard Modelgauge symmetries. After electroweak symmetry breaking the Higgsino becomes a pair of neutral Majorana fermions called neutralinos and a charged Dirac fermion called a chargino(plus and minus). These states finally mix with the neutralinos(photino and zino)and chargino (charged wino plus and minus)[1] to form the predicted particles which are four neutralinos and two charginos (plus and minus each). Such a linear combination of the Higgsino, bino and wino makes up thelightest supersymmetric particle(LSP), which is a particle physics candidate for thedark matterof the universe. In orderto be such a candidate,it must be neutral (i.e.a neutralino rather than chargino).

In natural scenarios of SUSY, top squarks, bottom squarks, gluinos, and higgsino-enriched neutralinos and charginos are expected to be relatively light, enhancing their production cross sections. Higgsino searches have been performed by both the ATLAS and CMS experiments at the Large Hadron Collider at CERN, where physicists have searched for the direct electroweak pair production of Higgsinos. As of March 2014, no experimental evidence for Higgsinos has been reported.[2][3]

30.1 Higgsino Mass

If dark matter is composed only of Higgsinos then the Higgsino mass is 1.1 TeV. On other hand if dark matter has multi-components then the Higgsino mass depends on the relevant multiverse distribution functions making the mass of the Higgsino lighter.

$$m_{\tilde{h}} \approx 1.1(\Omega_{\tilde{h}}/\Omega_{DM})^{1/2}\ \text{TeV}$$ [4]

30.2 Footnotes

[1] resulting from electroweak symmetry breaking of the bino and wino 0, 1, 2

[2] "ATLAS Supersymmetry Public Results". ATLAS, CERN. Retrieved 2014-03-25.

[3] "CMS Supersymmetry Public Results". CMS, CERN. Retrieved 2014-03-25.

[4] Lawrence J. Hall and Yasunori Nomura, "Spread Supersymmetry", *Berkeley Center for Theoretical Physics, Department of Physics,* and Theoretical Physics Group, Lawrence Berkeley National Laboratory, University of California, Berkeley, CA 94720, USA, *19 Nov 2011*

Chapter 31

Neutralino

In supersymmetry, the **neutralino**[1] is a hypothetical particle. There are four neutralinos that are fermions and are electrically neutral, the lightest of which is typically stable. They are typically labeled N0
1 (the lightest), N0
2, N0
3 and N0
4 (the heaviest) although sometimes $\tilde{\chi}_1^0, \ldots, \tilde{\chi}_4^0$ is also used when $\tilde{\chi}_i^\pm$ is used to refer to charginos. These four states are mixtures of the bino and the neutral wino (which are the neutral electroweak gauginos), and the neutral higgsinos. As the neutralinos are Majorana fermions, each of them is identical to its antiparticle. Because these particles only interact with the weak vector bosons, they are not directly produced at hadron colliders in copious numbers. They would primarily appear as particles in cascade decays of heavier particles (decays that happen in multiple steps) usually originating from colored supersymmetric particles such as squarks or gluinos.

In R-parity conserving models, the lightest neutralino is stable and all supersymmetric cascade-decays end up decaying into this particle which leaves the detector unseen and its existence can only be inferred by looking for unbalanced momentum in a detector.

The heavier neutralinos typically decay through a neutral Z boson to a lighter neutralino or through a charged W boson to a light chargino:[2]

The mass splittings between the different neutralinos will dictate which patterns of decays are allowed.

Up to present, neutralinos have never been observed or detected in an experiment.

31.1 Origins in supersymmetric theories

In supersymmetry models, all Standard Model particles have partner particles with the same quantum numbers except for the quantum number spin, which differs by 1/2 from its partner particle. Since the superpartners of the Z boson (zino), the photon (photino) and the neutral higgs (higgsino) have the same quantum numbers, they can mix to form four eigenstates of the mass operator called "neutralinos". In many models the lightest of the four neutralinos turns out to be the lightest supersymmetric particle (LSP), though other particles may also take on this role.

31.2 Phenomenology

The exact properties of each neutralino will depend on the details of the mixing[1] (e.g. whether they are more higgsino-like or gaugino-like), but they tend to have masses at the weak scale (100 GeV – 1 TeV) and couple to other particles with strengths characteristic of the weak interaction. In this way they are phenomenologically similar to neutrinos, and so are not directly observable in particle detectors at accelerators.

In models in which R-parity is conserved and the lightest of the four neutralinos is the LSP, the lightest neutralino is stable and is eventually produced in the decay chain of all other superpartners.[3] In such cases supersymmetric processes at accelerators are characterized by a large discrepancy in energy and momentum between the visible initial and final state particles, with this energy being carried off by a neutralino which departs the detector unnoticed.[4][5] This is an important signature to discriminate supersymmetry from Standard Model backgrounds.

31.3 Relationship to dark matter

As a heavy, stable particle, the lightest neutralino is an excellent candidate to form the universe's cold dark matter.[6][7][8] In many models the lightest neutralino can be produced thermally in the hot early universe and leave approximately the right relic abundance to account for the observed dark matter. A lightest neutralino of roughly 10–10000 GeV is the leading weakly interacting massive particle (WIMP) dark matter candidate.[9]

Neutralino dark matter could be observed experimentally in nature either indirectly or directly. For indirect observation, gamma ray and neutrino telescopes look for evidence of neutralino annihilation in regions of high dark matter density such as the galactic or solar centre.[4] For direct observation, special purpose experiments such as the Cryogenic Dark Matter Search (CDMS) seek to detect the rare impacts of WIMPs in terrestrial detectors. These experiments have begun to probe interesting supersymmetric parameter space, excluding some models for neutralino dark matter, and upgraded experiments with greater sensitivity are under development.

31.4 See also

- Lightest Supersymmetric Particle
- Real neutral particle

31.5 Notes

[1] Martin, pp. 71–74

[2] J.-F. Grivaz & the Particle Data Group (2010). "Supersymmetry, Part II (Experiment)" (PDF). *Journal of Physics G* **37** (7): 1309–1319.

[3] Martin, p. 83

[4] Feng, Jonathan L (2010). "Dark Matter Candidates from Particle Physics and Methods of Detection". *Annual Review of Astronomy and Astrophysics* **48**: 495–545. arXiv:1003.0904. Bibcode:2010ARA&A..48..495F. doi:10.1146/annurev-astro-082708-101659. |chapter= ignored (help)

[5] Ellis, John; Olive, Keith A. (2010). "Supersymmetric Dark Matter Candidates". arXiv:1001.3651 [astro-ph]. Also published as Chapter 8 in Bertone

[6] M. Drees; G. Gerbier & the Particle Data Group (2010). "Dark Matter" (PDF). *Journal of Physics G* **37** (7A): 255–260.

[7] Martin, p. 99

[8] Bertone, p. 8

[9] Martin, p. 124

31.6 References

- Martin, Stephen P. (2008). "A Supersymmetry Primer". v5. arXiv:hep-ph/9709356 [hep-ph]. Also published as Chapter 1 in Kane, Gordon L, ed. (2010). *Perspectives on Supersymmetry II*. World Scientific. p. 604. ISBN 978-981-4307-48-2.
- Bertone, Gianfranco, ed. (2010). *Particle Dark Matter: Observations, Models and Searches*. Cambridge University Press. p. 762. ISBN 978-0-521-76368-4.

Chapter 32

Chargino

In particle physics, the **chargino** is a hypothetical particle which refers to the mass eigenstates of a charged superpartner, i.e. any new electrically charged fermion (with spin 1/2) predicted by supersymmetry. They are linear combinations of the charged wino and charged higgsinos. There are two charginos that are fermions and are electrically charged, which are typically labeled C $\tilde{\chi}$ ±
1 (the lightest) and C $\tilde{\chi}$ ±
2 (the heaviest) although sometimes $\tilde{\chi}_1^{\pm}$ and $\tilde{\chi}_2^{\pm}$ is also used to refer to charginos, when $\tilde{\chi}_i^0$ is used to refer to neutralinos. The heavier chargino can decay through the neutral Z boson to the lighter chargino. Both can decay through a charged W boson to a neutralino:

C $\tilde{\chi}$ ±
2 → C $\tilde{\chi}$ ±
1 + Z0

C $\tilde{\chi}$ ±
2 → N0
2 + W±

C $\tilde{\chi}$ ±
1 → N0
1 + W±

32.1 External links

- http://lepsusy.web.cern.ch/lepsusy/www/inoslowdmsummer02/charginolowdm_pub.html
- http://arxiv.org/abs/1307.5073

Chapter 33

Axino

The **axino** is a hypothetical elementary particle predicted by some theories of particle physics. Peccei–Quinn theory attempts to explain the observed phenomenon known as the strong CP problem by introducing a hypothetical real scalar particle called the axion. Adding supersymmetry to the model predicts the existence of a fermionic superpartner for the axion, the axino, and a bosonic superpartner, the *saxion*. They are all bundled up in a chiral superfield.

The axino has been predicted to be the lightest supersymmetric particle in such a model.[1] In part due to this property, it is considered a candidate for the composition of dark matter.[2]

The supermultiplet containing an axion and axino has been suggested as the origin of supersymmetry breaking, where the supermultiplet gains an F-term expectation value. [3]

33.1 References

[1] Abe, Nobutaka; Moroi, Takeo; Yamaguchi, Masahiro (2002). "Anomaly-Mediated Supersymmetry Breaking with Axion". *Journal of High Energy Physics* **1**: 10. arXiv:hep-ph/0111155. Bibcode:2002JHEP...01..010A. doi:10.1088/1126-6708/2002/01/010.

[2] Hooper, Dan; Wang, Lian-Tao (2004). "Possible evidence for axino dark matter in the galactic bulge". *Physical Review D* **70** (6): 063506. arXiv:hep-ph/0402220. Bibcode:2004PhRvD..70f3506H. doi:10.1103/PhysRevD.70.063506.

[3] Baryakhtar, Masha; Hardy, Edward; March-Russell, John (2013). "Axion Mediation". *JHEP* **1307**: 096. arXiv:1301.0829. Bibcode:2013JHEP...07..096B. doi:10.1007/JHEP07(2013)096.

Chapter 34

Sfermion

In particle physics, a **sfermion** is the spin−0 superpartner particle (or *sparticle*) of its associated fermion. In supersymmetric extensions to the Standard Model (SM) each particle has a superpartner with spin that differs by ½. Fermions in the SM have spin-½ and therefore sfermions have spin 0.

In general the name sfermions is formed by prefixing an 's' to the name of its superpartner, denoting that it is a scalar particle with spin 0. For instance, the electron's superpartner is the selectron and the top quark's superpartner is the stop squark.

One corollary from supersymmetry is that sparticles have the same gauge numbers as their SM partners. This means that sparticle–particle pairs have the same color charge, weak isospin charge, and hypercharge (and consequently electric charge). Unbroken supersymmetry also implies that sparticle–particle pairs have the same mass. This is evidently not the case, since these sparticles would have already been detected. Thus, sparticles must have different masses from the particle partners and supersymmetry is said to be broken.

34.1 Fundamental sfermions

34.1.1 Squarks

Squarks are the superpartners of quarks. These include the sup squark, sdown squark, scharm squark, sstrange squark, stop squark, and sbottom squark.

34.1.2 Sleptons

Sleptons are the superpartners of leptons. These include the selectron, smuon, stau, and the sneutrinos.

34.2 See also

- Minimal Supersymmetric Standard Model (MSSM)

34.3 References

- Martin, Stephen, P. (2011). "A Supersymmetry Primer". arXiv:hep-ph/9709356 [hep-ph].

Chapter 35

Stop squark

In particle physics, a **stop squark** is the superpartner of the top quark as predicted by supersymmetry (SUSY). It is a sfermion, which means it is a spin-0 boson (scalar). While the top quark is the heaviest known quark, the stop squark is actually often the lightest squark in many supersymmetry models.[1]

The stop squark is a key ingredient of a wide range of SUSY models that address the hierarchy problem of the Standard Model (SM) in a natural way. A boson partner to the top quark would stabilize the Higgs boson mass against quadratically divergent quantum corrections, provided its mass is close to the electroweak symmetry breaking energy scale. If this was the case then the stop quark would be accessible at the Large Hadron Collider. In the generic R-parity conserving Minimal Supersymmetric Standard Model (MSSM) the scalar partners of right-handed and left-handed top quarks mix to form two stop mass eigenstates. Depending on the specific details of the SUSY model and the mass hierarchy of the sparticles, the stop might decay into a bottom quark and a chargino, with a subsequent decay of the chargino into the lightest neutralino (which is often the lightest supersymmetric particle).

Many searches for evidence of the stop squark have been performed by both the ATLAS and CMS experiments at the LHC but so far no signal has been discovered. [2][3]

35.1 References

[1] Search For Pair Production of Stop Quarks Mimicking Top Event Signatures

[2] "ATLAS Supersymmetry Public Results". ATLAS, CERN. Retrieved 2014-03-21.

[3] "CMS Supersymmetry Public Results". CMS, CERN. Retrieved 2014-03-21.

Chapter 36

Planck particle

A **Planck particle**, named after physicist Max Planck, is a hypothetical particle defined as a tiny black hole whose Compton wavelength is equal to its Schwarzschild radius.[1] Its mass is thus approximately the Planck mass, and its Compton wavelength and Schwarzschild radius are about the Planck length.[2] Planck particles are sometimes used as an exercise to define the Planck mass and Planck length.[3] They play a role in some models of the evolution of the universe during the Planck epoch.[4]

Compared to a proton, for example, the Planck particle would be extremely small (its radius being equal to the Planck length, which is about 10^{-20} times the proton's radius) and heavy (the Planck mass being 10^{19} times the proton's mass).[5]

It is thought that such a particle would vanish in Hawking radiation.

36.1 Derivation

While opinions vary as to its proper definition, the most common definition of a Planck particle is a particle whose Compton wavelength is equal to its Schwarzschild radius. This sets the relationship:

$$\lambda = \frac{h}{mc} = \frac{2Gm}{c^2}$$

Thus making the mass of such a particle:

$$m = \sqrt{\frac{hc}{2G}}$$

This mass will be $\sqrt{\pi}$ times larger than the Planck mass, making a Planck particle 1.772 times more massive than the Planck unit mass.

Its radius will be the Compton wavelength:

$$r = \frac{h}{mc} = \sqrt{\frac{2Gh}{c^3}}$$

36.2 Dimensions

Using the above derivations we can substitute the universal constants *h*, *G*, and *c*, and determine physical values for the particle's mass and radius. Assuming this radius represents a sphere of uniform density we can further determine the particle's volume and density.

It should be noted that the above dimensions do not correspond to any known physical entity or material.

36.3 See also

- Micro black hole
- Planck units
- Max Planck
- Black hole electron

36.4 References

[1] Michel M. Deza; Elena Deza. *Encyclopedia of Distances*. Springer; 1 June 2009. ISBN 978-3-642-00233-5. p. 433.

[2] "Light element synthesis in Planck fireballs" - SpringerLink

[3] B. Roy Frieden; Robert A. Gatenby. *Exploratory data analysis using Fisher information*. Springer; 2007. ISBN 978-1-84628-506-6. p. 163.

[4] Harrison, Edward Robert (2000), *Cosmology: the science of the universe*, Cambridge University Press, ISBN 978-0-521-66148-5 p. 424

[5] Harrison 2000, p. 478.

36.5 External links

- "The quasi-steady state cosmology: analytical solutions of field equations and their relationship to observations" - Astrophysics Data Systems
- "Mach's principle: from Newton's bucket to quantum gravity" - Google Books
- "Mysteries of Mass: Some Contrarian Views From an Experimenter"
- "The Gauge Hierarchy Problem and Planck Oscillators" - CERN Document Server
- "The First Turbulence and First Fossil Turbulence"
- "Lecture on Nuclear Physics for Plasma Engineers"
- The Planck Length

Chapter 37

Axion

For other uses, see Axion (disambiguation).

The **axion** is a hypothetical elementary particle postulated by the Peccei–Quinn theory in 1977 to resolve the strong CP problem in quantum chromodynamics (QCD). If axions exist and have low mass within a specific range, they are of interest as a possible component of cold dark matter.

37.1 History

37.1.1 Prediction

As shown by Gerardus 't Hooft, strong interactions of the standard model, QCD, possess a non-trivial vacuum structure that in principle permits violation of the combined symmetries of charge conjugation and parity, collectively known as CP. Together with effects generated by weak interactions, the effective periodic strong CP violating term, Θ, appears as a Standard Model input – its value is not predicted by the theory, but must be measured. However, large CP violating interactions originating from QCD would induce a large electric dipole moment (EDM) for the neutron. Experimental constraints on the currently unobserved EDM implies CP violation from QCD must be extremely tiny and thus Θ must itself be extremely small. Since a priori Θ could have any value between 0 and 2π, this presents a naturalness problem for the standard model. Why should this parameter find itself so close to 0? (Or, why should QCD find itself CP-preserving?) This question constitutes what is known as the strong CP problem.

One simple solution exists: if at least one of the quarks of the standard model is massless, Θ becomes unobservable. However, empirical evidence strongly suggests that none of the quarks are massless.

In 1977, Roberto Peccei and Helen Quinn postulated a more elegant solution to the strong CP problem, the Peccei–Quinn mechanism. The idea is to effectively promote Θ to a field. This is accomplished by adding a new global symmetry (called a Peccei–Quinn symmetry) that becomes spontaneously broken. This results in a new particle, as shown by Frank Wilczek and Steven Weinberg, that fills the role of Θ—naturally relaxing the CP violation parameter to zero. This hypothesized new particle is called the axion. (On a more technical note, the axion is the would-be Nambu–Goldstone boson that results from the spontaneously broken Peccei–Quinn symmetry. However, the non-trivial QCD vacuum effects (e.g., instantons) spoil the Peccei–Quinn symmetry explicitly and provide a small mass for the axion. Hence, the axion is actually a pseudo-Nambu–Goldstone boson.) The original Weinberg–Wilczek axion was ruled out. Current literature discusses the mechanism as the 'invisible axion' which has two forms: KSVZ [1][2] and DFSZ.[3][4]

37.1.2 Searches

It had been thought that the invisible axion solves the strong CP problem without being amenable to verification by experiment. Axion models choose coupling that does not appear in any of the prior experiments. The very weakly coupled axion is also very light because axion couplings and mass are proportional. The situation changed when it was shown that a very light axion is overproduced in the early universe and therefore excluded.[5][6][7] The critical mass is of order 10^{-11} times the electron mass, where axions may account for the dark matter. The axion is thus a dark matter candidate as well as a solution to the strong CP problem. Furthermore, in 1983, Pierre Sikivie wrote down the modification of Maxwell's equations from a light stable axion [8] and showed axions can be detected on Earth by converting them to photons with a strong magnetic field, the principle of the ADMX. Solar axions may be converted to x-rays, as in CAST. Many experiments are searching laser light for signs of axions.[9]

37.2 Experiments

The Italian PVLAS experiment searches for polarization changes of light propagating in a magnetic field. The concept was first put forward in 1986 by Luciano Maiani, Roberto Petronzio and Emilio Zavattini.[10] A rotation claim[11] in 2006 was excluded by an upgraded setup.[12] An optimized search began in 2014.

Another technique is so called "light shining through walls",[13] where light passes through an intense magnetic field to convert photons into axions, that pass through metal. Experiments by BFRS and a team led by Rizzo ruled out an axion cause.[14] GammeV saw no events in a 2008 PRL. ALPS-I conducted similar runs,[15] setting new constraints in 2010; ALPS-II will run in 2014. OSQAR found no signal, limiting coupling[16] and will continue.

Several experiments search for astrophysical axions by the Primakoff effect, which converts axions to photons and vice versa in electromagnetic fields. Axions can be produced in the Sun's core when x-rays scatter in strong electric fields. The CAST solar telescope is underway, and has set limits on coupling to photons and electrons. ADMX searches the galactic dark matter halo[17] for resonant axions with a cold microwave cavity and has excluded optimistic axion models in the 1.9-3.53 μeV range.[18][19][20] It is amidst a series of upgrades and is taking new data, including at 4.9-6.2 μeV.

Resonance effects may be evident in Josephson junctions[21] from a supposed high flux of axions from the galactic halo with mass of 0.11 meV and density $0.05 GeV/cm^3$ [22] compared to the implied dark matter density $(0.3{\pm}0.1) GeV/cm^3$, indicating said axions would only partially compose dark matter.

Dark matter cryogenic detectors have searched for electron recoils that would indicate axions. CDMS published in 2009 and EDELWEISS set coupling and mass limits in 2013. UORE and XMASS also set limits on solar axions in 2013. XENON100 used a 225-day run to set the best coupling limits to date and exclude some parameters.[23]

Axion-like bosons could have a signature in astrophysical settings. In particular, several recent works have proposed axion-like particles as a solution to the apparent transparency of the Universe to TeV photons.[24][25] It has also been demonstrated in a few recent works that, in the large magnetic fields threading the atmospheres of compact astrophysical objects (e.g., magnetars), photons will convert much more efficiently. This would in turn give rise to distinct absorption-like features in the spectra detectable by current telescopes.[26] A new promising means is looking for quasi-particle refraction in systems with strong magnetic gradients. In particular, the refraction will lead to beam splitting in the radio light curves of highly magnetized pulsars and allow much greater sensitivities than currently achievable.[27] The International Axion Observatory (IAXO) is a proposed fourth generation helioscope.[28]

37.3 Possible detection

Axions may have been detected through irregularities in X-ray emission due to interaction of the Earth's magnetic field with radiation streaming from the Sun. Studying 15 years of data by the European Space Agency's XMM-Newton observatory, a research group at Leicester University noticed a seasonal variation for which no conventional explanation could be found. One potential explanation for the variation, described as "plausible" by the senior author of the paper, was X-rays produced by axions from the Sun's core.[29]

A term analogous to the one that must be added to Maxwell's equations[30] also appears in recent theoretical models for topological insulators.[31] This term leads to several interesting predicted properties at the interface between topological and normal insulators.[32] In this situation the field θ describes something very different from its use in high-energy physics.[32] In 2013, Christian Beck suggested that axions might be detectable in Josephson junctions; and in 2014, he argued that a signature, consistent with a mass ~110μeV, had in fact been observed in several preexisting experiments.[33]

37.4 Properties

37.4.1 Predictions

One theory of axions relevant to cosmology had predicted that they would have no electric charge, a very small mass in the range from 10^{-6} to 1 eV/c^2, and very low interaction cross-sections for strong and weak forces. Because of their properties, axions would interact only minimally with ordinary matter. Axions would change to and from photons in magnetic fields.

Supersymmetry

In supersymmetric theories the axion has both a scalar and a fermionic superpartner. The fermionic superpartner of the axion is called the axino, the scalar superpartner is called the saxion or dilaton. They are all bundled up in a chiral superfield.

The axino has been predicted to be the lightest supersymmetric particle in such a model.[34] In part due to this property, it is considered a candidate for dark matter.[35]

37.4.2 Cosmological implications

Theory suggests that axions were created abundantly during the Big Bang.[36] Because of a unique coupling to the instanton field of the primordial universe (the "misalignment mechanism"), an effective dynamical friction is created during the acquisition of mass following cosmic inflation. This robs all such primordial axions of their kinetic energy.

If axions have low mass, thus preventing other decay modes, theories predict that the universe would be filled with a very cold Bose–Einstein condensate of primordial axions. Hence, axions could plausibly explain the dark matter problem of physical cosmology.[37] Observational studies are underway, but they are not yet sufficiently sensitive to probe the mass regions if they are the solution to the dark matter problem. High mass axions of the kind searched for by Jain and Singh (2007)[38] would not persist in the modern universe. Moreover, if axions exist, scatterings with other particles in the thermal bath of the early universe unavoidably produce a population of hot axions.[39]

Low mass axions could have additional structure at the galactic scale. As they continuously fell into a galaxy from the intergalactic medium, they would be denser in "caustic" rings, just as the stream of water in a continuously-flowing fountain is thicker at its peak.[40] The gravitational effects of these rings on galactic structure and rotation might then be observable.[41] Other cold dark matter theoretical candidates, such as WIMPs and MACHOs, could also form such rings, but because such candidates are fermionic and thus experience friction or scattering among themselves, the rings would be less pronounced.

Axions would also have stopped interaction with normal matter at a different moment than other more massive dark particles. The lingering effects of this difference could perhaps be calculated and observed astronomically. Axions may hold the key to the Solar corona heating problem.[42]

37.5 References

37.5.1 Notes

[1] Kim,J.E. (1979). "Weak-Interaction Singlet and Strong CP Invariance".*Phys.Rev.Lett.***43**(2): 103–107. Bibcode:1979PK. doi:10.1103/PhysRevLett.43.103.

[2] Shifman, M.; Vainshtein, A.; Zakharov, V. (1980). "Can confinement ensure natural CP invariance of strong interactions?". *Nucl. Phys.* **B166**: 493. Bibcode:1980NuPhB.166..493S. doi:10.1016/0550-3213(80)90209-6.

[3] Dine, M.; Fischler, W.; Srednicki, M. (1981). "A simple solution to the strong CP problem with a harmless axion". *Phys. Lett.* **B104**: 199. Bibcode:1981PhLB..104..199D. doi:10.1016/0370-2693(81)90590-6.

[4] Zhitnitsky, A. (1980). "On possible suppression of the axion-hadron interactions". *Sov. J. Nucl. Phys.* **31**: 260.

[5] Preskill,J.;Wise,M.;Wilczek,F. (1983). "Cosmology of the invisible axion".*Phys.Lett.***B120**: 127. Bibcode:1983PhP. doi:10.1016/0370-2693(83)90637-8.

[6] Abbott, L.; Sikivie, P. (1983). "A cosmological bound on the invisible axion". *Phys.Lett.* **B120**: 133.

[7] Dine,M.;Fischler,W. (1983). "The not-so-harmless axion".*Phys.Lett.***B120**: 137. Bibcode:1983PhLB..120..137D.do370-2693(83)90639-1.

[8] Sikivie,P. (1983). "Experimental Tests of the"Invisible"Axion".*Phys.Rev.Lett.***51**(16): 1413. Bibcode:1983PhR15S. doi:10.1103/physrevlett.51.1415.

[9] http://home.web.cern.ch/about/experiments/osqar

[10] Maiani, L.; Petronzio, R.; Zavattini, E. (1986). "Effects of nearly massless, spin-zero particles on light propagation in a magnetic field". *Phys. Lett.* **175** (3): 359–363. Bibcode:1986PhLB..175..359M. doi:10.1016/0370-2693(86)90869-5.

[11] Steve Reucroft, John Swain. Axion signature may be QED CERN Courier, 2006-10-05

[12] Zavattini, E.; Zavattini, G.; Ruoso, G.; Polacco, E.; Milotti, E.; Karuza, M.; Gastaldi, U.; Di Domenico, G.; Della Valle, F.; Cimino, R.; Carusotto, S.; Cantatore, G.; Bregant, M.; Pvlas, Collaboration (2006). "Experimental Observation of Optical Rotation Generated in Vacuum by a Magnetic Field". *Physical Review Letters* **96** (11): 110406. arXiv:hep-ex/0507107. Bibcode:2006PhRvL..96k0406Z. doi:10.1103/PhysRevLett.96.110406. PMID 16605804.

[13] Ringwald, A. (2003). "Electromagnetic Probes of Fundamental Physics - Proceedings of the Workshop". invited talk *"Fundamental Physics at an X-Ray Free Electron Laser"* at *Workshop on Electromagnetic Probes of Fundamental Physics*. The Science and Culture Series - Physics (Erice, Italy): 63–74. arXiv:hep-ph/0112254. doi:10.1142/9789812704214_0007. ISBN 9789812385666.

[14] Robilliard, C.; Battesti, R.; Fouche, M.; Mauchain, J.; Sautivet, A.-M.; Amiranoff, F.; Rizzo, C. (2007). "No "Light Shining through a Wall": Results from a Photoregeneration Experiment". *Physical Review Letters* **99** (19): 190403. arXiv:0707.1296. Bibcode:2007PhRvL..99s0403R. doi:10.1103/PhysRevLett.99.190403. PMID 18233050.

[15] Ehret, Klaus; Frede, Maik; Ghazaryan, Samvel; Hildebrandt, Matthias; Knabbe, Ernst-Axel; Kracht, Dietmar; Lindner, Axel; List, Jenny; Meier, Tobias; Meyer, Niels; Notz, Dieter; Redondo, Javier; Ringwald, Andreas; Wiedemann, Günter; Willke, Benno (May 2010). "New ALPS results on hidden-sector lightweights". *Phys Lett B* **689** (4–5): 149–155. arXiv:1004.1313. Bibcode:2010PhLB..689..149E. doi:10.1016/j.physletb.2010.04.066.

[16] Pugnat, P.; Ballou, R.; Schott, M.; Husek, T.; Sulc, M.; Deferne, G.; Duvillaret, L.; Finger, M.; Finger, M.; Flekova, L.; Hosek, J.; Jary, V.; Jost, R.; Kral, M.; Kunc, S.; MacUchova, K.; Meissner, K. A.; Morville, J.; Romanini, D.; Siemko, A.; Slunecka, M.; Vitrant, G.; Zicha, J. (Aug 2014). "Search for weakly interacting sub-eV particles with the OSQAR laser-based experiment: results and perspectives". *Eur Phys J C* **74** (8): 3027. arXiv:1306.0443. Bibcode:2014EPJC...74.3027P. doi:10.1140/epjc/s10052-014-3027-8.

[17] Duffy, L. D.; Sikivie, P.; Tanner, D. B.; Bradley, R. F.; Hagmann, C.; Kinion, D.; Rosenberg, L. J.; Van Bibber, K.; Yu, D. B.; Bradley, R. F. (2006). "High resolution search for dark-matter axions". *Physical Review D* **74**: 12006. arXiv:astro-ph/0603108. Bibcode:2006PhRvD..74a2006D. doi:10.1103/PhysRevD.74.012006.

[18] Asztalos, S. J.; Carosi, G.; Hagmann, C.; Kinion, D.; Van Bibber, K.; Hoskins, J.; Hwang, J.; Sikivie, P.; Tanner, D. B.; Hwang, J.; Sikivie, P.; Tanner, D. B.; Bradley, R.; Clarke, J.; ADMX Collaboration (2010). "SQUID-Based Microwave Cavity Search for Dark-Matter Axions". *Physical Review Letters* **104** (4): 41301. arXiv:0910.5914. Bibcode:2010PhRvL.104d1301A. doi:10.1103/PhysRevLett.104.041301.

[19] "ADMX | Axion Dark Matter eXperiment". Phys.washington.edu. Retrieved 2014-05-10.

[20] Phase 1 Results, dated 2006-03-04

[21] Beck, Christian (December 2, 2013). "Possible Resonance Effect of Axionic Dark Matter in Josephson Junctions". *Physical Review Letters* **111** (23): 1801. arXiv:1309.3790. Bibcode:2013PhRvL.111w1801B. doi:10.1103/PhysRevLett.111.231801.

[22] Moskvitch, Katia. "Hints of cold dark matter pop up in 10-year-old circuit". New Scientist magazine (Reed Business Information). Retrieved 3 December 2013.

[23] "First axion results from the XENON100experiment".*Phys.Rev.D90,062009*. 9September2014. Bibcode:2014PhRvD.. doi:10.1103/PhysRevD.90.062009.

[24] De Angelis, A.; Mansutti, O.; Roncadelli, M. (2007). "Evidence for a new light spin-zero boson from cosmological gamma-ray propagation?".*Physical Review D***76**(12): 121301. arXiv:0707.4312. Bibcode:2007PhRvD..76l1301D.doi:10.1103/PhysRev.

[25] De Angelis, A.; Mansutti, O.; Persic, M.; Roncadelli, M. (2009). "Photon propagation and the very high energy gamma-ray spectra of blazars: How transparent is the Universe?". *Monthly Notices of the Royal Astronomical Society: Letters* **394**: L21–L25. arXiv:0807.4246. Bibcode:2009MNRAS.394L..21D. doi:10.1111/j.1745-3933.2008.00602.x.

[26] Chelouche, Doron; Rabadan, Raul; Pavlov, Sergey S.; Castejon, Francisco (2009). "Spectral Signatures of Photon-Particle Oscillations from Celestial Objects". *The Astrophysical Journal Supplement Series* **180**: 1–29. arXiv:0806.0411. Bibcod .doi:10.1088/0067-0049/180/1/1.

[27] Chelouche, Doron; Guendelman, Eduardo I. (2009). "COSMIC ANALOGS OF THE STERN-GERLACH EXPERIMENT AND THE DETECTION OF LIGHT BOSONS". *The Astrophysical Journal* **699**: L5–L8. arXiv:0810.3002. Bibcode:2009 ApJ...699L...5C.doi:10.1088/0004-637X/699/1/L5.

[28] The International Axion Observatory (IAXO)

[29] Sample, Ian. "Dark matter may have been detected – streaming from sun's core". *www,theguardian.com*. The Guardian. Retrieved 16 October 2014.

[30] Wilczek, Frank (1987-05-04). "Two applications of axion electrodynamics". *Physical Review Letters* **58** (18): 1799–1802. Bibcode:1987PhRvL..58.1799W. doi:10.1103/PhysRevLett.58.1799. PMID 10034541.

[31] Qi, Xiao-Liang; Taylor L. Hughes, Shou-Cheng Zhang; Zhang, Shou-Cheng (2008-11-24). "Topological field theory of time-reversal invariant insulators". *Physical Review B* **78** (19): 195424. arXiv:0802.3537. Bibcode:2008PhRvB..78s5424Q. doi:10.1103/PhysRevB.78.195424.

[32] Franz,Marcel(2008-11-24). "High-energy physics in a new guise".*Physics***1**: 36. Bibcode:2008PhyOJ...1...36F.doi:10.1103.

[33]

[34] Abe, Nobutaka, Takeo Moroi and Masahiro Yamaguchi; Moroi; Yamaguchi (2002). "Anomaly-Mediated Supersymmetry Breaking with Axion". *Journal of High Energy Physics* **1**: 10. arXiv:hep-ph/0111155. Bibcode:2002JHEP...01..010A. doi:10. 6708/2002/01/010.

[35] Hooper, Dan and Lian-Tao Wang; Wang (2004). "Possible evidence for axino dark matter in the galactic bulge". *Physical Review D* **70** (6): 063506. arXiv:hep-ph/0402220. Bibcode:2004PhRvD..70f3506H. doi:10.1103/PhysRevD.70.063506.

[36] Redondo, J.; Raffelt, G.; Viaux Maira, N. (2012). "Journey at the axion meV mass frontier". *Journal of Physics: Conference Series.* **375** 022004. doi:10.1088/1742-6596/375/2/022004 (inactive 2015-03-30).

[37] P. Sikivie,*Dark matter axions*,arXiv.

[38] P. L. Jain, G. Singh, *Search for new particles decaying into electron pairs of mass below 100* MeV/c^2, J. Phys. G: Nucl. Part. Phys., **34**, 129–138, (2007); doi:10.1088/0954-3899/34/1/009, (possible early evidence of 7±1 and 19±1 MeV axions of less than 10^{-13} s lifetime).

[39] Alberto Salvio, Alessandro Strumia, Wei Xue, Alberto; Strumia, Alessandro; Xue, Wei (2014). "Thermal axion production". *Jcap 1401 (2014) 011* **2014**: 011. arXiv:1310.6982. Bibcode:2014JCAP...01..011S. doi:10.1088/1475-7516/2014/01/011.

[40] P. Sikivie, "Dark matter axions and caustic rings"

[41] P. Sikivie (personal website): pictures of alleged triangular structure in Milky Way; hypothetical flow diagram which could give rise to such a structure.

[42] The enigmatic Sun: a crucible for new physics

37.5.2 Journal entries

- Peccei, R. D.; Quinn, H. R. (1977). "*CP* Conservation in the Presence of Pseudoparticles". *Physical Review Letters* **38** (25): 1440–1443. Bibcode:1977PhRvL..38.1440P. doi:10.1103/PhysRevLett.38.1440.
- Peccei, R. D.; Quinn, H. R.; Quinn (1977). "Constraints imposed by *CP* conservation in the presence of pseudoparticles". *Physical Review* **D16** (6): 1791–1797. Bibcode:1977PhRvD..16.1791P. doi:10.1103/PhysRevD.16.1791.
- Weinberg,Steven(1978). "A New Light Boson?".*Physical Review Letters***40**(4): 223–226. Bibcode:1978Ph. doi:10.1103/PhysRevLett.40.223.
- Wilczek, Frank (1978). "Problem of Strong *P* and *T* Invariance in the Presence of Instantons". *Physical Review Letters* **40** (5): 279–282. Bibcode:1978PhRvL..40..279W. doi:10.1103/PhysRevLett.40.279.

37.6 External links

- November 24, 2008 article in APS Physics
- January 28, 2007 news article by newscientist.com
- December 06, 2006 news article by physorg.com
- July 17, 2006 news article from Scientific American
- March 27, 2006 news article by PhysicsWeb.org
- November 24, 2004 news article by PhysicsWeb.org
- CAST Experiment
- CAST at UNIZAR
- CAST at University of Technology Darmstadt
- ADMX at University of Washington

Chapter 38

Dilaton

In particle physics, the **dilaton** is a hypothetical particle that appears in theories with extra dimensions when the volume of the compactified dimensions is allowed to vary. It appears for instance in Kaluza–Klein theory's compactifications of extra dimensions. It is a particle of a scalar field Φ, a scalar field that always comes with gravity. For comparison, in standard general relativity, Newton's constant, or equivalently the Planck mass is a constant. If this constant is promoted to a dynamical field, the result is the dilaton.

In Kaluza–Klein theories, after dimensional reduction, the effective Planck mass varies as some power of the volume of compactified space. This is why volume can turn out as a dilaton in the lower-dimensional effective theory.

Although string theory naturally incorporates Kaluza–Klein theory (which first introduced the dilaton), perturbative string theories, such as type I string theory, type II string theory and heterotic string theory, already contain the dilaton in the maximal number of 10 dimensions. However, on the other hand, M-theory in 11 dimensions does not include the dilaton in its spectrum unless it is compactified. In fact, the dilaton in type IIA string theory is actually the radion of M-theory compactified over a circle, while the dilaton in $E_8 \times E_8$ string theory is the radion for the Hořava–Witten model. (For more on the M-theory origin of the dilaton, see .)

In string theory, there is also a dilaton in the worldsheet CFT(Conformal field theory). The exponential of its vacuum expectation value determines the coupling constant g, as $\int R = 2\pi\chi$ for compact worldsheets by the Gauss–Bonnet theorem and the Euler characteristic $\chi = 2 - 2g$, where g is the genus that counts the number of handles and thus the number of loops or string interactions described by a specific worldsheet.

$$g = \exp(\langle\phi\rangle)$$

Therefore the coupling constant is a dynamical variable in string theory, unlike the case of quantum field theory where it is constant. As long as supersymmetry is unbroken, such scalar fields can take arbitrary values (they are moduli). However, supersymmetry breaking usually creates a potential energy for the scalar fields and the scalar fields localize near a minimum whose position should in principle be calculable in string theory.

The dilaton acts like a Brans–Dicke scalar, with the effective Planck scale depending upon *both* the string scale and the dilaton field.

In supersymmetry, the superpartner of the dilaton is called the **dilatino**, and the dilaton combines with the axion to form a complex scalar field.

38.1 Dilaton action

The dilaton-gravity action is

$$\int d^D x \sqrt{-g} \left[\frac{1}{2\kappa} \left(\Phi R - \omega \left[\Phi \right] \frac{g^{\mu\nu} \partial_\mu \Phi \partial_\nu \Phi}{\Phi} \right) - V[\Phi] \right]$$

This is more general than Brans–Dicke in vacuum in that we have a dilaton potential.

38.2 See also

- CGHS model
- R=T model
- Quantum gravity

38.3 References

- Fujii, Y. (2003). "Mass of the dilaton and the cosmological constant". *Prog. Theor. Phys.* **110** (3): 433–439. arXiv:gr-qc/0212030. Bibcode:2003PThPh.110..433F. doi:10.1143/PTP.110.433.
- Hayashi, M.; Watanabe, T.; Aizawa, I. & Aketo, K. (2003). "Dilatonic Inflation and SUSY Breaking in String-inspired Supergravity". *Modern Physics Letters A* **18** (39): 2785–2793. arXiv:hep-ph/0303029. Bibcode: .doi:10.1142/S0217732303012465.
- Alvarenge, F.; Batista, A. & Fabris, J. (2005). "Does Quantum Cosmology Predict a Constant Dilatonic Field". *International Journal of Modern Physics D* **14** (2): 291–307. arXiv:gr-qc/0404034. Bibcode:2005IJMPD..14..291A. doi:10.1142/S0218271805005955.
- Lu, H.; Huang, Z.; Fang, W. & Zhang, K. (2004). "Dark Energy and Dilaton Cosmology". arXiv:hep-th/0409309 [hep-th].
- Wesson, Paul S. (1999). *Space-Time-Matter, Modern Kaluza-Klein Theory*. Singapore: World Scientific. p. 31. ISBN 981-02-3588-7.

Chapter 39

Graviton

This article is about the hypothetical particle. For other uses, see Graviton (disambiguation).

In physics, the **graviton** is a hypothetical elementary particle that mediates the force of gravitation in the framework of quantum field theory. If it exists, the graviton is expected to be massless (because the gravitational force appears to have unlimited range) and must be a spin−2 boson. The spin follows from the fact that the source of gravitation is the stress–energy tensor, a second-rank tensor (compared to electromagnetism's spin-1 photon, the source of which is the four-current, a first-rank tensor). Additionally, it can be shown that any massless spin-2 field would give rise to a force indistinguishable from gravitation, because a massless spin-2 field must couple to (interact with) the stress–energy tensor in the same way that the gravitational field does. Seeing as the graviton is hypothetical, its discovery would unite quantum theory with gravity.[4] This result suggests that, if a massless spin-2 particle is discovered, it must be the graviton, so that the only experimental verification needed for the graviton may simply be the discovery of a massless spin-2 particle.[5]

39.1 Theory

The four other known forces of nature are mediated by elementary particles: electromagnetism by the photon, the strong interaction by the gluons, the Higgs field by the Higgs Boson, and the weak interaction by the W and Z bosons. The hypothesis is that the gravitational interaction is likewise mediated by an – as yet undiscovered – elementary particle, dubbed as *the graviton*. In the classical limit, the theory would reduce to general relativity and conform to Newton's law of gravitation in the weak-field limit.[6][7][8]

39.1.1 Gravitons and renormalization

When describing graviton interactions, the classical theory (i.e., the tree diagrams) and semiclassical corrections (one-loop diagrams) behave normally, but Feynman diagrams with two (or more) loops lead to ultraviolet divergences; that is, infinite results that cannot be removed because the quantized general relativity is not renormalizable, unlike quantum electrodynamics. That is, the usual ways physicists calculate the probability that a particle will emit or absorb a graviton give nonsensical answers and the theory loses its predictive power. These problems, together with some conceptual puzzles, led many physicists to believe that a theory more complete than quantized general relativity must describe the behavior near the Planck scale.

39.1.2 Comparison with other forces

Unlike the force carriers of the other forces, gravitation plays a special role in general relativity in defining the spacetime in which events take place. In some descriptions, matter modifies the 'shape' of spacetime itself, and gravity is a result of this shape, an idea which at first glance may appear hard to match with the idea of a force acting between particles.[9]

Because the diffeomorphism invariance of the theory does not allow any particular space-time background to be singled out as the "true" space-time background, general relativity is said to be background independent. In contrast, the Standard Model is *not* background independent, with Minkowski space enjoying a special status as the fixed background space-time.[10] A theory of quantum gravity is needed in order to reconcile these differences.[11] Whether this theory should be background independent is an open question. The answer to this question will determine our understanding of what specific role gravitation plays in the fate of the universe.[12]

39.1.3 Gravitons in speculative theories

String theory predicts the existence of gravitons and their well-defined interactions. A graviton in perturbative string theory is a closed string in a very particular low-energy vibrational state. The scattering of gravitons in string theory can also be computed from the correlation functions in conformal field theory, as dictated by the AdS/CFT correspondence, or from matrix theory.

A feature of gravitons in string theory is that, as closed strings without endpoints, they would not be bound to branes and could move freely between them. If we live on a brane (as hypothesized by brane theories) this "leakage" of gravitons from the brane into higher-dimensional space could explain why gravitation is such a weak force, and gravitons from other branes adjacent to our own could provide a potential explanation for dark matter. However, if gravitons were to move completely freely between branes this would dilute gravity too much, causing a violation of Newton's inverse square law. To combat this, Lisa Randall found that a three-brane (such as ours) would have a gravitational pull of its own, preventing gravitons from drifting freely, possibly resulting in the diluted gravity we observe while roughly maintaining Newton's inverse square law.[13] See brane cosmology.

A theory by Ahmed Farag Ali and Saurya Das adds quantum mechanical corrections (using Bohm trajectories) to general relativistic geodesics. If gravitons are given a small but non-zero mass, it could explain the cosmological constant without need for dark energy and solve the smallness problem.[14]

39.2 Experimental observation

Unambiguous detection of individual gravitons, though not prohibited by any fundamental law, is impossible with any physically reasonable detector.[15] The reason is the extremely low cross section for the interaction of gravitons with matter. For example, a detector with the mass of Jupiter and 100% efficiency, placed in close orbit around a neutron star, would only be expected to observe one graviton every 10 years, even under the most favorable conditions. It would be impossible to discriminate these events from the background of neutrinos, since the dimensions of the required neutrino shield would ensure collapse into a black hole.[15]

However, experiments to detect gravitational waves, which may be viewed as coherent states of many gravitons, are underway (such as LIGO and VIRGO). Although these experiments cannot detect individual gravitons, they might provide information about certain properties of the graviton.[16] For example, if gravitational waves were observed to propagate slower than c (the speed of light in a vacuum), that would imply that the graviton has mass (however, gravitational waves must propagate slower than "c" in a region with non-zero mass density if they are to be detectable).[17] Astronomical observations of the kinematics of galaxies, especially the galaxy rotation problem and modified Newtonian dynamics, might point toward gravitons having non-zero mass.[18]

39.3 Difficulties and outstanding issues

Most theories containing gravitons suffer from severe problems. Attempts to extend the Standard Model or other quantum field theories by adding gravitons run into serious theoretical difficulties at high energies (processes involving energies close to or above the Planck scale) because of infinities arising due to quantum effects (in technical terms, gravitation is nonrenormalizable). Since classical general relativity and quantum mechanics seem to be incompatible at such energies, from a theoretical point of view, this situation is not tenable. One possible solution is to replace particles with strings.

String theories are quantum theories of gravity in the sense that they reduce to classical general relativity plus field theory at low energies, but are fully quantum mechanical, contain a graviton, and are believed to be mathematically consistent.[19]

39.4 See also

- Gravitomagnetism
- Gravitational wave
- Planck mass
- Gravitation
- Static forces and virtual-particle exchange
- Multiverse
- Gravitino

39.5 References

[1] G is used to avoid confusion with gluons (symbol g)

[2] Rovelli, C. (2001). "Notes for a brief history of quantum gravity". arXiv:gr-qc/0006061 [gr-qc].

[3] Blokhintsev, D. I.; Gal'perin, F. M. (1934). "Gipoteza neitrino i zakon sokhraneniya energii" [Neutrino hypothesis and conservation of energy]. *Pod Znamenem Marxisma* (in Russian) **6**: 147–157.

[4] Lightman, A. P.; Press, W. H.; Price, R. H.; Teukolsky, S. A. (1975). "Problem 12.16". *Problem book in Relativity and Gravitation*. Princeton University Press. ISBN 0-691-08162-X.

[5] For a comparison of the geometric derivation and the (non-geometric) spin-2 field derivation of general relativity, refer to box 18.1 (and also 17.2.5) of Misner, C. W.; Thorne, K. S.; Wheeler, J. A. (1973). *Gravitation*. W. H. Freeman. ISBN 0-7167-0344-0.

[6] Feynman, R. P.; Morinigo, F. B.; Wagner, W. G.; Hatfield, B. (1995). *Feynman Lectures on Gravitation*. Addison-Wesley. ISBN 0-201-62734-5.

[7] Zee, A. (2003). *Quantum Field Theory in a Nutshell*. Princeton University Press. ISBN 0-691-01019-6.

[8] Randall, L. (2005). *Warped Passages: Unraveling the Universe's Hidden Dimensions*. Ecco Press. ISBN 0-06-053108-8.

[9] See the other articles on General relativity, Gravitational field, Gravitational wave, etc

[10] Colosi, D. et al. (2005). "Background independence in a nutshell: The dynamics of a tetrahedron". *Classical and Quantum Gravity* **22** (14): 2971. arXiv:gr-qc/0408079. Bibcode:2005CQGra..22.2971C. doi:10.1088/0264-9381/22/14/008.

[11] Witten, E. (1993). "Quantum Background Independence In String Theory". arXiv:hep-th/9306122 [hep-th].

[12] Smolin, L. (2005). "The case for background independence". arXiv:hep-th/0507235 [hep-th].

[13] Kaku, Michio (2006). *Parallel Worlds - The science of alternative universes and our future in the Cosmos*. pp. 218–221.

[14] Ali, Ahmed Farang (2014). "Cosmology from quantum potential". *Physical Letters B* **741**: 276–279. arXiv:1404.3093v3. doi:10.1016/j.physletb.2014.12.057.

[15] Rothman, T.; Boughn, S. (2006). "Can Gravitons be Detected?". *Foundations of Physics* **36** (12): 1801–1825. arXiv:gr-qc/0601043. Bibcode:2006FoPh...36.1801R. doi:10.1007/s10701-006-9081-9.

[16] Dyson, Freeman (8 October 2013). "Is a graviton detectable?". *International Journal of Modern Physics A* **28** (25): 1330041–1–1330035–14. Bibcode:2013IJMPA..2830041D. doi:10.1142/S0217751X1330041X.

[17] Will, C. M. (1998). "Bounding the mass of the graviton using gravitational-wave observations of inspiralling compact binaries". *Physical Review D* **57** (4): 2061–2068. arXiv:gr-qc/9709011. Bibcode:1998PhRvD..57.2061W. doi:10.1103/PhysRevD.57.2061.

[18] Trippe, S. (2013), "A Simplified Treatment of Gravitational Interaction on Galactic Scales", J. Kor. Astron. Soc. **46**, 41. arXiv:1211.4692

[19] Sokal, A. (July 22, 1996). "Don't Pull the String Yet on Superstring Theory". *The New York Times*. Retrieved March 26, 2010.

39.6 External links

-
- Graviton on *In Our Time* at the BBC. (listen now)

Chapter 40

Majoron

Not to be confused with Majorana fermion.

In particle physics, **majorons** (named after Ettore Majorana) are a hypothetical type of Goldstone boson that are theorized to mediate the neutrino mass violation of lepton number or $B - L$ in certain high energy collisions such as

e− + e− → W− + W− + J

Where two electrons collide to form two W bosons and the majoron J. The U(1)B–L symmetry is assumed to be global so that the majoron isn't "eaten up" by the gauge boson and spontaneously broken. Majorons were originally formulated in four dimensions by Y. Chikashige, R. N. Mohapatra and R. D. Peccei to understand neutrino masses by the seesaw mechanism and are being searched for in the neutrino-less double beta decay process. There are theoretical extensions of this idea into supersymmetric theories and theories involving extra compactified dimensions. By propagating through the extra spatial dimensions the detectable number of majoron creation events vary accordingly. Mathematically, majorons may be modeled by allowing them to propagate through a material while all other Standard Model forces are fixed to an orbifold point.

40.1 Searches

Experiments studying double beta decay have set limits on decay modes that emit majorons.

NEMO[2] has observed a variety of elements . EXO [3] and Kamland-Zen [4] have set half-life limits for majoron decays in xenon.

40.2 References

[1] Lattanzi, M. (2008). "Decaying Majoron Dark Matter and Neutrino Masses". *AIP Conference Proceedings* **966** (1): 163–169. arXiv:0802.3155. doi:10.1063/1.2836988.

[2] Arnold, R.; Augier, C.; Baker, J. D.; Barabash, A. S.; Basharina-Freshville, A.; Blondel, S.; Blot, S.; Bongrand, M.; Brudanin, V.; Busto, J.; Caffrey, A. J.; Cerna, C.; Chapon, A.; Chauveau, E.; Duchesneau, D.; Durand, D.; Egorov, V.; Eurin, G.; Evans, J. J.; Flack, R.; Garrido, X.; Gómez, H.; Guillon, B.; Guzowski, P.; Hodák, R.; Hubert, P.; Hugon, C.; Jullian, S.; Klimenko, A.; Kochetov, O.; Konovalov, S. I.; Kovalenko, V.; Lalanne, D.; Lang, K.; Lemière, Y.; Liptak, Z.; Loaiza, P.; Lutter, G.; Mamedov, F.; Marquet, C.; Mauger, F.; Morgan, B.; Mott, J.; Nemchenok, I.; Nomachi, M.; Nova, F.; Nowacki, F.; Ohsumi, H.; Pahlka, R. B.; Perrot, F.; Piquemal, F.; Povinec, P.; Ramachers, Y. A.; Remoto, A.; Reyss, J. L.; Richards, B.; Riddle, C. L.; Rukhadze, E.; Saakyan, R.; Sarazin, X.; Shitov, Yu.; Simard, L.; Šimkovic, F.; Smetana, A.; Smolek, K.; Smolnikov, A.; Söldner-Rembold, S.; Soulé, B.; Štekl, I.; Suhonen, J.; Sutton, C. S.; Szklarz, G.; Thomas, J.; Timkin, V.; Torre, S.; Tretyak, Vl.

I.; Tretyak, V. I.; Umatov, V. I.; Vanushin, I.; Vilela, C.; Vorobel, V.; Waters, D.; Žukauskas, A. (12 June 2014). "Search for neutrinoless double-beta decay of with the NEMO-3 detector". *Physical Review D* **89** (11). Bibcode:2014PhRvD..89k1101A. doi:10.1103/PhysRevD.89.111101.

[3] Albert, J. B.; Auty, D. J.; Barbeau, P. S.; Beauchamp, E.; Beck, D.; Belov, V.; Benitez-Medina, C.; Breidenbach, M.; Brunner, T.; Burenkov, A.; Cao, G. F.; Chambers, C.; Chaves, J.; Cleveland, B.; Coon, M.; Craycraft, A.; Daniels, T.; Danilov, M.; Daugherty, S. J.; Davis, C. G.; Davis, J.; DeVoe, R.; Delaquis, S.; Didberidze, T.; Dolgolenko, A.; Dolinski, M. J.; Dunford, M.; Fairbank, W.; Farine, J.; Feldmeier, W.; Fierlinger, P.; Fudenberg, D.; Giroux, G.; Gornea, R.; Graham, K.; Gratta, G.; Hall, C.; Herrin, S.; Hughes, M.; Jewell, M. J.; Jiang, X. S.; Johnson, A.; Johnson, T. N.; Johnston, S.; Karelin, A.; Kaufman, L. J.; Killick, R.; Koffas, T.; Kravitz, S.; Kuchenkov, A.; Kumar, K. S.; Leonard, D. S.; Leonard, F.; Licciardi, C.; Lin, Y. H.; Ling, J.; MacLellan, R.; Marino, M. G.; Mong, B.; Moore, D.; Nelson, R.; Odian, A.; Ostrovskiy, I.; Ouellet, C.; Piepke, A.; Pocar, A.; Prescott, C. Y.; Rivas, A.; Rowson, P. C.; Rozo, M. P.; Russell, J. J.; Schubert, A.; Sinclair, D.; Smith, E.; Stekhanov, V.; Tarka, M.; Tolba, T.; Tosi, D.; Tsang, R.; Twelker, K.; Vogel, P.; Vuilleumier, J.-L.; Waite, A.; Walton, J.; Walton, T.; Weber, M.; Wen, L. J.; Wichoski, U.; Yang, L.; Yen, Y.-R.; Zeldovich, O. Ya. (10 November 2014). "Search for Majoron-emitting modes of double-beta decay of with EXO-200". *Physical Review D* **90** (9). Bibcode:2014PhRvD..90i2004A. doi:10.1103/PhysRevD.90.092004.

[4] Gando, A.; Gando, Y.; Hanakago, H.; Ikeda, H.; Inoue, K.; Kato, R.; Koga, M.; Matsuda, S.; Mitsui, T.; Nakada, T.; Nakamura, K.; Obata, A.; Oki, A.; Ono, Y.; Shimizu, I.; Shirai, J.; Suzuki, A.; Takemoto, Y.; Tamae, K.; Ueshima, K.; Watanabe, H.; Xu, B. D.; Yamada, S.; Yoshida, H.; Kozlov, A.; Yoshida, S.; Banks, T. I.; Detwiler, J. A.; Freedman, S. J.; Fujikawa, B. K.; Han, K.; O'Donnell, T.; Berger, B. E.; Efremenko, Y.; Karwowski, H. J.; Markoff, D. M.; Tornow, W.; Enomoto, S.; Decowski, M. P. (6 August 2012). "Limits on Majoron-emitting double- decays of Xe in the KamLAND-Zen experiment". *Physical Review C* **86** (2). arXiv:1205.6372. Bibcode:2012PhRvC..86b1601G. doi:10.1103/PhysRevC.86.021601.

40.3 Further reading

- Balysh, A. et al. (1996). "Bounds on new Majoron models from the Heidelberg-Moscow experiment". *Physical Review D***54**(5): 3641–3644. arXiv:nucl-ex/9511001. Bibcode:1996PhRvD..54.3641G.doi:10.1103/Phys.
- Mohapatra, R. N.; Pérez-Lorenzana, A.; de S. Pires, C. A. (2000). "Neutrino mass, bulk majoron and neutrinoless double beta decay". *Physics Letters B* **491** (1–2): 143–147. arXiv:hep-ph/0008158. Bibcode:2000PhLBM.doi:10.1016/S0370-2693(00)01031-5.
- Carone, C. D.; Conroy, J. M.; Kwee, H. J. (2002). "Bulk majorons at colliders". *Physics Letters B* **538** (1–2): 115–120. arXiv:hep-ph/0204045. Bibcode:2002PhLB..538..115C. doi:10.1016/S0370-2693(02)01943-3.
- Frampton, P. H.; Oh, M. C.; Yoshikawa, T. (2002). "Majoron mass zeros from Higgs triplet vacuum expectation values without a Majoron problem". *Physical Review D* **66** (3): 033007. arXiv:hep-ph/0204273. Bibco c3007F.doi:10.1103/PhysRevD.66.033007.
- Grossman, Y.; Haber, H. E. (2003). "The would-be Majoron in *R*-parity-violating supersymmetry". *Physical Review D***67**(3): 036002. arXiv:hep-ph/0210273. Bibcode:2003PhRvD..67c6002G.doi:10.1103/PhysRevD02.
- de S. Pires, C. A.; Rodrigues da Silva, P. S. (2004). "Spontaneous breaking of the lepton number and invisible majoron in a 3-3-1 model". *European Physical Journal C* **36**: 397–403. arXiv:hep-ph/0307253. Bibco 97D.doi:10.1140/epjc/s2004-01949-3.

Chapter 41

Majorana fermion

Not to be confused with Majoron.

A **Majorana fermion** (/maɪəˈrɒnə ˈfɛərmiːɒn/[1]), also referred to as a **Majorana particle**, is a fermion that is its own antiparticle. They were hypothesized by Ettore Majorana in 1937. The term is sometimes used in opposition to a Dirac fermion, which describes fermions that are not their own antiparticles.

All of the Standard Model fermions except the neutrino behave as Dirac fermions at low energy (after electroweak symmetry breaking), but the (massive) nature of the neutrino is not settled and it may be either Dirac or Majorana. In condensed matter physics, Majorana fermions exist as quasiparticle excitations in superconductors and can be used to form Majorana bound states governed by non-abelian statistics.

41.1 Theory

The concept goes back to Majorana's suggestion in 1937[2] that neutral spin$-1/2$ particles can be described by a real wave equation (the Majorana equation), and would therefore be identical to their antiparticle (because the wave functions of particle and antiparticle are related by complex conjugation).

The difference between Majorana fermions and Dirac fermions can be expressed mathematically in terms of the creation and annihilation operators of second quantization. The creation operator $\gamma_j^\dagger$ creates a fermion in quantum state j (described by a *real* wave function), whereas the annihilation operator γ_j annihilates it (or, equivalently, creates the corresponding antiparticle). For a Dirac fermion the operators $\gamma_j^\dagger$ and γ_j are distinct, whereas for a Majorana fermion they are identical. Majorana Fermions are the first exhibit of supersymmetry as their real and imaginary wave functions are the same. They could theoretically solve the Higgs boson mass problem.

41.2 Elementary particle

Because particles and antiparticles have opposite conserved charges, in order to be a Majorana fermion, namely, it is its own antiparticle, it is necessarily uncharged. All of the elementary fermions of the Standard Model have gauge charges, so they cannot have fundamental Majorana masses. However, the right-handed sterile neutrinos introduced to explain neutrino oscillation could have Majorana masses. If they do, then at low energy (after electroweak symmetry breaking), by the seesaw mechanism, the neutrino fields would naturally behave as six Majorana fields, with three expected to have very high masses (comparable to the GUT scale) and the other three expected to have very low masses (comparable to 1 eV). If right-handed neutrinos exist but do not have a Majorana mass, the neutrinos would instead behave as three Dirac fermions and their antiparticles with masses coming directly from the Higgs interaction, like the other Standard Model fermions.

The seesaw mechanism is appealing because it would naturally explain why the observed neutrino masses are so small. However, if the neutrinos are Majorana then they violate the conservation of lepton number and even B – L.

Ettore Majorana hypothesised the existence of Majorana fermions in 1937

Neutrinoless double beta decay, which can be viewed as two beta decay events with the produced antineutrinos immediately annihilating with one another, is only possible if neutrinos are their own antiparticles.[3] Experiments are underway to search for this type of decay.[4]

The high-energy analog of the neutrinoless double beta decay process is the production of same sign charged lepton pairs at hadron colliders;[5] it is being searched for by both the ATLAS and CMS experiments at the Large Hadron Collider. In theories based on left–right symmetry, there is a deep connection between these processes.[6] In the most accepted explanation of the smallness of neutrino mass, the seesaw mechanism, the neutrino is naturally a Majorana fermion.

Majorana fermions cannot possess intrinsic electric or magnetic moments, only toroidal moments.[7][8][9] Such minimal interaction with electromagnetic fields makes them potential candidates for cold dark matter.[10][11] The hypothetical neutralino of supersymmetric models is a Majorana fermion.

41.3 Majorana bound states

In superconducting materials, Majorana fermions can emerge as (non-fundamental) quasiparticles (which are more commonly referred as Bogoliubov quasiparticles in condensed matter.). This becomes possible because a quasiparticle in a superconductor is its own antiparticle. Majorana fermions (i.e. the Bogoliubov quasiparticles) in superconductors were observed by many experiments many years ago.

Mathematically, the superconductor imposes electron hole "symmetry" on the quasiparticle excitations, relating the creation operator $\gamma(E)$ at energy E to the annihilation operator $\gamma^{\dagger}(-E)$ at energy $-E$. Majorana fermions can be bound to a defect at zero energy, and then the combined objects are called Majorana bound states or Majorana zero modes.[12] This name is more appropriate than Majorana fermion (although the distinction is not always made in the literature), because the statistics of these objects is no longer fermionic. Instead, the Majorana bound states are an example of non-abelian anyons: interchanging them changes the state of the system in a way that depends only on the order in which the exchange was performed. The non-abelian statistics that Majorana bound states possess allows them to be used as a building block for a topological quantum computer.[13]

A quantum vortex in certain superconductors or superfluids can trap midgap states, so this is one source of Majorana bound states.[14][15][16] Shockley states at the end points of superconducting wires or line defects are an alternative, purely electrical, source.[17] An altogether different source uses the fractional quantum Hall effect as a substitute for the superconductor.[18]

41.3.1 Experiments in superconductivity

In 2008, Fu and Kane provided a groundbreaking development by theoretically predicting that Majorana bound states can appear at the interface between topological insulators and superconductors.[19][20] Many proposals of a similar spirit soon followed, where it was shown that Majorana bound states can appear even without any topological insulator. An intense search to provide experimental evidence of Majorana bound states in superconductors[21][22] first produced some positive results in 2012.[23][24] A team from the Kavli Institute of Nanoscience at Delft University of Technology in the Netherlands reported an experiment involving indium antimonide nanowires connected to a circuit with a gold contact at one end and a slice of superconductor at the other. When exposed to a moderately strong magnetic field the apparatus showed a peak electrical conductance at zero voltage that is consistent with the formation of a pair of Majorana bound states, one at either end of the region of the nanowire in contact with the superconductor.[25] This type of bounded state with zero energy was soon detected by several other groups in similar hybrid devices.[26][27][28][29]

This experiment from Delft marks a possible verification of independent 2010 theoretical proposals from two groups[30][31] predicting the solid state manifestation of Majorana bound states in semiconducting wires. However, it was also pointed out that some other trivial non-topological bounded states[32] could highly mimic the zero voltage conductance peak of Majorana bound state.

In 2014, evidence of Majorana bound states was observed using a low-temperature scanning tunneling microscope, by scientists at Princeton University.[33][34] It was suggested that Majorana bound states appeared at the edges of a chain of iron atoms formed on the surface of superconducting lead. Physicist Jason Alicea of California Institute of Technology, not involved in the research, said the study offered "compelling evidence" for Majorana fermions but that "we should keep in mind possible alternative explanations—even if there are no immediately obvious candidates".[35]

41.4 References

[1] "Quantum Computation possible with Majorana Fermions" on YouTube, uploaded 19 April 2013, retrieved 5 October 2014; and also based on the physicist's name's pronunciation.

[2] Majorana, Ettore; Maiani, Luciano (2006). "A symmetric theory of electrons and positrons". In Bassani, Giuseppe Franco. *Ettore Majorana Scientific Papers*. pp. 201–33. doi:10.1007/978-3-540-48095-2_10. ISBN 978-3-540-48091-4. Translated from: Majorana, Ettore (1937). "Teoria simmetrica dell'elettrone e del positrone". *Il Nuovo Cimento* (in Italian) **14** (4): 171–84. doi:10.1007/bf02961314.

[3] Schechter, J.; Valle, J.W.F. (1982). "Neutrinoless Double beta Decay in SU(2) x U(1) Theories". *Physical Review D* **25** (11): 2951. Bibcode:1982PhRvD..25.2951S. doi:10.1103/PhysRevD.25.2951. (subscription required (help)).

[4] Rodejohann, Werner (2011). "Neutrino-less Double Beta Decay and Particle Physics". *International Journal of Modern Physics* **E20** (9): 1833. arXiv:1106.1334. Bibcode:2011IJMPE..20.1833R. doi:10.1142/S0218301311020186. (registration required (help)).

[5] Keung, Wai-Yee; Senjanović, Goran (1983). "Majorana Neutrinos and the Production of the Right-Handed Charged Gauge Boson". *Physical Review Letters* **50** (19): 1427. Bibcode:1983PhRvL..50.1427K. doi:10.1103/PhysRevLett.50.1427. (subscription required (help)).

[6] Tello, Vladimir; Nemevšek, Miha; Nesti, Fabrizio; Senjanović, Goran; Vissani, Francesco (2011). "Left-Right Symmetry: from LHC to Neutrinoless Double Beta Decay". *Physical Review Letters* **106** (15): 151801. arXiv:1011.3522. Bibcode:2011PhR .doi:10.1103/PhysRevLett.106.151801. (subscription required(help)).

[7] Kayser, Boris; Goldhaber, Alfred S. (1983). "CPT and CP properties of Majorana particles, and the consequences". *Physical Review D* **28** (9): 2341–2344. Bibcode:1983PhRvD..28.2341K. doi:10.1103/PhysRevD.28.2341. (subscription required (help)).

[8] Radescu, E. E. (1985). "On the electromagnetic properties of Majorana fermions". *Physical Review D* **32** (5): 1266–1268. Bibcode:1985PhRvD..32.1266R. doi:10.1103/PhysRevD.32.1266. (subscription required (help)).

[9] Boudjema, F.; Hamzaoui, C.; Rahal, V.; Ren, H. C. (1989). "Electromagnetic Properties of Generalized Majorana Particles". *Physical Review Letters* **62** (8): 852–854. Bibcode:1989PhRvL..62..852B. doi:10.1103/PhysRevLett.62.852. (subscription required (help)).

[10] Pospelov, Maxim; ter Veldhuis, Tonnis (2000). "Direct and indirect limits on the electro-magnetic form factors of WIMPs". *Physics Letters B* **480**: 181–186. arXiv:hep-ph/0003010. Bibcode:2000PhLB..480..181P. doi:10.1016/S0370-2693(00)00358-0.

[11] Ho, Chiu Man; Scherrer, Robert J. (2013). "Anapole Dark Matter". *Physics Letters B* **722** (8): 341–346. arXiv:1211.0503. Bibcode:2013PhLB..722..341H. doi:10.1016/j.physletb.2013.04.039.

[12] Wilczek,Frank(2009). "Majorana returns"(PDF).*Nature Physics***5**(9): 614–618. Bibcode:2009NatPh...5..614W.doi:10.

[13] Nayak, Chetan; Simon, Steven H.; Stern, Ady; Freedman, Michael; Das Sarma, Sankar (2008). "Non-Abelian anyons and topological quantum computation". *Reviews of Modern Physics* **80** (3): 1083. arXiv:0707.1889. Bibcode:2008RvMP...80.1083N. doi:10.1103/RevModPhys.80.1083.

[14] N.B. Kopnin; M.M. Salomaa (1991). "Mutual friction in superfluid ^{3}He: Effects of bound states in the vortex core". *Physical Review B* **44** (17): 9667. Bibcode:1991PhRvB..44.9667K. doi:10.1103/PhysRevB.44.9667.

[15] Volovik, G. E. (1999). "Fermion zero modes on vortices in chiral superconductors". *JETP Letters* **70** (9): 609–614. arXiv:cond-mat/9909426. Bibcode:1999JETPL..70..609V. doi:10.1134/1.568223.

[16] Read, N.; Green, Dmitry (2000). "Paired states of fermions in two dimensions with breaking of parity and time-reversal symmetries and the fractional quantum Hall effect". *Physical Review B* **61** (15): 10267. arXiv:cond-mat/9906453. Bibcode:20 R.doi:10.1103/PhysRevB.61.10267.

[17] Kitaev, A. Yu (2001). "Unpaired Majorana fermions in quantum wires". *Physics-Uspekhi (supplement)* **44** (131): 131. arXiv:cond-mat/0010440. Bibcode:2001PhyU...44..131K. doi:10.1070/1063-7869/44/10S/S29.

[18] Moore, Gregory; Read, Nicholas (August 1991). "Nonabelions in the fractional quantum Hall effect". *Nuclear Physics B* **360** (2–3): 362. Bibcode:1991NuPhB.360..362M. doi:10.1016/0550-3213(91)90407-O.

[19] Fu, Liang; Kane, Charles L. (2008). "Superconducting Proximity Effect and Majorana Fermions at the Surface of a Topological Insulator".*Physical Review Letters***10**(9): 096407. arXiv:0707.1692. Bibcode:2008PhRvL.100i6407F.doi:10.1103/P407.

[20] Fu, Liang; Kane, Charles L. (2009). "Josephson current and noise at a superconductor/quantum-spin-Hall-insulator/superconductor junction". *Physical Review B* **79** (16): 161408. arXiv:0804.4469. Bibcode:2009PhRvB..79p1408F. doi:10.1103/Phy08.(subscription required(help)).

[21] Alicea, Jason (2012). "New directions in the pursuit of Majorana fermions in solid state systems". *Reports on Progress in Physics* **75** (7): 076501. arXiv:1202.1293. Bibcode:2012RPPh...75g6501A. doi:10.1088/0034-4885/75/7/076501. PMID 22790778. (subscription required (help)).

[22] Beenakker, C. W. J. (April 2013). "Search for Majorana fermions in superconductors". *Annual Review of Condensed Matter Physics* **4** (113): 113–136. arXiv:1112.1950. Bibcode:2013ARCMP...4..113B. doi:10.1146/annurev-conmatphys-030212-184337. (subscription required (help)).

[23] Reich, Eugenie Samuel (28 February 2012). "Quest for quirky quantum particles may have struck gold". *Nature News*. doi:10.1038/nature.2012.10124.

[24] Amos, Jonathan (13 April 2012). "Majorana particle glimpsed in lab". *BBC News*. Retrieved 15 April 2012.

[25] Mourik, V.; Zuo, K.; Frolov, S. M.; Plissard, S. R.; Bakkers, E. P. A. M.; Kouwenhoven, L. P. (12 April 2012). "Signatures of Majorana fermions in hybrid superconductor-semiconductor nanowire devices". *Science* **336** (6084): 1003–1007. arXiv:1204.2792. Bibcode:2012Sci...336.1003M. doi:10.1126/science.1222360.

[26] Deng, M.T.; Yu, C.L.; Huang, G.Y.; Larsson, M.; Caroff, P.; Xu, H.Q. (28 November 2012). "Anomalous zero-bias conductance peak in a Nb-InSb nanowire-Nb hybrid device". *Nano Letters* **12** (12): 6414–6419. Bibcode:2012NanoL..12.6414D. doi:10.1021/nl303758w.

[27] Das, A.; Ronen, Y.; Most, Y.; Oreg, Y.; Heiblum, M.; Shtrikman, H. (11 November 2012). "Zero-bias peaks and splitting in an Al-InAs nanowire topological superconductor as a signature of Majorana fermions.". *Nature Physics* **8** (12): 887–895. arXiv:1205.7073. Bibcode:2012NatPh...8..887D. doi:10.1038/nphys2479.

[28] Churchill, H. O. H.; Fatemi, V.; Grove-Rasmussen, K.; Deng, M.T.; Caroff, P.; Xu, H.Q.; Marcus, C.M. (6 June 2013). "Superconductor-nanowire devices from tunneling to the multichannel regime: Zero-bias oscillations and magnetoconductance crossover".*PHYSICAL REVIEW B***87**(24): 241401(R).arXiv:1303.2407. Bibcode:2013PhRvB..87x1401C.doi:10.1103/Ph1401.

[29] Deng, M.T.; Yu, C.L.; Huang, G.Y.; Larsson, Marcus; Caroff, P.; Xu, H.Q. (11 November 2014). "Parity independence of the zero-bias conductance peak in a nanowire based topological superconductor-quantum dot hybrid device". *Scientific Reports* **4**: 7261. arXiv:1406.4435. Bibcode:2014NatSR...4E7261D. doi:10.1038/srep07261.

[30] Lutchyn, Roman M.; Sau, Jay D.; Das Sarma, S. (August 2010). "Majorana Fermions and a Topological Phase Transition in Semiconductor-Superconductor Heterostructures". *Physical Review Letters* **105** (7): 077001. arXiv:1002.4033. Bibcod 001L.doi:10.1103/PhysRevLett.105.077001.

[31] Oreg, Yuval; Refael, Gil; von Oppen, Felix (October 2010). "Helical Liquids and Majorana Bound States in Quantum Wires". *Physical Review Letters***105**(17): 177002. arXiv:1003.1145. Bibcode:2010PhRvL.105q7002O.doi:10.1103/PhysRevLet2.

[32] Lee, E. J. H.; Jiang, X.; Houzet, M.; Aguado, R.; Lieber, C.M.; Franceschi, S.D. (15 December 2013). "Spin-resolved Andreev levels and parity crossings in hybrid superconductor–semiconductor nanostructures". *Nature Nanotechnology* **9**: 79–84. arXiv:1302.2611. Bibcode:2014NatNa...9...79L. doi:10.1038/nnano.2013.267.

[33] Nadj-Perge, Stevan; Drozdov, Ilya K.; Li, Jian; Chen, Hua; Jeon, Sangjun; Seo, Jungpil; MacDonald, Allan H.; Bernevig, B. Andrei; Yazdani, Ali (2 October 2014). "Observation of Majorana fermions in ferromagnetic atomic chains on a superconductor". Science. arXiv:1410.3453. Bibcode:2014Sci...346..602N. doi:10.1126/science.1259327. (subscription required (help)).

[34] "Majorana fermion: Physicists observe elusive particle that is its own antiparticle". Phys.org. October 2, 2014. Retrieved 3 October 2014.

[35] "New Particle Is Both Matter and Antimatter". *Scientific American*. October 2, 2014. Retrieved 3 October 2014.

41.5 Further reading

- Pal, Palash B. (2011) [12 October 2010]. "Dirac, Majorana and Weyl fermions". *American Journal of Physics* **79** (5): 485. arXiv:1006.1718. Bibcode:2011AmJPh..79..485P. doi:10.1119/1.3549729. (subscription required (help)).

Chapter 42

Magnetic monopole

A **magnetic monopole** is a hypothetical elementary particle in particle physics that is an isolated magnet with only one magnetic pole (a north pole without a south pole or vice versa).[1][2] In more technical terms, a magnetic monopole would have a net "magnetic charge". Modern interest in the concept stems from particle theories, notably the grand unified and superstring theories, which predict their existence.[3][4]

Magnetism in bar magnets and electromagnets does not arise from magnetic monopoles. There is no conclusive experimental evidence that magnetic monopoles exist at all in our universe.

Some condensed matter systems contain effective (non-isolated) magnetic monopole *quasi*-particles,[5] or contain phenomena that are mathematically analogous to magnetic monopoles.[6]

42.1 Historical background

42.1.1 Pre-twentieth century

Many early scientists attributed the magnetism of lodestones to two different "magnetic fluids" ("effluvia"), a north-pole fluid at one end and a south-pole fluid at the other, which attracted and repelled each other in analogy to positive and negative electric charge.[7][8] However, an improved understanding of electromagnetism in the nineteenth century showed that the magnetism of lodestones was properly explained by Ampère's circuital law, not magnetic monopole fluids. Gauss's law for magnetism, one of Maxwell's equations, is the mathematical statement that magnetic monopoles do not exist. Nevertheless, it was pointed out by Pierre Curie in 1894[9] that magnetic monopoles *could* conceivably exist, despite not having been seen so far.

42.1.2 Twentieth century

The *quantum* theory of magnetic charge started with a paper by the physicist Paul A.M. Dirac in 1931.[10] In this paper, Dirac showed that if *any* magnetic monopoles exist in the universe, then all electric charge in the universe must be quantized (Dirac quantization condition).[11] The electric charge *is*, in fact, quantized, which is consistent with (but does not prove) the existence of monopoles.[11]

Since Dirac's paper, several systematic monopole searches have been performed. Experiments in 1975[12] and 1982[13] produced candidate events that were initially interpreted as monopoles, but are now regarded as inconclusive.[14] Therefore, it remains an open question whether monopoles exist. Further advances in theoretical particle physics, particularly developments in grand unified theories and quantum gravity, have led to more compelling arguments (detailed below) that monopoles do exist. Joseph Polchinski, a string-theorist, described the existence of monopoles as "one of the safest bets that one can make about physics not yet seen".[15] These theories are not necessarily inconsistent with the experimental evidence. In some theoretical models, magnetic monopoles are unlikely to be observed, because they are too massive to

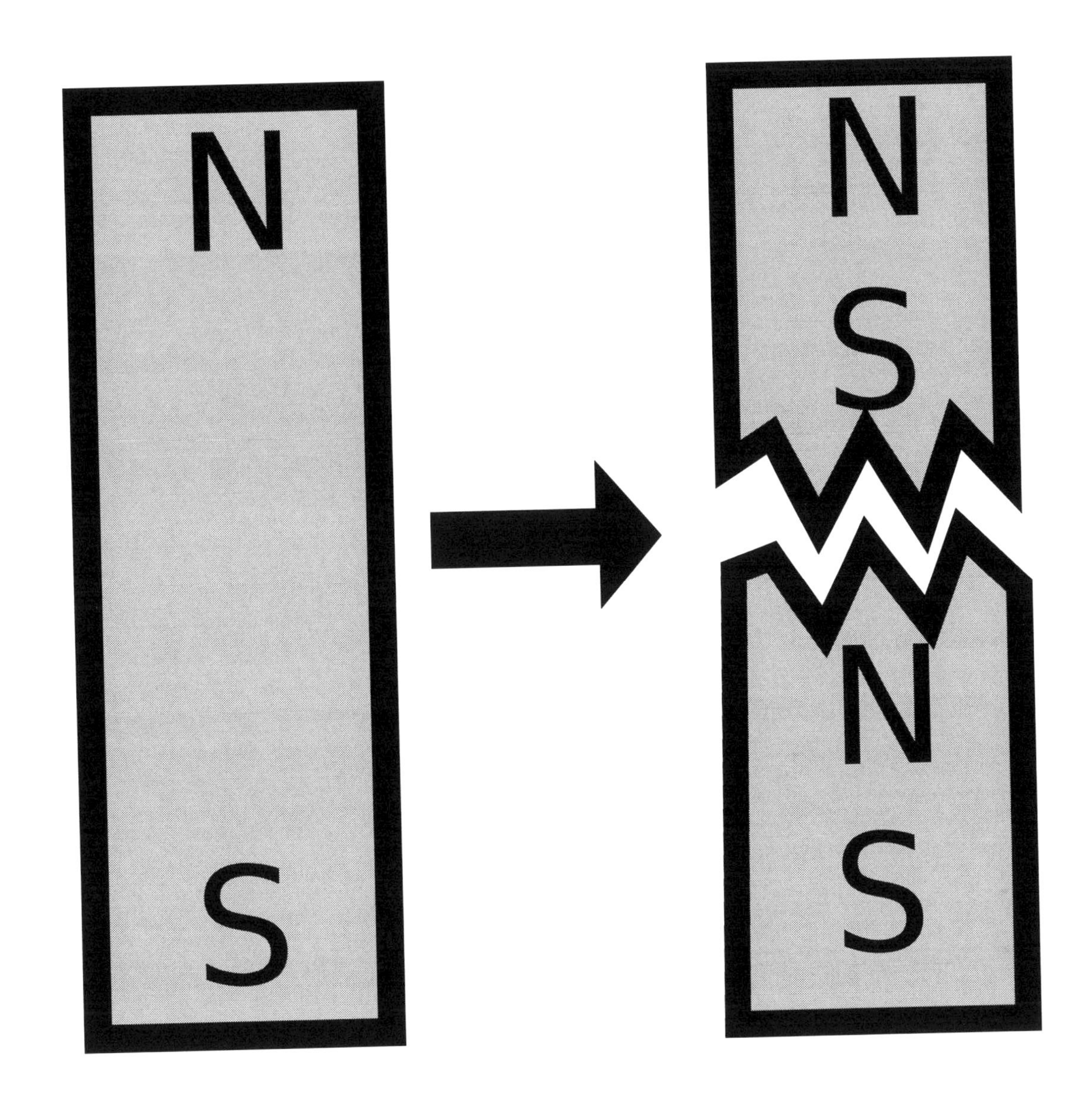

It is impossible to make ***magnetic monopoles*** *from a bar magnet. If a bar magnet is cut in half, it is* not *the case that one half has the north pole and the other half has the south pole. Instead, each piece has its own north and south poles. A magnetic monopole cannot be created from normal matter such as atoms and electrons, but would instead be a new elementary particle.*

be created in particle accelerators (see below), and also too rare in the Universe to enter a particle detector with much probability.[15]

Some condensed matter systems propose a structure superficially similar to a magnetic monopole, known as a flux tube. The ends of a flux tube form a magnetic dipole, but since they move independently, they can be treated for many purposes as independent magnetic monopole quasiparticles. Since 2009, numerous news reports from the popular media[16][17] have incorrectly described these systems as the long-awaited discovery of the magnetic monopoles, but the two phenomena are only superficially related to one another.[18][19] These condensed-matter systems continue to be an area of active research. (See "Monopoles" in condensed-matter systems below.)

42.2 Poles and magnetism in ordinary matter

Main article: Magnetism

All matter ever isolated to date—including every atom on the periodic table and every particle in the standard model—has zero magnetic monopole charge. Therefore, the ordinary phenomena of magnetism and magnets have nothing to do with magnetic monopoles.

Instead, magnetism in ordinary matter comes from two sources. First, electric currents create magnetic fields according to Ampère's law. Second, many elementary particles have an "intrinsic" magnetic moment, the most important of which is the electron magnetic dipole moment. (This magnetism is related to quantum-mechanical "spin".)

Mathematically, the magnetic field of an object is often described in terms of a multipole expansion. This is an expression of the field as the sum of component fields with specific mathematical forms. The first term in the expansion is called the "monopole" term, the second is called "dipole", then "quadrupole", then "octupole", and so on. Any of these terms can be present in the multipole expansion of an electric field, for example. However, in the multipole expansion of a *magnetic* field, the "monopole" term is always exactly zero (for ordinary matter). A magnetic monopole, if it exists, would have the defining property of producing a magnetic field whose "monopole" term is nonzero.

A magnetic dipole is something whose magnetic field is predominantly or exactly described by the magnetic dipole term of the multipole expansion. The term "dipole" means "two poles", corresponding to the fact that a dipole magnet typically contains a "north pole" on one side and a "south pole" on the other side. This is analogous to an electric dipole, which has positive charge on one side and negative charge on the other. However, an electric dipole and magnetic dipole are fundamentally quite different. In an electric dipole made of ordinary matter, the positive charge is made of protons and the negative charge is made of electrons, but a magnetic dipole does *not* have different types of matter creating the north pole and south pole. Instead, the two magnetic poles arise simultaneously from the aggregate effect of all the currents and intrinsic moments throughout the magnet. Because of this, the two poles of a magnetic dipole must always have equal and opposite strength, and the two poles cannot be separated from each other.

42.3 Maxwell's equations

Maxwell's equations of electromagnetism relate the electric and magnetic fields to each other and to the motions of electric charges. The standard equations provide for electric charges, but they posit no magnetic charges. Except for this difference, the equations are symmetric under the interchange of the electric and magnetic fields.[20] In fact, symmetric Maxwell's equations can be written when all charges (and hence electric currents) are zero, and this is how the electromagnetic wave equation is derived.

Fully symmetric Maxwell's equations can also be written if one allows for the possibility of "magnetic charges" analogous to electric charges.[21] With the inclusion of a variable for the density of these magnetic charges, say ϱ_m, there will also be a "magnetic current density" variable in the equations, $\mathbf{j}_m$.

If magnetic charges do not exist – or if they do exist but are not present in a region of space – then the new terms in Maxwell's equations are all zero, and the extended equations reduce to the conventional equations of electromagnetism such as $\nabla\cdot\mathbf{B} = 0$ (where $\nabla\cdot$ is divergence and $\mathbf{B}$ is the magnetic $\mathbf{B}$ field).

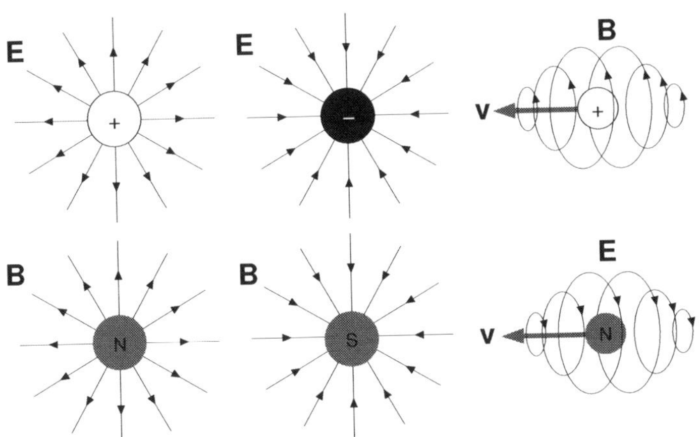

Left: Fields due to stationary electric and magnetic monopoles. **Right:** In motion (velocity **v**), an *electric* charge induces a **B** field while a *magnetic* charge induces an **E** field. Conventional current is used.

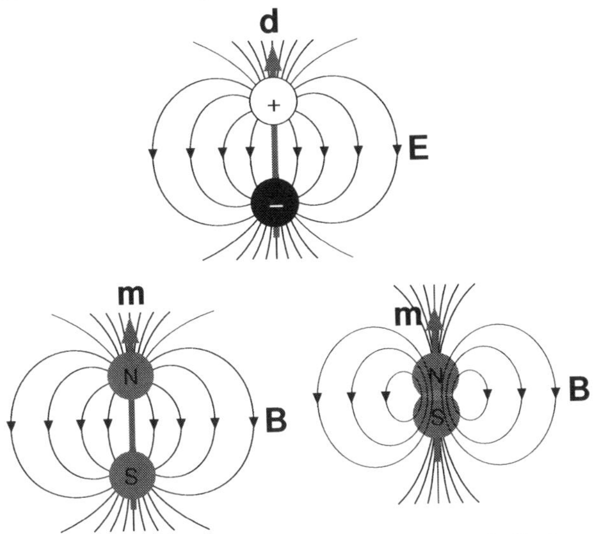

Top: **E** field due to an electric dipole moment **d**. **Bottom left:** **B** field due to a *mathematical* magnetic dipole **m** formed by two magnetic monopoles. **Bottom right:** **B** field due to a natural magnetic dipole moment **m** found in ordinary matter (*not* from monopoles).

The **E** fields and **B** fields due to electric charges (black/white) and magnetic poles (red/blue).[22][23]

42.3.1 In Gaussian cgs units

The extended Maxwell's equations are as follows, in Gaussian cgs units:[24]

In these equations ϱ_{m} is the *magnetic charge density*, $\mathbf{j}_{\mathrm{m}}$ is the *magnetic current density*, and q_{m} is the *magnetic charge* of a test particle, all defined analogously to the related quantities of electric charge and current; **v** is the particle's velocity and c is the speed of light. For all other definitions and details, see Maxwell's equations. For the equations in nondimensionalized form, remove the factors of c.

42.3.2 In SI units

In SI units, there are two conflicting units in use for magnetic charge q_m: webers (Wb) and ampere·meters (A·m). The conversion between them is $q_m(\text{Wb}) = \mu_0 q_m(\text{A·m})$, since the units are 1 Wb = 1 H·A = (1 H·m^{-1})·(1 A·m) by dimensional analysis (H is the henry – the SI unit of inductance).

Maxwell's equations then take the following forms (using the same notation above):[26]

42.3.3 Tensor formulation

Maxwell's equations in the language of tensors makes Lorentz covariance clear. The generalized equations are:[27][28]

where

- $F^{\alpha\beta}$ is the electromagnetic tensor, $^{\alpha\beta} = 1/2\varepsilon^{\alpha\beta\gamma\delta}F\gamma\delta$ is the dual electromagnetic tensor,
- for a particle with electric charge q_e and magnetic charge q_m; v is the four-velocity and p the four-momentum,
- for an electric and magnetic charge distribution; $J_e = (\varrho_e, \mathbf{j}_e)$ is the electric four-current and $J_m = (\varrho_m, \mathbf{j}_m)$ the magnetic four-current.

For a particle having only electric charge, one can express its field using a four-potential, according to the standard covariant formulation of classical electromagnetism:

$$F_{\alpha\beta} = \partial_\alpha A_\beta - \partial_\beta A_\gamma$$

However, this formula is inadequate for a particle that has both electric and magnetic charge, and we must add a term involving another potential P.[29][30]

$$F_{\alpha\beta} = \partial_\alpha A_\beta - \partial_\beta A_\alpha \ + \partial^\mu(\varepsilon_{\alpha\beta\mu\nu}P^\nu),$$

This formula for the fields is often called the Cabibbo-Ferrari relation, though Shanmugadhasan proposed it earlier.[30] The quantity $\varepsilon^{\alpha\beta\gamma\delta}$ is the Levi-Civita symbol, and the indices (as usual) behave according to the Einstein summation convention.

42.3.4 Duality transformation

The generalized Maxwell's equations possess a certain symmetry, called a *duality transformation*. One can choose any real angle ξ, and simultaneously change the fields and charges everywhere in the universe as follows (in Gaussian units):[31]

where the primed quantities are the charges and fields before the transformation, and the unprimed quantities are after the transformation. The fields and charges after this transformation still obey the same Maxwell's equations. The matrix is a two-dimensional rotation matrix.

Because of the duality transformation, one cannot uniquely decide whether a particle has an electric charge, a magnetic charge, or both, just by observing its behavior and comparing that to Maxwell's equations. For example, it is merely a convention, not a requirement of Maxwell's equations, that electrons have electric charge but not magnetic charge; after a $\xi = \pi/2$ transformation, it would be the other way around. The key empirical fact is that all particles ever observed have the same ratio of magnetic charge to electric charge.[31] Duality transformations can change the ratio to any arbitrary numerical value, but cannot change the fact that all particles have the same ratio. Since this is the case, a duality transformation can be made that sets this ratio to be zero, so that all particles have no magnetic charge. This choice underlies the "conventional" definitions of electricity and magnetism.[31]

42.4 Dirac's quantization

One of the defining advances in quantum theory was Paul Dirac's work on developing a relativistic quantum electromagnetism. Before his formulation, the presence of electric charge was simply "inserted" into the equations of quantum mechanics (QM), but in 1931 Dirac showed that a discrete charge naturally "falls out" of QM. That is to say, we can maintain the form of Maxwell's equations and still have magnetic charges.

Consider a system consisting of a single stationary electric monopole (an electron, say) and a single stationary magnetic monopole. Classically, the electromagnetic field surrounding them has a momentum density given by the Poynting vector, and it also has a total angular momentum, which is proportional to the product $q_e q_m$, and independent of the distance between them.

Quantum mechanics dictates, however, that angular momentum is quantized in units of $\hbar$, so therefore the product $q_e q_m$ must also be quantized. This means that if even a single magnetic monopole existed in the universe, and the form of Maxwell's equations is valid, all electric charges would then be quantized.

What are the units in which magnetic charge would be quantized? Although it would be possible simply to integrate over all space to find the total angular momentum in the above example, Dirac took a different approach. This led him to new ideas. He considered a point-like magnetic charge whose magnetic field behaves as q_m / r^2 and is directed in the radial direction, located at the origin. Because the divergence of **B** is equal to zero almost everywhere, except for the locus of the magnetic monopole at $r = 0$, one can locally define the vector potential such that the curl of the vector potential **A** equals the magnetic field **B**.

However, the vector potential cannot be defined globally precisely because the divergence of the magnetic field is proportional to the Dirac delta function at the origin. We must define one set of functions for the vector potential on the "northern hemisphere" (the half-space $z > 0$ above the particle), and another set of functions for the "southern hemisphere". These two vector potentials are matched at the "equator" (the plane $z = 0$ through the particle), and they differ by a gauge transformation. The wave function of an electrically-charged particle (a "probe charge") that orbits the "equator" generally changes by a phase, much like in the Aharonov–Bohm effect. This phase is proportional to the electric charge q_e of the probe, as well as to the magnetic charge q_m of the source. Dirac was originally considering an electron whose wave function is described by the Dirac equation.

Because the electron returns to the same point after the full trip around the equator, the phase φ of its wave function $e^{i\varphi}$ must be unchanged, which implies that the phase φ added to the wave function must be a multiple of 2π:

where ε_0 is the vacuum permittivity, $\hbar = h/2\pi$ is the reduced Planck's constant, c is the speed of light, and $\mathbb{Z}$ is the set of integers.

This is known as the **Dirac quantization condition**. The hypothetical existence of a magnetic monopole would imply that the electric charge must be quantized in certain units; also, the existence of the electric charges implies that the magnetic charges of the hypothetical magnetic monopoles, if they exist, must be quantized in units inversely proportional to the elementary electric charge.

At the time it was not clear if such a thing existed, or even had to. After all, another theory could come along that would explain charge quantization without need for the monopole. The concept remained something of a curiosity. However, in the time since the publication of this seminal work, no other widely accepted explanation of charge quantization has appeared. (The concept of local gauge invariance—see gauge theory below—provides a natural explanation of charge quantization, without invoking the need for magnetic monopoles; but only if the U(1) gauge group is compact, in which case we will have magnetic monopoles anyway.)

If we maximally extend the definition of the vector potential for the southern hemisphere, it will be defined everywhere except for a semi-infinite line stretched from the origin in the direction towards the northern pole. This semi-infinite line is called the Dirac string and its effect on the wave function is analogous to the effect of the solenoid in the Aharonov–Bohm effect. The quantization condition comes from the requirement that the phases around the Dirac string are trivial, which means that the Dirac string must be unphysical. The Dirac string is merely an artifact of the coordinate chart used and should not be taken seriously.

The Dirac monopole is a singular solution of Maxwell's equation (because it requires removing the worldline from spacetime); in more complicated theories, it is superseded by a smooth solution such as the 't Hooft–Polyakov monopole.

42.5 Topological interpretation

42.5.1 Dirac string

Main article: Dirac string

A gauge theory like electromagnetism is defined by a gauge field, which associates a group element to each path in space time. For infinitesimal paths, the group element is close to the identity, while for longer paths the group element is the successive product of the infinitesimal group elements along the way.

In electrodynamics, the group is U(1), unit complex numbers under multiplication. For infinitesimal paths, the group element is $1 + iA\mu dx^\mu$ which implies that for finite paths parametrized by s, the group element is:

$$\prod_s \left(1 + ieA_\mu \frac{dx^\mu}{ds} ds\right) = \exp\left(ie \int A \cdot dx\right).$$

The map from paths to group elements is called the Wilson loop or the holonomy, and for a U(1) gauge group it is the phase factor which the wavefunction of a charged particle acquires as it traverses the path. For a loop:

$$e \oint_{\partial D} A \cdot dx = e \int_D (\nabla \times A) dS = e \int_D B \, dS.$$

So that the phase a charged particle gets when going in a loop is the magnetic flux through the loop. When a small solenoid has a magnetic flux, there are interference fringes for charged particles which go around the solenoid, or around different sides of the solenoid, which reveal its presence.

But if all particle charges are integer multiples of e, solenoids with a flux of $2\pi/e$ have no interference fringes, because the phase factor for any charged particle is $e^{2\pi i} = 1$. Such a solenoid, if thin enough, is quantum-mechanically invisible. If such a solenoid were to carry a flux of $2\pi/e$, when the flux leaked out from one of its ends it would be indistinguishable from a monopole.

Dirac's monopole solution in fact describes an infinitesimal line solenoid ending at a point, and the location of the solenoid is the singular part of the solution, the Dirac string. Dirac strings link monopoles and antimonopoles of opposite magnetic charge, although in Dirac's version, the string just goes off to infinity. The string is unobservable, so you can put it anywhere, and by using two coordinate patches, the field in each patch can be made nonsingular by sliding the string to where it cannot be seen.

42.5.2 Grand unified theories

Main article: 't Hooft–Polyakov monopole

In a U(1) gauge group with quantized charge, the group is a circle of radius $2\pi/e$. Such a U(1) gauge group is called compact. Any U(1) which comes from a Grand Unified Theory is compact – because only compact higher gauge groups make sense. The size of the gauge group is a measure of the inverse coupling constant, so that in the limit of a large-volume gauge group, the interaction of any fixed representation goes to zero.

The case of the U(1) gauge group is a special case because all its irreducible representations are of the same size – the charge is bigger by an integer amount, but the field is still just a complex number – so that in U(1) gauge field theory it is possible to take the decompactified limit with no contradiction. The quantum of charge becomes small, but each charged particle has a huge number of charge quanta so its charge stays finite. In a non-compact U(1) gauge group theory, the charges of particles are generically not integer multiples of a single unit. Since charge quantization is an experimental certainty, it is clear that the U(1) gauge group of electromagnetism is compact.

GUTs lead to compact U(1) gauge groups, so they explain charge quantization in a way that seems to be logically independent from magnetic monopoles. However, the explanation is essentially the same, because in any GUT which breaks down into a U(1) gauge group at long distances, there are magnetic monopoles.

The argument is topological:

1. The holonomy of a gauge field maps loops to elements of the gauge group. Infinitesimal loops are mapped to group elements infinitesimally close to the identity.

2. If you imagine a big sphere in space, you can deform an infinitesimal loop which starts and ends at the north pole as follows: stretch out the loop over the western hemisphere until it becomes a great circle (which still starts and ends at the north pole) then let it shrink back to a little loop while going over the eastern hemisphere. This is called *lassoing the sphere*.

3. Lassoing is a sequence of loops, so the holonomy maps it to a sequence of group elements, a continuous path in the gauge group. Since the loop at the beginning of the lassoing is the same as the loop at the end, the path in the group is closed.

4. If the group path associated to the lassoing procedure winds around the U(1), the sphere contains magnetic charge. During the lassoing, the holonomy changes by the amount of magnetic flux through the sphere.

5. Since the holonomy at the beginning and at the end is the identity, the total magnetic flux is quantized. The magnetic charge is proportional to the number of windings N, the magnetic flux through the sphere is equal to $2\pi N/e$. This is the Dirac quantization condition, and it is a topological condition which demands that the long distance U(1) gauge field configurations be consistent.

6. When the U(1) gauge group comes from breaking a compact Lie group, the path which winds around the U(1) group enough times is topologically trivial in the big group. In a non-U(1) compact Lie group, the covering space is a Lie group with the same Lie algebra, but where all closed loops are contractible. Lie groups are homogenous, so that any cycle in the group can be moved around so that it starts at the identity, then its lift to the covering group ends at P, which is a lift of the identity. Going around the loop twice gets you to P^2, three times to P^3, all lifts of the identity. But there are only finitely many lifts of the identity, because the lifts can't accumulate. This number of times one has to traverse the loop to make it contractible is small, for example if the GUT group is SO(3), the covering group is SU(2), and going around any loop twice is enough.

7. This means that there is a continuous gauge-field configuration in the GUT group allows the U(1) monopole configuration to unwind itself at short distances, at the cost of not staying in the U(1). In order to do this with as little energy as possible, you should leave only the U(1) gauge group in the neighborhood of one point, which is called the **core** of the monopole. Outside the core, the monopole has only magnetic field energy.

Hence, the Dirac monopole is a topological defect in a compact U(1) gauge theory. When there is no GUT, the defect is a singularity – the core shrinks to a point. But when there is some sort of short-distance regulator on space time, the monopoles have a finite mass. Monopoles occur in lattice U(1), and there the core size is the lattice size. In general, they are expected to occur whenever there is a short-distance regulator.

42.5.3 String theory

In the universe, quantum gravity provides the regulator. When gravity is included, the monopole singularity can be a black hole, and for large magnetic charge and mass, the black hole mass is equal to the black hole charge, so that the mass of the magnetic black hole is not infinite. If the black hole can decay completely by Hawking radiation, the lightest charged particles cannot be too heavy.[33] The lightest monopole should have a mass less than or comparable to its charge in natural units.

So in a consistent holographic theory, of which string theory is the only known example, there are always finite-mass monopoles. For ordinary electromagnetism, the mass bound is not very useful because it is about same size as the Planck mass.

42.5.4 Mathematical formulation

In mathematics, a (classical) gauge field is defined as a connection over a principal G-bundle over spacetime. G is the gauge group, and it acts on each fiber of the bundle separately.

A *connection* on a G bundle tells you how to glue fibers together at nearby points of M. It starts with a continuous symmetry group G which acts on the fiber F, and then it associates a group element with each infinitesimal path. Group multiplication along any path tells you how to move from one point on the bundle to another, by having the G element associated to a path act on the fiber F.

In mathematics, the definition of bundle is designed to emphasize topology, so the notion of connection is added on as an afterthought. In physics, the connection is the fundamental physical object. One of the fundamental observations in the theory of characteristic classes in algebraic topology is that many homotopical structures of nontrivial principal bundles may be expressed as an integral of some polynomial over **any** connection over it. Note that a connection over a trivial bundle can never give us a nontrivial principal bundle.

If space time is $\mathbf{R}^4$ the space of all possible connections of the G-bundle is connected. But consider what happens when we remove a timelike worldline from spacetime. The resulting spacetime is homotopically equivalent to the topological sphere S^2.

A principal G-bundle over S^2 is defined by covering S^2 by two charts, each homeomorphic to the open 2-ball such that their intersection is homeomorphic to the strip $S^1 \times I$. 2-balls are homotopically trivial and the strip is homotopically equivalent to the circle S^1. So a topological classification of the possible connections is reduced to classifying the transition functions. The transition function maps the strip to G, and the different ways of mapping a strip into G are given by the first homotopy group of G.

So in the G-bundle formulation, a gauge theory admits Dirac monopoles provided G is not simply connected, whenever there are paths that go around the group that cannot be deformed to a constant path (a path whose image consists of a single point). U(1), which has quantized charges, is not simply connected and can have Dirac monopoles while **R**, its universal covering group, **is** simply connected, doesn't have quantized charges and does not admit Dirac monopoles. The mathematical definition is equivalent to the physics definition provided that, following Dirac, gauge fields are allowed which are defined only patch-wise and the gauge field on different patches are glued after a gauge transformation.

The total magnetic flux is none other than the first Chern number of the principal bundle, and depends only upon the choice of the principal bundle, and not the specific connection over it. In other words, it's a topological invariant.

This argument for monopoles is a restatement of the lasso argument for a pure U(1) theory. It generalizes to $d + 1$ dimensions with $d \geq 2$ in several ways. One way is to extend everything into the extra dimensions, so that U(1) monopoles become sheets of dimension $d - 3$. Another way is to examine the type of topological singularity at a point with the homotopy group πd_{-2}(G).

42.6 Grand unified theories

In more recent years, a new class of theories has also suggested the existence of magnetic monopoles.

During the early 1970s, the successes of quantum field theory and gauge theory in the development of electroweak theory and the mathematics of the strong nuclear force led many theorists to move on to attempt to combine them in a single theory known as a Grand Unified Theory (GUT). Several GUTs were proposed, most of which implied the presence of a real magnetic monopole particle. More accurately, GUTs predicted a range of particles known as dyons, of which the most basic state was a monopole. The charge on magnetic monopoles predicted by GUTs is either 1 or 2 gD, depending on the theory.

The majority of particles appearing in any quantum field theory are unstable, and they decay into other particles in a variety of reactions that must satisfy various conservation laws. Stable particles are stable because there are no lighter particles into which they can decay and still satisfy the conservation laws. For instance, the electron has a lepton number of one and an electric charge of one, and there are no lighter particles that conserve these values. On the other hand, the muon, essentially a heavy electron, can decay into the electron plus two quanta of energy, and hence it is not stable.

The dyons in these GUTs are also stable, but for an entirely different reason. The dyons are expected to exist as a side effect of the "freezing out" of the conditions of the early universe, or a symmetry breaking. In this scenario, the dyons arise due to the configuration of the vacuum in a particular area of the universe, according to the original Dirac theory. They remain stable not because of a conservation condition, but because there is no simpler *topological* state into which they can decay.

The length scale over which this special vacuum configuration exists is called the *correlation length* of the system. A correlation length cannot be larger than causality would allow, therefore the correlation length for making magnetic monopoles must be at least as big as the horizon size determined by the metric of the expanding universe. According to that logic, there should be at least one magnetic monopole per horizon volume as it was when the symmetry breaking took place.

Cosmological models of the events following the big bang make predictions about what the horizon volume was, which lead to predictions about present-day monopole density. Early models predicted an enormous density of monopoles, in clear contradiction to the experimental evidence.[34][35] This was called the "monopole problem". Its widely accepted resolution was not a change in the particle-physics prediction of monopoles, but rather in the cosmological models used to infer their present-day density. Specifically, more recent theories of cosmic inflation drastically reduce the predicted number of magnetic monopoles, to a density small enough to make it unsurprising that humans have never seen one.[36] This resolution of the "monopole problem" was regarded as a success of cosmic inflation theory. (However, of course, it is only a noteworthy success if the particle-physics monopole prediction is correct.[37]) For these reasons, monopoles became a major interest in the 1970s and 80s, along with the other "approachable" predictions of GUTs such as proton decay.

Many of the other particles predicted by these GUTs were beyond the abilities of current experiments to detect. For instance, a wide class of particles known as the X and Y bosons are predicted to mediate the coupling of the electroweak and strong forces, but these particles are extremely heavy and well beyond the capabilities of any reasonable particle accelerator to create.

42.7 Searches for magnetic monopoles

A number of attempts have been made to detect magnetic monopoles. One of the simpler ones is to use a loop of superconducting wire to look for even tiny magnetic sources, a so-called "superconducting quantum interference device", or SQUID. Given the predicted density, loops small enough to fit on a lab bench would expect to see about one monopole event per year. Although there have been tantalizing events recorded, in particular the event recorded by Blas Cabrera on the night of February 14, 1982 (thus, sometimes referred to as the "Valentine's Day Monopole"[38]), there has never been reproducible evidence for the existence of magnetic monopoles.[13] The lack of such events places a limit on the number of monopoles of about one monopole per 10^{29} nucleons.

Another experiment in 1975 resulted in the announcement of the detection of a moving magnetic monopole in cosmic rays by the team led by P. Buford Price.[12] Price later retracted his claim, and a possible alternative explanation was offered by Alvarez.[39] In his paper it was demonstrated that the path of the cosmic ray event that was claimed to be due to a magnetic monopole could be reproduced by the path followed by a platinum nucleus decaying first to osmium, and then to tantalum.

Other experiments rely on the strong coupling of monopoles with photons, as is the case for any electrically-charged particle as well. In experiments involving photon exchange in particle accelerators, monopoles should be produced in reasonable numbers, and detected due to their effect on the scattering of the photons. The probability of a particle being created in such experiments is related to their mass – with heavier particles being less likely to be created – so by examining the results of such experiments, limits on the mass of a magnetic monopole can be calculated. The most recent such experiments suggest that monopoles with masses below 600 GeV/c^2 do not exist, while upper limits on their mass due to the very existence of the universe – which would have collapsed by now if they were too heavy – are about 10^{17} GeV/c^2.

The MoEDAL experiment, installed at the Large Hadron Collider, is currently searching for magnetic monopoles and large supersymmetric particles using layers of special plastic sheets attached to the walls around LHCb's VELO detector. The particles it is looking for will damage the sheets along their path, with various identifying features.

The Russian astrophysicist Igor Novikov claims the fields of macroscopic black holes to be potential magnetic monopoles, representing the entrance to an Einstein–Rosen bridge.[40]

42.8 "Monopoles" in condensed-matter systems

Since around 2003, various condensed-matter physics groups have used the term "magnetic monopole" to describe a different and largely unrelated phenomenon.[18][19]

A true magnetic monopole would be a new elementary particle, and would violate the law $\nabla\cdot\mathbf{B} = 0$. A monopole of this kind, which would help to explain the law of charge quantization as formulated by Paul Dirac in 1931,[41] has never been observed in experiments.

The monopoles studied by condensed-matter groups have none of these properties. They are not a new elementary particle, but rather are an emergent phenomenon in systems of everyday particles (protons, neutrons, electrons, photons); in other words, they are quasi-particles. They are not sources for the **B**-field (i.e., they do not violate $\nabla\cdot\mathbf{B} = 0$); instead, they are sources for other fields, for example the **H**-field,[5] or the "B*-field" (related to superfluid vorticity).[6] They are not directly relevant to grand unified theories or other aspects of particle physics, and do not help explain charge quantization—except insofar as studies of analogous situations can help confirm that the mathematical analyses involved are sound.[42]

There are a number of examples in condensed-matter physics where collective behavior leads to emergent phenomena that resemble magnetic monopoles in certain respects,[17][43][44][45] including most prominently the spin ice materials.[5][46] While these should not be confused with hypothetical elementary monopoles existing in the vacuum, they nonetheless have similar properties and can be probed using similar techniques.

Some researchers use the term **magnetricity** to describe the manipulation of magnetic monopole quasiparticles in spin ice,[46][47] in analogy to the word "electricity".

One example of the work on magnetic monopole quasiparticles is a paper published in the journal *Science* in September 2009, in which researchers Jonathan Morris and Alan Tennant from the Helmholtz-Zentrum Berlin für Materialien und Energie (HZB) along with Santiago Grigera from Instituto de Física de Líquidos y Sistemas Biológicos (IFLYSIB, CONICET) and other colleagues from Dresden University of Technology, University of St. Andrews and Oxford University described the observation of quasiparticles resembling magnetic monopoles. A single crystal of the spin ice material dysprosium titanate was cooled to a temperature between 0.6 kelvin and 2.0 kelvin. Using observations of neutron scattering, the magnetic moments were shown to align into interwoven tubelike bundles resembling Dirac strings. At the defect formed by the end of each tube, the magnetic field looks like that of a monopole. Using an applied magnetic field to break the symmetry of the system, the researchers were able to control the density and orientation of these strings. A contribution to the heat capacity of the system from an effective gas of these quasiparticles was also described.[16][48]

This research went on to win the 2012 Europhysics Prize for condensed matter physics.

Another example is a paper in the February 11, 2011 issue of *Nature Physics* which describes creation and measurement of long-lived magnetic monopole quasiparticle currents in spin ice. By applying a magnetic-field pulse to crystal of dysprosium titanate at 0.36 K, the authors created a relaxing magnetic current that lasted for several minutes. They measured the current by means of the electromotive force it induced in a solenoid coupled to a sensitive amplifier, and quantitatively described it using a chemical kinetic model of point-like charges obeying the Onsager–Wien mechanism of carrier dissociation and recombination. They thus derived the microscopic parameters of monopole motion in spin ice and identified the distinct roles of free and bound magnetic charges.[47]

In superfluids, there is a field **B***, related to superfluid vorticity, which is mathematically analogous to the magnetic **B**-field. Because of the similarity, the field **B*** is called a "synthetic magnetic field". In January 2014, it was reported that monopole quasiparticles[49] for the **B*** field were created and studied in a spinor Bose–Einstein condensate.[6] This constitutes the first example of a quasi-magnetic monopole observed within a system governed by quantum field theory.[42]

42.9 Further descriptions in particle physics

In physics the phrase "magnetic monopole" usually denoted a Yang–Mills potential A and Higgs field ϕ whose equations of motion are determined by the Yang–Mills action

$$\int (F_A, F_A) + (D_A\phi, D_A\phi) - \lambda(1 - \|\phi\|^2)^2.$$

In mathematics, the phrase customarily refers to a static solution to these equations in the Bogomolny–Parasad–Sommerfeld limit $\lambda \to \phi$ which realizes, within topological class, the absolutes minimum of the functional

$$\int_{R^3} (F_A, F_A) + (D_A\phi, D_A\phi).$$

This means that it in a connection A on a principal G-bundle over $\mathbf{R}^3$ (c.f. also Connections on a manifold; principal G-object) and a section ϕ of the associated adjoint bundle of Lie algebras such that the curvature FA and covariant derivative $DA\ \phi$ satisfy the Bogomolny equations

$$F_A = *D_A\phi$$

and the boundary conditions.

$$\|\phi\| = 1 - \frac{m}{r} + \theta(r^2), \quad \|D_A\phi\| = \mathcal{O}(r^2)$$

Pure mathematical advances in the theory of monopoles from the 1980s onwards have often proceeded on the basis of physically motived questions.

The equations themselves are invariant under gauge transformation and orientation-preserving symmetries. When γ is large, $\phi/\|\phi\|$ defines a mapping from a 2-sphere of radius γ in $\mathbf{R}^3$ to an adjoint orbit G/k and the homotopy class of this mapping is called the magnetic charge. Most work has been done in the case G = SU(2), where the charge is a positive integer k. The absolute minimum value of the functional is then $8\pi k$ and the coefficient m in the asymptotic expansion of $\phi/\|\phi\|$ is $k/2$.

The first SU(2) solution was found by E. B. Bogomolny, J. K. Parasad and C. M. Sommerfield in 1975. It is spherically symmetric of charge 1 and has the form

$$A = \left(\frac{1}{\sinh\gamma} - \frac{1}{\gamma}\right) \epsilon_{ijk}\frac{x_j}{\gamma}\sigma_k\, dx_i,$$

$$\phi = \left(\frac{1}{\tanh\gamma} - \frac{1}{\gamma}\right) \frac{x_j}{\gamma}\sigma_i$$

In 1980, C.H.Taubes[50] showed by a gluing construction that there exist solutions for all large k and soon after explicit axially-symmetric solutions were found. The first exact solution in the general case was given in 1981 by R.S.Ward for k = 2 in terms of elliptic functions.

There are two ways of solving the Bogomolny equations. The first is by twistor methods. In the formulation of N.J. Hitchin,[51] an arbitrary solution corresponds to a holomorphic vector bundle over the complex surface TP^1, the tangent bundle of the projective line. This is naturally isomorphic to the space of oriented straight lines in $\mathbf{R}^3$.

The boundary condition show that the holomorphic bundle is an extension of line bundles determined by a compact algebraic curve of genus $(k-1)^2$ (the spectral curve) in TP^1, satisfying certain constraints.

The second method, due to W.Nahm,[52] involves solving an eigen value problem for the coupled Dirac operator and transforming the equations with their boundary conditions into a system of ordinary differential equations, the Nahm equations.

$$\frac{dT_1}{ds} = [T_2, T_3],\ \frac{dT_2}{ds} = [T_3, T_1],\ \frac{dT_3}{ds} = [T_1, T_2]$$

where $Ti(s)$ is a $k \times k$ -matrix valued function on (0,2).

Both constructions are based on analogous procedures for instantons, the key observation due to N.S.Manton being of the self-dual Yang–Mills equations (c.f. also Yang–Mills field) in $\mathbf{R}^4$.

The equivalence of the two methods for SU(2) and their general applicability was established in[53] (see also[54]). Explicit formulas for A and ϕ are difficult to obtain by either method, despite some exact solutions of Nahm's equations in symmetric situations.[55]

Maximally imbedded spherically symmetric magnetic monopole solutions in the Bogolomony-Parasad-Sommerfield limit for the gauge group SU(n) were exhibited by Bais.[56][57] Gannoulis, Goddard and Olive,[58] and Farwell and Minami [59] showed that maximally imbedded spherically symmetric magnetic monopole solutions in the Bogolomony-Parasad-Sommerfield limit for an arbitrary simple gauge group G corresponding to a Lie Algebra with Cartan matrix K and level vector[60] R, are solutions to the Toda molecule[61][62] equation:

$$\frac{d^2\theta_i}{dr^2} = \exp K_{ij}\theta_j,\ where: B_r(r,\hat{k}) \ = \frac{1}{2}\{\frac{d^2\theta_i}{dr^2} - \frac{\bar{R}_i}{r^2}\}H_i,\ \phi(r,\hat{k}) = \frac{1}{2}\{\frac{d\theta_i}{dr} + \frac{\bar{R}_i}{r}\}H_i$$

Non-singular solutions have a magnetic field vanishes at the origin. Explicit finite energy solutions for the Lie Algebras A_n, B_n and C_n have been obtained using this method.

The case of a more general Lie group G, where the stabilizer of ϕ at infinity is a maximal torus, was treated by M.K.Murray from the twistor point of view,where the single spectral curve of an SU(2)-monopole is replaced by a collection of curves indexed by the vertices of theDynkin diagramof$G$.The corresponding Nahm construction was designed by J.Hustubise and Murray.[64]

The moduli space (c.f. also Moduli theory) of all SU(2) monopoles of charge k up to gauge equivalence was shown by Taubes[65] to be a smooth non-compact manifold of dimension $4k-1$. Restricting to gauge transformations that preserve the connection at infinity gives a $4k$-dimensional manifold Mk, which is a circle bundle over the true moduli space and carries a natural complete hyper-Kähler metric[66] (c.f. also Kähler–Einstein manifold). With suspected to any of the complex structures of the hyper-Kähler family, this manifold is holomorphically equivalent to the space of based rational mapping of degree k from P_1 to itself.[67]

The metric is known in twistor terms,[66] and its Kähler potential can be written using the Riemann theta functions of the spectral curve,[54] but only the case $k = 2$ is known in a more conventional and usable form[66] (as of 2000). This Atiyah–Hitchin manifold, the Einstein Taub-NUT metric and $\mathbf{R}^4$ are the only 4-dimensional complete hyper-Kähler manifolds with a non-triholomorphic SU(2) action. Its geodesics have been studied and a programme of Manton concerning monopole dynamics put into effect. Further dynamical features have been elucidated by numerical and analytical techniques.

A cyclic k-fold conering of Mk splits isometrically is a product $Mk \times S^1 \times \mathbf{R}^3$, where Mk is the space of strongly centred monopoles. This space features in an application of S-duality in theoretical physics, and in[68] G.B.Segal and A.Selby studied its topology and the L^2 harmonic forms defined on it, partially confirming the physical prediction.

Magnetic monopole on hyperbolic three-space were investigated from the twistor point of view by M. F. Atiyah[69] (replacing the complex surface TP^1 by the complement of the anti-diagonal in $P^1 \times P^1$) and in terms of discrete Nahm equations by Murray and M. A. Singer.[70]

42.10 See also

- Horizon problem
- Flatness problem
- Bogomolny equations
- Dirac string
- Dyon
- Felix Ehrenhaft
- Gauss's law for magnetism
- Halbach array
- Instanton
- Meron
- Soliton
- 't Hooft–Polyakov monopole
- Wu–Yang monopole

42.11 Notes

[1] Dark Cosmos: In Search of Our Universe's Missing Mass and Energy, by Dan Hooper, p192

[2] Particle Data Group summary of magnetic monopole search

[3] Wen, Xiao-Gang; Witten, Edward, *Electric and magnetic charges in superstring models*, Nuclear Physics B, Volume 261, p. 651–677

[4] S. Coleman, *The Magnetic Monopole 50 years Later*, reprinted in *Aspects of Symmetry*

[5] C. Castelnovo, R. Moessner and S. L. Sondhi (January 3, 2008). "Magnetic monopoles in spin ice". *Nature* **451**: 42–45. arXiv:0710.5515. Bibcode:2008Natur.451...42C. doi:10.1038/nature06433.

[6] Ray, M.W.; Ruokokoski, E.; Kandel, S.; Möttönen, M.; Hall, D. S. (2014). "Observation of Dirac monopoles in a synthetic magnetic field". *Nature* **505** (7485): 657–660. arXiv:1408.3133. Bibcode:2014Natur.505..657R. doi:10.1038/nature12954. ISSN 0028-0836.

[7] The encyclopædia britannica, Volume 17, p352

[8] Principles of Physics by William Francis Magie, p424

[9] Pierre Curie, *Sur la possibilité d'existence de la conductibilité magnétique et du magnétisme libre* (*On the possible existence of magnetic conductivity and free magnetism*), Séances de la Société Française de Physique (Paris), p76 (1894). (French)Free access online copy.

[10] Paul Dirac, "Quantised Singularities in the Electromagnetic Field". Proc. Roy. Soc. (London) **A 133**, 60 (1931). Journal Site, Free Access .

[11] Lecture notes by Robert Littlejohn, University of California, Berkeley, 2007–8

[12] P. B. Price; E. K. Shirk; W. Z. Osborne; L. S. Pinsky (August 25, 1975). "Evidence for Detection of a Moving Magnetic Monopole". *Physical Review Letters* (American Physical Society) **35** (8): 487–490. Bibcode:1975PhRvL..35..487P. doi:10.1103/PhysRevLett.35.487.

[13] Blas Cabrera (May 17, 1982). "First Results from a Superconductive Detector for Moving Magnetic Monopoles". *Physical Review Letters*(American Physical Society)**48**(20): 1378–1381. Bibcode:1982PhRvL..48.1378C.doi:10.1103/PhysRev378.

[14] Milton p.60

[15] Polchinski, arXiv 2003

[16] "Magnetic Monopoles Detected in a Real Magnet for the First Time". Science Daily. September 4, 2009. Retrieved September 4, 2009.

[17] Making magnetic monopoles, and other exotica, in the lab, Symmetry Breaking, January 29, 2009. Retrieved January 31, 2009.

[18] Magnetic monopoles spotted in spin ices, September 3, 2009. "Oleg Tchernyshyov at Johns Hopkins University [a researcher in this field] cautions that the theory and experiments are specific to spin ices, and are not likely to shed light on magnetic monopoles as predicted by Dirac."

[19] Elizabeth Gibney(29January2014). "Quantum cloud simulates magnetic monopole".*Nature(news section)*. doi:10.1038/4612. "This is not the first time that physicists have created monopole analogues. In 2009, physicists observed magnetic monopoles in a crystalline material called spin ice, which, when cooled to near-absolute zero, seems to fill with atom-sized, classical monopoles. These are magnetic in a true sense, but cannot be studied individually. Similar analogues have also been seen in other materials, such as in superfluid helium.... Steven Bramwell, a physicist at University College London who pioneered work on monopoles in spin ices, says that the [2014 experiment led by David Hall] is impressive, but that what it observed is not a Dirac monopole in the way many people might understand it. "There's a mathematical analogy here, a neat and beautiful one. But they're not magnetic monopoles."

[20] The fact that the electric and magnetic fields can be written in a symmetric way is specific to the fact that space is three-dimensional. When the equations of electromagnetism are extrapolated to other dimensions, the magnetic field is described as being a rank-two antisymmetric tensor, whereas the electric field remains a true vector. In dimensions other than three, these two mathematical objects do not have the same number of components.

[21] http://www.ieeeghn.org/wiki/index.php/STARS:Maxwell%27s_Equations

[22] Parker, C.B. (1994). *McGraw-Hill Encyclopaedia of Physics* (2nd ed.). McGraw-Hill. ISBN 0-07-051400-3.

[23] M. Mansfield, C. O'Sullivan (2011). *Understanding Physics* (4th ed.). John Wiley & Sons. ISBN 978-0-47-0746370.

[24] F. Moulin (2001). "Magnetic monopoles and Lorentz force". *Nuovo Cimento B* **116** (8): 869–877. arXiv:math-ph/0203043. Bibcode:2001NCimB.116..869M.

[25] Wolfgang Rindler (November 1989). "Relativity and electromagnetism: The force on a magnetic monopole". *American Journal of Physics* (American Journal of Physics) **57** (11): 993–994. Bibcode:1989AmJPh..57..993R. doi:10.1119/1.15782.

[26] For the convention where magnetic charge has units of webers, see Jackson 1999. In particular, for Maxwell's equations, see section 6.11, equation (6.150), page 273, and for the Lorentz force law, see page 290, exercise 6.17(a). For the convention where magnetic charge has units of ampere-meters, see (for example) arXiv:physics/0508099v1, eqn (4).

[27] J.A. Heras, G. Baez (2009). "The covariant formulation of Maxwell's equations expressed in a form independent of specific units". arXiv:0901.0194.

[28] F. Moulin (2002). "Magnetic monopoles and Lorentz force". arXiv:math-ph/0203043.

[29] Shanmugadhasan, S (1952). "The Dynamical Theory of Magnetic Monopoles".*Canadian Journal of Physics***30**: 218. Bibcode doi:10.1139/p52-021.

[30] Fryberger, D (1989). "On Generalized Electromagnetism and Dirac Algebra"(PDF).*Foundations of Physics***19**: 125. Bibcode: doi:10.1007/bf00734522.

[31] Jackson 1999, section 6.11.

[32] Jackson 1999, section 6.11, equation (6.153), page 275

[33] Nima Arkani-Hamed, Lubos Motl, Alberto Nicolis, Cumrun Vafa: The String Landscape, Black Holes and Gravity as the Weakest Force(arXiv:hep-th/0601001, JHEP 0706:060,2007)

[34] Zel'dovich, Ya. B.; Khlopov, M. Yu. (1978). "On the concentration of relic monopoles in the universe". *Phys. Lett.* **B79** (3): 239–41. Bibcode:1978PhLB...79..239Z. doi:10.1016/0370-2693(78)90232-0.

[35] Preskill, John (1979). "Cosmological production of superheavy magnetic monopoles". *Phys. Rev. Lett.* **43** (19): 1365. Bibcode:1979PhRvL..43.1365P. doi:10.1103/PhysRevLett.43.1365.

[36] Preskill, John (1984). "Magnetic Monopoles". *Ann. Rev. Nucl. Part. Sci.* **34**: 461. Bibcode:1984ARNPS..34..461P. doi:10.1146/annurev.ns.34.120184.002333.

[37] Rees, Martin. (1998). *Before the Beginning* (New York: Basic Books) p. 185 ISBN 0-201-15142-1

[38] http://www.nature.com/nature/journal/v429/n6987/full/429010a.html

[39] Alvarez, Luis W. "Analysis of a Reported Magnetic Monopole". In ed. Kirk, W. T. *Proceedings of the 1975 international symposium on lepton and photon interactions at high energies*. International symposium on lepton and photon interactions at high energies, Aug 21, 1975. p. 967.

[40] „*If the structures of the magnetic fields appear to be magnetic monopoles, that are macroscopic in size, then this is a wormhole.*" Taken from All About Space, issue No. 24, April 2014, item „Could wormholes really exist?"

[41] "Quantised Singularities in the Electromagnetic Field" Paul Dirac, *Proceedings of the Royal Society*, May 29, 1931. Retrieved February 1, 2014.

[42] Elizabeth Gibney(29January2014). "Quantum cloud simulates magnetic monopole".*Nature(news section)*. doi:10.1034612.

[43] Zhong, Fang; Nagosa, Naoto; Takahashi, Mei S.; Asamitsu, Atsushi; Mathieu, Roland; Ogasawara, Takeshi; Yamada, Hiroyuki; Kawasaki, Masashi; Tokura, Yoshinori; Terakura, Kiyoyuki (2003). "The Anomalous Hall Effect and Magnetic Monopoles in Momentum Space". *Science* **302** (5642): 92–95. arXiv:cond-mat/0310232. Bibcode:2003Sci...302...92F. doi:10.1126/science.1089408.

[44] Inducing a Magnetic Monopole with Topological Surface States, American Association for the Advancement of Science (AAAS) *Science Express* magazine, Xiao-Liang Qi, Rundong Li, Jiadong Zang, Shou-Cheng Zhang, January 29, 2009. Retrieved January 31, 2009.

[45] *Artificial Magnetic Monopoles Discovered*

[46] S. T. Bramwell, S. R. Giblin, S. Calder, R. Aldus, D. Prabhakaran, T. Fennell (15 October 2009). "Measurement of the charge and current of magnetic monopoles in spin ice". *Nature* **461** (7266): 956–959. arXiv:0907.0956. Bibcode:2009Natur.461..956B. doi:10.1038/nature08500. PMID 19829376.

[47] S. R. Giblin, S. T. Bramwell, P. C. W. Holdsworth, D. Prabhakaran & I. Terry (February 13, 2011). "Creation and measurement of long-lived magnetic monopole currents in spin ice" **7** (3). Nature Physics. Bibcode:2011NatPh...7..252G. doi:10.10896. Retrieved February28, 2011.

[48] D.J.P. Morris, D.A. Tennant, S.A. Grigera, B. Klemke, C. Castelnovo, R. Moessner, C. Czter-nasty, M. Meissner, K.C. Rule, J.-U. Hoffmann, K. Kiefer, S. Gerischer, D. Slobinsky, and R.S. Perry (September 3, 2009) [2009-07-09]. "Dirac Strings and Magnetic Monopoles in Spin Ice $Dy_2Ti_2O_7$". *Science* **326** (5951): 411–4. arXiv:1011.1174. Bibcode:2009Sci...326..411M. doi:10.1126/science.1178868. PMID 19729617.

[49] Pietilä, Ville; Möttönen, Mikko (2009). "Creation of Dirac Monopoles in Spinor Bose–Einstein Condensates". *Phys. Rev. Lett.* **103**: 030401. arXiv:0903.4732. Bibcode:2009PhRvL.103c0401P. doi:10.1103/physrevlett.103.030401.

[50] A.Jaffe, C.H.Taubes (1980). *Vortices and monopoles.*

[51] N.J. Hitchin (1982). *Monopoles and geodesics.*

[52] W.Nahm (1982). *The construction of all self-dual monopoles by the ADHM method.*

[53] N.J. Hitchin (1983). *On the construction of monopoles.*

[54] N.J. Hitchin (1999). *Integrable sustems in Riemannian geometry* (K.Uhlenbeck ed.). C-L.Terng (ed.).

[55] N.J. Hitchin, N.S. Manton, M.K. Murray (1995). *Symmetric Monopoles.*

[56] F.A. Bais and H. Weldon, (1978). *Exact Monopole Solutions in SU(N) Gauge Theory*, Phys. Rev. Let. 41, 601.

[57] D. Wilkinson and F.A. Bais, (1979). *Exact SU(N) monopole solutions with spherical symmetry*, Phys. Rev D. 19, 2410

[58] N. Ganoulis, P. Goddard, D. Olive, (1982).*Self dual Monopoles and Toda Molecules*, Nucl. Phys. B205, 601

[59] Farwell, Ruth and Minami, Masatsugu, (1983). *One-dimensional Toda Molecule. 2. The Solutions Applied To Bogomolny Monopoles With Spherical Symmetry*, Prog. Theor. Phys. 70 710.

[60] R. Slansky,(1981). *Group theory for unified model building", Physics Reports, 79, 1. (See table 10 pg. 84 of http://citeseerx.ist.psu.edu/viewdoc/download?doi=10.1.1.126.1581&rep=rep1&type=pdf)*

[61] M. Toda, (1975). *Studies of a non-linear lattice*, Phys. Rep., 8, 1.

[62] B. Kostant, (1979).*The solution to a generalized Toda lattice and representation theory.*, Adv. in Math. 34, 195.

[63] M.K.Murray (1983). *Monopoles and spectral curves for arbitrary Lie groups.*

[64] Hurtubise, Jacques; Murray, Michael K. (1989). "On the construction of monopoles for the classical groups". *Communications in Mathematical Physics* **122** (1): 35–89. Bibcode:1989CMaPh.122...35H. doi:10.1007/bf01221407. MR 994495.

[65] C.H.Taubes (1983). *Stability in Yang–Mills theories.*

[66] M.F. Atiyah; N.J. Hitchin (1988). *The geometry and dynamics of magnetic monopoles.* Princeton Univ.Press.

[67] S.K.Donaldson (1984). *Nahm's equations and the classification of monopoles.*

[68] G.B.Segal, A.Selby (1996). *The cohomology of the space of magnetic monopoles.*

[69] M.F.Atiyah (1987). *Magnetic monopoles in hyperbolic space, Vector bundles on algebraic varieties.* Oxford University Press.

[70] M.K.Murray (2000). *On the complete integrability of the discrete Nahm equations.*

42.12 References

- Brau, Charles A. (2004). *Modern Problems in Classical Electrodynamics.* Oxford University Press. ISBN 0-19-514665-4.
- Hitchin, N.J.; Murray, M.K. (1988). *Spectral curves and the ADHM method.*
- Jackson, John David (1999). *Classical Electrodynamics* (3rd ed.). New York: Wiley. ISBN 0-471-30932-X.
- Milton, Kimball A. (June 2006). "Theoretical and experimental status of magnetic monopoles". *Reports on Progress in Physics***69**(6): 1637–1711. arXiv:hep-ex/0602040. Bibcode:2006RPPh...69.1637M.doi:10.1088/0034/R02.
- Shnir, Yakov M. (2005). *Magnetic Monopoles.* Springer-Verlag. ISBN 3-540-25277-0.
- Sutcliffe, P.M. (1997). *BPS monopoles.*
- Vonsovsky, Sergey V. (1975). *Magnetism of Elemetary Particles.* Mir Publishers.

42.13 External links

- Magnetic Monopole Searches (lecture notes)
- Particle Data Group summary of magnetic monopole search
- 'Race for the Pole' Dr David Milstead Freeview 'Snapshot' video by the Vega Science Trust and the BBC/OU.
- Interview with Jonathan Morris about magnetic monopoles and magnetic monopole quasiparticles. Drillingsraum, April 16, 2010
- *Nature*, 2009
- *Sciencedaily*, 2009
- H. Kadowaki, N. Doi, Y. Aoki, Y.Tabata, T.J. Sato, J.W. Lynn, K. Matsuhira, Z. Hiroi (2009). “Observation of Magnetic Monopoles in Spin Ice”. arXiv:0908.3568.
- *Video of lecture by Paul Dirac on magnetic monopoles*, 1975 on YouTube

Chapter 43

Tachyon

This article is about hypothetical faster-than-light particles. For quantum fields with imaginary mass, see Tachyonic field. For other uses, see Tachyon (disambiguation).

A **tachyon** /ˈtæki.ɒn/ or **tachyonic particle** is a hypothetical particle that always moves faster than light. The word

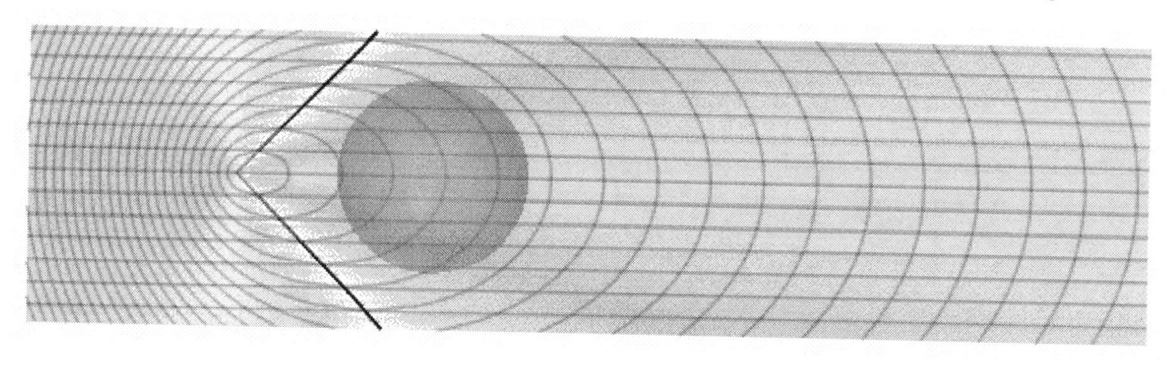

Because a tachyon would always move faster than light, it would not be possible to see it approaching. After a tachyon has passed nearby, we would be able to see two images of it, appearing and departing in opposite directions. The black line is the shock wave of Cherenkov radiation, shown only in one moment of time. This double image effect is most prominent for an observer located directly in the path of a superluminal object (in this example a sphere, shown in grey). The right hand bluish shape is the image formed by the blue-doppler shifted light arriving at the observer—who is located at the apex of the black Cherenkov lines—from the sphere as it approaches. The left-hand reddish image is formed from red-shifted light that leaves the sphere after it passes the observer. Because the object arrives before the light, the observer sees nothing until the sphere starts to pass the observer, after which the image-as-seen-by-the-observer splits into two—one of the arriving sphere (to the right) and one of the departing sphere (to the left).

comes from the Greek: ταχύ pronounced *tachy /ˈtaːxi/*, meaning rapid. It was coined in 1967 by Gerald Feinberg.[1] The complementary particle types are called luxon (always moving at the speed of light) and bradyon (always moving slower than light), which both exist. The possibility of particles moving faster than light was first proposed by O. M. P. Bilaniuk, V. K. Deshpande, and E. C. G. Sudarshan in 1962, although the term they used for it was "meta-particle".[2]

Most physicists think that faster-than-light particles cannot exist because they are not consistent with the known laws of physics.[3][4] If such particles did exist, they could be used to build a tachyonic antitelephone and send signals faster than light, which (according to special relativity) would lead to violations of causality.[4] Potentially consistent theories that allow faster-than-light particles include those that break Lorentz invariance, the symmetry underlying special relativity, so that the speed of light is not a barrier.

In the 1967 paper that coined the term,[1] Feinberg proposed that tachyonic particles could be quanta of a quantum field with negative squared mass. However, it was soon realized that excitations of such imaginary mass fields do *not* in fact propagate faster than light,[5] and instead represent an instability known as tachyon condensation.[3] Nevertheless, negative squared mass fields are commonly referred to as "tachyons",[6] and in fact have come to play an important role in modern physics.

Despite theoretical arguments against the existence of faster-than-light particles, experiments have been conducted to search for them. No compelling evidence for their existence has been found. In September 2011, it was reported that a tau neutrino had travelled faster than the speed of light in a major release by CERN; however, later updates from CERN on the OPERA project indicate that the faster-than-light readings were resultant from “a faulty element of the experiment’s fibre optic timing system”.[7]

43.1 Tachyons in relativistic theory

In special relativity, a faster-than-light particle would have space-like four-momentum,[1] in contrast to ordinary particles that have time-like four-momentum. It would also have imaginary mass and proper time. Being constrained to the spacelike portion of the energy–momentum graph, it could not slow down to subluminal speeds.[1]

43.1.1 Mass

Main articles: Mass § Tachyonic particles and imaginary (complex) mass and Tachyonic field

In a Lorentz invariant theory, the same formulas that apply to ordinary slower-than-light particles (sometimes called "bradyons" in discussions of tachyons) must also apply to tachyons. In particular the energy–momentum relation:

$$E^2 = p^2c^2 + m^2c^4$$

(where **p** is the relativistic momentum of the bradyon and **m** is its rest mass) should still apply, along with the formula for the total energy of a particle:

$$E = \frac{mc^2}{\sqrt{1 - \frac{v^2}{c^2}}}.$$

This equation shows that the total energy of a particle (bradyon or tachyon) contains a contribution from its rest mass (the “rest mass–energy”) and a contribution from its motion, the kinetic energy. When v is larger than c, the denominator in the equation for the energy is “imaginary”, as the value under the radical is negative. Because the total energy must be real, the numerator must *also* be imaginary: i.e. the rest mass **m** must be imaginary, as a pure imaginary number divided by another pure imaginary number is a real number.

43.1.2 Speed

One curious effect is that, unlike ordinary particles, the speed of a tachyon *increases* as its energy decreases. In particular, E approaches zero when v approaches infinity. (For ordinary bradyonic matter, E increases with increasing speed, becoming arbitrarily large as v approaches c, the speed of light). Therefore, just as bradyons are forbidden to break the light-speed barrier, so too are tachyons forbidden from slowing down to below c, because infinite energy is required to reach the barrier from either above or below.

As noted by Einstein, Tolman, and others, special relativity implies that faster-than-light particles, if they existed, could be used to communicate backwards in time.[8]

43.1.3 Neutrinos

In 1985 Chodos et al. proposed that neutrinos can have a tachyonic nature.[9] The possibility of standard model particles moving at superluminal speeds can be modeled using Lorentz invariance violating terms, for example in the Standard-Model Extension.[10][11][12] In this framework, neutrinos experience Lorentz-violating oscillations and can travel faster than light at high energies. This proposal was strongly criticized.[13]

43.1.4 Cherenkov radiation

A tachyon with an electric charge would lose energy as Cherenkov radiation[14]—just as ordinary charged particles do when they exceed the local speed of light in a medium. A charged tachyon traveling in a vacuum therefore undergoes a constant proper time acceleration and, by necessity, its worldline forms a hyperbola in space-time. However reducing a tachyon's energy *increases* its speed, so that the single hyperbola formed is of *two* oppositely charged tachyons with opposite momenta (same magnitude, opposite sign) which annihilate each other when they simultaneously reach infinite speed at the same place in space. (At infinite speed, the two tachyons have no energy each and finite momentum of opposite direction, so no conservation laws are violated in their mutual annihilation. The time of annihilation is frame dependent.)

Even an electrically neutral tachyon would be expected to lose energy via gravitational Cherenkov radiation, because it has a gravitational mass, and therefore increase in speed as it travels, as described above. If the tachyon interacts with any other particles, it can also radiate Cherenkov energy into those particles. Neutrinos interact with the other particles of the Standard Model, and Andrew Cohen and Sheldon Glashow recently used this to argue that the faster-than-light neutrino anomaly cannot be explained by making neutrinos propagate faster than light, and must instead be due to an error in the experiment.[15]

43.1.5 Causality

Causality is a fundamental principle of physics. If tachyons can transmit information faster than light, then according to relativity they violate causality, leading to logical paradoxes of the "kill your own grandfather" type. This is often illustrated with thought experiments such as the "tachyon telephone paradox"[8] or "logically pernicious self-inhibitor."[16]

The problem can be understood in terms of the relativity of simultaneity in special relativity, which says that different inertial reference frames will disagree on whether two events at different locations happened "at the same time" or not, and they can also disagree on the order of the two events (technically, these disagreements occur when spacetime interval between the events is 'space-like', meaning that neither event lies in the future light cone of the other).[17]

If one of the two events represents the sending of a signal from one location and the second event represents the reception of the same signal at another location, then as long as the signal is moving at the speed of light or slower, the mathematics of simultaneity ensures that all reference frames agree that the transmission-event happened before the reception-event.[17] However, in the case of a hypothetical signal moving faster than light, there would always be some frames in which the signal was received before it was sent, so that the signal could be said to have moved backwards in time. Because one of the two fundamental postulates of special relativity says that the laws of physics should work the same way in every inertial frame, if it is possible for signals to move backwards in time in any one frame, it must be possible in all frames. This means that if observer A sends a signal to observer B which moves faster than light in A's frame but backwards in time in B's frame, and then B sends a reply which moves faster than light in B's frame but backwards in time in A's frame, it could work out that A receives the reply before sending the original signal, challenging causality in *every* frame and opening the door to severe logical paradoxes.[18] Mathematical details can be found in the tachyonic antitelephone article, and an illustration of such a scenario using spacetime diagrams can be found in *Baker, R. (2003)*[19]

Reinterpretation principle

The **reinterpretation principle**[1][2][18] asserts that a tachyon sent *back* in time can always be *reinterpreted* as a tachyon traveling *forward* in time, because observers cannot distinguish between the emission and absorption of tachyons. The attempt to *detect* a tachyon *from* the future (and violate causality) would actually *create* the same tachyon and send it *forward* in time (which is causal).

However, this principle is not widely accepted as resolving the paradoxes.[8][18][20] Instead, what would be required to avoid paradoxes is that unlike any known particle, tachyons do not interact in any way and can never be detected or observed, because otherwise a tachyon beam could be modulated and used to create an anti-telephone[8] or a "logically pernicious self-inhibitor".[16] All forms of energy are believed to interact at least gravitationally, and many authors state that superluminal propagation in Lorentz invariant theories always leads to causal paradoxes.[21][22]

43.2 Fundamental models

In modern physics, all fundamental particles are regarded as excitations of quantum fields. There are several distinct ways in which tachyonic particles could be embedded into a field theory.

43.2.1 Fields with imaginary mass

Main article: Tachyonic field

In the paper that coined the term "tachyon", Gerald Feinberg studied Lorentz invariant quantum fields with imaginary mass.[1] Because the group velocity for such a field is superluminal, naively it appears that its excitations propagate faster than light. However, it was quickly understood that the superluminal group velocity does not correspond to the speed of propagation of any localized excitation (like a particle). Instead, the negative mass represents an instability to tachyon condensation, and all excitations of the field propagate subluminally and are consistent with causality.[5] Despite having no faster-than-light propagation, such fields are referred to simply as "tachyons" in many sources.[3][6][23][24][25][26]

Tachyonic fields play an important role in modern physics. Perhaps the most famous is the Higgs boson of the Standard Model of particle physics, which—in its uncondensed phase—has an imaginary mass. In general, the phenomenon of spontaneous symmetry breaking, which is closely related to tachyon condensation, plays a very important role in many aspects of theoretical physics, including the Ginzburg–Landau and BCS theories of superconductivity. Another example of a tachyonic field is the tachyon of bosonic string theory.[23][25][27]

Tachyons are predicted by bosonic string theory and also the NS (which is the open bosonic sector) and NS-NS (which is the closed bosonic sector) sectors of RNS Superstring theory before GSO projection. However, due to the Sen conjecture—also known as tachyon condensation—this is not possible. This resulted in the necessity for the GSO projection.

43.2.2 Lorentz-violating theories

In theories that do not respect Lorentz invariance the speed of light is not (necessarily) a barrier, and particles can travel faster than the speed of light without infinite energy or causal paradoxes.[21] A class of field theories of that type are the so-called Standard Model extensions. However, the experimental evidence for Lorentz invariance is extremely good, so such theories are very tightly constrained.[28][29]

43.2.3 Fields with non-canonical kinetic term

By modifying the kinetic energy of the field, it is possible to produce Lorentz invariant field theories with excitations that propagate superluminally.[5][22] However, such theories in general do not have a well-defined Cauchy problem (for reasons related to the issues of causality discussed above), and are probably inconsistent quantum mechanically.

43.3 History

As mentioned above, the term "tachyon" was coined by Gerald Feinberg in a 1967 paper titled "Possibility of Faster-Than-Light Particles".[1] He had been inspired by the science-fiction story "Beep" by James Blish.[30] Feinberg studied the kinematics of such particles according to special relativity. In his paper he also introduced fields with imaginary mass (now also referred to as "tachyons") in an attempt to understand the microphysical origin such particles might have.

The first hypothesis regarding faster-than-light particles is sometimes attributed to German physicist Arnold Sommerfeld in 1904,[31] and more recent discussions happened in 1962[2] and 1969.[32]

43.4 In fiction

Main article: Tachyons in fiction

Tachyons have appeared in many works of fiction. They have been used as a standby mechanism upon which many science fiction authors rely to establish faster-than-light communication, with or without reference to causality issues. The word *tachyon* has become widely recognized to such an extent that it can impart a science-fictional connotation even if the subject in question has no particular relation to superluminal travel (a form of technobabble, akin to *positronic brain*).

- In the limited series *Watchmen*, Dr. Manhattan's ability to see into the future is blocked by tachyons generated by Adrian Veidt.
- In the book series *Koban*, by Stephen W. Bennett, tachyons are used to generate power and allow for faster-than-light travel.
- In the book series The Missing by Margaret Peterson Haddix, tachyons are used for transporting “doomed” children across time.
- In the movie, *K-PAX*, Kevin Spacey’s character claims to have traveled to Earth at Tachyon speeds.
- Tachyons figure prominently in the *Star Trek* universe where they are often associated with time travel scenarios.[33]
- In *The Flash* TV series, tachyons play a significant role. Eobard Thawne uses a stolen tachyon prototype to temporarily restore his super-speed.
- In the Disney film *Tomorrowland*, tachyons are said to power the Thinking Machine, rockets, and the Monitor, among other technologies.
- In the film *Land of The Lost*, Rick Marshall, played by Will Ferrell, builds a tachyon amplifier which enables the user to travel “sideways in time.”
- In the film *Fantastic Four: Rise of the Silver Surfer*,a tachyon pulse is used by the Fantastic Four to remove the Silver Surfer from his board so they can study it.
- In the novel *Timescape*, author Gregory Benford employs the tachyon to allow communication backwards in time.

43.5 See also

- Massive particle (bradyon)
- Massless particle (luxon)
- Lorentz-violating neutrino oscillations
- Retrocausality
- Tachyonic antitelephone
- Wheeler–Feynman absorber theory

43.6 References

[1] Feinberg,G. (1967). “Possibility of Faster-Than-Light Particles”.*Physical Review***159**(5): 1089–1105. Bibcode:1967P. doi:10.1103/PhysRev.159.1089. See also Feinberg’s later paper: Phys. Rev. D 17, 1651 (1978)

[2] Bilaniuk, O.-M. P.; Deshpande, V. K.; Sudarshan, E. C. G. (1962). "'Meta' Relativity". *American Journal of Physics* **30** (10): 718. Bibcode:1962AmJPh..30..718B. doi:10.1119/1.1941773.

[3] Lisa Randall, *Warped Passages: Unraveling the Mysteries of the Universe's Hidden Dimensions*, p.286: "People initially thought of tachyons as particles travelling faster than the speed of light...But we now know that a tachyon indicates an instability in a theory that contains it. Regrettably for science fiction fans, tachyons are not real physical particles that appear in nature."

[4] Tipler, Paul A.; Llewellyn, Ralph A. (2008). *Modern Physics* (5th ed.). New York: W.H. Freeman & Co. p. 54. ISBN 978-0-7167-7550-8. ... so existence of particles v > c ... Called tachyons ... would present relativity with serious ... problems of infinite creation energies and causality paradoxes.

[5] Aharonov, Y.; Komar, A.; Susskind, L. (1969). "Superluminal Behavior, Causality, and Instability". *Phys. Rev.* (American Physical Society) **182** (5): 1400–1403. Bibcode:1969PhRv..182.1400A. doi:10.1103/PhysRev.182.1400.

[6] A. Sen, "Rolling tachyon," JHEP **0204**, 048 (2002). Cited 720 times as of 2/2012.

[7] "Neutrinos sent from CERN to Gran Sasso respect the cosmic speed limit" (Press release). CERN. 8 June 2012. Retrieved 2012-06-08.

[8] Benford,G.;Book,D.;Newcomb,W. (1970). "The Tachyonic Antitelephone".*Physical Review D***2**(2): 263–265. Bibc3B. doi:10.1103/PhysRevD.2.263.

[9] Chodos, A. (1985). "The Neutrino as a Tachyon". *Physics Letters B* **150** (6): 431–435. Bibcode:1985PhLB..150..431C. doi:10.1016/0370-2693(85)90460-5.

[10] Colladay, D.; Kostelecky, V. A. (1997). "CPT Violation and the Standard Model". *Physical Review D* **55** (11): 6760–6774. arXiv:hep-ph/9703464. Bibcode:1997PhRvD..55.6760C. doi:10.1103/PhysRevD.55.6760.

[11] Colladay, D.; Kostelecky, V. A. (1998). "Lorentz-Violating Extension of the Standard Model". *Physical Review D* **58** (11): 116002. arXiv:hep-ph/9809521. Bibcode:1998PhRvD..58k6002C. doi:10.1103/PhysRevD.58.116002.

[12] Kostelecky, V. A. (2004). "Gravity, Lorentz Violation, and the Standard Model". *Physical Review D* **69** (10): 105009. arXiv:hep-th/0312310. Bibcode:2004PhRvD..69j5009K. doi:10.1103/PhysRevD.69.105009.

[13] R. J. Hughes and G. J. Stephenson, Jr., *Against tachyonic neutrinos*, Phys. Lett. B 244, 95–100 (1990).

[14] Bock, R. K. (9 April 1998). "Cherenkov Radiation". *The Particle Detector BriefBook*. CERN. Retrieved 2011-09-23.

[15] Cohen, Andrew G. & Glashow, Sheldon L. (2011). "Pair Creation Constrains Superluminal Neutrino Propagation". *Phys.Rev.Lett.* **107**: 181803. arXiv:1109.6562. Bibcode:2011PhRvL.107r1803C. doi:10.1103/PhysRevLett.107.181803.

[16] P. Fitzgerald, "Tachyons, Backward Casuation, and Freedom", PSA: Proceedings of the Biennial Meeting of the Philosophy of Science Association, Vol. 1970 (1970), pp. 425–426: "A more powerful argument to show that retrocausal tachyons involve an intolerable conceptual difficulty is illustrated by the Case of the Logically Pernicious Self-Inhibitor..."

[17] Mark, J. "The Special Theory of Relativity" (PDF). University of Cincinnati. pp. 7–11. Archived from the original (PDF) on 2006-09-13. Retrieved 2006-10-27.

[18] Grøn, Ø.; Hervik, S. (2007). *Einstein's General Theory of Relativity: With Modern Applications in Cosmology*. Springer. p. 39. ISBN 978-0-387-69199-2. The tachyon telephone paradox cannot be resolved by means of the reinterpretation principle.

[19] . Baker, R. (12 September 2003). "Relativity, FTL and causality". *Sharp Blue*. Retrieved 2011-09-23.

[20] Erasmo Recami, Flavio Fontana, Roberto Garavaglia, "About Superluminal motions and Special Relativity: A Discussion of some recent Experiments, and the solution of the Causal Paradoxes", International Journal of Modern Physics A15 (2000) 2793–2812, abstract: "it is possible...to solve also the known causal paradoxes, devised for "faster than light" motion, although *this is not widely recognized yet.*" [emphasis added].

[21] Carlos Barceló, Stefano Finazzi, Stefano Liberati, "On the impossibility of superluminal travel: the warp drive lesson", Second prize of the 2009 FQXi essay contest "What is Ultimately Possible in Physics?", p.8: "As a matter of fact, any mechanism for superluminal travel can be easily turned into a time machine and hence lead to the typical causality paradoxes..."

[22] Allan Adams, Nima Arkani-Hamed, Sergei Dubovsky, Alberto Nicolis, Riccardo Rattazzi, "Causality, Analyticity and an IR Obstruction to UV Completion", JHEP 0610 (2006) 014 .

[23] Brian Greene, *The Elegant Universe*, Vintage Books (2000)

[24] Kutasov, David; Marino, Marcos & Moore, Gregory W. (2000). "Some exact results on tachyon condensation in string field theory". *JHEP* **0010**: 045. arXiv EFI-2000-32, RUNHETC-2000-34.

[25] NOVA, "The Elegant Universe", PBS television special, http://www.pbs.org/wgbh/nova/elegant/

[26] G. W. Gibbons, "Cosmological evolution of the rolling tachyon," Phys. Lett. B **537**, 1 (2002)

[27] J. Polchinski, *String Theory*, Cambridge University Press, Cambridge, UK (1998)

[28] Glashow, Sheldon Lee (2004). "Atmospheric neutrino constraints on Lorentz violation".

[29] Coleman, Sidney R. & Glashow, Sheldon L. (1999). "High-energy tests of Lorentz invariance". *Phys.Rev.* **D59**: 116008. arXiv:hep-ph/9812418. Bibcode:1999PhRvD..59k6008C. doi:10.1103/PhysRevD.59.116008.

[30] "He told me years later that he had begun thinking about tachyons because he was inspired by James Blish's [1954] short story, "Beep". In it, a faster-than-light communicator plays a crucial role in a future society, but has an annoying final *beep* at the end of every message. The communicator necessarily allows sending of signals backward in time, even when that's not your intention. Eventually the characters discover that all future messages are compressed into that *beep*, so the future is known, more or less by accident. Feinberg had set out to see if such a gadget was theoretically possible." pg276 of Gregory Benford's "Old Legends"

[31] Sommerfeld, A. (1904). "Simplified deduction of the field and the forces of an electron moving in any given way". *Knkl. Acad. Wetensch* **7**: 345–367.

[32] Bilaniuk,O.-M.P.;Sudarshan,E.C.G. (1969). "Particles beyond the Light Barrier".*Physics Today***22**(5): 43–51. Bibco43B. doi:10.1063/1.3035574.

[33] "Tachyon". *Memory Alpha, the Star Trek Wiki*. 2012-08-09. Retrieved 2014-05-09.

43.7 External links

- The Faster Than Light (FTL) FAQ (from the Internet Archive)
- Weisstein, Eric W., Tachyon from ScienceWorld.
- Tachyon entry from the *Physics FAQ*
-
- Tachyon at Memory Alpha (a *Star Trek* wiki)

Chapter 44

Leptoquark

Leptoquarks are hypothetical particles that carry information between quarks and leptons of a given generation that allow quarks and leptons to interact. They are color-triplet bosons that carry both lepton and baryon numbers. They are encountered in various extensions of the Standard Model, such as technicolor theories or GUTs based on Pati–Salam model, SU(5) or E_6, etc. Their quantum numbers like spin, (fractional) electric charge and weak isospin vary among theories.

Leptoquarks, predicted to be nearly as heavy as an atom of lead, could only be created at high energies, and would decay rapidly. A third generation leptoquark, for example, might decay into a bottom quark and a tau lepton. Some theorists propose that the 'leptoquark' observed by HERA and DESY could be a new force that bonds positrons and quarks or be examples of preons found at high energies.[1] Leptoquarks could explain the reason for the three generations of matter. Furthermore, leptoquarks could explain why the same number of quarks and leptons exist and many other similarities between the quark and the lepton sectors. At high energies, when leptons that do not feel the strong force and quarks that cannot be separately observed because of the strong force become one, it could form a more fundamental particle and describe a higher symmetry. There would be three kinds of leptoquarks made of the leptons and quarks of each generation.

The LHeC project to add an electron ring to collide bunches with the existing LHC proton ring is proposed as a project to look for higher-generation leptoquarks.[2]

44.1 Existence

In 1997, an excess of events at the HERA accelerator created a stir in the particle physics community, because one possible explanation of the excess was the involvement of leptoquarks. However, more recent studies performed both at HERA and at the Tevatron with larger samples of data ruled out this possibility for masses of the leptoquark up to 275-325 GeV.[3] Second generation leptoquarks were also looked for and not found.[4] For leptoquarks to be proven to exist, the missing energy in particle collisions attributed to neutrinos would have to be excessively energetic. It is likely that the creation of leptoquarks would mimic the creation of massive quarks.[5]

44.2 See also

- Quark–lepton complementarity
- X and Y bosons

44.3 References

[1] Scientific American

[2] Birmingham LHeC project page

[3] H1 Collaboration; Andreev, V.; Anthonis, T.; Aplin, S.; Asmone, A.; Astvatsatourov, A.; Babaev, A.; Backovic, S.; Bähr, J.; Baghdasaryan, A.; Baranov, P.; Barrelet, E.; Bartel, W.; Baudrand, S.; Baumgartner, S.; Becker, J.; Beckingham, M.; Behnke, O.; Behrendt, O.; Belousov, A.; Berger, Ch.; Berger, N.; Bizot, J.C.; Boenig, M.-O.; Boudry, V.; Bracinik, J.; Brandt, G.; Brisson, V.; Brown, D.P. et al. (2005). “Search for Leptoquark Bosons in ep Collisions at HERA”. *Physics Letters B* **629**: 9–19. arXiv:hep-ex/0506044. Bibcode:2005PhLB..629....9H. doi:10.1016/j.physletb.2005.09.048.

[4] The Search for Leptoquarks.

[5] Search for Third Generation Leptoquarks

Chapter 45

X and Y bosons

In particle physics, the **X and Y bosons** (sometimes collectively called "**X bosons**"[1]) are hypothetical elementary particles analogous to the W and Z bosons, but corresponding to a new type of force predicted by the Georgi–Glashow model, a grand unified theory.

45.1 Details

The X and Y bosons couple quarks to leptons, allowing violation of the conservation of baryon number, and thus permitting proton decay.

An X boson would have the following decay modes:[2]

X → u + u

X → e+ + d

where the two decay products in each process have opposite chirality, u is an up quark, d is a down quark and e+ is a positron.

A Y boson would have the following decay modes:[2]

Y → e+ + u

Y → d + u

Y → d + ν
e

where the first decay product in each process has left-handed chirality and the second has right-handed chirality and ν
e is an electron antineutrino.

Similar decay products exist for the other quark-lepton generations

In these reactions, neither the lepton number (L) nor the baryon number (B) is conserved, but $B-L$ is. Different branching ratios between the X boson and its antiparticle (as is the case with the K-meson) would explain baryogenesis.

45.2 See also

- $B-L$

- Grand unification theory
- Proton decay
- W' and Z' bosons
- Leptoquark

45.3 References

[1] Ta-Pei Cheng; Ling-Fong Li (1983). *Gauge Theory of Elementary Particle Physics*. Oxford University Press. p. 437. ISBN 0-19-851961-3.

[2] Ta-Pei Cheng; Ling-Fong Li (1983). *Gauge Theory of Elementary Particle Physics*. Oxford University Press. p. 442. ISBN 0-19-851961-3.

Chapter 46

W′ and Z′ bosons

In particle physics, **W′ and Z′ bosons** (or **W-prime and Z-prime bosons**) refer to hypothetical new gauge bosons that arise from extensions of the electroweak symmetry of the Standard Model. They are named in analogy with the Standard Model W and Z bosons.

46.1 Types

46.1.1 Types of W′ bosons

W′ bosons often arise in models with an extra SU(2) gauge group. SU(2) × SU(2) is spontaneously broken to the diagonal subgroup SU(2)W which corresponds to the electroweak SU(2). More generally, we might have n copies of SU(2), which are then broken down to a diagonal SU(2)W. This gives rise to n^2-1 $W^{+\prime}$, $W^{-\prime}$ and Z′ bosons. Such models might arise from quiver diagram, for example. In order for the W′ bosons to couple to weak isospin, the extra SU(2) and the Standard Model SU(2) must mix; one copy of SU(2) must break around the TeV scale (to get W′ bosons with a TeV mass) leaving a second SU(2) for the Standard Model. This happens in Little Higgs models that contain more than one copy of SU(2). Because the W′ comes from the breaking of an SU(2), it is generically accompanied by a Z′ boson of (almost) the same mass and with couplings related to the W′ couplings.

Another model with W′ bosons but without an additional SU(2) factor is the so-called 331 model with $\beta = \pm\, ^1/\sqrt{3}$. The symmetry breaking chain SU(3)L × U(1)W → SU(2)W × U(1)Y leads to a pair of $W'^{\pm}$ bosons and three Z′ bosons.

W′ bosons also arise in Kaluza–Klein theories with SU(2) in the bulk.

46.1.2 Types of Z′ bosons

Various models of physics beyond the Standard Model predict different kinds of Z′ bosons.

- Models with a new U(1) gauge symmetry. The Z′ is the gauge boson of the (broken) U(1) symmetry.
- E_6 models. This type of model contains two Z′ bosons, which can mix in general.
- Topcolor and Top Seesaw Models of Dynamical Electroweak Symmetry Breaking have Z′ bosons to select the formation of particular condensates.
- Little Higgs models. These models typically include an enlarged gauge sector, which is broken down to the Standard Model gauge symmetry around the TeV scale. In addition to one or more Z′ bosons, these models often contain W′ bosons.
- Kaluza–Klein models. The Z′ boson are the excited modes of a neutral bulk gauge symmetry.

- Stueckelberg Extensions (see Stueckelberg action). The Z' boson is sourced from couplings found in string theories with intersecting D-branes.

46.2 Searches

46.2.1 Direct searches

A W' boson could be detected at hadron colliders through its decay to lepton plus neutrino or top quark plus bottom quark, after being produced in quark–antiquark annihilation. The LHC reach for W' discovery is expected to be a few TeV.

Direct searches for Z' bosons are carried out at hadron colliders, since these give access to the highest energies available. The search looks for high-mass dilepton resonances: the Z' boson would be produced by quark–antiquark annihilation and decay to an electron-positron pair or a pair of opposite-charged muons. The most stringent current limits come from the Fermilab Tevatron, and depend on the couplings of the Z' boson (which control the production cross section); as of 2006, the Tevatron excludes Z' bosons up to masses of about 800 GeV for "typical" cross sections predicted in various models.[2]

The above statements apply to "wide width" models. Recent classes of models have emerged that naturally provide cross section signatures that fall on the edge, or slightly below the 95 confidence level limits set by the Tevatron, and hence can produce detectable cross section signals for a Z' boson in a mass range much closer to the Z pole mass than the "wide width" models discussed above.

These "narrow width" models which fall into this category are those that predict a Stückelberg Z' as well as a Z' from a universal extra dimension (see the Z' Hunter's Guide for links to these papers).

On April 7, 2011, the CDF collaboration at the Tevatron reported an excess in proton–antiproton collision events that produce a W boson accompanied by two hadronic jets. This could possibly be interpreted in terms of a Z' boson.[3][4]

On June 2, 2015, the ATLAS experiment at the LHC reported evidence for W' bosons at significance 3.4 sigma, still too low to claim a formal discovery.[5] Researchers at the CMS experiment also independently reported signals that corroborate ATLAS's findings.

46.2.2 Indirect searches

The most stringent limits on new W' bosons are set by their indirect effects on low-energy processes like muon decay, where they can substitute for the Standard Model W boson exchange.

Indirect searches for Z' bosons are carried out at electron-positron colliders, since these give access to high-precision measurements of the properties of the Standard Model Z boson. The constraints come from mixing between the Z' and the Z, and are model dependent because they depend not only on the Z' mass but also its mixing with the Z. The current most stringent limits are from the CERN LEP collider, which constrains Z' bosons to be heavier than a few hundred GeV, for typical model parameters. The ILC will extend this reach up to 5 to 10 TeV depending on the model under consideration, providing complementarity with the LHC because it will offer measurements of additional properties of the Z' boson.

46.3 Z'–Y mixings

We might have gauge kinetic mixings between the $U(1)'$ of the Z' boson and $U(1)Y$ of hypercharge. This mixing leads to a tree level modification of the Peskin–Takeuchi parameters.

46.4 See also

- X and Y bosons

46.5 References

[1] J. Beringer et al. (Particle Data Group) (2012). "Notes in the Gauge and Higgs Boson Listings". *Physical Review D* **86**: 010001. Bibcode:2012PhRvD..86a0001B. doi:10.1103/PhysRevD.86.010001.

[2] A. Abulencia et al. (CDF collaboration) (2006). "Search for $Z' \to e^+e^-$ using dielectron mass and angular distribution". *Physical Review Letters* **96**(21): 211801. arXiv:hep-ex/0602045. Bibcode:2006PhRvL..96u1801A.doi:10.1103/PhysRevLett.96.211.

[3] Emma Woollacott (2011-04-07). "Tevatron data indicates unknown new particle". TG Daily.

[4] "Fermilab's data peak that causes excitement". Fermilab/SLAC. 2011-04-07.

[5] Michael Slezak (22 August 2015). "Possible new particle hints that universe may not be left-handed". *New Scientist*.

46.6 Further reading

- T.G. Rizzo (2006). "Z′ Phenomenology and the LHC". arXiv:hep-ph/0610104 [hep-ph]., a pedagogical overview of Z′ phenomenology (TASI 2006 lectures)
- P. Rincon (17 May 2010). "LHC particle search 'nearing', says physicist". BBC News.

More advanced:

- Abulencia, A.; CDF Collaboration et al. (2006). "Search for $Z' \to e^+e^-$ using dielectron mass and angular distribution". *Physical Review Letters* **96** (211801). arXiv:hep-ex/0602045. Bibcode:2006PhRvL..96u1801A. doi:10.1103/PhysRevLett.96.211801.
- Amini, Hassib (2003). "Radiative corrections to Higgs masses in Z′ models". *New Journal of Physics* **5** (49). arXiv:hep-ph/0210086. Bibcode:2003NJPh....5...49A. doi:10.1088/1367-2630/5/1/349.
- Aoki, Mayumi; Oshimo, Noriyuki (2000). "Supersymmetric extension of the standard model with naturally stable proton". *Physical Review D* **62** (055013): 55013. arXiv:hep-ph/0003286. Bibcode:2000PhRvD..62e5013A. doi:10.1103/PhysRevD.62.055013.
- Aoki, Mayumi; Oshimo, Noriyuki (2000). "A supersymmetric model with an extra U(1) gauge symmetry". *Physical Review Letters* **84**(23): 5269–5272. arXiv:hep-ph/9907481. Bibcode:2000PhRvL..84.5269A.doi 69.PMID10990921.
- Appelquist, Thomas; Dobrescu, Bogdan A.; Hopper, Adam R. (2003). "Nonexotic neutral gauge bosons". *Physical Review D* **68**(035012): 35012. arXiv:hep-ph/0212073. Bibcode:2003PhRvD..68c5012A.doivD.68.035012.
- Babu, K. S.; Kolda, Christopher F.; March-Russell, John (1996). "Leptophobic U(1)s and the R_b–R_c crisis" **54**. pp. 4635–4647. arXiv:hep-ph/9603212. Bibcode:1996PhRvD..54.4635B. doi:10.1103/PhysRevD.54.4635.
- Barger, Vernon D.; Whisnant, K. (1987). "Use of Z lepton asymmetry to determine mixing between Z boson and Z′ boson of E_6 superstrings". *Physical Review D* **36** (3): 979–82. Bibcode:1987PhRvD..36..979B. doi:10.1103/PhysRevD.36.979.
- Barr, S.M.; Dorsner, I. (2005). "The origin of a peculiar extra U(1)". *Physical Review D* **72** (015011). arXiv:hep-ph/0503186. Bibcode:2005PhRvD..72a5011B. doi:10.1103/PhysRevD.72.015011.

- Batra, Puneet; Dobrescu, Bogdan A.; Spivak, David (2006). "Anomaly-free sets of fermions". *Journal of Mathematical Physics***47**(082301): 2301. arXiv:hep-ph/0510181. Bibcode:2006JMP....47h2301B.doi:10.1063/11.
- Carena, Marcela S.; Daleo, Alejandro; Dobrescu, Bogdan A.; Tait, Tim M.P. (2004). "Z′ gauge bosons at the Tevatron".*Physical Review D***70**(093009). arXiv:hep-ph/0408098. Bibcode:2004PhRvD..70i3009C.doi:10.113009.
- Demir, Durmus A.; Kane, Gordon L.; Wang, Ting T. (2005). "The Minimal U(1)′ extension of the MSSM". *Physical Review D***72**(015012). arXiv:hep-ph/0503290. Bibcode:2005PhRvD..72a5012D.doi:10.1103/PhysRevD2.
- Dittmar, Michael; Nicollerat, Anne-Sylvie; Djouadi, Abdelhak (2004). "Z′ studies at the LHC: an update". *Physical Letters B***583**: 111–120. arXiv:hep-ph/0307020. Bibcode:2004PhLB..583..111D.doi:10.1016/j.physletb.03.
- Emam, W.; Khalil, S. (2007). "Higgs and Z′ phenomenology in *B–L* extension of the standard model at LHC". *European Physical Journal C* **522**: 625–633. arXiv:0704.1395. Bibcode:2007EPJC...52..625E. doi:10.1140/e 2-007-0411-7.
- Erler, Jens (2000). "Chiral models of weak scale supersymmetry". *Nuclear Physics B* **586**: 73–91. arXiv:hep-ph/0006051. Bibcode:2000NuPhB.586...73E. doi:10.1016/S0550-3213(00)00427-2.
- Everett, Lisa L.; Langacker, Paul; Plumacher, Michael; Wang, Jing (2000). "Alternative supersymmetric spectra". *Physics Letters B* **477**: 233–241. arXiv:hep-ph/0001073. Bibcode:2000PhLB..477..233E. doi:10.1016/S0370-2693(00)00187-8.
- Fajfer, S.; Singer, P. (2002). "Constraints on heavy Z′ couplings from $\Delta S = 2$ $B^- \to K^-K^-\pi^+$ decay". *Physical Review D* **65** (017301). arXiv:hep-ph/0110233. Bibcode:2002PhRvD..65a7301F. doi:10.1103/PhysRevD.65.017301.
- Ferroglia, A.; Lorca, A.; van der Bij, J. J. (2007). "The Z′ reconsidered". *Annalen der Physik* **16**: 563–578. arXiv:hep-ph/0611174. Bibcode:2007AnP...519..563F. doi:10.1002/andp.200710249.
- Hayreter, Alper (2007). "Dilepton signatures of family non-universal U(1)′". *Physical Letters B* **649** (2–3): 191–196. arXiv:hep-ph/0703269. Bibcode:2007PhLB..649..191H. doi:10.1016/j.physletb.2007.03.049.
- Kang, Junhai; Langacker, Paul (2005). "Z′ discovery limits for supersymmetric E_6 models". *Physical Review D* **71** (035014). arXiv:hep-ph/0412190. Bibcode:2005PhRvD..71c5014K. doi:10.1103/PhysRevD.71.035014.
- Morrissey, David E.; Wells, James D. (2006). "The tension between gauge coupling unification, the Higgs boson mass, and a gauge-breaking origin of the supersymmetric μ-term". *Physical Review D* **74** (015008): 15008. arXiv:hep-ph/0512019. Bibcode:2006PhRvD..74a5008M. doi:10.1103/PhysRevD.74.015008.

46.7 External links

- The Z′ Hunter's Guide, a collection of papers and talks regarding Z′ physics
- Z′ physics on arxiv.org

Chapter 47

Sterile neutrino

Sterile neutrinos (or **inert neutrinos**) are hypothetical particles (neutral leptons – neutrinos) that interact only via gravity and do not interact via any of the fundamental interactions of the Standard Model. The term *sterile neutrino* is used to distinguish them from the known *active neutrinos* in the Standard Model, which are charged under the weak interaction.

This term usually refers to neutrinos with right-handed chirality (see right-handed neutrino), which may be added to the Standard Model. Occasionally it is used in a more general sense for any neutral fermion.

The existence of right-handed neutrinos is theoretically well-motivated, as all other known fermions have been observed with left and right chirality, and they can explain the observed active neutrino masses in a natural way. The mass of the right-handed neutrinos themselves is unknown and could have any value between 10^{15} GeV and less than one eV.[1]

The number of sterile neutrino types is unknown. This is in contrast to the number of active neutrino types, which has to equal that of charged leptons and quark generations to ensure the anomaly freedom of the electroweak interaction.

The search for sterile neutrinos is an active area of particle physics. If they exist and their mass is smaller than the energies of particles in the experiment, they can be produced in the laboratory, either by mixing between active and sterile neutrinos or in high energy particle collisions. If they are heavier, the only directly observable consequence of their existence would be the observed active neutrino masses. They may, however, be responsible for a number of unexplained phenomena in physical cosmology and astrophysics, including dark matter, baryogenesis or dark radiation.[1]

Sterile neutrinos may be Neutral Heavy Leptons (NHLs, or Heavy Neutral Leptons, HNLs).

47.1 Motivation

See also: Neutrino: Chirality and Neutrino oscillation

Experimental results show that all produced and observed neutrinos have left-handed helicities (spins antiparallel to momenta), and all antineutrinos have right-handed helicities, within the margin of error. In the massless limit, it means that only one of two possible chiralities is observed for either particle. These are the only helicities (and chiralities) included in the Standard Model of particle interactions.

Recent experiments such as neutrino oscillation, however, have shown that neutrinos have a non-zero mass, which is not predicted by the Standard Model and suggests new, unknown physics. This unexpected mass explains neutrinos with right-handed helicity and antineutrinos with left-handed helicity: since they do not move at the speed of light, their helicity is not relativistic invariant (it is possible to move faster than them and observe the opposite helicity). Yet all neutrinos have been observed with left-handed *chirality*, and all antineutrinos right-handed. Chirality is a fundamental property of particles and *is* relativistic invariant: it is the same regardless of the particle's speed and mass in every reference frame. The question, thus, remains: can neutrinos and antineutrinos be differentiated only by chirality? Or do right-handed neutrinos and left-handed antineutrinos exist as separate particles?

47.2 Properties

Such particles would belong to a singlet representation with respect to the strong interaction and the weak interaction, having zero electric charge, zero weak hypercharge, zero weak isospin, and, as with the other leptons, no color, although they do have a B-L of −1. If the standard model is embedded in a hypothetical SO(10) grand unified theory, they can be assigned an X charge of −5. The left-handed anti-neutrino has a B-L of 1 and an X charge of 5.

Due to the lack of charge, sterile neutrinos would not interact electromagnetically, weakly, or strongly, making them extremely difficult to detect. They have Yukawa interactions with ordinary leptons and Higgs bosons, which via the Higgs mechanism lead to mixing with ordinary neutrinos. In experiments involving energies larger than their mass they would participate in all processes in which ordinary neutrinos take part, but with a quantum mechanical probability that is suppressed by the small mixing angle. That makes it possible to produce them in experiments if they are light enough. They would also interact gravitationally due to their mass, however, and if they are heavy enough, they could explain cold dark matter or warm dark matter. In some grand unification theories, such as SO(10), they also interact via gauge interactions which are extremely suppressed at ordinary energies because their gauge boson is extremely massive. They do not appear at all in some other GUTs, such as the Georgi–Glashow model (i.e. all its SU(5) charges or quantum numbers are zero).

47.2.1 Mass

All particles are initially massless under the Standard Model, since there are no Dirac mass terms in the Standard Model's Lagrangian. The only mass terms are generated by the Higgs mechanism, which produces non-zero Yukawa couplings between the left-handed components of fermions, the Higgs field, and their right-handed components. This occurs when the **SU**(2) doublet Higgs field ϕ acquires its non-zero vacuum expectation value, ν , spontaneously breaking its SU(2)L × U(1) symmetry, and thus yielding non-zero Yukawa couplings:

$$\mathcal{L}(\psi) = \bar{\psi}(i\partial\!\!\!/)\psi - G\bar{\psi}_L\phi\psi_R$$

Such is the case for charged leptons, like the electron; but within the standard model, the right-handed neutrino does not exist, so even with a Yukawa coupling neutrinos remain massless. In other words, there are no mass terms for neutrinos under the Standard Model: the model only contains a left-handed neutrino and its antiparticle, a right-handed antineutrino, for each generation, produced in weak eigenstates during weak interactions. See neutrino masses in the Standard Model for a detailed explanation.

In the seesaw mechanism, one eigenvector of the neutrino mass matrix, which includes sterile neutrinos, is predicted to be significantly heavier than the other.

A sterile neutrino would have the same weak hypercharge, weak isospin, and mass as its antiparticle. For any charged particle, for example the electron, this is not the case: its antiparticle, the positron, has opposite electric charge, among other opposite charges. Similarly, an up quark has a charge of + ⅔ and (for example) a color charge of red, while its antiparticle has an electric charge of - ⅔ and a color charge of anti-red.

Dirac and Majorana terms

Sterile neutrinos allow the introduction of a **Dirac mass** term as usual. This can yield the observed neutrino mass, but it requires that the strength of the Yukawa coupling be much weaker for the electron neutrino than the electron, without explanation. Similar problems (although less severe) are observed in the quark sector, where the top and bottom masses differ by a factor 40.

Unlike for the left-handed neutrino, a **Majorana mass** term can be added for a sterile neutrino without violating local symmetries (weak isospin and weak hypercharge) since it has no weak charge. However, this would still violate total lepton number.

It is possible to include **both** Dirac and Majorana terms: this is done in the seesaw mechanism (below). In addition to satisfying the Majorana equation, if the neutrino were also its own antiparticle, then it would be the first Majorana fermion.

In that case, it could annihilate with another neutrino, allowing neutrinoless double beta decay. The other case is that it is a Dirac fermion, which is not its own antiparticle.

To put this in mathematical terms, we have to make use of the transformation properties of particles. For free fields, a Majorana field is defined as an eigenstate of charge conjugation. However, neutrinos interact only via the weak interactions, which are not invariant under charge conjugation (C), so an interacting Majorana neutrino cannot be an eigenstate of C. The generalized definition is: "a Majorana neutrino field is an eigenstate of the CP transformation". Consequently, Majorana and Dirac neutrinos would behave differently under CP transformations (actually Lorentz and CPT transformations). Also, a massive Dirac neutrino would have nonzero magnetic and electric dipole moments, whereas a Majorana neutrino would not. However, the Majorana and Dirac neutrinos are different only if their rest mass is not zero. For Dirac neutrinos, the dipole moments are proportional to mass and would vanish for a massless particle. Both Majorana and Dirac mass terms however can appear in the mass Lagrangian.

47.2.2 Seesaw mechanism

Main article: Seesaw mechanism

In addition to the left-handed neutrino, which couples to its family charged lepton in weak charged currents, if there is also a right-handed sterile neutrino partner, a weak isosinglet with no charge, then it is possible to add a Majorana mass term without violating electroweak symmetry. Both neutrinos have mass and handedness is no longer preserved (thus "left or right-handed neutrino" means that the state is mostly left or right-handed). To get the neutrino mass eigenstates, we have to diagonalize the general mass matrix **M**:

$$m_\nu = \begin{pmatrix} 0 & m_D \\ m_D & M_{NHL} \end{pmatrix}$$

where M_{NHL} is big and m_D is of intermediate size terms.

Apart from empirical evidence, there is also a theoretical justification for the seesaw mechanism in various extensions to the Standard Model. Both Grand Unification Theories (GUTs) and left-right symmetrical models predict the following relation:

$$m_\nu \ll m_D \ll M_{NHL}$$

According to GUTs and left-right models, the right-handed neutrino is extremely heavy: *MNHL* $\approx 10^5$—10^{12} GeV, while the smaller eigenvalue is approximately equal to

$$m_\nu \approx \frac{m_D^2}{M_{NHL}}$$

This is the seesaw mechanism: as the sterile right-handed neutrino gets heavier, the normal left-handed neutrino gets lighter. The left-handed neutrino is a mixture of two Majorana neutrinos, and this mixing process is how sterile neutrino mass is generated.

47.3 Detection attempts

The production and decay of sterile neutrinos could happen through the mixing with virtual ("off mass shell") neutrinos. There were several experiments set up to discover or observe NHLs, for example the NuTeV (E815) experiment at Fermilab or LEP-l3 at CERN. They all lead to establishing limits to observation, rather than actual observation of those particles. If they are indeed a constituent of dark matter, sensitive X-ray detectors would be needed to observe the radiation emitted by their decays.[2]

Sterile neutrinos may mix with ordinary neutrinos via a Dirac mass after electroweak symmetry breaking, in analogy to quarks and charged leptons. Sterile neutrinos and (in more-complicated models) ordinary neutrinos may also have Majorana masses. In type 1 seesaw mechanism both Dirac and Majorana masses are used to drive ordinary neutrino masses down and make the sterile neutrinos much heavier than the Standard Model's interacting neutrinos. In some models the heavy neutrinos can be as heavy as the GUT scale ($\approx 10^{15}$ GeV). In other models they could be lighter than the weak gauge bosons W and Z as in the so-called νMSM model where their masses are between GeV and keV. A light (with the mass ≈ 1 eV) sterile neutrino was suggested as a possible explanation of the results of the Liquid Scintillator Neutrino Detector experiment. On April 11, 2007, researchers at the MiniBooNE experiment at Fermilab announced that they had not found any evidence supporting the existence of such a sterile neutrino.[3] More-recent results and analysis have provided some support for the existence of the sterile neutrino.[4][5] Two separate detectors near a nuclear reactor in France found 3% of anti-neutrinos missing. They suggested the existence of a 4th neutrino with a mass of 0.7 keV.[6] Sterile neutrinos are also candidates for dark radiation. Daya Bay has also searched for a light sterile neutrino and excluded some mass regions.[7]

The number of neutrinos and the masses of the particles can have large-scale effects that shape the appearance of the cosmic microwave background. The total number of neutrino species, for instance, affects the rate at which the cosmos expanded in its earliest epochs: more neutrinos means a faster expansion. The Planck Satellite 2013 data release found no evidence of additional neutrino-like particles.[8]

47.4 See also

- MiniBooNE at Fermilab

47.5 References

Notes

References

[1] Marco Drewes (2013). "The Phenomenology of Right Handed Neutrinos". *International Journal of Modern Physics E* **22** (8): 1330019. arXiv:1303.6912. Bibcode:2013IJMPE..2230019D. doi:10.1142/S0218301313300191.

[2] Battison, Leila (2011-09-16). "Dwarf galaxies suggest dark matter theory may be wrong". BBC News. Retrieved 2011-09-18.

[3] First_Results (PDF)

[4] Scientific American: "Dimensional Shortcuts", August 2007

[5] Bulbul, E.; Markevitch, M.; Foster, A.; Smith, R.K.; Loewenstein, M.; Randall, S.W. (2014). "Detection of an Unidentified Emission Line in the Stacked X-ray Spectrum of Galaxy Clusters". *The Astrophysical Journal* **789** (1): 13. arXiv:1402.2301v2. Bibcode:2014ApJ...789...13B. doi:10.1088/0004-637X/789/1/13.

[6] The Reactor Antineutrino Anomaly

[7] "Search for a Light Sterile Neutrino at Daya Bay". *Phys. Rev. Lett. 113, 141802*. 1 October 2014. arXiv:1407.7259. Bibcode:2014PhRvL.113n1802A. doi:10.1103/PhysRevLett.113.141802.

[8] Ade, P.A.R.; et al. (Planck Collaboration) (2013). "Planck 2013 results. XVI. Cosmological parameters". arXiv:1303.5076 [astro-ph.CO].

Bibliography

- M. Drewes (2013). "The Phenomenology of Right Handed Neutrinos". *International Journal of Modern Physics E*. arXiv:1303.6912. Bibcode:2013IJMPE..2230019D. doi:10.1142/S0218301313300191.

- A. Merle (2013). "keV Neutrino Model Building". *International Journal of Modern Physics D* **22** (10): 1330020. arXiv:1302.2625. Bibcode:2013IJMPD..2230020M. doi:10.1142/S0218271813300206.
- A. G. Vaitaitis et al. (1999). "Search for Neutral Heavy Leptons in a High-Energy Neutrino Beam". *Physical Review Letters***83**(24): 4943–4946. arXiv:hep-ex/9908011. Bibcode:1999PhRvL..83.4943V.doi:10.1103/Phys943.
- J. A. Formaggio; J. Conrad; M. Shaevitz; A. Vaitaitis (1998). "Helicity effects in neutral heavy lepton decays". *Physical Review D* **57** (11): 7037–7040. Bibcode:1998PhRvD..57.7037F. doi:10.1103/PhysRevD.57.7037.
- K. Nakamura; Particle Data Group (2010). "Review of Particle Physics". *Journal of Physics G* **37** (75021): 075021. Bibcode:2010JPhG...37g5021N. doi:10.1088/0954-3899/37/7A/075021.

47.6 External links

- The NuTeV experiment at Fermilab
- The L3 Experiment at CERN
- Experiment Nixes Fourth Neutrino (April 2007 Scientific American)

Chapter 48

Preon

For the protein diseases, see Prion. For the Freon trade name, see Chlorofluorocarbon.

In particle physics, **preons** are "point-like" particles, conceived to be subcomponents of quarks and leptons.[1] The word was coined by Jogesh Pati and Abdus Salam in 1974. Interest in preon models peaked in the 1980s but has slowed as the Standard Model of particle physics continues to describe the physics mostly successfully, and no direct experimental evidence for lepton and quark compositeness has been found.

Note that in the hadronic sector there are some intriguing open questions and some effects considered anomalies within the Standard Model. For example, four very important open questions are the proton spin puzzle, the EMC effect, the distributions of electric charges inside the nucleons as found by Hofstadter in 1956, and the ad hoc CKM matrix elements.

48.1 Background

Before the Standard Model (SM) was developed in the 1970s (the key elements of the Standard Model known as quarks were proposed by Murray Gell-Mann and George Zweig in 1964), physicists observed hundreds of different kinds of particles in particle accelerators. These were organized into relationships on their physical properties in a largely ad-hoc system of hierarchies, not entirely unlike the way taxonomy grouped animals based on their physical features. Not surprisingly, the huge number of particles was referred to as the "particle zoo".

The Standard Model, which is now the prevailing model of particle physics, dramatically simplified this picture by showing that most of the observed particles were mesons, which are combinations of two quarks, or baryons which are combinations of three quarks, plus a handful of other particles. The particles being seen in the ever-more-powerful accelerators were, according to the theory, typically nothing more than combinations of these quarks.

Within the Standard Model, there are several different types of particles. One of these, the quarks, has six different types, of which there are three varieties in each (dubbed "colors", red, green, and blue, giving rise to quantum chromodynamics). Additionally, there are six different types of what are known as leptons. Of these six leptons, there are three charged particles: the electron, muon, and tau. The neutrinos comprise the other three leptons, and for each neutrino there is a corresponding member from the other set of three leptons. In the Standard Model, there are also bosons, including the photons; W^+, W^-, and Z bosons; gluons and the Higgs boson; and an open space left for the graviton. Almost all of these particles come in "left-handed" and "right-handed" versions (see *chirality*). The quarks, leptons and W boson all have antiparticles with opposite electric charge.

The Standard Model also has a number of problems which have not been entirely solved. In particular, no successful theory of gravitation based on a particle theory has yet been proposed. Although the Model assumes the existence of a graviton, all attempts to produce a consistent theory based on them have failed. Additionally, mass remains a mystery in the Standard Model. Additionally Kalman [2] notes that according to the concept of atomism, the fundamental building blocks of nature are invisible and indivisible bits of matter that are ungenerated and indestructible. Quarks are not indestructible, some can decay into other quarks. Thus on fundamental grounds- quarks must be composed of fundamental quantities-preons. Although the mass of each successive particle follows certain patterns, predictions of the rest mass of most particles cannot be made precisely, except for the masses of almost all baryons which have been recently described very well by the model of de Souza.[3] The Higgs boson explains why particles show inertial mass (but does not explain rest mass).

The Standard Model also has problems predicting the large scale structure of the universe. For instance, the SM generally predicts equal amounts of matter and antimatter in the universe, something that is observably not the case. A number of attempts have been made to "fix" this through a variety of mechanisms, but to date none have won widespread support. Likewise, basic adaptations of the Model suggest the presence of proton decay, which has not yet been observed.

Preon theory is motivated by a desire to replicate the achievements of the periodic table, and the later Standard Model which tamed the "particle zoo", by finding more fundamental answers to the huge number of arbitrary constants present in the Standard Model. It is one of several models to have been put forward in an attempt to provide a more fundamental explanation of the results in experimental and theoretical particle physics. The preon model has attracted comparatively little interest to date among the particle physics community.

48.2 Motivations

Preon research is motivated by the desire to explain already known facts (retrodiction), which include

- To reduce the large number of particles, many that differ only in charge, to a smaller number of more fundamental particles. For example, the electron and positron are identical except for charge, and preon research is motivated by explaining that electrons and positrons are composed of similar preons with the relevant difference accounting for charge. The hope is to reproduce the reductionist strategy that has worked for the periodic table of elements.
- To explain the three generations of fermions.
- To calculate parameters that are currently unexplained by the Standard Model, such as particle masses, electric charges, and color charges, and reduce the number of experimental input parameters required by the Standard Model.
- To provide reasons for the very large differences in energy-masses observed in supposedly fundamental particles, from the electron neutrino to the top quark.
- To provide alternative explanations for the electro-weak symmetry breaking without invoking a Higgs field, which in turn possibly needs a supersymmetry to correct the theoretical problems involved with the Higgs field. Supersymmetry itself has theoretical problems.
- To account for neutrino oscillation and mass.
- The desire to make new nontrivial predictions, for example, to provide possible cold dark matter candidates.
- To explain why there exists only the observed variety of particle species and not something else and to reproduce only these observed particles (since the prediction of non-observed particles is one of the major theoretical problems, as, for example, with supersymmetry).

48.3 History

A number of physicists have attempted to develop a theory of "pre-quarks" (from which the name *preon* derives) in an effort to justify theoretically the many parts of the Standard Model that are known only through experimental data.

Other names which have been used for these proposed fundamental particles (or particles intermediate between the most fundamental particles and those observed in the Standard Model) include *prequarks*, *subquarks*, *maons*,[4] *alphons*, *quinks*, *rishons*, *tweedles*, *helons*, *haplons*, *Y-particles*,[5] and *primons*.[6] *Preon* is the leading name in the physics community.

Efforts to develop a substructure date at least as far back as 1974 with a paper by Pati and Salam in *Physical Review*.[7] Other attempts include a 1977 paper by Terazawa, Chikashige and Akama,[8] similar, but independent, 1979 papers by Ne'eman,[9] Harari,[10] and Shupe,[11] a 1981 paper by Fritzsch and Mandelbaum,[12] and a 1992 book by D'Souza and Kalman.[1] None of these has gained wide acceptance in the physics world. However, in a recent work[13] de Souza has shown that his model describes well all weak decays of hadrons according to selection rules dictated by a quantum number derived from his compositeness model. In his model leptons are elementary particles and each quark is composed of two *primons*, and thus, all quarks are described by four *primons*. Therefore, there is no need for the Standard Model Higgs boson and each quark mass is derived from the interaction between each pair of *primons* by means of three Higgs-like bosons. In his 1989 Nobel Prize acceptance lecture, Hans Dehmelt described a most fundamental elementary particle, with definable properties, which he called the *cosmon*, as the likely end result of a long but finite chain of increasingly more elementary particles.[14]

Each of the preon models postulates a set of fewer fundamental particles than those of the Standard Model, together with the rules governing how those fundamental particles operate. Based on these rules, the preon models try to explain the Standard Model, often predicting small discrepancies with this model and generating new particles and certain phenomena, which do not belong to the Standard Model. The Rishon model illustrates some of the typical efforts in the field.

Many of the preon models theorize that the apparent imbalance of matter and antimatter in the universe is in fact illusory, with large quantities of preon level antimatter confined within more complex structures.

Many preon models either do not account for the Higgs boson or rule it out, and propose that electro-weak symmetry is broken not by a scalar Higgs field but by composite preons.[15] For example, Fredriksson preon theory does not need the Higgs boson, and explains the electro-weak breaking as the rearrangement of preons, rather than a Higgs-mediated field. In fact, Fredriksson preon model and de Souza model predict that the Standard Model Higgs boson does not exist.

When the term "preon" was coined, it was primarily to explain the two families of spin-1/2 fermions: leptons and quarks. More-recent preon models also account for spin-1 bosons, and are still called "preons".

48.4 Rishon model

Main article: Rishon model

The *rishon model* (RM) is the earliest effort to develop a preon model to explain the phenomenon appearing in the Standard Model (SM) of particle physics. It was first developed by Haim Harari and Michael A. Shupe (independently of each other), and later expanded by Harari and his then-student Nathan Seiberg.

The model has two kinds of fundamental particles called **rishons** (which means "primary" in Hebrew). They are **T** ("Third" since it has an electric charge of ⅓ *e*, or Tohu which means "unformed" in Hebrew Genesis) and **V** ("Vanishes", since it is electrically neutral, or Vohu which means "void" in Hebrew Genesis). All leptons and all flavours of quarks are three-rishon ordered triplets. These groups of three rishons have spin-½.

48.5 Criticisms

48.5.1 The mass paradox

One preon model started as an internal paper at the Collider Detector at Fermilab (CDF) around 1994. The paper was written after an unexpected and inexplicable excess of jets with energies above 200 GeV were detected in the 1992–1993 running period. However, scattering experiments have shown that quarks and leptons are "pointlike" down to distance scales of less than 10^{-18} m (or 1/1000 of a proton diameter). The momentum uncertainty of a preon (of whatever mass) confined to a box of this size is about 200 GeV/c, 50,000 times larger than the rest mass of an up-quark and 400,000 times larger than the rest mass of an electron.

Heisenberg's uncertainty principle states that $\Delta x \Delta p \geq \hbar/2$ and thus anything confined to a box smaller than Δx would have a momentum uncertainty proportionally greater. Thus, the preon model proposed particles smaller than the elementary particles they make up, since the momentum uncertainty Δp should be greater than the particles themselves. And so the preon model represents a mass paradox: How could quarks or electrons be made of smaller particles that would have many orders of magnitude greater mass-energies arising from their enormous momenta? This paradox is resolved by postulating a large binding force between preons cancelling their mass-energies.

48.5.2 Constraints

Any candidate preon theory must address particle chirality and the 't Hooft Chiral anomaly constraints, and would ideally have simpler theoretical structure than the Standard Model itself.

48.6 Conflicts with observed physics

Preon models propose additional unobserved forces or dynamics to account for the observed properties of elementary particles, which may have implications in conflict with observation.

For example, now that the LHC's observation of a Higgs boson is confirmed, the observation contradicts the predictions of many preon models that did not include it.

Preon theories require that quarks and electrons should have a finite size. It is possible that the Large Hadron Collider will observe this when raised to higher energies.

48.7 Popular culture

- In the 1948 reprint/edit of his 1930 novel *Skylark Three*, E. E. Smith postulated a series of 'subelectrons of the first and second type' with the latter being fundamental particles that were associated with the gravitation force. While this may not have been an element of the original novel (the scientific basis of some of the other novels in the series was revised extensively due to the additional eighteen years of scientific development), even the edited publication may be the first, or one of the first, mentions of the possibility that electrons are not fundamental particles.
- In the novelized version of the 1982 motion picture *Star Trek II: The Wrath of Khan*, written by Vonda McIntyre, two of Dr. Carol Marcus' Genesis project team, Vance Madison and Delwyn March, have studied sub-elementary particles they've named “boojums” and “snarks”, in a field they jokingly call “kindergarten physics” because it is lower than “elementary” (analogy to school levels).
- James P. Hogan's novel *Voyage from Yesteryear* discussed preons (called *tweedles*), the physics of which became central to the plot. Hogan's “tweedle” physics was patently derived from the Rishon model.

48.8 See also

- Technicolor (physics)
- Preon star
- Preon-degenerate matter

48.9 Notes

[1] D'Souza, I.A.; Kalman, C.S. (1992). *Preons: Models of Leptons, Quarks and Gauge Bosons as Composite Objects*. World Scientific. ISBN 978-981-02-1019-9.

[2] Kalman, C. S. (2005). *Nuclear Physics B (Proc. Suppl.)* **142**: 235–237. Missing or empty |title= (help)

[3] de Souza,M.E. (2010). "Calculation of almost all energy levels of baryons".*Papers in Physics***3**: 030003–1. doi:10.4279/3.

[4] Overbye, D. (5 December 2006). "China Pursues Major Role in Particle Physics". *The New York Times*. Retrieved 2011-09-12.

[5] Yershov,V.N. (2005). "Equilibrium Configurations of Tripolar Charges".*Few-Body Systems***37**(1–2): 79–106. arXiv:p85. Bibcode:2005FBS....37...79Y. doi:10.1007/s00601-004-0070-2.

[6] de Souza, M.E. (2005). "The Ultimate Division of Matter". *Scientia Plena* **1** (4): 83.

[7] Pati,J.C.;Salam,A. (1974). "Lepton number as the fourth"color"".*Physical Review D***10**: 275–289. Bibcode:19.275P. doi:10.1103/PhysRevD.10.275.

with erratum published as *Physical Review D* **11** (3): 703. 1975. Bibcode:1975PhRvD..11..703P. doi:10.110303.2. Missing or empty|title=(help)

[8] Terazawa, H.; Chikashige, Y.; Akama, K. (1977). "Unified model of the Nambu-Jona-Lasinio type for all elementary particles". *Physical Review D* **15** (2): 480–487. Bibcode:1977PhRvD..15..480T. doi:10.1103/PhysRevD.15.480.

[9] Ne'eman, Y. (1979). "Irreducible gauge theory of a consolidated Weinberg-Salam model". *Physics Letters B* **81** (2): 190–194. Bibcode:1979PhLB...81..190N. doi:10.1016/0370-2693(79)90521-5.

[10] Harari, H. (1979). "A schematic model of quarks and leptons" (PDF). *Physics Letters B* **86**: 83–6. Bibcode:1979PhLB...86...83H. doi:10.1016/0370-2693(79)90626-9.

[11] Shupe, M.A. (1979). "A composite model of leptons and quarks". *Physics Letters B* **86**: 87–92. Bibcode:1979PhLB...86...87S. doi:10.1016/0370-2693(79)90627-0.

[12] Fritzsch, H.; Mandelbaum, G. (1981). "Weak interactions as manifestations of the substructure of leptons and quarks". *Physics Letters B* **102** (5): 319. Bibcode:1981PhLB..102..319F. doi:10.1016/0370-2693(81)90626-2.

[13] de Souza, M.E. (2008). "Weak decays of hadrons reveal compositeness of quarks". *Scientia Plena* **4** (6): 064801–1.

[14] Dehmelt, H.G. (1989). "Experiments with an Isolated Subatomic Particle at Rest". *Nobel Lecture*. The Nobel Foundation. See also references therein.

[15] Dugne, J.-J.; Fredriksson, S.; Hansson, J.; Predazzi, E. (1997). "Higgs pain? Take a preon!". arXiv:hep-ph/9709227 [hep-ph].

48.10 Further reading

- Ball, P. (2007). "Splitting the quark". *Nature*. doi:10.1038/news.2007.292.
- Have We Hit Bottom Yet?- an article about preons and minuteness

Chapter 49

Bound state

In quantum physics, a **bound state** describes a system where a particle is subject to a potential such that the particle has a tendency to remain localised in one or more regions of space. The potential may be either an external potential, or may be the result of the presence of another particle.

In quantum mechanics (where the number of particles is conserved), a bound state is a state in Hilbert space representing two or more particles whose interaction energy is less than the total energy of each separate particle, and therefore these particles cannot be separated unless energy is added from outside. The energy spectrum of the set of bound states is discrete, unlike the continuous spectrum of free particles.

(Actually, it is possible to have unstable "bound states", which aren't really bound states in the strict sense, with a net positive interaction energy, provided that there is an "energy barrier" that has to be tunnelled through in order to decay. This is true for some radioactive nuclei and for some electret materials able to carry electric charge for rather long periods.)

For a given potential, a bound state is represented by a stationary square-integrable wavefunction. The energy of such a wavefunction is negative.

In relativistic quantum field theory, a stable bound state of n particles with masses $m_1, \ldots, mn$ shows up as a pole in the S-matrix with a center of mass energy which is less than $m_1 + \ldots + mn$. An unstable bound state (see resonance) shows up as a pole with a complex center of mass energy.

49.1 Examples

- A proton and an electron can move separately; when they do, the total center-of-mass energy is positive, and such a pair of particles can be described as an ionized atom. Once the electron starts to "orbit" the proton, the energy becomes negative, and a bound state – namely the hydrogen atom – is formed. Only the lowest-energy bound state, the ground state, is stable. The other excited states are unstable and will decay into bound states with less energy by emitting a photon.
- A nucleus is a bound state of protons and neutrons (nucleons).
- A positronium "atom" is an unstable bound state of an electron and a positron. It decays into photons.
- The proton itself is a bound state of three quarks (two up and one down; one red, one green and one blue). However, unlike the case of the hydrogen atom, the individual quarks can never be isolated. See confinement.
- The eigenstates of the Hubbard model and Jaynes-Cummings-Hubbard model (JCH) Hamiltonian in the two-excitation subspace are also examples of bound states. In Hubbard model, two repulsive bosonic atoms can form a bound pair in an optical lattice.[1][2][3] The JCH Hamiltonian also supports two-polariton bound states when the photon-atom interaction is sufficiently strong. In particular, the two polaritons associated with the bound states exhibit a strong correlation such that they stay close to each other in position space. The results discussed has been published in Ref.[4]

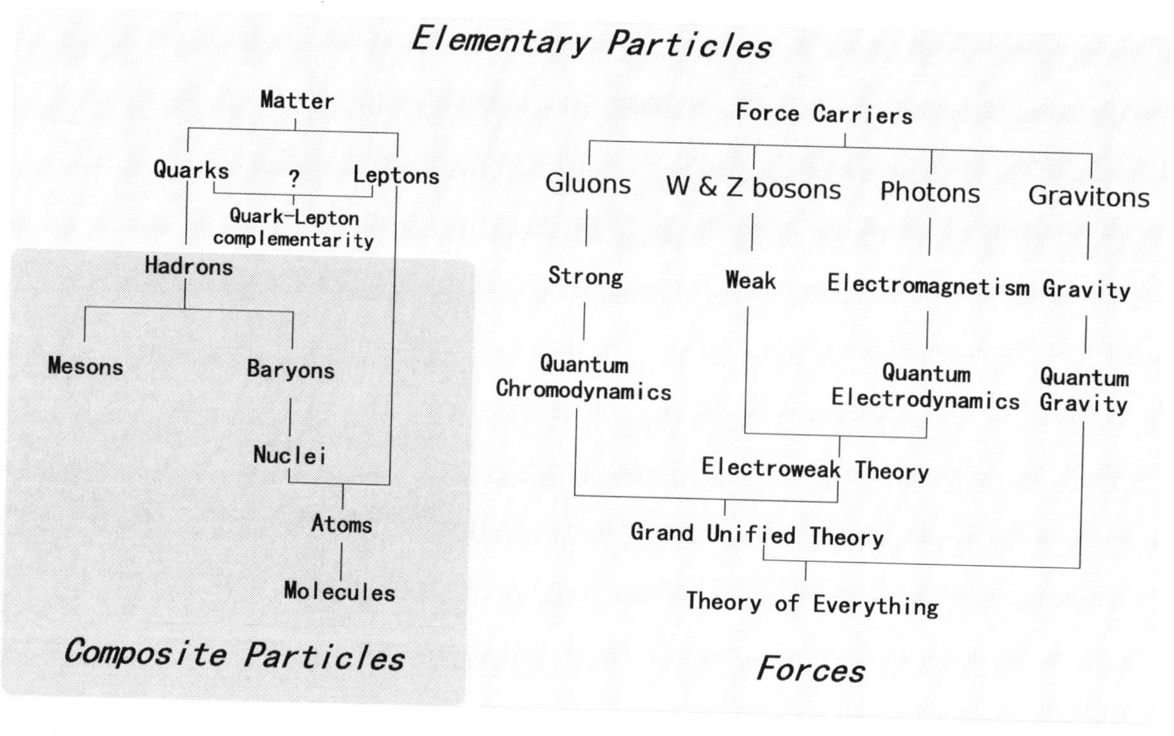

An overview of the various families of elementary and composite particles, and the theories describing their interactions

49.2 In mathematical quantum physics

Let H be a complex separable Hilbert space, $U = \{U(t) \mid t \in \mathbb{R}\}$ be a one-parametric group of unitary operators on H and $\rho = \rho(t_0)$ be a statistical operator on H. Let A be an observable on H and let $\mu(A, \rho)$ be the induced probability distribution of A with respect to ρ on the Borel σ-algebra on $\mathbb{R}$. Then the evolution of ρ induced by U is said to be **bound** with respect to A if $\lim_{R\to\infty} \sum_{t\geq t_0} \mu(A, \rho(t))(\mathbb{R}_{>R}) = 0$, where $\mathbb{R}_{>R} = \{x \in \mathbb{R} \mid x > R\}$.

Example: Let $H = L^2(\mathbb{R})$ and let A be the position observable. Let $\rho = \rho(0) \in H$ have compact support and $[-1, 1] \subseteq \text{Supp}(\rho)$.

- If the state evolution of ρ "moves this wave package constantly to the right", e.g. if $[t-1, t+1] \in \text{Supp}(\rho(t))$ for all $t \geq 0$, then ρ is not a bound state with respect to the position.
- If ρ does not change in time, i.e. $\rho(t) = \rho$ for all $t \geq 0$, then ρ is a bound state with respect to position.
- More generally: If the state evolution of ρ "just moves ρ inside a bounded domain", then ρ is also a bound state with respect to position.

It should be emphasized that a bound state can have its energy located in the continuum spectrum. This fact was first pointed out by John von Neumann and Eugene Wigner in 1929.[5] Autoionization states are discrete states located in a continuum, but not referred to as bound states because they are short lived.

49.3 See also

- Composite field
- Resonance
- Bethe–Salpeter equation

49.4 References

[1] K. Winkler, G. Thalhammer, F. Lang, R. Grimm, J. H. Denschlag, A. J. Daley, A. Kantian, H. P. Buchler and P. Zoller (2006). "Repulsively bound atom pairs in an optical lattice". *Nature* **441**: 853. arXiv:cond-mat/0605196. Bibcode:2006Natur.441..853W. doi:10.1038/nature04918.

[2] Javanainen, Juha and Odong, Otim and Sanders, Jerome C. (Apr 2010). "Dimer of two bosons in a one-dimensional optical lattice". *Phys. Rev. A* (American Physical Society) **81** (4): 043609. arXiv:1004.5118. Bibcode:2010PhRvA..81d3609J. doi:10.1103/PhysRevA.81.043609.

[3] M. Valiente and D. Petrosyan (2008). "Two-particle states in the Hubbard model". *J. Phys. B: At. Mol. Opt. Phys.* **41**: 161002. Bibcode:2008JPhB...41p1002V. doi:10.1088/0953-4075/41/16/161002.

[4] Max T. C. Wong and C. K. Law (May 2011). "Two-polariton bound states in the Jaynes-Cummings-Hubbard model". *Phys. Rev. A*(American Physical Society)**83**(5): 055802. arXiv:1101.1366. Bibcode:2011PhRvA..83e5802W.doi:10.1103/PhysR02.

[5] von Neumann, John; Wigner, Eugene (1929). "Über merkwürdige diskrete Eigenwerte". *Physikalische Zeitschrift* **30**: 465–467.

Chapter 50

Hadron

In particle physics, a **hadron** /ˈhædrɒn/ (Greek: ἁδρός, *hadrós*, "stout, thick") is a composite particle made of quarks held together by the strong force (in a similar way as molecules are held together by the electromagnetic force).

Hadrons are categorized into two families: baryons, made of three quarks, and mesons, made of one quark and one antiquark. Protons and neutrons are examples of baryons; pions are an example of a meson. Hadrons containing more than three valence quarks (exotic hadrons) have been discovered in recent years. A tetraquark state (an exotic meson), named the Z(4430)$^{-}$, was discovered in 2007 by the Belle Collaboration [1] and confirmed as a resonance in 2014 by the LHCb collaboration.[2] Two pentaquark states (exotic baryons), named P+
c(4380) and P+
c(4450), were discovered in 2015 by the LHCb collaboration.[3] There are several more exotic hadron candidates, and other colour-singlet quark combinations may also exist.

Of the hadrons, protons are stable, and neutrons bound within atomic nuclei are stable. Other hadrons are unstable under ordinary conditions; free neutrons decay with a half-life of about 611 seconds. Experimentally, hadron physics is studied by colliding protons or nuclei of heavy elements such as lead, and detecting the debris in the produced particle showers.

50.1 Etymology

The term "hadron" was introduced by Lev B. Okun in a plenary talk at the 1962 International Conference on High Energy Physics.[4] In this talk he said:

> Not withstanding the fact that this report deals with weak interactions, we shall frequently have to speak of strongly interacting particles. These particles pose not only numerous scientific problems, but also a terminological problem. The point is that "strongly interacting particles" is a very clumsy term which does not yield itself to the formation of an adjective. For this reason, to take but one instance, decays into strongly interacting particles are called non-leptonic. This definition is not exact because "non-leptonic" may also signify "photonic". In this report I shall call strongly interacting particles "hadrons", and the corresponding decays "hadronic" (the Greek ἁδρός signifies "large", "massive", in contrast to λεπτός which means "small", "light"). I hope that this terminology will prove to be convenient. — Lev B. Okun, 1962

50.2 Properties

According to the quark model,[5] the properties of hadrons are primarily determined by their so-called *valence quarks*. For example, a proton is composed of two up quarks (each with electric charge $+2/3$, for a total of $+4/3$ together) and one down quark (with electric charge $-1/3$). Adding these together yields the proton charge of +1. Although quarks also carry color charge, hadrons must have zero total color charge because of a phenomenon called color confinement. That is, hadrons must be “colorless” or “white”. These are the simplest of the two ways: three quarks of different colors, or a quark of one color and an antiquark carrying the corresponding anticolor. Hadrons with the first arrangement are called baryons, and those with the second arrangement are mesons.

Hadrons, however, are not composed of just three or two quarks, because of the strength of the strong force. More accurately, strong force gluons have enough energy (E) to have resonances composed of massive (m) quarks ($E > mc^2$) . Thus, virtual quarks and antiquarks, in a 1:1 ratio, form the majority of massive particles inside a hadron. The two or three quarks are the excess of quarks vs. antiquarks in hadrons, and vice versa in anti-hadrons. Because the virtual quarks are not stable wave packets (quanta), but irregular and transient phenomena, it is not meaningful to ask which quark is real and which virtual; only the excess is apparent from the outside. Massless virtual gluons compose the numerical majority of particles inside hadrons.

Like all subatomic particles, hadrons are assigned quantum numbers corresponding to the representations of the Poincaré group: $J^{PC}(m)$, where J is the spin quantum number, P the intrinsic parity (or P-parity), and C, the charge conjugation (or C-parity), and the particle’s mass, m. Note that the mass of a hadron has very little to do with the mass of its valence quarks; rather, due to mass–energy equivalence, most of the mass comes from the large amount of energy associated with the strong interaction. Hadrons may also carry flavor quantum numbers such as isospin (or G parity), and strangeness. All quarks carry an additive, conserved quantum number called a baryon number (B), which is $+\frac{1}{3}$ for quarks and $-\frac{1}{3}$ for antiquarks. This means that baryons (groups of three quarks) have $B = 1$ whereas mesons have $B = 0$.

Hadrons have excited states known as resonances. Each ground state hadron may have several excited states; several hundreds of resonances have been observed in particle physics experiments. Resonances decay extremely quickly (within about 10^{-24} seconds) via the strong nuclear force.

In other phases of matter the hadrons may disappear. For example, at very high temperature and high pressure, unless there are sufficiently many flavors of quarks, the theory of quantum chromodynamics (QCD) predicts that quarks and gluons will no longer be confined within hadrons, “because the strength of the strong interaction diminishes with energy". This property, which is known as asymptotic freedom, has been experimentally confirmed in the energy range between 1 GeV (gigaelectronvolt) and 1 TeV (teraelectronvolt).[6]

All free hadrons except the proton (and antiproton) are unstable.

50.3 Baryons

Main article: Baryon

All known baryons are made of three valence quarks, so they are fermions, *i.e.*, they have odd half-integral spin, because they have an odd number of quarks. As quarks possess baryon number $B = \frac{1}{3}$, baryons have baryon number $B = 1$. The best-known baryons are the proton and the neutron.

One can hypothesise baryons with further quark-antiquark pairs in addition to their three quarks. Hypothetical baryons with one extra quark-antiquark pair (5 quarks in all) are called pentaquarks. As of August 2015, there are two known pentaquarks, P+
c(4380) and P+
c(4450), both discovered in 2015 by the LHCb collaboration.[3]

Each type of baryon has a corresponding antiparticle (antibaryon) in which quarks are replaced by their corresponding antiquarks. For example, just as a proton is made of two up-quarks and one down-quark, its corresponding antiparticle, the antiproton, is made of two up-antiquarks and one down-antiquark.

50.4 Mesons

Main article: Meson

Mesons are hadrons composed of a quark-antiquark pair. They are bosons, meaning they have integral spin, *i.e.*, 0, 1, or −1, as they have an even number of quarks. They have baryon number $B = 0$. Examples of mesons commonly produced in particle physics experiments include pions and kaons. Pions also play a role in holding atomic nuclei together via the residual strong force.

In principle, mesons with more than one quark-antiquark pair may exist; a hypothetical meson with two pairs is called a tetraquark. Several tetraquark candidates were found in the 2000s, but their status is under debate.[7] Several other hypothetical "exotic" mesons lie outside the quark model of classification. These include glueballs and hybrid mesons (mesons bound by excited gluons).

50.5 See also

- Hadronization, the formation of hadrons out of quarks and gluons
- Large Hadron Collider (LHC)
- List of particles
- Standard model
- Subatomic particles
- Hadron therapy, a.k.a. particle therapy
- Exotic hadrons

50.6 References

[1] Choi, S.-K.; Belle Collaboration et al. (2007). "Observation of a resonance-like structure in the π±Ψ′ mass distribution in exclusive B→Kπ±Ψ′ decays". *Physical Review Letters* **100** (14). arXiv:0708.1790. Bibcode:2008PhRvL.100n2001C. doi:10.1103/PhysRevLett.100.142001.

[2] LHCb collaboration (2014): Observation of the resonant character of the Z(4430)⁻ state

[3] R. Aaij et al. (LHCb collaboration) (2015). "Observation of J/ψp resonances consistent with pentaquark states in Λ0
b→J/ψK−
p decays". *Physical Review Letters* **115** (7). doi:10.1103/PhysRevLett.115.072001.

[4] Lev B. Okun (1962). "The Theory of Weak Interaction". *Proceedings of 1962 International Conference on High-Energy Physics at CERN*. Geneva. p. 845. Bibcode:1962hep..conf..845O.

[5] C. Amsler *et al.* (Particle Data Group) (2008). "Review of Particle Physics – Quark Model" (PDF). *Physics Letters B* **667**: 1. Bibcode:2008PhLB..667....1P. doi:10.1016/j.physletb.2008.07.018.

[6] S. Bethke (2007). "Experimental tests of asymptotic freedom". *Progress in Particle and Nuclear Physics* **58** (2): 351. arXiv:hep-ex/0606035. Bibcode:2007PrPNP..58..351B. doi:10.1016/j.ppnp.2006.06.001.

[7] Mysterious Subatomic Particle May Represent Exotic New Form of Matter

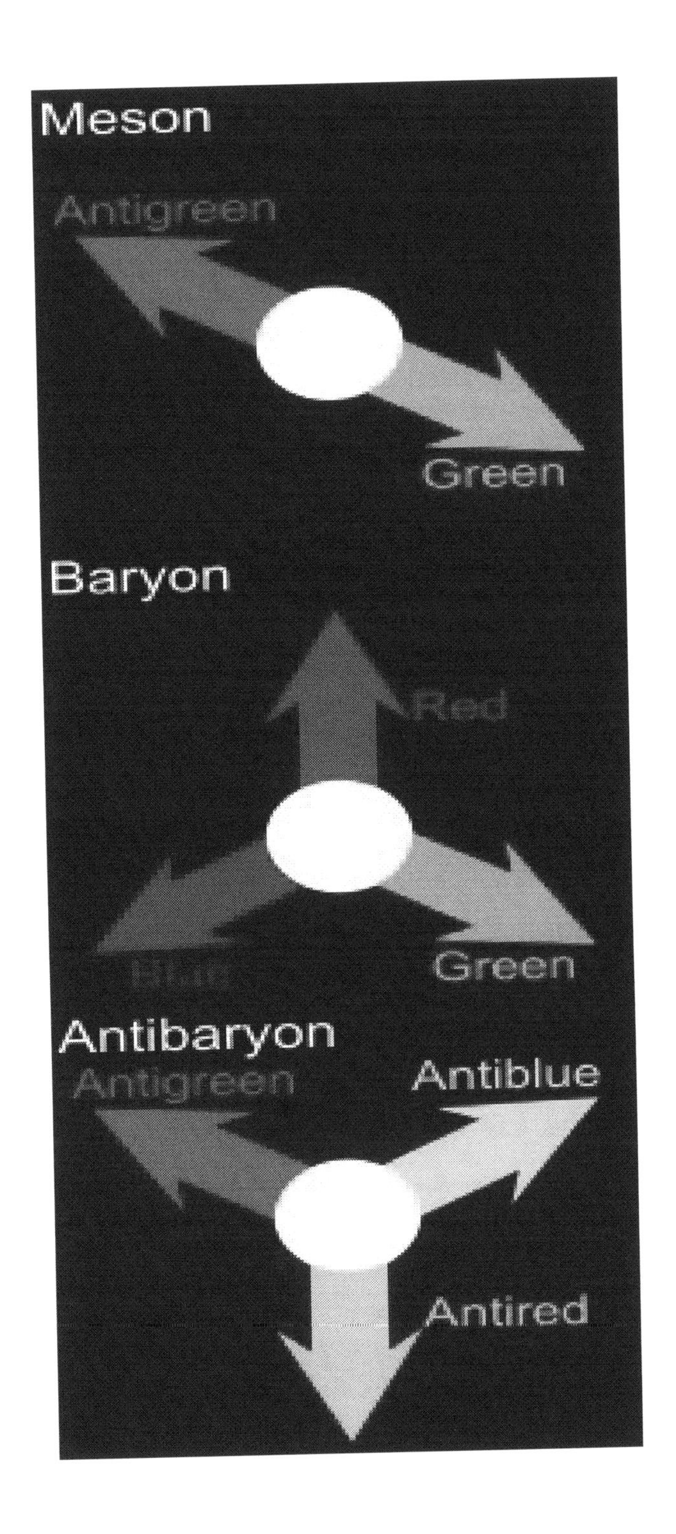

All types of hadrons have zero total color charge. (three examples shown)

Chapter 51

Baryon

Not to be confused with Baryonyx.

A **baryon** is a composite subatomic particle made up of three quarks (as distinct from mesons, which are composed of one quark and one antiquark). Baryons and mesons belong to the hadron family of particles, which are the quark-based particles. The name "baryon" comes from the Greek word for "heavy" (βαρύς, *barys*), because, at the time of their naming, most known elementary particles had lower masses than the baryons.

As quark-based particles, baryons participate in the strong interaction, whereas leptons, which are not quark-based, do not. The most familiar baryons are the protons and neutrons that make up most of the mass of the visible matter in the universe. Electrons (the other major component of the atom) are leptons.

Each baryon has a corresponding antiparticle (antibaryon) where quarks are replaced by their corresponding antiquarks. For example, a proton is made of two up quarks and one down quark; and its corresponding antiparticle, the antiproton, is made of two up antiquarks and one down antiquark.

51.1 Background

Baryons are strongly interacting fermions that is, they experience the strong nuclear force and are described by Fermi–Dirac statistics, which apply to all particles obeying the Pauli exclusion principle. This is in contrast to the bosons, which do not obey the exclusion principle.

Baryons, along with mesons, are hadrons, meaning they are particles composed of quarks. Quarks have baryon numbers of $B = 1/3$ and antiquarks have baryon number of $B = -1/3$. The term "baryon" usually refers to *triquarks*—baryons made of three quarks ($B = 1/3 + 1/3 + 1/3 = 1$).

Other exotic baryons have been proposed, such as pentaquarks—baryons made of four quarks and one antiquark ($B = 1/3 + 1/3 + 1/3 + 1/3 - 1/3 = 1$), but their existence is not generally accepted. In theory, heptaquarks (5 quarks, 2 antiquarks), nonaquarks (6 quarks, 3 antiquarks), etc. could also exist. Until recently, it was believed that some experiments showed the existence of pentaquarks—baryons made of four quarks and one antiquark.[1][2] The particle physics community as a whole did not view their existence as likely in 2006,[3] and in 2008, considered evidence to be overwhelmingly against the existence of the reported pentaquarks.[4] However, in July 2015, the LHCb experiment observed two resonances consistent with pentaquark states in the Λ0
b → J/ψK−
p decay, with a combined statistical significance of 15σ.[5][6]

51.2 Baryonic matter

Nearly all matter that may be encountered or experienced in everyday life is baryonic matter, which includes atoms of any sort, and provides those with the quality of mass. Non-baryonic matter, as implied by the name, is any sort of matter that is not composed primarily of baryons. Those might include neutrinos or free electrons, dark matter, such as supersymmetric particles, axions, or black holes.

The very existence of baryons is also a significant issue in cosmology because it is assumed that the Big Bang produced a state with equal amounts of baryons and antibaryons. The process by which baryons came to outnumber their antiparticles is called baryogenesis.

51.3 Baryogenesis

Main article: Baryogenesis

Experiments are consistent with the number of quarks in the universe being a constant and, to be more specific, the number of baryons being a constant ; in technical language, the total baryon number appears to be *conserved*. Within the prevailing Standard Model of particle physics, the number of baryons may change in multiples of three due to the action of sphalerons, although this is rare and has not been observed under experiment. Some grand unified theories of particle physics also predict that a single proton can decay, changing the baryon number by one; however, this has not yet been observed under experiment. The excess of baryons over antibaryons in the present universe is thought to be due to non-conservation of baryon number in the very early universe, though this is not well understood.

51.4 Properties

51.4.1 Isospin and charge

Main article: Isospin

The concept of isospin was first proposed by Werner Heisenberg in 1932 to explain the similarities between protons and neutrons under the strong interaction.[7] Although they had different electric charges, their masses were so similar that physicists believed they were actually the same particle. The different electric charges were explained as being the result of some unknown excitation similar to spin. This unknown excitation was later dubbed *isospin* by Eugene Wigner in 1937.[8]

This belief lasted until Murray Gell-Mann proposed the quark model in 1964 (containing originally only the u, d, and s quarks).[9] The success of the isospin model is now understood to be the result of the similar masses of the u and d quarks. Since the u and d quarks have similar masses, particles made of the same number then also have similar masses. The exact specific u and d quark composition determines the charge, as u quarks carry charge $+\frac{2}{3}$ while d quarks carry charge $-\frac{1}{3}$. For example the four Deltas all have different charges (Δ++ (uuu), Δ+ (uud), Δ0 (udd), Δ− (ddd)), but have similar masses (~1,232 MeV/c^2) as they are each made of a combination of three u and d quarks. Under the isospin model, they were considered to be a single particle in different charged states.

The mathematics of isospin was modeled after that of spin. Isospin projections varied in increments of 1 just like those of spin, and to each projection was associated a "charged state". Since the "Delta particle" had four “charged states”, it was said to be of isospin $I = \frac{3}{2}$. Its “charged states” Δ++, Δ+, Δ0, and Δ−, corresponded to the isospin projections $I_3 = +\frac{3}{2}$, $I_3 = +\frac{1}{2}$, $I_3 = -\frac{1}{2}$, and $I_3 = -\frac{3}{2}$, respectively. Another example is the “nucleon particle”. As there were two nucleon “charged states”, it was said to be of isospin $\frac{1}{2}$. The positive nucleon N+ (proton) was identified with $I_3 = +\frac{1}{2}$ and the neutral nucleon N0 (neutron) with $I_3 = -\frac{1}{2}$.[10] It was later noted that the isospin projections were related to the up and down quark content of particles by the relation:

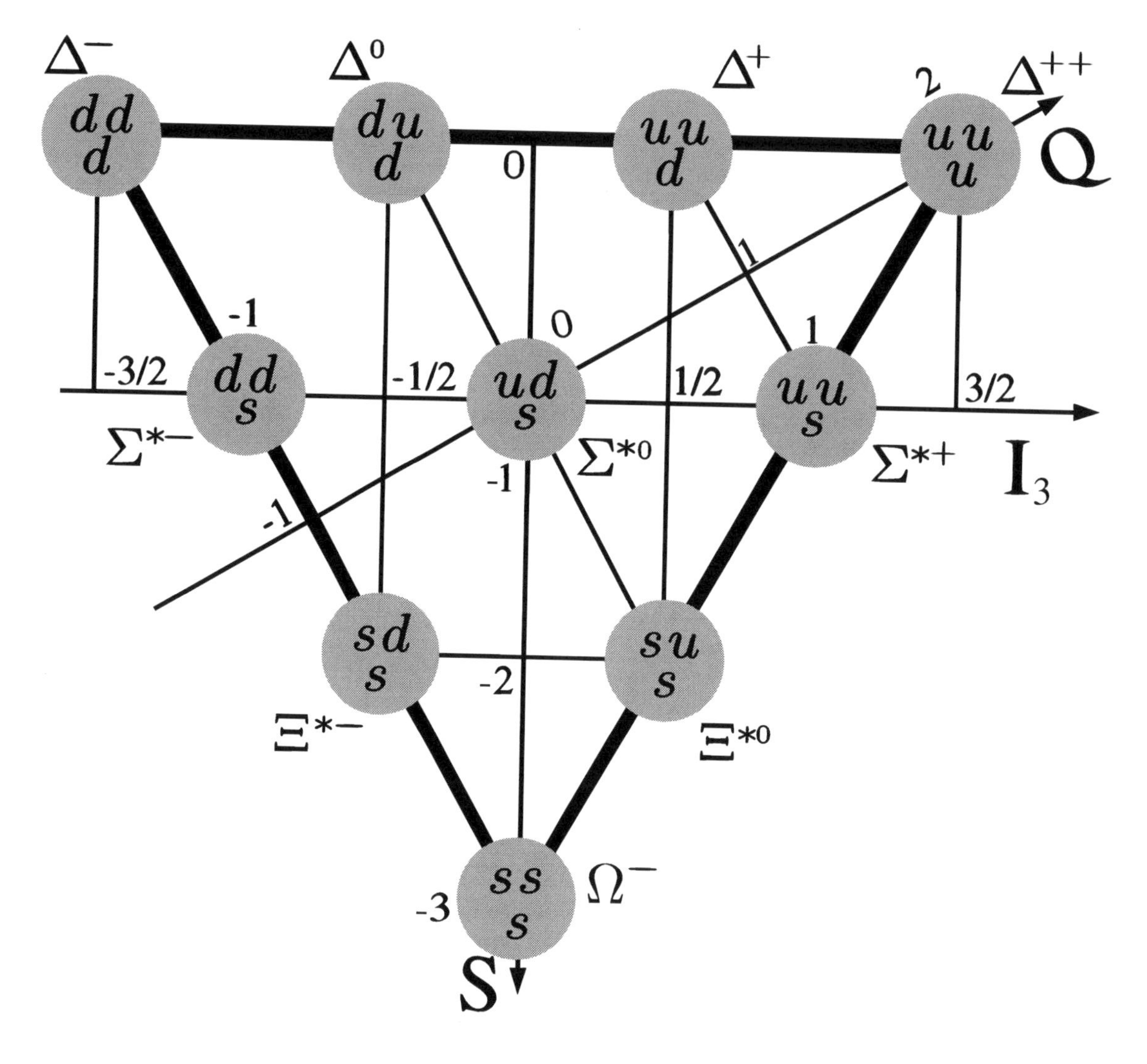

Combinations of three **u, d** *or s quarks forming baryons with a spin-*$^3\!/_2$ *form the* uds baryon decuplet

$$I_3 = \frac{1}{2}[(n_{\mathrm{u}} - n_{\bar{\mathrm{u}}}) - (n_{\mathrm{d}} - n_{\bar{\mathrm{d}}})],$$

where the *n*'s are the number of up and down quarks and antiquarks.

In the "isospin picture", the four Deltas and the two nucleons were thought to be the different states of two particles. However in the quark model, Deltas are different states of nucleons (the N^{++} or N^- are forbidden by Pauli's exclusion principle). Isospin, although conveying an inaccurate picture of things, is still used to classify baryons, leading to unnatural and often confusing nomenclature.

51.4.2 Flavour quantum numbers

Main article: Flavour (particle physics) § Flavour quantum numbers

The strangeness flavour quantum number *S* (not to be confused with spin) was noticed to go up and down along with particle mass. The higher the mass, the lower the strangeness (the more s quarks). Particles could be described with

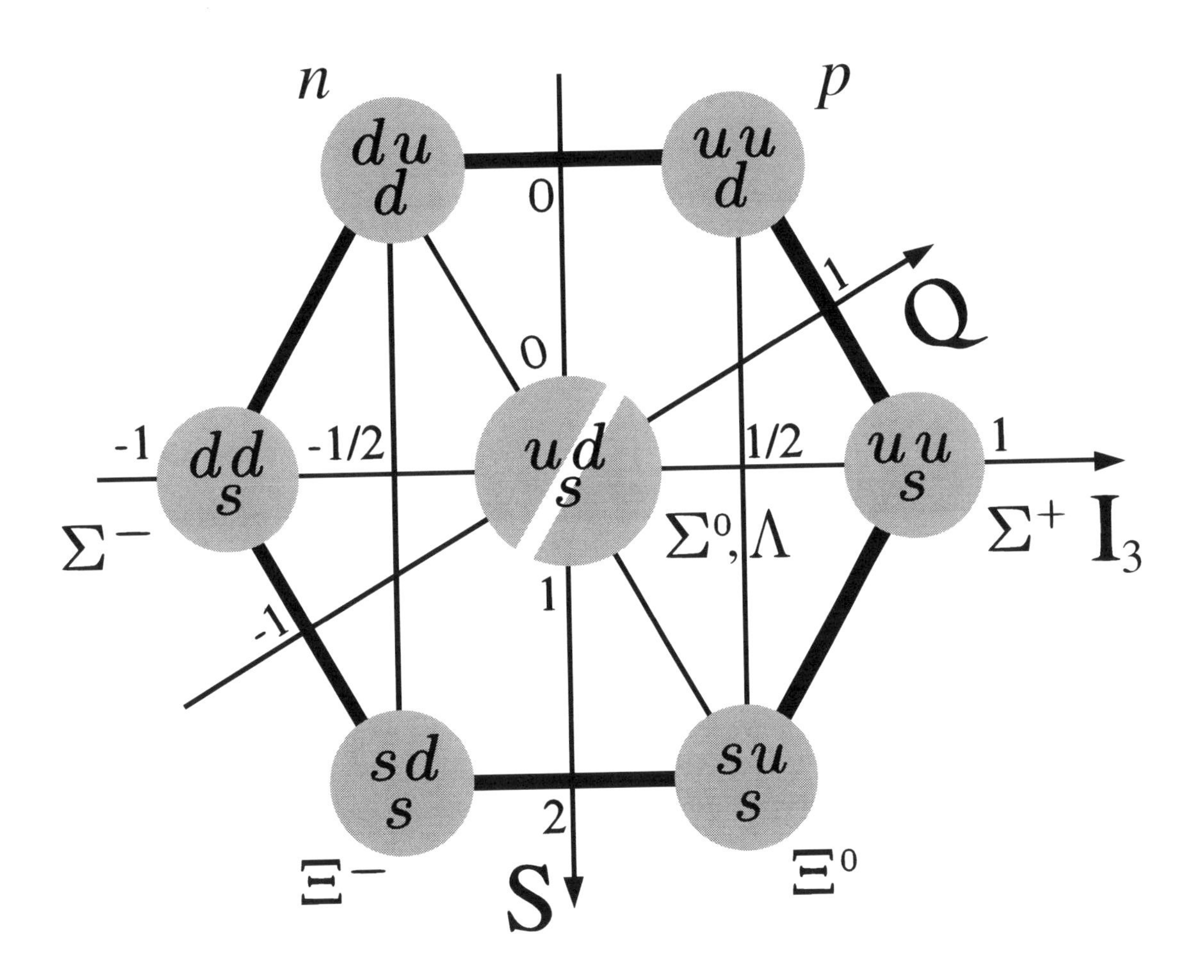

*Combinations of three **u**, **d** or s quarks forming baryons with a spin-½ form the* uds baryon octet

isospin projections (related to charge) and strangeness (mass) (see the uds octet and decuplet figures on the right). As other quarks were discovered, new quantum numbers were made to have similar description of udc and udb octets and decuplets. Since only the u and d mass are similar, this description of particle mass and charge in terms of isospin and flavour quantum numbers works well only for octet and decuplet made of one u, one d, and one other quark, and breaks down for the other octets and decuplets (for example, ucb octet and decuplet). If the quarks all had the same mass, their behaviour would be called *symmetric*, as they would all behave in exactly the same way with respect to the strong interaction. Since quarks do not have the same mass, they do not interact in the same way (exactly like an electron placed in an electric field will accelerate more than a proton placed in the same field because of its lighter mass), and the symmetry is said to be broken.

It was noted that charge (Q) was related to the isospin projection (I_3), the baryon number (B) and flavour quantum numbers (S, C, B', T) by the Gell-Mann–Nishijima formula:[10]

$$Q = I_3 + \frac{1}{2}(B + S + C + B' + T),$$

where S, C, B', and T represent the strangeness, charm, bottomness and topness flavour quantum numbers, respectively. They are related to the number of strange, charm, bottom, and top quarks and antiquark according to the relations:

$$S = -(n_s - n_{\bar{s}}),$$

$$C = +(n_c - n_{\bar{c}}),$$

$$B' = -(n_b - n_{\bar{b}}),$$

$$T = +(n_t - n_{\bar{t}}),$$

meaning that the Gell-Mann–Nishijima formula is equivalent to the expression of charge in terms of quark content:

$$Q = \frac{2}{3}[(n_u - n_{\bar{u}}) + (n_c - n_{\bar{c}}) + (n_t - n_{\bar{t}})] - \frac{1}{3}[(n_d - n_{\bar{d}}) + (n_s - n_{\bar{s}}) + (n_b - n_{\bar{b}})].$$

51.4.3 Spin, orbital angular momentum, and total angular momentum

Main articles: Spin (physics), Angular momentum operator, Quantum numbers and Clebsch–Gordan coefficients

Spin (quantum number *S*) is a vector quantity that represents the "intrinsic" angular momentum of a particle. It comes in increments of $\frac{1}{2}$ ħ (pronounced "h-bar"). The ħ is often dropped because it is the "fundamental" unit of spin, and it is implied that "spin 1" means "spin 1 ħ". In some systems of natural units, ħ is chosen to be 1, and therefore does not appear anywhere.

Quarks are fermionic particles of spin $\frac{1}{2}$ ($S = \frac{1}{2}$). Because spin projections vary in increments of 1 (that is 1 ħ), a single quark has a spin vector of length $\frac{1}{2}$, and has two spin projections ($S_z = +\frac{1}{2}$ and $S_z = -\frac{1}{2}$). Two quarks can have their spins aligned, in which case the two spin vectors add to make a vector of length $S = 1$ and three spin projections ($S_z = +1$, $S_z = 0$, and $S_z = -1$). If two quarks have unaligned spins, the spin vectors add up to make a vector of length $S = 0$ and has only one spin projection ($S_z = 0$), etc. Since baryons are made of three quarks, their spin vectors can add to make a vector of length $S = \frac{3}{2}$, which has four spin projections ($S_z = +\frac{3}{2}$, $S_z = +\frac{1}{2}$, $S_z = -\frac{1}{2}$, and $S_z = -\frac{3}{2}$), or a vector of length $S = \frac{1}{2}$ with two spin projections ($S_z = +\frac{1}{2}$, and $S_z = -\frac{1}{2}$).[11]

There is another quantity of angular momentum, called the orbital angular momentum, (azimuthal quantum number *L*), that comes in increments of 1 ħ, which represent the angular moment due to quarks orbiting around each other. The total angular momentum (total angular momentum quantum number *J*) of a particle is therefore the combination of intrinsic angular momentum (spin) and orbital angular momentum. It can take any value from $J = |L - S|$ to $J = |L + S|$, in increments of 1.

Particle physicists are most interested in baryons with no orbital angular momentum ($L = 0$), as they correspond to ground states—states of minimal energy. Therefore the two groups of baryons most studied are the $S = \frac{1}{2}$; $L = 0$ and $S = \frac{3}{2}$; $L = 0$, which corresponds to $J = \frac{1}{2}^+$ and $J = \frac{3}{2}^+$, respectively, although they are not the only ones. It is also possible to obtain $J = \frac{3}{2}^+$ particles from $S = \frac{1}{2}$ and $L = 2$, as well as $S = \frac{3}{2}$ and $L = 2$. This phenomenon of having multiple particles in the same total angular momentum configuration is called *degeneracy*. How to distinguish between these degenerate baryons is an active area of research in baryon spectroscopy.[12][13]

51.4.4 Parity

Main article: Parity (physics)

If the universe were reflected in a mirror, most of the laws of physics would be identical—things would behave the same way regardless of what we call "left" and what we call "right". This concept of mirror reflection is called *intrinsic parity* or *parity* (*P*). Gravity, the electromagnetic force, and the strong interaction all behave in the same way regardless of whether or not the universe is reflected in a mirror, and thus are said to conserve parity (P-symmetry). However, the weak interaction *does* distinguish "left" from "right", a phenomenon called parity violation (P-violation).

Based on this, one might think that, if the wavefunction for each particle (in more precise terms, the quantum field for each particle type) were simultaneously mirror-reversed, then the new set of wavefunctions would perfectly satisfy the laws of physics (apart from the weak interaction). It turns out that this is not quite true: In order for the equations to be satisfied, the wavefunctions of certain types of particles have to be multiplied by −1, in addition to being mirror-reversed. Such particle types are said to have *negative* or *odd* parity ($P = -1$, or alternatively $P = -$), while the other particles are said to have *positive* or *even* parity ($P = +1$, or alternatively $P = +$).

For baryons, the parity is related to the orbital angular momentum by the relation:[14]

$$P = (-1)^L.$$

As a consequence, baryons with no orbital angular momentum ($L = 0$) all have even parity ($P = +$).

51.5 Nomenclature

Baryons are classified into groups according to their isospin (I) values and quark (q) content. There are six groups of baryons—nucleon (N), Delta (Δ), Lambda (Λ), Sigma (Σ), Xi (Ξ), and Omega (Ω). The rules for classification are defined by the Particle Data Group. These rules consider the up (u), down (d) and strange (s) quarks to be *light* and the charm (c), bottom (b), and top (t) quarks to be *heavy*. The rules cover all the particles that can be made from three of each of the six quarks, even though baryons made of t quarks are not expected to exist because of the t quark's short lifetime. The rules do not cover pentaquarks.[15]

- Baryons with three u and/or d quarks are N's ($I = 1/2$) or Δ's ($I = 3/2$).
- Baryons with two u and/or d quarks are Λ's ($I = 0$) or Σ's ($I = 1$). If the third quark is heavy, its identity is given by a subscript.
- Baryons with one u or d quark are Ξ's ($I = 1/2$). One or two subscripts are used if one or both of the remaining quarks are heavy.
- Baryons with no u or d quarks are Ω's ($I = 0$), and subscripts indicate any heavy quark content.
- Baryons that decay strongly have their masses as part of their names. For example, Σ^0 does not decay strongly, but $\Delta^{++}(1232)$ does.

It is also a widespread (but not universal) practice to follow some additional rules when distinguishing between some states that would otherwise have the same symbol.[10]

- Baryons in total angular momentum $J = 3/2$ configuration that have the same symbols as their $J = 1/2$ counterparts are denoted by an asterisk (*).
- Two baryons can be made of three different quarks in $J = 1/2$ configuration. In this case, a prime (′) is used to distinguish between them.
  - *Exception*: When two of the three quarks are one up and one down quark, one baryon is dubbed Λ while the other is dubbed Σ.

Quarks carry charge, so knowing the charge of a particle indirectly gives the quark content. For example, the rules above say that a Λ+
c contains a c quark and some combination of two u and/or d quarks. The c quark has a charge of ($Q = +2/3$), therefore the other two must be a u quark ($Q = +2/3$), and a d quark ($Q = -1/3$) to have the correct total charge ($Q = +1$).

51.6 See also

- Eightfold way
- List of baryons
- List of particles
- Meson
- Timeline of particle discoveries

51.7 Notes

[1] H. Muir (2003)

[2] K. Carter (2003)

[3] W.-M. Yao *et al.* (2006): Particle listings – Θ^+

[4] C. Amsler *et al.* (2008): Pentaquarks

[5] LHCb (14 July 2015). "Observation of particles composed of five quarks, pentaquark-charmonium states, seen in $\Lambda_b^0 \to J/\psi pK^-$ decays.". *CERN website*. Retrieved 2015-07-14.

[6] R. Aaij et al. (LHCb collaboration) (2015). "Observation of J/ψp resonances consistent with pentaquark states in Λ0 b→J/ψK–
p decays". *Physical Review Letters* **115** (7). Bibcode:2015PhRvL.115g2001A. doi:10.1103/PhysRevLett.115.072001.

[7] W. Heisenberg (1932)

[8] E. Wigner (1937)

[9] M. Gell-Mann (1964)

[10] S.S.M. Wong (1998a)

[11] R. Shankar (1994)

[12] H. Garcilazo *et al.* (2007)

[13] D.M. Manley (2005)

[14] S.S.M. Wong (1998b)

[15] C. Amsler *et al.* (2008): Naming scheme for hadrons

51.8 References

- C. Amsler *et al.* (Particle Data Group) (2008). "Review of Particle Physics". *Physics Letters B* **667** (1): 1–1340. Bibcode:2008PhLB..667....1P. doi:10.1016/j.physletb.2008.07.018.
- H. Garcilazo, J. Vijande, and A. Valcarce (2007). "Faddeev study of heavy-baryon spectroscopy". *Journal of Physics G* **34** (5): 961–976. doi:10.1088/0954-3899/34/5/014.
- K. Carter (2006). "The rise and fall of the pentaquark". Fermilab and SLAC. Retrieved 2008-05-27.
- W.-M. Yao *et al.*(Particle Data Group) (2006). "Review of Particle Physics". *Journal of Physics G* **33**: 1–1232. arXiv:astro-ph/0601168. Bibcode:2006JPhG...33....1Y. doi:10.1088/0954-3899/33/1/001.

- D.M. Manley (2005). "Status of baryon spectroscopy". *Journal of Physics: Conference Series* **5**: 230–237. Bibcode:2005JPhCS...9..230M. doi:10.1088/1742-6596/9/1/043.
- H. Muir (2003). "Pentaquark discovery confounds sceptics". New Scientist. Retrieved 2008-05-27.
- S.S.M. Wong (1998a). "Chapter 2—Nucleon Structure". *Introductory Nuclear Physics* (2nd ed.). New York (NY): John Wiley & Sons. pp. 21–56. ISBN 0-471-23973-9.
- S.S.M. Wong (1998b). "Chapter 3—The Deuteron". *Introductory Nuclear Physics* (2nd ed.). New York (NY): John Wiley & Sons. pp. 57–104. ISBN 0-471-23973-9.
- R. Shankar (1994). *Principles of Quantum Mechanics* (2nd ed.). New York (NY): Plenum Press. ISBN 0-306-44790-8.
- E. Wigner (1937). "On the Consequences of the Symmetry of the Nuclear Hamiltonian on the Spectroscopy of Nuclei". *Physical Review* **51** (2): 106–119. Bibcode:1937PhRv...51..106W. doi:10.1103/PhysRev.51.106.
- M.Gell-Mann(1964). "A Schematic of Baryons and Mesons".*Physics Letters***8**(3): 214–215. Bibcode:1964PhL... doi:10.1016/S0031-9163(64)92001-3.
- W.Heisenberg(1932). "Über den Bau der Atomkerne I".*Zeitschrift für Physik*(in German)**77**: 1–11. Bibcode1H. doi:10.1007/BF01342433.
- W. Heisenberg (1932). "Über den Bau der Atomkerne II". *Zeitschrift für Physik* (in German) **78** (3–4): 156–164. Bibcode:1932ZPhy...78..156H. doi:10.1007/BF01337585.
- W. Heisenberg (1932). "Über den Bau der Atomkerne III". *Zeitschrift für Physik* (in German) **80** (9–10): 587–596. Bibcode:1933ZPhy...80..587H. doi:10.1007/BF01335696.

51.9 External links

- Particle Data Group—Review of Particle Physics (2008).
- Georgia State University—HyperPhysics
- Baryons made thinkable, an interactive visualisation allowing physical properties to be compared

Chapter 52

Hyperon

Not to be confused with Hyperion (disambiguation).

In particle physics, a **hyperon** is any baryon containing one or more strange quarks, but no charm, bottom, or top quark.

52.1 Properties and behavior of hyperons

Being baryons, all hyperons are fermions. That is, they have half-integer spin and obey Fermi–Dirac statistics. They all interact via the strong nuclear force, making them types of hadron. They are composed of three light quarks, at least one of which is a strange quark, which makes them strange baryons. Hyperons decay weakly with non-conserved parity.

52.2 List of hyperons

Notes:

- Since strangeness is conserved by the strong interactions, the ground-state hyperons cannot decay strongly. However, they do participate in strong interactions.
- Λ^0 may also decay on rare occurrences via these processes:

  $\Lambda^0 \rightarrow p^+ + e^- + \nu_e$

  $\Lambda^0 \rightarrow p^+ + \mu^- + \nu_\mu$

- Ξ^0 and Ξ^- are also known as “cascade” hyperons, since they go through a two-step cascading decay into a nucleon.
- The Ω^- has a baryon number of +1 and hypercharge of −2, giving it strangeness of −3.

It takes multiple flavor-changing weak decays for it to decay into a proton or neutron. Murray Gell-Mann's and Yuval Ne'eman's SU(3) model (sometimes called the Eightfold Way) predicted this hyperon’s existence, mass and that it will only undergo weak decay processes. Experimental evidence for its existence was discovered in 1964 at Brookhaven National Laboratory. Further examples of its formation and observation using particle accelerators confirmed the SU(3) model.

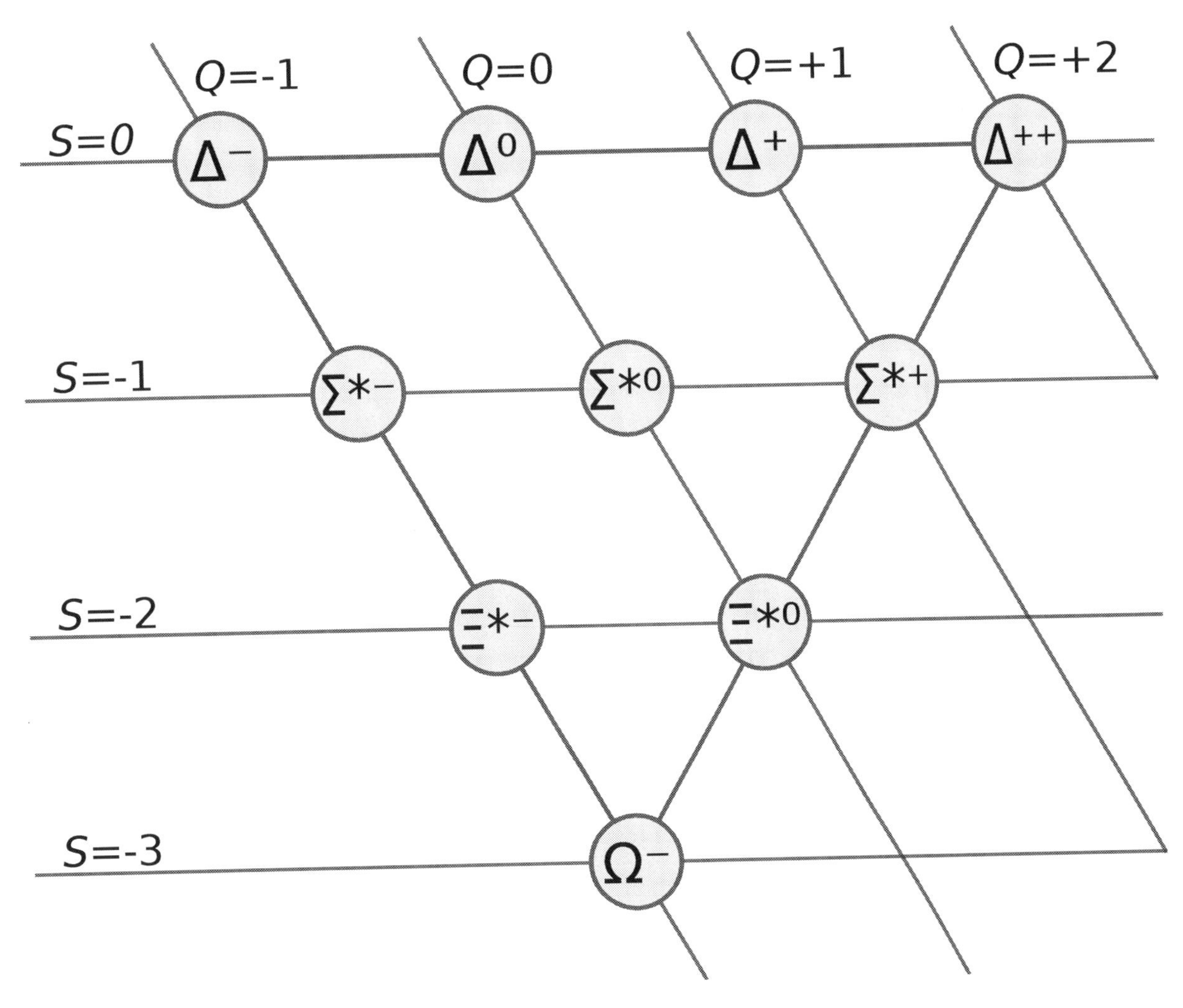

A combination of three u, d or s-quarks with a total spin of 3/2 form the so-called ***baryon decuplet****. The lower six are hyperons.*

52.3 Hyperon research

The first research into hyperons happened in the 1950s, and spurred physicists on to the creation of an organized classification of particles. Today, research in this area is carried out on data taken at many facilities around the world, including CERN, Fermilab, SLAC, JLAB, Brookhaven National Laboratory, KEK, and others. Physics topics include searches for CP violation, measurements of spin, studies of excited states (commonly referred to as *spectroscopy*), and hunts for exotic states such as pentaquarks and dibaryons.

52.4 See also

- Delta baryon
- Hypernucleus
- List of mesons
- List of particles
- Nucleon

- Physics portal
- Timeline of particle discoveries

52.5 References

[1] "Particle Data Groups: 2006 Review of Particle Physics – Lambda" (PDF). Retrieved 2008-04-20.

[2] "Physics Particle Overview – Baryons". Retrieved 2008-04-20.

[3] "Particle Data Groups: 2006 Review of Particle Physics – Sigma+" (PDF). Retrieved 2008-04-20.

[4] "Particle Data Groups: 2006 Review of Particle Physics – Sigma0" (PDF). Retrieved 2008-04-20.

[5] "Particle Data Groups: 2006 Review of Particle Physics – Sigma-" (PDF). Retrieved 2008-04-20.

[6] "Particle Data Groups: 2006 Review of Particle Physics – Sigma(1385)" (PDF). Retrieved 2008-04-20.

[7] "Particle Data Groups: 2006 Review of Particle Physics – Xi0" (PDF). Retrieved 2008-04-20.

[8] "Particle Data Groups: 2006 Review of Particle Physics – Xi-" (PDF). Retrieved 2008-04-20.

[9] "Particle Data Groups: 2006 Review of Particle Physics – Xi(1530)" (PDF). Retrieved 2008-04-20.

[10] "Particle Data Groups: 2006 Review of Particle Physics – Omega-" (PDF). Retrieved 2008-04-20.

- Henry Semat, John R. Albright (1984). *Introduction to atomic and nuclear physics*. Chapman and Hall. ISBN 0-412-15670-9.

Chapter 53

Nucleon

For the Ford concept car, see Ford Nucleon. For the fictional power source in the Transformers universe, see Nucleon (Transformers).

In chemistry and physics, a **nucleon** is one of the particles that makes up the atomic nucleus. Each atomic nucleus consists of one or more nucleons, and each atom in turn consists of a cluster of nucleons surrounded by one or more electrons. There are two known kinds of nucleon: the neutron and the proton. The mass number of a given atomic isotope is identical to its number of nucleons. Thus the term nucleon number may be used in place of the more common terms mass number or atomic mass number.

Until the 1960s, nucleons were thought to be elementary particles, each of which would not then have been made up of smaller parts. Now they are known to be composite particles, made of three quarks bound together by the so-called strong interaction. The interaction between two or more nucleons is called internucleon interactions or nuclear force, which is also ultimately caused by the strong interaction. (Before the discovery of quarks, the term "strong interaction" referred to just internucleon interactions.)

Nucleons sit at the boundary where particle physics and nuclear physics overlap. Particle physics, particularly quantum chromodynamics, provides the fundamental equations that explain the properties of quarks and of the strong interaction. These equations explain quantitatively how quarks can bind together into protons and neutrons (and all the other hadrons). However, when multiple nucleons are assembled into an atomic nucleus (nuclide), these fundamental equations become too difficult to solve directly (see lattice QCD). Instead, nuclides are studied within nuclear physics, which studies nucleons and their interactions by approximations and models, such as the nuclear shell model. These models can successfully explain nuclide properties, for example, whether or not a certain nuclide undergoes radioactive decay.

The proton and neutron are both baryons and both fermions. They are quite similar. One carries a non-zero net charge and the other carries a zero net charge; the proton's mass is only 0.1% less than the neutron's. Thus, they can be viewed as two states of the same nucleon. They together form the isospin doublet ($\mathbf{I} = \frac{1}{2}$). In isospin space, neutrons can be rotationally transformed into protons, and vice versa. These nucleons are acted upon equally by the strong interaction. This implies that strong interaction is invariant when doing rotation transformation in isospin space. According to the Noether theorem, isospin is conserved with respect to the strong interaction.[1]:129–130

53.1 Overview

Main articles: Proton and Neutron

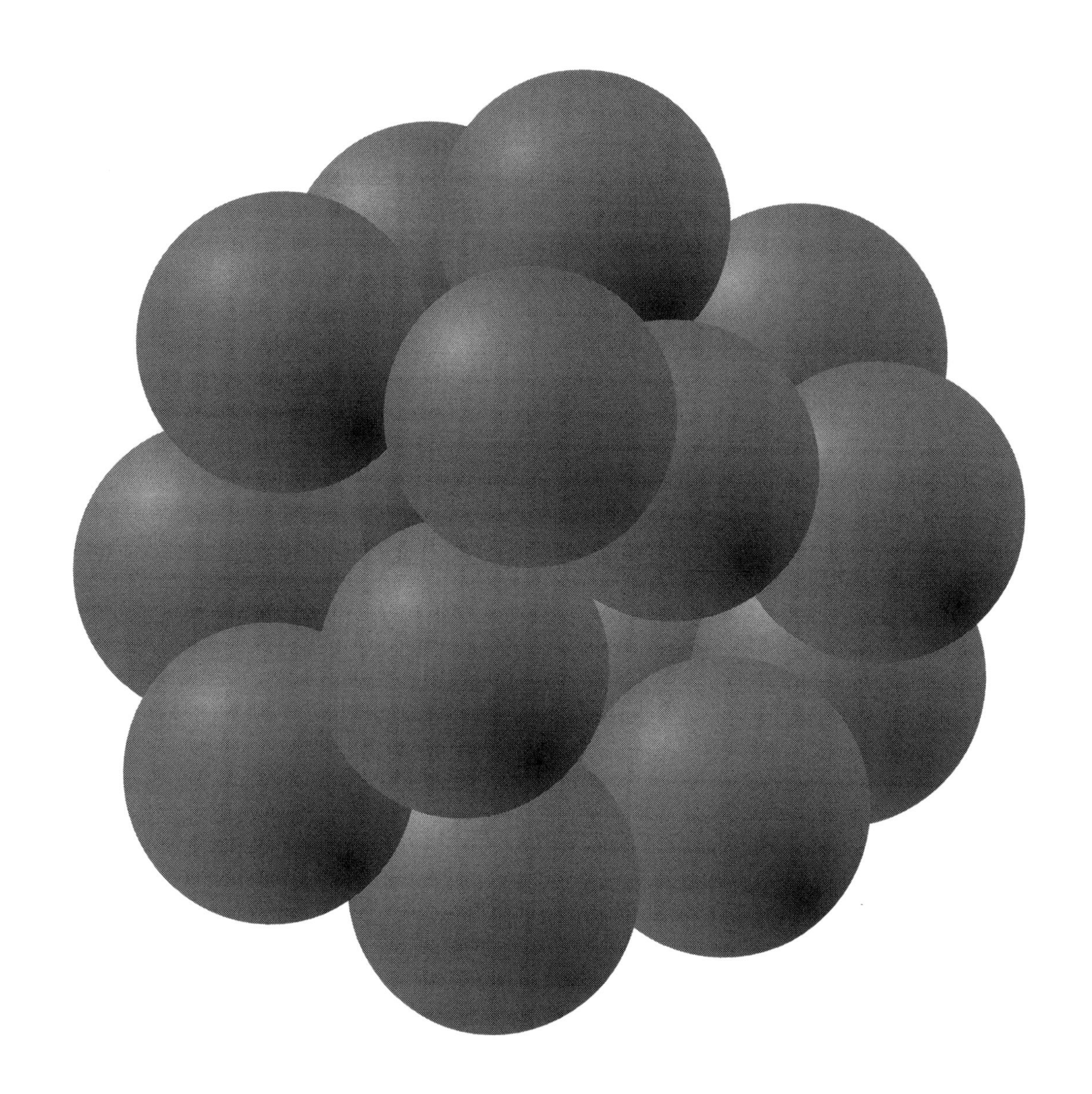

An atomic nucleus is a compact bundle of the two types of nucleons: Protons (red) and neutrons (blue). In this picture, the protons and neutrons look like little balls stuck together, but an actual nucleus, as understood by modern nuclear physics, does not look like this. An actual nucleus can only be accurately described using quantum mechanics. For example, in a real nucleus, each nucleon is in multiple locations at once, spread throughout the nucleus.

53.1.1 Properties

Nucleon quark composition

Proton (p): uud

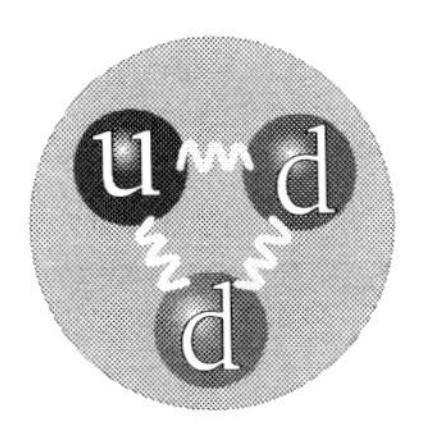

Neutron (n): udd

Antiproton (p): uud

Antineutron (n): udd

Protons and neutrons are most important and best known for constituting atomic nuclei, but they can also be found on their own, not part of a larger nucleus. A proton on its own is the nucleus of the hydrogen-1 atom (^{1}H). A neutron on its own is unstable (see below), but they can be found in nuclear reactions (see neutron radiation) and are used in scientific analysis (see neutron scattering).

Both the proton and neutron are made of three quarks. The proton is made of two up quarks and one down quark, while the neutron is one up quark and two down quarks. The quarks are held together by the strong force. It is also said that the quarks are held together by gluons, but this is just a different way to say the same thing (gluons mediate the strong force).

An up quark has electric charge $+\frac{2}{3}\ e$, and a down quark has charge $-\frac{1}{3}\ e$, so the total electric charge of the proton and neutron are $+e$ and 0, respectively. The word "neutron" comes from the fact that it is electrically "neutral".

The mass of the proton and neutron is quite similar: The proton is 1.6726×10^{-27} kg or 938.27 MeV/c^2, while the neutron is 1.6749×10^{-27} kg or 939.57 MeV/c^2. The neutron is roughly 0.1% heavier. The similarity in mass can be explained roughly by the slight difference in mass of up quark and down quark composing the nucleons. However, detailed explanation remains an unsolved problem in particle physics.[1]:135–136

The spin of both protons and neutrons is ½. This means they are fermions not bosons, and therefore, like electrons, they are subject to the Pauli exclusion principle. This is a very important fact in nuclear physics: Protons and neutrons in an atomic nucleus cannot all be in the same quantum state, but instead they spread out into nuclear shells analogous to electron shells in chemistry. Another reason that the spin of the proton and neutron is important is because it is the source of nuclear spin in larger nuclei. Nuclear spin is best known for its crucial role in the NMR/MRI technique for chemistry and biochemistry analysis.

The magnetic moment of a proton, denoted μ_p, is 2.79 nuclear magnetons (μN), while the magnetic moment of a neutron is $\mu_n = -1.91$ μN. These parameters are also important in NMR/MRI.

53.1.2 Stability

A neutron by itself is an unstable particle: It undergoes β− decay (a type of radioactive decay) by turning into a proton, electron, and electron antineutrino, with a half-life around ten minutes. (See the Neutron article for further discussion of neutron decay.) A proton by itself is thought to be stable, or at least its lifetime is too long to measure. (This is an important issue in particle physics, see Proton decay.)

Inside a nucleus, on the other hand, both protons and neutrons can be stable or unstable, depending on the nuclide. Inside some nuclides, a neutron can turn into a proton (plus other particles) as described above; inside other nuclides the reverse can happen, where a proton turns into a neutron (plus other particles) through β+ decay or electron capture; and inside still other nuclides, both protons and neutrons are stable and do not change form.

53.1.3 Antinucleons

Main articles: Antineutron, Antiproton and Antimatter

Both of the nucleons have corresponding antiparticles: The antiproton and the antineutron. These antimatter particles have the same mass and opposite charge as the proton and neutron respectively, and they interact in the same way. (This is generally believed to be *exactly* true, due to CPT symmetry. If there is a difference, it is too small to measure in all experiments to date.) In particular, antinucleons can bind into an "antinucleus". So far, scientists have created antideuterium[2][3] and antihelium-3[4] nuclei.

53.2 Tables of detailed properties

53.2.1 Nucleons

^a The masses of the proton and neutron are known with far greater precision in atomic mass units (u) than in MeV/c^2, due to the relatively poorly known value of the elementary charge. The conversion factor used is 1 u = 931.494028±0.000023 MeV/c^2. The masses of their antiparticles are assumed to be identical, and no experiments have refuted this to date. Current experiments show any percent difference between the masses of the proton and antiproton must be less than 2×10^{-9}[PDG 1] and the difference between the neutron and antineutron masses is on the order of $(9\pm6)\times10^{-5}$ MeV/c^2.[PDG 2]

^b At least 10^{35} years. See proton decay.

^c For free neutrons; in most common nuclei, neutrons are stable.

53.2.2 Nucleon resonances

Nucleon resonances are excited states of nucleon particles, often corresponding to one of the quarks having a flipped spin state, or with different orbital angular momentum when the particle decays. Only resonances with a 3 or 4 star rating at the Particle Data Group (PDG) are included in this table. Due to their extraordinarily short lifetimes, many properties of these particles are still under investigation.

The symbol format is given as N(M) L_2I_2J, where M is the particle's approximate mass, L is the orbital angular momentum of the Nucleon-meson pair produced when it decays, and I and J are the particle's isospin and total angular momentum respectively. Since nucleons are defined as having ½ isospin, the first number will always be 1, and the second number will always be odd. When discussing nucleon resonances, sometimes the N is omitted and the order is reversed, giving L_2I_2J (M). For example, a proton can be symbolized as "N(939) S_{11}" or "S_{11} (939)".

The table below lists only the base resonance; each individual entry represents 4 baryons: 2 nucleon resonances particles, as well as their 2 antiparticles. Each resonance exists in a form with a positive electric charge (Q), with a quark composition of uud like the proton, and a neutral form, with a quark composition of udd like the neutron, as well as the corresponding antiparticles with antiquark compositions of uud and udd respectively. Since they contain no strange, charm, bottom, or top quarks, these particles do not possess strangeness, etc. The table only lists the resonances with an isospin of ½. For resonances with 3⁄2 isospin, see the Delta baryon article.

† *The P_{11}(939) nucleon represents the excited state of a normal proton or neutron, for example, within the nucleus of an atom. Such particles are usually stable within the nucleus, i.e. Lithium-6.*

53.3 Quark model classification

In the quark model with SU(2) flavour, the two nucleons are part of the ground state doublet. The proton has quark content of *uud*, and the neutron, *udd*. In SU(3) flavour, they are part of the ground state octet (**8**) of spin ½ baryons, known as the Eightfold way. The other members of this octet are the hyperons strange isotriplet Σ+, Σ0, Σ−, the Λ and the strange isodoublet Ξ0, Ξ−. One can extend this multiplet in SU(4) flavour (with the inclusion of the charm quark) to the ground state **20**-plet, or to SU(6) flavour (with the inclusion of the top and bottom quarks) to the ground state **56**-plet.

The article on isospin provides an explicit expression for the nucleon wave functions in terms of the quark flavour eigenstates.

53.4 Models

Although it is known that the nucleon is made from three quarks, as of 2006, it is not known how to solve the equations of motion for quantum chromodynamics. Thus, the study of the low-energy properties of the nucleon are performed by means of models. The only first-principles approach available is to attempt to solve the equations of QCD numerically, using lattice QCD. This requires complicated algorithms and very powerful supercomputers. However, several analytic models also exist:

The Skyrmion models the nucleon as a topological soliton in a non-linear SU(2) pion field. The topological stability of the Skyrmion is interpreted as the conservation of baryon number, that is, the non-decay of the nucleon. The local topological winding number density is identified with the local baryon number density of the nucleon. With the pion isospin vector field oriented in the shape of a hedgehog space, the model is readily solvable, and is thus sometimes called the **hedgehog model**. The hedgehog model is able to predict low-energy parameters, such as the nucleon mass, radius and axial coupling constant, to approximately 30% of experimental values.

The MIT bag model confines three non-interacting quarks to a spherical cavity, with the boundary condition that the quark vector current vanish on the boundary. The non-interacting treatment of the quarks is justified by appealing to the idea of asymptotic freedom, whereas the hard boundary condition is justified by quark confinement. Mathematically, the model vaguely resembles that of a radar cavity, with solutions to the Dirac equation standing in for solutions to the Maxwell equations and the vanishing vector current boundary condition standing for the conducting metal walls of the radar cavity. If the radius of the bag is set to the radius of the nucleon, the bag model predicts a nucleon mass that is within 30% of the actual mass. Although the basic bag model does not provide a pion-mediated interaction, it describes excellently the nucleon-nucleon forces through the 6-quark bag s-channel mechanism using the P matrix.[5] [6]

The **chiral bag model**[7] merges the MIT bag model and the Skyrmion model. In this model, a hole is punched out of the middle of the Skyrmion, and replaced with a bag model. The boundary condition is provided by the requirement of continuity of the axial vector current across the bag boundary. Very curiously, the missing part of the topological winding number (the baryon number) of the hole punched into the Skyrmion is exactly made up by the non-zero vacuum expectation value (or spectral asymmetry) of the quark fields inside the bag. As of 2006, this remarkable trade-off between topology and the spectrum of an operator does not have any grounding or explanation in the mathematical theory of Hilbert spaces and their relationship to geometry. Several other properties of the chiral bag are notable: it provides a better fit to the low energy nucleon properties, to within 5–10%, and these are almost completely independent of the chiral bag radius (as long as the radius is less than the nucleon radius). This independence of radius is referred to as the **Cheshire Cat principle**, after the fading to a smile of Lewis Carroll's Cheshire Cat. It is expected that a first-principles solution of the equations of QCD will demonstrate a similar duality of quark-pion descriptions.

53.5 See also

- Hadrons
- Electroweak interaction

53.6 Further reading

- A.W. Thomas and W.Weise, *The Structure of the Nucleon*, (2001) Wiley-WCH, Berlin, ISBN ISBN 3-527-40297-7
- YAN Kun. Equation of average binding energy per nucleon. doi:10.3969/j.issn.1004-2903.2011.01.018
- Brown, G. E.; Jackson, A. D. (1976). *The Nucleon–Nucleon Interaction*. North-Holland Publishing. ISBN 0-7204-0335-9.
- Vepstas, L.; Jackson, A.D.; Goldhaber, A.S. (1984). "Two-phase models of baryons and the chiral Casimir effect". *Physics Letters B* **140** (5–6): 280–284. Bibcode:1984PhLB..140..280V. doi:10.1016/0370-2693(84)90753-6.
- Vepstas,L.;Jackson,A.D. (1990). "Justifying the chiral bag".*Physics Reports***187**(3): 109–143. Bibcode:. doi:10.1016/0370-1573(90)90056-8.
- Nakamura, N.; Particle Data Group et al. (2011). "Review of Particle Physics". *Journal of Physics G* **37** (7): 075021. Bibcode:2010JPhG...37g5021N. doi:10.1088/0954-3899/37/7A/075021.

53.7 References

[1] Griffiths, David J. (2008), *Introduction to Elementary Particles* (2nd revised ed.), WILEY-VCH, ISBN 978-3-527-40601-2

[2] Massam, T; Muller, Th.; Righini, B.; Schneegans, M.; Zichichi, A. (1965). "Experimental observation of antideuteron production". *Il Nuovo Cimento* **39**: 10–14. Bibcode:1965NCimS..39...10M. doi:10.1007/BF02814251.

[3] Dorfan, D. E; Eades, J.; Lederman, L. M.; Lee, W.; Ting, C. C. (June 1965). "Observation of Antideuterons". *Phys. Rev. Lett.* **14** (24): 1003–1006. Bibcode:1965PhRvL..14.1003D. doi:10.1103/PhysRevLett.14.1003.

[4] R. Arsenescu et al. (2003). "Antihelium-3 production in lead-lead collisions at 158 *A* GeV/*c*". *New Journal of Physics* **5**: 1. Bibcode:2003NJPh....5....1A. doi:10.1088/1367-2630/5/1/301.

[5] Jaffe, R. L.; Low, F. E. (1979). "Connection between quark-model eigenstates and low-energy scattering". *Phys. Rev. D* **19**: 2105. Bibcode:1979PhRvD..19.2105J. doi:10.1103/PhysRevD.19.2105.

[6] Yu;Simonov,A. (1981). "The quark compound bag model and the Jaffe-Low P matrix".*Phys.Lett.B***107**: 1. Bibcode:1981Ph1S. doi:10.1016/0370-2693(81)91133-3.

[7] Gerald E. Brown and Mannque Rho (March 1979). "The little bag". *Phys. Lett. B* **82** (2): 177–180. Bibcode:1979PhLB...82..177B. doi:10.1016/0370-2693(79)90729-9.

53.7.1 Particle listings

[1] Particle listings – p

[2] Particle listings – n

[3] Particle listings — Note on N and Delta Resonances

[4] Particle listings — N(1440)

[5] Particle listings — N(1520)

[6] Particle listings — N(1535)

[7] Particle listings — N(1650)

[8] Particle listings — N(1675)

[9] Particle listings — N(1680)

[10] Particle listings — N(1700)

[11] Particle listings — N(1710)

[12] Particle listings — N(1720)

[13] Particle listings — N(2190)

[14] Particle listings — N(2220)

[15] Particle listings — N(2250)

Chapter 54

Proton

This article is about the proton as a subatomic particle. For other uses, see Proton (disambiguation).

The **proton** is an elementary subatomic particle, symbol p or p+, with a positive electric charge of +1e elementary charge and mass slightly less than that of a neutron. Protons and neutrons, each with mass approximately one atomic mass unit, are collectively referred to as "nucleons". One or more protons are present in the nucleus of an atom. The number of protons in the nucleus is referred to as its atomic number. Since each element has a unique number of protons, each element has its own unique atomic number. The word *proton* is Greek for "first", and this name was given to the hydrogen nucleus by Ernest Rutherford in 1920. In previous years Rutherford had discovered that the hydrogen nucleus (known to be the lightest nucleus) could be extracted from the nuclei of nitrogen by collision. The proton was therefore a candidate to be a fundamental particle and a building block of nitrogen and all other heavier atomic nuclei.

In the modern Standard Model of particle physics, the proton is a hadron, and like the neutron, the other nucleon (particle present in atomic nuclei), is composed of three quarks. Although the proton was originally considered a fundamental particle, it is composed of three valence quarks: two up quarks and one down quark. The rest masses of the quarks contribute only about 1% of the proton's mass, however.[2] The remainder of the proton mass is due to the kinetic energy of the quarks and to the energy of the gluon fields that bind the quarks together. Because the proton is not a fundamental particle, it possesses a physical size; the radius of the proton is about 0.84–0.87 fm.[3]

At sufficiently low temperatures, free protons will bind to electrons. However, the character of such bound protons does not change, and they remain protons. A fast proton moving through matter will slow by interactions with electrons and nuclei, until it is captured by the electron cloud of an atom. The result is a protonated atom, which is a chemical compound of hydrogen. In vacuum, when free electrons are present, a sufficiently slow proton may pick up a single free electron, becoming a neutral hydrogen atom, which is chemically a free radical. Such "free hydrogen atoms" tend to react chemically with many other types of atoms at sufficiently low energies. When free hydrogen atoms react with each other, they form neutral hydrogen molecules (H_2), which are the most common molecular component of molecular clouds in interstellar space. Such molecules of hydrogen on Earth may then serve (among many other uses) as a convenient source of protons for accelerators (as used in proton therapy) and other hadron particle physics experiments that require protons to accelerate, with the most powerful and noted example being the Large Hadron Collider.

54.1 Description

Protons are spin-½ fermions and are composed of three valence quarks,[4] making them baryons (a sub-type of hadrons). The two up quarks and one down quark of the proton are held together by the strong force, mediated by gluons.[5]:21–22 A modern perspective has the proton composed of the valence quarks (up, up, down), the gluons, and transitory pairs of sea quarks. The proton has an approximately exponentially decaying positive charge distribution with a mean square radius of about 0.8 fm.[6]

Protons and neutrons are both nucleons, which may be bound together by the nuclear force to form atomic nuclei. The nucleus of the most common isotope of the hydrogen atom (with the chemical symbol "H") is a lone proton. The nuclei of the heavy hydrogen isotopes deuterium and tritium contain one proton bound to one and two neutrons, respectively. All other types of atomic nuclei are composed of two or more protons and various numbers of neutrons.

54.2 History

The concept of a hydrogen-like particle as a constituent of other atoms was developed over a long period. As early as 1815, William Prout proposed that all atoms are composed of hydrogen atoms (which he called "protyles"), based on a simplistic interpretation of early values of atomic weights (see Prout's hypothesis), which was disproved when more accurate values were measured.[7]:39–42

In 1886, Eugen Goldstein discovered canal rays (also known as anode rays) and showed that they were positively charged particles (ions) produced from gases. However, since particles from different gases had different values of charge-to-mass ratio (e/m), they could not be identified with a single particle, unlike the negative electrons discovered by J. J. Thomson.

Following the discovery of the atomic nucleus by Ernest Rutherford in 1911, Antonius van den Broek proposed that the place of each element in the periodic table (its atomic number) is equal to its nuclear charge. This was confirmed experimentally by Henry Moseley in 1913 using X-ray spectra.

In 1917, (in experiments reported in 1919) Rutherford proved that the hydrogen nucleus is present in other nuclei, a result usually described as the discovery of the proton.[8] Rutherford had earlier learned to produce hydrogen nuclei as a type of radiation produced as a product of the impact of alpha particles on nitrogen gas, and recognize them by their unique penetration signature in air and their appearance in scintillation detectors. These experiments were begun when Rutherford had noticed that, when alpha particles were shot into air (mostly nitrogen), his scintillation detectors showed the signatures of typical hydrogen nuclei as a product. After experimentation Rutherford traced the reaction to the nitrogen in air, and found that when alphas were produced into pure nitrogen gas, the effect was larger. Rutherford determined that this hydrogen could have come only from the nitrogen, and therefore nitrogen must contain hydrogen nuclei. One hydrogen nucleus was being knocked off by the impact of the alpha particle, producing oxygen-17 in the process. This was the first reported nuclear reaction, $^{14}N + \alpha \rightarrow {}^{17}O + p$. (This reaction would later be observed happening directly in a cloud chamber in 1925).

Rutherford knew hydrogen to be the simplest and lightest element and was influenced by Prout's hypothesis that hydrogen was the building block of all elements. Discovery that the hydrogen nucleus is present in all other nuclei as an elementary particle, led Rutherford to give the hydrogen nucleus a special name as a particle, since he suspected that hydrogen, the lightest element, contained only one of these particles. He named this new fundamental building block of the nucleus the *proton,* after the neuter singular of the Greek word for "first", πρῶτον. However, Rutherford also had in mind the word *protyle* as used by Prout. Rutherford spoke at the British Association for the Advancement of Science at its Cardiff meeting beginning 24 August 1920.[9] Rutherford was asked by Oliver Lodge for a new name for the positive hydrogen nucleus to avoid confusion with the neutral hydrogen atom. He initially suggested both *proton* and *prouton* (after Prout).[10] Rutherford later reported that the meeting had accepted his suggestion that the hydrogen nucleus be named the "proton", following Prout's word "protyle".[11] The first use of the word "proton" in the scientific literature appeared in 1920.[12]

54.3 Stability

Main article: Proton decay

The free proton (a proton not bound to nucleons or electrons) is a stable particle that has not been observed to break down spontaneously to other particles. Free protons are found naturally in a number of situations in which energies or temperatures are high enough to separate them from electrons, for which they have some affinity. Free protons exist in plasmas in which temperatures are too high to allow them to combine with electrons. Free protons of high energy and velocity make up 90% of cosmic rays, which propagate in vacuum for interstellar distances. Free protons are emitted directly from atomic nuclei in some rare types of radioactive decay. Protons also result (along with electrons and antineutrinos) from the radioactive decay of free neutrons, which are unstable.

The spontaneous decay of free protons has never been observed, and the proton is therefore considered a stable particle. However, some grand unified theories of particle physics predict that proton decay should take place with lifetimes of the order of 10^{36} years, and experimental searches have established lower bounds on the mean lifetime of the proton for various assumed decay products.[13][14][15]

Experiments at the Super-Kamiokande detector in Japan gave lower limits for proton mean lifetime of 6.6×10^{33} years for decay to an antimuon and a neutral pion, and 8.2×10^{33} years for decay to a positron and a neutral pion.[16] Another experiment at the Sudbury Neutrino Observatory in Canada searched for gamma rays resulting from residual nuclei resulting from the decay of a proton from oxygen-16. This experiment was designed to detect decay to any product, and established a lower limit to the proton lifetime of 2.1×10^{29} years.[17]

However, protons are known to transform into neutrons through the process of electron capture (also called inverse beta decay). For free protons, this process does not occur spontaneously but only when energy is supplied. The equation is:

p+ + e− → n + ν
e

The process is reversible; neutrons can convert back to protons through beta decay, a common form of radioactive decay. In fact, a free neutron decays this way, with a mean lifetime of about 15 minutes.

54.4 Quarks and the mass of the proton

In quantum chromodynamics, the modern theory of the nuclear force, most of the mass of the proton and the neutron is explained by special relativity. The mass of the proton is about 80–100 times greater than the sum of the rest masses of the quarks that make it up, while the gluons have zero rest mass. The extra energy of the quarks and gluons in a region within a proton, as compared to the rest energy of the quarks alone in the QCD vacuum, accounts for almost 99% of the mass. The rest mass of the proton is, thus, the invariant mass of the system of moving quarks and gluons that make up the particle, and, in such systems, even the energy of massless particles is still measured as part of the rest mass of the system.

Two terms are used in referring to the mass of the quarks that make up protons: *current quark mass* refers to the mass of a quark by itself, while *constituent quark mass* refers to the current quark mass plus the mass of the gluon particle field surrounding the quark.[18]:285–286 [19]:150–151 These masses typically have very different values. As noted, most of a proton's mass comes from the gluons that bind the current quarks together, rather than from the quarks themselves. While gluons are inherently massless, they possess energy—to be more specific, quantum chromodynamics binding energy (QCBE)—and it is this that contributes so greatly to the overall mass of the proton (see mass in special relativity). A proton has a mass of approximately 938 MeV/c^2, of which the rest mass of its three valence quarks contributes only about 9.4 MeV/c^2; much of the remainder can be attributed to the gluons' QCBE.[20][21][22]

The internal dynamics of the proton are complicated, because they are determined by the quarks' exchanging gluons, and interacting with various vacuum condensates. Lattice QCD provides a way of calculating the mass of the proton directly from the theory to any accuracy, in principle. The most recent calculations[23][24] claim that the mass is determined to better than 4% accuracy, even to 1% accuracy (see Figure S5 in Dürr *et al.*[24]). These claims are still controversial, because the calculations cannot yet be done with quarks as light as they are in the real world. This means that the predictions are found by a process of extrapolation, which can introduce systematic errors.[25] It is hard to tell whether these errors are controlled properly, because the quantities that are compared to experiment are the masses of the hadrons, which are known in advance.

These recent calculations are performed by massive supercomputers, and, as noted by Boffi and Pasquini: "a detailed description of the nucleon structure is still missing because ... long-distance behavior requires a nonperturbative and/or numerical treatment..."[26] More conceptual approaches to the structure of the proton are: the topological soliton approach originally due to Tony Skyrme and the more accurate AdS/QCD approach that extends it to include a string theory of gluons,[27] various QCD-inspired models like the bag model and the constituent quark model, which were popular in the 1980s, and the SVZ sum rules, which allow for rough approximate mass calculations.[28] These methods do not have the same accuracy as the more brute-force lattice QCD methods, at least not yet.

54.5 Charge radius

Main article: Charge radius

The internationally accepted value of the proton's charge radius is 0.8768 fm (see orders of magnitude for comparison to other sizes). This value is based on measurements involving a proton and an electron.

However, since 5 July 2010, an international research team has been able to make measurements involving an exotic atom made of a proton and a negatively charged muon. After a long and careful analysis of those measurements, the team concluded that the root-mean-square charge radius of a proton is "0.84184(67) fm, which differs by 5.0 standard deviations from the CODATA value of 0.8768(69) fm".[29] In January 2013, an updated value for the charge radius of a proton—0.84087(39) fm—was published. The precision was improved by 1.7 times, but the difference with CODATA value persisted at 7σ significance.[30]

The international research team that obtained this result at the Paul Scherrer Institut (PSI) in Villigen (Switzerland) includes scientists from the Max Planck Institute of Quantum Optics (MPQ) in Garching, the Ludwig-Maximilians-Universität (LMU) Munich and the Institut für Strahlwerkzeuge (IFWS) of the Universität Stuttgart (both from Germany), and the University of Coimbra, Portugal.[31][32] They are now attempting to explain the discrepancy, and re-examining the results of both previous high-precision measurements and complicated calculations. If no errors are found in the measurements or calculations, it could be necessary to re-examine the world's most precise and best-tested fundamental theory: quantum electrodynamics.[31] The proton radius remains a puzzle as of early 2015.[33]

54.6 Interaction of free protons with ordinary matter

Main article: Proton therapy

Although protons have affinity for oppositely charged electrons, free protons must lose sufficient velocity (and kinetic energy) in order to become closely associated and bound to electrons, since this is a relatively low-energy interaction. High energy protons, in traversing ordinary matter, lose energy by collisions with atomic nuclei, and by ionization of atoms (removing electrons) until they are slowed sufficiently to be captured by the electron cloud in a normal atom.

However, in such an association with an electron, the character of the bound proton is not changed, and it remains a proton. The attraction of low-energy free protons to any electrons present in normal matter (such as the electrons in normal atoms) causes free protons to stop and to form a new chemical bond with an atom. Such a bond happens at any sufficiently "cold" temperature (i.e., comparable to temperatures at the surface of the Sun) and with any type of atom. Thus, in interaction with any type of normal (non-plasma) matter, low-velocity free protons are attracted to electrons in any atom or molecule with which they come in contact, causing the proton and molecule to combine. Such molecules are then said to be "protonated", and chemically they often, as a result, become so-called Bronsted acids.

54.7 Proton in chemistry

54.7.1 Atomic number

In chemistry, the number of protons in the nucleus of an atom is known as the atomic number, which determines the chemical element to which the atom belongs. For example, the atomic number of chlorine is 17; this means that each chlorine atom has 17 protons and that all atoms with 17 protons are chlorine atoms. The chemical properties of each atom are determined by the number of (negatively charged) electrons, which for neutral atoms is equal to the number of (positive) protons so that the total charge is zero. For example, a neutral chlorine atom has 17 protons and 17 electrons, whereas a Cl^- anion has 17 protons and 18 electrons for a total charge of −1.

All atoms of a given element are not necessarily identical, however, as the number of neutrons may vary to form different isotopes, and energy levels may differ forming different nuclear isomers. For example, there are two stable isotopes of

chlorine: 35
17Cl with 35 − 17 = 18 neutrons and 37
17Cl with 37 − 17 = 20 neutrons.

54.7.2 Hydrogen ion

See also: Hydron (chemistry)
In chemistry, the term proton refers to the hydrogen ion, H+
. Since the atomic number of hydrogen is 1, a hydrogen ion has no electrons and corresponds to a bare nucleus, consisting of a proton (and 0 neutrons for the most abundant isotope *protium* 1
1H). The proton is a "bare charge" with only about **1/64,000** of the radius of a hydrogen atom, and so is extremely reactive chemically. The free proton, thus, has an extremely short lifetime in chemical systems such as liquids and it reacts immediately with the electron cloud of any available molecule. In aqueous solution, it forms the hydronium ion, H_3O^+, which in turn is further solvated by water molecules in clusters such as $[H_5O_2]^+$ and $[H_9O_4]^+$.[34]

The transfer of H+
in an acid–base reaction is usually referred to as "proton transfer". The acid is referred to as a proton donor and the base as a proton acceptor. Likewise, biochemical terms such as proton pump and proton channel refer to the movement of hydrated H+
ions.

The ion produced by removing the electron from a deuterium atom is known as a deuteron, not a proton. Likewise, removing an electron from a tritium atom produces a triton.

54.7.3 Proton nuclear magnetic resonance (NMR)

Also in chemistry, the term "proton NMR" refers to the observation of hydrogen-1 nuclei in (mostly organic) molecules by nuclear magnetic resonance. This method uses the spin of the proton, which has the value one-half. The name refers to examination of protons as they occur in protium (hydrogen-1 atoms) in compounds, and does not imply that free protons exist in the compound being studied.

54.8 Human exposure

Main article: Effect of spaceflight on the human body

The Apollo Lunar Surface Experiments Packages (ALSEP) determined that more than 95% of the particles in the solar wind are electrons and protons, in approximately equal numbers.[35][36]

> Because the Solar Wind Spectrometer made continuous measurements, it was possible to measure how the Earth's magnetic field affects arriving solar wind particles. For about two-thirds of each orbit, the Moon is outside of the Earth's magnetic field. At these times, a typical proton density was 10 to 20 per cubic centimeter, with most protons having velocities between 400 and 650 kilometers per second. For about five days of each month, the Moon is inside the Earth's geomagnetic tail, and typically no solar wind particles were detectable. For the remainder of each lunar orbit, the Moon is in a transitional region known as the magnetosheath, where the Earth's magnetic field affects the solar wind but does not completely exclude it. In this region, the particle flux is reduced, with typical proton velocities of 250 to 450 kilometers per second. During the lunar night, the spectrometer was shielded from the solar wind by the Moon and no solar wind particles were measured.[35]

Protons also occur in from extrasolar origin in space, from galactic cosmic rays, where they make up about 90% of the total particle flux. These protons often have higher energy than solar wind protons, but their intensity is far more uniform and less variable than protons coming from the Sun, the production of which is heavily affected by solar proton events such as coronal mass ejections.

Research has been performed on the dose-rate effects of protons, as typically found in space travel, on human health.[36][37] To be more specific, there are hopes to identify what specific chromosomes are damaged, and to define the damage, during cancer development from proton exposure.[36] Another study looks into determining "the effects of exposure to proton irradiation on neurochemical and behavioral endpoints, including dopaminergic functioning, amphetamine-induced conditioned taste aversion learning, and spatial learning and memory as measured by the Morris water maze.[37] Electrical charging of a spacecraft due to interplanetary proton bombardment has also been proposed for study.[38] There are many more studies that pertain to space travel, including galactic cosmic rays and their possible health effects, and solar proton event exposure.

The American Biostack and Soviet Biorack space travel experiments have demonstrated the severity of molecular damage induced by heavy ions on micro organisms including Artemia cysts.[39]

54.9 Antiproton

Main article: Antiproton

CPT-symmetry puts strong constraints on the relative properties of particles and antiparticles and, therefore, is open to stringent tests. For example, the charges of the proton and antiproton must sum to exactly zero. This equality has been tested to one part in 10^8. The equality of their masses has also been tested to better than one part in 10^8. By holding antiprotons in a Penning trap, the equality of the charge to mass ratio of the proton and the antiproton has been tested to one part in 6×10^9.[40] The magnetic moment of the antiproton has been measured with error of 8×10^{-3} nuclear Bohr magnetons, and is found to be equal and opposite to that of the proton.

54.10 See also

- Fermion field
- Hydrogen
- Hydron (chemistry)
- List of particles
- Proton-proton chain reaction
- Quark model
- Proton spin crisis

54.11 References

[1] Mohr, P.J.; Taylor, B.N. and Newell, D.B. (2011), "The 2010 CODATA Recommended Values of the Fundamental Physical Constants", National Institute of Standards and Technology, Gaithersburg, MD, US.

[2] Cho, Adiran (2 April 2010). "Mass of the Common Quark Finally Nailed Down". *http://news.sciencemag.org". American Association for the Advancement of Science. Retrieved 27 September 2014.*

[3] "Proton size puzzle reinforced!". Paul Shearer Institute. 25 January 2013.

[4] Adair, R.K. (1989). *The Great Design: Particles, Fields, and Creation*. Oxford University Press. p. 214.

[5] Cottingham,W.N.;Greenwood,D.A. (1986).*An Introduction to Nuclear Physics*. Cambridge University Press. ISBN9784.

[6] Basdevant, J.-L.; Rich, J.; M. Spiro (2005). *Fundamentals in Nuclear Physics*. Springer. p. 155. ISBN 0-387-01672-4.

[7] Department of Chemistry and Biochemistry UCLA Eric R. Scerri Lecturer. *The Periodic Table : Its Story and Its Significance: Its Story and Its Significance*. Oxford University Press. ISBN 978-0-19-534567-4.

[8] Petrucci, R.H.; Harwood, W.S.; Herring, F.G. (2002). *General Chemistry* (8th ed.). p. 41.

[9] See meeting report and announcement

[10] Romer A(1997). "Proton or prouton? Rutherford and the depths of the atom".*Amer.J.Phys.***65**(8): 707. Bibcode:1997A7R. doi:10.1119/1.18640.

[11] Rutherford reported acceptance by the *British Association* in a footnote to Masson, O. (1921). "XXIV.The constitution of atoms". *Philosophical Magazine Series 6* **41** (242): 281. doi:10.1080/14786442108636219.

[12] Pais, A. (1986) *Inward Bound*, Oxford Press, ISBN 0198519974, p. 296. Pais believed the first science literature use of the word *proton* occurs in "Physics at the British Association". *Nature* **106** (2663): 357. 1920. doi:10.1038/106357a0.

[13] Buccella, F.; Miele, G.; Rosa, L.; Santorelli, P.; Tuzi, T. (1989). "An upper limit for the proton lifetime in SO(10)". *Physics Letters B* **233**: 178. doi:10.1016/0370-2693(89)90637-0.

[14] Lee, D. G.; Mohapatra, R.; Parida, M.; Rani, M. (1995). "Predictions for the proton lifetime in minimal nonsupersymmetric SO(10) models: An update". *Physical Review D* **51**: 229. arXiv:hep-ph/9404238. doi:10.1103/PhysRevD.51.229.

[15] "Proton lifetime is longer than 1034 years". Kamioka Observatory. November 2009.

[16] Nishino, H.; Clark, S.; Abe, K.; Hayato, Y.; Iida, T.; Ikeda, M.; Kameda, J.; Kobayashi, K.; Koshio, Y.; Miura, M.; Moriyama, S.; Nakahata, M.; Nakayama, S.; Obayashi, Y.; Ogawa, H.; Sekiya, H.; Shiozawa, M.; Suzuki, Y.; Takeda, A.; Takenaga, Y.; Takeuchi, Y.; Ueno, K.; Ueshima, K.; Watanabe, H.; Yamada, S.; Hazama, S.; Higuchi, I.; Ishihara, C.; Kajita, T. et al. (2009). "Search for Proton Decay via $p \rightarrow e^+\pi^0$ and $p \rightarrow \mu^+\pi^0$ in a Large Water Cherenkov Detector". *Physical Review Letters* **102** (14). arXiv:0903.0676. Bibcode:2009PhRvL.102n1801N. doi:10.1103/PhysRevLett.102.141801.

[17] Ahmed, S.; Anthony, A.; Beier, E.; Bellerive, A.; Biller, S.; Boger, J.; Boulay, M.; Bowler, M.; Bowles, T.; Brice, S.; Bullard, T.; Chan, Y.; Chen, M.; Chen, X.; Cleveland, B.; Cox, G.; Dai, X.; Dalnoki-Veress, F.; Doe, P.; Dosanjh, R.; Doucas, G.; Dragowsky, M.; Duba, C.; Duncan, F.; Dunford, M.; Dunmore, J.; Earle, E.; Elliott, S.; Evans, H. et al. (2004). "Constraints on Nucleon Decay via Invisible Modes from the Sudbury Neutrino Observatory". *Physical Review Letters* **92** (10). arXiv:hep-ex/0310030. Bibcode:2004PhRvL..92j2004A. doi:10.1103/PhysRevLett.92.102004. PMID 15089201.

[18] Watson, A. (2004). *The Quantum Quark*. Cambridge University Press. pp. 285–286. ISBN 0-521-82907-0.

[19] Timothy Paul Smith (2003). *Hidden Worlds: Hunting for Quarks in Ordinary Matter*. Princeton University Press. ISBN 0-691-05773-7.

[20] Weise, W.; Green, A.M. (1984). *Quarks and Nuclei*. World Scientific. pp. 65–66. ISBN 9971-966-61-1.

[21] Ball, Philip (Nov 20, 2008). "Nuclear masses calculated from scratch". Nature. doi:10.1038/news.2008.1246. Retrieved Aug 27, 2014.

[22] Reynolds, Mark (Apr 2009). "Calculating the Mass of a Proton". *CNRS international magazine* (CNRS) (13). ISSN 2270-5317. Retrieved Aug 27, 2014.

[23] See this news report and links

[24] Durr, S.; Fodor, Z.; Frison, J.; Hoelbling, C.; Hoffmann, R.; Katz, S. D.; Krieg, S.; Kurth, T.; Lellouch, L.; Lippert, T.; Szabo, K. K.; Vulvert, G. (2008). "Ab Initio Determination of Light Hadron Masses". *Science* **322** (5905): 1224–7. arXiv:0906.3599. doi:10.1126/science.1163233. PMID 19023076.

[25] Perdrisat, C. F.; Punjabi, V.; Vanderhaeghen, M. (2007). "Nucleon electromagnetic form factors". *Progress in Particle and Nuclear Physics* **59** (2): 694. arXiv:hep-ph/0612014. Bibcode:2007PrPNP..59..694P. doi:10.1016/j.ppnp.2007.05.001.

[26] Boffi, Sigfrido; Pasquini, Barbara (2007). "Generalized parton distributions and the structure of the nucleon". *Rivista del Nuovo Cimento* **30**. arXiv:0711.2625. Bibcode:2007NCimR..30..387B. doi:10.1393/ncr/i2007-10025-7.

[27] Joshua, Erlich (December 2008). "Recent Results in AdS/QCD". *Proceedings, 8th Conference on Quark Confinement and the Hadron Spectrum, September 1–6, 2008, Mainz, Germany*. arXiv:0812.4976.

[28] Pietro, Colangelo; Alex, Khodjamirian (October 2000). “QCD Sum Rules, a Modern Perspective”. In Shifman, M. *At the Frontier of Particle Physics / Handbook of QCD*. World Scientific. arXiv:hep-ph/0010175.

[29] Pohl, R.; Antognini, A.; Nez, F. O.; Amaro, F. D.; Biraben, F. O.; Cardoso, J. O. M. R.; Covita, D. S.; Dax, A.; Dhawan, S.; Fernandes, L. M. P.; Giesen, A.; Graf, T.; Hänsch, T. W.; Indelicato, P.; Julien, L.; Kao, C. Y.; Knowles, P.; Le Bigot, E. O.; Liu, Y. W.; Lopes, J. A. M.; Ludhova, L.; Monteiro, C. M. B.; Mulhauser, F. O.; Nebel, T.; Rabinowitz, P.; Dos Santos, J. M. F.; Schaller, L. A.; Schuhmann, K.; Schwob, C. et al. (2010). “The size of the proton”. *Nature* **466** (7303): 213–6. doi:10.1038/nature09250. PMID 20613837.

[30] Antognini, A.; Nez, F.; Schuhmann, K.; Amaro, F. D.; Biraben, F.; Cardoso, J. M. R.; Covita, D. S.; Dax, A.; Dhawan, S.; Diepold, M.; Fernandes, L. M. P.; Giesen, A.; Gouvea, A. L.; Graf, T.; Hänsch, T. W.; Indelicato, P.; Julien, L.; Kao, C. -Y.; Knowles, P.; Kottmann, F.; Le Bigot, E. -O.; Liu, Y. -W.; Lopes, J. A. M.; Ludhova, L.; Monteiro, C. M. B.; Mulhauser, F.; Nebel, T.; Rabinowitz, P.; Dos Santos, J. M. F.; Schaller, L. A. (2013). “Proton Structure from the Measurement of 2S-2P Transition Frequencies of Muonic Hydrogen”. *Science* **339** (6118): 417–420. doi:10.1126/science.1230016. PMID 23349284.

[31] Researchers Observes Unexpectedly Small Proton Radius in a Precision Experiment. *Azonano*. July 9, 2010

[32] “The Proton Just Got Smaller”. *Photonics.Com*. 12 July 2010. Retrieved 2010-07-19.

[33] Carlson, Carl E. (February 19, 2015), *The Proton Radius Puzzle*, arXiv:1502.05314

[34] Headrick, J.M.; Diken, E.G.; Walters, R. S.; Hammer, N. I.; Christie, R.A.; Cui, J.; Myshakin, E.M.; Duncan, M.A.; Johnson, M.A.; Jordan, K.D. (2005). “Spectral Signatures of Hydrated Proton Vibrations in Water Clusters”. *Science* **308** (5729): 1765–69. Bibcode:2005Sci...308.1765H. doi:10.1126/science.1113094. PMID 15961665.

[35] “Apollo 11 Mission”. Lunar and Planetary Institute. 2009. Retrieved 2009-06-12.

[36] “Space Travel and Cancer Linked? Stony Brook Researcher Secures NASA Grant to Study Effects of Space Radiation”. Brookhaven National Laboratory. 12 December 2007. Retrieved 2009-06-12.

[37] Shukitt-Hale, B.; Szprengiel, A.; Pluhar, J.; Rabin, B.M.; Joseph, J.A. “The effects of proton exposure on neurochemistry and behavior”. Elsevier/COSPAR. Retrieved 2009-06-12.

[38] Green, N.W.; Frederickson, A.R. “A Study of Spacecraft Charging due to Exposure to Interplanetary Protons” (PDF). Jet Propulsion Laboratory. Retrieved 2009-06-12.

[39] Planel, H. (2004). *Space and life: an introduction to space biology and medicine*. CRC Press. pp. 135–138. ISBN 0-415-31759-2.

[40] Gabrielse, G. (2006). “Antiproton mass measurements”. *International Journal of Mass Spectrometry* **251** (2–3): 273–280. Bibcode:2006IJMSp.251..273G. doi:10.1016/j.ijms.2006.02.013.

54.12 External links

- Particle Data Group
- Large Hadron Collider
- Eaves, Laurence; Copeland, Ed; Padilla, Antonio (Tony) (2010). “The shrinking proton”. *Sixty Symbols*. Brady Haran for the University of Nottingham.

Ernest Rutherford at the first Solvay Conference, 1911

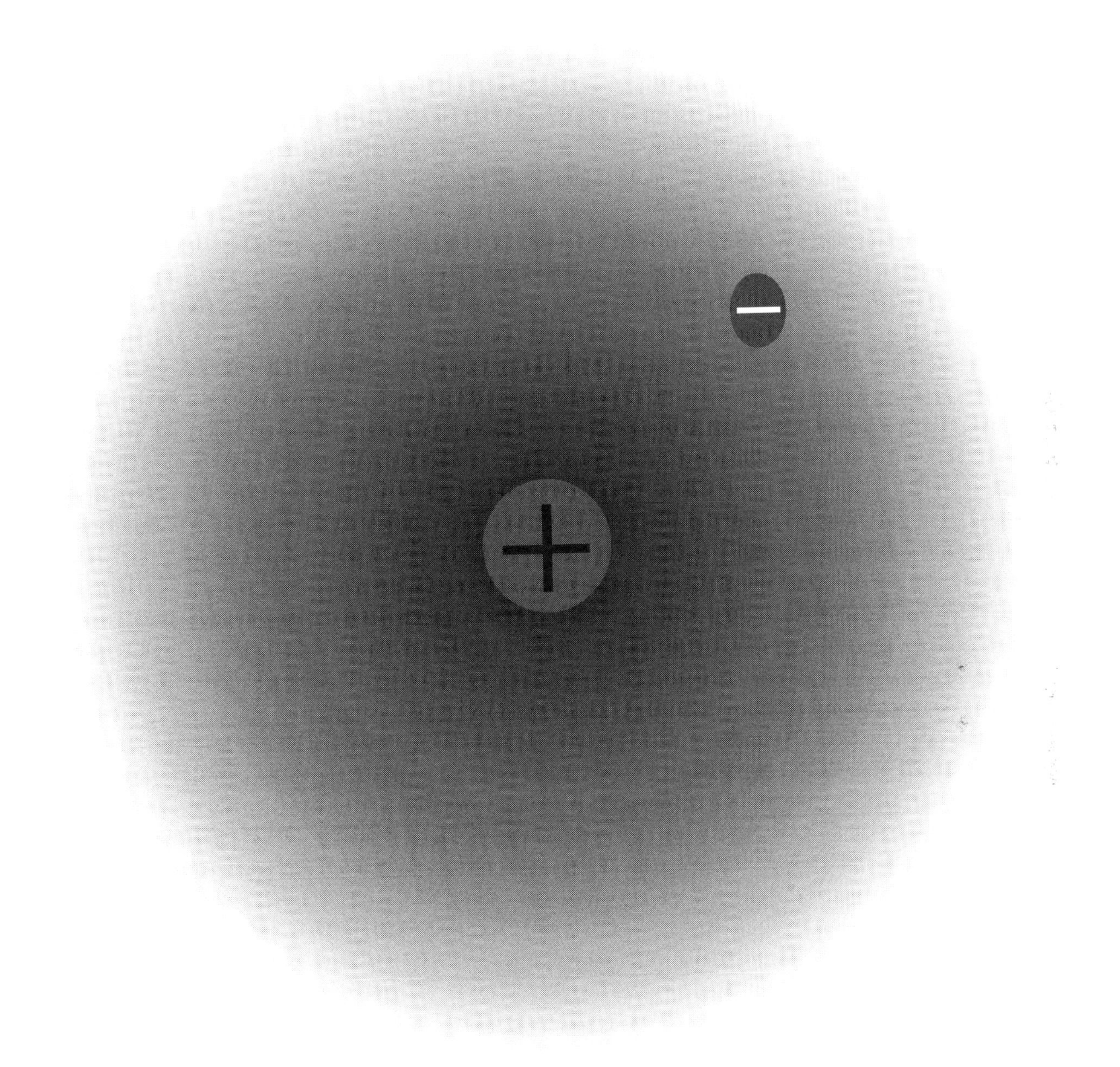

*Protium, the most common isotope of hydrogen, consists of one proton and one electron (it has no neutrons). The term "hydrogen ion" (H+
) implies that that H-atom has lost its one electron, causing only a proton to remain. Thus, in chemistry, the terms "proton" and "hydrogen ion" (for the protium isotope) are used synonymously*

Chapter 55

Neutron

This article is about the subatomic particle. For other uses, see Neutron (disambiguation).

The **neutron** is a subatomic particle, symbol n or n0, with no net electric charge and a mass slightly larger than that of a proton. Protons and neutrons, each with mass approximately one atomic mass unit, constitute the nucleus of an atom, and they are collectively referred to as nucleons.[4] Their properties and interactions are described by nuclear physics.

The nucleus consists of Z protons, where Z is called the atomic number, and N neutrons, where N is the neutron number. The atomic number defines the chemical properties of the atom, and the neutron number determines the isotope or nuclide.[5] The terms isotope and nuclide are often used synonymously, but they refer to chemical and nuclear properties, respectively. The atomic mass number, symbol A, equals Z+N. For example, carbon has atomic number 6, and its abundant carbon-12 isotope has 6 neutrons, whereas its rare carbon-13 isotope has 7 neutrons. Some elements occur in nature with only one stable isotope, such as fluorine (see stable nuclide). Other elements occur as many stable isotopes, such as tin with ten stable isotopes. Even though it is not a chemical element, the neutron is included in the table of nuclides.[6]

Within the nucleus, protons and neutrons are bound together through the nuclear force, and neutrons are required for the stability of nuclei. Neutrons are produced copiously in nuclear fission and fusion. They are a primary contributor to the nucleosynthesis of chemical elements within stars through fission, fusion, and neutron capture processes.

The neutron is essential to the production of nuclear power. In the decade after the neutron was discovered in 1932,[7] neutrons were used to effect many different types of nuclear transmutations. With the discovery of nuclear fission in 1938,[8] it was quickly realized that, if a fission event produced neutrons, each of these neutrons might cause further fission events, etc., in a cascade known as a nuclear chain reaction.[5] These events and findings led to the first self-sustaining nuclear reactor (Chicago Pile-1, 1942) and the first nuclear weapon (Trinity, 1945).

Free neutrons, or individual neutrons free of the nucleus, are effectively a form of ionizing radiation, and as such, are a biological hazard, depending upon dose.[5] A small natural “neutron background” flux of free neutrons exists on Earth, caused by cosmic ray muons, and by the natural radioactivity of spontaneously fissionable elements in the Earth’s crust.[9] Dedicated neutron sources like neutron generators, research reactors and spallation sources produce free neutrons for use in irradiation and in neutron scattering experiments.

55.1 Description

Neutrons and protons are both nucleons, which are attracted and bound together by the nuclear force to form atomic nuclei. The nucleus of the most common isotope of the hydrogen atom (with the chemical symbol “H”) is a lone proton. The nuclei of the heavy hydrogen isotopes deuterium and tritium contain one proton bound to one and two neutrons, respectively. All other types of atomic nuclei are composed of two or more protons and various numbers of neutrons. The most common nuclide of the common chemical element lead, ^{208}Pb has 82 protons and 126 neutrons, for example.

The free neutron has a mass of about 1.675×10^{-27} kg (equivalent to 939.6 MeV/c^2, or 1.0087 u).[3] The neutron has a mean square radius of about 0.8×10^{-15} m, or 0.8 fm,[10] and it is a spin-½ fermion.[11] The neutron has a magnetic moment with a negative value, because its orientation is opposite to the neutron's spin.[12] The neutron's magnetic moment causes its motion to be influenced by magnetic fields. Although the neutron has no net electric charge, it does have a slight distribution of charge within it. With its positive electric charge, the proton is directly influenced by electric fields, whereas the response of the neutron to this force is much weaker.

A free neutron is unstable, decaying to a proton, electron and antineutrino with a mean lifetime of just under 15 minutes (881.5±1.5 s). This radioactive decay, known as beta decay,[13] is possible since the mass of the neutron is slightly greater than the proton. The free proton is stable. Neutrons or protons bound in a nucleus can be stable or unstable, however, depending on the nuclide. Beta decay, in which neutrons decay to protons, or vice versa, is governed by the weak force, and it requires the emission or absorption of electrons and neutrinos, or their antiparticles.

Protons and neutrons behave almost identically under the influence of the nuclear force within the nucleus. The concept of isospin, in which the proton and neutron are viewed as two quantum states of the same particle, is used to model the interactions of nucleons by the nuclear or weak forces. Because of the strength of the nuclear force at short distances, the binding energy of nucleons is more than seven orders of magnitude larger than the electromagnetic energy binding electrons in atoms. Nuclear reactions (such as nuclear fission) therefore have an energy density that is more than ten million times that of chemical reactions. Because of the mass–energy equivalence, nuclear binding energies add or subtract from the mass of nuclei. Ultimately, the ability of the nuclear force to store energy arising from the electromagnetic repulsion of nuclear components is the basis for most of the energy that makes nuclear reactors or bombs possible. In nuclear fission, the absorption of a neutron by a heavy nuclide (e.g., uranium-235) causes the nuclide to become unstable and break into light nuclides and additional neutrons. The positively charged light nuclides then repel, releasing electromagnetic potential energy.

The neutron is classified as a hadron, since it is composed of quarks, and as a baryon, since it is composed of three quarks.[14] The finite size of the neutron and its magnetic moment indicate the neutron is a composite, rather than elementary, particle. The neutron consists of two down quarks with charge −⅓ *e* and one up quark with charge +⅔ *e*, although this simple model belies the complexities of the Standard Model for nuclei.[15] The masses of the three quarks sum to only about 12 MeV/c^2, whereas the neutron's mass is about 940 MeV/c^2, for example.[15] Like the proton, the quarks of the neutron are held together by the strong force, mediated by gluons.[16] The nuclear force results from secondary effects of the more fundamental strong force.

55.2 Discovery

Main article: Discovery of the neutron

The story of the discovery of the neutron and its properties is central to the extraordinary developments in atomic physics that occurred in the first half of the 20th century, leading ultimately to the atomic bomb in 1945. In the 1911 Rutherford model, the atom consisted of a small positively charged massive nucleus surrounded by a much larger cloud of negatively charged electrons. In 1920 Rutherford suggested the nucleus consisted of positive protons and neutrally-charged particles, suggested to be a proton and an electron bound in some way.[17] Electrons were assumed to reside within the nucleus because it was known that beta radiation consisted of electrons emitted from the nucleus.[17] Rutherford called these uncharged particles *neutrons*, by the Latin root for *neutralis* (neuter) and the Greek suffix *-on* (a suffix used in the names of subatomic particles, i.e. *electron* and *proton*).[18][19] References to the word *neutron* in connection with the atom can be found in the literature as early as 1899, however.[20]

Throughout the 1920s, physicists assumed that the atomic nucleus was composed of protons and "nuclear electrons"[21][22] but there were obvious problems. It was difficult to reconcile the proton–electron model for nuclei with the Heisenberg uncertainty relation of quantum mechanics.[23][24] The Klein paradox,[25] discovered by Oskar Klein in 1928, presented further quantum mechanical objections to the notion of an electron confined within a nucleus.[23] Observed properties of atoms and molecules were inconsistent with the nuclear spin expected from proton–electron hypothesis. Since both protons and electrons carry an intrinsic spin of ½ $\hbar$, there is no way to arrange an odd number of spins ±½ $\hbar$ to give a spin integer multiple of $\hbar$. Nuclei with integer spin are common, e.g., ^{14}N.

In 1931, Walther Bothe and Herbert Becker found that if alpha particle radiation from polonium fell on beryllium, boron, or lithium, an unusually penetrating radiation was produced. The radiation was not influenced by an electric field, so Bothe and Becker assumed it was gamma radiation.[26][27] The following year Irène Joliot-Curie and Frédéric Joliot in Paris showed that if this "gamma" radiation fell on paraffin, or any other hydrogen-containing compound, it ejected protons of very high energy.[28] Neither Rutherford nor James Chadwick at the Cavendish Laboratory in Cambridge were convinced by the gamma ray interpretation.[21] Chadwick quickly performed a series of experiments that showed that the new radiation consisted of uncharged particles with about the same mass as the proton.[7][29][30] These particles were neutrons. Chadwick won the Nobel Prize in Physics for this discovery in 1935.[2]

Models for atomic nucleus consisting of protons and neutrons were quickly developed by Werner Heisenberg[31][32][33] and others.[34][35] The proton–neutron model explained the puzzle of nuclear spins. The origins of beta radiation were explained by Enrico Fermi in 1934 by the process of beta decay, in which the neutron decays to a proton by *creating* an electron and a (as yet undiscovered) neutrino.[36] In 1935 Chadwick and his doctoral student Maurice Goldhaber, reported the first accurate measurement of the mass of the neutron.[37][38]

By 1934, Fermi had bombarded heavier elements with neutrons to induce radioactivity in elements of high atomic number. In 1938, Fermi received the Nobel Prize in Physics *"for his demonstrations of the existence of new radioactive elements produced by neutron irradiation, and for his related discovery of nuclear reactions brought about by slow neutrons"*.[39] In 1938 Otto Hahn, Lise Meitner, and Fritz Strassmann discovered nuclear fission, or the fractionation of uranium nuclei into light elements, induced by neutron bombardment.[40][41][42] In 1945 Hahn received the 1944 Nobel Prize in Chemistry *"for his discovery of the fission of heavy atomic nuclei."* [43][44][45] The discovery of nuclear fission would lead to the development of nuclear power and the atomic bomb by the end of World War II.

55.3 Beta decay and the stability of the nucleus

Under the Standard Model of particle physics, the only possible decay mode for the neutron that conserves baryon number is for one of the neutron's quarks to change flavour via the weak interaction. The decay of one of the neutron's down quarks into a lighter up quark can be achieved by the emission of a W boson. By this process, the Standard Model description of beta decay, the neutron decays into a proton (which contains one down and two up quarks), an electron, and an electron antineutrino.

Since interacting protons have a mutual electromagnetic repulsion that is stronger than their attractive nuclear interaction, neutrons are a necessary constituent of any atomic nucleus that contains more than one proton (see diproton and neutron–proton ratio).[46] Neutrons bind with protons and one another in the nucleus via the nuclear force, effectively moderating the repulsive forces between the protons and stabilizing the nucleus.

See also: Beta-decay stable isobars and Neutron emission

55.3.1 Free neutron decay

Outside the nucleus, free neutrons are unstable and have a mean lifetime of 881.5±1.5 s (about 14 minutes, 42 seconds); therefore the half-life for this process (which differs from the mean lifetime by a factor of ln(2) = 0.693) is 611.0±1.0 s (about 10 minutes, 11 seconds).[13] Beta decay of the neutron, described above, can be denoted by the radioactive decay:[47]

n0 → p+ + e− + ν
e

where p+, e−, and ν
e denote the proton, electron and electron antineutrino, respectively. For the free neutron the decay energy for this process (based on the masses of the neutron, proton, and electron) is 0.782343 MeV. The maximal energy of the beta decay electron (in the process wherein the neutrino receives a vanishingly small amount of kinetic energy) has been measured at 0.782 ± .013 MeV.[48] The latter number is not well-enough measured to determine the comparatively tiny rest mass

of the neutrino (which must in theory be subtracted from the maximal electron kinetic energy) as well as neutrino mass is constrained by many other methods.

A small fraction (about one in 1000) of free neutrons decay with the same products, but add an extra particle in the form of an emitted gamma ray:

$$n^0 \rightarrow p^+ + e^- + \nu_e + \gamma$$

This gamma ray may be thought of as a sort of "internal bremsstrahlung" that arises as the emitted beta particle interacts with the charge of the proton in an electromagnetic way. Internal bremsstrahlung gamma ray production is also a minor feature of beta decays of bound neutrons (as discussed below).

A very small minority of neutron decays (about four per million) are so-called "two-body (neutron) decays", in which a proton, electron and antineutrino are produced as usual, but the electron fails to gain the 13.6 eV necessary energy to escape the proton, and therefore simply remains bound to it, as a neutral hydrogen atom (one of the "two bodies"). In this type of free neutron decay, in essence all of the neutron decay energy is carried off by the antineutrino (the other "body").

The transformation of a free proton to a neutron (plus a positron and a neutrino) is energetically impossible, since a free neutron has a greater mass than a free proton.

55.3.2 Bound neutron decay

Main article: Atomic nucleus

While a free neutron has a half life of about 10.2 min, most neutrons within nuclei are stable. According to the nuclear shell model, the protons and neutrons of a nuclide are a quantum mechanical system organized into discrete energy levels with unique quantum numbers. For a neutron to decay, the resulting proton requires an available state at lower energy than the initial neutron state. In stable nuclei the possible lower energy states are all filled, meaning they are each occupied by two protons with spin up and spin down. The Pauli exclusion principle therefore disallows the decay of a neutron to a proton within stable nuclei. The situation is similar to electrons of an atom, where electrons have distinct atomic orbitals and are prevented from decaying to lower energy states, with the emission of a photon, by the exclusion principle.

Neutrons in unstable nuclei can decay by beta decay as described above. In this case, an energetically allowed quantum state is available for the proton resulting from the decay. One example of this decay is carbon-14 (6 protons, 8 neutrons) that decays to nitrogen-14 (7 protons, 7 neutrons) with a half-life of about 5,730 years.

Inside a nucleus, a proton can transform into a neutron via inverse beta decay, if an energetically allowed quantum state is available for the neutron. This transformation occurs by emission of an antielectron (also called positron) and an electron neutrino:

$$p^+ \rightarrow n^0 + e^+ + \nu_e$$

The transformation of a proton to a neutron inside of a nucleus is also possible through electron capture:

$$p^+ + e^- \rightarrow n^0 + \nu_e$$

Positron capture by neutrons in nuclei that contain an excess of neutrons is also possible, but is hindered because positrons are repelled by the positive nucleus, and quickly annihilate when they encounter electrons.

55.3.3 Competition of beta decay types

Three types of beta decay in competition are illustrated by the single isotope copper-64 (29 protons, 35 neutrons), which has a half-life of about 12.7 hours. This isotope has one unpaired proton and one unpaired neutron, so either the proton or the neutron can decay. This particular nuclide (though not all nuclides in this situation) is almost equally likely to decay through proton decay by positron emission (18%) or electron capture (43%), as through neutron decay by electron emission (39%).

55.4 Intrinsic properties

55.4.1 Electric charge

The total electric charge of the neutron is 0 *e*. This zero value has been tested experimentally, and the present experimental limit for the charge of the neutron is $-2(8)\times10^{-22}$ *e*,[49] or $-3(13)\times10^{-41}$ C. This value is consistent with zero, given the experimental uncertainties (indicated in parentheses). By comparison, the charge of the proton is, of course, +1 *e*.

55.4.2 Electric dipole moment

Main article: Neutron electric dipole moment

The Standard Model of particle physics predicts a tiny separation of positive and negative charge within the neutron leading to a permanent electric dipole moment.[50] The predicted value is, however, well below the current sensitivity of experiments. From several unsolved puzzles in particle physics, it is clear that the Standard Model is not the final and full description of all particles and their interactions. New theories going beyond the Standard Model generally lead to much larger predictions for the electric dipole moment of the neutron. Currently, there are at least four experiments trying to measure for the first time a finite neutron electric dipole moment, including:

- Cryogenic neutron EDM experiment being set up at the Institut Laue–Langevin[51]
- nEDM experiment under construction at the new UCN source at the Paul Scherrer Institute[52]
- nEDM experiment being envisaged at the Spallation Neutron Source[53]
- nEDM experiment being built at the Institut Laue–Langevin[54]

55.4.3 Magnetic moment

Main article: Neutron magnetic moment

Even though the neutron is a neutral particle, the magnetic moment of a neutron is not zero. Since the neutron is a neutral particle, it is not affected by electric fields, but with its magnetic moment it is affected by magnetic fields. The magnetic moment of the neutron is an indication of its quark substructure and internal charge distribution.[55] The value for the neutron's magnetic moment was first directly measured by Luis Alvarez and Felix Bloch at Berkeley, California in 1940,[56] using an extension of the magnetic resonance methods developed by Rabi. Alvarez and Bloch determined the magnetic moment of the neutron to be $\mu_n = -1.93(2)\,\mu N$, where μN is the nuclear magneton.

55.4.4 Structure and geometry of charge distribution

An article published in 2007 featuring a model-independent analysis concluded that the neutron has a negatively charged exterior, a positively charged middle, and a negative core.[57] In a simplified classical view, the negative "skin" of the neutron assists it to be attracted to the protons with which it interacts in the nucleus. (However, the main attraction between neutrons and protons is via the nuclear force, which does not involve charge.)

The simplified classical view of the neutron's charge distribution also "explains" the fact that the neutron magnetic dipole points in the opposite direction from its spin angular momentum vector (as compared to the proton). This gives the neutron, in effect, a magnetic moment which resembles a negatively charged particle. This can be reconciled classically with a neutral neutron composed of a charge distribution in which the negative sub-parts of the neutron have a larger average radius of distribution, and therefore contribute more to the particle's magnetic dipole moment, than do the positive parts that are, on average, nearer the core.

55.4.5 Mass

The mass of a neutron cannot be directly determined by mass spectrometry due to lack of electric charge. However, since the mass of protons and deuterons can be measured by mass spectrometry, the mass of a neutron can be deduced by subtracting proton mass from deuteron mass, with the difference being the mass of the neutron plus the binding energy of deuterium (expressed as a positive emitted energy). The latter can be directly measured by measuring the energy (B_d) of the single 0.7822 MeV gamma photon emitted when neutrons are captured by protons (this is exothermic and happens with zero-energy neutrons), plus the small recoil kinetic energy (E_{rd}) of the deuteron (about 0.06% of the total energy).

$$m_n = m_d - m_p + B_d - E_{rd}$$

The energy of the gamma ray can be measured to high precision by X-ray diffraction techniques, as was first done by Bell and Elliot in 1948. The best modern (1986) values for neutron mass by this technique are provided by Greene, et al.[58] These give a neutron mass of:

$$m_{\text{neutron}} = 1.008644904(14)\ \text{u}$$

The value for the neutron mass in MeV is less accurately known, due to less accuracy in the known conversion of u to MeV:[59]

$$m_{\text{neutron}} = 939.56563(28)\ \text{MeV}/c^2.$$

Another method to determine the mass of a neutron starts from the beta decay of the neutron, when the momenta of the resulting proton and electron are measured.

55.4.6 Anti-neutron

Main article: Antineutron

The antineutron is the antiparticle of the neutron. It was discovered by Bruce Cork in the year 1956, a year after the antiproton was discovered. CPT-symmetry puts strong constraints on the relative properties of particles and antiparticles, so studying antineutrons yields provide stringent tests on CPT-symmetry. The fractional difference in the masses of the neutron and antineutron is $(9\pm6)\times10^{-5}$. Since the difference is only about two standard deviations away from zero, this does not give any convincing evidence of CPT-violation.[13]

55.5 Neutron compounds

55.5.1 Dineutrons and tetraneutrons

Main articles: Dineutron and Tetraneutron

The existence of stable clusters of 4 neutrons, or tetraneutrons, has been hypothesised by a team led by Francisco-Miguel Marqués at the CNRS Laboratory for Nuclear Physics based on observations of the disintegration of beryllium−14 nuclei. This is particularly interesting because current theory suggests that these clusters should not be stable.

The dineutron is another hypothetical particle. In 2012, Artemis Spyrou from Michigan State University and coworkers reported that they observed, for the first time, the dineutron emission in the decay of ^{16}Be. The dineutron character is evidenced by a small emission angle between the two neutrons. The authors measured the two-neutron separation energy to be 1.35(10) MeV, in good agreement with shell model calculations, using standard interactions for this mass region.[60]

55.5.2 Neutronium and neutron stars

Main articles: Neutronium and Neutron star

At extremely high pressures and temperatures, nucleons and electrons are believed to collapse into bulk neutronic matter, called neutronium. This is presumed to happen in neutron stars.

The extreme pressure inside a neutron star may deform the neutrons into a cubic symmetry, allowing tighter packing of neutrons.[61]

55.6 Detection

Main article: Neutron detection

The common means of detecting a charged particle by looking for a track of ionization (such as in a cloud chamber) does not work for neutrons directly. Neutrons that elastically scatter off atoms can create an ionization track that is detectable, but the experiments are not as simple to carry out; other means for detecting neutrons, consisting of allowing them to interact with atomic nuclei, are more commonly used. The commonly used methods to detect neutrons can therefore be categorized according to the nuclear processes relied upon, mainly neutron capture or elastic scattering. A good discussion on neutron detection is found in chapter 14 of the book *Radiation Detection and Measurement* by Glenn F. Knoll (John Wiley & Sons, 1979).

55.6.1 Neutron detection by neutron capture

A common method for detecting neutrons involves converting the energy released from neutron capture reactions into electrical signals. Certain nuclides have a high neutron capture cross section, which is the probability of absorbing a neutron. Upon neutron capture, the compound nucleus emits more easily detectable radiation, for example an alpha particle, which is then detected. The nuclides 3He, 6Li, 10B, 233U, 235U, 237Np and 239Pu are useful for this purpose.

55.6.2 Neutron detection by elastic scattering

Neutrons can elastically scatter off nuclei, causing the struck nucleus to recoil. Kinematically, a neutron can transfer more energy to light nuclei such as hydrogen or helium than to heavier nuclei. Detectors relying on elastic scattering are called fast neutron detectors. Recoiling nuclei can ionize and excite further atoms through collisions. Charge and/or scintillation light produced in this way can be collected to produce a detected signal. A major challenge in fast neutron detection is discerning such signals from erroneous signals produced by gamma radiation in the same detector.

Fast neutron detectors have the advantage of not requiring a moderator, and therefore being capable of measuring the neutron's energy, time of arrival, and in certain cases direction of incidence.

55.7 Sources and production

Main articles: Neutron source, neutron generator and research reactor

Free neutrons are unstable, although they have the longest half-life of any unstable sub-atomic particle by several orders of magnitude. Their half-life is still only about 10 minutes, however, so they can be obtained only from sources that produce them freshly.

Natural neutron background. A small natural background flux of free neutrons exists everywhere on Earth. In the atmosphere and deep into the ocean, the "neutron background" is caused by muons produced by cosmic ray interaction with the atmosphere. These high energy muons are capable of penetration to considerable depths in water and soil. There, in striking atomic nuclei, among other reactions they induce spallation reactions in which a neutron is liberated from the nucleus. Within the Earth's crust a second source is neutrons produced primarily by spontaneous fission of uranium and thorium present in crustal minerals. The neutron background is not strong enough to be a biological hazard, but it is of importance to very high resolution particle detectors that are looking for very rare events, such as (hypothesized) interactions that might be caused by particles of dark matter.[9] Recent research has shown that even thunderstorms can produce neutrons with energies of up to several tens of MeV.[62]

Even stronger neutron background radiation is produced at the surface of Mars, where the atmosphere is thick enough to generate neutrons from cosmic ray muon production and neutron-spallation, but not thick enough to provide significant protection from the neutrons produced. These neutrons not only produce a Martian surface neutron radiation hazard from direct downward-going neutron radiation but may also produce a significant hazard from reflection of neutrons from the Martian surface, which will produce reflected neutron radiation penetrating upward into a Martian craft or habitat from the floor.[63]

Sources of neutrons for research. These include certain types of radioactive decay (spontaneous fission and neutron emission), and from certain nuclear reactions. Convenient nuclear reactions include tabletop reactions such as natural alpha and gamma bombardment of certain nuclides, often beryllium or deuterium, and induced nuclear fission, such as occurs in nuclear reactors. In addition, high-energy nuclear reactions (such as occur in cosmic radiation showers or accelerator collisions) also produce neutrons from disintigration of target nuclei. Small (tabletop) particle accelerators optimized to produce free neutrons in this way, are called neutron generators.

In practice, the most commonly used small laboratory sources of neutrons use radioactive decay to power neutron production. One noted neutron-producing radioisotope, californium−252 decays (half-life 2.65 years) by spontaneous fission 3% of the time with production of 3.7 neutrons per fission, and is used alone as a neutron source from this process. Nuclear reaction sources (that involve two materials) powered by radioisotopes use an alpha decay source plus a beryllium target, or else a source of high-energy gamma radiation from a source that undergoes beta decay followed by gamma decay, which produces photoneutrons on interaction of the high energy gamma ray with ordinary stable beryllium, or else with the deuterium in heavy water. A popular source of the latter type is radioactive antimony-124 plus beryllium, a system with a half-life of 60.9 days, which can be constructed from natural antimony (which is 42.8% stable antimony-123) by activating it with neutrons in a nuclear reactor, then transported to where the neutron source is needed.[64]

Nuclear fission reactors naturally produce free neutrons; their role is to sustain the energy-producing chain reaction. The intense neutron radiation can also be used to produce various radioisotopes through the process of neutron activation, which is a type of neutron capture.

Experimental nuclear fusion reactors produce free neutrons as a waste product. However, it is these neutrons that possess most of the energy, and converting that energy to a useful form has proved a difficult engineering challenge. Fusion reactors that generate neutrons are likely to create radioactive waste, but the waste is composed of neutron-activated lighter isotopes, which have relatively short (50–100 years) decay periods as compared to typical half-lives of 10,000 years for fission waste, which is long due primarily to the long half-life of alpha-emitting transuranic actinides.[65]

55.7.1 Neutron beams and modification of beams after production

Free neutron beams are obtained from neutron sources by neutron transport. For access to intense neutron sources, researchers must go to a specialist neutron facility that operates a research reactor or a spallation source.

The neutron's lack of total electric charge makes it difficult to steer or accelerate them. Charged particles can be accelerated, decelerated, or deflected by electric or magnetic fields. These methods have little effect on neutrons. However, some effects may be attained by use of inhomogeneous magnetic fields because of the neutron's magnetic moment. Neutrons can be controlled by methods that include moderation, reflection, and velocity selection. Thermal neutrons can be polarized by transmission through magnetic materials in a method analogous to the Faraday effect for photons. Cold neutrons of wavelengths of 6–7 angstroms can be produced in beams of a high degree of polarization, by use of magnetic mirrors and magnetized interference filters.[66]

55.8 Applications

The neutron plays an important role in many nuclear reactions. For example, neutron capture often results in neutron activation, inducing radioactivity. In particular, knowledge of neutrons and their behavior has been important in the development of nuclear reactors and nuclear weapons. The fissioning of elements like uranium-235 and plutonium-239 is caused by their absorption of neutrons.

Cold, *thermal* and *hot* neutron radiation is commonly employed in neutron scattering facilities, where the radiation is used in a similar way one uses X-rays for the analysis of condensed matter. Neutrons are complementary to the latter in terms of atomic contrasts by different scattering cross sections; sensitivity to magnetism; energy range for inelastic neutron spectroscopy; and deep penetration into matter.

The development of “neutron lenses” based on total internal reflection within hollow glass capillary tubes or by reflection from dimpled aluminum plates has driven ongoing research into neutron microscopy and neutron/gamma ray tomography.[67][68][69]

A major use of neutrons is to excite delayed and prompt gamma rays from elements in materials. This forms the basis of neutron activation analysis (NAA) and prompt gamma neutron activation analysis (PGNAA). NAA is most often used to analyze small samples of materials in a nuclear reactor whilst PGNAA is most often used to analyze subterranean rocks around bore holes and industrial bulk materials on conveyor belts.

Another use of neutron emitters is the detection of light nuclei, in particular the hydrogen found in water molecules. When a fast neutron collides with a light nucleus, it loses a large fraction of its energy. By measuring the rate at which slow neutrons return to the probe after reflecting off of hydrogen nuclei, a neutron probe may determine the water content in soil.

55.9 Medical therapies

Main articles: Fast neutron therapy and Neutron capture therapy of cancer

Because neutron radiation is both penetrating and ionizing, it can be exploited for medical treatments. Neutron radiation can have the unfortunate side-effect of leaving the affected area radioactive, however. Neutron tomography is therefore not a viable medical application.

Fast neutron therapy utilizes high energy neutrons typically greater than 20 MeV to treat cancer. Radiation therapy of cancers is based upon the biological response of cells to ionizing radiation. If radiation is delivered in small sessions to damage cancerous areas, normal tissue will have time to repair itself, while tumor cells often cannot.[70] Neutron radiation can deliver energy to a cancerous region at a rate an order of magnitude larger than gamma radiation[71]

Beams of low energy neutrons are used in boron capture therapy to treat cancer. In boron capture therapy, the patient is given a drug that contains boron and that preferentially accumulates in the tumor to be targeted. The tumor is then bombarded with very low energy neutrons (although often higher than thermal energy) which are captured by the boron-10 isotope in the boron, which produces an excited state of boron-11 that then decays to produce lithium-7 and an alpha particle that have sufficient energy to kill the malignant cell, but insufficient range to damage nearby cells. For such a therapy to be applied to the treatment of cancer, a neutron source having an intensity of the order of billion (10^9) neutrons per second per cm^2 is preferred. Such fluxes require a research nuclear reactor.

55.10 Protection

Exposure to free neutrons can be hazardous, since the interaction of neutrons with molecules in the body can cause disruption to molecules and atoms, and can also cause reactions that give rise to other forms of radiation (such as protons). The normal precautions of radiation protection apply: Avoid exposure, stay as far from the source as possible, and keep exposure time to a minimum. Some particular thought must be given to how to protect from neutron exposure, however. For other types of radiation, e.g. alpha particles, beta particles, or gamma rays, material of a high atomic number and with high density make for good shielding; frequently, lead is used. However, this approach will not work with neutrons, since the absorption of neutrons does not increase straightforwardly with atomic number, as it does with alpha, beta, and gamma radiation. Instead one needs to look at the particular interactions neutrons have with matter (see the section on detection above). For example, hydrogen-rich materials are often used to shield against neutrons, since ordinary hydrogen both scatters and slows neutrons. This often means that simple concrete blocks or even paraffin-loaded plastic blocks afford better protection from neutrons than do far more dense materials. After slowing, neutrons may then be absorbed with an isotope that has high affinity for slow neutrons without causing secondary capture radiation, such as lithium-6.

Hydrogen-rich ordinary water affects neutron absorption in nuclear fission reactors: Usually, neutrons are so strongly absorbed by normal water that fuel enrichment with fissionable isotope is required. The deuterium in heavy water has a very much lower absorption affinity for neutrons than does protium (normal light hydrogen). Deuterium is, therefore, used in CANDU-type reactors, in order to slow (moderate) neutron velocity, to increase the probability of nuclear fission compared to neutron capture.

55.11 Neutron temperature

Main article: Neutron temperature

55.11.1 Thermal neutrons

A *thermal neutron* is a free neutron that is Boltzmann distributed with kT = 0.0253 eV (4.0×10^{-21} J) at room temperature. This gives characteristic (not average, or median) speed of 2.2 km/s. The name 'thermal' comes from their energy being that of the room temperature gas or material they are permeating. (see *kinetic theory* for energies and speeds of molecules). After a number of collisions (often in the range of 10–20) with nuclei, neutrons arrive at this energy level, provided that they are not absorbed.

In many substances, thermal neutron reactions show a much larger effective cross-section than reactions involving faster neutrons, and thermal neutrons can therefore be absorbed more readily (i.e., with higher probability) by any atomic nuclei that they collide with, creating a heavier — and often unstable — isotope of the chemical element as a result.

Most fission reactors use a neutron moderator to slow down, or *thermalize* the neutrons that are emitted by nuclear fission so that they are more easily captured, causing further fission. Others, called fast breeder reactors, use fission energy neutrons directly.

55.11.2 Cold neutrons

Cold neutrons are thermal neutrons that have been equilibrated in a very cold substance such as liquid deuterium. Such a *cold source* is placed in the moderator of a research reactor or spallation source. Cold neutrons are particularly valuable for neutron scattering experiments.

55.11.3 Ultracold neutrons

Ultracold neutrons are produced by inelastically scattering cold neutrons in substances with a temperature of a few kelvins, such as solid deuterium or superfluid helium. An alternative production method is the mechanical deceleration of cold neutrons.

55.11.4 Fission energy neutrons

Main article: nuclear fission

A *fast neutron* is a free neutron with a kinetic energy level close to 1 MeV (1.6×10^{-13} J), hence a speed of ~14000 km/s (~ 5% of the speed of light). They are named *fission energy* or *fast* neutrons to distinguish them from lower-energy thermal neutrons, and high-energy neutrons produced in cosmic showers or accelerators. Fast neutrons are produced by nuclear processes such as nuclear fission. Neutrons produced in fission, as noted above, have a Maxwell–Boltzmann distribution of kinetic energies from 0 to ~14 MeV, a mean energy of 2 MeV (for U-235 fission neutrons), and a mode of only 0.75 MeV, which means that more than half of them do not qualify as fast (and thus have almost no chance of initiating fission in fertile materials, such as U-238 and Th-232).

Fast neutrons can be made into thermal neutrons via a process called moderation. This is done with a neutron moderator. In reactors, typically heavy water, light water, or graphite are used to moderate neutrons.

55.11.5 Fusion neutrons

For more details on this topic, see Nuclear fusion § Criteria and candidates for terrestrial reactions.

D–T (deuterium–tritium) fusion is the fusion reaction that produces the most energetic neutrons, with 14.1 MeV of kinetic energy and traveling at 17% of the speed of light. D–T fusion is also the easiest fusion reaction to ignite, reaching near-peak rates even when the deuterium and tritium nuclei have only a thousandth as much kinetic energy as the 14.1 MeV that will be produced.

14.1 MeV neutrons have about 10 times as much energy as fission neutrons, and are very effective at fissioning even non-fissile heavy nuclei, and these high-energy fissions produce more neutrons on average than fissions by lower-energy neutrons. This makes D–T fusion neutron sources such as proposed tokamak power reactors useful for transmutation of transuranic waste. 14.1 MeV neutrons can also produce neutrons by knocking them loose from nuclei.

On the other hand, these very high energy neutrons are less likely to simply be captured without causing fission or spallation. For these reasons, nuclear weapon design extensively utilizes D–T fusion 14.1 MeV neutrons to cause more fission. Fusion neutrons are able to cause fission in ordinarily non-fissile materials, such as depleted uranium (uranium-238), and these materials have been used in the jackets of thermonuclear weapons. Fusion neutrons also can cause fission in substances that are unsuitable or difficult to make into primary fission bombs, such as reactor grade plutonium. This physical fact thus causes ordinary non-weapons grade materials to become of concern in certain nuclear proliferation discussions and treaties.

Other fusion reactions produce much less energetic neutrons. D–D fusion produces a 2.45 MeV neutron and helium-3 half of the time, and produces tritium and a proton but no neutron the other half of the time. D–^{3}He fusion produces no neutron.

55.11.6 Intermediate-energy neutrons

A fission energy neutron that has slowed down but not yet reached thermal energies is called an epithermal neutron.

Cross sections for both capture and fission reactions often have multiple resonance peaks at specific energies in the epithermal energy range. These are of less significance in a fast neutron reactor, where most neutrons are absorbed before slowing down to this range, or in a well-moderated thermal reactor, where epithermal neutrons interact mostly with moderator nuclei, not with either fissile or fertile actinide nuclides. However, in a partially moderated reactor with more interactions of epithermal neutrons with heavy metal nuclei, there are greater possibilities for transient changes in reactivity that might make reactor control more difficult.

Ratios of capture reactions to fission reactions are also worse (more captures without fission) in most nuclear fuels such as plutonium-239, making epithermal-spectrum reactors using these fuels less desirable, as captures not only waste the one neutron captured but also usually result in a nuclide that is not fissile with thermal or epithermal neutrons, though still fissionable with fast neutrons. The exception is uranium-233 of the thorium cycle, which has good capture-fission ratios at all neutron energies.

55.11.7 High-energy neutrons

These neutrons have much more energy than fission energy neutrons and are generated as secondary particles by particle accelerators or in the atmosphere from cosmic rays. They can have energies as high as tens of joules per neutron. These neutrons are extremely efficient at ionization and far more likely to cause cell death than X-rays or protons.[72][73]

55.12 See also

- Ionizing radiation
- Isotope
- List of particles
- Neutronium
- Neutron magnetic moment
- Neutron radiation and the Sievert radiation scale
- Nuclear reaction
- Thermal reactor
- Nucleosynthesis
    - Neutron capture nucleosynthesis
    - R-process
    - S-process

55.12.1 Neutron sources

- Neutron generator
- Neutron sources

55.12.2 Processes involving neutrons

- Neutron bomb
- Neutron diffraction
- Neutron flux
- Neutron transport

55.13 References

[1] Ernest Rutherford. Chemed.chem.purdue.edu. Retrieved on 2012-08-16.

[2] 1935 Nobel Prize in Physics. Nobelprize.org. Retrieved on 2012-08-16.

[3] Mohr, P.J.; Taylor, B.N. and Newell, D.B. (2011), "The 2010 CODATA Recommended Values of the Fundamental Physical Constants" (Web Version 6.0). The database was developed by J. Baker, M. Douma, and S. Kotochigova. (2011-06-02). National Institute of Standards and Technology, Gaithersburg, Maryland 20899.

[4] Thomas, A.W.; Weise, W. (2001), *The Structure of the Nucleon*, Wiley-WCH, Berlin, ISBN 3-527-40297-7

[5] Glasstone, Samuel; Dolan, Philip J., eds. (1977), *The Effects of Nuclear Weapons, Third Edition*, U.S. Dept. of Defense and Energy Research and Development Administration, U.S. Government Printing Office, ISBN 1-60322-016-X

[6] Nudat 2. Nndc.bnl.gov. Retrieved on 2010-12-04.

[7] Chadwick, James (1932). "Possible Existence of a Neutron". *Nature* **129** (3252): 312. Bibcode:1932Natur.129Q.312C. doi:10.1038/129312a0.

[8] O. Hahn and F. Strassmann (1939). "Über den Nachweis und das Verhalten der bei der Bestrahlung des Urans mittels Neutronen entstehenden Erdalkalimetalle ("On the detection and characteristics of the alkaline earth metals formed by irradiation of uranium with neutrons")". *Naturwissenschaften* **27** (1): 11–15. Bibcode:1939NW.....27...11H. doi:10.1007/BF01488241.. The authors were identified as being at the Kaiser-Wilhelm-Institut für Chemie, Berlin-Dahlem. Received 22 December 1938.

[9] M. J. Carson et al. (2004). "Neutron background in large-scale xenon detectors for dark matter searches". *Astroparticle Physics* **21** (6): 667–687. doi:10.1016/j.astropartphys.2004.05.001.

[10] Povh, B.; Rith, K.; Scholz, C.; Zetsche, F. (2002). *Particles and Nuclei: An Introduction to the Physical Concepts*. Berlin: Springer-Verlag. p. 73. ISBN 978-3-540-43823-6.

[11] J.-L. Basdevant, J. Rich, M. Spiro (2005). *Fundamentals in Nuclear Physics*. Springer. p. 155. ISBN 0-387-01672-4.

[12] Paul Allen Tipler, Ralph A. Llewellyn (2002). *Modern Physics* (4 ed.). Macmillan. p. 310. ISBN 0-7167-4345-0.

[13] Nakamura, K (2010). "Review of Particle Physics". *Journal of Physics G: Nuclear and Particle Physics* **37** (7A): 075021. Bibcode:2010JPhG...37g5021N. doi:10.1088/0954-3899/37/7A/075021. PDF with 2011 partial update for the 2012 edition The exact value of the mean lifetime is still uncertain, due to conflicting results from experiments. The Particle Data Group reports values up to six seconds apart (more than four standard deviations), commenting that "our 2006, 2008, and 2010 Reviews stayed with 885.7±0.8 s; but we noted that in light of SEREBROV 05 our value should be regarded as suspect until further experiments clarified matters. Since our 2010 Review, PICHLMAIER 10 has obtained a mean life of 880.7±1.8 s, closer to the value of SEREBROV 05 than to our average. And SEREBROV 10B[...] claims their values should be lowered by about 6 s, which would bring them into line with the two lower values. However, those reevaluations have not received an enthusiastic response from the experimenters in question; and in any case the Particle Data Group would have to await published changes (by those experimenters) of published values. At this point, we can think of nothing better to do than to average the seven best but discordant measurements, getting 881.5±1.5s. Note that the error includes a scale factor of 2.7. This is a jump of 4.2 old (and 2.8 new) standard deviations. This state of affairs is a particularly unhappy one, because the value is so important. We again call upon the experimenters to clear this up."

[14] R.K. Adair (1989). *The Great Design: Particles, Fields, and Creation*. Oxford University Press. p. 214.

[15] Cho, Adiran (2 April 2010). "Mass of the Common Quark Finally Nailed Down". *http://news.sciencemag.org". American Association for the Advancement of Science. Retrieved 27 September 2014.*

[16] W.N.Cottingham,D.A.Greenwood(1986).*An Introduction to Nuclear Physics*. Cambridge University Press. ISBN978052.

[17] E.Rutherford(1920). "Nuclear Constitution of Atoms".*Proceedings of the Royal Society* A**97**(686): 374–400. Bibcode:1R. doi:10.1098/rspa.1920.0040.

[18] "Wolfgang Pauli". *Sources in the History of Mathematics and Physical Sciences*. Sources in the History of Mathematics and Physical Sciences **6**: 105–144. 1985. doi:10.1007/978-3-540-78801-0_3. ISBN 978-3-540-13609-5. |chapter= ignored (help)

[19] Hendry, John, ed. (1984), *Cambridge Physics in the Thirties*, Adam Hilger Ltd, Bristol, ISBN 0852747616

[20] N.Feather(1960). “A history of neutrons and nuclei. Part1”.*Contemporary Physics***1**(3): 191–203. doi:10.1080/00107516.

[21] Brown,Laurie M. (1978). “The idea of the neutrino”.*Physics Today***31**(9): 23. Bibcode:1978PhT....31i..23B.doi:10.10631.

[22] Friedlander G., Kennedy J.W. and Miller J.M. (1964) *Nuclear and Radiochemistry* (2nd edition), Wiley, pp. 22–23 and 38–39

[23] Stuewer, Roger H. (1985). “Niels Bohr and Nuclear Physics”. In French, A. P.; Kennedy, P. J. *Niels Bohr: A Centenary Volume.* Harvard University Press. pp. 197–220. ISBN 0674624165.

[24] Pais, Abraham (1986). *Inward Bound.* Oxford: Oxford University Press. p. 299. ISBN 0198519974.

[25] Klein, O. (1929). “Die Reflexion von Elektronen an einem Potentialsprung nach der relativistischen Dynamik von Dirac”. *Zeitschrift für Physik* **53** (3–4): 157–165. Bibcode:1929ZPhy...53..157K. doi:10.1007/BF01339716.

[26] Bothe, W.; Becker, H. (1930). “Künstliche Erregung von Kern-γ-Strahlen” [Artificial excitation of nuclear γ-radiation]. *Zeitschrift für Physik* **66** (5–6): 289–306. Bibcode:1930ZPhy...66..289B. doi:10.1007/BF01390908.

[27] Becker, H.; Bothe, W. (1932). “Die in Bor und Beryllium erregten γ-Strahlen” [Γ-rays excited in boron and beryllium]. *Zeitschrift für Physik* **76** (7–8): 421–438. Bibcode:1932ZPhy...76..421B. doi:10.1007/BF01336726.

[28] Joliot-Curie, Irène and Joliot, Frédéric (1932). "Émission de protons de grande vitesse par les substances hydrogénées sous l'influence des rayons γ très pénétrants” [Emission of high-speed protons by hydrogenated substances under the influence of very penetrating γ-rays]. *Comptes Rendus* **194**: 273.

[29] “Atop the Physics Wave: Rutherford Back in Cambridge, 1919–1937”. *Rutherford's Nuclear World.* American Institute of Physics. 2011–2014. Retrieved 19 August 2014.

[30] Chadwick, J. (1933). “Bakerian Lecture. The Neutron”. *Proceedings of the Royal Society A: Mathematical, Physical and Engineering Sciences* **142** (846): 1–25. Bibcode:1933RSPSA.142....1C. doi:10.1098/rspa.1933.0152.

[31] Heisenberg, W. (1932). "Über den Bau der Atomkerne. I”. *Z. Phys.* **77**: 1–11. doi:10.1007/BF01342433.

[32] Heisenberg, W. (1932). "Über den Bau der Atomkerne. II”. *Z. Phys.* **78** (3–4): 156–164. doi:10.1007/BF01337585.

[33] Heisenberg, W. (1933). "Über den Bau der Atomkerne. III”. *Z. Phys.* **80** (9–10): 587–596. doi:10.1007/BF01335696.

[34] Iwanenko, D.D., The neutron hypothesis, Nature **129** (1932) 798.

[35] Miller A. I. *Early Quantum Electrodynamics: A Sourcebook*, Cambridge University Press, Cambridge, 1995, ISBN 0521568919, pp. 84–88.

[36] Wilson, Fred L. (1968). “Fermi's Theory of Beta Decay”. *Am. J. Phys.* **36** (12): 1150–1160. Bibcode:1968AmJPh..36.1150W. doi:10.1119/1.1974382.

[37] Chadwick, J.; Goldhaber, M. (1934). “A nuclear photo-effect: disintegration of the diplon by gamma rays”. *Nature* **134**: 237–238. doi:10.1038/134237a0.

[38] Chadwick,J.;Goldhaber,M. (1935). “A nuclear photoelectric effect”(PDF).*Proc.R.Soc.Lond***151**: 479–493. doi:10.1098/rspa.

[39] Cooper, Dan (1999). *Enrico Fermi: And the Revolutions in Modern physics.* New York: Oxford University Press. ISBN 0-19-511762-X. OCLC 39508200.

[40] Hahn, O. (1958). “The Discovery of Fission”. *Scientific American* **198** (2): 76. doi:10.1038/scientificamerican0258-76.

[41] Rife, Patricia (1999). *Lise Meitner and the dawn of the nuclear age.* Basel, Switzerland: Birkhäuser. ISBN 0-8176-3732-X.

[42] Hahn, O.; Strassmann, F. (10 February 1939). “Proof of the Formation of Active Isotopes of Barium from Uranium and Thorium Irradiated with Neutrons; Proof of the Existence of More Active Fragments Produced by Uranium Fission”. *Die Naturwissenschaften* **27**: 89–95.

[43] “The Nobel Prize in Chemistry 1944”. Nobel Foundation. Retrieved 2007-12-17.

[44] Bernstein, Jeremy (2001). *Hitler's uranium club: the secret recordings at Farm Hall.* New York: Copernicus. p. 281. ISBN 0-387-95089-3.

[45] “The Nobel Prize in Chemistry 1944: Presentation Speech”. Nobel Foundation. Retrieved 2008-01-03.

[46] Sir James Chadwick's Discovery of Neutrons. ANS Nuclear Cafe. Retrieved on 2012-08-16.

[47] Particle Data Group Summary Data Table on Baryons. lbl.gov (2007). Retrieved on 2012-08-16.

[48] Basic Ideas and Concepts in Nuclear Physics: An Introductory Approach, Third Edition K. Heyde Taylor & Francis 2004. Print ISBN 978-0-7503-0980-6. eBook ISBN 978-1-4200-5494-1. DOI: 10.1201/9781420054941.ch5. full text

[49] Olive, K.A. et al. (2014). "Review of Particle Physics". *Chin. Phys. C* **38**: 090001. doi:10.1088/1674-1137/38/9/090001. |first2= missing |last2= in Authors list (help)

[50] "Pear-shaped particles probe big-bang mystery" (Press release). University of Sussex. 20 February 2006. Retrieved 2009-12-14.

[51] A cryogenic experiment to search for the EDM of the neutron. Hepwww.rl.ac.uk. Retrieved on 2012-08-16.

[52] Search for the neutron electric dipole moment: nEDM. Nedm.web.psi.ch (2001-09-12). Retrieved on 2012-08-16.

[53] SNS Neutron EDM Experiment. P25ext.lanl.gov. Retrieved on 2012-08-16.

[54] Measurement of the Neutron Electric Dipole Moment. Nrd.pnpi.spb.ru. Retrieved on 2012-08-16.

[55] Gell, Y.; Lichtenberg, D. B. (1969). "Quark model and the magnetic moments of proton and neutron". *Il Nuovo Cimento A*. Series 10 **61**: 27–40. Bibcode:1969NCimA..61...27G. doi:10.1007/BF02760010.

[56] Alvarez, L. W; Bloch, F. (1940). "A quantitative determination of the neutron magnetic moment in absolute nuclear magnetons". *Physical Review* **57**: 111–122. doi:10.1103/physrev.57.111.

[57] Miller,G.A. (2007). "Charge Densities of the Neutron and Proton".*Physical Review Letters***99**(11): 112001. Bibcode:201M. doi:10.1103/PhysRevLett.99.112001.

[58] Greene, GL et al. (1986). "New determination of the deuteron binding energy and the neutron mass". *Phys. Rev. Lett.* **56**: 819–822. Bibcode:1986PhRvL..56..819G. doi:10.1103/PhysRevLett.56.819.

[59] Byrne, J. *Neutrons, Nuclei, and Matter*, Dover Publications, Mineola, New York, 2011, ISBN 0486482383, pp. 18–19

[60] Spyrou, A. et al. (2012). "First Observation of Ground State Dineutron Decay: 16Be". *Physical Review Letters* **108** (10): 102501. Bibcode:2012PhRvL.108j2501S. doi:10.1103/PhysRevLett.108.102501. PMID 22463404.

[61] Llanes-Estrada, Felipe J.; Moreno Navarro, Gaspar (2011). "Cubic neutrons". arXiv:1108.1859v1 [nucl-th].

[62] Köhn, C., Ebert, U., Calculation of beams of positrons, neutrons and protons associated with terrestrial gamma-ray flashes, J. Geophys. Res. Atmos. (2015), vol. 23, doi:10.1002/2014JD022229

[63] Clowdsley, MS; Wilson, JW; Kim, MH; Singleterry, RC; Tripathi, RK; Heinbockel, JH; Badavi, FF; Shinn, JL (2001). "Neutron Environments on the Martian Surface" (PDF). *Physica Medica* **17** (Suppl 1): 94–6. PMID 11770546.

[64] Byrne, J. *Neutrons, Nuclei, and Matter*, Dover Publications, Mineola, New York, 2011, ISBN 0486482383, pp. 32–33.

[65] Science/Nature | Q&A: Nuclear fusion reactor. BBC News (2006-02-06). Retrieved on 2010-12-04.

[66] Byrne, J. *Neutrons, Nuclei, and Matter*, Dover Publications, Mineola, New York, 2011, ISBN 0486482383, p. 453.

[67] Kumakhov, M. A.; Sharov, V. A. (1992). "A neutron lens". *Nature* **357** (6377): 390–391. Bibcode:1992Natur.357..390K. doi:10.1038/357390a0.

[68] Physorg.com, "New Way of 'Seeing': A 'Neutron Microscope'". Physorg.com (2004-07-30). Retrieved on 2012-08-16.

[69] "NASA Develops a Nugget to Search for Life in Space". NASA.gov (2007-11-30). Retrieved on 2012-08-16.

[70] Hall EJ. Radiobiology for the Radiologist. Lippincott Williams & Wilkins; 5th edition (2000)

[71] Johns HE and Cunningham JR. The Physics of Radiology. Charles C Thomas 3rd edition 1978

[72] Tami Freeman (May 23, 2008). "Facing up to secondary neutrons". Medical Physics Web. Retrieved 2011-02-08.

[73] Heilbronn, L.; Nakamura, T; Iwata, Y; Kurosawa, T; Iwase, H; Townsend, LW (2005). "Expand+Overview of secondary neutron production relevant to shielding in space". *Radiation Protection Dosimetry* **116** (1–4): 140–143. doi:10.1093/rpd/nci033. PMID 16604615.

55.14 Further reading

- Annotated bibliography for neutrons from the Alsos Digital Library for Nuclear Issues
- Abraham Pais, *Inward Bound*, Oxford: Oxford University Press, 1986. ISBN 0198519974.
- Sin-Itiro Tomonaga, *The Story of Spin*, The University of Chicago Press, 1997
- Herwig Schopper, *Weak interactions and nuclear beta decay*, Publisher, North-Holland Pub. Co., 1966.

55.15 External links

- neutron properties at Particle Data Group, Lawrence Berkeley National Laboratory in Berkeley, CA. (pdgLive)

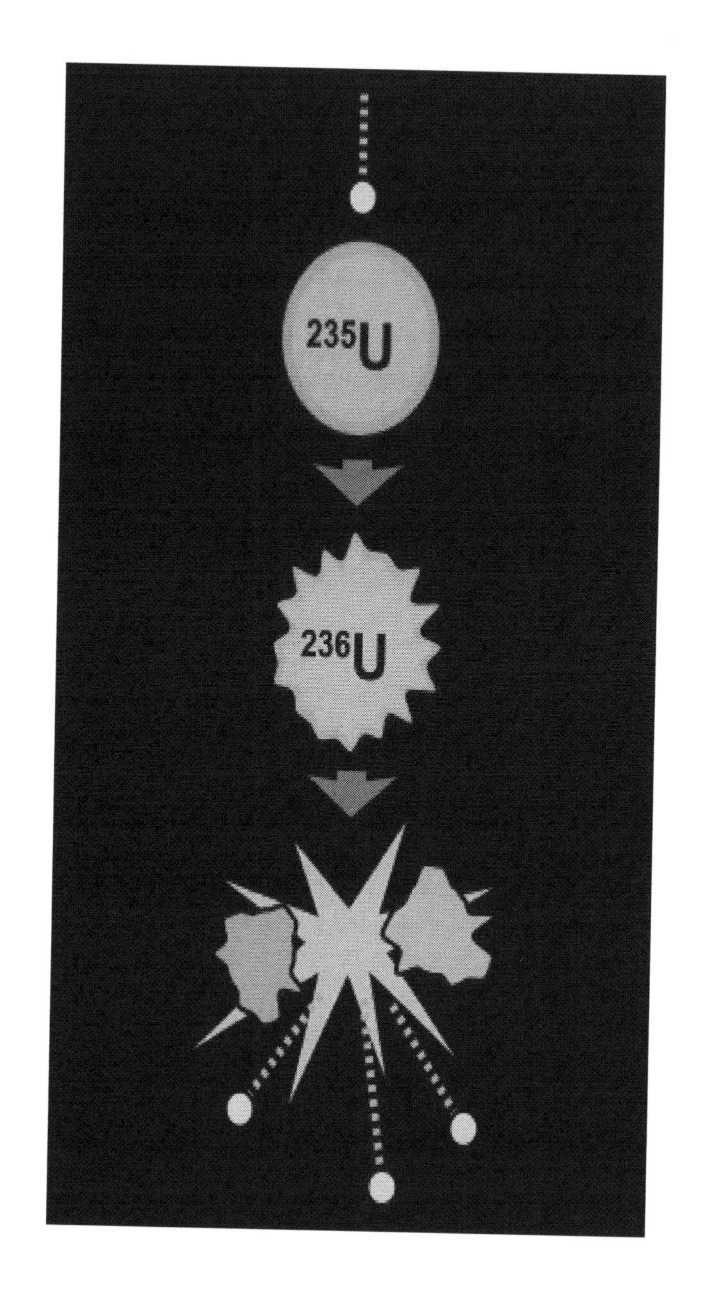

Nuclear fission caused by absorption of a neutron by uranium-235. The heavy nuclide fragments into lighter components and additional neutrons.

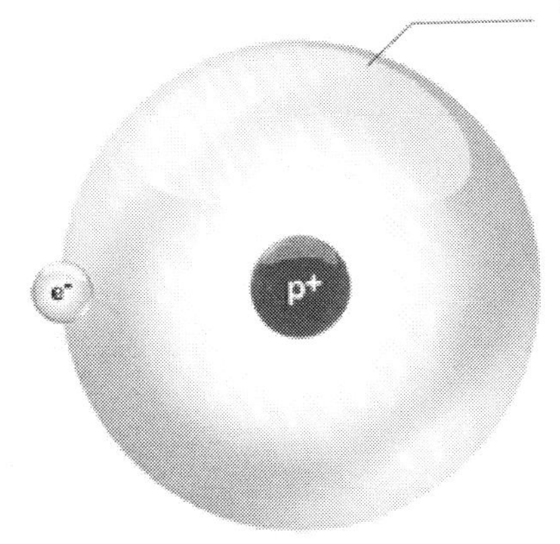

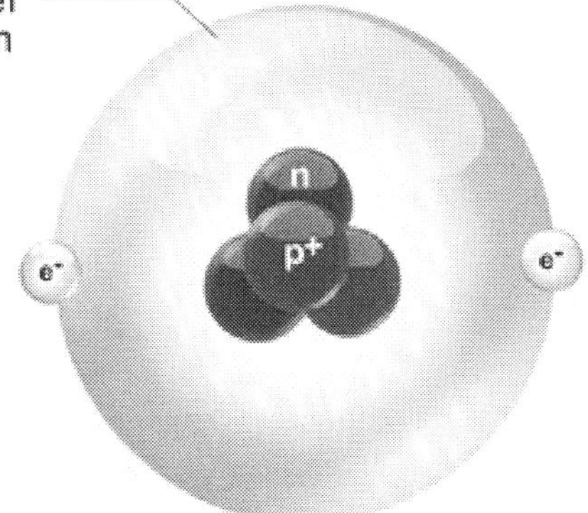

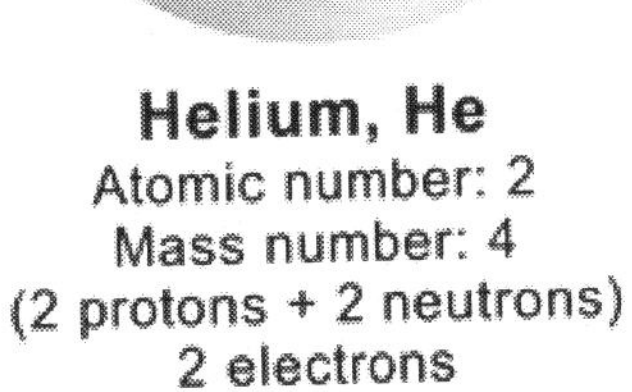

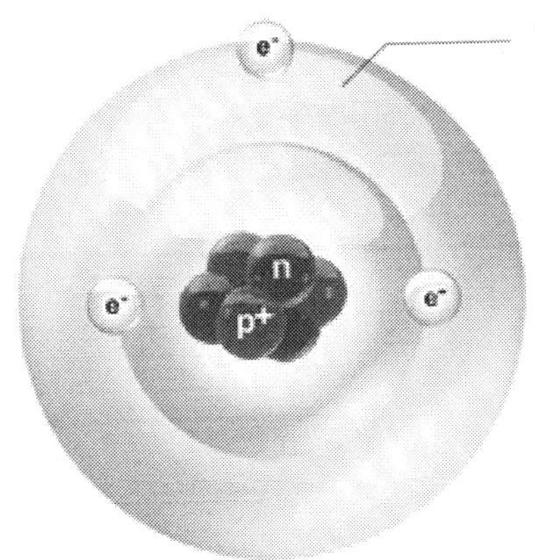

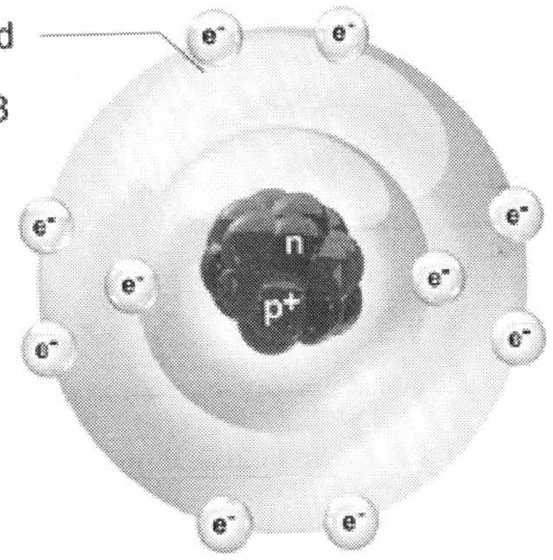

Models depicting the nucleus and electron energy levels in hydrogen, helium, lithium, and neon atoms. In reality, the diameter of the nucleus is about 100,000 times smaller than the diameter of the atom.

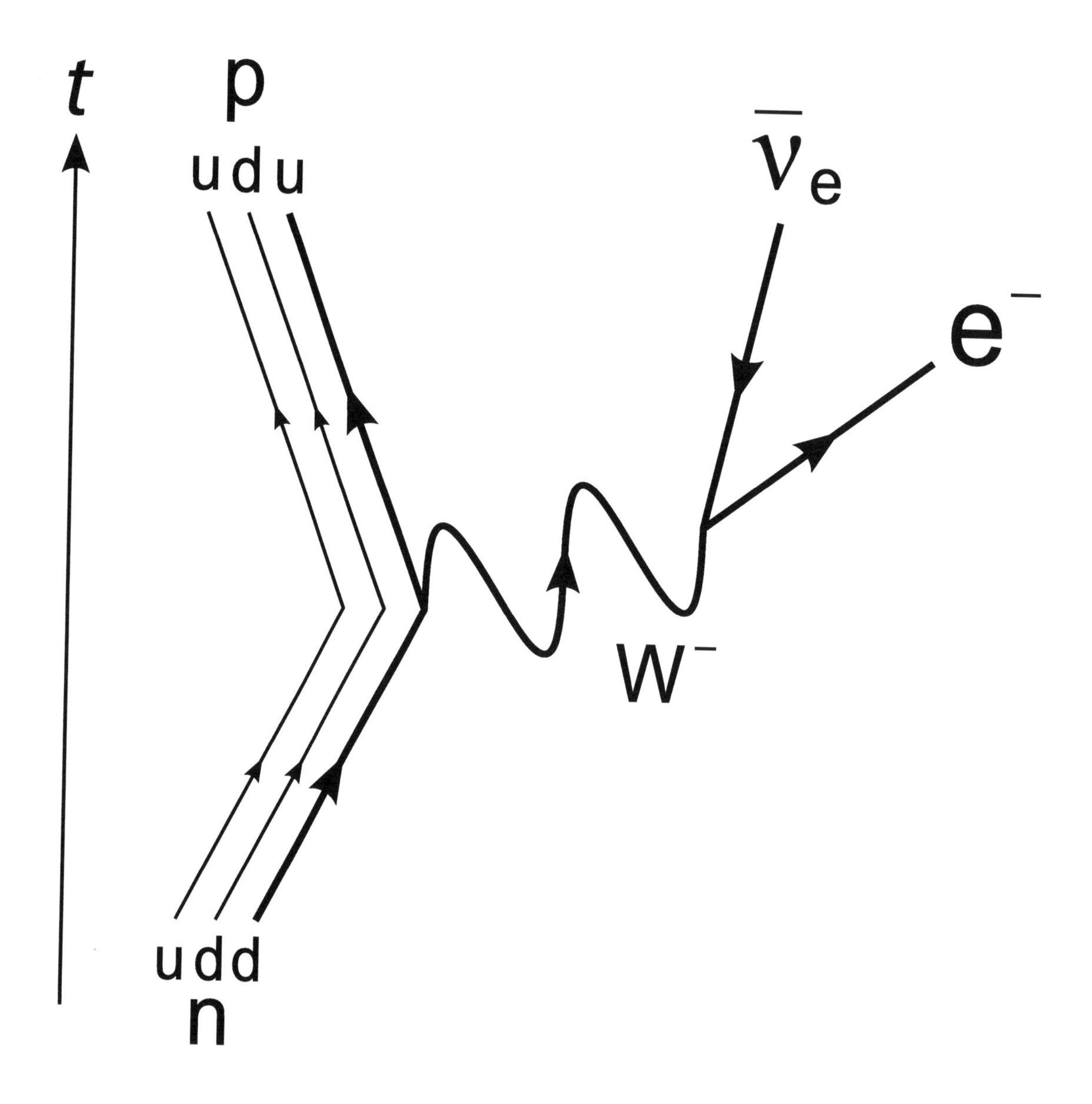

The Feynman diagram for beta decay of a neutron into a proton, electron, and electron antineutrino via an intermediate heavy W boson

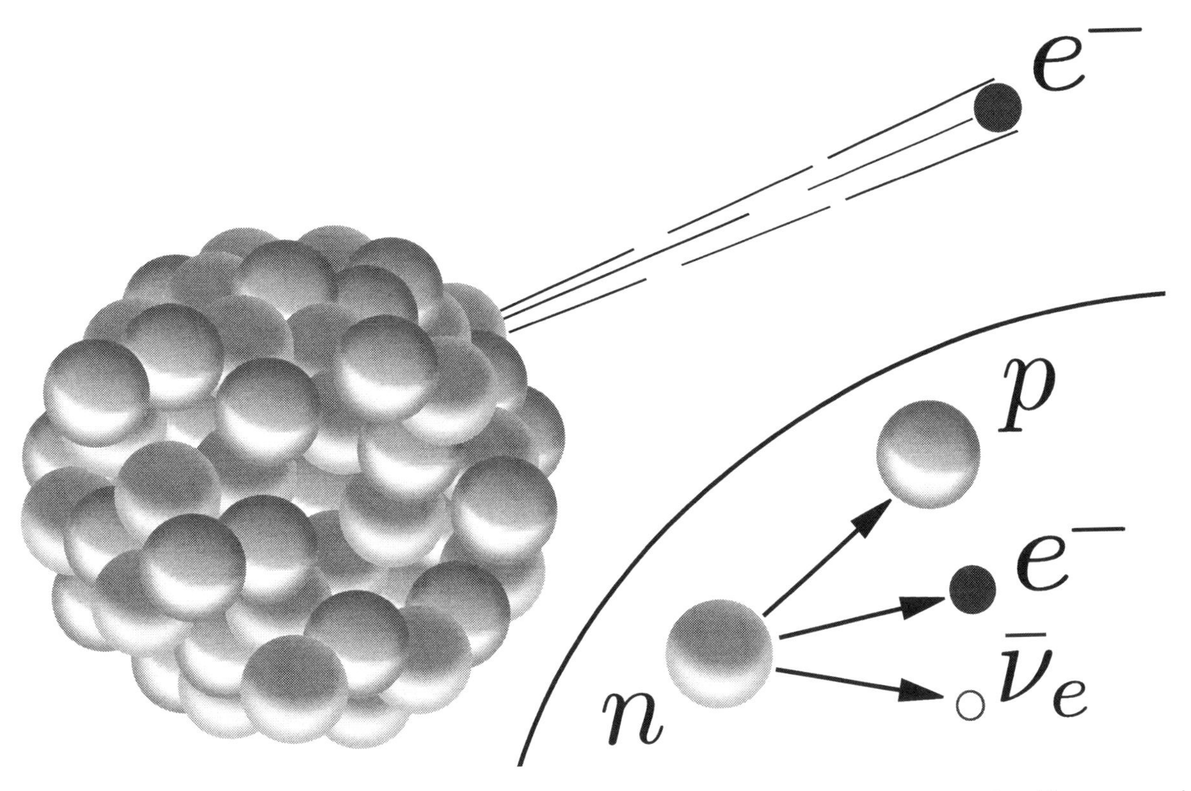

A schematic of the nucleus of an atom indicating β− radiation, the emission of a fast electron from the nucleus (the accompanying antineutrino is omitted). In the Rutherford model for the nucleus, red spheres were protons with positive charge and blue spheres were protons tightly bound to an electron with no net charge.
*The **inset** shows beta decay of a free neutron as it is understood today; an electron and antineutrino are created in this process.*

Institut Laue–Langevin (ILL) in Grenoble, France – a major neutron research facility.

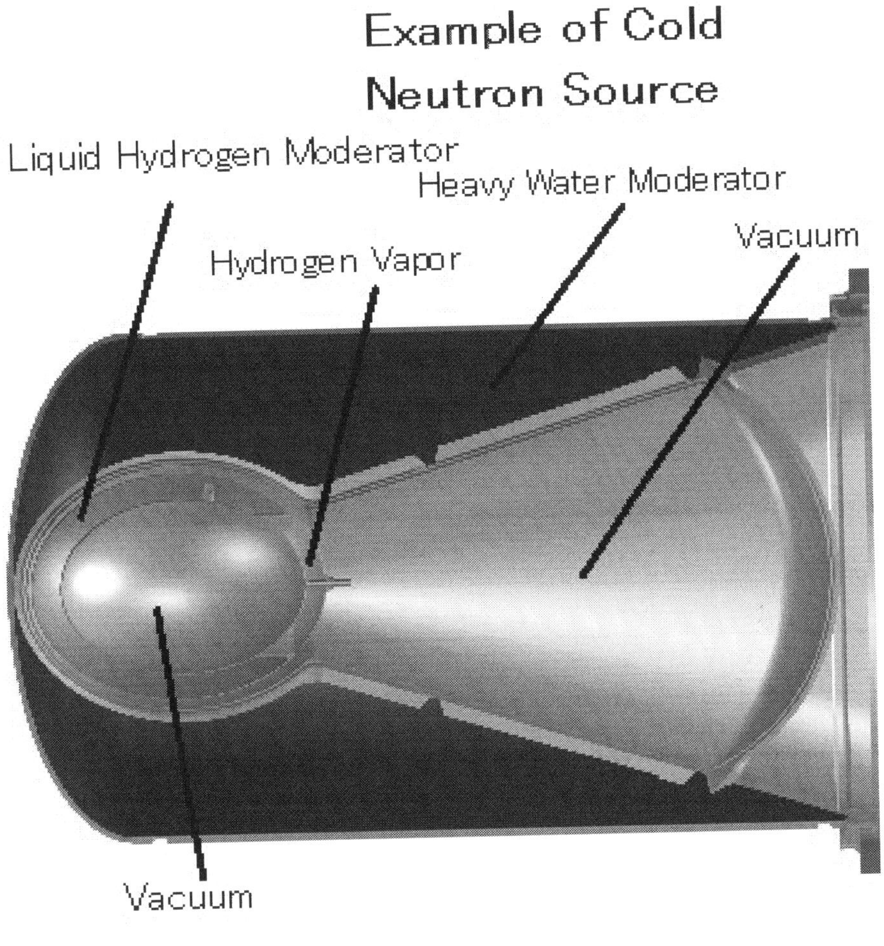

Example of Cold
Neutron Source

Liquid Hydrogen Moderator
Heavy Water Moderator
Vacuum
Hydrogen Vapor
Vacuum

Cold neutron source providing neutrons at about the temperature of liquid hydrogen

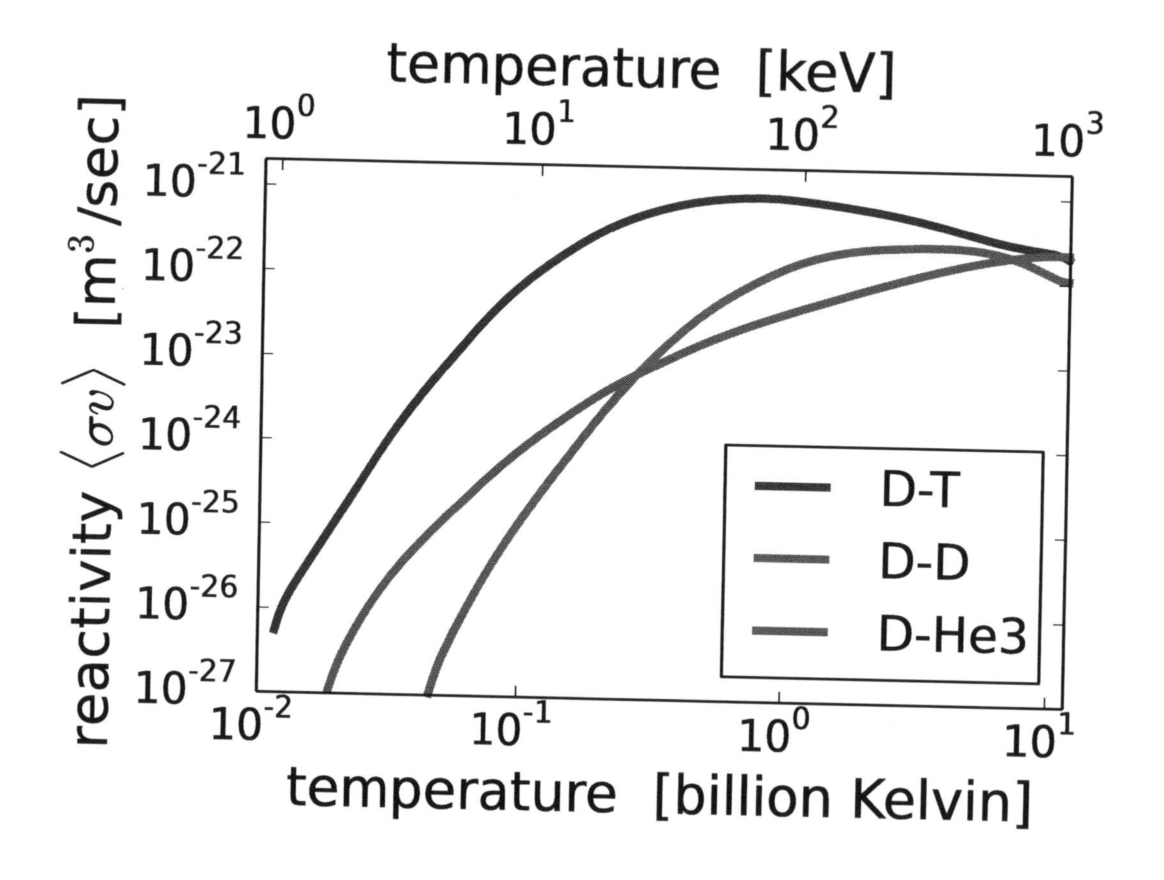

The fusion reaction rate increases rapidly with temperature until it maximizes and then gradually drops off. The DT rate peaks at a lower temperature (about 70 keV, or 800 million kelvins) and at a higher value than other reactions commonly considered for fusion energy.

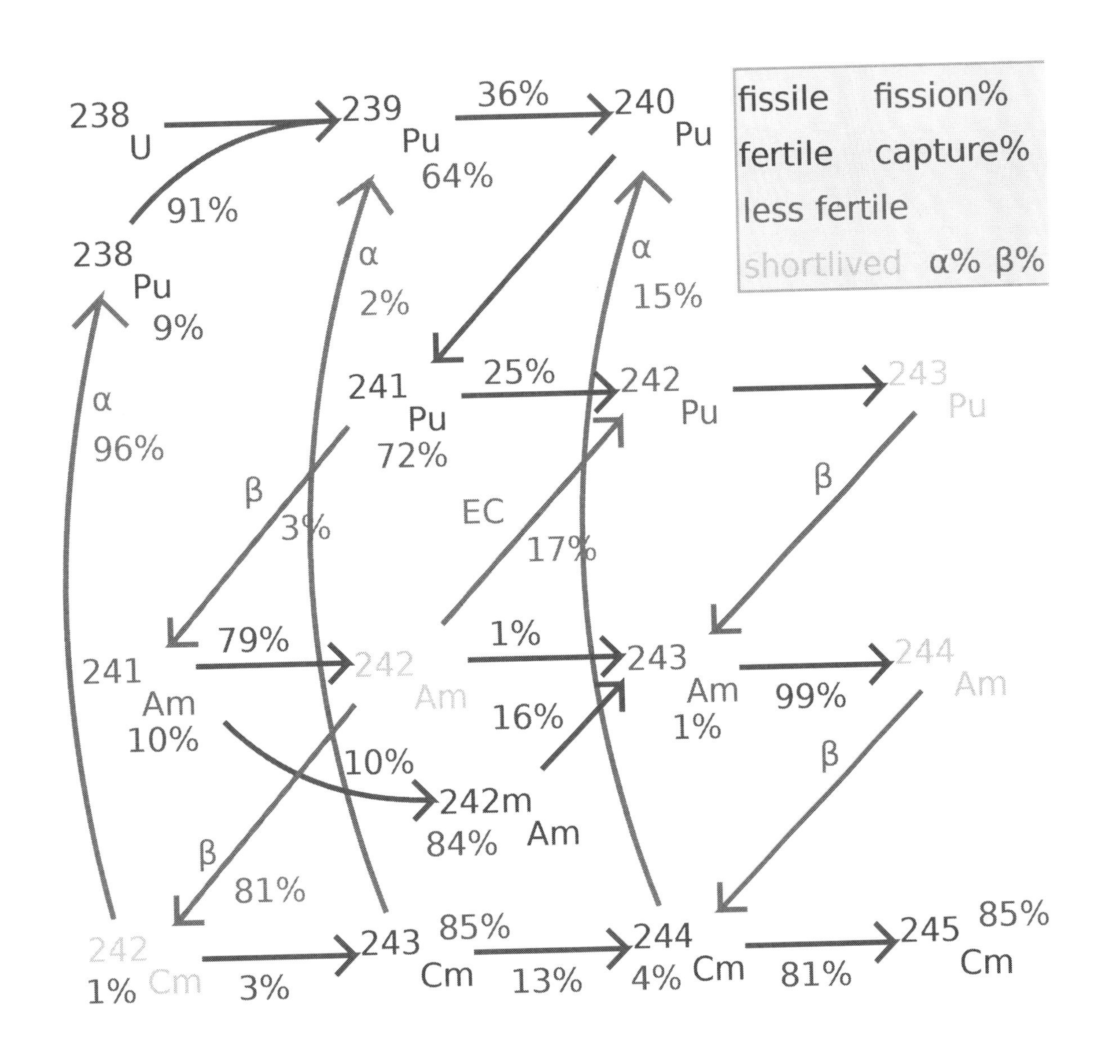

Transmutation flow in light water reactor, which is a thermal-spectrum reactor

Chapter 56

Delta baryon

The **Delta baryons** (or **Δ baryons**, also called **Delta resonances**) are a family of subatomic particle made of three up or down quarks (u or d quarks).

Four Δ baryons exist: Δ++ (constituent quarks: uuu), Δ+ (uud), Δ0 (udd), and Δ– (ddd), which respectively carry an electric charge of +2 *e*, +1 *e*, 0 *e*, and −1 *e*.

The Δ baryons have a mass of about 1232 MeV/c^2, a spin of $^3/_2$, and an isospin of $^3/_2$. In many ways, Δ baryons are 'excited' nucleons (symbol N), which are made of the same constituent quarks in a lower-energy spin configuration (spin $^1/_2$). The Δ+ (uud), Δ0 (udd) are the higher-energy equivalent of the proton (N+, uud) and neutron (N0, udd). However, the Δ++ and Δ– have no nucleon equivalent.

56.1 Composition

The four Δ baryons are distinguished by their electrical charges, which is the sum of the charges of the quarks from which they are composed. There are also four antiparticles with opposite charges, made up of the corresponding antiquarks. The existence of the Δ++, with its unusual +2 charge, was a crucial clue in the development of the quark model.

56.2 Decay

All varieties of Δ baryons quickly decay via the strong force into a nucleon (proton or neutron) and a pion of appropriate charge. The amplitudes of various final charge states given by their respective isospin couplings. More rarely and more slowly, the Δ+ can decay into a proton and a photon and the Δ0 can decay into a neutron and a photon.

56.3 List

[a] ^ PDG reports the resonance width (Γ). Here the conversion $\tau = \hbar/\Gamma$ is given instead.

56.4 References

[1] J. Beringer *et al.* (2013): Particle listings – Δ(1232)

56.4.1 Bibliography

- C. Amsler et al. (Particle Data Group) (2008). "Review of Particle Physics". *Physics Letters B* **667** (1): 1. Bibcode:2008PhLB..667....1P. doi:10.1016/j.physletb.2008.07.018.

Chapter 57

Lambda baryon

The **Lambda baryons** are a family of subatomic hadron particles that have the symbols Λ0, Λ+
c, Λ0
b, and Λ+
t and have +1 elementary charge or are neutral. They are baryons containing three different quarks: one up, one down, and one third quark, which can be a strange (Λ0), a charm (Λ+
c), a bottom (Λ0
b), or a top (Λ+
t) quark. The top Lambda is not expected to be observed as the Standard Model predicts the mean lifetime of top quarks to be roughly 5×10^{-25} s.[1] This is about one-twentieth the timescale for strong interactions, and, therefore it does not form hadrons.

The Lambda baryon Λ0 was first discovered in October 1950, by V. D. Hopper and S. Biswas of the University of Melbourne, as a neutral V particle with a proton as a decay product, thus correctly distinguishing it as a baryon, rather than a meson[2] *i.e.*, different in kind from the K meson discovered in 1947 by Rochester and Butler;[3] they were produced by cosmic rays and detected in photographic emulsions flown in a balloon at 70,000 feet (21,000 m).[4] Though the particle was expected to live for $\sim1\times10^{-23}$ s,[5] it actually survived for $\sim1\times10^{-10}$ s.[6] The property that caused it to live so long was dubbed *strangeness* and led to the discovery of the strange quark.[5] Furthermore, these discoveries led to a principle known as the *conservation of strangeness*, wherein lightweight particles do not decay as quickly if they exhibit strangeness (because non-weak methods of particle decay must preserve the strangeness of the decaying baryon).[5]

The Lambda baryon has also been observed in atomic nuclei called hypernuclei. These nuclei contain the same number of protons and neutrons as a known nucleus, but also contains one or in rare cases two Lambda particles.[7] In such a scenario, the Lambda slides into the center of the nucleus (it is not a proton or a neutron, and thus is not affected by the Pauli exclusion principle), and it binds the nucleus more tightly together due to its interaction via the strong force. In a lithium isotope (Λ7Li), it made the nucleus 19% smaller.[8]

57.1 List

The symbols encountered in this list are: I (*isospin*), J (*total angular momentum quantum number*), P (*parity*), Q (*charge*), S (*strangeness*), C (*charmness*), B′ (*bottomness*), T (*topness*), B (*baryon number*), u (*up quark*), d (*down quark*), s (*strange quark*), c (*charm quark*), b (*bottom quark*), t (*top quark*), as well as other subatomic particles (hover for name).

Antiparticles are not listed in the table; however, they simply would have all quarks changed to antiquarks, and Q, B, S, C, B′, T, would be of opposite signs. I, J, and P values in red have not been firmly established by experiments, but are predicted by the quark model and are consistent with the measurements.[9][10] The top lambda (Λ+
t) is listed for comparison, but is not expected to be observed, because top quarks decay before they have time to hadronize.[11]

† ^ Particle unobserved, because the top-quark decays before it hadronizes.

57.2 See also

- List of baryons

57.3 References

[1] A.Quadt(2006). "Top quark physics at hadron colliders".*European Physical Journal C***48**(3): 835–1000. Bibcode:2006EPJC. doi:10.1140/epjc/s2006-02631-6.

[2] Hopper, V.D.; Biswas, S. (1950). "Evidence Concerning the Existence of the New Unstable Elementary Neutral Particle". *Phys. Rev.* **80**: 1099. Bibcode:1950PhRv...80.1099H. doi:10.1103/physrev.80.1099.

[3] Rochester, G.D.; Butler, C.C. (1947). "Evidence for the Existence of New Unstable Elementary Particles". *Nature* **160**: 855. Bibcode:1947Natur.160..855R. doi:10.1038/160855a0.

[4] Pais, Abraham (1986). *Inward Bound*. Oxford University Press, p 21, 511-517.

[5] The Strange Quark

[6] C. Amsler et al. (2008): Particle listings – Λ

[7] "Media Advisory: The Heaviest Known Antimatter". bnl.gov.

[8] Brumfiel, Geoff. "Focus: The Incredible Shrinking Nucleus".

[9] C. Amsler et al. (2008): Particle summary tables – Baryons

[10] J. G. Körner et al. (1994)

[11] Ho-Kim, Quang; Pham, Xuan Yem (1998). "Quarks and SU(3) Symmetry". *Elementary Particles and Their Interactions: Concepts and Phenomena*. Berlin: Springer-Verlag. p. 262. ISBN 3-540-63667-6. OCLC 38965994. Because the top quark decays before it can be hadronized, there are no bound $t\bar{t}$ states and no top-flavored mesons or baryons[...].

[12] C. Amsler et al. (2008): Particle listings – Λ
c

[13] C. Amsler et al. (2008): Particle listings – Λ
b

57.3.1 Bibliography

- C.Amsler*et al.* (Particle Data Group) (2008). "Review of Particle Physics".*Physics Letters B***667**: 1. BibcodP. doi:10.1016/j.physletb.2008.07.018.
- C. Caso *et al.* (Particle Data Group) (1998). "Review of Particle Physics". *European Physical Journal C* **3**: 1. Bibcode:1998EPJC....3....1P. doi:10.1007/s10052-998-0104-x.
- J. G. Körner, M. Krämer, and D. Pirjol (1994). "Heavy Baryons". *Progress in Particle and Nuclear Physics* **33**: 787–868. arXiv:hep-ph/9406359. Bibcode:1994PrPNP..33..787K. doi:10.1016/0146-6410(94)90053-1.
- R. Nave (12 April 2005). "The Lambda Baryon". *HyperPhysics*. Retrieved 2010-07-14.

Chapter 58

Sigma baryon

The **Sigma baryons** are a family of subatomic hadron particles which have a +2, +1 or −1 elementary charge or are neutral. They are baryons containing three quarks: two up and/or down quarks, and one third quark, which can be either a strange (symbols Σ+, Σ0, Σ−), a charm (symbols Σ++
c, Σ+
c, Σ0
c), a bottom (symbols Σ+
b, Σ0
b, Σ−
b) or a top (symbols Σ++
t, Σ+
t, Σ0
t) quark. However, the top Sigmas are not expected to be observed as the Standard Model predicts the mean lifetime of top quarks to be roughly 5×10^{-25} s.[1] This is about 20 times shorter than the timescale for strong interactions, and therefore it does not form hadrons.

58.1 List

The symbols encountered in these lists are: I (*isospin*), J (*total angular momentum*), P (*parity*), u (*up quark*), d (*down quark*), s (*strange quark*), c (*charm quark*), t (*top quark*), b (*bottom quark*), Q (*charge*), B (*baryon number*), S (*strangeness*), C (*charmness*), B′ (*bottomness*), T (*topness*), as well as other subatomic particles (hover for name).

Antiparticles are not listed in the table; however, they simply would have all quarks changed to antiquarks, and Q, B, S, C, B′, T, would be of opposite signs. I, J, and P values in red have not been firmly established by experiments, but are predicted by the quark model and are consistent with the measurements.[2][3]

58.1.1 $J^P = \frac{1}{2}^+$ Sigma baryons

† ^ Particle currently unobserved, but predicted by the standard model.
[a] ^ PDG reports the resonance width (Γ). Here the conversion $\tau = \hbar/\Gamma$ is given instead.
[b] ^ The specific values of the name has not been decided yet, but will likely be close to Σ
b(5810).

58.1.2 $J^P = \frac{3}{2}^+$ Sigma baryons

† ^ Particle currently unobserved, but predicted by the standard model.
[c] ^ PDG reports the resonance width (Γ). Here the conversion $\tau = \hbar/\Gamma$ is given instead.

58.2 See also

- Delta baryon
- Hyperon
- Lambda baryon
- List of mesons
- List of particles
- Nucleon
- Omega baryon
- Physics portal
- Timeline of particle discoveries
- Xi baryon

58.3 References

[1] A.Quadt(2006). "Top quark physics at hadron colliders".*European Physical Journal C***48**(3): 835–1000. Bibcode:2005Q. doi:10.1140/epjc/s2006-02631-6.

[2] C. Amsler et al. (2008): Particle summary tables – Baryons

[3] J. G. Körner et al. (1994)

[4] C. Amsler et al. (2008): Particle listings – Σ+

[5] C. Amsler et al. (2008): Particle listings – Σ0

[6] C. Amsler et al. (2008): Particle listings – Σ–

[7] C. Amsler et al. (2008): Particle listings – Σ
c(2455)

[8] T. Aaltonen et al. (2007a)

[9] C. Amsler et al. (2008): Particle listings – Σ(1385)

[10] C. Amsler et al. (2008): Particle listings – Σ
c(2520)

58.4 Bibliography

- C. Amsler *et al.* (Particle Data Group) (2008). "Review of Particle Physics". *Physics Letters B* **667** (1): 1. Bibcode:2008PhLB..667....1P. doi:10.1016/j.physletb.2008.07.018.
- J. G. Körner, M. Krämer, and D. Pirjol (1994). "Heavy Baryons". *Progress in Particle and Nuclear Physics* **33**: 787–868. arXiv:hep-ph/9406359. Bibcode:1994PrPNP..33..787K. doi:10.1016/0146-6410(94)90053-1.
- T. Aaltonen *et al.* (CDF Collaboration) (2007a). "First Observation of Heavy Baryons Σ
b and Σ∗
b".*Physical Review Letters***99**(20): 202001. arXiv:0706.3868. Bibcode:2007PhRvL..99t2001A.doi:9.202001.

Chapter 59

Xi baryon

The **Xi baryons** or *cascade particles* are a family of subatomic hadron particles which have the symbol Ξ and may have an elementary charge (*Q*) of +2, +1, 0, or −1. Like all baryons, they contain three quarks: one up or down quark, and two more massive quarks. They are historically called the *cascade particles* because of their unstable state; they decay rapidly into lighter particles through a chain of decays.[1] The first discovery of a charged Xi baryon was in cosmic ray experiments by the Manchester group in 1952.[2] The first discovery of the neutral Xi particle was at Lawrence Berkeley Laboratory in 1959.[3] It was also observed as a daughter product from the decay of the omega baryon (Ω−) observed at Brookhaven National Laboratory in 1964.[1] The Xi spectrum is important to nonperturbative quantum chromodynamics (QCD).

The Ξ−
b particle is also known as the **cascade B** particle and contains quarks from all three families. It was discovered by D0 and CDF experiments at Fermilab. The discovery was announced on 12 June 2007. It was the first known particle made of quarks from all three quark generations – namely, a down quark, a strange quark, and a bottom quark. The D0 and CDF collaborations reported the consistent masses of the new state. The Particle Data Group world average mass is 5.7924±0.0030 GeV/c^2.

Unless specified, the non-up/down quark content of Xi baryons is strange (i.e. there is one up or down quark and two strange quarks). However a Ξ0
b contains one up, one strange, and one bottom quark, while a Ξ0
bb contains one up and two bottom quarks.

In 2012, the CMS experiment at the Large Hadron Collider detected a Ξ∗0
b baryon (reported mass 5945±2.8 MeV/c^2).[4][5] LHCb discovered two new Xi baryons in 2014: Ξ′−
b and Ξ∗−
b [6]

59.1 Xi baryons

[1] Particle (or quantity, i.e. spin) has neither been observed nor indicated

[2] Some controversy exists about this data. See references

[3] This is actually a measurement of the average lifetime of b-baryons that decay to a jet containing a same sign $\Xi^{\pm}l^{\pm}$ pair. Presumably the mix is mainly Ξ
b, with some Λ
b.

59.2 See also

- Delta baryon
- Hyperon
- Lambda baryon
- List of baryons
- List of mesons
- List of particles
- Nucleon
- Omega baryon
- Physics portal
- Sigma baryon
- Timeline of particle discoveries

59.3 References

[1] R. Nave. "The Xi Baryon". *HyperPhysics*. Retrieved 4 April 2013.

[2] R.Armenteros et al. (1952). "The properties of charged V-particles".*Philosophical Magazine***43**(341): 597. doi:10.1080/147816.

[3] L. W. Alvarez et al. (1959). "Neutral Cascade Hyperon Event". *Physical Review Letters* **2** (5): 215. Bibcode:1959PhRvL...2..215A. doi:10.1103/PhysRevLett.2.215.

[4] C. Simpson (28 April 2013). "New Particle Discovered with 'Higgs Boson' Machine". *The Atlantic Wire*. Retrieved 2013-04-04.

[5] S. Chatrchyan et al. (CMS Collaboration) (2012). "Observation of an excite excited Ξ_b baryon". *Physical Review Letters* **108** (25): 252002. arXiv:1204.5955. Bibcode:2012PhRvL.108y2002C. doi:10.1103/PhysRevLett.108.252002.

[6] "LHCb experiment observes two new baryon particles never seen before". 19 Nov 2014.

[7] V. Abazov et al. (D0 Collaboration) (2007). "Direct observation of the strange b baryon Ξ− b".*Physical Review Letters***99**(5): 52001. arXiv:0706.1690. Bibcode:2007PhRvL..99e2001A.doi:10.1103/PhysRevLett001.

[8] "Fermilab physicists discover "triple-scoop" baryon" (Press release). Fermilab. 13 June 2007. Retrieved 2007-06-14.

[9] "Back-to-Back b Baryons in Batavia" (Press release). Fermilab. 25 June 2007. Retrieved 2007-06-25.

[10] T. Aaltonen et al. (CDF Collaboration) (2007). "Observation and mass measurement of the baryon Xi(b)-". *Physical Review Letters* **99** (5): 52002. arXiv:0707.0589. Bibcode:2007PhRvL..99e2002A. doi:10.1103/PhysRevLett.99.052002.

[11] W.-M. Yao et al. (Particle Data Group) (2006). "Particle listings – Xi^0" (PDF). *Journal of Physics G* **33** (1): 1. arXiv:astro-ph/0601168. Bibcode:2006JPhG...33....1Y. doi:10.1088/0954-3899/33/1/001.

[12] W.-M. Yao et al. (Particle Data Group) (2006). "Particle listings – Xi^-" (PDF). *Journal of Physics G* **33** (1): 1. arXiv:astro-ph/0601168. Bibcode:2006JPhG...33....1Y. doi:10.1088/0954-3899/33/1/001.

[13] W.-M. Yao et al. (Particle Data Group) (2006). "Particle Listings – Xi(1530)" (PDF). *Journal of Physics G* **33** (1): 1. arXiv:astro-ph/0601168. Bibcode:2006JPhG...33....1Y. doi:10.1088/0954-3899/33/1/001.

[14] W.-M. Yao et al. (Particle Data Group) (2006). "Particle listings – Charmed baryons" (PDF). *Journal of Physics G* **33** (1): 1. arXiv:astro-ph/0601168. Bibcode:2006JPhG...33....1Y. doi:10.1088/0954-3899/33/1/001.

[15] W.-M. Yao et al. (Particle Data Group) (2006). "Particle listings – Ξ+ cc"(PDF).*Journal of Physics G***33**(1): 1. arXiv:astro-ph/0601168. Bibcode:2006JPhG...33....1Y.doi:10.1088/0954-1.

[16] W.-M. Yao et al. (Particle Data Group) (2006). "Particle listings – Xi_b" (PDF). *Journal of Physics G* **33** (1): 1. arXiv:astro-ph/0601168. Bibcode:2006JPhG...33....1Y. doi:10.1088/0954-3899/33/1/001.

[17] T. Aaltonen et al. (CDF Collaboration) (2007). "Observation and mass measurement of the baryon Ξ– b".*Physical Review Letters***99**(5): 052002. arXiv:0707.0589. Bibcode:2007PhRvL..99e2002A.doi:10.1103/PhysR.

[18] "Observation of two new Ξ–b baryon resonances". *submitted to PRL*. 18 Nov 2014.

59.4 External links

- "Listings for Xi particles". Particle Data Group. Retrieved 2007-07-31.
- "Direct observation of the strange b baryon Xi(b)-"
- "Fermilab physicists discover "triple-scoop" baryon" (Press release). Fermilab. 2007-06-18. Retrieved 2007-06-18.
- "Back-to-Back b Baryons in Batavia" (Press release). Fermilab. 2007-06-25. Retrieved 2007-06-25.
- "Observation and mass measurement of the baryon Xi(b)-"

Chapter 60

Omega baryon

"Omega particle" redirects here. For the *Star Trek* episode, see The Omega Directive.

The **Omega baryons** are a family of subatomic hadron particles that are represented by the symbol Ω and are either neutral or have a +2, +1 or −1 elementary charge. They are baryons containing no up or down quarks.[1] Omega baryons containing top quarks are not expected to be observed. This is because the Standard Model predicts the mean lifetime of top quarks to be roughly 5×10^{-25} s,[2] which is about a twentieth of the timescale for strong interactions, and therefore that they do not form hadrons.

The first Omega baryon discovered was the Ω−, made of three strange quarks, in 1964.[3] The discovery was a great triumph in the study of quark processes, since it was found only after its existence, mass, and decay products had been predicted in 1962 by the American physicist Murray Gell-Mann and, independently, by the Israeli physicist Yuval Ne'eman. Besides the Ω−, a charmed Omega particle (Ω0
c) was discovered, in which a strange quark is replaced by a charm quark. The Ω− decays only via the weak interaction and has therefore a relatively long lifetime.[4] Spin (*J*) and parity (*P*) values for unobserved baryons are predicted by the quark model.[5]

Since Omega baryons do not have any up or down quarks, they all have isospin 0.

60.1 Omega baryons

† Particle (or quantity, i.e. spin) has neither been observed nor indicated.

60.2 Recent discoveries

The Ω−
b particle is a "doubly strange" baryon containing two strange quarks and a bottom quark. A discovery of this particle was first claimed in September 2008 by physicists working on the DØ experiment at the Tevatron facility of the Fermi National Accelerator Laboratory.[9][10] However, the reported mass, 6165±16 MeV/c^2, was significantly higher than expected in the quark model. The apparent discrepancy from the Standard Model has since been dubbed the "Ω
b puzzle". In May 2009, the CDF collaboration made public their results on the search for the Ω−
b based on analysis of a data sample roughly four times the size of the one used by the DØ experiment.[8] CDF measured the mass to be 6054.4±6.8 MeV/c^2, which was in excellent agreement with the Standard Model prediction. No signal has been observed at the DØ reported value. The two results differ by 111±18 MeV/c^2, which is equivalent to 6.2 standard deviations and are therefore inconsistent. Excellent agreement between the CDF measured mass and theoretical expectations is a strong indication that the particle discovered by CDF is indeed the Ω−
b.

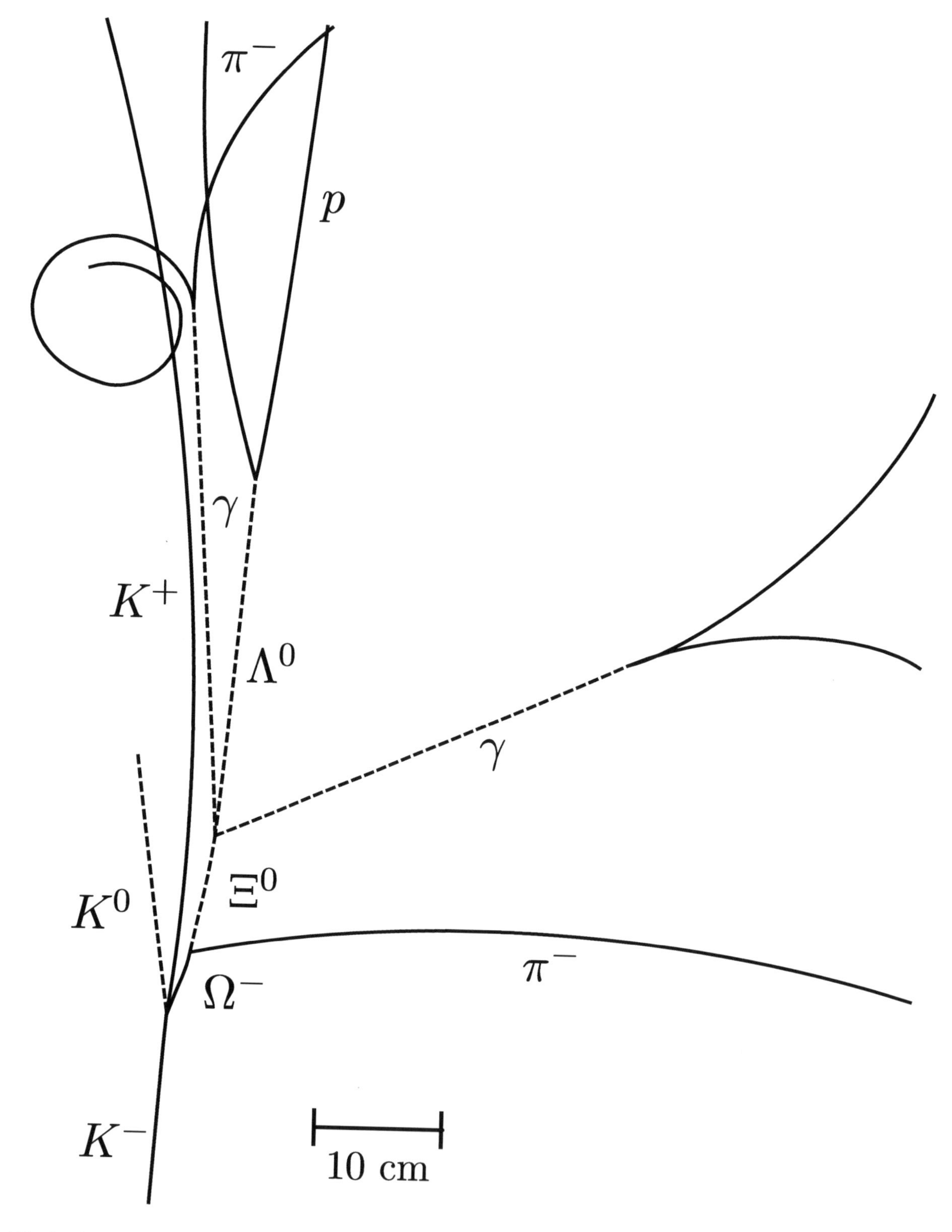

Bubble chamber trace of the first observed Ω baryon event at Brookhaven National Laboratory

60.3 See also

- Delta baryon

- Hyperon
- Lambda baryon
- List of mesons
- List of particles
- Nucleon
- Physics portal
- Sigma baryon
- Timeline of particle discoveries
- Xi baryon

60.4 References

[1] Particle Data Group. "2010 Review of Particle Physics – Naming scheme for hadrons" (PDF). Retrieved 2011-12-26.

[2] A.Quadt(2006). "Top quark physics at hadron colliders".*European Physical Journal* C**48**(3): 835–1000. Bibcode:2006EPJC. doi:10.1140/epjc/s2006-02631-6.

[3] V. E. Barnes et al. (1964). "Observation of a Hyperon with Strangeness Minus Three" (PDF). *Physical Review Letters* **12** (8): 204. Bibcode:1964PhRvL..12..204B. doi:10.1103/PhysRevLett.12.204.

[4] R. Nave. "The Omega baryon". *HyperPhysics*. Retrieved 2009-11-26.

[5] J. G. Körner, M. Krämer, and D. Pirjol (1994). "Heavy Baryons". *Progress in Particle and Nuclear Physics* **33**: 787–868. arXiv:hep-ph/9406359. Bibcode:1994PrPNP..33..787K. doi:10.1016/0146-6410(94)90053-1.

[6] Particle Data Group. "2006 Review of Particle Physics – Ω–" (PDF). Retrieved 2008-04-20.

[7] Particle Data Group. "2006 Review of Particle Physics – Ω0
c" (PDF). Retrieved 2008-04-20.

[8] T. Aaltonen *et al.* (CDF Collaboration) (2009). "Observation of the Ω–
b and Measurement of the Properties of the Ξ–
b and Ω–
b". *Physical Review D* **80** (7). arXiv:0905.3123. Bibcode:2009PhRvD..80g2003A. doi:10.1103/PhysRevD.80.072003.

[9] "Fermilab physicists discover "doubly strange" particle". Fermilab. 3 September 2008. Retrieved 2008-09-04.

[10] V. Abazov *et al.* (DØ Collaboration) (2008). "Observation of the doubly strange b baryon Ω–
b".*Physical Review Letters***101**(23): 232002. arXiv:0808.4142. Bibcode:2008PhRvL.101w2002A.doi:10.1103/Phy.

60.5 External links

- Picture of the first event containing the Ω–, which happens to contain the complete decay chain of the Ω–.
- Science Daily – Discovery of the Ω–
  b
- Strangeness Minus Three - BBC Horizon 1964

Chapter 61

Meson

In particle physics, **mesons** (/ˈmiːzɒnz/ or /ˈmɛzɒnz/) are hadronic subatomic particles composed of one quark and one antiquark, bound together by the strong interaction. Because mesons are composed of sub-particles, they have a physical size, with a diameter of roughly one fermi, which is about ⅔ the size of a proton or neutron. All mesons are unstable, with the longest-lived lasting for only a few hundredths of a microsecond. Charged mesons decay (sometimes through intermediate particles) to form electrons and neutrinos. Uncharged mesons may decay to photons.

Mesons are not produced by radioactive decay, but appear in nature only as short-lived products of very high-energy interactions in matter, between particles made of quarks. In cosmic ray interactions, for example, such particles are ordinary protons and neutrons. Mesons are also frequently produced artificially in high-energy particle accelerators that collide protons, anti-protons, or other particles.

In nature, the importance of lighter mesons is that they are the associated quantum-field particles that transmit the nuclear force, in the same way that photons are the particles that transmit the electromagnetic force. The higher energy (more massive) mesons were created momentarily in the Big Bang, but are not thought to play a role in nature today. However, such particles are regularly created in experiments, in order to understand the nature of the heavier types of quark that compose the heavier mesons.

Mesons are part of the hadron particle family, defined simply as particles composed of two quarks. The other members of the hadron family are the baryons: subatomic particles composed of three quarks rather than two. Some experiments show evidence of exotic mesons, which don't have the conventional valence quark content of one quark and one antiquark.

Because quarks have a spin of ½, the difference in quark-number between mesons and baryons results in conventional two-quark mesons being bosons, whereas baryons are fermions.

Each type of meson has a corresponding antiparticle (antimeson) in which quarks are replaced by their corresponding antiquarks and vice versa. For example, a positive pion ($\pi+$) is made of one up quark and one down antiquark; and its corresponding antiparticle, the negative pion ($\pi-$), is made of one up antiquark and one down quark.

Because mesons are composed of quarks, they participate in both the weak and strong interactions. Mesons with net electric charge also participate in the electromagnetic interaction. They are classified according to their quark content, total angular momentum, parity and various other properties, such as C-parity and G-parity. Although no meson is stable, those of lower mass are nonetheless more stable than the most massive mesons, and are easier to observe and study in particle accelerators or in cosmic ray experiments. They are also typically less massive than baryons, meaning that they are more easily produced in experiments, and thus exhibit certain higher energy phenomena more readily than baryons composed of the same quarks would. For example, the charm quark was first seen in the J/Psi meson (J/ψ) in 1974,[1][2] and the bottom quark in the upsilon meson (Υ) in 1977.[3]

61.1 History

From theoretical considerations, in 1934 Hideki Yukawa[4][5] predicted the existence and the approximate mass of the "meson" as the carrier of the nuclear force that holds atomic nuclei together. If there were no nuclear force, all nuclei with two or more protons would fly apart because of the electromagnetic repulsion. Yukawa called his carrier particle the meson, from μέσος *mesos*, the Greek word for "intermediate," because its predicted mass was between that of the electron and that of the proton, which has about 1,836 times the mass of the electron. Yukawa had originally named his particle the "mesotron", but he was corrected by the physicist Werner Heisenberg (whose father was a professor of Greek at the University of Munich). Heisenberg pointed out that there is no "tr" in the Greek word "mesos".[6]

The first candidate for Yukawa's meson, now known in modern terminology as the muon, was discovered in 1936 by Carl David Anderson and others in the decay products of cosmic ray interactions. The mu meson had about the right mass to be Yukawa's carrier of the strong nuclear force, but over the course of the next decade, it became evident that it was not the right particle. It was eventually found that the "mu meson" did not participate in the strong nuclear interaction at all, but rather behaved like a heavy version of the electron, and was eventually classed as a lepton like the electron, rather than a meson. Physicists in making this choice decided that properties other than particle mass should control their classification.

There were years of delays in the subatomic particle research during World War II in 1939–45, with most physicists working in applied projects for wartime necessities. When the war ended in August 1945, many physicists gradually returned to peacetime research. The first true meson to be discovered was what would later be called the "pi meson" (or pion). This discovery was made in 1947, by Cecil Powell, César Lattes, and Giuseppe Occhialini, who were investigating cosmic ray products at the University of Bristol in England, based on photographic films placed in the Andes mountains. Some mesons in these films had about the same mass as the already-known meson, yet seemed to decay into it, leading physicist Robert Marshak to hypothesize in 1947 that it was actually a new and different meson. Over the next few years, more experiments showed that the pion was indeed involved in strong interactions. The pion (as a virtual particle) is the primary force carrier for the nuclear force in atomic nuclei. Other mesons, such as the rho mesons are involved in mediating this force as well, but to lesser extents. Following the discovery of the pion, Yukawa was awarded the 1949 Nobel Prize in Physics for his predictions.

The word *meson* has at times been used to mean *any* force carrier, such as the "Z^0 meson", which is involved in mediating the weak interaction.[7] However, this spurious usage has fallen out of favor. Mesons are now defined as particles composed of pairs of quarks and antiquarks.

61.2 Overview

61.2.1 Spin, orbital angular momentum, and total angular momentum

Main articles: Spin (physics), angular momentum operator, Total angular momentum and Quantum numbers

Spin (quantum number S) is a vector quantity that represents the "intrinsic" angular momentum of a particle. It comes in increments of $\frac{1}{2}$ ħ. The ħ is often dropped because it is the "fundamental" unit of spin, and it is implied that "spin 1" means "spin 1 ħ". (In some systems of natural units, ħ is chosen to be 1, and therefore does not appear in equations).

Quarks are fermions—specifically in this case, particles having spin $\frac{1}{2}$ ($S = \frac{1}{2}$). Because spin projections vary in increments of 1 (that is 1 ħ), a single quark has a spin vector of length $\frac{1}{2}$, and has two spin projections ($S_z = +\frac{1}{2}$ and $S_z = -\frac{1}{2}$). Two quarks can have their spins aligned, in which case the two spin vectors add to make a vector of length $S = 1$ and three spin projections ($S_z = +1$, $S_z = 0$, and $S_z = -1$), called the spin-1 triplet. If two quarks have unaligned spins, the spin vectors add up to make a vector of length S = 0 and only one spin projection ($S_z = 0$), called the spin-0 singlet. Because mesons are made of one quark and one antiquark, they can be found in triplet and singlet spin states.

There is another quantity of quantized angular momentum, called the orbital angular momentum (quantum number L), that comes in increments of 1 ħ, which represent the angular momentum due to quarks orbiting around each other. The total angular momentum (quantum number J) of a particle is therefore the combination of intrinsic angular momentum (spin) and orbital angular momentum. It can take any value from $J = |L - S|$ to $J = |L + S|$, in increments of 1.

Particle physicists are most interested in mesons with no orbital angular momentum ($L = 0$), therefore the two groups of mesons most studied are the $S = 1$; $L = 0$ and $S = 0$; $L = 0$, which corresponds to $J = 1$ and $J = 0$, although they are not the only ones. It is also possible to obtain $J = 1$ particles from $S = 0$ and $L = 1$. How to distinguish between the $S = 1$, $L = 0$ and $S = 0$, $L = 1$ mesons is an active area of research in meson spectroscopy.

61.2.2 Parity

Main article: Parity (physics)

If the universe were reflected in a mirror, most of the laws of physics would be identical—things would behave the same way regardless of what we call "left" and what we call "right". This concept of mirror reflection is called parity (*P*). Gravity, the electromagnetic force, and the strong interaction all behave in the same way regardless of whether or not the universe is reflected in a mirror, and thus are said to conserve parity (P-symmetry). However, the weak interaction does distinguish "left" from "right", a phenomenon called parity violation (P-violation).

Based on this, one might think that, if the wavefunction for each particle (more precisely, the quantum field for each particle type) were simultaneously mirror-reversed, then the new set of wavefunctions would perfectly satisfy the laws of physics (apart from the weak interaction). It turns out that this is not quite true: In order for the equations to be satisfied, the wavefunctions of certain types of particles have to be multiplied by −1, in addition to being mirror-reversed. Such particle types are said to have *negative* or *odd* parity ($P = -1$, or alternatively $P = -$), whereas the other particles are said to have *positive* or *even* parity ($P = +1$, or alternatively $P = +$).

For mesons, the parity is related to the orbital angular momentum by the relation:[8]

$$P = (-1)^{L+1}$$

where the L is a result of the parity of the corresponding spherical harmonic of the wavefunction. The '+1' in the exponent comes from the fact that, according to the Dirac equation, a quark and an antiquark have opposite intrinsic parities. Therefore, the intrinsic parity of a meson is the product of the intrinsic parities of the quark (+1) and antiquark (−1). As these are different, their product is −1, and so it contributes a +1 in the exponent.

As a consequence, mesons with no orbital angular momentum ($L = 0$) all have odd parity ($P = -1$).

61.2.3 C-parity

Main article: C-parity

C-parity is only defined for mesons that are their own antiparticle (i.e. neutral mesons). It represents whether or not the wavefunction of the meson remains the same under the interchange of their quark with their antiquark.[9] If

$$|q\bar{q}\rangle = |\bar{q}q\rangle$$

then, the meson is "C even" (C = +1). On the other hand, if

$$|q\bar{q}\rangle = -|\bar{q}q\rangle$$

then the meson is "C odd" (C = −1).

C-parity rarely is studied on its own, but more commonly in combination with P-parity into CP-parity. CP-parity was thought to be conserved, but was later found to be violated in weak interactions.[10][11][12]

61.2.4 G-parity

Main article: G-parity

G parity is a generalization of the C-parity. Instead of simply comparing the wavefunction after exchanging quarks and antiquarks, it compares the wavefunction after exchanging the meson for the corresponding antimeson, regardless of quark content.[13] In the case of neutral meson, G-parity is equivalent to C-parity because neutral mesons are their own antiparticles.

If

$$|q_1\bar{q}_2\rangle = |\bar{q}_1 q_2\rangle$$

then, the meson is “G even” (G = +1). On the other hand, if

$$|q_1\bar{q}_2\rangle = -|\bar{q}_1 q_2\rangle$$

then the meson is “G odd” (G = −1).

61.2.5 Isospin and charge

Main article: Isospin

The concept of isospin was first proposed by Werner Heisenberg in 1932 to explain the similarities between protons and neutrons under the strong interaction.[14] Although they had different electric charges, their masses were so similar that physicists believed that they were actually the same particle. The different electric charges were explained as being the result of some unknown excitation similar to spin. This unknown excitation was later dubbed *isospin* by Eugene Wigner in 1937.[15] When the first mesons were discovered, they too were seen through the eyes of isospin and so the three pions were believed to be the same particle, but in different isospin states.

This belief lasted until Murray Gell-Mann proposed the quark model in 1964 (containing originally only the u, d, and s quarks).[16] The success of the isospin model is now understood to be the result of the similar masses of the u and d quarks. Because the u and d quarks have similar masses, particles made of the same number of them also have similar masses. The exact specific u and d quark composition determines the charge, because u quarks carry charge $+\frac{2}{3}$ whereas d quarks carry charge $-\frac{1}{3}$. For example the three pions all have different charges (π+ (ud), π0 (a quantum superposition of uu and dd states), π− (du)), but have similar masses (~140 MeV/c^2) as they are each made of a same number of total of up and down quarks and antiquarks. Under the isospin model, they were considered to be a single particle in different charged states.

The mathematics of isospin was modeled after that of spin. Isospin projections varied in increments of 1 just like those of spin, and to each projection was associated a "charged state". Because the “pion particle” had three “charged states”, it was said to be of isospin $I = 1$. Its “charged states” π+, π0, and π−, corresponded to the isospin projections $I_3 = +1$, $I_3 = 0$, and $I_3 = -1$ respectively. Another example is the "rho particle", also with three charged states. Its “charged states” ρ+, ρ0, and ρ−, corresponded to the isospin projections $I_3 = +1$, $I_3 = 0$, and $I_3 = -1$ respectively. It was later noted that the isospin projections were related to the up and down quark content of particles by the relation

$$I_3 = \frac{1}{2}[(n_u - n_{\bar{u}}) - (n_d - n_{\bar{d}})],$$

where the *n*'s are the number of up and down quarks and antiquarks.

In the “isospin picture”, the three pions and three rhos were thought to be the different states of two particles. However, in the quark model, the rhos are excited states of pions. Isospin, although conveying an inaccurate picture of things, is still used to classify hadrons, leading to unnatural and often confusing nomenclature. Because mesons are hadrons, the isospin classification is also used, with $I_3 = +\frac{1}{2}$ for up quarks and down antiquarks, and $I_3 = -\frac{1}{2}$ for up antiquarks and down quarks.

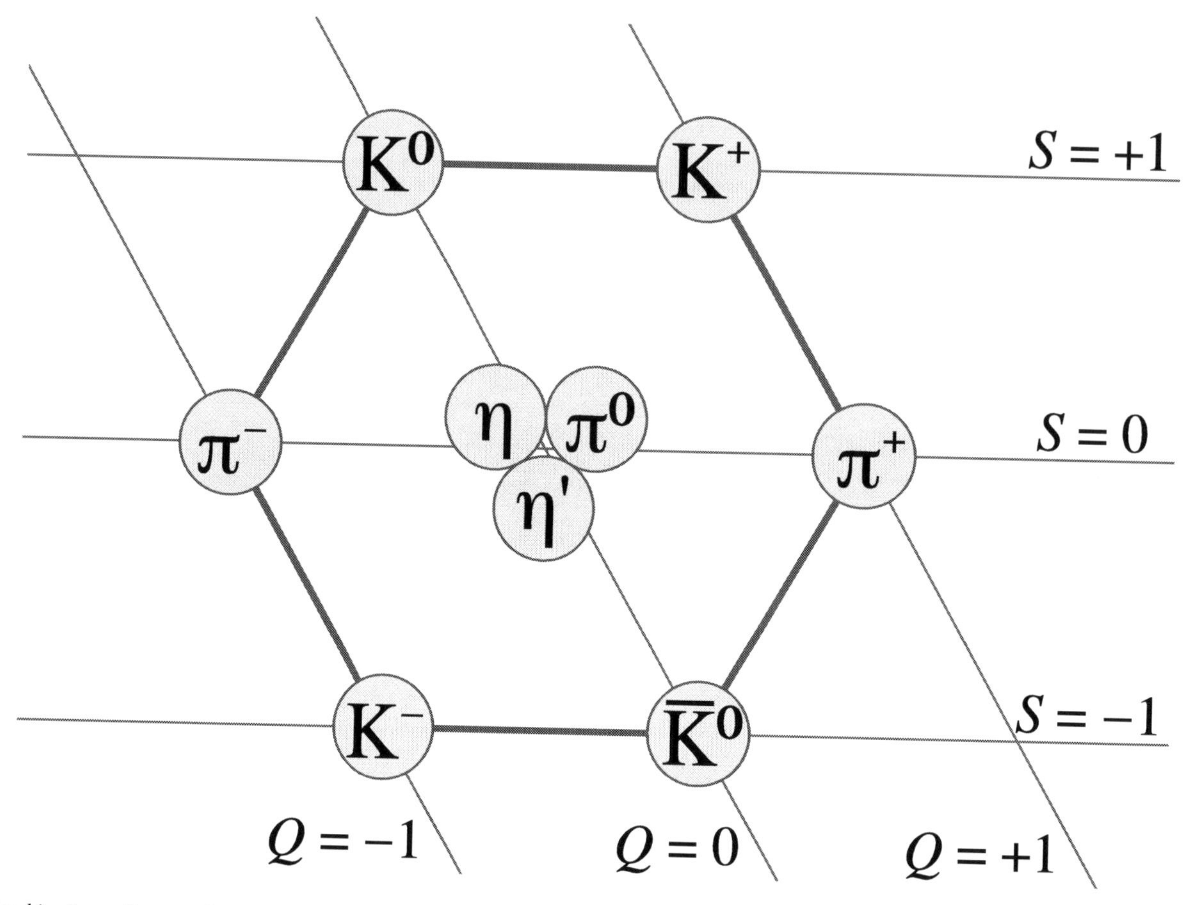

Combinations of one u, d or s quarks and one u, d, or s antiquark in $J^P = 0^-$ configuration form a nonet.

61.2.6 Flavour quantum numbers

Main article: Flavour (particle physics) § Flavour quantum numbers

The strangeness quantum number S (not to be confused with spin) was noticed to go up and down along with particle mass. The higher the mass, the lower the strangeness (the more s quarks). Particles could be described with isospin projections (related to charge) and strangeness (mass) (see the uds nonet figures). As other quarks were discovered, new quantum numbers were made to have similar description of udc and udb nonets. Because only the u and d mass are similar, this description of particle mass and charge in terms of isospin and flavour quantum numbers only works well for the nonets made of one u, one d and one other quark and breaks down for the other nonets (for example ucb nonet). If the quarks all had the same mass, their behaviour would be called *symmetric*, because they would all behave in exactly the same way with respect to the strong interaction. However, as quarks do not have the same mass, they do not interact in the same way (exactly like an electron placed in an electric field will accelerate more than a proton placed in the same field because of its lighter mass), and the symmetry is said to be broken.

It was noted that charge (Q) was related to the isospin projection (I_3), the baryon number (B) and flavour quantum numbers (S, C, B', T) by the Gell-Mann–Nishijima formula:[17]

$$Q = I_3 + \frac{1}{2}(B + S + C + B' + T),$$

where S, C, B', and T represent the strangeness, charm, bottomness and topness flavour quantum numbers respectively. They are related to the number of strange, charm, bottom, and top quarks and antiquark according to the relations:

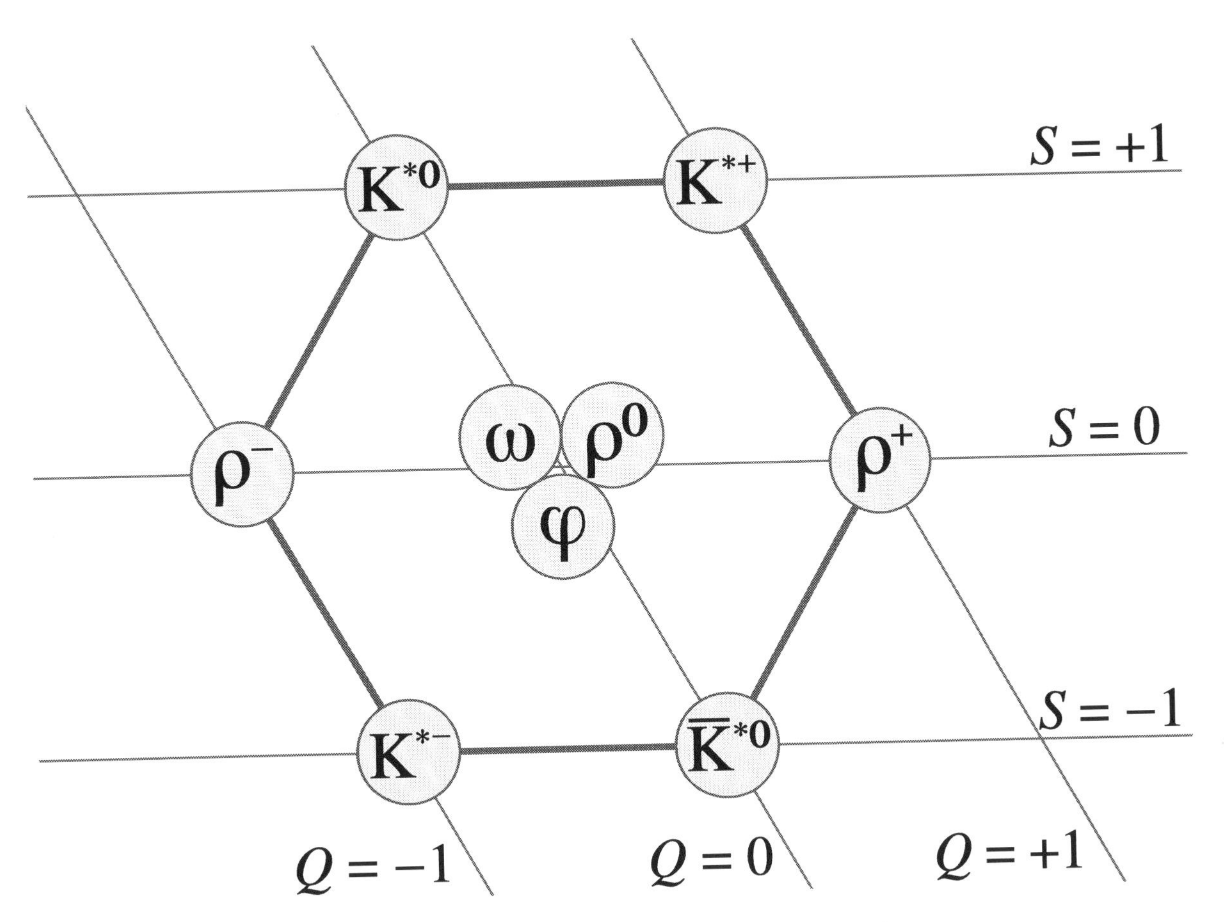

Combinations of one u, d or s quarks and one u, d, or s antiquark in $J^P = 1^-$ configuration also form a nonet.

$S = -(n_s - n_{\bar{s}})$

$C = +(n_c - n_{\bar{c}})$

$B' = -(n_b - n_{\bar{b}})$

$T = +(n_t - n_{\bar{t}})$,

meaning that the Gell-Mann–Nishijima formula is equivalent to the expression of charge in terms of quark content:

$$Q = \frac{2}{3}[(n_u - n_{\bar{u}}) + (n_c - n_{\bar{c}}) + (n_t - n_{\bar{t}})] - \frac{1}{3}[(n_d - n_{\bar{d}}) + (n_s - n_{\bar{s}}) + (n_b - n_{\bar{b}})].$$

61.3 Classification

Mesons are classified into groups according to their isospin (*I*), total angular momentum (*J*), parity (*P*), G-parity (*G*) or C-parity (*C*) when applicable, and quark (q) content. The rules for classification are defined by the Particle Data Group, and are rather convoluted.[18] The rules are presented below, in table form for simplicity.

61.3.1 Types of meson

Mesons are classified into types according to their spin configurations. Some specific configurations are given special names based on the mathematical properties of their spin configuration.

61.3.2 Nomenclature

Flavourless mesons

Flavourless mesons are mesons made of pair of quark and antiquarks of the same flavour (all their flavour quantum numbers are zero: $S=0$, $C=0$, $B'=0$, $T=0$).[20] The rules for flavourless mesons are:[18]

† ^ The C parity is only relevant to neutral mesons.
†† ^ For $J^{PC}=1^{--}$, the ψ is called the J/ψ

In addition:

- When the spectroscopic state of the meson is known, it is added in parentheses.
- When the spectroscopic state is unknown, mass (in MeV/c^2) is added in parentheses.
- When the meson is in its ground state, nothing is added in parentheses.

Flavoured mesons

Flavoured mesons are mesons made of pair of quark and antiquarks of different flavours. The rules are simpler in this case: the main symbol depends on the heavier quark, the superscript depends on the charge, and the subscript (if any) depends on the lighter quark. In table form, they are:[18]

In addition:

- If J^P is in the "normal series" (i.e., $J^P = 0^+, 1^-, 2^+, 3^-, \ldots$), a superscript $*$ is added.
- If the meson is not pseudoscalar ($J^P = 0^-$) or vector ($J^P = 1^-$), J is added as a subscript.
- When the spectroscopic state of the meson is known, it is added in parentheses.
- When the spectroscopic state is unknown, mass (in MeV/c^2) is added in parentheses.
- When the meson is in its ground state, nothing is added in parentheses.

61.4 Exotic mesons

Main article: Exotic meson

There is experimental evidence for particles that are hadrons (i.e., are composed of quarks) and are color-neutral with zero baryon number, and thus by conventional definition are mesons. Yet, these particles do not consist of a single quark-antiquark pair, as all the other conventional mesons discussed above do. A tentative category for these particles is exotic mesons.

There are at least five exotic meson resonances that have been experimentally confirmed to exist by two or more independent experiments. The most statistically significant of these is the Z(4430), discovered by the Belle experiment in 2007 and confirmed by LHCb in 2014. It is a candidate for being a tetraquark: a particle composed of two quarks and two antiquarks.[21] See the main article above for other particle resonances that are candidates for being exotic mesons.

61.5 List

Main article: List of mesons

61.6 See also

- Standard Model

61.7 Notes

[1] J.J. Aubert *et al.* (1974)

[2] J.E. Augustin *et al.* (1974)

[3] S.W. Herb *et al.* (1977)

[4] The Noble Foundation (1949) Nobel Prize in Physics 1949 – Presentation Speech

[5] H. Yukawa (1935)

[6] G. Gamow (1961)

[7] J. Steinberger (1998)

[8] C. Amsler *et al.* (2008): Quark Model

[9] M.S. Sozzi (2008b)

[10] J.W. Cronin (1980)

[11] V.L. Fitch (1980)

[12] M.S. Sozzi (2008c)

[13] K. Gottfried, V.F. Weisskopf (1986)

[14] W. Heisenberg (1932)

[15] E. Wigner (1937)

[16] M. Gell-Mann (1964)

[17] S.S.M Wong (1998)

[18] C. Amsler *et al.* (2008): Naming scheme for hadrons

[19] W.E. Burcham, M. Jobes (1995)

[20] For the purpose of nomenclature, the isospin projection I_3 isn't considered a flavour quantum number. This means that the charged pion-like mesons ($\pi^{\pm}$, $a^{\pm}$, $b^{\pm}$, and $\rho^{\pm}$ mesons) follow the rules of flavourless mesons, even if they aren't truly "flavourless".

[21] LHCb collaborators (2014): Observation of the resonant character of the Z(4430)– state

61.8 References

- M.S. Sozzi (2008a). “Parity”. *Discrete Symmetries and CP Violation: From Experiment to Theory*. Oxford University Press. pp. 15–87. ISBN 0-19-929666-9.
- M.S. Sozzi (2008b). “Charge Conjugation”. *Discrete Symmetries and CP Violation: From Experiment to Theory*. Oxford University Press. pp. 88–120. ISBN 0-19-929666-9.
- M.S. Sozzi (2008c). “CP-Symmetry”. *Discrete Symmetries and CP Violation: From Experiment to Theory*. Oxford University Press. pp. 231–275. ISBN 0-19-929666-9.
- C. Amsler *et al.* (Particle Data Group) (2008). “Review of Particle Physics”. *Physics Letters B* **667** (1): 1–1340. Bibcode:2008PhLB..667....1P. doi:10.1016/j.physletb.2008.07.018.
- S.S.M. Wong (1998). “Nucleon Structure”. *Introductory Nuclear Physics* (2nd ed.). New York (NY): John Wiley & Sons. pp. 21–56. ISBN 0-471-23973-9.
- W.E. Burcham, M. Jobes (1995). *Nuclear and Particle Physics* (2nd ed.). Longman Publishing. ISBN 0-582-45088-8.
- R. Shankar (1994). *Principles of Quantum Mechanics* (2nd ed.). New York (NY): Plenum Press. ISBN 0-306-44790-8.
- J. Steinberger (1989). “Experiments with high-energy neutrino beams”. *Reviews of Modern Physics* **61** (3): 533–545. Bibcode:1989RvMP...61..533S. doi:10.1103/RevModPhys.61.533.
- K. Gottfried, V.F. Weisskopf (1986). “Hadronic Spectroscopy: G-parity”. *Concepts of Particle Physics* **2**. Oxford University Press. pp. 303–311. ISBN 0-19-503393-0.
- J.W. Cronin (1980). “CP Symmetry Violation—The Search for its origin” (PDF). The Nobel Foundation.
- V.L. Fitch (1980). “The Discovery of Charge—Conjugation Parity Asymmetry” (PDF). The Nobel Foundation.
- S.W. Herb; Hom, D.; Lederman, L.; Sens, J.; Snyder, H.; Yoh, J.; Appel, J.; Brown, B. et al. (1977). “Observation of a Dimuon Resonance at 9.5 Gev in 400-GeV Proton-Nucleus Collisions”. *Physical Review Letters* **39** (5): 252–255. Bibcode:1977PhRvL..39..252H. doi:10.1103/PhysRevLett.39.252.
- J.J. Aubert; Becker, U.; Biggs, P.; Burger, J.; Chen, M.; Everhart, G.; Goldhagen, P.; Leong, J. et al. (1974). “Experimental Observation of a Heavy Particle *J*". *Physical Review Letters* **33** (23): 1404–1406. Bibcode:1974 .doi:10.1103/PhysRevLett.33.1404.
- J.E. Augustin; Boyarski, A.; Breidenbach, M.; Bulos, F.; Dakin, J.; Feldman, G.; Fischer, G.; Fryberger, D. et al. (1974). “Discovery of a Narrow Resonance in e^+e^- Annihilation”. *Physical Review Letters* **33** (23): 1406–1408. Bibcode:1974PhRvL..33.1406A. doi:10.1103/PhysRevLett.33.1406.
- M.Gell-Mann(1964). “A Schematic of Baryons and Mesons”.*Physics Letters***8**(3): 214–215. Bibcode:1.....8..214G. doi:10.1016/S0031-9163(64)92001-3.
- Ishfaq Ahmad (1965). “the Interactions of 200 MeV $\pi\pm$ -Mesons with Complex Nuclei Proposal to Study the Interactions of 200 MeV $\pi\pm$ -Mesons with Complex Nuclei” (PDF). *CERN documents* **3** (5).
- G. Gamow (1988) [1961]. *The Great Physicists from Galileo to Einstein* (Reprint ed.). Dover Publications. p. 315. ISBN 978-0-486-25767-9.
- E. Wigner (1937). “On the Consequences of the Symmetry of the Nuclear Hamiltonian on the Spectroscopy of Nuclei”. *Physical Review* **51** (2): 106–119. Bibcode:1937PhRv...51..106W. doi:10.1103/PhysRev.51.106.
- H. Yukawa (1935). “On the Interaction of Elementary Particles” (PDF). *Proc. Phys. Math. Soc. Jap.* **17** (48).
- W.Heisenberg(1932). "Über den Bau der Atomkerne I".*Zeitschrift für Physik*(in German)**77**: 1–11. Bibcode....1H. doi:10.1007/BF01342433.

- W. Heisenberg (1932). "Über den Bau der Atomkerne II". *Zeitschrift für Physik* (in German) **78** (3–4): 156–164. Bibcode:1932ZPhy...78..156H. doi:10.1007/BF01337585.
- W. Heisenberg (1932). "Über den Bau der Atomkerne III". *Zeitschrift für Physik* (in German) **80** (9–10): 587–596. Bibcode:1933ZPhy...80..587H. doi:10.1007/BF01335696.

61.9 External links

- A table of some mesons and their properties
- *Particle Data Group*—Compiles authoritative information on particle properties
- hep-ph/0211411: The light scalar mesons within quark models
- Naming scheme for hadrons (a PDF file)
- Mesons made thinkable, an interactive visualisation allowing physical properties to be compared

61.9.1 Recent findings

- What Happened to the Antimatter? Fermilab's DZero Experiment Finds Clues in Quick-Change Meson
- CDF experiment's definitive observation of matter-antimatter oscillations in the Bs meson

Chapter 62

Quarkonium

In particle physics, **quarkonium** (from quark + onium, pl. **quarkonia**) designates a flavorless meson whose constituents are a quark and its own antiquark. Examples of quarkonia are the J/ψ meson (an example of **charmonium**, cc) and the ϒ meson (**bottomonium**, bb). Because of the high mass of the top quark, **toponium** does not exist, since the top quark decays through the electroweak interaction before a bound state can form. Usually quarkonium refers only to charmonium and bottomonium, and not to any of the lighter quark–antiquark states. This usage is because the lighter quarks (up, down, and strange) are much less massive than the heavier quarks, and so the physical states actually seen in experiments (η, η′, and π^0 mesons) are quantum mechanical mixtures of the light quark states. The much larger mass differences between the charm and bottom quarks and the lighter quarks results in states that are well defined in terms of a quark–antiquark pair of a given flavor.

62.1 Charmonium states

See also: J/ψ meson

In the following table, the same particle can be named with the spectroscopic notation or with its mass. In some cases excitation series are used: Ψ' is the first excitation of Ψ (for historical reasons, this one is called *J/ψ* particle); Ψ" is a second excitation, and so on. That is, names in the same cell are synonymous.

Some of the states are predicted, but have not been identified; others are unconfirmed. The quantum numbers of the X(3872) particle have been measured recently by the LHCb experiment at CERN[1] . This measurement shed some light on its identity, excluding the third option among the three envised, which are :

- a candidate for the 1^1D_2 state;
- a charmonium hybrid state;
- a $D^0\bar{D}^{*0}$ molecule.

In 2005, the BaBar experiment announced the discovery of a new state: Y(4260).[2][3] CLEO and Belle have since corroborated these observations. At first, Y(4260) was thought to be a charmonium state, but the evidence suggests more exotic explanations, such as a D “molecule”, a 4-quark construct, or a hybrid meson.

Notes:

* Needs confirmation.

† Predicted, but not yet identified.

† Interpretation as a 1^{--} charmonium state not favored.

62.2 Bottomonium states

See also: Upsilon meson

In the following table, the same particle can be named with the spectroscopic notation or with its mass.

Some of the states are predicted, but have not been identified; others are unconfirmed.

Notes:

* Preliminary results. Confirmation needed.

The χ_b (3P) state was the first particle discovered in the Large Hadron Collider. The article about this discovery was first submitted to arXiv on 21 December 2011.[4][5] On April 2012, Tevatron's DØ experiment confirms the result in a paper published in *Phys. Rev. D*.[6][7]

62.3 QCD and quarkonia

The computation of the properties of mesons in Quantum chromodynamics (QCD) is a fully non-perturbative one. As a result, the only general method available is a direct computation using lattice QCD (LQCD) techniques. However, other techniques are effective for heavy quarkonia as well.

The light quarks in a meson move at relativistic speeds, since the mass of the bound state is much larger than the mass of the quark. However, the speed of the charm and the bottom quarks in their respective quarkonia is sufficiently smaller, so that relativistic effects affect these states much less. It is estimated that the speed, **v**, is roughly 0.3 times the speed of light for charmonia and roughly 0.1 times the speed of light for bottomonia. The computation can then be approximated by an expansion in powers of **v/c** and $\mathbf{v}^2/\mathbf{c}^2$. This technique is called non-relativistic QCD (NRQCD).

NRQCD has also been quantized as a lattice gauge theory, which provides another technique for LQCD calculations to use. Good agreement with the bottomonium masses has been found, and this provides one of the best non-perturbative tests of LQCD. For charmonium masses the agreement is not as good, but the LQCD community is actively working on improving their techniques. Work is also being done on calculations of such properties as widths of quarkonia states and transition rates between the states.

An early, but still effective, technique uses models of the *effective* potential to calculate masses of quarkonia states. In this technique, one uses the fact that the motion of the quarks that comprise the quarkonium state is non-relativistic to assume that they move in a static potential, much like non-relativistic models of the hydrogen atom. One of the most popular potential models is the so-called *Cornell potential*

$V(r) = -\frac{a}{r} + br$ [8]

where r is the effective radius of the quarkonium state, a and b are parameters. This potential has two parts. The first part, a/r corresponds to the potential induced by one-gluon exchange between the quark and its anti-quark, and is known as the *Coulombic* part of the potential, since its $1/r$ form is identical to the well-known Coulombic potential induced by the electromagnetic force. The second part, br , is known as the *confinement* part of the potential, and parameterizes the poorly understood non-perturbative effects of QCD. Generally, when using this approach, a convenient form for the wave function of the quarks is taken, and then a and b are determined by fitting the results of the calculations to the masses of well-measured quarkonium states. Relativistic and other effects can be incorporated into this approach by adding extra terms to the potential, much in the same way that they are for the hydrogen atom in non-relativistic quantum mechanics. This form has been derived from QCD up to $\mathcal{O}(\Lambda^3_{\text{QCD}}r^2)$ by Y. Sumino in 2003.[9] It is popular because it allows for accurate predictions of quarkonia parameters without a lengthy lattice computation, and provides a separation between the short-distance *Coulombic* effects and the long-distance *confinement* effects that can be useful in understanding the quark/anti-quark force generated by QCD.

Quarkonia have been suggested as a diagnostic tool of the formation of the quark–gluon plasma: both disappearance and enhancement of their formation depending on the yield of heavy quarks in plasma can occur.

62.4 See also

- Onium
- OZI Rule
- J/ψ meson
- Phi meson
- Upsilon meson
- Theta meson
- Non-relativistic QCD
- Lattice QCD
- Quantum chromodynamics

62.5 References

[1] LHCb collaboration; Aaij, R.; Abellan Beteta, C.; Adeva, B.; Adinolfi, M.; Adrover, C.; Affolder, A.; Ajaltouni, Z. et al. (February 2013). "Determination of the X(3872) meson quantum numbers". *Physical Review Letters* **1302** (22): 6269. arXiv:1302.6269. Bibcode:2013PhRvL.110v2001A. doi:10.1103/PhysRevLett.110.222001.

[2] "A new particle discovered by BaBar experiment". Istituto Nazionale di Fisica Nucleare. 6 July 2005. Retrieved 2010-03-06.

[3] B. Aubert *et al.* (BaBar Collaboration) (2005). "Observation of a broad structure in the $\pi^+\pi^-$J/ψ mass spectrum around 4.26

[4] ATLAS Collaboration (2012). "Observation of a new χ
b state in radiative transitions to ϒ(1S) and ϒ(2S) at ATLAS". arXiv:1112.5154v4 [hep-ex].

[5] Jonathan Amos (2011-12-22). "LHC reports discovery of its first new particle". BBC.

[6] *Tevatron experiment confirms LHC discovery of Chi-b (P3) particle*

[7] *Observation of a narrow mass state decaying into ϒ(1S) + γ in pp collisions at 1.96 TeV*

[8] Hee Sok Chung; Jungil Lee; Daekyoung Kang (2008). "Cornell Potential Parameters for S-wave Heavy Quarkonia". *Journal of the Korean Physical Society* **52** (4): 1151. arXiv:0803.3116. Bibcode:2008JKPS...52.1151C. doi:10.3938/jkps.52.1151.

[9] Y. Sumino (2003). "QCD potential as a "Coulomb-plus-linear" potential". *Phys. Lett. B* **571**: 173–183. arXiv:hep-ph/0303120. doi:10.1016/j.physletb.2003.05.010.

Chapter 63

Pion

In particle physics, a **pion** (or a **pi meson**, denoted with the Greek letter pi: π) is any of three subatomic particles: π0, π+, and π−. Each pion consists of a quark and an antiquark and is therefore a meson. Pions are the lightest mesons (and, more generally, the lightest hadrons), because they are composed of the lightest quarks (the u and d quarks). They are unstable, with the charged pions π+ and π− decaying with a mean lifetime of 26 nanoseconds (2.6×10^{-8} seconds), and the neutral pion π0 decaying with a much shorter lifetime of 8.4×10^{-17} seconds. Charged pions most often decay into muons and muon neutrinos, and neutral pions into gamma rays.

The exchange of virtual pions, along with the vector, rho and omega mesons, provides an explanation for the residual strong force between nucleons. Pions are not produced in radioactive decay, but are produced commonly in high energy accelerators in collisions between hadrons. All types of pions are also produced in natural processes when high energy cosmic ray protons and other hadronic cosmic ray components interact with matter in the Earth's atmosphere. Recently, detection of characteristic gamma rays originating from decay of neutral pions in two supernova remnant stars has shown that pions are produced copiously in supernovas, most probably in conjunction with production of high energy protons that are detected on Earth as cosmic rays.[1]

The concept of mesons as the carrier particles of the nuclear force was first proposed in 1935 by Hideki Yukawa. While the muon was first proposed to be this particle after its discovery in 1936, later work found that it did not participate in the strong nuclear interaction. The pions, which turned out to be examples of Yukawa's proposed mesons, were discovered later: the charged pions in 1947, and the neutral pion in 1950.

63.1 History

Theoretical work by Hideki Yukawa in 1935 had predicted the existence of mesons as the carrier particles of the strong nuclear force. From the range of the strong nuclear force (inferred from the radius of the atomic nucleus), Yukawa predicted the existence of a particle having a mass of about 100 MeV. Initially after its discovery in 1936, the muon (initially called the "mu meson") was thought to be this particle, since it has a mass of 106 MeV. However, later particle physics experiments showed that the muon did not participate in the strong nuclear interaction. In modern terminology, this makes the muon a lepton, and not a true meson. However, some communities of nuclear physicists, continue to call the muon a "mu-meson."

In 1947, the first true mesons, the charged pions, were found by the collaboration of Cecil Powell, César Lattes, Giuseppe Occhialini, *et al.*, at the University of Bristol, in England. Since the advent of particle accelerators had not yet come, high-energy subatomic particles were only obtainable from atmospheric cosmic rays. Photographic emulsions, which used the gelatin-silver process, were placed for long periods of time in sites located at high altitude mountains, first at Pic du Midi de Bigorre in the Pyrenees, and later at Chacaltaya in the Andes Mountains, where they were impacted by cosmic rays.

After the development of the photographic plates, microscopic inspection of the emulsions revealed the tracks of charged subatomic particles. Pions were first identified by their unusual "double meson" tracks, which were left by their decay into another "meson". (It was actually the muon, which is not classified as a meson in modern particle physics.) In

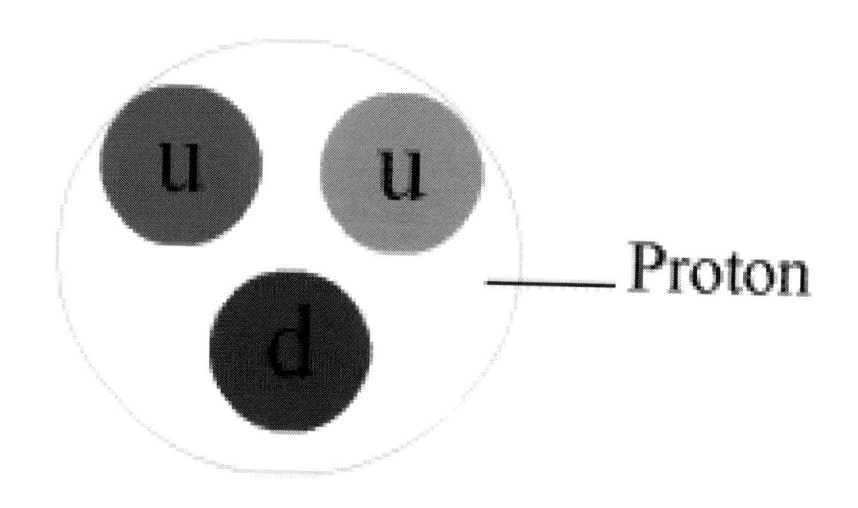

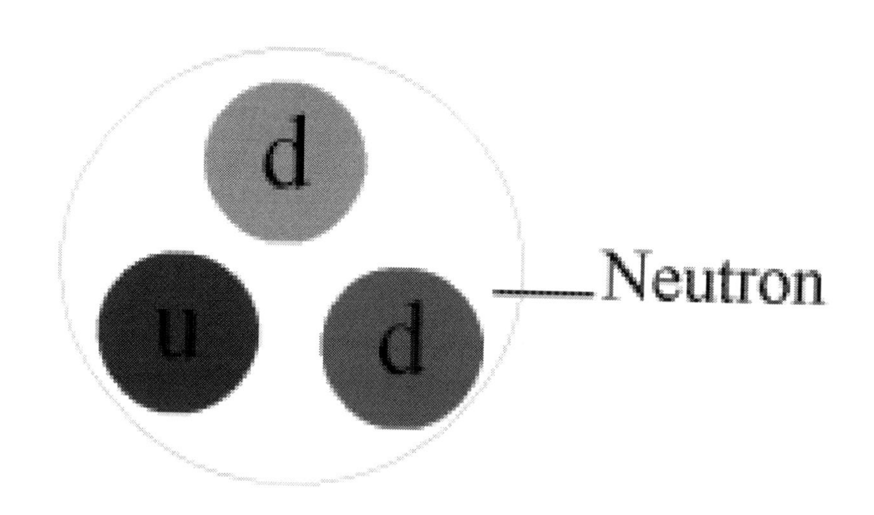

An animation of the nuclear force (or residual strong force) interaction. The small colored double disks are gluons. Anticolors are shown as per this diagram (larger version).

1948, Lattes, Eugene Gardner, and their team first artificially produced pions at the University of California's cyclotron in Berkeley, California, by bombarding carbon atoms with high-speed alpha particles. Further advanced theoretical work was carried out by Riazuddin, who in 1959, used the dispersion relation for Compton scattering of virtual photons on pions to analyze their charge radius.[2]

Nobel Prizes in Physics were awarded to Yukawa in 1949 for his theoretical prediction of the existence of mesons, and to Cecil Powell in 1950 for developing and applying the technique of particle detection using photographic emulsions.

Since the neutral pion is not electrically charged, it is more difficult to detect and observe than the charged pions are.

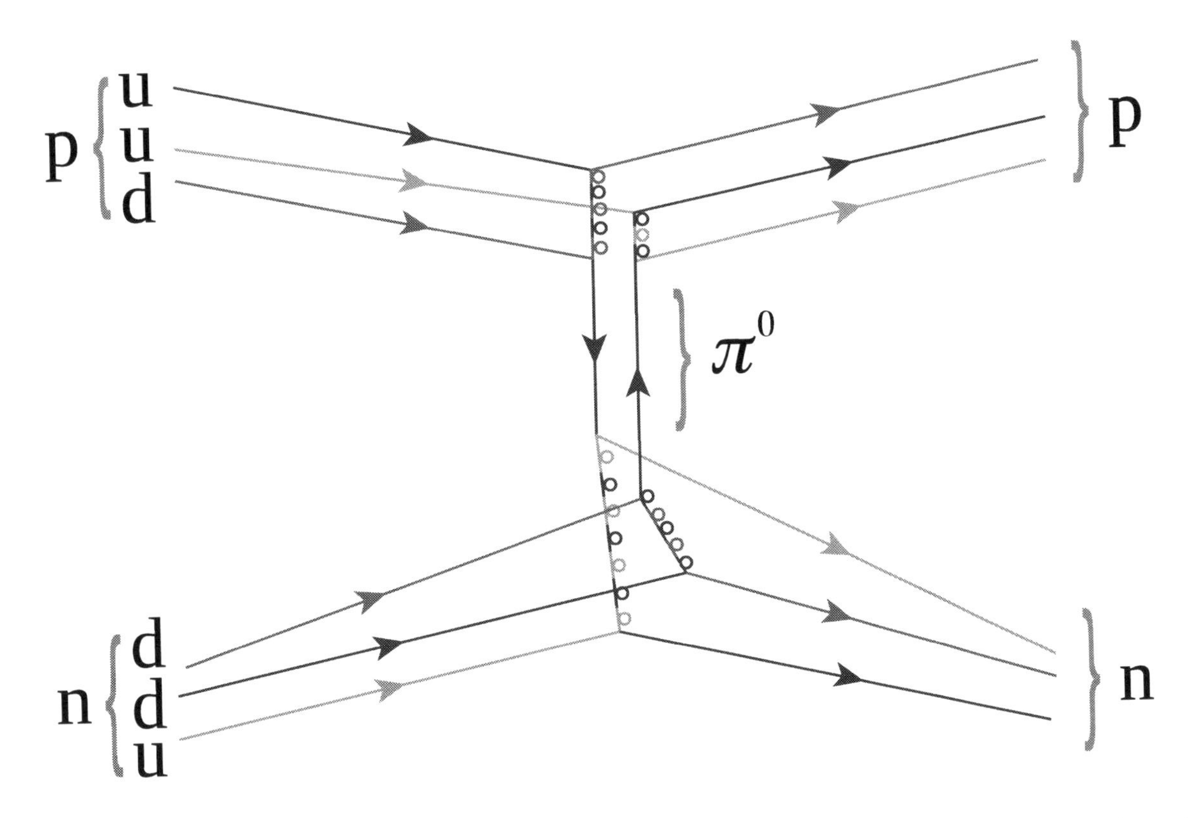

The same process as in the animation with the individual quark constituents shown, to illustrate how the fundamental *strong interaction gives rise to the* ***nuclear force****. Straight lines are quarks, while multi-colored loops are gluons (the carriers of the fundamental force). Other gluons, which bind together the proton, neutron, and pion "in-flight," are not shown.*

Neutral pions do not leave tracks in photographic emulsions, and neither do they in Wilson cloud chambers. The existence of the neutral pion was inferred from observing its decay products from cosmic rays, a so-called "soft component" of slow electrons with photons. The π0 was identified definitively at the University of California's cyclotron in 1950 by observing its decay into two photons.[3] Later in the same year, they were also observed in cosmic-ray balloon experiments at Bristol University.

The pion also plays a crucial role in cosmology, by imposing an upper limit on the energies of cosmic rays surviving collisions with the cosmic microwave background, through the Greisen–Zatsepin–Kuzmin limit.

In the standard understanding of the strong force interaction (called QCD, "quantum chromodynamics"), pions are understood to be the pseudo-Nambu-Goldstone bosons of spontaneously broken chiral symmetry. This explains why the three kinds of pions' masses are considerably less than the masses of the other mesons, such as the scalar or vector mesons. If their current quarks were massless particles, hypothetically, making the chiral symmetry exact, then the Goldstone theorem would dictate that all pions have zero masses. In reality, since the light quarks actually have minuscule nonzero masses, the pions also have nonzero rest masses, albeit *almost an order of magnitude smaller* than that of the nucleons, roughly[4] $m\pi \approx \sqrt{v\, m_q} / f\pi \approx \sqrt{m_q}$ 45 MeV, where m are the relevant current quark masses in MeV, 5–10 MeVs.

The use of pions in medical radiation therapy, such as for cancer, was explored at a number of research institutions, including the Los Alamos National Laboratory's Meson Physics Facility, which treated 228 patients between 1974 and 1981 in New Mexico,[5] and the TRIUMF laboratory in Vancouver, British Columbia.

63.2 Theoretical overview

The pion can be thought of as one of the particles that mediate the interaction between a pair of nucleons. This interaction is attractive: it pulls the nucleons together. Written in a non-relativistic form, it is called the Yukawa potential. The pion, being spinless, has kinematics described by the Klein–Gordon equation. In the terms of quantum field theory, the effective field theory Lagrangian describing the pion-nucleon interaction is called the Yukawa interaction.

The nearly identical masses of $\pi\pm$ and $\pi0$ imply that there must be a symmetry at play; this symmetry is called the SU(2) flavour symmetry or isospin. The reason that there are three pions, $\pi+$, $\pi-$ and $\pi0$, is that these are understood to belong to the triplet representation or the adjoint representation **3** of SU(2). By contrast, the up and down quarks transform according to the fundamental representation **2** of SU(2), whereas the anti-quarks transform according to the conjugate representation **2***.

With the addition of the strange quark, one can say that the pions participate in an SU(3) flavour symmetry, belonging to the adjoint representation **8** of SU(3). The other members of this octet are the four kaons and the eta meson.

Pions are pseudoscalars under a parity transformation. Pion currents thus couple to the axial vector current and pions participate in the chiral anomaly.

63.3 Basic properties

Pions are mesons with zero spin, and they are composed of first-generation quarks. In the quark model, an up quark and an anti-down quark make up a $\pi+$, whereas a down quark and an anti-up quark make up the $\pi-$, and these are the antiparticles of one another. The neutral pion $\pi0$ is a combination of an up quark with an anti-up quark or a down quark with an anti-down quark. The two combinations have identical quantum numbers, and hence they are only found in superpositions. The lowest-energy superposition of these is the $\pi0$, which is its own antiparticle. Together, the pions form a triplet of isospin. Each pion has isospin ($I = 1$) and third-component isospin equal to its charge ($I_z = +1, 0$ or -1).

63.3.1 Charged pion decays

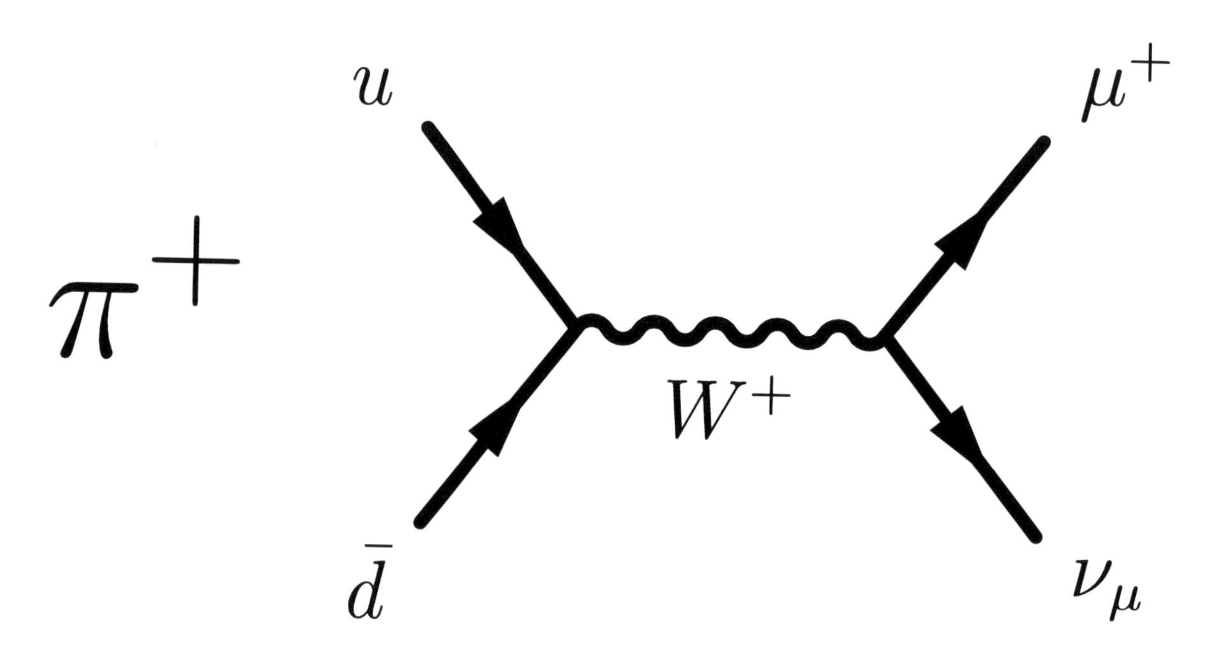

Feynman diagram of the dominating leptonic pion decay.

The $\pi\pm$ mesons have a mass of 139.6 MeV/c^2 and a mean lifetime of 2.6×10^{-8} s. They decay due to the weak interaction. The primary decay mode of a pion, with probability 0.999877, is a purely leptonic decay into an anti-muon and a muon neutrino:

The second most common decay mode of a pion, with probability 0.000123, is also a leptonic decay into an electron and the corresponding electron antineutrino. This "electronic mode" was discovered at CERN in 1958:[6]

The suppression of the electronic mode, with respect to the muonic one, is given approximately (to within radiative corrections) by the ratio of the half-widths of the pion–electron and the pion–muon decay reactions:

$$R_\pi = (m_e/m_\mu)^2 \left(\frac{m_\pi^2 - m_e^2}{m_\pi^2 - m_\mu^2}\right)^2 = 1.283 \times 10^{-4}$$

and is a spin effect known as the helicity suppression. Its mechanism is as follows: The negative pion has spin zero, therefore the lepton and antineutrino must be emitted with opposite spins (and opposite linear momenta) to preserve net zero spin (and conserve linear momentum). However, the antineutrino, due to very high speed, is always right-handed, so this implies that the lepton must be emitted with spin in the direction of its linear momentum (i.e., also right-handed). If, however, leptons were massless, they would only exist in the left-handed form, just as the neutrino does (due to parity violation), and this decay mode would be prohibited. Therefore, suppression of the electron decay channel comes from the fact that the electron's mass is much smaller than the muon's. The electron is thus relatively massless compared with the muon, and thus the electronic mode is *almost* prohibited.[7]

Hence, electronic mode decay favors the left-handed symmetry and inhibits this decay channel. Measurements of the above ratio have been considered for decades to be tests of the *V* – *A structure* (vector minus axial vector or left-handed lagrangian) of the charged weak current and of lepton universality. Experimentally this ratio is $1.230(4)\times10^{-4}$.[8]

Besides the purely leptonic decays of pions, some structure-dependent radiative leptonic decays (that is, decay to the usual leptons plus a gamma ray) have also been observed.

Also observed, for charged pions only, is the very rare "pion beta decay" (with probability of about 10^{-8}) into a neutral pion plus an electron and electron antineutrino (or for positive pions, a neutral pion, positron, and electron neutrino).

The rate at which pions decay is a prominent quantity in many sub-fields of particle physics, such as chiral perturbation theory. This rate is parametrized by the pion decay constant ($f\pi$), related to the wave function overlap of the quark and antiquark, which is about 130 MeV.[9]

63.3.2 Neutral pion decays

The π0 meson has a mass of 135.0 MeV/c^2 and a mean lifetime of 8.4×10^{-17} s. It decays via the electromagnetic force, which explains why its mean lifetime is much smaller than that of the charged pion (which can only decay via the weak force). The main π^0 decay mode, with a branching ratio of BR=0.98823, is into two photons:

The decay $\pi^0\rightarrow3\gamma$ (as well as decays into any odd number of photons) is forbidden by the C-symmetry of the electromagnetic interaction. The intrinsic C-parity of the π^0 is **+1**, while the C-parity of a system of **n** photons is $\mathbf{(-1)^n}$.

The second largest π^0 decay mode (BR=0.01174) is the Dalitz decay (named after Richard Dalitz), which is a two-photon decay with an internal photon conversion resulting a photon and an electron-positron pair in the final state:

The third largest established decay mode (BR=3.34×10^{-5}) is the double Dalitz decay, with both photons undergoing internal conversion which leads to further suppression of the rate:

The fourth largest established decay mode is the loop-induced and therefore suppressed (and additionally helicity-suppressed) leptonic decay mode (BR=6.46×10^{-8}):

The neutral pion has also been observed to decay into positronium with a branching fraction of the order of 10^{-9}. No other decay modes have been established experimentally. The branching fractions above are the PDG central values, and their uncertainties are not quoted.

[a] ^ Make-up inexact due to non-zero quark masses.[12]

63.4 See also

- Pionium
- List of particles
- Quark model
- Static forces and virtual-particle exchange
- César Lattes

63.5 References

[1] M. Ackermann et al. (2013). "Detection of the Characteristic Pion-Decay Signature in Supernova Remnants". *Science* **339** (6424): 807–811. arXiv:1302.3307. Bibcode:2013Sci...339..807A. doi:10.1126/science.1231160.

[2] Riazuddin(1959). "Charge Radius of Pion".*Physical Review***114**(4): 1184–1186. Bibcode:1959PhRv..114.1184R.doi:84.

[3] R. Bjorklund; W. E. Crandall; B. J. Moyer; H. F. York (1950). "High Energy Photons from Proton-Nucleon Collisions". *Physical Review* **77** (2): 213–218. Bibcode:1950PhRv...77..213B. doi:10.1103/PhysRev.77.213.

[4] Gell-Mann, M.; Renner, B. (1968). "Behavior of Current Divergences under SU_{3}×SU_{3}". *Physical Review* **175** (5): 2195. Bibcode:1968PhRv..175.2195G. doi:10.1103/PhysRev.175.2195.

[5] von Essen, C. F.; Bagshaw, M. A.; Bush, S. E.; Smith, A. R.; Kligerman, M. M. (1987). "Long-term results of pion therapy at Los Alamos". *International Journal of Radiation Oncology*Biology*Physics* **13** (9): 1389–98. doi:10.1016/0360-3016(87)90235-5. PMID 3114189.

[6] Fazzini, T.; Fidecaro, G.; Merrison, A.; Paul, H.; Tollestrup, A. (1958). "Electron Decay of the Pion". *Physical Review Letters* **1** (7): 247. doi:10.1103/PhysRevLett.1.247.

[7] Mesons at Hyperphysics

[8] C. Amsler *et al.*. (2008): Particle listings – π±

[9] LEPTONIC DECAYS OF CHARGED PSEUDO- SCALAR MESONS J. L. Rosner and S. Stone. Particle Data Group. December 18, 2013

[10] C. Amsler *et al.*. (2008): Quark Model

[11] C. Amsler *et al.*. (2008): Particle listings – π0

[12] D. J. Griffiths (1987). *Introduction to Elementary Particles*. John Wiley & Sons. ISBN 0-471-60386-4.

63.6 Further reading

- Gerald Edward Brown and A. D. Jackson, *The Nucleon-Nucleon Interaction*, (1976) North-Holland Publishing, Amsterdam ISBN 0-7204-0335-9

63.7 External links

- Mesons at the Particle Data Group

Chapter 64

Rho meson

In particle physics, a **rho meson** is a short-lived hadronic particle that is an isospin triplet whose three states are denoted as ρ+, ρ0 and ρ–. After the pions and kaons, the rho mesons are the lightest strongly interacting particle with a mass of roughly 770 MeV for all three states. There should be a small mass difference between the ρ+ and the ρ0 that can be attributed to the electromagnetic self-energy of the particle as well as a small effect due to isospin breaking arising from the light quark masses; however, the current experimental limit is that this mass difference is less than 0.7 MeV.

The rho mesons have a very short lifetime and their decay width is about 145 MeV with the peculiar feature that the decay widths are not described by a Breit-Wigner form. The principal decay route of the rho mesons is to a pair of pions with a branching rate of 99.9%. Neutral rho mesons can decay to a pair of electrons or muons which occurs with a branching ratio of 5×10^{-5}. This decay of the neutral rho to leptons can be interpreted as a mixing between the photon and rho. In principle the charged rho mesons mix with the weak vector bosons and can lead to decay to an electron or muon plus a neutrino; however, this has never been observed.

In the De Rujula–Georgi–Glashow description of hadrons,[1] the rho mesons can be interpreted as a bound state of a quark and an anti-quark and is an excited version of the pion. Unlike the pion, the rho meson has spin $j = 1$ (a vector meson) and a much higher value of the mass. This mass difference between the pions and rho mesons is attributed to a large hyperfine interaction between the quark and anti-quark. The main objection with the De Rujula–Georgi–Glashow description is that it attributes the lightness of the pions as an accident rather than a result of chiral symmetry breaking.

The rho mesons can be thought of as the gauge bosons of a spontaneously broken gauge symmetry whose local character is emergent (arising from QCD); Note that this broken gauge symmetry (sometimes called hidden local symmetry) is distinct from the global chiral symmetry acting on the flavors. This was described by Howard Georgi in a paper titled "The Vector Limit of Chiral Symmetry" where he ascribed much of the literature of hidden local symmetry to a non-linear sigma model.[2]

[a] ^ PDG reports the resonance width (Γ). Here the conversion $\tau = \hbar/\Gamma$ is given instead.
[b] ^ The exact value depends on the method used. See the given reference for detail.

64.1 References

[1] Rujula, Georgi, Glashow (1975) "Hadron Masses in Gauge Theory." Physical Review D12, p.147

[2] H. Georgi. (1990) "Vector Realization of Chiral Symmetry." inSPIRE Record

[3] C. Amsler *et al.* (2008): Quark Model

[4] C. Amsler *et al.* (2008): Particle listings – ρ

Chapter 65

Eta meson

The **eta** (η) and **eta prime meson** (η′) are mesons made of a mixture of up, down and strange quarks and their antiquarks. The charmed eta meson (η
c) and bottom eta meson (η
b) are forms of quarkonium; they have the same spin and parity as the light eta but are made of charm quarks and bottom quarks respectively. The top quark is too heavy to form a similar meson, due to its very fast decay.

65.1 General

The eta was discovered in pion-nucleon collisions at the Bevatron in 1961 by A. Pevsner et al. at a time when the proposal of the Eightfold Way was leading to predictions and discoveries of new particles from symmetry considerations.[2]

The difference between the mass of the η and that of the η' is larger than the quark model can naturally explain. This "η-η' puzzle" can be resolved[3][4][5] by the 't Hooft instanton mechanism,[6] whose 1/N realization is also known as Witten-Veneziano mechanism.[7][8]

65.2 Quark composition

The η particles belong to the "pseudo-scalar" nonet of mesons which have spin $J = 0$ and negative parity,[9][10] and η and η′ have zero total isospin, I, and zero strangeness and hypercharge. Each quark which appears in an η particle is accompanied by its antiquark (the particle overall is "flavourless") and all the main quantum numbers are zero.

The basic SU(3) symmetry theory of quarks for the three lightest quarks, which only takes into account the strong force, predicts corresponding particles

$$\eta_1 = \frac{u\bar{u}+d\bar{d}+s\bar{s}}{\sqrt{3}}$$

$$\eta_8 = \frac{u\bar{u}+d\bar{d}-2s\bar{s}}{\sqrt{6}}$$

The subscripts refer to the fact that η_1 belongs to a singlet (which is fully antisymmetrical) and η_8 is part of an octet. However in this case the weak and electromagnetic forces, which can transform one flavour of quark into another, cause a significant, though small, amount of "mixing" of the eigenstates (with mixing angle $\theta P = -11.5$ degrees),[11] so that the actual quark composition is a linear combination of these formulae. That is:

$$\begin{pmatrix} \cos\theta_P & -\sin\theta_P \\ \sin\theta_P & \cos\theta_P \end{pmatrix} \begin{pmatrix} \eta_8 \\ \eta_1 \end{pmatrix} = \begin{pmatrix} \eta \\ \eta' \end{pmatrix}$$

The unsubscripted name η refers to the real particle which is actually observed and which is close to the η_8. The η′ is the observed particle close to η_1.[10]

The η and η′ particles are closely related to the better-known neutral pion π0, where

$$\pi^0 = \frac{u\bar{u} - d\bar{d}}{\sqrt{2}}$$

In fact π^0, η_1 and η_8 are three mutually orthogonal linear combinations of the quark pairs *uu*, *dd* and *ss*; they are at the centre of the pseudo-scalar nonet of mesons[9][10] with all the main quantum numbers equal to zero.

65.3 Eta Prime Meson

The Eta Prime Meson is essentially a superposition of the Eta Meson, the only significant differences being a higher mass, a different decay state, and a shorter lifetime.

65.4 See also

- List of mesons
- Special Unitary Group

65.5 External links

- Eta Meson at the Particle Data Group

65.6 References

[1] Light Unflavored Mesons as appearing in Olive, K. A.; et al. (PDG) (2014). "Review of Particle Physics". *Chinese Physics C* **38**: 090001.

[2] Kupść, A. (2007). "What is interesting in η and η′ Meson Decays?". *AIP Conference Proceedings* **950**: 165–179. arXiv:0709.0603. Bibcode:2007AIPC..950..165K. doi:10.1063/1.2819029.

[3] Del Debbio, L.; Giusti, L.; Pica, C. (2005). "Topological Susceptibility in SU(3) Gauge Theory". *Physical Review Letters* **94** (3): 032003. arXiv:hep-th/0407052. Bibcode:2005PhRvL..94c2003D. doi:10.1103/PhysRevLett.94.032003.

[4] Lüscher, M.; Palombi, F. (2010). "Universality of the topological susceptibility in the SU(3) gauge theory". *Journal of High Energy Physics* **2010** (9): 110. arXiv:1008.0732. Bibcode:2010JHEP...09..110L. doi:10.1007/JHEP09(2010)110.

[5] Cè, M.; Consonni, C.; Engel, G.; Giusti, L. (2014). *Testing the Witten-Veneziano mechanism with the Yang-Mills gradient flow on the lattice*. 32nd International Symposium on Lattice Field Theory. arXiv:1410.8358.

[6] 't Hooft,G. (1976). "Symmetry Breaking through Bell-Jackiw Anomalies".*Physical Review Letters***37**(1): 8–11. Bibcode:8T. doi:10.1103/PhysRevLett.37.8.

[7] Witten,E. (1979). "Current algebra theorems for the U(1) "Goldstone boson"".*Nuclear Physics B***156**(2): 269–28369W. doi:10.1016/0550-3213(79)90031-2.

[8] Veneziano, G. (1979). "U(1) without instantons". *Nuclear Physics B* **159** (1–2): 213–224. Bibcode:1979NuPhB.159..213V. doi:10.1016/0550-3213(79)90332-8.

[9] The Wikipedia meson article describes the SU(3) pseudo-scalar nonet of mesons including η and η′.

[10] Jones, H. F. (1998). *Groups, Representations and Physics*. IOP Publishing. ISBN 0-7503-0504-5. Page 150 describes the SU(3) pseudo-scalar nonet of mesons including η and η′. Page 154 defines η_1 and η_8 and explains the mixing (leading to η and η′).

[11] Quark Model Review as appearing in Beringer, J.; et al. (PDG) (2012). "Review of Particle Physics" (PDF). *Physical Review D* **86** (1): 010001. Bibcode:2012PhRvD..86a0001B. doi:10.1103/PhysRevD.86.010001.

Chapter 66

Phi meson

Not to be confused with the Φ^{--}, formerly thought to be a pentaquark.

In particle physics, the **phi meson** is a vector meson formed of a strange quark and a strange antiquark. It has a mass of 1019.445±0.020 MeV/c^2.

66.1 References

[1] C. Amsler *et al.* (2008): Particle listings – φ

66.2 See also

- Charmonium
- List of mesons
- List of particles
- Quark model

Chapter 67

List of mesons

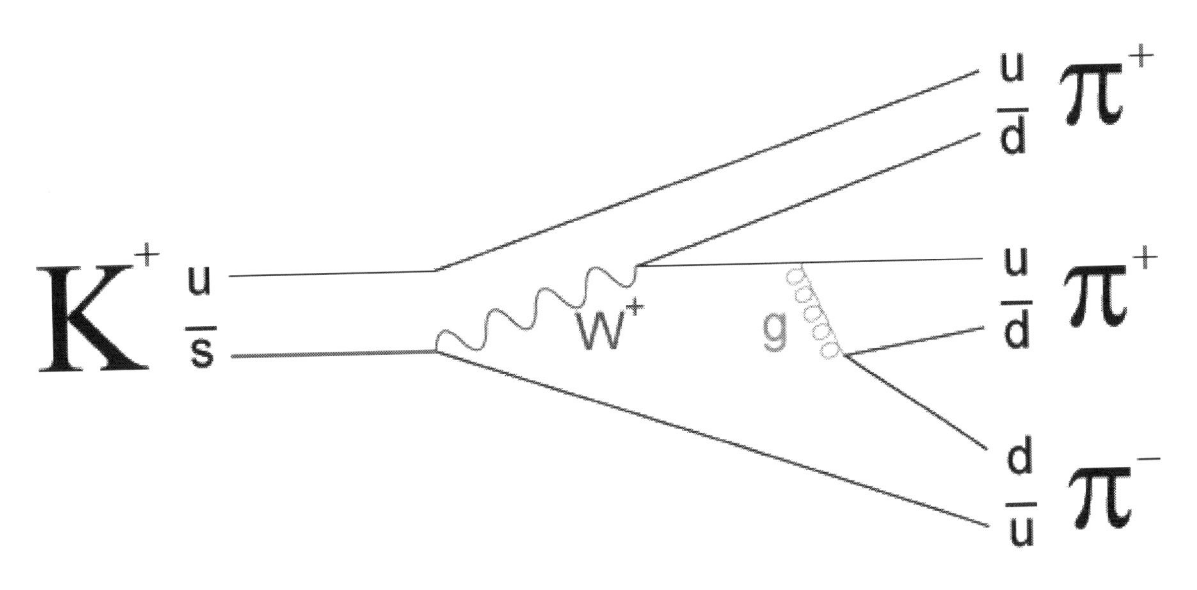

The decay of a kaon (K+) into three pions (2 π+, 1 π−) is a process that involves both weak and strong interactions.
Weak interactions*: The strange antiquark (s) of the kaon transmutes into an up antiquark (u) by the emission of a W+ boson; the W+ boson subsequently decays into a down antiquark (d) and an up quark (u).*
Strong interactions*: An up quark (u) emits a gluon (g) which decays into a down quark (d) and a down antiquark (d).*

This list is of all known and predicted scalar, pseudoscalar and vector mesons. See list of particles for a more detailed list of particles found in particle physics.

Mesons are unstable subatomic particles composed of one quark and one antiquark. They are part of the hadron particle family – particles made of quarks. The other members of the hadron family are the baryons – subatomic particles composed of three quarks. The main difference between mesons and baryons is that mesons have integer spin (thus are bosons) while baryons are fermions (half-integer spin). Because mesons are bosons, the Pauli exclusion principle does not apply to them. Because of this, they can act as force mediating particles on short distances, and thus play a part in processes such as the nuclear interaction.

Since mesons are composed of quarks, they participate in both the weak and strong interactions. Mesons with net electric charge also participate in the electromagnetic interaction. They are classified according to their quark content, total angular momentum, parity, and various other properties such as C-parity and G-parity. While no meson is stable, those of lower mass are nonetheless more stable than the most massive mesons, and are easier to observe and study in particle accelerators or in cosmic ray experiments. They are also typically less massive than baryons, meaning that they are more easily produced in experiments, and will exhibit higher-energy phenomena sooner than baryons would. For example, the charm quark was first seen in the J/Psi meson (J/ψ) in 1974,[1][2] and the bottom quark in the upsilon meson (ϒ) in 1977.[3]

Each meson has a corresponding antiparticle (antimeson) where quarks are replaced by their corresponding antiquarks and vice versa. For example, a positive pion (π+) is made of one up quark and one down antiquark; and its corresponding antiparticle, the negative pion (π−), is made of one up antiquark and one down quark. Some experiments show the evidence of *tetraquarks* – "exotic" mesons made of two quarks and two antiquarks, but the particle physics community as a whole does not view their existence as likely, although still possible.[4]

The symbols encountered in these lists are: I (*isospin*), J (*total angular momentum*), P (*parity*), C (*C-parity*), G (*G-parity*), u (*up quark*), d (*down quark*), s (*strange quark*), c (*charm quark*), b (*bottom quark*), Q (*charge*), B (*baryon number*), S (*strangeness*), C (*charm*), and B′ (*bottomness*), as well as a wide array of subatomic particles (hover for name).

67.1 Summary table

Because this table was initially derived from published results and many of those results were preliminary, as many as 64 of the mesons in the following table may not exist or have the wrong mass or quantum numbers.

67.2 Meson properties

The following lists detail all known and predicted pseudoscalar ($J^P = 0^-$) and vector ($J^P = 1^-$) mesons.

The properties and quark content of the particles are tabulated below; for the corresponding antiparticles, simply change quarks into antiquarks (and vice versa) and flip the sign of Q, B, S, C, and B′. Particles with † next to their names have been predicted by the standard model but not yet observed. Values in red have not been firmly established by experiments, but are predicted by the quark model and are consistent with the measurements.

67.2.1 Pseudoscalar mesons

[a] ^ Makeup inexact due to non-zero quark masses.
[b] ^ PDG reports the resonance width (Γ). Here the conversion $\tau = ^{\hbar}\!/\Gamma$ is given instead.
[c] ^ Strong eigenstate. No definite lifetime (see kaon notes below)
[d] ^ The mass of the K0
L and K0
S are given as that of the K0. However, it is known that a difference between the masses of the K0
L and K0
S on the order of $2.2{\times}10^{-11}$ MeV/c^2 exists.[15]
[e] ^ Weak eigenstate. Makeup is missing small CP–violating term (see notes on neutral kaons below).

67.2.2 Vector mesons

[f] ^ PDG reports the resonance width (Γ). Here the conversion $\tau = ^{\hbar}\!/\Gamma$ is given instead.
[g] ^ The exact value depends on the method used. See the given reference for detail.

67.2.3 Notes on neutral kaons

There are two complications with neutral kaons:[34]

- Due to neutral kaon mixing, the K0
  S and K0

L are not eigenstates of strangeness. However, they *are* eigenstates of the weak force, which determines how they decay, so these are the particles with definite lifetime.

- The linear combinations given in the table for the K0
  S and K0
  L are not exactly correct, since there is a small correction due to CP violation. See CP violation in kaons.

Note that these issues also exist in principle for other neutral flavored mesons; however, the weak eigenstates are considered separate particles only for kaons because of their dramatically different lifetimes.[34]

67.3 See also

- List of baryons
- List of particles
- Timeline of particle discoveries

67.4 References

[1] J.J. Aubert *et al.* (1974)

[2] J.E. Augustin *et al.* (1974)

[3] S.W. Herb *et al.* (1977)

[4] C. Amsler *et al.* (2008): Charmonium States

[5] K.A. Olive *et al.* (2014): Meson Summary Table

[6] K.A. Olive *et al.* (2014): Particle listings – π±

[7] K.A. Olive *et al.* (2014): Particle listings – π0

[8] K.A. Olive *et al.* (2014): Particle listings – η

[9] K.A. Olive *et al.* (2014): Particle listings – η′

[10] K.A. Olive *et al.* (2014): Particle listings – η
c

[11] K.A. Olive *et al.* (2014): Particle listings – η
b

[12] K.A. Olive *et al.* (2014): Particle listings – K±

[13] K.A. Olive *et al.* (2014): Particle listings – K0

[14] K.A. Olive *et al.* (2014): Particle listings – K0
S

[15] K.A. Olive *et al.* (2014): Particle listings – K0
L

[16] K.A. Olive *et al.* (2014): Particle listings – D±

[17] K.A. Olive *et al.* (2014): Particle listings – D0

[18] K.A. Olive *et al.* (2014): Particle listings – D±
s

[19] K.A. Olive *et al.* (2014): Particle listings – $B^{\pm}$

[20] K.A. Olive *et al.* (2014): Particle listings – B^0

[21] K.A. Olive *et al.* (2014): Particle listings – B^0_s

[22] K.A. Olive *et al.* (2014): Particle listings – $B^{\pm}_c$

[23] K.A. Olive *et al.* (2014): Particle listings – ρ

[24] K.A. Olive *et al.* (2014): Particle listings – $\omega(782)$

[25] K.A. Olive *et al.* (2014): Particle listings – ϕ

[26] K.A. Olive *et al.* (2014): Particle listings – J/Ψ

[27] K.A. Olive *et al.* (2014): Particle listings – $\Upsilon(1S)$

[28] K.A. Olive *et al.* (2014): Particle listings – $K^*(892)$

[29] K.A. Olive *et al.* (2014): Particle listings – $D^{*\pm}(2010)$

[30] K.A. Olive *et al.* (2014): Particle listings – $D^{*0}(2007)$

[31] K.A. Olive *et al.* (2014): Particle listings – $D^{*\pm}_s$

[32] K.A. Olive *et al.* (2014): Particle listings – B^*

[33] K.A. Olive *et al.* (2014): Particle listings – B^*_s

[34] J.W. Cronin (1980)

67.4.1 Bibliography

- K.A. Olive *et al.* (Particle Data Group) (2014). “Review of Particle Physics”. *Chinese Physics C* **38** (9): 090001.
- M.S. Sozzi (2008a). “Parity”. *Discrete Symmetries and CP Violation: From Experiment to Theory*. Oxford University Press. pp. 15–87. ISBN 0-19-929666-9.
- M.S. Sozzi (2008a). “Charge Conjugation”. *Discrete Symmetries and CP Violation: From Experiment to Theory*. Oxford University Press. pp. 88–120. ISBN 0-19-929666-9.
- M.S. Sozzi (2008c). “CP-Symmetry”. *Discrete Symmetries and CP Violation: From Experiment to Theory*. Oxford University Press. pp. 231–275. ISBN 0-19-929666-9.
- C. Amsler *et al.* (Particle Data Group); Amsler; Doser; Antonelli; Asner; Babu; Baer; Band; Barnett; Bergren; Beringer; Bernardi; Bertl; Bichsel; Biebel; Bloch; Blucher; Blusk; Cahn; Carena; Caso; Ceccucci; Chakraborty; Chen; Chivukula; Cowan; Dahl; d'Ambrosio; Damour et al. (2008). “Review of Particle Physics”. *Physics Letters B* **667** (1): 1–1340. Bibcode:2008PhLB..667....1P. doi:10.1016/j.physletb.2008.07.018.
- S.S.M. Wong (1998). “Nucleon Structure”. *Introductory Nuclear Physics* (2nd ed.). John Wiley & Sons. pp. 21–56. ISBN 0-471-23973-9.
- R. Shankar (1994). *Principles of Quantum Mechanics* (2nd ed.). Plenum Press. ISBN 0-306-44790-8.
- K. Gottfried, V.F. Weisskopf (1986). “Hadronic Spectroscopy: G-parity”. *Concepts of Particle Physics* **2**. Oxford University Press. pp. 303–311. ISBN 0-19-503393-0.

- J.W. Cronin (1980). "CP Symmetry Violation – The Search for its origin" (PDF). *Nobel Lecture*. The Nobel Foundation.
- V.L. Fitch (1980). "The Discovery of Charge – Conjugation Parity Asymmetry" (PDF). *Nobel Lecture*. The Nobel Foundation.
- S.W. Herb; Hom, D.; Lederman, L.; Sens, J.; Snyder, H.; Yoh, J.; Appel, J.; Brown, B. et al. (1977). "Observation of a Dimuon Resonance at 9.5 Gev in 400-GeV Proton-Nucleus Collisions". *Physical Review Letters* **39** (5): 252–255. Bibcode:1977PhRvL..39..252H. doi:10.1103/PhysRevLett.39.252.
- J.J. Aubert; Becker, U.; Biggs, P.; Burger, J.; Chen, M.; Everhart, G.; Goldhagen, P.; Leong, J. et al. (1974). "Experimental Observation of a Heavy Particle J". *Physical Review Letters* **33** (23): 1404–1406. Bibcode:1974 A.doi:10.1103/PhysRevLett.33.1404.
- J.E. Augustin; Boyarski, A.; Breidenbach, M.; Bulos, F.; Dakin, J.; Feldman, G.; Fischer, G.; Fryberger, D. et al. (1974). "Discovery of a Narrow Resonance in e^+e^- Annihilation". *Physical Review Letters* **33** (23): 1406–1408. Bibcode:1974PhRvL..33.1406A. doi:10.1103/PhysRevLett.33.1406.
- M.Gell-Mann(1964). "A Schematic of Baryons and Mesons".*Physics Letters***8**(3): 214–215. Bibcode:1964PhL doi:10.1016/S0031-9163(64)92001-3.
- E. Wigner (1937). "On the Consequences of the Symmetry of the Nuclear Hamiltonian on the Spectroscopy of Nuclei". *Physical Review* **51** (2): 106–119. Bibcode:1937PhRv...51..106W. doi:10.1103/PhysRev.51.106.
- W. Heisenberg (1932). "Über den Bau der Atomkerne I". *Zeitschrift für Physik* (in German) **77** (1–2): 1–11. Bibcode:1932ZPhy...77....1H. doi:10.1007/BF01342433.
- W. Heisenberg (1932). "Über den Bau der Atomkerne II". *Zeitschrift für Physik* (in German) **78** (3–4): 156–164. Bibcode:1932ZPhy...78..156H. doi:10.1007/BF01337585.
- W. Heisenberg (1932). "Über den Bau der Atomkerne III". *Zeitschrift für Physik* (in German) **80** (9–10): 587–596. Bibcode:1933ZPhy...80..587H. doi:10.1007/BF01335696.

67.5 External links

- Particle Data Group – The Review of Particle Physics (2008)
- Mesons made thinkable, an interactive visualisation allowing physical properties to be compared

Chapter 68

J/psi meson

The **J/ψ (J/Psi) meson** or **psion** [1] is a subatomic particle, a flavor-neutral meson consisting of a charm quark and a charm antiquark. Mesons formed by a bound state of a charm quark and a charm anti-quark are generally known as "charmonium". The J/ψ is the first excited state of charmonium (i.e. the form of the charmonium with the second-smallest rest mass). The J/ψ has a rest mass of 3.0969 GeV/c^2, and a mean lifetime of 7.2×10^{-21} s. This lifetime was about a thousand[2] times longer than expected.

Its discovery was made independently by two research groups, one at the Stanford Linear Accelerator Center, headed by Burton Richter, and one at the Brookhaven National Laboratory, headed by Samuel Ting of MIT. They discovered they had actually found the same particle, and both announced their discoveries on 11 November 1974. The importance of this discovery is highlighted by the fact that the subsequent, rapid changes in high-energy physics at the time have become collectively known as the "**November Revolution**". Richter and Ting were rewarded for their shared discovery with the 1976 Nobel Prize in Physics.

68.1 Background to discovery

The background to the discovery of the J/ψ was both theoretical and experimental. In the 1960s, the first quark models of elementary particle physics were proposed, which said that protons, neutrons and all other baryons, and also all mesons, are made from three kinds of fractionally-charged particles, the "quarks", that come in three different types or "flavors", called *up*, *down*, and *strange*. Despite the impressive ability of quark models to bring order to the "elementary particle zoo", their status was considered something like mathematical fiction at the time, a simple artifact of deeper physical reasons.

Starting in 1969, deep inelastic scattering experiments at SLAC revealed surprising experimental evidence for particles inside of protons. Whether these were quarks or something else was not known at first. Many experiments were needed to fully identify the properties of the subprotonic components. To a first approximation, they were indeed the already-described quarks.

On the theoretical front, gauge theories with broken symmetry became the first fully viable contenders for explaining the weak interaction after Gerardus 't Hooft discovered in 1971 how to calculate with them beyond tree level. The first experimental evidence for these electroweak unification theories was the discovery of the weak neutral current in 1973. Gauge theories with quarks became a viable contender for the strong interaction in 1973 when the concept of asymptotic freedom was identified.

However, a naive mixture of electroweak theory and the quark model led to calculations about known decay modes that contradicted observation: in particular, it predicted Z boson-mediated *flavor-changing* decays of a strange quark into a down quark, which were not observed. A 1970 idea of Sheldon Glashow, John Iliopoulos, and Luciano Maiani, known as the GIM mechanism, showed that the flavor-changing decays would be eliminated if there were a fourth quark, *charm*, that paired with the strange quark. This work led, by the summer of 1974, to theoretical predictions of what a charm/anticharm meson would be like. These predictions were ignored. The work of Richter and Ting was done for other reasons, mostly

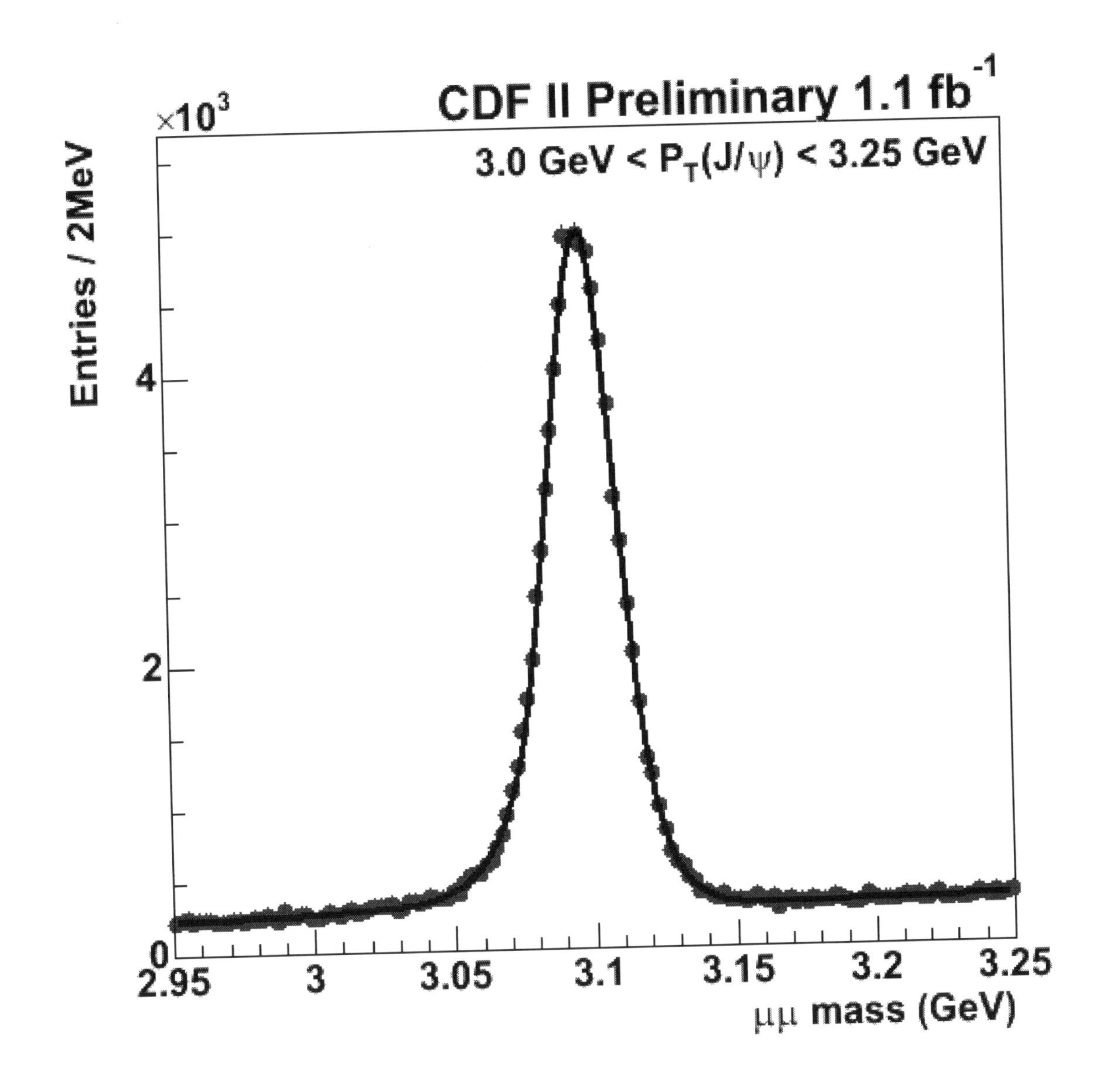

J/Ψ Production at Fermilab

to explore new energy regimes.

68.2 The name

Because of the nearly simultaneous discovery, the J/ψ is the only particle to have a two-letter name. Richter named it "SP", after the SPEAR accelerator used at SLAC; however, none of his coworkers liked that name. After consulting with Greek-born Leo Resvanis to see which Greek letters were still available, and rejecting "iota" because its name implies insignificance, Richter chose "psi" – a name which, as Gerson Goldhaber pointed out, contains the original name "SP", but in reverse order.[3] Coincidentally, later spark chamber pictures often resembled the psi shape. Ting assigned the name "J" to it, which is one letter removed from "K", the name of the already-known strange meson; possibly by coincidence, "J" strongly resembles the Chinese character for Ting's name (?). (Cf. the naming of Gallium.) J is also the first letter of Ting's oldest daughter's name, Jeanne.

Since the scientific community considered it unjust to give one of the two discoverers priority, most subsequent publications have referred to the particle as the "J/ψ".

The first excited state of the J/ψ was called the ψ'; it is now called the ψ(2S), indicating its quantum state. The next excited state was called the ψ"; it is now called ψ(3770), indicating mass in MeV. Other vector charm-anticharm states are denoted similarly with ψ and the quantum state (if known) or the mass.[4] The "J" is not used, since Richter's group alone first found excited states.

The name *charmonium* is used for the J/ψ and other charm-anticharm bound states. This is by analogy with positronium, which also consists of a particle and its antiparticle (an electron and positron in the case of positronium).

68.3 J/ψ melting

In a hot QCD medium, when the temperature is raised well beyond the Hagedorn temperature, the J/ψ and its excitations are expected to melt.[5] This is one of the predicted signals of the formation of the quark–gluon plasma. Heavy-ion experiments at CERN's Super Proton Synchrotron and at BNL's Relativistic Heavy Ion Collider have studied this phenomenon without a conclusive outcome as of 2009. This is due to the requirement that the disappearance of J/ψ mesons is evaluated with respect to the baseline provided by the total production of all charm quark-containing subatomic particles, and because it is widely expected that some J/ψ are produced and/or destroyed at time of QGP hadronization. Thus, there is uncertainty in the prevailing conditions at the initial collisions.

In fact, instead of suppression, enhanced production of J/ψ is expected[6] in heavy ion experiments at LHC where the quark-combinant production mechanism should be dominant given the large abundance of charm quarks in the QGP. Aside of J/ψ, charmed B mesons (B
c), offer a signature that indicates that quarks move freely and bind at-will when combining to form hadrons.[7][8]

68.4 Decay modes

Hadronic decay modes of J/ψ are strongly suppressed because of the OZI Rule. This effect strongly increases the lifetime of the particle and thereby gives it its very narrow decay width of just 93.2±2.1 keV. Because of this strong suppression, electromagnetic decays begin to compete with hadronic decays. This is why the J/ψ has a significant branching fraction to leptons.

68.5 See also

- OZI Rule
- List of multiple discoveries

68.6 Notes

[1] http://books.google.com.au/books?id=8AD3GDoVaMkC&pg=PA462&dq=psion+meson+-wikipedia&XpioC4Bw&ved=0CC8Q6AEwAA#v=onepage&q=psion%20meson%20-wikipedia&f=false retrieved 25 September 2014

[2] "Shared Physics prize for elementary particle" (Press release). The Royal Swedish Academy of Sciences. 18 October 1976. Retrieved 2012-04-23.

[3] Zielinski, L (8 August 2006). "Physics Folklore". QuarkNet. Retrieved 2009-04-13.

[4] Roos, M; Wohl, CG; (Particle Data Group) (2004). "Naming schemes for hadrons" (PDF). Retrieved 2009-04-13.

[5] Matsui,T;Satz,H(1986). "J/ψsuppression by quark-gluon plasma formation".*Physics Letters B***178**(4): 416–42216M. doi:10.1016/0370-2693(86)91404-8.

[6] Thews, R. L.; Schroedter, M.; Rafelski, J. (2001). "Enhanced J/ψ production in deconfined quark matter". *Physical Review C* **63** (5): 054905. arXiv:hep-ph/0007323. Bibcode:2001PhRvC..63e4905T. doi:10.1103/PhysRevC.63.054905.

[7] Schroedter, M.; Thews, R. L.; Rafelski, J. (2000). "B_c-meson production in ultrarelativistic nuclear collisions". *Physical Review C* **62** (2): 024905. arXiv:hep-ph/0004041. Bibcode:2000PhRvC..62b4905S. doi:10.1103/PhysRevC.62.024905.

[8] Fulcher, L. P.; Rafelski, J.; Thews, R. L. (1999). "B_c mesons as a signal of deconfinement". arXiv:hep-ph/9905201 [hep-ph].

68.7 References

- Glashow, S. L.; Iliopoulos, J.; Maiani, L. (1970). "Weak Interactions with Lepton-Hadron Symmetry". *Physical Review D* **2** (7): 1285–1292. Bibcode:1970PhRvD...2.1285G. doi:10.1103/PhysRevD.2.1285.
- Aubert, J. et al. (1974). "Experimental Observation of a Heavy Particle J". *Physical Review Letters* **33** (23): 1404–1406. Bibcode:1974PhRvL..33.1404A. doi:10.1103/PhysRevLett.33.1404.
- Augustin, J. et al. (1974). "Discovery of a Narrow Resonance in e^+e^- Annihilation". *Physical Review Letters* **33** (23): 1406–1408. Bibcode:1974PhRvL..33.1406A. doi:10.1103/PhysRevLett.33.1406.
- Bobra, M. (2005). "Logbook: J/ψ particle". *Symmetry Magazine* **2** (7): 34.
- Yao, W.-M. (Particle Data Group) et al. (2006). "Review of Particle Physics: Naming Scheme for Hadrons" (PDF). *Journal of Physics G* **33**: 108. doi:10.1088/0954-3899/33/1/001.

Chapter 69

Upsilon meson

The **Upsilon meson** (ϒ) is a quarkonium state (i.e. flavourless meson) formed from a bottom quark and its antiparticle. It was discovered by the E288 collaboration, headed by Leon Lederman, at Fermilab in 1977, and was the first particle containing a bottom quark to be discovered because it is the lightest that can be produced without additional massive particles. It has a lifetime of 1.21×10^{-20} s and a mass about 9.46 GeV/c^2 in the ground state.

69.1 See also

- Oops-Leon, an erroneously-claimed discovery of a similar particle at a lower mass in 1976.
- The φ particle is the analogous state made from strange quarks.
- The J/ψ particle is the analogous state made from charm quarks.
- List of mesons

69.2 References

- D.C. Hom et al. (1977). "Observation of a Dimuon Resonance at 9.5 Gev in 400-GeV Proton-Nucleus Collisions" (PDF). *Physical Review Letters* **39**: 252–255. Bibcode:1977PhRvL..39..252H. doi:10.1103/PhysRevLett.39.252.
- J. Yoh (1998). "The Discovery of the *b* Quark at Fermilab in 1977: The Experiment Coordinator's Story" (PDF). *AIP Conference Proceedings* **424**: 29–42.
- S. Eidelman *et al.* (Particle Data Group) (2004). "Review of Particle Physics – ϒ meson" (PDF). *Physics Letters B* **592**: 1. arXiv:astro-ph/0406663. Bibcode:2004PhLB..592....1P. doi:10.1016/j.physletb.2004.06.001.

Chapter 70

Theta meson

Not to be confused with the θ^+ of the τ–θ puzzle, which is now identified as kaon, or with the Θ^+ and Θ+ c, formerly thought to be pentaquarks.

The **theta meson** (θ) is a hypothetical form of quarkonium (i.e. a flavourless meson) formed by a top quark and top antiquark. It is the equivalent of the phi (strange quark, strange antiquark), psi (charm quark, charm antiquark) and upsilon (bottom quark, bottom antiquark) mesons. Due to the top quark's shortlifetime, the theta meson is not expected to be observed in nature.

70.1 See also

- List of mesons

Chapter 71

Kaon

For other uses, see Kaon (disambiguation).

In particle physics, a **kaon** /ˈkeɪ.ɒn/, also called a **K meson** and denoted K,[nb 1] is any of a group of four mesons distinguished by a quantum number called strangeness. In the quark model they are understood to be bound states of a strange quark (or antiquark) and an up or down antiquark (or quark).

Kaons have proved to be a copious source of information on the nature of fundamental interactions since their discovery in cosmic rays in 1947. They were essential in establishing the foundations of the Standard Model of particle physics, such as the quark model of hadrons and the theory of quark mixing (the latter was acknowledged by a Nobel Prize in Physics in 2008). Kaons have played a distinguished role in our understanding of fundamental conservation laws: CP violation, a phenomenon generating the observed matter-antimatter asymmetry of the universe, was discovered in the kaon system in 1964 (which was acknowledged by a Nobel prize in 1980). Moreover, direct CP violation was also discovered in the kaon decays in the early 2000s.

71.1 Basic properties

The four kaons are :

1. K−, negatively charged (containing a strange quark and an up antiquark) has mass 493.667±0.013 MeV and mean lifetime (1.2384±0.0024)×10^{-8} s.
2. K+ (antiparticle of above) positively charged (containing an up quark and a strange antiquark) must (by CPT invariance) have mass and lifetime equal to that of K−. The mass difference is 0.032±0.090 MeV, consistent with zero. The difference in lifetime is (0.11±0.09)×10^{-8} s.
3. K0, neutrally charged (containing a down quark and a strange antiquark) has mass 497.648±0.022 MeV. It has mean squared charge radius of −0.076±0.01 fm^2.
4. K0, neutrally charged (antiparticle of above) (containing a strange quark and a down antiquark) has the same mass.

It is clear from the quark model assignments that the kaons form two doublets of isospin; that is, they belong to the fundamental representation of SU(2) called the **2**. One doublet of strangeness +1 contains the K+ and the K0. The antiparticles form the other doublet (of strangeness −1).

[a] ^ Strong eigenstate. No definite lifetime (see kaon notes below)
[b] ^ Weak eigenstate. Makeup is missing small CP–violating term (see notes on neutral kaons below).
[c] ^ The mass of the K0
L and K0

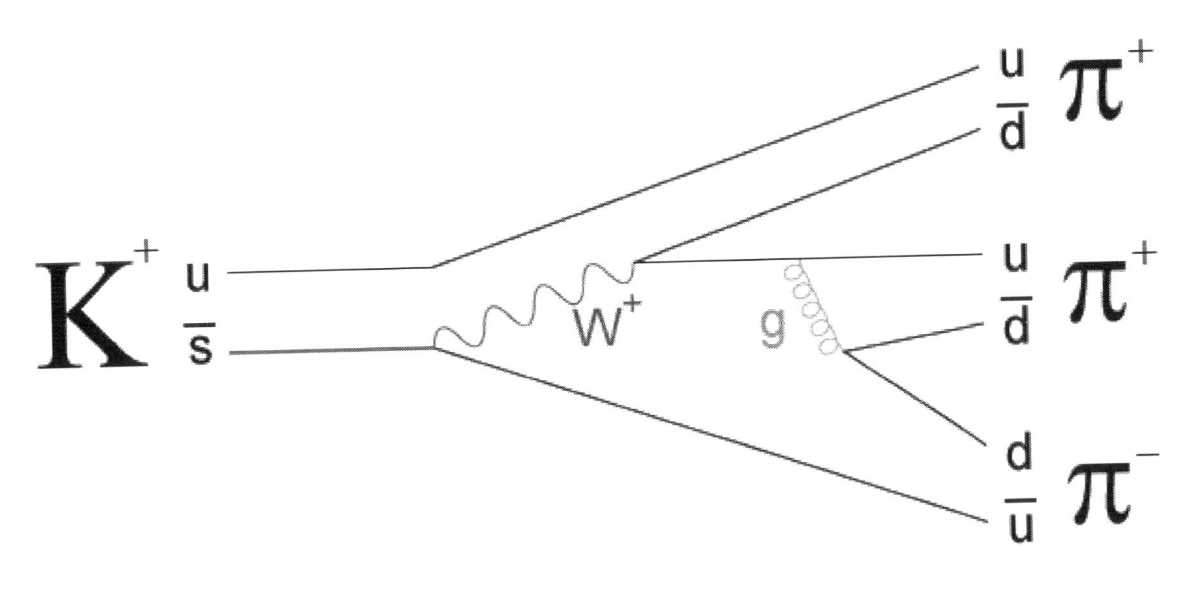

The decay of a kaon (K+) into three pions (2 π+, 1 π−) is a process that involves both weak and strong interactions.
Weak interactions : *The strange antiquark (s) of the kaon transmutes into an up antiquark (u) by the emission of a W+ boson; the W+ boson subsequently decays into a down antiquark (d) and an up quark (u).*
Strong interactions : *An up quark (u) emits a gluon (g) which decays into a down quark (d) and a down antiquark (d).*

S are given as that of the K0. However, it is known that a difference between the masses of the K0
L and K0
S on the order of 3.5×10^{-12} MeV/c^2 exists.[4]

Although the K0 and its antiparticle K0 are usually produced via the strong force, they decay weakly. Thus, once created the two are better thought of as superpositions of two weak eigenstates which have vastly different lifetimes:

1. The long-lived neutral kaon is called the K
   L ("K-long"), decays primarily into three pions, and has a mean lifetime of 5.18×10^{-8} s.
2. The short-lived neutral kaon is called the K
   S ("K-short"), decays primarily into two pions, and has a mean lifetime 8.958×10^{-11} s.

(See discussion of neutral kaon mixing below.)

An experimental observation made in 1964 that K-longs rarely decay into two pions was the discovery of CP violation (see below).

Main decay modes for K+:

Decay modes for the K− are charge conjugates of the ones above.

71.2 Strangeness

Main article: Strangeness

> The discovery of hadrons with the internal quantum number "strangeness" marks the beginning of a most exciting epoch in particle physics that even now, fifty years later, has not yet found its conclusion ... by

and large experiments have driven the development, and that major discoveries came unexpectedly or even against expectations expressed by theorists. — I.I. Bigi and A.I. Sanda, *CP violation*, (ISBN 0-521-44349-0)

In 1947, G. D. Rochester and Clifford Charles Butler of the University of Manchester published two cloud chamber photographs of cosmic ray-induced events, one showing what appeared to be a neutral particle decaying into two charged pions, and one which appeared to be a charged particle decaying into a charged pion and something neutral. The estimated mass of the new particles was very rough, about half a proton's mass. More examples of these "V-particles" were slow in coming.

The first breakthrough was obtained at Caltech, where a cloud chamber was taken up Mount Wilson, for greater cosmic ray exposure. In 1950, 30 charged and 4 neutral V-particles were reported. Inspired by this, numerous mountaintop observations were made over the next several years, and by 1953, the following terminology was adopted: "L-meson" meant muon or pion. "K meson" meant a particle intermediate in mass between the pion and nucleon. "Hyperon" meant any particle heavier than a nucleon.

The decays were extremely slow; typical lifetimes are of the order of 10^{-10} s. However, production in pion-proton reactions proceeds much faster, with a time scale of 10^{-23} s. The problem of this mismatch was solved by Abraham Pais who postulated the new quantum number called "strangeness" which is conserved in strong interactions but violated by the weak interactions. Strange particles appear copiously due to "associated production" of a strange and an antistrange particle together. It was soon shown that this could not be a multiplicative quantum number, because that would allow reactions which were never seen in the new synchrotrons which were commissioned in Brookhaven National Laboratory in 1953 and in the Lawrence Berkeley Laboratory in 1955.

71.3 Parity violation

Two different decays were found for charged strange mesons:

The intrinsic parity of a pion is P = −1, and parity is a multiplicative quantum number. Therefore, the two final states have different parity (P = +1 and P = −1, respectively). It was thought that the initial states should also have different parities, and hence be two distinct particles. However, with increasingly precise measurements, no difference was found between the masses and lifetimes of each, respectively, indicating that they are the same particle. This was known as the **τ–θ puzzle**. It was resolved only by the discovery of parity violation in weak interactions. Since the mesons decay through weak interactions, parity is not conserved, and the two decays are actually decays of the same particle,[5] now called the K+.

71.4 CP violation in neutral meson oscillations

Initially it was thought that although parity was violated, CP (charge parity) symmetry was conserved. In order to understand the discovery of CP violation, it is necessary to understand the mixing of neutral kaons; this phenomenon does not require CP violation, but it is the context in which CP violation was first observed.

71.4.1 Neutral kaon mixing

Since neutral kaons carry strangeness, they cannot be their own antiparticles. There must be then two different neutral kaons, differing by two units of strangeness. The question was then how to establish the presence of these two mesons. The solution used a phenomenon called **neutral particle oscillations**, by which these two kinds of mesons can turn from one into another through the weak interactions, which cause them to decay into pions (see the adjacent figure).

These oscillations were first investigated by Murray Gell-Mann and Abraham Pais together. They considered the CP-invariant time evolution of states with opposite strangeness. In matrix notation one can write

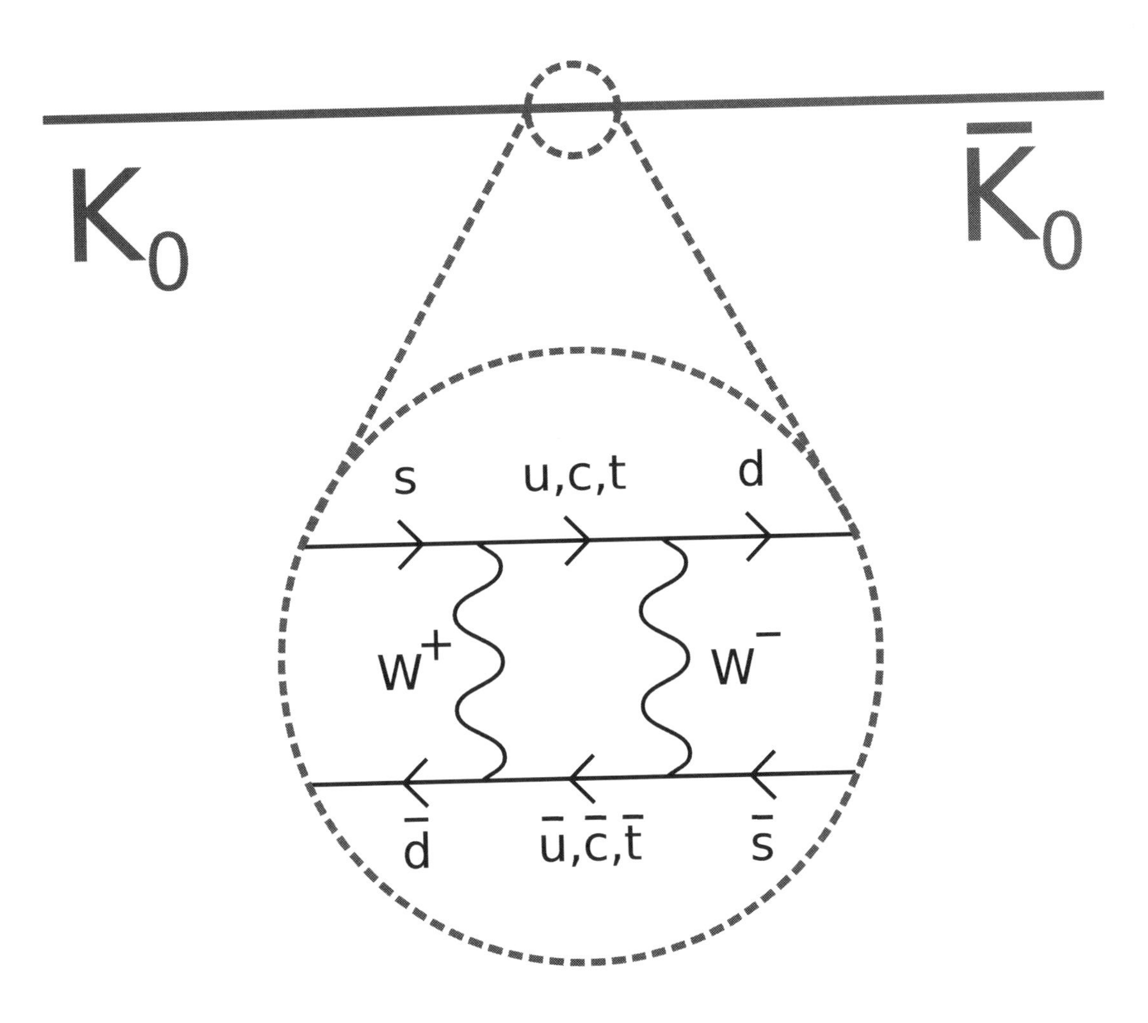

Two different neutral K mesons, carrying different strangeness, can turn from one into another through the weak interactions, since these interactions do not conserve strangeness. The strange quark in the K0 turns into a down quark by successively emitting two W-bosons of opposite charge. The down antiquark in the K0 turns into a strange antiquark by absorbing them.

$$\psi(t) = U(t)\psi(0) = e^{iHt}\begin{pmatrix} a \\ b \end{pmatrix}, \qquad H = \begin{pmatrix} M & \Delta \\ \Delta & M \end{pmatrix}$$

where ψ is a quantum state of the system specified by the amplitudes of being in each of the two basis states (which are a and b at time $t = 0$). The diagonal elements (M) of the Hamiltonian are due to strong interaction physics which conserves strangeness. The two diagonal elements must be equal, since the particle and antiparticle have equal masses in the absence of the weak interactions. The off-diagonal elements, which mix opposite strangeness particles, are due to weak interactions; CP symmetry requires them to be real.

The consequence of the matrix H being real is that the probabilities of the two states will forever oscillate back and forth. However, if any part of the matrix were imaginary, as is forbidden by CP symmetry, then part of the combination will diminish over time. The diminishing part can be either one component (a) or the other (b), or a mixture of the two.

Mixing

The eigenstates are obtained by diagonalizing this matrix. This gives new eigenvectors, which we can call $\mathbf{K}_1$ which is the difference of the two states of opposite strangeness, and $\mathbf{K}_2$, which is the sum. The two are eigenstates of **CP** with opposite eigenvalues; $\mathbf{K}_1$ has $\mathbf{CP} = +1$, and $\mathbf{K}_2$ has $\mathbf{CP} = -1$ Since the two-pion final state also has $\mathbf{CP} = +1$, only the $\mathbf{K}_1$ can decay this way. The $\mathbf{K}_2$ must decay into three pions. Since the mass of $\mathbf{K}_2$ is just a little larger than the sum of the masses of three pions, this decay proceeds very slowly, about 600 times slower than the decay of $\mathbf{K}_1$ into two pions. These two different modes of decay were observed by Leon Lederman and his coworkers in 1956, establishing the existence of the two weak eigenstates (states with definite lifetimes under decays via the weak force) of the neutral kaons.

These two weak eigenstates are called the K
L (K-long) and K
S (K-short). CP symmetry, which was assumed at the time, implies that K
S = $\mathbf{K}_1$ and K
L = $\mathbf{K}_2$.

Oscillation

Main article: Neutral particle oscillation

An initially pure beam of K0 will turn into its antiparticle while propagating, which will turn back into the original particle, and so on. This is called particle oscillation. On observing the weak decay *into leptons*, it was found that a K0 always decayed into an electron, whereas the antiparticle K0 decayed into the positron. The earlier analysis yielded a relation between the rate of electron and positron production from sources of pure K0 and its antiparticle K0. Analysis of the time dependence of this semileptonic decay showed the phenomenon of oscillation, and allowed the extraction of the mass splitting between the K
S and K
L. Since this is due to weak interactions it is very small, 10^{-15} times the mass of each state.

Regeneration

A beam of neutral kaons decays in flight so that the short-lived K
S disappears, leaving a beam of pure long-lived K
L. If this beam is shot into matter, then the K0 and its antiparticle K0 interact differently with the nuclei. The K0 undergoes quasi-elastic scattering with nucleons, whereas its antiparticle can create hyperons. Due to the different interactions of the two components, quantum coherence between the two particles is lost. The emerging beam then contains different linear superpositions of the K0 and K0. Such a superposition is a mixture of K
L and K
S; the K
S is regenerated by passing a neutral kaon beam through matter. Regeneration was observed by Oreste Piccioni and his collaborators at Lawrence Berkeley National Laboratory. Soon thereafter, Robert Adair and his coworkers reported excess K
S regeneration, thus opening a new chapter in this history.

71.4.2 CP violation

While trying to verify Adair's results, J. Christenson, James Cronin, Val Fitch and Rene Turlay of Princeton University found decays of K
L into two pions ($\mathbf{CP} = +1$) in an experiment performed in 1964 at the Alternating Gradient Synchrotron at the Brookhaven laboratory.[6] As explained in an earlier section, this required the assumed initial and final states to have different values of $\mathbf{CP}$, and hence immediately suggested CP violation. Alternative explanations such as non-linear quantum mechanics and a new unobserved particle were soon ruled out, leaving CP violation as the only possibility. Cronin and Fitch received the Nobel Prize in Physics for this discovery in 1980.

It turns out that although the K
L and K

S are weak eigenstates (because they have definite lifetimes for decay by way of the weak force), they are *not quite* **CP** eigenstates. Instead, for small ε (and up to normalization),

K
L = $\mathbf{K}_2 + \varepsilon\mathbf{K}_1$

and similarly for K
S. Thus occasionally the K
L decays as a $\mathbf{K}_1$ with **CP** = +1, and likewise the K
S can decay with **CP** = −1. This is known as **indirect CP violation**, CP violation due to mixing of K0 and its antiparticle. There is also a **direct CP violation** effect, in which the CP violation occurs during the decay itself. Both are present, because both mixing and decay arise from the same interaction with the W boson and thus have CP violation predicted by the CKM matrix.

71.5 See also

- Hadrons, mesons, hyperons and flavour
- Strange quark and the quark model
- Parity (physics), charge conjugation, time reversal symmetry, CPT invariance and CP violation
- Neutrino oscillation
- Neutral particle oscillation

71.6 Notes and references

Notes

[1] The positively charged kaon used to be called τ^+ and θ^+, as it was supposed to be two different particles until the 1960s. See the parity violation section.

References

[1] J. Beringer *et al.* (2012): Particle listings – K±

[2] J. Beringer *et al.* (2012): Particle listings – K0

[3] J. Beringer *et al.* (2012): Particle listings – K0
S

[4] J. Beringer *et al.* (2012): Particle listings – K0
L

[5] Lee, T. D.; Yang, C. N. (1 October 1956). "Question of Parity Conservation in Weak Interactions". *Physical Review* **104** (1): 254. Bibcode:1956PhRv..104..254L. doi:10.1103/PhysRev.104.254. One way out of the difficulty is to assume that parity is not strictly conserved, so that Θ+ and τ+ are two different decay modes of the same particle, which necessarily has a single mass value and a single lifetime.

[6] http://journals.aps.org/prl/pdf/10.1103/PhysRevLett.13.138

71.6.1 Bibliography

- C.Amsler; Doser, M; Antonelli, M; Asner, D; Babu, K; Baer, H; Band, H; Barnett, R; Bergren, E; Bergren, E.; Beringer, J.; Bernardi, G.; Bertl, W.; Bichsel, H.; Biebel, O.; Bloch, P.; Blucher, E.; Blusk, S.; Cahn, R. N.; Carena, M.; Caso, C.; Ceccucci, A.; Chakraborty, D.; Chen, M.-C.; Chivukula, R. S.; Cowan, G.; Dahl, O.; d'Ambrosio, G.; Damour, T. et al. (2008). "Review of Particle Physics". *Physics Letters B* (Particle Data Group) **667** (1): 1–1340. Bibcode:2008PhLB..667....1P. doi:10.1016/j.physletb.2008.07.018.
- S. Eidelman et al. (2004). "Review of Particle Physics 2004 – Strange Mesons". Particle Data Group.

  Particle Data Group; Eidelman, S.; Hayes, K. G.; Olive, K. A.; Aguilar-Benitez, M.; Amsler, C.; Asner, D.; Babu, K. S.; Barnett, R. M.; Beringer, J.; Burchat, P. R.; Carone, C. D.; Caso, S.; Conforto, G.; Dahl, O.; d'Ambrosio, G.; Doser, M.; Feng, J. L.; Gherghetta, T.; Gibbons, L.; Goodman, M.; Grab, C.; Groom, D. E.; Gurtu, A.; Hagiwara, K.; Hernández-Rey, J. J.; Hikasa, K.; Honscheid, K.; Jawahery, H. et al. (2004). "Review of Particle Physics*1". *Physics Letters B* **592** (1): 1. arXiv:astro-ph/0406663. Bibcode:2004PhLB..592....1P. doi:10.1016/j.physletb.2004.06.001.

- *The quark model*, by J.J.J. Kokkedee
- M.S. Sozzi (2008). *Discrete symmetries and CP violation*. Oxford University Press. ISBN 978-0-19-929666-8.
- I.I. Bigi, A.I. Sanda (2000). *CP violation*. Cambridge University Press. ISBN 0-521-44349-0.
- D.J. Griffiths (1987). *Introduction to Elementary Particle*. John Wiley & Sons. ISBN 0-471-60386-4.

Chapter 72

B meson

In particle physics, **B mesons** are mesons composed of a bottom antiquark and either an up (B+), down (B0), strange (B0
s) or charm quark (B+
c). The combination of a bottom antiquark and a top quark is not thought to be possible because of the top quark's short lifetime. The combination of a bottom antiquark and a bottom quark is not a B meson, but rather *bottomonium*.

Each B meson has an antiparticle that is composed of a bottom quark and an up (B–), down (B0), strange (B0
s) or charm antiquark (B–
c) respectively.

72.1 List of B mesons

72.2 B–B oscillations

Main article: B–B oscillation

The neutral B mesons, B0 and B0
s, spontaneously transform into their own antiparticles and back. This phenomenon is called flavor oscillation. The existence of neutral B meson oscillations is a fundamental prediction of the Standard Model of particle physics. It has been measured in the B0–B0 system to be about 0.496 ps^{-1},[1] and in the B0
s–B0
s system to be $\Delta m_s = 17.77 \pm 0.10$ (stat) ± 0.07 (syst) ps^{-1} measured by CDF experiment at Fermilab.[2] A first estimation of the lower and upper limit of the B0
s–B0
s system value have been made by the DØ experiment also at Fermilab.[3]

On 25 September 2006, Fermilab announced that they had claimed discovery of previously-only-theorized B_s meson oscillation.[4] According to Fermilab's press release:

> This first major discovery of Run 2 continues the tradition of particle physics discoveries at Fermilab, where the bottom (1977) and top (1995) quarks were discovered. Surprisingly, the bizarre behavior of the B_s (pronounced "B sub s") mesons is actually predicted by the Standard Model of fundamental particles and forces. The discovery of this oscillatory behavior is thus another reinforcement of the Standard Model's durability... CDF physicists have previously measured the rate of the matter-antimatter transitions for the B_s meson, which consists of the heavy bottom quark bound by the strong nuclear interaction to a strange

antiquark. Now they have achieved the standard for a discovery in the field of particle physics, where the probability for a false observation must be proven to be less than about 5 in 10 million (5/10,000,000). For CDF's result the probability is even smaller, at 8 in 100 million (8/100,000,000).

Ronald Kotulak, writing for the Chicago Tribune, called the particle "bizarre" and stated that the meson "may open the door to a new era of physics" with its proven interactions with the "spooky realm of antimatter".[5]

On 14 May 2010, physicists at the Fermi National Accelerator Laboratory reported that the oscillations decayed into matter 1% more often than into antimatter, which may help explain the abundance of matter over antimatter in the observed Universe.[6] However, more recent results at LHCb with larger data samples have suggested no significant deviation from the Standard Model.[7]

72.3 See also

- B–B oscillation

72.4 References

[1] http://repository.ubn.ru.nl/bitstream/2066/26242/

[2] A. Abulencia *et al.* (CDF Collaboration) (2006). "Observation of B0
s–B0
s Oscillations".*Physical Review Letters***97**(24): 242003. arXiv:hep-ex/0609040. Bibcode:2006PhRvL..97x297.242003.

[3] V.M. Abazov *et al.* (D0 Collaboration) (2006). "Direct Limits on the $B_s{}^0$ Oscillation Frequency" (PDF). *Physical Review Letters* **97** (2): 021802. arXiv:hep-ex/0603029. Bibcode:2006PhRvL..97b1802A. doi:10.1103/PhysRevLett.97.021802.

[4] "It might be…It could be…It is!!!" (Press release). Fermilab. 25 September 2006. Retrieved 2007-12-08.

[5] R. Kotulak (26 September 2006). "Antimatter discovery could alter physics: Particle tracked between real world, spooky realm". *Deseret News*. Archived from the original on 29 November 2007. Retrieved 2007-12-08.

[6] A New Clue to Explain Existence

[7] Article on LHCb results

72.5 External links

- W.-M. Yao *et al.* (Particle Data Group), J. Phys. G 33, 1 (2006) and 2007 partial update for edition 2008 (URL: http://pdg.lbl.gov)
- V. Jamieson (18 March 2008). "Flipping particle could explain missing antimatter". *New Scientist*. Retrieved 2010-01-23.

Chapter 73

D meson

The **D mesons** are the lightest particle containing charm quarks. They are often studied to gain knowledge on the weak interaction.[1] The strange D mesons (D_s) were called the "F mesons" prior to 1986.

73.1 Overview

The D mesons were discovered in 1976 by the Mark I detector at the Stanford Linear Accelerator Center.[2]

Since the D mesons are the lightest mesons containing a single charm quark (or antiquark), they must change the charm (anti)quark into an (anti)quark of another type to decay. Such transitions violate the internal charm quantum number, and can take place only via the weak interaction. In D mesons, the charm quark preferentially changes into a strange quark via an exchange of a W particle, therefore the D meson preferentially decays into Ks and πs.[1]

In November 2011, researchers at the LHCb experiment at CERN reported (3.5 sigma significance) that they have observed a direct CP violation in the neutral D meson decay, possibly beyond the Standard Model.[3]

73.2 List of D mesons

[a] ^ PDG reports the resonance width (Γ). Here the conversion $\tau = \hbar/\Gamma$ is given instead.

73.3 See also

- List of mesons
- List of baryons
- List of particles
- Timeline of particle discoveries

73.4 References

[1] D Meson

[2] http://www.kudryavtsev.staff.shef.ac.uk/phy466/charmed-mesons_files/charmed-mesons.ppt

[3] New Physics at LHC? An Anomaly in CP Violation : Cosmic Variance

[4] C. Amsler *et al.*. (2008): Quark Model

[5] C. Amsler *et al.*. (2008): Particle listings – $D^{\pm}$

[6] C. Amsler *et al.*. (2008): Particle listings – D^0

[7] N. Nakamura *et al.* (2010): Particle listings – $D_s^{\pm}$

[8] C. Amsler *et al.*. (2008): Particle listings – $D^{*\pm}(2010)$

[9] C. Amsler *et al.*. (2008): Particle listings – $D^{*0}(2007)$

Chapter 74

T meson

T mesons are hypothetical mesons composed of a top quark and either an up (T0), down (T+), strange (T+
s) or charm antiquark (T0
c).[1] Because of the top quark's short lifetime, T mesons are not expected to be found in nature. The combination of a top quark and top antiquark is not a T meson, but rather toponium. Each T meson has an antiparticle that is composed of a top antiquark and an up (T0), down (T−), strange (T−
s) or charm quark (T0
c) respectively.

74.1 References

[1] C. Amsler et al. (2008). "Review of Particle Physics: Naming Scheme for Hadrons" (PDF). *Physics Letters B* (Particle Data Group) **667** (1). Bibcode:2008PhLB..667....1P. doi:10.1016/j.physletb.2008.07.018.

74.2 External links

- W.-M. Yao *et al.* (Particle Data Group), J. Phys. G 33, 1 (2006) and 2007 partial update for edition 2008 (URL: http://pdg.lbl.gov)

Chapter 75

Atomic nucleus

The **nucleus** is the small, dense region consisting of protons and neutrons at the center of an atom. The atomic nucleus was discovered in 1911 by Ernest Rutherford based on the 1909 Geiger–Marsden gold foil experiment. After the discovery of the neutron in 1932, models for a nucleus composed of protons and neutrons were quickly developed by Dmitri Ivanenko[1] and Werner Heisenberg.[2][3][4][5][6] Almost all of the mass of an atom is located in the nucleus, with a very small contribution from the electron cloud. Protons and neutrons are bound together to form a nucleus by the nuclear force.

The diameter of the nucleus is in the range of 1.75 fm (1.75×10^{-15} m) for hydrogen (the diameter of a single proton)[7] to about 15 fm for the heaviest atoms, such as uranium. These dimensions are much smaller than the diameter of the atom itself (nucleus + electron cloud), by a factor of about 23,000 (uranium) to about 145,000 (hydrogen).

The branch of physics concerned with the study and understanding of the atomic nucleus, including its composition and the forces which bind it together, is called nuclear physics.

75.1 Introduction

75.1.1 History

Main article: Rutherford model

The nucleus was discovered in 1911, as a result of Ernest Rutherford's efforts to test Thomson's "plum pudding model" of the atom.[8] The electron had already been discovered earlier by J.J. Thomson himself. Knowing that atoms are neutral, Thomson postulated that there must be a positive charge as well. In his plum pudding model, Thomson stated that an atom consisted of negative electrons randomly scattered within a sphere of positive charge. Ernest Rutherford later devised an experiment, performed by Hans Geiger and Ernest Marsden under Rutherford's direction, that involved the deflection of alpha particles directed at a thin sheet of metal foil. He reasoned that if Thomson's model were correct, the positively charged alpha nuclei would easily pass through the foil with very little deviation in their paths as the foil should act in a manner as to be neutrally charged if the negative and positive charges are so intimately mixed as to make it appear neutral. To his surprise, many of the particles were deflected at very large angles. Because the mass of alpha particles is about 8000 times that of an electron, it became apparent that a very strong force must be present if it could deflect the massive and fast moving helium nuclei. He realized that the plum pudding model could not be accurate and that the deflections of the alpha particles could only be explained if the positive and negatives charges were in fact separated from each other and that the mass of the atom was a concentrated point of positive charge. Thus, the idea of a nuclear atom with a dense center of positive charge and mass became justified.

A model of the atomic nucleus showing it as a compact bundle of the two types of nucleons: protons (red) and neutrons (blue). In this diagram, protons and neutrons look like little balls stuck together, but an actual nucleus (as understood by modern nuclear physics) cannot be explained like this, but only by using quantum mechanics. In a nucleus which occupies a certain energy level (for example, the ground state), each nucleon can be said to occupy a range of locations.

75.1.2 Etymology

The term **nucleus** is from the Latin word *nucleus*, a diminutive of *nux* ("nut"), meaning the kernel (i.e., the "small nut") inside a watery type of fruit (like a peach). In 1844, Michael Faraday used the term to refer to the "central point of an atom". The modern atomic meaning was proposed by Ernest Rutherford in 1912.[9] The adoption of the term "nucleus" to atomic theory, however, was not immediate. In 1916, for example, Gilbert N. Lewis stated, in his famous article *The Atom and the Molecule*, that "the atom is composed of the *kernel* and an outer atom or *shell*"[10]

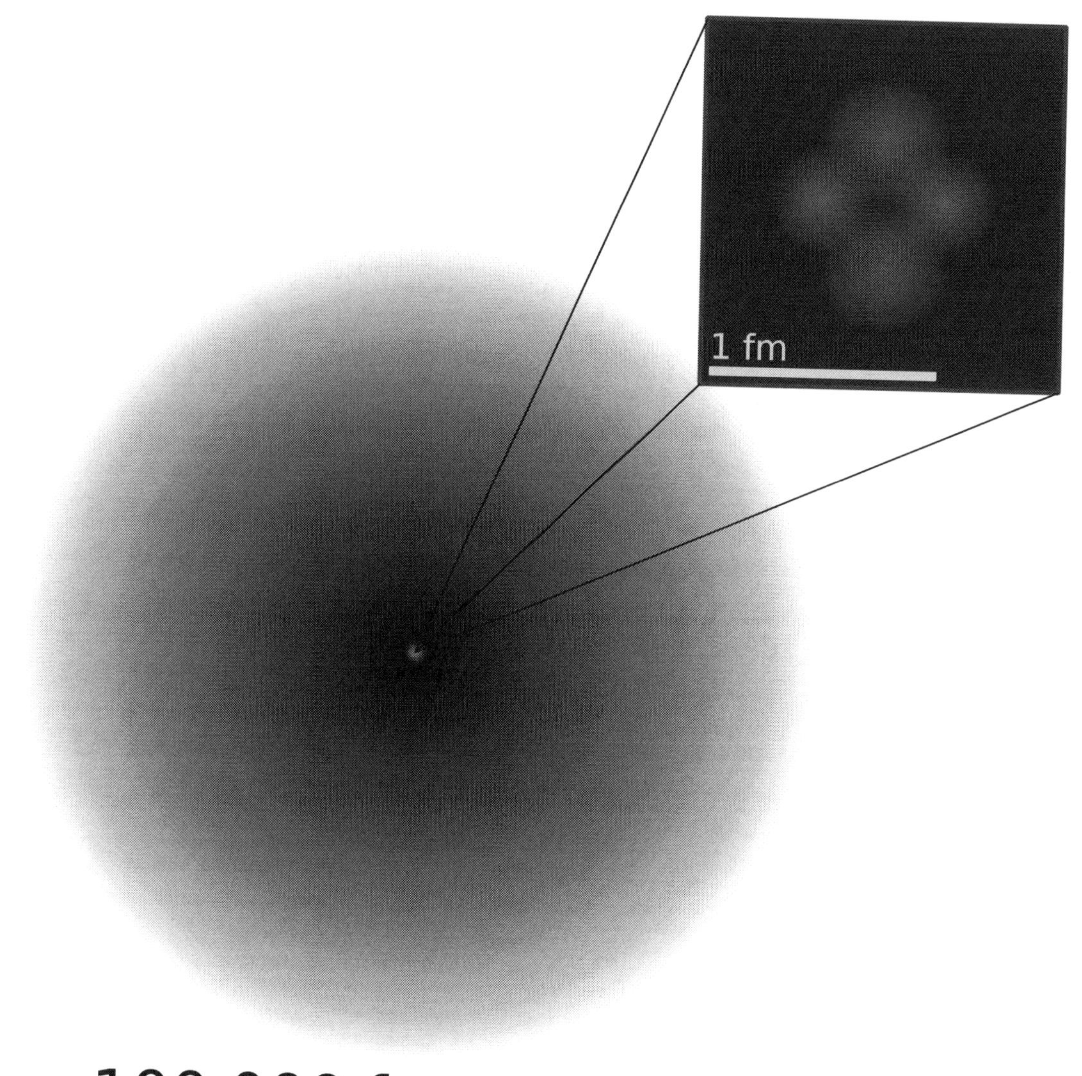

A figurative depiction of the helium–4 atom with the electron cloud in shades of gray. In the nucleus, the two protons and two neutrons are depicted in red and blue. This depiction shows the particles as separate, whereas in an actual helium atom, the protons are superimposed in space and most likely found at the very center of the nucleus, and the same is true of the two neutrons. Thus, all four particles are most likely found in exactly the same space, at the central point. Classical images of separate particles fail to model known charge distributions in very small nuclei. A more accurate image is that the spatial distribution of nucleons in helium's nucleus, although on a far smaller scale, is much closer to the helium ***electron cloud*** *shown here, than to the fanciful nucleus image.*

75.1.3 Nuclear makeup

The nucleus of an atom consists of neutrons and protons, which in turn are the manifestation of fundamental particles, called quarks, that are held in association by the nuclear strong force in certain stable combinations of hadrons, called baryons. The nuclear strong force extends far enough from each baryon so as to bind the neutrons and protons together against the repulsive force of the positively charged protons. The nuclear strong force has a very short range and essentially drops to zero just beyond the edge of the nucleus. The collective action of the positively charged nucleus is to hold the electrically negative charged electrons in their orbits about the nucleus. The collection of negatively charged electrons orbiting the nucleus display an affinity for certain configurations and numbers of electrons that make their orbits stable. Which chemical element an atom represents is determined by the number of protons in the nucleus; the atom will have an equal number of electrons orbiting that nucleus. Individual chemical elements can create more stable electron configurations by combining to share their electrons. It is that sharing of electrons to create stable electronic orbits about the nucleus that appears to us as the chemistry of our macro world.

While protons define the entire charge of a nucleus and, hence, its chemical identity, neutrons are electrically neutral, but contribute to the mass of a nucleus to nearly the same extent as the protons. Neutrons explain the phenomenon of isotopes – varieties of the same chemical element which differ only in their atomic mass, not their chemical action.

75.2 Protons and neutrons

Protons and neutrons are fermions, with different values of the strong isospin quantum number, so two protons and two neutrons can share the same space wave function since they are not identical quantum entities. They sometimes are viewed as two different quantum states of the same particle, the *nucleon*.[11][12] Two fermions, such as two protons, or two neutrons, or a proton + neutron (the deuteron) can exhibit bosonic behavior when they become loosely bound in pairs.

In the rare case of a hypernucleus, a third baryon called a hyperon, with a different value of the strangeness quantum number can also share the wave function. However, the latter type of nuclei are extremely unstable and are not found on Earth except in high energy physics experiments.

The neutron has a positively charged core of radius $\approx$ 0.3 fm surrounded by a compensating negative charge of radius between 0.3 fm and 2 fm. The proton has an approximately exponentially decaying positive charge distribution with a mean square radius of about 0.8 fm.[13]

75.3 Forces

Nuclei are bound together by the residual strong force (nuclear force). The residual strong force is a minor residuum of the strong interaction which binds quarks together to form protons and neutrons. This force is much weaker *between* neutrons and protons because it is mostly neutralized within them, in the same way that electromagnetic forces *between* neutral atoms (such as van der Waals forces that act between two inert gas atoms) are much weaker than the electromagnetic forces that hold the parts of the atoms internally together (for example, the forces that hold the electrons in an inert gas atom bound to its nucleus).

The nuclear force is highly attractive at the distance of typical nucleon separation, and this overwhelms the repulsion between protons which is due to the electromagnetic force, thus allowing nuclei to exist. However, because the residual strong force has a limited range because it decays quickly with distance (see Yukawa potential), only nuclei smaller than a certain size can be completely stable. The largest known completely stable (e.g., stable to alpha, beta, and gamma decay) nucleus is lead-208 which contains a total of 208 nucleons (126 neutrons and 82 protons). Nuclei larger than this maximal size of 208 particles are unstable and (as a trend) become increasingly short-lived with larger size, as the number of neutrons and protons which compose them increases beyond this number. However, bismuth-209 is also stable to beta decay and has the longest half-life to alpha decay of any known isotope, estimated at a billion times longer than the age of the universe.

The residual strong force is effective over a very short range (usually only a few fermis; roughly one or two nucleon diameters) and causes an attraction between any pair of nucleons. For example, between protons and neutrons to form [NP] deuteron, and also between protons and protons, and neutrons and neutrons.

75.4 Halo nuclei and strong force range limits

The effective absolute limit of the range of the strong force is represented by halo nuclei such as lithium-11 or boron-14, in which dineutrons, or other collections of neutrons, orbit at distances of about ten fermis (roughly similar to the 8 fermi radius of the nucleus of uranium-238). These nuclei are not maximally dense. Halo nuclei form at the extreme edges of the chart of the nuclides—the neutron drip line and proton drip line—and are all unstable with short half-lives, measured in milliseconds; for example, lithium-11 has a half-life of 8.8 milliseconds.

Halos in effect represent an excited state with nucleons in an outer quantum shell which has unfilled energy levels "below" it (both in terms of radius and energy). The halo may be made of either neutrons [NN, NNN] or protons [PP, PPP]. Nuclei which have a single neutron halo include ^{11}Be and ^{19}C. A two-neutron halo is exhibited by ^{6}He, ^{11}Li, ^{17}B, ^{19}B and ^{22}C. Two-neutron halo nuclei break into three fragments, never two, and are called *Borromean nuclei* because of this behavior (referring to a system of three interlocked rings in which breaking any ring frees both of the others). ^{8}He and ^{14}Be both exhibit a four-neutron halo. Nuclei which have a proton halo include ^{8}B and ^{26}P. A two-proton halo is exhibited by ^{17}Ne and ^{27}S. Proton halos are expected to be more rare and unstable than the neutron examples, because of the repulsive electromagnetic forces of the excess proton(s).

75.5 Nuclear models

Although the standard model of physics is widely believed to completely describe the composition and behavior of the nucleus, generating predictions from theory is much more difficult than for most other areas of particle physics. This is due to two reasons:

- In principle, the physics within a nuclei can be derived entirely from quantum chromodynamics (QCD). In practice however, current computational and mathematical approaches for solving QCD in low-energy systems such as the nuclei are extremely limited. This is due to the phase transition that occurs between high-energy quark matter and low-energy hadronic matter, which renders perturbative techniques unusable, making it difficult to construct an accurate QCD-derived model of the forces between nucleons. Current approaches are limited to either phenomenological models such as the Argonne v18 potential or chiral effective field theory.[14]

- Even if the nuclear force is well constrained, a significant amount of computational power is required to accurately compute the properties of nuclei *ab initio*. Developments in many-body theory have made this possible for many low mass and relatively stable nuclei, but further improvements in both computational power and mathematical approaches are required before heavy nuclei or highly unstable nuclei can be tackled.

Historically, experiments have been compared to relatively crude models that are necessarily imperfect. None of these models can completely explain experimental data on nuclear structure.[15]

The nuclear radius (R) is considered to be one of the basic quantities that any model must predict. For stable nuclei (not halo nuclei or other unstable distorted nuclei) the nuclear radius is roughly proportional to the cube root of the mass number (A) of the nucleus, and particularly in nuclei containing many nucleons, as they arrange in more spherical configurations:

The stable nucleus has approximately a constant density and therefore the nuclear radius R can be approximated by the following formula,

$$R = r_0 A^{1/3}$$

where A = Atomic mass number (the number of protons Z, plus the number of neutrons N) and r_0 = 1.25 fm = 1.25 × 10^{-15} m. In this equation, the constant r_0 varies by 0.2 fm, depending on the nucleus in question, but this is less than 20% change from a constant.[16]

In other words, packing protons and neutrons in the nucleus gives *approximately* the same total size result as packing hard spheres of a constant size (like marbles) into a tight spherical or almost spherical bag (some stable nuclei are not quite spherical, but are known to be prolate).

75.5.1 Liquid drop model

Main article: Semi-empirical mass formula

Early models of the nucleus viewed the nucleus as a rotating liquid drop. In this model, the trade-off of long-range electromagnetic forces and relatively short-range nuclear forces, together cause behavior which resembled surface tension forces in liquid drops of different sizes. This formula is successful at explaining many important phenomena of nuclei, such as their changing amounts of binding energy as their size and composition changes (see semi-empirical mass formula), but it does not explain the special stability which occurs when nuclei have special "magic numbers" of protons or neutrons.

The terms in the semi-empirical mass formula, which can be used to approximate the binding energy of many nuclei, are considered as the sum of five types of energies (see below). Then the picture of a nucleus as a drop of incompressible liquid roughly accounts for the observed variation of binding energy of the nucleus:

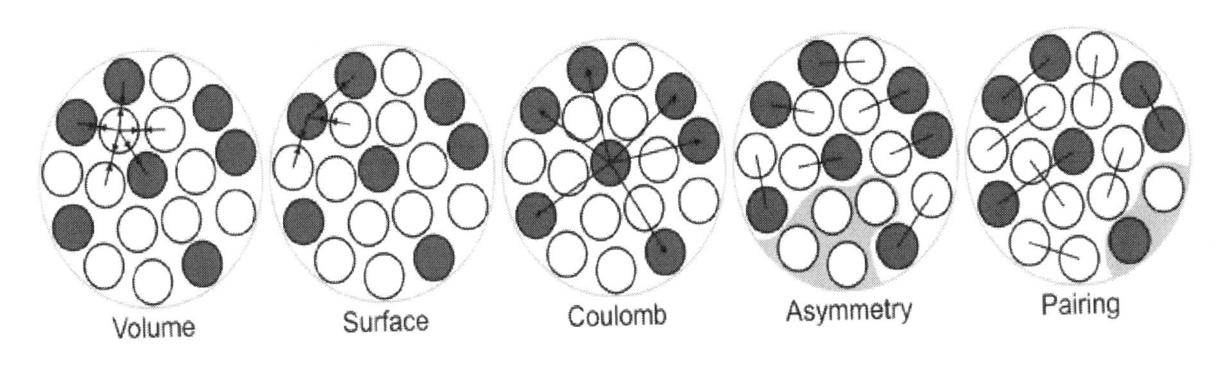

Volume energy. When an assembly of nucleons of the same size is packed together into the smallest volume, each interior nucleon has a certain number of other nucleons in contact with it. So, this nuclear energy is proportional to the volume.

Surface energy. A nucleon at the surface of a nucleus interacts with fewer other nucleons than one in the interior of the nucleus and hence its binding energy is less. This surface energy term takes that into account and is therefore negative and is proportional to the surface area.

Coulomb Energy. The electric repulsion between each pair of protons in a nucleus contributes toward decreasing its binding energy.

Asymmetry energy (also called Pauli Energy). An energy associated with the Pauli exclusion principle. Were it not for the Coulomb energy, the most stable form of nuclear matter would have the same number of neutrons as protons, since unequal numbers of neutrons and protons imply filling higher energy levels for one type of particle, while leaving lower energy levels vacant for the other type.

Pairing energy. An energy which is a correction term that arises from the tendency of proton pairs and neutron pairs to occur. An even number of particles is more stable than an odd number.

75.5.2 Shell models and other quantum models

Main article: Nuclear shell model

A number of models for the nucleus have also been proposed in which nucleons occupy orbitals, much like the atomic orbitals in atomic physics theory. These wave models imagine nucleons to be either sizeless point particles in potential wells, or else probability waves as in the "optical model", frictionlessly orbiting at high speed in potential wells.

In the above models, the nucleons may occupy orbitals in pairs, due to being fermions, which allows to explain even/odd Z and N effects well-known from experiments. The exact nature and capacity of nuclear shells differs from those of electrons in atomic orbitals, primarily because the potential well in which the nucleons move (especially in larger nuclei) is quite different from the central electromagnetic potential well which binds electrons in atoms. Some resemblance to atomic orbital models may be seen in a small atomic nucleus like that of helium-4, in which the two protons and two neutrons separately occupy 1s orbitals analogous to the 1s orbital for the two electrons in the helium atom, and achieve unusual stability for the same reason. Nuclei with 5 nucleons are all extremely unstable and short-lived, yet, helium-3, with 3 nucleons, is very stable even with lack of a closed 1s orbital shell. Another nucleus with 3 nucleons, the triton hydrogen-3 is unstable and will decay into helium-3 when isolated. Weak nuclear stability with 2 nucleons {NP} in the 1s orbital is found in the deuteron hydrogen-2, with only one nucleon in each of the proton and neutron potential wells. While each nucleon is a fermion, the {NP} deuteron is a boson and thus does not follow Pauli Exclusion for close packing within shells. Lithium-6 with 6 nucleons is highly stable without a closed second 1p shell orbital. For light nuclei with total nucleon numbers 1 to 6 only those with 5 do not show some evidence of stability. Observations of beta-stability of light nuclei outside closed shells indicate that nuclear stability is much more complex than simple closure of shell orbitals with magic numbers of protons and neutrons.

For larger nuclei, the shells occupied by nucleons begin to differ significantly from electron shells, but nevertheless, present nuclear theory does predict the magic numbers of filled nuclear shells for both protons and neutrons. The closure of the stable shells predicts unusually stable configurations, analogous to the noble group of nearly-inert gases in chemistry. An example is the stability of the closed shell of 50 protons, which allows tin to have 10 stable isotopes, more than any other element. Similarly, the distance from shell-closure explains the unusual instability of isotopes which have far from stable numbers of these particles, such as the radioactive elements 43 (technetium) and 61 (promethium), each of which is preceded and followed by 17 or more stable elements.

There are however problems with the shell model when an attempt is made to account for nuclear properties well away from closed shells. This has led to complex *post hoc* distortions of the shape of the potential well to fit experimental data, but the question remains whether these mathematical manipulations actually correspond to the spatial deformations in real nuclei. Problems with the shell model have led some to propose realistic two-body and three-body nuclear force effects involving nucleon clusters and then build the nucleus on this basis. Two such cluster models are the Close-Packed Spheron Model of Linus Pauling and the 2D Ising Model of MacGregor.[15]

75.5.3 Consistency between models

Main article: Nuclear structure

As with the case of superfluid liquid helium, atomic nuclei are an example of a state in which both (1) "ordinary" particle physical rules for volume and (2) non-intuitive quantum mechanical rules for a wave-like nature apply. In superfluid helium, the helium atoms have volume, and essentially "touch" each other, yet at the same time exhibit strange bulk properties, consistent with a Bose–Einstein condensation. The latter reveals that they also have a wave-like nature and do not exhibit standard fluid properties, such as friction. For nuclei made of hadrons which are fermions, the same type of condensation does not occur, yet nevertheless, many nuclear properties can only be explained similarly by a combination of properties of particles with volume, in addition to the frictionless motion characteristic of the wave-like behavior of objects trapped in Erwin Schrödinger's quantum orbitals.

75.6 See also

- Giant resonance
- List of particles
- Nuclear medicine
- Radioactivity
- Semi-empirical mass formula

75.7 References

[1] Iwanenko, D.D., The neutron hypothesis, Nature **129** (1932) 798.

[2] Heisenberg, W. (1932). "Über den Bau der Atomkerne. I". *Z. Phys.* **77**: 1–11. Bibcode:1932ZPhy...77....1H. doi:1

3.[3] Heisenberg, W. (1932). "Über den Bau der Atomkerne. II".*Z.Phys.* **78** (3–4): 156–164. Bibcode:1932ZPhy...78..156H. doi:10.1007/BF01337585.

[4] Heisenberg, W. (1933). "Über den Bau der Atomkerne. III". *Z. Phys.* **80** (9–10): 587–596. Bibcode:1933ZPhy...80..587H. doi:10.1007/BF01335696.

[5] Miller A. I. *Early Quantum Electrodynamics: A Sourcebook*, Cambridge University Press, Cambridge, 1995, ISBN 0521568919, pp. 84–88.

[6] Bernard Fernandez and Georges Ripka (2012). "Nuclear Theory After the Discovery of the Neutron". *Unravelling the Mystery of the Atomic Nucleus: A Sixty Year Journey 1896 — 1956*. Springer. p. 263. ISBN 9781461441809. Retrieved 15 February 2013.

[7] Geoff Brumfiel (July 7, 2010). "The proton shrinks in size". *Nature*. doi:10.1038/news.2010.337.

[8] *Rutgers University*. "The Rutherford Experiment". physics.rutgers.edu. Retrieved February 26, 2013.

[9] D. Harper. "Nucleus". *Online Etymology Dictionary*. Retrieved 2010-03-06.

[10] G.N. Lewis (1916). "The Atom and the Molecule". *Journal of the American Chemical Society* **38** (4): 4. doi:10.1021/ja02261a002.

[11] A.G. Sitenko, V.K. Tartakovskiĭ (1997). *Theory of Nucleus: Nuclear Structure and Nuclear Interaction*. Kluwer Academic. p. 3. ISBN 0-7923-4423-5.

[12] M.A. Srednicki (2007). *Quantum Field Theory*. Cambridge University Press. pp. 522–523. ISBN 978-0-521-86449-7.

[13] J.-L. Basdevant, J. Rich, M. Spiro (2005). *Fundamentals in Nuclear Physics*. Springer. p. 155. ISBN 0-387-01672-4.

[14] Machleidt,R.;Entem,D.R. (2011). "Chiral effectivefield theory and nuclear forces".*Physics Reports***503**(1): 1–75. arXi1. Bibcode:2011PhR...503....1M. doi:10.1016/j.physrep.2011.02.001.

[15] N.D. Cook (2010). *Models of the Atomic Nucleus* (2nd ed.). Springer. p. 57 ff. ISBN 978-3-642-14736-4.

[16] K.S. Krane (1987). *Introductory Nuclear Physics*. Wiley-VCH. ISBN 0-471-80553-X.

75.8 External links

- The Nucleus – a chapter from an online textbook
- The LIVEChart of Nuclides – IAEA in Java or HTML
- Article on the "nuclear shell model," giving nuclear shell filling for the various elements. Accessed Sept. 16, 2009.
- Timeline: Subatomic Concepts, Nuclear Science & Technology.

Chapter 76

Atom

For other uses, see Atom (disambiguation).

An **atom** is the smallest constituent unit of ordinary matter that has the properties of a chemical element.[1] Every solid, liquid, gas, and plasma is made up of neutral or ionized atoms. Atoms are very small; typical sizes are around 100 pm (a ten-billionth of a meter, in the short scale).[2] However, atoms do not have well defined boundaries, and there are different ways to define their size which give different but close values.

Atoms are small enough that classical physics give noticeably incorrect results. Through the development of physics, atomic models have incorporated quantum principles to better explain and predict the behavior.

Every atom is composed of a nucleus and one or more electrons bound to the nucleus. The nucleus is made of one or more protons and typically a similar number of neutrons (none in hydrogen-1). Protons and neutrons are called nucleons. Over 99.94% of the atom's mass is in the nucleus. The protons have a positive electric charge, the electrons have a negative electric charge, and the neutrons have no electric charge. If the number of protons and electrons are equal, that atom is electrically neutral. If an atom has more or fewer electrons than protons, then it has an overall negative or positive charge, respectively, and it is called an ion.

Electrons of an atom are attracted to the protons in an atomic nucleus by this electromagnetic force. The protons and neutrons in the nucleus are attracted to each other by a different force, the nuclear force, which is usually stronger than the electromagnetic force repelling the positively charged protons from one another. Under certain circumstances the repelling electromagnetic force becomes stronger than the nuclear force, and nucleons can be ejected from the nucleus, leaving behind a different element: nuclear decay resulting in nuclear transmutation.

The number of protons in the nucleus defines to what chemical element the atom belongs: for example, all copper atoms contain 29 protons. The number of neutrons defines the isotope of the element.[3] The number of electrons influences the magnetic properties of an atom. Atoms can attach to one or more other atoms by chemical bonds to form chemical compounds such as molecules. The ability of atoms to associate and dissociate is responsible for most of the physical changes observed in nature, and is the subject of the discipline of chemistry.

Not all the matter of the universe is composed of atoms. Dark matter comprises more of the Universe than matter, and is composed not of atoms, but of particles of a currently unknown type.

76.1 History of atomic theory

Main article: Atomic theory

76.1.1 Atoms in philosophy

Main article: Atomism

The idea that matter is made up of discrete units is a very old one, appearing in many ancient cultures such as Greece and India. The word “atom”, in fact, was coined by ancient Greek philosophers. However, these ideas were founded in philosophical and theological reasoning rather than evidence and experimentation. As a result, their views on what atoms look like and how they behave were incorrect. They also could not convince everybody, so atomism was but one of a number of competing theories on the nature of matter. It was not until the 19th century that the idea was embraced and refined by scientists, when the blossoming science of chemistry produced discoveries that only the concept of atoms could explain.

76.1.2 First evidence-based theory

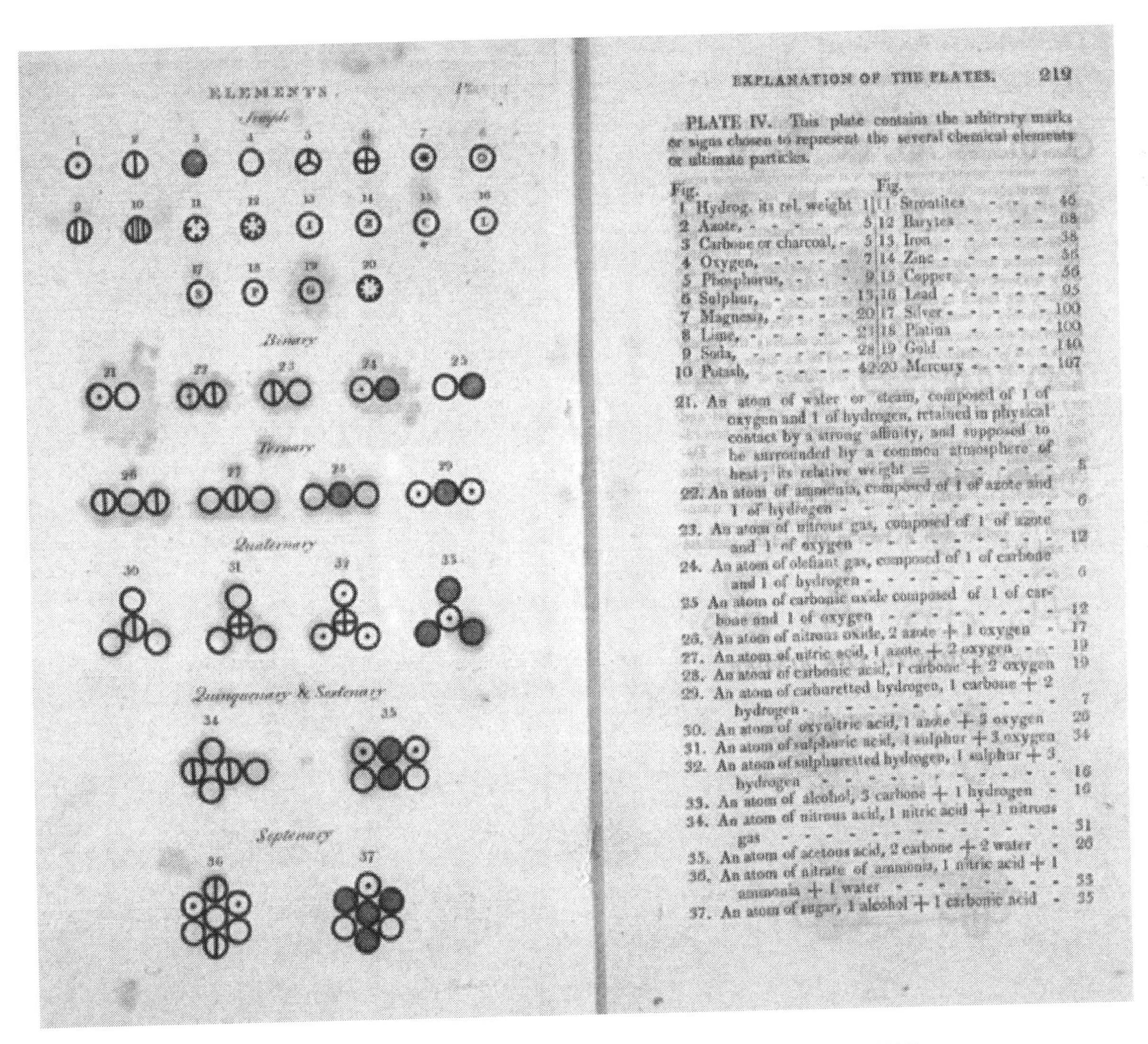

EXPLANATION OF THE PLATES. 219

PLATE IV. This plate contains the arbitrary marks or signs chosen to represent the several chemical elements or ultimate particles.

Fig.			Fig.		
1	Hydrog. its rel. weight	1	11	Strontites	46
2	Azote,	5	12	Barytes	68
3	Carbone or charcoal,	5	13	Iron	38
4	Oxygen,	7	14	Zinc	56
5	Phosphorus,	9	15	Copper	56
6	Sulphur,	13	16	Lead	95
7	Magnesia,	20	17	Silver	100
8	Lime,	24	18	Platina	100
9	Soda,	28	19	Gold	140
10	Potash,	42	20	Mercury	167

21. An atom of water or steam, composed of 1 of oxygen and 1 of hydrogen, retained in physical contact by a strong affinity, and supposed to be surrounded by a common atmosphere of heat; its relative weight = 8
22. An atom of ammonia, composed of 1 of azote and 1 of hydrogen 6
23. An atom of nitrous gas, composed of 1 of azote and 1 of oxygen 12
24. An atom of olefiant gas, composed of 1 of carbone and 1 of hydrogen 6
25 An atom of carbonic oxide composed of 1 of carbone and 1 of oxygen 12
26. An atom of nitrous oxide, 2 azote + 1 oxygen 17
27. An atom of nitric acid, 1 azote + 2 oxygen 19
28. An atom of carbonic acid, 1 carbone + 2 oxygen 19
29. An atom of carburetted hydrogen, 1 carbone + 2 hydrogen 7
30. An atom of oxynitric acid, 1 azote + 3 oxygen 26
31. An atom of sulphuric acid, 1 sulphur + 3 oxygen 34
32. An atom of sulphuretted hydrogen, 1 sulphur + 3 hydrogen 16
33. An atom of alcohol, 3 carbone + 1 hydrogen 16
34. An atom of nitrous acid, 1 nitric acid + 1 nitrous gas 31
35. An atom of acetous acid, 2 carbone + 2 water 26
36. An atom of nitrate of ammonia, 1 nitric acid + 1 ammonia + 1 water 33
37. An atom of sugar, 1 alcohol + 1 carbonic acid 35

Various atoms and molecules as depicted in John Dalton's A New System of Chemical Philosophy *(1808).*

In the early 1800s, John Dalton used the concept of atoms to explain why elements always react in ratios of small whole numbers (the law of multiple proportions). For instance, there are two types of tin oxide: one is 88.1% tin and 11.9% oxygen and the other is 78.7% tin and 21.3% oxygen (tin(II) oxide and tin dioxide respectively). This means that 100g of tin will combine either with 13.5g or 27g of oxygen. 13.5 and 27 form a ratio of 1:2, a ratio of small whole numbers. This common pattern in chemistry suggested to Dalton that elements react in whole number multiples of discrete units—in other words, atoms. In the case of tin oxides, one tin atom will combine with either one or two oxygen atoms.[4]

Dalton also believed atomic theory could explain why water absorbs different gases in different proportions. For example, he found that water absorbs carbon dioxide far better than it absorbs nitrogen.[5] Dalton hypothesized this was due to the differences in mass and complexity of the gases' respective particles. Indeed, carbon dioxide molecules (CO_2) are heavier and larger than nitrogen molecules (N_2).

76.1.3 Brownian motion

In 1827, botanist Robert Brown used a microscope to look at dust grains floating in water and discovered that they moved about erratically, a phenomenon that became known as "Brownian motion". This was thought to be caused by water molecules knocking the grains about. In 1905 Albert Einstein produced the first mathematical analysis of the motion.[6][7][8] French physicist Jean Perrin used Einstein's work to experimentally determine the mass and dimensions of atoms, thereby conclusively verifying Dalton's atomic theory.[9]

76.1.4 Discovery of the electron

The physicist J. J. Thomson measured the mass of cathode rays, showing they were made of particles, but were around 1800 times lighter than the lightest atom, hydrogen. Therefore, they were not atoms, but a new particle, the first *subatomic* particle to be discovered, which he originally called "*corpuscle*" but was later named *electron*, after particles postulated by George Johnstone Stoney in 1874. He also showed they were identical to particles given off by photoelectric and radioactive materials.[10] It was quickly recognized that they are the particles that carry electric currents in metal wires, and carry the negative electric charge within atoms. Thomson was given the 1906 Nobel Prize in Physics for this work. Thus he overturned the belief that atoms are the indivisible, ultimate particles of matter.[11] Thomson also incorrectly postulated that the low mass, negatively charged electrons were distributed throughout the atom in a uniform sea of positive charge. This became known as the plum pudding model.

76.1.5 Discovery of the nucleus

Main article: Geiger-Marsden experiment

In 1909, Hans Geiger and Ernest Marsden, under the direction of Ernest Rutherford, bombarded a metal foil with alpha particles to observe how they scattered. They expected all the alpha particles to pass straight through with little deflection, because Thomson's model said that the charges in the atom are so diffuse that their electric fields could not affect the alpha particles much. However, Geiger and Marsden spotted alpha particles being deflected by angles greater than 90°, which was supposed to be impossible according to Thomson's model. To explain this, Rutherford proposed that the positive charge of the atom is concentrated in a tiny nucleus at the center of the atom.[12]

76.1.6 Discovery of isotopes

While experimenting with the products of radioactive decay, in 1913 radiochemist Frederick Soddy discovered that there appeared to be more than one type of atom at each position on the periodic table.[13] The term isotope was coined by Margaret Todd as a suitable name for different atoms that belong to the same element. J.J. Thomson created a technique for separating atom types through his work on ionized gases, which subsequently led to the discovery of stable isotopes.[14]

76.1.7 Bohr model

Main article: Bohr model

In 1913 the physicist Niels Bohr proposed a model in which the electrons of an atom were assumed to orbit the nucleus but could only do so in a finite set of orbits, and could jump between these orbits only in discrete changes of energy corresponding to absorption or radiation of a photon.[15] This quantization was used to explain why the electrons orbits are stable (given that normally, charges in acceleration, including circular motion, lose kinetic energy which is emitted as electromagnetic radiation, see *synchrotron radiation*) and why elements absorb and emit electromagnetic radiation in discrete spectra.[16]

Later in the same year Henry Moseley provided additional experimental evidence in favor of Niels Bohr's theory. These results refined Ernest Rutherford's and Antonius Van den Broek's model, which proposed that the atom contains in its nucleus a number of positive nuclear charges that is equal to its (atomic) number in the periodic table. Until these experiments, atomic number was not known to be a physical and experimental quantity. That it is equal to the atomic nuclear charge remains the accepted atomic model today.[17]

76.1.8 Chemical bonding explained

Chemical bonds between atoms were now explained, by Gilbert Newton Lewis in 1916, as the interactions between their constituent electrons.[18] As the chemical properties of the elements were known to largely repeat themselves according to the periodic law,[19] in 1919 the American chemist Irving Langmuir suggested that this could be explained if the electrons in an atom were connected or clustered in some manner. Groups of electrons were thought to occupy a set of electron shells about the nucleus.[20]

76.1.9 Further developments in quantum physics

The Stern–Gerlach experiment of 1922 provided further evidence of the quantum nature of the atom. When a beam of silver atoms was passed through a specially shaped magnetic field, the beam was split based on the direction of an atom's angular momentum, or spin. As this direction is random, the beam could be expected to spread into a line. Instead, the beam was split into two parts, depending on whether the atomic spin was oriented up or down.[21]

In 1924, Louis de Broglie proposed that all particles behave to an extent like waves. In 1926, Erwin Schrödinger used this idea to develop a mathematical model of the atom that described the electrons as three-dimensional waveforms rather than point particles. A consequence of using waveforms to describe particles is that it is mathematically impossible to obtain precise values for both the position and momentum of a particle at the same time; this became known as the uncertainty principle, formulated by Werner Heisenberg in 1926. In this concept, for a given accuracy in measuring a position one could only obtain a range of probable values for momentum, and vice versa. This model was able to explain observations of atomic behavior that previous models could not, such as certain structural and spectral patterns of atoms larger than hydrogen. Thus, the planetary model of the atom was discarded in favor of one that described atomic orbital zones around the nucleus where a given electron is most likely to be observed.[22][23]

76.1.10 Discovery of the neutron

The development of the mass spectrometer allowed the mass of atoms to be measured with increased accuracy. The device uses a magnet to bend the trajectory of a beam of ions, and the amount of deflection is determined by the ratio of an atom's mass to its charge. The chemist Francis William Aston used this instrument to show that isotopes had different masses. The atomic mass of these isotopes varied by integer amounts, called the whole number rule.[24] The explanation for these different isotopes awaited the discovery of the neutron, an uncharged particle with a mass similar to the proton, by the physicist James Chadwick in 1932. Isotopes were then explained as elements with the same number of protons, but different numbers of neutrons within the nucleus.[25]

76.1.11 Fission, high-energy physics and condensed matter

In 1938, the German chemist Otto Hahn, a student of Rutherford, directed neutrons onto uranium atoms expecting to get transuranium elements. Instead, his chemical experiments showed barium as a product.[26] A year later, Lise Meitner and her nephew Otto Frisch verified that Hahn's result were the first experimental *nuclear fission*.[27][28] In 1944, Hahn received the Nobel prize in chemistry. Despite Hahn's efforts, the contributions of Meitner and Frisch were not recognized.[29]

In the 1950s, the development of improved particle accelerators and particle detectors allowed scientists to study the impacts of atoms moving at high energies.[30] Neutrons and protons were found to be hadrons, or composites of smaller particles called quarks. The standard model of particle physics was developed that so far has successfully explained the properties of the nucleus in terms of these sub-atomic particles and the forces that govern their interactions.[31]

76.2 Structure

76.2.1 Subatomic particles

Main article: Subatomic particle

Though the word *atom* originally denoted a particle that cannot be cut into smaller particles, in modern scientific usage the atom is composed of various subatomic particles. The constituent particles of an atom are the electron, the proton and the neutron; all three are fermions. However, the hydrogen-1 atom has no neutrons and the hydron ion has no electrons.

The electron is by far the least massive of these particles at 9.11×10^{-31} kg, with a negative electrical charge and a size that is too small to be measured using available techniques.[32] It is the lightest particle with a positive rest mass measured. Under ordinary conditions, electrons are bound to the positively charged nucleus by the attraction created from opposite electric charges. If an atom has more or fewer electrons than its atomic number, then it becomes respectively negatively or positively charged as a whole; a charged atom is called an ion. Electrons have been known since the late 19th century, mostly thanks to J.J. Thomson; see history of subatomic physics for details.

Protons have a positive charge and a mass 1,836 times that of the electron, at 1.6726×10^{-27} kg. The number of protons in an atom is called its atomic number. Ernest Rutherford (1919) observed that nitrogen under alpha-particle bombardment ejects what appeared to be hydrogen nuclei. By 1920 he had accepted that the hydrogen nucleus is a distinct particle within the atom and named it proton.

Neutrons have no electrical charge and have a free mass of 1,839 times the mass of the electron,[33] or 1.6929×10^{-27} kg, the heaviest of the three constituent particles, but it can be reduced by the nuclear binding energy. Neutrons and protons (collectively known as nucleons) have comparable dimensions—on the order of 2.5×10^{-15} m—although the 'surface' of these particles is not sharply defined.[34] The neutron was discovered in 1932 by the English physicist James Chadwick.

In the Standard Model of physics, electrons are truly elementary particles with no internal structure. However, both protons and neutrons are composite particles composed of elementary particles called quarks. There are two types of quarks in atoms, each having a fractional electric charge. Protons are composed of two up quarks (each with charge $+\frac{2}{3}$) and one down quark (with a charge of $-\frac{1}{3}$). Neutrons consist of one up quark and two down quarks. This distinction accounts for the difference in mass and charge between the two particles.[35][36]

The quarks are held together by the strong interaction (or strong force), which is mediated by gluons. The protons and neutrons, in turn, are held to each other in the nucleus by the nuclear force, which is a residuum of the strong force that has somewhat different range-properties (see the article on the nuclear force for more). The gluon is a member of the family of gauge bosons, which are elementary particles that mediate physical forces.[35][36]

76.2.2 Nucleus

Main article: Atomic nucleus

All the bound protons and neutrons in an atom make up a tiny atomic nucleus, and are collectively called nucleons. The radius of a nucleus is approximately equal to $1.07 \sqrt[3]{A}$ fm, where A is the total number of nucleons.[37] This is much smaller than the radius of the atom, which is on the order of 10^5 fm. The nucleons are bound together by a short-ranged attractive potential called the residual strong force. At distances smaller than 2.5 fm this force is much more powerful than the electrostatic force that causes positively charged protons to repel each other.[38]

Atoms of the same element have the same number of protons, called the atomic number. Within a single element, the number of neutrons may vary, determining the isotope of that element. The total number of protons and neutrons determine the nuclide. The number of neutrons relative to the protons determines the stability of the nucleus, with certain isotopes undergoing radioactive decay.[39]

The proton, the electron, and the neutron are classified as fermions. Fermions obey the Pauli exclusion principle which prohibits *identical* fermions, such as multiple protons, from occupying the same quantum state at the same time. Thus, every proton in the nucleus must occupy a quantum state different from all other protons, and the same applies to all neutrons of the nucleus and to all electrons of the electron cloud. However, a proton and a neutron are allowed to occupy the same quantum state.[40]

For atoms with low atomic numbers, a nucleus that has more neutrons than protons tends to drop to a lower energy state through radioactive decay so that the neutron–proton ratio is closer to one. However, as the atomic number increases, a higher proportion of neutrons is required to offset the mutual repulsion of the protons. Thus, there are no stable nuclei with equal proton and neutron numbers above atomic number $Z = 20$ (calcium) and as Z increases, the neutron–proton ratio of stable isotopes increases.[40] The stable isotope with the highest proton–neutron ratio is lead-208 (about 1.5).

The number of protons and neutrons in the atomic nucleus can be modified, although this can require very high energies because of the strong force. Nuclear fusion occurs when multiple atomic particles join to form a heavier nucleus, such as through the energetic collision of two nuclei. For example, at the core of the Sun protons require energies of 3–10 keV to overcome their mutual repulsion—the coulomb barrier—and fuse together into a single nucleus.[41] Nuclear fission is the opposite process, causing a nucleus to split into two smaller nuclei—usually through radioactive decay. The nucleus can also be modified through bombardment by high energy subatomic particles or photons. If this modifies the number of protons in a nucleus, the atom changes to a different chemical element.[42][43]

If the mass of the nucleus following a fusion reaction is less than the sum of the masses of the separate particles, then the difference between these two values can be emitted as a type of usable energy (such as a gamma ray, or the kinetic energy of a beta particle), as described by Albert Einstein's mass–energy equivalence formula, $E = mc^2$, where m is the mass loss and c is the speed of light. This deficit is part of the binding energy of the new nucleus, and it is the non-recoverable loss of the energy that causes the fused particles to remain together in a state that requires this energy to separate.[44]

The fusion of two nuclei that create larger nuclei with lower atomic numbers than iron and nickel—a total nucleon number of about 60—is usually an exothermic process that releases more energy than is required to bring them together.[45] It is this energy-releasing process that makes nuclear fusion in stars a self-sustaining reaction. For heavier nuclei, the binding energy per nucleon in the nucleus begins to decrease. That means fusion processes producing nuclei that have atomic numbers higher than about 26, and atomic masses higher than about 60, is an endothermic process. These more massive nuclei can not undergo an energy-producing fusion reaction that can sustain the hydrostatic equilibrium of a star.[40]

76.2.3 Electron cloud

Main articles: Atomic orbital and Electron configuration

The electrons in an atom are attracted to the protons in the nucleus by the electromagnetic force. This force binds the electrons inside an electrostatic potential well surrounding the smaller nucleus, which means that an external source of energy is needed for the electron to escape. The closer an electron is to the nucleus, the greater the attractive force. Hence electrons bound near the center of the potential well require more energy to escape than those at greater separations.

Electrons, like other particles, have properties of both a particle and a wave. The electron cloud is a region inside the potential well where each electron forms a type of three-dimensional standing wave—a wave form that does not move relative to the nucleus. This behavior is defined by an atomic orbital, a mathematical function that characterises the probability that an electron appears to be at a particular location when its position is measured.[46] Only a discrete (or quantized) set of these orbitals exist around the nucleus, as other possible wave patterns rapidly decay into a more stable form.[47] Orbitals can have one or more ring or node structures, and they differ from each other in size, shape and orientation.[48]

Each atomic orbital corresponds to a particular energy level of the electron. The electron can change its state to a higher energy level by absorbing a photon with sufficient energy to boost it into the new quantum state. Likewise, through spontaneous emission, an electron in a higher energy state can drop to a lower energy state while radiating the excess energy as a photon. These characteristic energy values, defined by the differences in the energies of the quantum states, are responsible for atomic spectral lines.[47]

The amount of energy needed to remove or add an electron—the electron binding energy—is far less than the binding energy of nucleons. For example, it requires only 13.6 eV to strip a ground-state electron from a hydrogen atom,[49] compared to 2.23 *million* eV for splitting a deuterium nucleus.[50] Atoms are electrically neutral if they have an equal number of protons and electrons. Atoms that have either a deficit or a surplus of electrons are called ions. Electrons that are farthest from the nucleus may be transferred to other nearby atoms or shared between atoms. By this mechanism, atoms are able to bond into molecules and other types of chemical compounds like ionic and covalent network crystals.[51]

76.3 Properties

76.3.1 Nuclear properties

Main articles: Isotope, Stable isotope, List of nuclides and List of elements by stability of isotopes

By definition, any two atoms with an identical number of *protons* in their nuclei belong to the same chemical element. Atoms with equal numbers of protons but a different number of *neutrons* are different isotopes of the same element. For example, all hydrogen atoms admit exactly one proton, but isotopes exist with no neutrons (hydrogen-1, by far the most common form,[52] also called protium), one neutron (deuterium), two neutrons (tritium) and more than two neutrons. The known elements form a set of atomic numbers, from the single proton element hydrogen up to the 118-proton element ununoctium.[53] All known isotopes of elements with atomic numbers greater than 82 are radioactive.[54][55]

About 339 nuclides occur naturally on Earth,[56] of which 254 (about 75%) have not been observed to decay, and are referred to as "stable isotopes". However, only 90 of these nuclides are stable to all decay, even in theory. Another 164 (bringing the total to 254) have not been observed to decay, even though in theory it is energetically possible. These are also formally classified as “stable”. An additional 34 radioactive nuclides have half-lives longer than 80 million years, and are long-lived enough to be present from the birth of the solar system. This collection of 288 nuclides are known as primordial nuclides. Finally, an additional 51 short-lived nuclides are known to occur naturally, as daughter products of primordial nuclide decay (such as radium from uranium), or else as products of natural energetic processes on Earth, such as cosmic ray bombardment (for example, carbon-14).[57][note 1]

For 80 of the chemical elements, at least one stable isotope exists. As a rule, there is only a handful of stable isotopes for each of these elements, the average being 3.2 stable isotopes per element. Twenty-six elements have only a single stable isotope, while the largest number of stable isotopes observed for any element is ten, for the element tin. Elements 43, 61, and all elements numbered 83 or higher have no stable isotopes.[58]

Stability of isotopes is affected by the ratio of protons to neutrons, and also by the presence of certain “magic numbers” of neutrons or protons that represent closed and filled quantum shells. These quantum shells correspond to a set of energy levels within the shell model of the nucleus; filled shells, such as the filled shell of 50 protons for tin, confers unusual stability on the nuclide. Of the 254 known stable nuclides, only four have both an odd number of protons *and* odd number of neutrons: hydrogen-2 (deuterium), lithium-6, boron-10 and nitrogen-14. Also, only four naturally occurring, radioactive odd–odd nuclides have a half-life over a billion years: potassium-40, vanadium-50, lanthanum-138 and tantalum-180m. Most odd–odd nuclei are highly unstable with respect to beta decay, because the decay products are even–even, and are therefore more strongly bound, due to nuclear pairing effects.[58]

76.3.2 Mass

Main articles: Atomic mass and mass number

The large majority of an atom’s mass comes from the protons and neutrons that make it up. The total number of these particles (called “nucleons”) in a given atom is called the mass number. It is a positive integer and dimensionless (instead of having dimension of mass), because it expresses a count. An example of use of a mass number is “carbon-12,” which has 12 nucleons (six protons and six neutrons).

The actual mass of an atom at rest is often expressed using the unified atomic mass unit (u), also called dalton (Da). This unit is defined as a twelfth of the mass of a free neutral atom of carbon-12, which is approximately 1.66×10^{-27} kg.[59] Hydrogen-1 (the lightest isotope of hydrogen which is also the nuclide with the lowest mass) has an atomic weight of 1.007825 u.[60] The value of this number is called the atomic mass. A given atom has an atomic mass approximately equal (within 1%) to its mass number times the atomic mass unit (for example the mass of a nitrogen-14 is roughly 14 u). However, this number will not be exactly an integer except in the case of carbon-12 (see below).[61] The heaviest stable atom is lead-208,[54] with a mass of 207.9766521 u.[62]

As even the most massive atoms are far too light to work with directly, chemists instead use the unit of moles. One mole of atoms of any element always has the same number of atoms (about 6.022×10^{23}). This number was chosen so that if an element has an atomic mass of 1 u, a mole of atoms of that element has a mass close to one gram. Because of the definition of the unified atomic mass unit, each carbon-12 atom has an atomic mass of exactly 12 u, and so a mole of carbon-12 atoms weighs exactly 0.012 kg.[59]

76.3.3 Shape and size

Main article: Atomic radius

Atoms lack a well-defined outer boundary, so their dimensions are usually described in terms of an atomic radius. This is a measure of the distance out to which the electron cloud extends from the nucleus.[2] However, this assumes the atom to exhibit a spherical shape, which is only obeyed for atoms in vacuum or free space. Atomic radii may be derived from the distances between two nuclei when the two atoms are joined in a chemical bond. The radius varies with the location of an atom on the atomic chart, the type of chemical bond, the number of neighboring atoms (coordination number) and a quantum mechanical property known as spin.[63] On the periodic table of the elements, atom size tends to increase when moving down columns, but decrease when moving across rows (left to right).[64] Consequently, the smallest atom is helium with a radius of 32 pm, while one of the largest is caesium at 225 pm.[65]

When subjected to external forces, like electrical fields, the shape of an atom may deviate from spherical symmetry. The deformation depends on the field magnitude and the orbital type of outer shell electrons, as shown by group-theoretical considerations. Aspherical deviations might be elicited for instance in crystals, where large crystal-electrical fields may occur at low-symmetry lattice sites. Significant ellipsoidal deformations have recently been shown to occur for sulfur ions[66] and chalcogen ions[67] in pyrite-type compounds.

Atomic dimensions are thousands of times smaller than the wavelengths of light (400–700 nm) so they cannot be viewed using an optical microscope. However, individual atoms can be observed using a scanning tunneling microscope. To visualize the minuteness of the atom, consider that a typical human hair is about 1 million carbon atoms in width.[68] A single drop of water contains about 2 sextillion (2×10^{21}) atoms of oxygen, and twice the number of hydrogen atoms.[69] A single carat diamond with a mass of 2×10^{-4} kg contains about 10 sextillion (10^{22}) atoms of carbon.[note 2] If an apple were magnified to the size of the Earth, then the atoms in the apple would be approximately the size of the original apple.[70]

76.3.4 Radioactive decay

Main article: Radioactive decay

Every element has one or more isotopes that have unstable nuclei that are subject to radioactive decay, causing the nucleus to emit particles or electromagnetic radiation. Radioactivity can occur when the radius of a nucleus is large compared with the radius of the strong force, which only acts over distances on the order of 1 fm.[71]

The most common forms of radioactive decay are:[72][73]

- Alpha decay: this process is caused when the nucleus emits an alpha particle, which is a helium nucleus consisting of two protons and two neutrons. The result of the emission is a new element with a lower atomic number.
- Beta decay (and electron capture): these processes are regulated by the weak force, and result from a transformation of a neutron into a proton, or a proton into a neutron. The neutron to proton transition is accompanied by the emission of an electron and an antineutrino, while proton to neutron transition (except in electron capture) causes

the emission of a positron and a neutrino. The electron or positron emissions are called beta particles. Beta decay either increases or decreases the atomic number of the nucleus by one. Electron capture is more common than positron emission, because it requires less energy. In this type of decay, an electron is absorbed by the nucleus, rather than a positron emitted from the nucleus. A neutrino is still emitted in this process, and a proton changes to a neutron.

- Gamma decay: this process results from a change in the energy level of the nucleus to a lower state, resulting in the emission of electromagnetic radiation. The excited state of a nucleus which results in gamma emission usually occurs following the emission of an alpha or a beta particle. Thus, gamma decay usually follows alpha or beta decay.

Other more rare types of radioactive decay include ejection of neutrons or protons or clusters of nucleons from a nucleus, or more than one beta particle. An analog of gamma emission which allows excited nuclei to lose energy in a different way, is internal conversion— a process that produces high-speed electrons that are not beta rays, followed by production of high-energy photons that are not gamma rays. A few large nuclei explode into two or more charged fragments of varying masses plus several neutrons, in a decay called spontaneous nuclear fission.

Each radioactive isotope has a characteristic decay time period—the half-life—that is determined by the amount of time needed for half of a sample to decay. This is an exponential decay process that steadily decreases the proportion of the remaining isotope by 50% every half-life. Hence after two half-lives have passed only 25% of the isotope is present, and so forth.[71]

76.3.5 Magnetic moment

Main articles: Electron magnetic moment and Nuclear magnetic moment

Elementary particles possess an intrinsic quantum mechanical property known as spin. This is analogous to the angular momentum of an object that is spinning around its center of mass, although strictly speaking these particles are believed to be point-like and cannot be said to be rotating. Spin is measured in units of the reduced Planck constant (ħ), with electrons, protons and neutrons all having spin ½ ħ, or "spin-½". In an atom, electrons in motion around the nucleus possess orbital angular momentum in addition to their spin, while the nucleus itself possesses angular momentum due to its nuclear spin.[74]

The magnetic field produced by an atom—its magnetic moment—is determined by these various forms of angular momentum, just as a rotating charged object classically produces a magnetic field. However, the most dominant contribution comes from electron spin. Due to the nature of electrons to obey the Pauli exclusion principle, in which no two electrons may be found in the same quantum state, bound electrons pair up with each other, with one member of each pair in a spin up state and the other in the opposite, spin down state. Thus these spins cancel each other out, reducing the total magnetic dipole moment to zero in some atoms with even number of electrons.[75]

In ferromagnetic elements such as iron, cobalt and nickel, an odd number of electrons leads to an unpaired electron and a net overall magnetic moment. The orbitals of neighboring atoms overlap and a lower energy state is achieved when the spins of unpaired electrons are aligned with each other, a spontaneous process known as an exchange interaction. When the magnetic moments of ferromagnetic atoms are lined up, the material can produce a measurable macroscopic field. Paramagnetic materials have atoms with magnetic moments that line up in random directions when no magnetic field is present, but the magnetic moments of the individual atoms line up in the presence of a field.[75][76]

The nucleus of an atom will have no spin when it has even numbers of both neutrons and protons, but for other cases of odd numbers, the nucleus may have a spin. Normally nuclei with spin are aligned in random directions because of thermal equilibrium. However, for certain elements (such as xenon-129) it is possible to polarize a significant proportion of the nuclear spin states so that they are aligned in the same direction—a condition called hyperpolarization. This has important applications in magnetic resonance imaging.[77][78]

76.3.6 Energy levels

The potential energy of an electron in an atom is negative, its dependence of its position reaches the minimum (the most absolute value) inside the nucleus, and vanishes when the distance from the nucleus goes to infinity, roughly in an inverse proportion to the distance. In the quantum-mechanical model, a bound electron can only occupy a set of states centered on the nucleus, and each state corresponds to a specific energy level; see time-independent Schrödinger equation for theoretical explanation. An energy level can be measured by the amount of energy needed to unbind the electron from the atom, and is usually given in units of electronvolts (eV). The lowest energy state of a bound electron is called the ground state, i.e. stationary state, while an electron transition to a higher level results in an excited state.[79] The electron's energy raises when n increases because the (average) distance to the nucleus increases. Dependence of the energy on ℓ is caused not by electrostatic potential of the nucleus, but by interaction between electrons.

For an electron to transition between two different states, e.g. grounded state to first excited level (ionization), it must absorb or emit a photon at an energy matching the difference in the potential energy of those levels, according to Niels Bohr model, what can be precisely calculated by the Schrödinger equation. Electrons jump between orbitals in a particle-like fashion. For example, if a single photon strikes the electrons, only a single electron changes states in response to the photon; see Electron properties.

The energy of an emitted photon is proportional to its frequency, so these specific energy levels appear as distinct bands in the electromagnetic spectrum.[80] Each element has a characteristic spectrum that can depend on the nuclear charge, subshells filled by electrons, the electromagnetic interactions between the electrons and other factors.[81]

When a continuous spectrum of energy is passed through a gas or plasma, some of the photons are absorbed by atoms, causing electrons to change their energy level. Those excited electrons that remain bound to their atom spontaneously emit this energy as a photon, traveling in a random direction, and so drop back to lower energy levels. Thus the atoms behave like a filter that forms a series of dark absorption bands in the energy output. (An observer viewing the atoms from a view that does not include the continuous spectrum in the background, instead sees a series of emission lines from the photons emitted by the atoms.) Spectroscopic measurements of the strength and width of atomic spectral lines allow the composition and physical properties of a substance to be determined.[82]

Close examination of the spectral lines reveals that some display a fine structure splitting. This occurs because of spin–orbit coupling, which is an interaction between the spin and motion of the outermost electron.[83] When an atom is in an external magnetic field, spectral lines become split into three or more components; a phenomenon called the Zeeman effect. This is caused by the interaction of the magnetic field with the magnetic moment of the atom and its electrons. Some atoms can have multiple electron configurations with the same energy level, which thus appear as a single spectral line. The interaction of the magnetic field with the atom shifts these electron configurations to slightly different energy levels, resulting in multiple spectral lines.[84] The presence of an external electric field can cause a comparable splitting and shifting of spectral lines by modifying the electron energy levels, a phenomenon called the Stark effect.[85]

If a bound electron is in an excited state, an interacting photon with the proper energy can cause stimulated emission of a photon with a matching energy level. For this to occur, the electron must drop to a lower energy state that has an energy difference matching the energy of the interacting photon. The emitted photon and the interacting photon then move off in parallel and with matching phases. That is, the wave patterns of the two photons are synchronized. This physical property is used to make lasers, which can emit a coherent beam of light energy in a narrow frequency band.[86]

76.3.7 Valence and bonding behavior

Main articles: Valence (chemistry) and Chemical bond

Valency is the combining power of an element. It is equal to number of hydrogen atoms that atom can combine or displace in forming compounds.[87] The outermost electron shell of an atom in its uncombined state is known as the valence shell, and the electrons in that shell are called valence electrons. The number of valence electrons determines the bonding behavior with other atoms. Atoms tend to chemically react with each other in a manner that fills (or empties) their outer valence shells.[88] For example, a transfer of a single electron between atoms is a useful approximation for bonds that form between atoms with one-electron more than a filled shell, and others that are one-electron short of a full shell, such as occurs in the compound sodium chloride and other chemical ionic salts. However, many elements display multiple valences, or tendencies to share differing numbers of electrons in different compounds. Thus, chemical bonding between these elements takes many forms of electron-sharing that are more than simple electron transfers. Examples include the element carbon and the organic compounds.[89]

The chemical elements are often displayed in a periodic table that is laid out to display recurring chemical properties, and elements with the same number of valence electrons form a group that is aligned in the same column of the table. (The horizontal rows correspond to the filling of a quantum shell of electrons.) The elements at the far right of the table have their outer shell completely filled with electrons, which results in chemically inert elements known as the noble gases.[90][91]

76.3.8 States

Main articles: State of matter and Phase (matter)

Quantities of atoms are found in different states of matter that depend on the physical conditions, such as temperature and pressure. By varying the conditions, materials can transition between solids, liquids, gases and plasmas.[92] Within a state, a material can also exist in different allotropes. An example of this is solid carbon, which can exist as graphite or diamond.[93] Gaseous allotropes exist as well, such as dioxygen and ozone.

At temperatures close to absolute zero, atoms can form a Bose–Einstein condensate, at which point quantum mechanical effects, which are normally only observed at the atomic scale, become apparent on a macroscopic scale.[94][95] This super-cooled collection of atoms then behaves as a single super atom, which may allow fundamental checks of quantum mechanical behavior.[96]

76.4 Identification

The scanning tunneling microscope is a device for viewing surfaces at the atomic level. It uses the quantum tunneling phenomenon, which allows particles to pass through a barrier that would normally be insurmountable. Electrons tunnel through the vacuum between two planar metal electrodes, on each of which is an adsorbed atom, providing a tunneling-current density that can be measured. Scanning one atom (taken as the tip) as it moves past the other (the sample) permits plotting of tip displacement versus lateral separation for a constant current. The calculation shows the extent to which scanning-tunneling-microscope images of an individual atom are visible. It confirms that for low bias, the microscope images the space-averaged dimensions of the electron orbitals across closely packed energy levels—the Fermi level local density of states.[97][98]

An atom can be ionized by removing one of its electrons. The electric charge causes the trajectory of an atom to bend when it passes through a magnetic field. The radius by which the trajectory of a moving ion is turned by the magnetic field is determined by the mass of the atom. The mass spectrometer uses this principle to measure the mass-to-charge ratio of ions. If a sample contains multiple isotopes, the mass spectrometer can determine the proportion of each isotope in the sample by measuring the intensity of the different beams of ions. Techniques to vaporize atoms include inductively coupled plasma atomic emission spectroscopy and inductively coupled plasma mass spectrometry, both of which use a plasma to vaporize samples for analysis.[99]

A more area-selective method is electron energy loss spectroscopy, which measures the energy loss of an electron beam within a transmission electron microscope when it interacts with a portion of a sample. The atom-probe tomograph has sub-nanometer resolution in 3-D and can chemically identify individual atoms using time-of-flight mass spectrometry.[100]

Spectra of excited states can be used to analyze the atomic composition of distant stars. Specific light wavelengths contained in the observed light from stars can be separated out and related to the quantized transitions in free gas atoms. These colors can be replicated using a gas-discharge lamp containing the same element.[101] Helium was discovered in this way in the spectrum of the Sun 23 years before it was found on Earth.[102]

76.5 Origin and current state

Atoms form about 4% of the total energy density of the observable Universe, with an average density of about 0.25 atoms/m^3.[103] Within a galaxy such as the Milky Way, atoms have a much higher concentration, with the density of matter in the interstellar medium (ISM) ranging from 10^5 to 10^9 atoms/m^3.[104] The Sun is believed to be inside the Local Bubble, a region of highly ionized gas, so the density in the solar neighborhood is only about 10^3 atoms/m^3.[105] Stars form from dense clouds in the ISM, and the evolutionary processes of stars result in the steady enrichment of the ISM with elements more massive than hydrogen and helium. Up to 95% of the Milky Way's atoms are concentrated inside stars and the total mass of atoms forms about 10% of the mass of the galaxy.[106] (The remainder of the mass is an unknown dark matter.)[107]

76.5.1 Formation

Electrons are thought to exist in the Universe since early stages of the Big Bang. Atomic nuclei forms in nucleosynthesis reactions. In about three minutes Big Bang nucleosynthesis produced most of the helium, lithium, and deuterium in the Universe, and perhaps some of the beryllium and boron.[108][109][110]

Ubiquitousness and stability of atoms relies on their binding energy, which means that an atom has a lower energy than an unbound system of the nucleus and electrons. Where the temperature is much higher than ionization potential, the matter exists in the form of plasma—a gas of positively charged ions (possibly, bare nuclei) and electrons. When the temperature drops below the ionization potential, atoms become statistically favorable. Atoms (complete with bound electrons) became to dominate over charged particles 380,000 years after the Big Bang—an epoch called recombination, when the expanding Universe cooled enough to allow electrons to become attached to nuclei.[111]

Since the Big Bang, which produced no carbon or heavier elements, atomic nuclei have been combined in stars through the process of nuclear fusion to produce more of the element helium, and (via the triple alpha process) the sequence of elements from carbon up to iron;[112] see stellar nucleosynthesis for details.

Isotopes such as lithium-6, as well as some beryllium and boron are generated in space through cosmic ray spallation.[113] This occurs when a high-energy proton strikes an atomic nucleus, causing large numbers of nucleons to be ejected.

Elements heavier than iron were produced in supernovae through the r-process and in AGB stars through the s-process, both of which involve the capture of neutrons by atomic nuclei.[114] Elements such as lead formed largely through the radioactive decay of heavier elements.[115]

76.5.2 Earth

Most of the atoms that make up the Earth and its inhabitants were present in their current form in the nebula that collapsed out of a molecular cloud to form the Solar System. The rest are the result of radioactive decay, and their relative proportion can be used to determine the age of the Earth through radiometric dating.[116][117] Most of the helium in the crust of the Earth (about 99% of the helium from gas wells, as shown by its lower abundance of helium-3) is a product of alpha decay.[118]

There are a few trace atoms on Earth that were not present at the beginning (i.e., not "primordial"), nor are results of radioactive decay. Carbon-14 is continuously generated by cosmic rays in the atmosphere.[119] Some atoms on Earth have been artificially generated either deliberately or as by-products of nuclear reactors or explosions.[120][121] Of the transuranic elements—those with atomic numbers greater than 92—only plutonium and neptunium occur naturally on Earth.[122][123] Transuranic elements have radioactive lifetimes shorter than the current age of the Earth[124] and thus identifiable quantities of these elements have long since decayed, with the exception of traces of plutonium-244 possibly deposited by cosmic dust.[125] Natural deposits of plutonium and neptunium are produced by neutron capture in uranium ore.[126]

The Earth contains approximately 1.33×10^{50} atoms.[127] Although small numbers of independent atoms of noble gases exist, such as argon, neon, and helium, 99% of the atmosphere is bound in the form of molecules, including carbon dioxide and diatomic oxygen and nitrogen. At the surface of the Earth, an overwhelming majority of atoms combine to form various compounds, including water, salt, silicates and oxides. Atoms can also combine to create materials that do not consist of discrete molecules, including crystals and liquid or solid metals.[128][129] This atomic matter forms networked arrangements that lack the particular type of small-scale interrupted order associated with molecular matter.[130]

76.5.3 Rare and theoretical forms

Superheavy elements

Main article: Transuranium element

While isotopes with atomic numbers higher than lead (82) are known to be radioactive, an "island of stability" has been proposed for some elements with atomic numbers above 103. These superheavy elements may have a nucleus that is relatively stable against radioactive decay.[131] The most likely candidate for a stable superheavy atom, unbihexium, has 126 protons and 184 neutrons.[132]

Exotic matter

Main article: Exotic matter

Each particle of matter has a corresponding antimatter particle with the opposite electrical charge. Thus, the positron is a positively charged antielectron and the antiproton is a negatively charged equivalent of a proton. When a matter and corresponding antimatter particle meet, they annihilate each other. Because of this, along with an imbalance between the number of matter and antimatter particles, the latter are rare in the universe. The first causes of this imbalance are not yet fully understood, although theories of baryogenesis may offer an explanation. As a result, no antimatter atoms have been discovered in nature.[133][134] However, in 1996 the antimatter counterpart of the hydrogen atom (antihydrogen) was synthesized at the CERN laboratory in Geneva.[135][136]

Other exotic atoms have been created by replacing one of the protons, neutrons or electrons with other particles that have the same charge. For example, an electron can be replaced by a more massive muon, forming a muonic atom. These types of atoms can be used to test the fundamental predictions of physics.[137][138][139]

76.6 See also

- History of quantum mechanics
- Infinite divisibility
- List of basic chemistry topics
- Timeline of atomic and subatomic physics
- Vector model of the atom
- Nuclear model
- Radioactive isotope

76.7 Notes

[1] For more recent updates see Interactive Chart of Nuclides (Brookhaven National Laboratory).

[2] A carat is 200 milligrams. By definition, carbon-12 has 0.012 kg per mole. The Avogadro constant defines 6×10^{23} atoms per mole.

76.8 References

[1] "Atom". *Compendium of Chemical Terminology (IUPAC Gold Book)* (2nd ed.). IUPAC. Retrieved 2015-04-25.

[2] Ghosh, D. C.; Biswas, R. (2002). "Theoretical calculation of Absolute Radii of Atoms and Ions. Part 1. The Atomic Radii". *Int. J. Mol. Sci.* **3**: 87–113. doi:10.3390/i3020087.

[3] Leigh, G. J., ed. (1990). *International Union of Pure and Applied Chemistry, Commission on the Nomenclature of Inorganic Chemistry, Nomenclature of Organic Chemistry – Recommendations 1990*. Oxford: Blackwell Scientific Publications. p. 35. ISBN 0-08-022369-9. An atom is the smallest unit quantity of an element that is capable of existence whether alone or in chemical combination with other atoms of the same or other elements.

[4] Andrew G. van Melsen (1952). *From Atomos to Atom*. Mineola, N.Y.: Dover Publications. ISBN 0-486-49584-1.

[5] Dalton, John. "On the Absorption of Gases by Water and Other Liquids", in *Memoirs of the Literary and Philosophical Society of Manchester*. 1803. Retrieved on August 29, 2007.

[6] Einstein, Albert (1905). "Über die von der molekularkinetischen Theorie der Wärme geforderte Bewegung von in ruhenden Flüssigkeiten suspendierten Teilchen" (PDF). *Annalen der Physik* (in German) **322** (8): 549–560. Bibcode:1905AnP...322..549E. doi:10.1002/andp.19053220806. Retrieved 4 February 2007.

[7] Mazo, Robert M. (2002). *Brownian Motion: Fluctuations, Dynamics, and Applications*. Oxford University Press. pp. 1–7. ISBN 0-19-851567-7. OCLC 48753074.

[8] Lee, Y.K.; Hoon, K. (1995). "Brownian Motion". Imperial College. Archived from the original on 18 December 2007. Retrieved 18 December 2007.

[9] Patterson,G. (2007). "Jean Perrin and the triumph of the atomic doctrine".*Endeavour***31**(2): 50–53. doi:10.1016/j.. PMID 17602746.

[10] Thomson, J. J. (August 1901). "On bodies smaller than atoms". *The Popular Science Monthly* (Bonnier Corp.): 323–335. Retrieved 2009-06-21.

[11] "J.J. Thomson". Nobel Foundation. 1906. Retrieved 20 December 2007.

[12] Rutherford, E. (1911). "The Scattering of α and β Particles by Matter and the Structure of the Atom" (PDF). *Philosophical Magazine* **21** (125): 669–88. doi:10.1080/14786440508637080.

[13] "Frederick Soddy, The Nobel Prize in Chemistry 1921". Nobel Foundation. Retrieved 18 January 2008.

[14] Thomson,Joseph John(1913). "Rays of positive electricity".*Proceedings of the Royal Society*. A**89**(607): 1–20. Bibcode:19T. doi:10.1098/rspa.1913.0057.

[15] Stern, David P. (16 May 2005). "The Atomic Nucleus and Bohr's Early Model of the Atom". NASA/Goddard Space Flight Center. Retrieved 20 December 2007.

[16] Bohr, Niels (11 December 1922). "Niels Bohr, The Nobel Prize in Physics 1922, Nobel Lecture". Nobel Foundation. Retrieved 16 February 2008.

[17] Pais, Abraham (1986). *Inward Bound: Of Matter and Forces in the Physical World*. New York: Oxford University Press. pp. 228–230. ISBN 0-19-851971-0.

[18] Lewis, Gilbert N. (1916). "The Atom and the Molecule". *Journal of the American Chemical Society* **38** (4): 762–786. doi:10.1021/ja02261a002.

[19] Scerri, Eric R. (2007). *The periodic table: its story and its significance*. Oxford University Press US. pp. 205–226. ISBN 0-19-530573-6.

[20] Langmuir, Irving (1919). "The Arrangement of Electrons in Atoms and Molecules". *Journal of the American Chemical Society* **41** (6): 868–934. doi:10.1021/ja02227a002.

[21] Scully, Marlan O.; Lamb, Willis E.; Barut, Asim (1987). "On the theory of the Stern-Gerlach apparatus". *Foundations of Physics* **17** (6): 575–583. Bibcode:1987FoPh...17..575S. doi:10.1007/BF01882788.

[22] Brown, Kevin (2007). "The Hydrogen Atom". MathPages. Retrieved 21 December 2007.

[23] Harrison, David M. (2000). "The Development of Quantum Mechanics". University of Toronto. Archived from the original on 25 December 2007. Retrieved 21 December 2007.

[24] Aston,Francis W. (1920). "The constitution of atmospheric neon".*Philosophical Magazine***39**(6): 449–55. doi:10.1080/147.

[25] Chadwick, James (12 December 1935). "Nobel Lecture: The Neutron and Its Properties". Nobel Foundation. Retrieved 21 December 2007.

[26] "Otto Hahn, Lise Meitner and Fritz Strassmann". *Chemical Achievers: The Human Face of the Chemical Sciences*. Chemical Heritage Foundation. Archived from the original on 24 October 2009. Retrieved 15 September 2009.

[27] Meitner, Lise; Frisch, Otto Robert (1939). "Disintegration of uranium by neutrons: a new type of nuclear reaction". *Nature* **143** (3615): 239–240. Bibcode:1939Natur.143..239M. doi:10.1038/143239a0.

[28] Schroeder, M. "Lise Meitner – Zur 125. Wiederkehr Ihres Geburtstages" (in German). Retrieved 4 June 2009.

[29] Crawford, E.; Sime, Ruth Lewin; Walker, Mark (1997). "A Nobel tale of postwar injustice". *Physics Today* **50** (9): 26–32. Bibcode:1997PhT....50i..26C. doi:10.1063/1.881933.

[30] Kullander, Sven (28 August 2001). "Accelerators and Nobel Laureates". Nobel Foundation. Retrieved 31 January 2008.

[31] "The Nobel Prize in Physics 1990". Nobel Foundation. 17 October 1990. Retrieved 31 January 2008.

[32] Demtröder, Wolfgang (2002). *Atoms, Molecules and Photons: An Introduction to Atomic- Molecular- and Quantum Physics* (1st ed.). Springer. pp. 39–42. ISBN 3-540-20631-0. OCLC 181435713.

[33] Woan, Graham (2000). *The Cambridge Handbook of Physics*. Cambridge University Press. p. 8. ISBN 0-521-57507-9. OCLC 224032426.

[34] MacGregor, Malcolm H. (1992). *The Enigmatic Electron*. Oxford University Press. pp. 33–37. ISBN 0-19-521833-7. OCLC 223372888.

[35] Particle Data Group (2002). "The Particle Adventure". Lawrence Berkeley Laboratory. Archived from the original on 4 January 2007. Retrieved 3 January 2007.

[36] Schombert, James (18 April 2006). "Elementary Particles". University of Oregon. Retrieved 3 January 2007.

[37] Jevremovic, Tatjana (2005). *Nuclear Principles in Engineering*. Springer. p. 63. ISBN 0-387-23284-2. OCLC 228384008.

[38] Pfeffer, Jeremy I.; Nir, Shlomo (2000). *Modern Physics: An Introductory Text*. Imperial College Press. pp. 330–336. ISBN 1-86094-250-4. OCLC 45900880.

[39] Wenner, Jennifer M. (10 October 2007). "How Does Radioactive Decay Work?". Carleton College. Retrieved 9 January 2008.

[40] Raymond, David (7 April 2006). "Nuclear Binding Energies". New Mexico Tech. Archived from the original on 11 December 2006. Retrieved 3 January 2007.

[41] Mihos, Chris (23 July 2002). "Overcoming the Coulomb Barrier". Case Western Reserve University. Retrieved 13 February 2008.

[42] Staff (30 March 2007). "ABC's of Nuclear Science". Lawrence Berkeley National Laboratory. Archived from the original on 5 December 2006. Retrieved 3 January 2007.

[43] Makhijani, Arjun; Saleska, Scott (2 March 2001). "Basics of Nuclear Physics and Fission". Institute for Energy and Environmental Research. Archived from the original on 16 January 2007. Retrieved 3 January 2007.

[44] Shultis, J. Kenneth; Faw, Richard E. (2002). *Fundamentals of Nuclear Science and Engineering*. CRC Press. pp. 10–17. ISBN 0-8247-0834-2. OCLC 123346507.

[45] Fewell, M. P. (1995). "The atomic nuclide with the highest mean binding energy". *American Journal of Physics* **63** (7): 653–658. Bibcode:1995AmJPh..63..653F. doi:10.1119/1.17828.

[46] Mulliken,Robert S. (1967). "Spectroscopy,Molecular Orbitals,and Chemical Bonding".*Science***157**(3784): 13–24. Bib3M. doi:10.1126/science.157.3784.13. PMID 5338306.

[47] Brucat, Philip J. (2008). "The Quantum Atom". University of Florida. Archived from the original on 7 December 2006. Retrieved 4 January 2007.

[48] Manthey, David (2001). "Atomic Orbitals". Orbital Central. Archived from the original on 10 January 2008. Retrieved 21 January 2008.

[49] Herter, Terry (2006). "Lecture 8: The Hydrogen Atom". Cornell University. Retrieved 14 February 2008.

[50] Bell, R. E.; Elliott, L. G. (1950). "Gamma-Rays from the Reaction $H^1(n,\gamma)D^2$ and the Binding Energy of the Deuteron". *Physical Review* **79** (2): 282–285. Bibcode:1950PhRv...79..282B. doi:10.1103/PhysRev.79.282.

[51] Smirnov, Boris M. (2003). *Physics of Atoms and Ions*. Springer. pp. 249–272. ISBN 0-387-95550-X.

[52] Matis, Howard S. (9 August 2000). "The Isotopes of Hydrogen". *Guide to the Nuclear Wall Chart*. Lawrence Berkeley National Lab. Archived from the original on 18 December 2007. Retrieved 21 December 2007.

[53] Weiss, Rick (17 October 2006). "Scientists Announce Creation of Atomic Element, the Heaviest Yet". Washington Post. Retrieved 21 December 2007.

[54] Sills, Alan D. (2003). *Earth Science the Easy Way*. Barron's Educational Series. pp. 131–134. ISBN 0-7641-2146-4. OCLC 51543743.

[55] Dumé, Belle (23 April 2003). "Bismuth breaks half-life record for alpha decay". Physics World. Archived from the original on 14 December 2007. Retrieved 21 December 2007.

[56] Lindsay, Don (30 July 2000). "Radioactives Missing From The Earth". Don Lindsay Archive. Archived from the original on 28 April 2007. Retrieved 23 May 2007.

[57] Tuli, Jagdish K. (April 2005). "Nuclear Wallet Cards". National Nuclear Data Center, Brookhaven National Laboratory. Retrieved 16 April 2011.

[58] CRC Handbook (2002).

[59] Mills, Ian; Cvitaš, Tomislav; Homann, Klaus; Kallay, Nikola; Kuchitsu, Kozo (1993). *Quantities, Units and Symbols in Physical Chemistry* (PDF) (2nd ed.). Oxford: International Union of Pure and Applied Chemistry, Commission on Physiochemical Symbols Terminology and Units, Blackwell Scientific Publications. p. 70. ISBN 0-632-03583-8. OCLC 27011505.

[60] Chieh, Chung (22 January 2001). "Nuclide Stability". University of Waterloo. Retrieved 4 January 2007.

[61] "Atomic Weights and Isotopic Compositions for All Elements". National Institute of Standards and Technology. Archived from the original on 31 December 2006. Retrieved 4 January 2007.

[62] Audi, G.; Wapstra, A.H.; Thibault, C. (2003). "The Ame2003 atomic mass evaluation (II)" (PDF). *Nuclear Physics A* **729** (1): 337–676. Bibcode:2003NuPhA.729..337A. doi:10.1016/j.nuclphysa.2003.11.003.

[63] Shannon, R. D. (1976). "Revised effective ionic radii and systematic studies of interatomic distances in halides and chalcogenides". *Acta Crystallographica A* **32** (5): 751–767. Bibcode:1976AcCrA..32..751S. doi:10.1107/S0567739476001551.

[64] Dong, Judy (1998). "Diameter of an Atom". The Physics Factbook. Archived from the original on 4 November 2007. Retrieved 19 November 2007.

[65] Zumdahl, Steven S. (2002). *Introductory Chemistry: A Foundation* (5th ed.). Houghton Mifflin. ISBN 0-618-34342-3. OCLC 173081482. Archived from the original on 4 March 2008. Retrieved 5 February 2008.

[66] Birkholz, M.; Rudert, R. (2008). "Interatomic distances in pyrite-structure disulfides – a case for ellipsoidal modeling of sulfur ions]" (PDF). *phys. stat. sol. b* **245**: 1858–1864. Bibcode:2008PSSBR.245.1858B. doi:10.1002/pssb.200879532.

[67] Birkholz, M. (2014). "Modeling the Shape of Ions in Pyrite-Type Crystals". *Crystals* **4**: 390–403. doi:10.3390/cryst4030390.

[68] Staff (2007). "Small Miracles: Harnessing nanotechnology". Oregon State University. Retrieved 7 January 2007.—describes the width of a human hair as 10^5 nm and 10 carbon atoms as spanning 1 nm.

[69] Padilla, Michael J.; Miaoulis, Ioannis; Cyr, Martha (2002). *Prentice Hall Science Explorer: Chemical Building Blocks*. Upper Saddle River, New Jersey USA: Prentice-Hall, Inc. p. 32. ISBN 0-13-054091-9. OCLC 47925884. There are 2,000,00 000(that's2 sextillion)atoms of oxygen in one drop of water—and twice as many atoms of hydrogen.

[70] Feynman, Richard (1995). *Six Easy Pieces*. The Penguin Group. p. 5. ISBN 978-0-14-027666-4. OCLC 40499574.

[71] "Radioactivity". Splung.com. Archived from the original on 4 December 2007. Retrieved 19 December 2007.

[72] L'Annunziata, Michael F. (2003). *Handbook of Radioactivity Analysis*. Academic Press. pp. 3–56. ISBN 0-12-436603-1. OCLC 16212955.

[73] Firestone, Richard B. (22 May 2000). "Radioactive Decay Modes". Berkeley Laboratory. Retrieved 7 January 2007.

[74] Hornak, J. P. (2006). "Chapter 3: Spin Physics". *The Basics of NMR*. Rochester Institute of Technology. Archived from the original on 3 February 2007. Retrieved 7 January 2007.

[75] Schroeder, Paul A. (25 February 2000). "Magnetic Properties". University of Georgia. Archived from the original on 29 April 2007. Retrieved 7 January 2007.

[76] Goebel, Greg (1 September 2007). "[4.3] Magnetic Properties of the Atom". *Elementary Quantum Physics*. In The Public Domain website. Retrieved 7 January 2007.

[77] Yarris, Lynn (Spring 1997). "Talking Pictures". *Berkeley Lab Research Review*. Archived from the original on 13 January 2008. Retrieved 9 January 2008.

[78] Liang, Z.-P.; Haacke, E. M. (1999). Webster, J. G., ed. *Encyclopedia of Electrical and Electronics Engineering: Magnetic Resonance Imaging*. vol. 2. John Wiley & Sons. pp. 412–426. ISBN 0-471-13946-7.

[79] Zeghbroeck, Bart J. Van (1998). "Energy levels". Shippensburg University. Archived from the original on 15 January 2005. Retrieved 23 December 2007.

[80] Fowles, Grant R. (1989). *Introduction to Modern Optics*. Courier Dover Publications. pp. 227–233. ISBN 0-486-65957-7. OCLC 18834711.

[81] Martin, W. C.; Wiese, W. L. (May 2007). "Atomic Spectroscopy: A Compendium of Basic Ideas, Notation, Data, and Formulas". National Institute of Standards and Technology. Archived from the original on 8 February 2007. Retrieved 8 January 2007.

[82] "Atomic Emission Spectra — Origin of Spectral Lines". Avogadro Web Site. Retrieved 10 August 2006.

[83] Fitzpatrick, Richard (16 February 2007). "Fine structure". University of Texas at Austin. Retrieved 14 February 2008.

[84] Weiss, Michael (2001). "The Zeeman Effect". University of California-Riverside. Archived from the original on 2 February 2008. Retrieved 6 February 2008.

[85] Beyer, H. F.; Shevelko, V. P. (2003). *Introduction to the Physics of Highly Charged Ions*. CRC Press. pp. 232–236. ISBN 0-7503-0481-2. OCLC 47150433.

[86] Watkins, Thayer. "Coherence in Stimulated Emission". San José State University. Archived from the original on 12 January 2008. Retrieved 23 December 2007.

[87] oxford dictionary – valency

[88] Reusch, William (16 July 2007). "Virtual Textbook of Organic Chemistry". Michigan State University. Retrieved 11 January 2008.

[89] "Covalent bonding – Single bonds". chemguide. 2000.

[90] Husted, Robert et al. (11 December 2003). "Periodic Table of the Elements". Los Alamos National Laboratory. Archived from the original on 10 January 2008. Retrieved 11 January 2008.

[91] Baum, Rudy (2003). "It's Elemental: The Periodic Table". Chemical & Engineering News. Retrieved 11 January 2008.

[92] Goodstein, David L. (2002). *States of Matter*. Courier Dover Publications. pp. 436–438. ISBN 0-13-843557-X.

[93] Brazhkin, Vadim V. (2006). "Metastable phases, phase transformations, and phase diagrams in physics and chemistry". *Physics-Uspekhi* **49** (7): 719–24. Bibcode:2006PhyU...49..719B. doi:10.1070/PU2006v049n07ABEH006013.

[94] Myers, Richard (2003). *The Basics of Chemistry*. Greenwood Press. p. 85. ISBN 0-313-31664-3. OCLC 50164580.

[95] Staff (9 October 2001). "Bose-Einstein Condensate: A New Form of Matter". National Institute of Standards and Technology. Archived from the original on 3 January 2008. Retrieved 16 January 2008.

[96] Colton, Imogen; Fyffe, Jeanette (3 February 1999). "Super Atoms from Bose-Einstein Condensation". The University of Melbourne. Archived from the original on 29 August 2007. Retrieved 6 February 2008.

[97] Jacox, Marilyn; Gadzuk, J. William (November 1997). "Scanning Tunneling Microscope". National Institute of Standards and Technology. Archived from the original on 7 January 2008. Retrieved 11 January 2008.

[98] "The Nobel Prize in Physics 1986". The Nobel Foundation. Retrieved 11 January 2008.—in particular, see the Nobel lecture by G. Binnig and H. Rohrer.

[99] Jakubowski, N.; Moens, Luc; Vanhaecke, Frank (1998). "Sector field mass spectrometers in ICP-MS". *Spectrochimica Acta Part B: Atomic Spectroscopy* **53** (13): 1739–63. Bibcode:1998AcSpe..53.1739J. doi:10.1016/S0584-8547(98)00222-5.

[100] Müller, Erwin W.; Panitz, John A.; McLane, S. Brooks (1968). "The Atom-Probe Field Ion Microscope". *Review of Scientific Instruments* **39** (1): 83–86. Bibcode:1968RScI...39...83M. doi:10.1063/1.1683116.

[101] Lochner, Jim; Gibb, Meredith; Newman, Phil (30 April 2007). "What Do Spectra Tell Us?". NASA/Goddard Space Flight Center. Archived from the original on 16 January 2008. Retrieved 3 January 2008.

[102] Winter, Mark (2007). "Helium". WebElements. Archived from the original on 30 December 2007. Retrieved 3 January 2008.

[103] Hinshaw, Gary (10 February 2006). "What is the Universe Made Of?". NASA/WMAP. Archived from the original on 31 December 2007. Retrieved 7 January 2008.

[104] Choppin, Gregory R.; Liljenzin, Jan-Olov; Rydberg, Jan (2001). *Radiochemistry and Nuclear Chemistry*. Elsevier. p. 441. ISBN 0-7506-7463-6. OCLC 162592180.

[105] Davidsen, Arthur F. (1993). "Far-Ultraviolet Astronomy on the Astro-1 Space Shuttle Mission". *Science* **259** (5093): 327–34. Bibcode:1993Sci...259..327D. doi:10.1126/science.259.5093.327. PMID 17832344.

[106] Lequeux, James (2005). *The Interstellar Medium*. Springer. p. 4. ISBN 3-540-21326-0. OCLC 133157789.

[107] Smith, Nigel (6 January 2000). "The search for dark matter". Physics World. Archived from the original on 16 February 2008. Retrieved 14 February 2008.

[108] Croswell, Ken (1991). "Boron, bumps and the Big Bang: Was matter spread evenly when the Universe began? Perhaps not; the clues lie in the creation of the lighter elements such as boron and beryllium". *New Scientist* (1794): 42. Archived from the original on 7 February 2008. Retrieved 14 January 2008.

[109] Copi, Craig J.; Schramm, DN; Turner, MS (1995). "Big-Bang Nucleosynthesis and the Baryon Density of the Universe". *Science* **267** (5195): 192–99. arXiv:astro-ph/9407006. Bibcode:1995Sci...267..192C. doi:10.1126/science.7809624. PMID 7809624.

[110] Hinshaw, Gary (15 December 2005). "Tests of the Big Bang: The Light Elements". NASA/WMAP. Archived from the original on 17 January 2008. Retrieved 13 January 2008.

[111] Abbott, Brian (30 May 2007). "Microwave (WMAP) All-Sky Survey". Hayden Planetarium. Retrieved 13 January 2008.

[112] Hoyle, F. (1946). "The synthesis of the elements from hydrogen". *Monthly Notices of the Royal Astronomical Society* **106**: 343–83. Bibcode:1946MNRAS.106..343H. doi:10.1093/mnras/106.5.343.

[113] Knauth, D. C.; Knauth, D. C.; Lambert, David L.; Crane, P. (2000). "Newly synthesized lithium in the interstellar medium". *Nature* **405** (6787): 656–58. doi:10.1038/35015028. PMID 10864316.

[114] Mashnik, Stepan G. (2000). "On Solar System and Cosmic Rays Nucleosynthesis and Spallation Processes". arXiv:astro-ph/0008382 [astro-ph].

[115] Kansas Geological Survey (4 May 2005). "Age of the Earth". University of Kansas. Retrieved 14 January 2008.

[116] Manuel 2001, pp. 407–430, 511–519.

[117] Dalrymple, G. Brent (2001). "The age of the Earth in the twentieth century: a problem (mostly) solved". *Geological Society, London, Special Publications* **190** (1): 205–21. Bibcode:2001GSLSP.190..205D. doi:10.1144/GSL.SP.2001.190.01.14. Retrieved 14 January 2008.

[118] Anderson, Don L.; Foulger, G. R.; Meibom, Anders (2 September 2006). "Helium: Fundamental models". MantlePlumes.org. Archived from the original on 8 February 2007. Retrieved 14 January 2007.

[119] Pennicott, Katie (10 May 2001). "Carbon clock could show the wrong time". PhysicsWeb. Archived from the original on 15 December 2007. Retrieved 14 January 2008.

[120] Yarris, Lynn (27 July 2001). "New Superheavy Elements 118 and 116 Discovered at Berkeley Lab". Berkeley Lab. Archived from the original on 9 January 2008. Retrieved 14 January 2008.

[121] Diamond, H et al. (1960). "Heavy Isotope Abundances in Mike Thermonuclear Device". *Physical Review* **119** (6): 2000–04. Bibcode:1960PhRv..119.2000D. doi:10.1103/PhysRev.119.2000.

[122] Poston Sr., John W. (23 March 1998). "Do transuranic elements such as plutonium ever occur naturally?". Scientific American.

[123] Keller, C. (1973). "Natural occurrence of lanthanides, actinides, and superheavy elements". *Chemiker Zeitung* **97** (10): 522–30. OSTI 4353086.

[124] Zaider, Marco; Rossi, Harald H. (2001). *Radiation Science for Physicians and Public Health Workers*. Springer. p. 17. ISBN 0-306-46403-9. OCLC 44110319.

[125] Manuel 2001, pp. 407–430,511–519.

[126] "Oklo Fossil Reactors". Curtin University of Technology. Archived from the original on 18 December 2007. Retrieved 15 January 2008.

[127] Weisenberger, Drew. "How many atoms are there in the world?". Jefferson Lab. Retrieved 16 January 2008.

[128] Pidwirny, Michael. "Fundamentals of Physical Geography". University of British Columbia Okanagan. Archived from the original on 21 January 2008. Retrieved 16 January 2008.

[129] Anderson, Don L. (2002). "The inner inner core of Earth". *Proceedings of the National Academy of Sciences* **99** (22): 13966–68. Bibcode:2002PNAS...9913966A. doi:10.1073/pnas.232565899. PMC 137819. PMID 12391308.

[130] Pauling, Linus (1960). *The Nature of the Chemical Bond*. Cornell University Press. pp. 5–10. ISBN 0-8014-0333-2. OCLC 17518275.

[131] Anonymous (2 October 2001). "Second postcard from the island of stability". *CERN Courier*. Archived from the original on 3 February 2008. Retrieved 14 January 2008.

[132] Jacoby, Mitch (2006). "As-yet-unsynthesized superheavy atom should form a stable diatomic molecule with fluorine". *Chemical & Engineering News* **84** (10): 19. doi:10.1021/cen-v084n010.p019a.

[133] Koppes, Steve (1 March 1999). "Fermilab Physicists Find New Matter-Antimatter Asymmetry". University of Chicago. Retrieved 14 January 2008.

[134] Cromie, William J. (16 August 2001). "A lifetime of trillionths of a second: Scientists explore antimatter". Harvard University Gazette. Retrieved 14 January 2008.

[135] Hijmans, Tom W. (2002). "Particle physics: Cold antihydrogen". *Nature* **419** (6906): 439–40. Bibcode:2002Natur.419..439H. doi:10.1038/419439a. PMID 12368837.

[136] Staff (30 October 2002). "Researchers 'look inside' antimatter". BBC News. Retrieved 14 January 2008.

[137] Barrett, Roger (1990). "The Strange World of the Exotic Atom". *New Scientist* (1728): 77–115. Archived from the original on 21 December 2007. Retrieved 4 January 2008.

[138] Indelicato,Paul(2004). "Exotic Atoms".*Physica Scripta***T112**(1): 20–26. arXiv:physics/0409058. Bibcode:2004PhST..112. doi:10.1238/Physica.Topical.112a00020.

[139] Ripin, Barrett H. (July 1998). "Recent Experiments on Exotic Atoms". American Physical Society. Retrieved 15 February 2008.

76.9 Sources

- Manuel, Oliver (2001). *Origin of Elements in the Solar System: Implications of Post-1957 Observations*. Springer. ISBN 0-306-46562-0. OCLC 228374906.

76.10 Further reading

- Dalton, J. (1808). *A New System of Chemical Philosophy, Part 1*. London and Manchester: S. Russell.
- Gangopadhyaya, Mrinalkanti (1981). *Indian Atomism: History and Sources*. Atlantic Highlands, New Jersey: Humanities Press. ISBN 0-391-02177-X. OCLC 10916778.
- Harrison, Edward Robert (2003). *Masks of the Universe: Changing Ideas on the Nature of the Cosmos*. Cambridge University Press. ISBN 0-521-77351-2. OCLC 50441595.
- Iannone, A. Pablo (2001). *Dictionary of World Philosophy*. Routledge. ISBN 0-415-17995-5. OCLC 44541769.
- King, Richard (1999). *Indian philosophy: an introduction to Hindu and Buddhist thought*. Edinburgh University Press. ISBN 0-7486-0954-7.
- Levere, Trevor, H. (2001). *Transforming Matter – A History of Chemistry for Alchemy to the Buckyball*. The Johns Hopkins University Press. ISBN 0-8018-6610-3.
- Liddell, Henry George; Scott, Robert. "A Greek-English Lexicon". Perseus Digital Library.
- Liddell, Henry George; Scott, Robert. "ἄτομος". *A Greek-English Lexicon*. Perseus Digital Library. Retrieved 21 June 2010.
- McEvilley, Thomas (2002). *The shape of ancient thought: comparative studies in Greek and Indian philosophies*. Allworth Press. ISBN 1-58115-203-5.
- Moran, Bruce T. (2005). *Distilling Knowledge: Alchemy, Chemistry, and the Scientific Revolution*. Harvard University Press. ISBN 0-674-01495-2.
- Ponomarev, Leonid Ivanovich (1993). *The Quantum Dice*. CRC Press. ISBN 0-7503-0251-8. OCLC 26853108.
- Roscoe, Henry Enfield (1895). *John Dalton and the Rise of Modern Chemistry*. Century science series. New York: Macmillan. Retrieved 3 April 2011.
- Siegfried, Robert (2002). *From Elements to Atoms: A History of Chemical Composition*. DIANE. ISBN 0-87169-924-9. OCLC 186607849.
- Teresi, Dick (2003). *Lost Discoveries: The Ancient Roots of Modern Science*. Simon & Schuster. pp. 213–214. ISBN 0-7432-4379-X.
- Various (2002). Lide, David R., ed. *Handbook of Chemistry & Physics* (88th ed.). CRC. ISBN 0-8493-0486-5. OCLC 179976746. Archived from the original on 23 May 2008. Retrieved 23 May 2008.
- Wurtz, Charles Adolphe (1881). *The Atomic Theory*. New York: D. Appleton and company. ISBN 0-559-43636-X.

76.11 External links

- "Quantum Mechanics and the Structure of Atoms" on YouTube
- Freudenrich, Craig C. "How Atoms Work". How Stuff Works. Archived from the original on 8 January 2007. Retrieved 9 January 2007.
- "The Atom". *Free High School Science Texts: Physics*. Wikibooks. Retrieved 10 July 2010.
- Anonymous (2007). "The atom". Science aid+. Retrieved 10 July 2010.—a guide to the atom for teens.
- Anonymous (3 January 2006). "Atoms and Atomic Structure". BBC. Archived from the original on 2 January 2007. Retrieved 11 January 2007.
- Various (3 January 2006). "Physics 2000, Table of Contents". University of Colorado. Archived from the original on 14 January 2008. Retrieved 11 January 2008.
- Various (3 February 2006). "What does an atom look like?". University of Karlsruhe. Retrieved 12 May 2008.

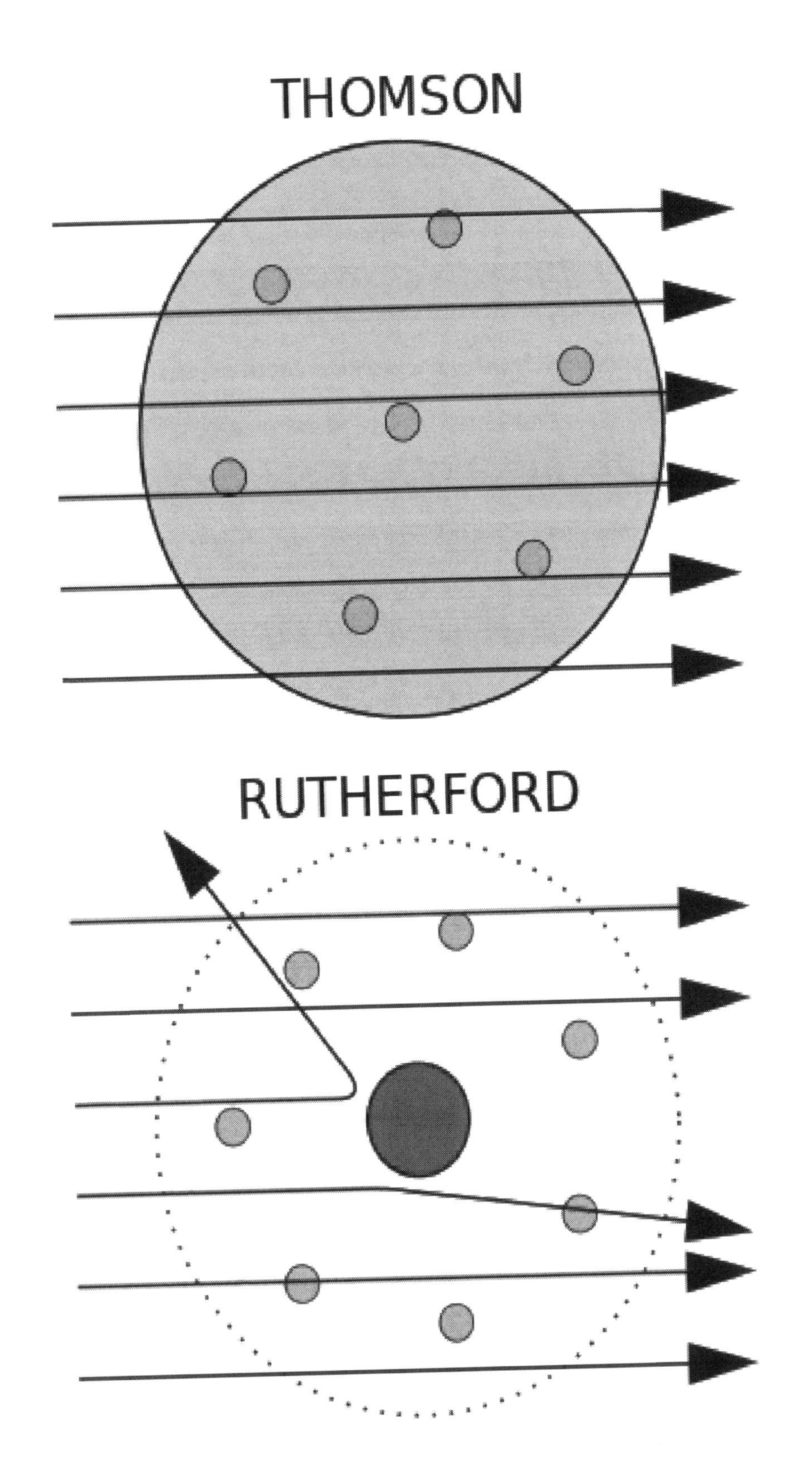

The Geiger–Marsden experiment Top: Expected results: alpha particles passing through the plum pudding model of the atom with negligible deflection. Bottom: Observed results: a small portion of the particles were deflected by the concentrated positive charge of the nucleus.

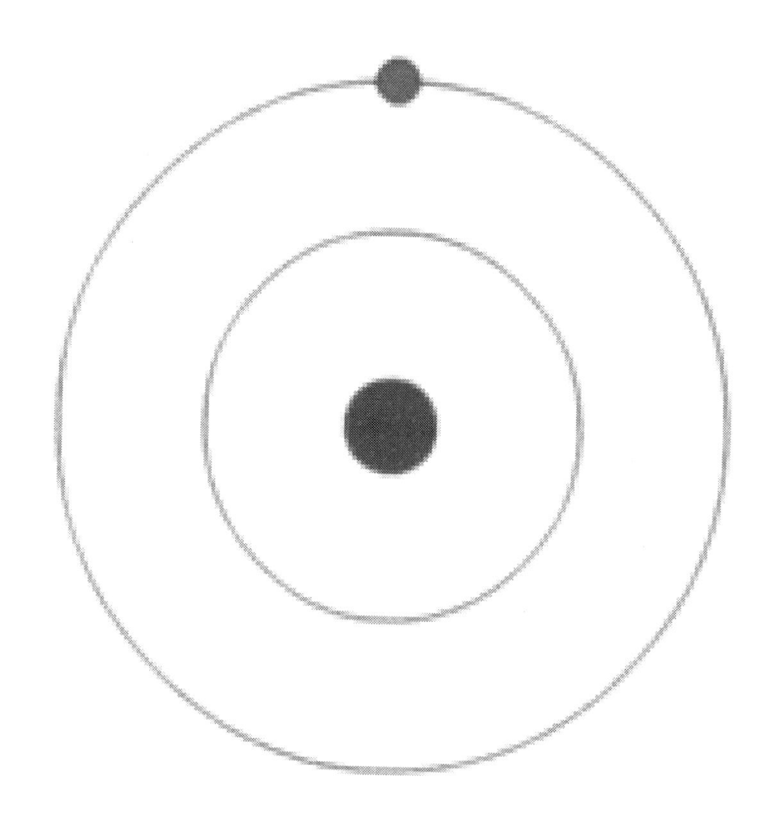

The Bohr model of the atom, with an electron making instantaneous "quantum leaps" from one orbit to another. This model is obsolete.

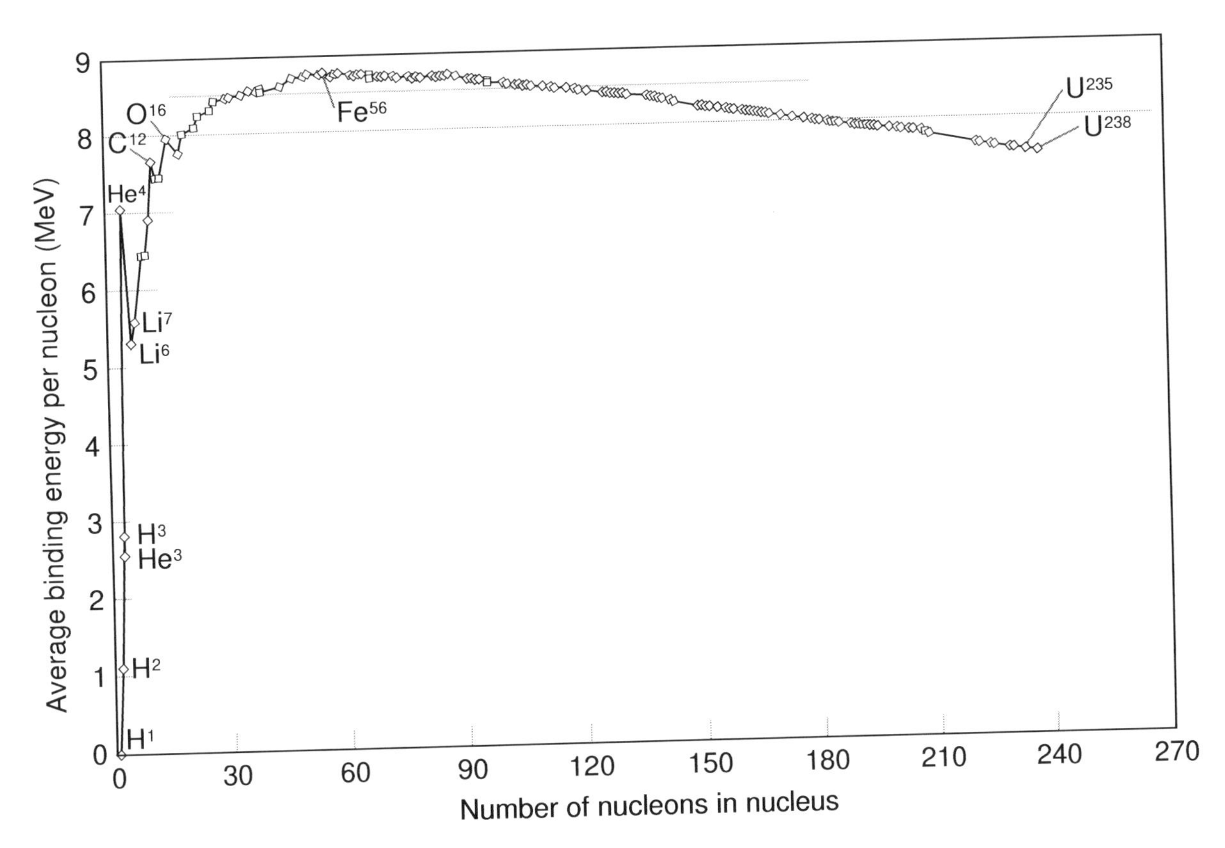

The binding energy needed for a nucleon to escape the nucleus, for various isotopes

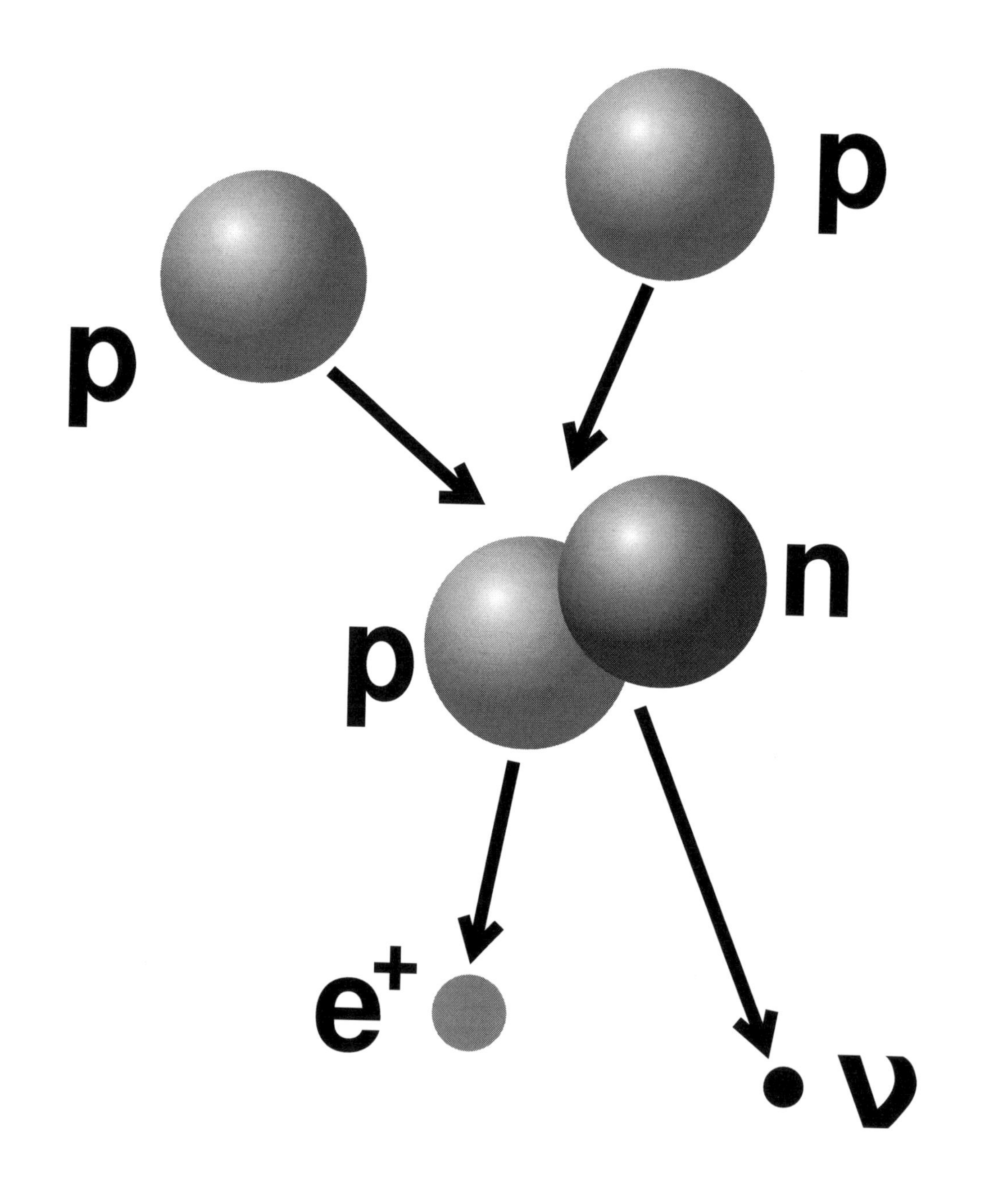

Illustration of a nuclear fusion process that forms a deuterium nucleus, consisting of a proton and a neutron, from two protons. A positron (e^+)—an antimatter electron—is emitted along with an electron neutrino.

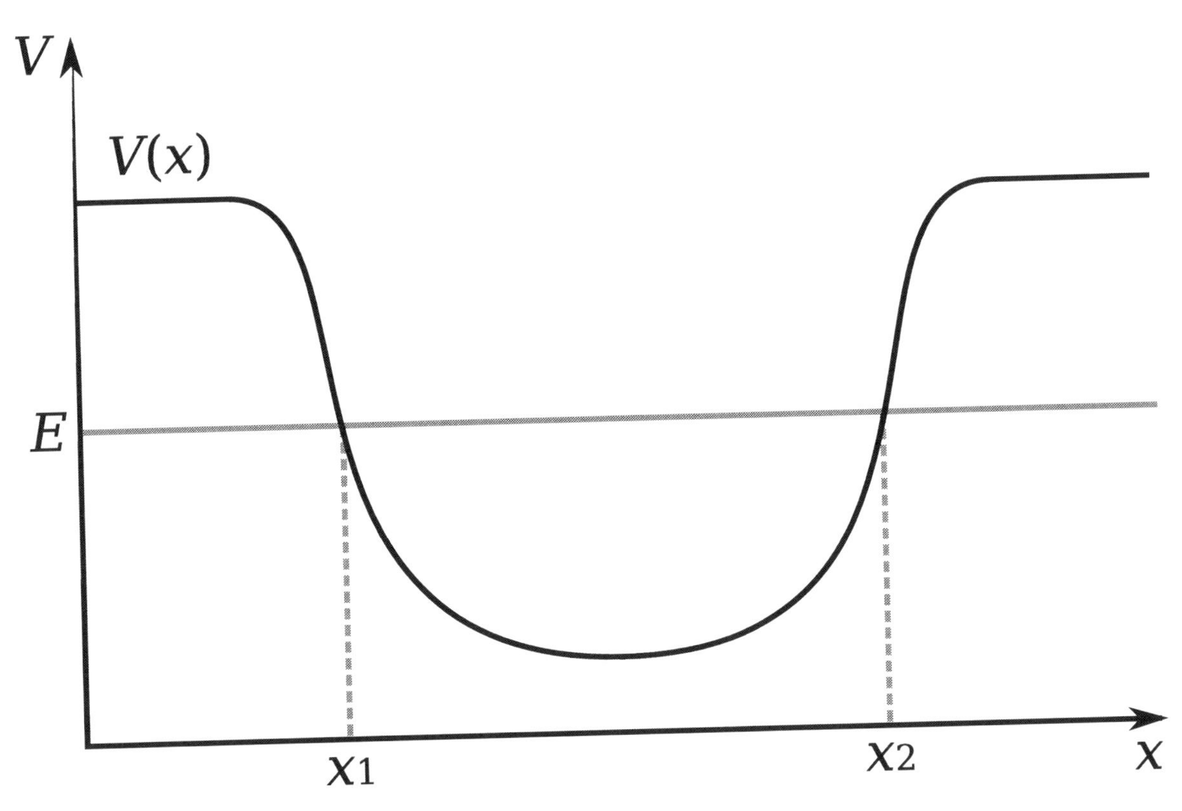

A potential well, showing, according to classical mechanics, the minimum energy V(x) *needed to reach each position* x. *Classically, a particle with energy* E *is constrained to a range of positions between* x_1 *and* x_2.

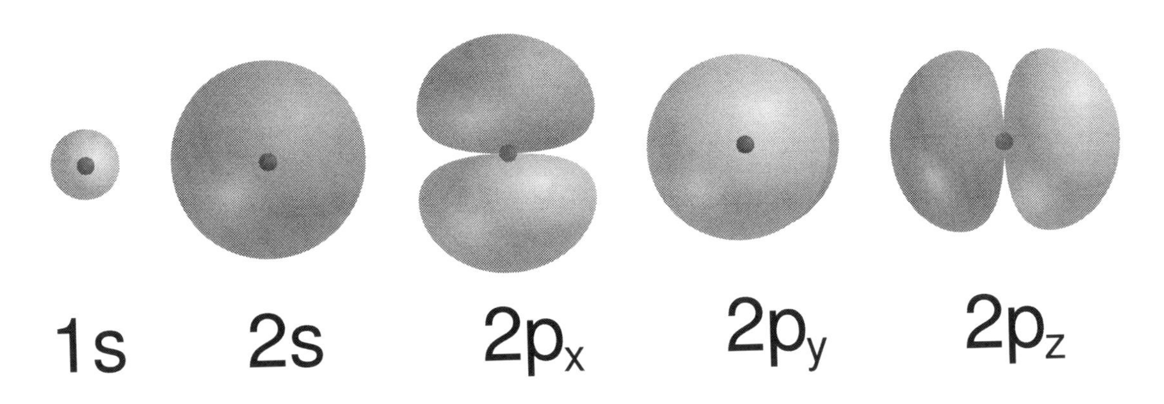

Wave functions of the first five atomic orbitals. The three 2p orbitals each display a single angular node that has an orientation and a minimum at the center.

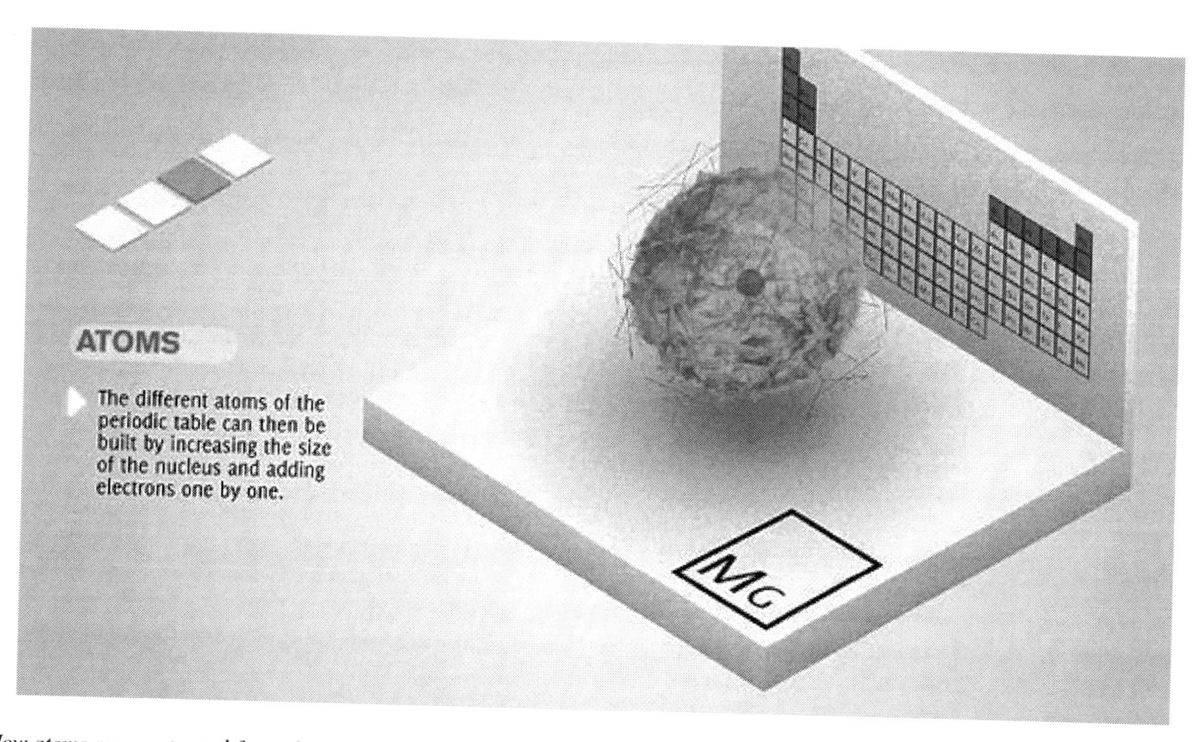

How atoms are constructed from electron orbitals and link to the periodic table

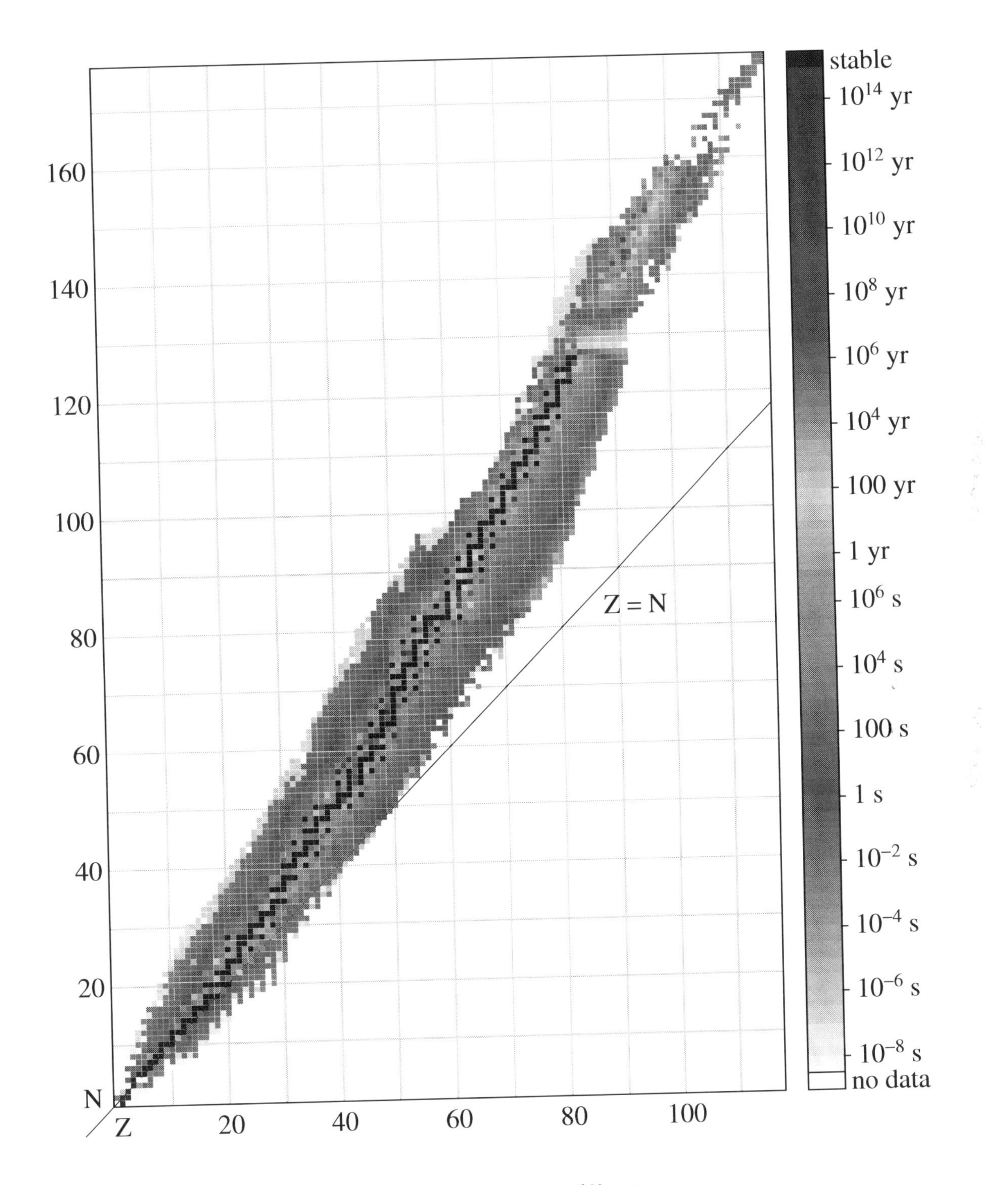

This diagram shows the half-life (T½) of various isotopes with Z protons and N neutrons.

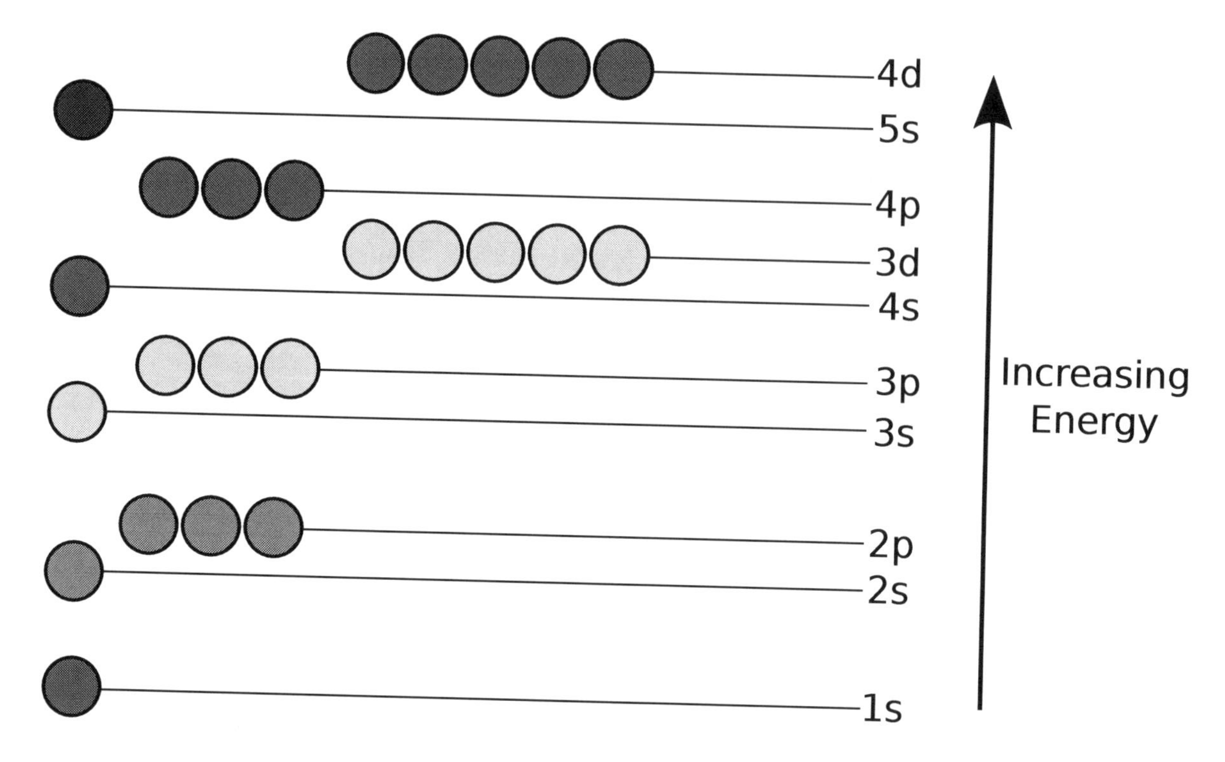

These electron's energy levels (not to scale) are sufficient for ground states of atoms up to cadmium ($5s^2$ $4d^{10}$) inclusively. Do not forget that even the top of the diagram is lower than an unbound electron state.

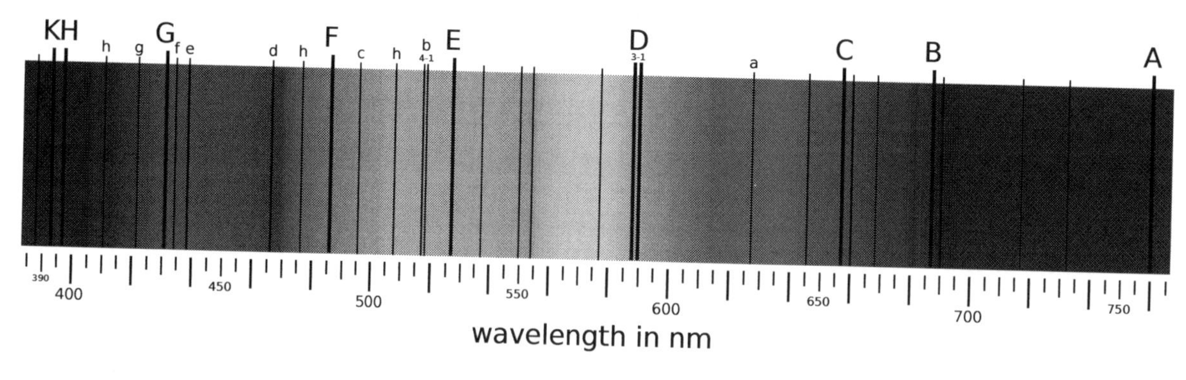

An example of absorption lines in a spectrum

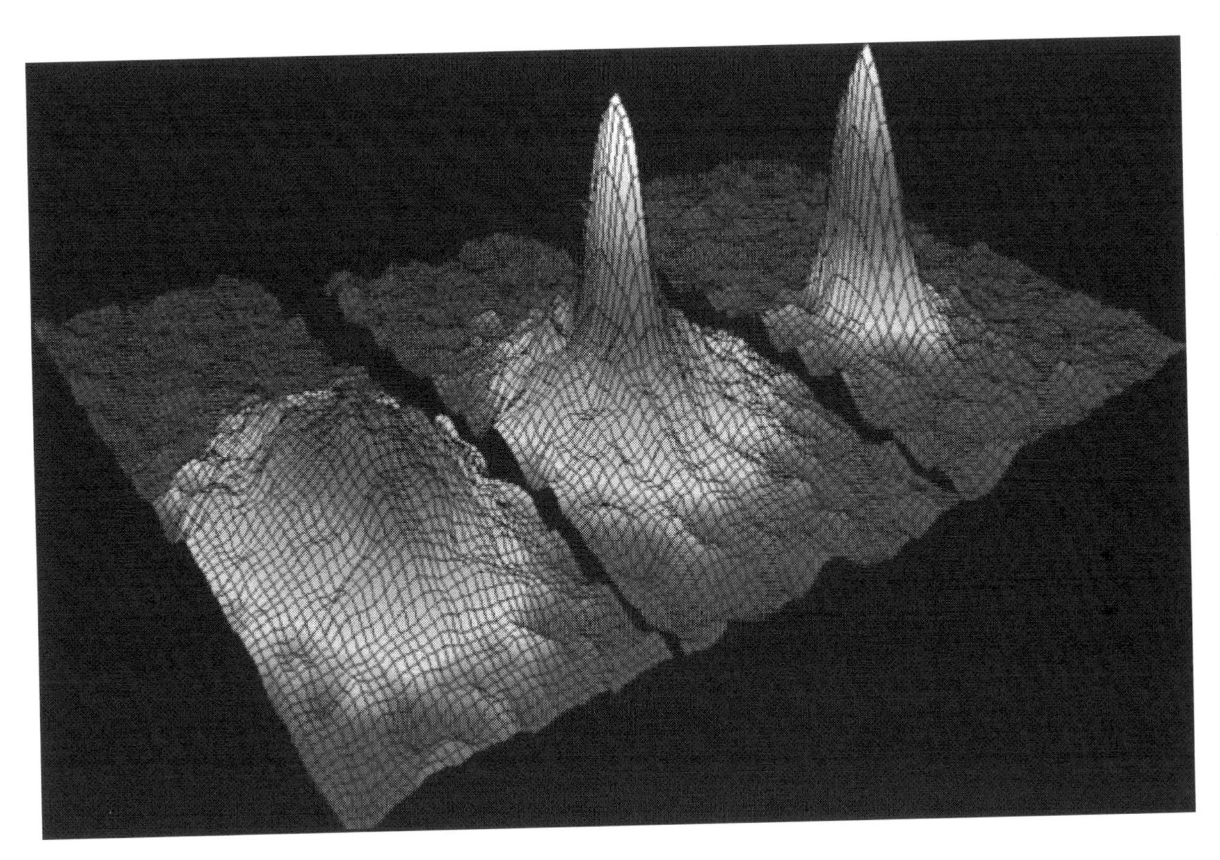

Snapshots illustrating the formation of a Bose–Einstein condensate

Scanning tunneling microscope image showing the individual atoms making up this gold (100) surface. The surface atoms deviate from the bulk crystal structure and arrange in columns several atoms wide with pits between them (See surface reconstruction).

Chapter 77

Diquark

Not to be confused with meson.

In particle physics, a **diquark**, or **diquark correlation/clustering**, is a hypothetical state of two quarks grouped inside a baryon (that consists of three quarks) (Lichtenberg 1982). Corresponding models of baryons are referred to as **quark–diquark models**. The diquark is often treated as a single subatomic particle with which the third quark interacts via the strong interaction. The existence of diquarks inside the nucleons is a disputed issue, but it helps to explain some nucleon properties and to reproduce experimental data sensitive to the nucleon structure. Diquark–antidiquark pairs have also been advanced for anomalous particles such as the X(3872).

77.1 Formation

The forces between the two quarks in a diquark is attractive when both the colors and spins are antisymmetric. When both quarks are correlated in this way they tend to form a very low energy configuration. This low energy configuration has become known as a diquark.

77.2 Controversy

Many scientists theorize that a diquark should not be considered a particle. Even though they may contain two quarks they are not colour neutral, and therefore cannot exist as isolated bound states. So instead they tend to float freely inside hadrons as composite entities; while free-floating they have a size of about 1 fm. This also happens to be the same size as the hadron itself.

77.3 Uses

Diquarks are the conceptual building blocks, and as such give scientists an ordering principle for the most important states in the hadronic spectrum. There are many different pieces of evidence that prove diquarks are fundamental in the structure of hadrons. One of the most compelling pieces of evidence comes from a recent study of baryons. In this study the baryon had one heavy and two light quarks. Since the heavy quark is inert, the scientists were able to discern the properties of the different quark configurations in the hadronic spectrum.

77.4 Λ and Σ baryon experiment

An experiment was conducted using diquarks in an attempt to study the Λ and Σ baryons that are produced in the creation of hadrons created by fast-moving quarks. In the experiment the quarks ionized the vacuum area. This produced the quark–antiquark pairs, which then converted themselves into mesons. When generating a baryon by assembling quarks, it is helpful if the quarks first form a stable two-quark state. The Λ and the Σ are created as a result of up, down and strange quarks. Scientists found that the Λ contains the [ud] diquark, however the Σ does not. From this experiment scientists inferred that Λ baryons are more common than Σ baryons, and indeed they are more common by a factor of 10.

77.5 References

- D. B. Lichtenberg, W. Namgung, E. Predazzi, J. G. Wills (1982). “Baryon Masses In A Relativistic Quark-Diquark Model”.*Physical Review Letters***48**(24): 1653–1656. Bibcode:1982PhRvL..48.1653L.doi:10.1103/PhysRev.
- R. Rapp, T. Schëfer, E. Shuryak, M. Velkovsky (1998). “Diquark bose condensates in high density matter and instantons”. *Physical Review Letters* **81** (1): 53–56. arXiv:hep-ph/9711396. Bibcode:1998PhRvL..81...53R. doi:10.1103/PhysRevLett.81.53.

Chapter 78

Exotic atom

An **exotic atom** is an otherwise normal atom in which one or more sub-atomic particles have been replaced by other particles of the same charge. For example, electrons may be replaced by other negatively charged particles such as muons (muonic atoms) or pions (pionic atoms).[1][2] Because these substitute particles are usually unstable, exotic atoms typically have very short lifetimes.

78.1 Muonic atoms

In a *muonic atom* (also called a *mu-mesic* atom),[3] an electron is replaced by a muon, which, like the electron, is a lepton. Since leptons are only sensitive to weak, electromagnetic and gravitational forces, muonic atoms are governed to very high precision by the electromagnetic interaction. The description of these atoms is not complicated by strong forces between the lepton and the nucleus.

Since a muon is more massive than an electron, the Bohr orbits are closer to the nucleus in a muonic atom than in an ordinary atom, and corrections due to quantum electrodynamics are more important. Study of muonic atoms' energy levels as well as transition rates from excited states to the ground state therefore provide experimental tests of quantum electrodynamics.

Muon-catalyzed fusion is a technical application of muonic atoms.

78.2 Hadronic atoms

A *hadronic atom* is an atom in which one or more of the orbital electrons is replaced by a charged hadron.[4] Possible hadrons include mesons such as the pion or kaon, yielding a *mesonic atom*; antiprotons, yielding an *antiprotonic atom*; and the $\Sigma-$ particle, yielding a $\Sigma-$ or *sigmaonic atom*.[5][6][7]

Unlike leptons, hadrons can interact via the strong force, so the orbitals of hadronic atoms are influenced by nuclear forces between the nucleus and the hadron. Since the strong force is a short-range interaction, these effects are strongest if the atomic orbital involved is close to the nucleus, when the energy levels involved may broaden or disappear because of the absorption of the hadron by the nucleus.[2][6] Hadronic atoms, such as pionic hydrogen and kaonic hydrogen, thus provide experimental probes of the theory of strong interactions, quantum chromodynamics.[8]

78.3 Onium

Main article: Onium

An *onium* (plural: *onia*) is the bound state of a particle and its antiparticle. The classic onium is positronium, which consists of an electron and a positron bound together as a long-lived metastable state. Positronium has been studied since the 1950s to understand bound states in quantum field theory. A recent development called non-relativistic quantum electrodynamics (NRQED) used this system as a proving ground.

Pionium, a bound state of two oppositely-charged pions, is useful for exploring the strong interaction. This should also be true of protonium. The true analogs of positronium in the theory of strong interactions, however, are not exotic atoms but certain mesons, the *quarkonium states*, which are made of a heavy quark such as the charm or bottom quark and its antiquark. (Top quarks are so heavy that they decay through the weak force before they can form bound states.) Exploration of these states through non-relativistic quantum chromodynamics (NRQCD) and lattice QCD are increasingly important tests of quantum chromodynamics.

Muonium, despite its name, is *not* an onium containing a muon and an antimuon, because IUPAC assigned that name to the system of an antimuon bound with an electron. However, the production of a muon/antimuon bound state, which is an onium, has been theorized.[9]

Understanding bound states of hadrons such as pionium and protonium is also important in order to clarify notions related to exotic hadrons such as mesonic molecules and pentaquark states.

78.4 Hypernuclear atoms

Main article: Hypernucleus

Atoms may be composed of electrons orbiting a hypernucleus that includes strange particles called hyperons. Such hypernuclear atoms are generally studied for their nuclear behaviour, falling into the realm of nuclear physics rather than atomic physics.

78.5 Quasiparticle atoms

In condensed matter systems, specifically in some semiconductors, there are states called excitons which are bound states of an electron and an electron hole.

78.6 See also

- Antihydrogen
- Antiprotonic helium
- Di-positronium
- Kaonic hydrogen
- Lattice QCD
- Muonium
- Neutronium
- Positronium
- Quantum chromodynamics
- Quantum electrodynamics
- Quarkonium

78.7 References

[1] §1.8, *Constituents of Matter: Atoms, Molecules, Nuclei and Particles*, Ludwig Bergmann, Clemens Schaefer, and Wilhelm Raith, Berlin: Walter de Gruyter, 1997, ISBN 3-11-013990-1.

[2] Exotic atoms, AccessScience, McGraw-Hill. Accessed on line September 26, 2007.

[3] Dr. Richard Feynman's Douglas Robb Memorial Lectures

[4] p. 3, *Fundamentals in Hadronic Atom Theory*, A. Deloff, River Edge, New Jersey: World Scientific, 2003. ISBN 981-238-371-9.

[5] p. 8, §16.4, §16.5, Deloff.

[6] The strange world of the exotic atom, Roger Barrett, Daphne Jackson and Habatwa Mweene, *New Scientist*, August 4, 1990. Accessed on line September 26, 2007.

[7] p. 180, *Quantum Mechanics*, B. K. Agarwal and Hari Prakash, New Delhi: Prentice-Hall of India Private Ltd., 1997. ISBN 81-203-1007-1.

[8] Exotic atoms cast light on fundamental questions, *CERN Courier*, November 1, 2006. Accessed on line September 26, 2007.

[9] DOE/SLAC National Accelerator Laboratory (2009, June 4). *Theorists Reveal Path To True Muonium – Never-seen Atom.* ScienceDaily. Retrieved June 7, 2009.

Chapter 79

Positronium

Positronium (Ps) is a system consisting of an electron and its anti-particle, a positron, bound together into an *exotic atom*, specifically an *onium*. The system is unstable: the two particles annihilate each other to predominantly produce two or three gamma-rays, depending on the relative spin states. The orbit and energy levels of the two particles are similar to that of the hydrogen atom (which is a bound state of a proton and an electron). However, because of the reduced mass, the frequencies of the spectral lines are less than half of the corresponding hydrogen lines.

79.1 States

The ground state of positronium, like that of hydrogen, has two possible configurations depending on the relative orientations of the spins of the electron and the positron.

The *singlet* state, 1S
0, with antiparallel spins ($S = 0$, $Ms = 0$) is known as *para-positronium* (*p*-Ps). It has a mean lifetime of 125 picoseconds and decays preferentially into two gamma rays with energy of 511 keV each (in the center-of-mass frame). Detection of these photons allows to reconstruct the vertex of the decay and is used in the positron-emission tomography. Para-positronium can decay into any even number of photons (2, 4, 6, ...), but the probability quickly decreases with the number: the branching ratio for decay into 4 photons is $1.439(2)\times10^{-6}$.[1]

Para-positronium lifetime in vacuum is approximately[1]

$$t_0 = \frac{2\hbar}{m_e c^2 \alpha^5} = 1.244 \times 10^{-10} \text{ s.}$$

The *triplet* state, $^3S_1$, with parallel spins ($S = 1$, $Ms = -1, 0, 1$) is known as *ortho-positronium* (*o*-Ps). It has a mean lifetime of 142.05 ± 0.02 ns,[2] and the leading decay is three gammas. Other modes of decay are negligible; for instance, the five-photons mode has branching ratio of $\sim1.0\times10^{-6}$.[3]

Ortho-positronium lifetime in vacuum can be calculated approximately as:[1]

$$t_1 = \frac{\frac{1}{2}9h}{2m_e c^2 \alpha^6 (\pi^2 - 9)} = 1.386 \times 10^{-7} \text{ s.}$$

However more accurate calculations with corrections to order $O(\alpha^2)$ yield a value of 7.04 μs^{-1} for the decay rate, corresponding to a lifetime of 1.42×10^{-7} s.[4][5]

Positronium in the 2S state is metastable having a lifetime of 1.1 μs against annihilation. The positronium created in such an excited state will quickly cascade down to the ground state, where annihilation will occur more quickly.

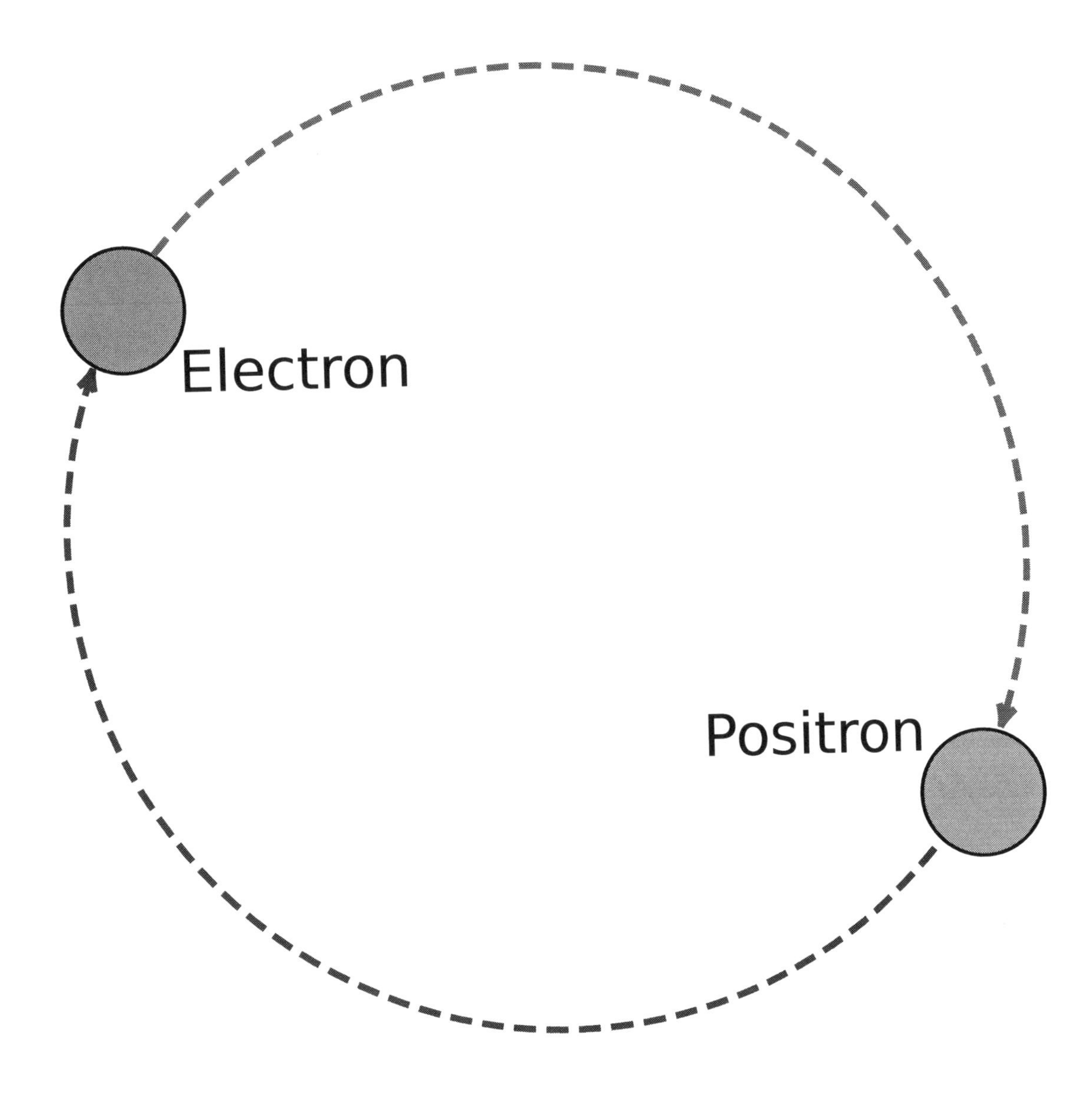

An electron and positron orbiting around their common centre of mass. This is a bound quantum state known as ***positronium****.*

79.1.1 Measurements

Measurements of these lifetimes and energy levels have been used in precision tests of quantum electrodynamics, confirming QED predictions to high precision.[1][6][7] Annihilation can proceed via a number of channels, each producing gamma rays with total energy of 1022 keV (sum of the electron and positron mass-energy), usually 2 or 3, with up to 5 recorded.

The annihilation into a neutrino–antineutrino pair is also possible, but the probability is predicted to be negligible. The branching ratio for *o*-Ps decay for this channel is 6.2×10^{-18} (electron neutrino–antineutrino pair) and 9.5×10^{-21} (for other flavour)[3] in predictions based on the Standard Model, but it can be increased by non-standard neutrino properties, like relatively high magnetic moment. The experimental upper limits on branching ratio for this decay (as well as for a decay into any "invisible" particles) are $<4.3\times10^{-7}$ for *p*-Ps and $<4.2\times10^{-7}$ for *o*-Ps.[2]

79.2 Energy levels

Main article: Bohr model § Electron energy levels

While precise calculation of positronium energy levels uses the Bethe–Salpeter equation or the Breit equation, the similarity between positronium and hydrogen allows a rough estimate. In this approximation, the energy levels are different because of a different effective mass, *m**, in the energy equation (see electron energy levels for a derivation):

$$E_n = -\frac{\mu q_e^4}{8h^2 \varepsilon_0^2} \frac{1}{n^2},$$

where:

q_e is the charge magnitude of the electron (same as the positron),

h is Planck's constant,

ε_0 is the electric constant (otherwise known as the permittivity of free space),

μ is the reduced mass:

$$\mu = \frac{m_e m_p}{m_e + m_p} = \frac{m_e^2}{2m_e} = \frac{m_e}{2},$$

where m_e and m_p are, respectively, the mass of the electron and the positron (which are *the same* by definition as antiparticles).

Thus, for positronium, its reduced mass only differs from the electron by a factor of 2. This causes the energy levels to also roughly be half of what they are for the hydrogen atom.

So finally, the energy levels of positronium are given by

$$E_n = -\frac{1}{2} \frac{m_e q_e^4}{8h^2 \varepsilon_0^2} \frac{1}{n^2} = \frac{-6.8 \text{ eV}}{n^2}.$$

The lowest energy level of positronium ($n = 1$) is −6.8 electronvolts (eV). The next level is −1.7 eV. The negative sign implies a bound state. Positronium can also be considered by a particular form of the two-body Dirac equation; Two point particles with a Coulomb interaction can be exactly separated in the (relativistic) center-of-momentum frame and the resulting ground-state energy has been obtained very accurately using finite element methods of J. Shertzer.[8] The Dirac equation whose Hamiltonian comprises two Dirac particles and a static Coulomb potential is not relativistically invariant. But if one adds the $1/c^{2n}$ (or α^{2n}, where α is the fine-structure constant) terms, where $n = 1,2\ldots$, then the result is relativistically invariant. Only the leading term is included. The α^2 contribution is the Breit term; workers rarely go to α^4 because at α^3 one has the Lamb shift, which requires quantum electrodynamics.[8]

79.3 History

Croatian scientist Stjepan Mohorovičić predicted the existence of positronium in a 1934 article published in *Astronomische Nachrichten*, in which he called it the "electrum".[9] Other sources credit Carl Anderson as having predicted its existence in 1932 while at Caltech.[10] It was experimentally discovered by Martin Deutsch at MIT in 1951 and became known as

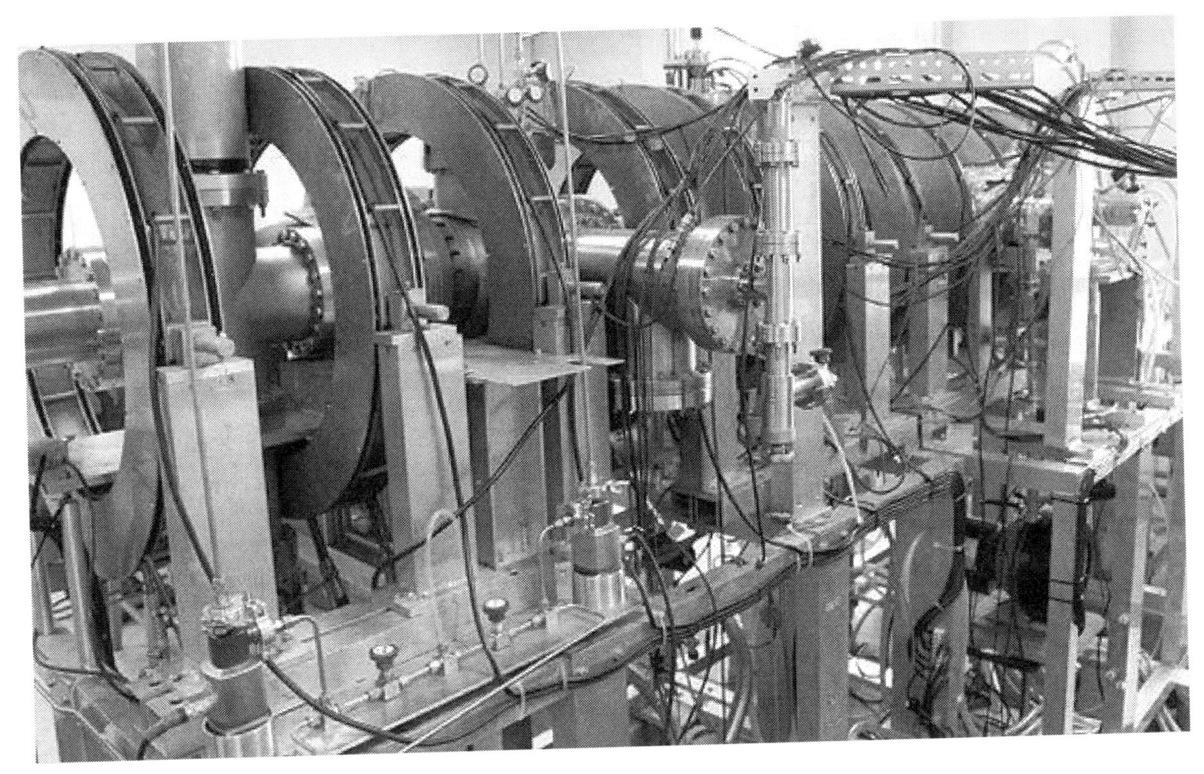

The Positronium Beam at University College London, a lab used to study the properties of positronium

positronium.[10] Many subsequent experiments have precisely measured its properties and verified predictions of quantum electrodynamics (QED). There was a discrepancy known as the ortho-positronium lifetime puzzle that persisted for some time, but was eventually resolved with further calculations and measurements.[11] Measurements were in error because of the lifetime measurement of unthermalised positronium, which was only produced at a small rate. This had yielded lifetimes that were too long. Also calculations using relativistic QED are difficult to perform, so they had been done to only the first order. Corrections that involved higher orders were then calculated in a non-relativistic QED.[4]

79.4 Exotic compounds

Molecular bonding was predicted for positronium.[12] Molecules of positronium hydride (PsH) can be made.[13] Positronium can also form a cyanide and can form bonds with halogens or lithium.[14]

The first observation of di-positronium molecules—molecules consisting of two positronium atoms—was reported on 12 September 2007 by David Cassidy and Allen Mills from University of California, Riverside.[15][16]

79.5 Natural occurrence

Positronium in high energy states has been predicted to be the dominant form of atomic matter in the universe in the far future, if proton decay is a reality.[17]

79.6 See also

- Breit equation
- Antiprotonic helium

- Quantum electrodynamics
- Protonium
- Two-body Dirac equations

79.7 References

[1] Karshenboim, Savely G. (2003). "Precision Study of Positronium: Testing Bound State QED Theory". *International Journal of Modern Physics A [Particles and Fields; Gravitation; Cosmology; Nuclear Physics]* **19** (23): 3879–3896. arXiv:hep-ph/0310099. Bibcode:2004IJMPA..19.3879K. doi:10.1142/S0217751X04020142.

[2] A. Badertscher et al. (2007). "An Improved Limit on Invisible Decays of Positronium". *Physical Review D* **75** (3): 032004. arXiv:hep-ex/0609059. Bibcode:2007PhRvD..75c2004B. doi:10.1103/PhysRevD.75.032004.

[3] Czarnecki, Andrzej; Karshenboim, Savely G. (2000). Levchenko, B.B.; Savrin, V.I., eds. "Decays of Positronium". *Proceedings of the International Workshop on High Energy Physics and Quantum Field Theory (QFTHEP)* (Moscow: MSU-Press) **14** (99): 538–544. arXiv:hep-ph/9911410. Bibcode:1999hep.ph...11410C.

[4] Kataoka, Y.; Asai, S.; Kobayashi, t. (9 September 2008). "First Test of O(α^2) Correction of the Orthopositronium Decay Rate" (PDF). International Center for Elementary Particle Physics.

[5] Adkins, G. S.; Fell, R. N.; Sapirstein, J. (29 May 2000). "Order α^2 Corrections to the Decay Rate of Orthopositronium". *Physical Review Letters***84**(22): 5086–5089. arXiv:hep-ph/0003028. Bibcode:2000PhRvL..84.5086A.doi:10.1103/PhysRevLett6.

[6] Rubbia, A. (2004). "Positronium as a probe for new physics beyond the standard model". *International Journal of Modern Physics A [Particles and Fields; Gravitation; Cosmology; Nuclear Physics]* **19** (23): 3961–3985. arXiv:hep-ph/0402151. Bibcode:2004IJMPA..19.3961R. doi:10.1142/S0217751X0402021X.

[7] Vetter, P.A.; Freedman, S.J. (2002). "Branching-ratio measurements of multiphoton decays of positronium". *Physical Review A* **66** (5): 052505. Bibcode:2002PhRvA..66e2505V. doi:10.1103/PhysRevA.66.052505.

[8] Scott, T.C.; Shertzer, J.; Moore, R.A. (1992). "Accurate finite element solutions of the two-body Dirac equation". *Physical Review A* **45** (7): 4393–4398. Bibcode:1992PhRvA..45.4393S. doi:10.1103/PhysRevA.45.4393. PMID 9907514.

[9] Mohorovičić, S. (1934). "Möglichkeit neuer Elemente und ihre Bedeutung für die Astrophysik". *Astronomische Nachrichten* **253** (4): 94. Bibcode:1934AN....253...93M. doi:10.1002/asna.19342530402.

[10] "Martin Deutsch, MIT physicist who discovered positronium, dies at 85" (Press release). MIT. 2002.

[11] Dumé, Belle (May 23, 2003). "Positronium puzzle is solved". *Physics World*.

[12] Usukura, J.; Varga, K.; Suzuki, Y. (1998). "Signature of the existence of the positronium molecule". arXiv:physics/9804023v1 [physics.atom-ph].

[13] ""Out of This World" Chemical Compound Observed" (PDF). p. 9.

[14] Saito, Shiro L. (2000). "Is Positronium Hydride Atom or Molecule?". *Nuclear Instruments and Methods in Physics Research B* **171**: 60–66. Bibcode:2000NIMPB.171...60S. doi:10.1016/s0168-583x(00)00005-7.

[15] Cassidy,D.B.;Mills,A.P. (Jr.) (2007). "The production of molecular positronium".*Nature***449**(7159): 195–197. Bibcode:2007. doi:10.1038/nature06094. PMID 17851519. Lay summary.

[16] "Molecules of positronium observed in the lab for the first time". Physorg.com. Retrieved 2007-09-07.

[17] A dying universe: the long-term fate and evolution of astrophysical objects, Fred C. Adams and Gregory Laughlin, *Reviews of Modern Physics* **69**, #2 (April 1997), pp. 337–372. Bibcode: 1997RvMP...69..337A. doi:10.1103/RevModPhys.69.337 arXiv:astro-ph/9701131.

79.8 External links

- The Search for Positronium
- Obituary of Martin Deutsch, discoverer of Positronium
- Website about positrons, positronium and antihydrogen. Positron Laboratory, Como, Italy

Chapter 80

Muonium

For atoms where muons have replaced one or more electrons, see muonic atom.

Muonium is an exotic atom made up of an antimuon and an electron,[1] which was discovered in 1960[2] and is given the chemical symbol Mu. During the muon's 2.2 μs lifetime, muonium can enter into compounds such as muonium chloride (MuCl) or sodium muonide (NaMu).[3] Due to the mass difference between the antimuon and the electron, muonium (μ+e−) is more similar to atomic hydrogen (p+e−) than positronium (e+e−). Its Bohr radius and ionization energy are within 0.5% of hydrogen, deuterium, and tritium.[4]

Although muonium is short-lived, physical chemists study it using Muon spin spectroscopy (μSR),[5] a magnetic resonance technique analogous to nuclear magnetic resonance (NMR) or electron spin resonance (ESR) spectroscopy. Like ESR, μSR is useful for the analysis of chemical transformations and the structure of compounds with novel or potentially valuable electronic properties. Muonium is usually studied by muon spin rotation, in which the Mu atom's spin precesses in a magnetic field applied transverse to the muon spin direction (since muons are typically produced in a spin-polarized state from the decay of pions), and by avoided level crossing (ALC), which is also called level crossing resonance (LCR).[5] The latter employs a magnetic field applied longitudinally to the polarization direction, and monitors the relaxation of muon spins caused by "flip/flop" transitions with other magnetic nuclei.

Because the muon is a lepton, the atomic energy levels of muonium can be calculated with great precision from quantum electrodynamics (QED), unlike in the case of hydrogen, where the precision is limited by uncertainties related to the internal structure of the proton. For this reason, muonium is an ideal system for studying bound-state QED and also for searching for physics beyond the standard model.[6]

80.1 Nomenclature

Normally in the nomenclature of particle physics, an atom composed of a positively charged particle bound to an electron is named after the positive particle with "-ium" appended, in this case "muium". The suffix "-onium" is mostly used for bound states of a particle with its own antiparticle. The exotic atom consisting of a muon and an antimuon is known as "true muonium". It is yet to be observed, but it may have been generated in the collision of electron and positron beams.[7][8]

80.2 References

[1] IUPAC (1997). "Muonium". In A.D. McNaught, A. Wilkinson. *Compendium of Chemical Terminology* (2nd ed.). Blackwell Scientific Publications. doi:10.1351/goldbook.M04069. ISBN 0-86542-684-8.

[2] V.W Hughes et al. (1960). "Formation of Muonium and Observation of its Larmor Precession". *Physical Review Letters* **5** (2): 63–65. Bibcode:1960PhRvL...5...63H. doi:10.1103/PhysRevLett.5.63.

[3] W.H. Koppenol (IUPAC) (2001). "Names for muonium and hydrogen atoms and their ions" (PDF). *Pure and Applied Chemistry* **73** (2): 377–380. doi:10.1351/pac200173020377.

[4] Walker, David C (1983-09-08). *Muon and Muonium Chemistry*. p. 4. ISBN 978-0-521-24241-7.

[5] J.H. Brewer (1994). "Muon Spin Rotation/Relaxation/Resonance". *Encyclopedia of Applied Physics* (VCH Publishers, Inc.) **11**: 23–53.

[6] K.P. Jungmann (2004). "Past, Present and Future of Muonium". *Proc. of Memorial Symp. in Honor of V. W. Hughes, New Haven, Connecticut, 14–15 Nov 2003*: 134. arXiv:nucl-ex/0404013. Bibcode:2004shvw.conf..134J. doi:10.1142/9789812702425_0009.ISBN978-981-256-050-6.

[7] S.J. Brodsky, R.F. Lebed (2009). "Production of the smallest QED atom: True muonium ($\mu^+\mu^-$)". *Physical Review Letters* **102** (21): 213401. arXiv:0904.2225. Bibcode:2009PhRvL.102u3401B. doi:10.1103/PhysRevLett.102.213401.

[8] H. Lamm, R.F. Lebed (2013). "True Muonium ($\mu^+\mu^-$) on the Light Front: A Toy Model". arXiv:1311.3245.

Chapter 81

Onium

Not to be confused with onium compounds, ions such as ammonium.

An **onium** (plural: **onia**) is a bound state of a particle and its antiparticle. They are usually named by adding the suffix *-onium* to the name of the constituting particle except for Muonium which, despite its name, not a bound muon–antimuon onium, but an electron–antimuon bound state, and whose name was assigned by IUPAC. A muon–antimuon onium would be named muononium.

81.1 Examples

Positronium is an onium which consists of an electron and a positron bound together as a long-lived metastable state. Positronium has been studied since the 1950s to understand bound states in quantum field theory. A recent development called non-relativistic quantum electrodynamics (NRQED) used this system as a proving ground.

Pionium, a bound state of two oppositely-charged pions, is interesting for exploring the strong interaction. This should also be true of protonium. The true analogs of positronium in the theory of strong interactions, however, are not exotic atoms but certain mesons, the *quarkonium states*, which are made of a heavy quark such as the charm or bottom quark and its antiquark. (Top quarks are so heavy that they decay through the weak force before they can form bound states.) Exploration of these states through non-relativistic quantum chromodynamics (NRQCD) and lattice QCD are increasingly important tests of quantum chromodynamics.

Understanding bound states of hadrons such as pionium and protonium is also important in order to clarify notions related to exotic hadrons such as mesonic molecules and pentaquark states.

81.2 See also

- Exotic atom

Chapter 82

Superatom

A **superatom** is any cluster of atoms that seem to exhibit some of the properties of elemental atoms.

Sodium atoms, when cooled from vapor, naturally condense into clusters, preferentially containing a magic number of atoms (2, 8, 20, 40, 58, etc.). The first two of these can be recognized as the numbers of electrons needed to fill the first and second shells, respectively. The superatom suggestion is that free electrons in the cluster occupy a new set of orbitals that are defined by the entire group of atoms, i.e. cluster, rather than each individual atom separately (non-spherical or doped clusters show deviations in the number of electrons that form a closed shell as the potential is defined by the shape of the positive nuclei.) Superatoms tend to behave chemically in a way that will allow them to have a closed shell of electrons, in this new counting scheme. Therefore, a superatom with one more electron than a full shell should give up that electron very easily, similar to an alkali metal, and a cluster with one electron short of full shell should have a large electron affinity, such as a halogen.

82.1 Aluminium clusters

Certain aluminium clusters have superatom properties. These aluminium clusters are generated as anions (Aln^- with n = 1, 2, 3, …) in helium gas and reacted with a gas containing iodine. When analyzed by mass spectrometry one main reaction product turns out to be $Al_{13}I^-$.[1] These clusters of 13 aluminium atoms with an extra electron added do not appear to react with oxygen when it is introduced in the same gas stream. Assuming each atom liberates its 3 valence electrons, this means that there are 40 electrons present, which is one of the magic numbers noted above for sodium, and implies that these numbers are a reflection of the noble gases. Calculations show that the additional electron is located in the aluminium cluster at the location directly opposite from the iodine atom. The cluster must therefore have a higher electron affinity for the electron than iodine and therefore the aluminium cluster is called a **superhalogen**. The cluster component in $Al_{13}I^-$ ion is similar to an iodide ion or better still a bromide ion. The related $Al_{13}I_2{}^-$ cluster is expected to behave chemically like the triiodide ion.

Similarly it has been noted that Al_{14} clusters with 42 electrons (2 more than the magic numbers) appear to exhibit the properties of an alkaline earth metal which typically adopt +2 valence states. This is only known to occur when there are at least 3 iodine atoms attached to an $Al_{14}{}^-$ cluster, $Al_{14}I_3{}^-$. The anionic cluster has a total of 43 itinerant electrons, but the three iodine atoms each remove one of the itinerant electrons to leave 40 electrons in the jellium shell.[2][3]

It is particularly easy and reliable to study atomic clusters of inert gas atoms by computer simulation because interaction between two atoms can be approximated very well by the Lennard-Jones potential. Other methods are readily available and it has been established that the magic numbers are 13, 19, 23, 26, 29, 32, 34, 43, 46, 49, 55, etc.[4]

- Al_7 = the property is similar to germanium atoms.
- Al_{13} = the property is similar to halogen atoms, more specifically, chlorine.
  - $Al_{13}Ix^-$, where x = 1–13.[5]

- Al_{14} = the property is similar to alkaline earth metals.
  - $Al_{14}Ix^-$, where x = 1–14.[5]
- Al_{23}
- Al_{37}

82.2 Other clusters

- $Li(HF)_3Li$ = the $(HF)_3$ interior causes 2 valence electrons from the Li to orbit the entire molecule as if it were an atom's nucleus.[6]
- $VSi_{16}F$ = has ionic bonding.[7]
- A cluster of 13 platinum becomes paramagnetic.[8]
- A cluster of 2000 rubidium atoms.[9]

82.3 Superatom complexes

Superatom complexes are a special group of superatoms that incorporate a metal core which is stabilized by organic ligands. In thiolate-protected gold cluster complexes a simple electron counting rule can be used to determine the total number of electrons (n_e) which correspond to a magic number via,

$$n_e = N\nu_A - M - z$$

where N is the number of metal atoms (A) in the core, v is the atomic valence, M is the number of electron withdrawing ligands, and z is the overall charge on the complex.[10] For example the Au_{102}(p-MBA)$_{44}$ has 58 electrons and corresponds to a closed shell magic number.[11]

82.3.1 Gold superatom complexes

- $Au_{25}(SMe)_{18}^-$ [12]
- Au_{102}(p-MBA)$_{44}$
- $Au_{144}(SR)_{60}$ [13]

82.3.2 Other superatom complexes

- $Ga_{23}(N(Si(CH_3)_3)_2)_{11}$[14]
- $Al_{50}(C_5(CH_3)_5)_{12}$[15]

82.4 See also

- Bose–Einstein condensate

82.5 References

[1] *Formation of $Al_{13}I^-$: Evidence for the Superhalogen Character of Al_{13}* D. E. Bergeron, A.W. Castleman Jr., T. Morisato, S. N. Khanna Science, Vol 304, Issue 5667, 84–87 , 2 April **2004** Abstract MS spectra

[2] Philip Ball, "A New Kind of Alchemy", *New Scientist* Issue dated 2005-04-16.

[3] *Al Cluster Superatoms as Halogens in Polyhalides and as Alkaline Earths in Iodide Salts* D. E. Bergeron, P. J. Roach, A.W. Castleman Jr., N.O. Jones, S. N. Khanna Science, Vol 307, Issue 5707, 231–235 , 14 January **2005** Abstract MS spectrum

[4] I. A. Harris *et al.* Phys. Rev. Lett. Vol. 53, 2390–94 (1984).

[5] Naiche Owen Jones, 2006.

[6] Extraordinary superatom containing double shell nucleus: Li(HF)3Li connected mainly by intermolecular interactions, Sun, Xiao-Ying, Li, Zhi-Ru, Wu, Di, & Sun, Chia-Chung, 2007.

[7] Electronic and geometric stabilities of clusters with transition metal encapsulated by silicon, Kiichirou Koyasu et al.

[8] Platinum nanoclusters go magnetic, nanotechweb.org, 2007

[9] Ultra Cold Trap Yields Superatom, NIST, 1995

[10] M. Walter, J. Akola, O. Lopez-Acevedo, P. D. Jadzinsky, G. Calero, C. J. Ackerson, R. L. Whetten, H. Grönbeck, H. Häkkinen, Gold Superatom Complexes"A unified view of ligand-protected gold clusters as superatom complexes ", PNAS 105, 9157 (2008)

[11] P.D. Jadzinsky, G. Calero, C.J. Ackerson, D.A. Bushnell, R.D. Kornberg, Gold Superatom Complexes Structure of a thiol monolayer-protected gold nanoparticle at 1.1 Å resolution" Science 318, 430–433 (2007)

[12] J. Akola, M. Walter, R.L. Whetten, H. Häkkinen and H. Grönbeck, "On the structure of thiolate-protected Au_{25}", JACS 130, 3756–3757 (2008)

[13] O. Lopez-Acevedo, J. Akola, R.L. Whetten, H. Grönbeck, H. Häkkinen, "Structure and Bonding in the Ubiquitous Icosahedral Metallic Gold Cluster $Au_{144}(SR)_{60}$", JPCC 130, 3756–3757 (2009)

[14] J. Hartig, A. Stösser, H. Schnöckel, "A metalloid $(Ga_{23}\{N(SiMe_3)_2\}_{11})$ cluster: The jellium model put to test" Angew. Chemie. Int. Ed. 46, 1658–1662 (2007).

[15] P.A. Clayborne, O. Lopez-Acevedo, R.L. Whetten, H. Grönbeck and H. Häkkinen, "$Al_{50}Cp^*_{12}$ Cluster: A 138-electron (L=6) Superatom", Eur. J. Inorg. Chem. 2011.

82.6 External links

- Designer Magnetic Superatoms, J.U. Reveles, et al. 2009
- Gold Superatom Complexes, M. Walter et al. 2008
- Gold Superatom Complexes P.D. Jadzinsky et al. 2007
- Multiple Valence Superatoms, J.U. Reveles, S.N. Khanna, P.J. Roach, and A.W. Castleman Jr., 2006
- On the Aluminum Cluster Superatoms acting as Halogens and Alkaline-earth Metals, Bergeron, Dennis E et al., 2006
- Clusters of Aluminum Atoms Found to Have Properties of Other Elements Reveal a New Form of Chemistry, innovations report, 2005. Have a picture of Al_{14}.
- Clusters of Aluminum Atoms Found to Have Properties of Other Elements Reveal a New Form of Chemistry, Penn State, Eberly College of Science, 2005
- Research Reveals Halogen Characteristics innovations report, 2004. Have pictures of Al_{13}.

Chapter 83

Molecule

For the scientific journal, see Molecules (journal).

A **molecule** (/ˈmɒlɪkjuːl/ from Latin moles "mass"[1]) is an electrically neutral group of two or more atoms held together

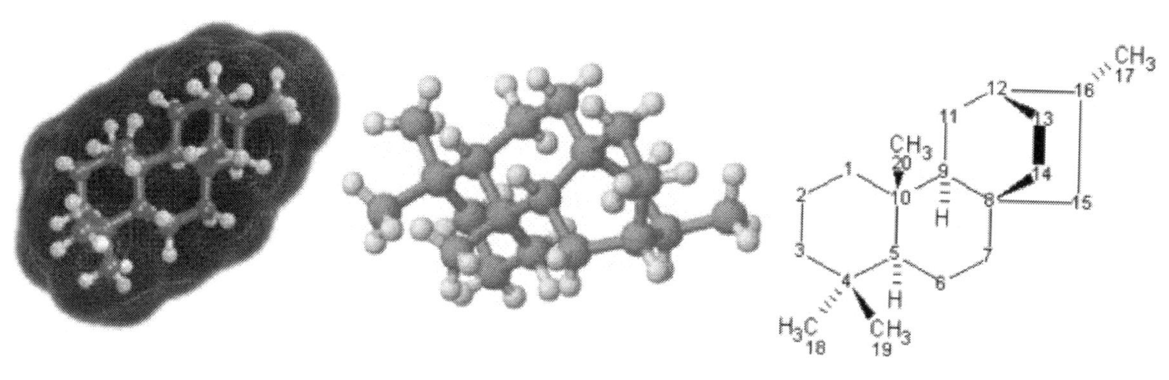

3D (left and center) and 2D (right) representations of the terpenoid molecule atisane

by chemical bonds.[2][3][4][5][6] Molecules are distinguished from ions by their lack of electrical charge. However, in quantum physics, organic chemistry, and biochemistry, the term *molecule* is often used less strictly, also being applied to polyatomic ions.

In the kinetic theory of gases, the term *molecule* is often used for any gaseous particle regardless of its composition. According to this definition, noble gas atoms are considered molecules despite being composed of a single non-bonded atom.[7]

A molecule may be homonuclear, that is, it consists of atoms of a single chemical element, as with oxygen (O_2); or it may be heteronuclear, a chemical compound composed of more than one element, as with water (H_2O). Atoms and complexes connected by non-covalent bonds such as hydrogen bonds or ionic bonds are generally not considered single molecules.[8]

Molecules as components of matter are common in organic substances (and therefore biochemistry). They also make up most of the oceans and atmosphere. However, the majority of familiar solid substances on Earth, including most of the minerals that make up the crust, mantle, and core of the Earth, contain many chemical bonds, but are *not* made of identifiable molecules. Also, no typical molecule can be defined for ionic crystals (salts) and covalent crystals (network solids), although these are often composed of repeating unit cells that extend either in a plane (such as in graphene) or three-dimensionally (such as in diamond, quartz, or sodium chloride). The theme of repeated unit-cellular-structure also holds for most condensed phases with metallic bonding, which means that solid metals are also not made of molecules. In glasses (solids that exist in a vitreous disordered state), atoms may also be held together by chemical bonds without presence of any definable molecule, but also without any of the regularity of repeating units that characterizes crystals.

83.1 Molecular science

The science of molecules is called *molecular chemistry* or *molecular physics*, depending on whether the focus is on chemistry or physics. Molecular chemistry deals with the laws governing the interaction between molecules that results in the formation and breakage of chemical bonds, while molecular physics deals with the laws governing their structure and properties. In practice, however, this distinction is vague. In molecular sciences, a molecule consists of a stable system (bound state) composed of two or more atoms. Polyatomic ions may sometimes be usefully thought of as electrically charged molecules. The term *unstable molecule* is used for very reactive species, i.e., short-lived assemblies (resonances) of electrons and nuclei, such as radicals, molecular ions, Rydberg molecules, transition states, van der Waals complexes, or systems of colliding atoms as in Bose–Einstein condensate.

83.2 History and etymology

Main article: History of molecular theory

According to Merriam-Webster and the Online Etymology Dictionary, the word "molecule" derives from the Latin "moles" or small unit of mass.

- **Molecule** (1794) – "extremely minute particle", from Fr. *molécule* (1678), from modern Latin. *molecula*, diminutive of Latin *moles* "mass, barrier". A vague meaning at first; the vogue for the word (used until the late 18th century only in Latin form) can be traced to the philosophy of Descartes.

The definition of the molecule has evolved as knowledge of the structure of molecules has increased. Earlier definitions were less precise, defining molecules as the smallest particles of pure chemical substances that still retain their composition and chemical properties.[9] This definition often breaks down since many substances in ordinary experience, such as rocks, salts, and metals, are composed of large crystalline networks of chemically bonded atoms or ions, but are not made of discrete molecules.

83.3 Molecular size

Most molecules are far too small to be seen with the naked eye, but there are exceptions. DNA, a macromolecule, can reach macroscopic sizes, as can molecules of many polymers. Molecules commonly used as building blocks for organic synthesis have a dimension of a few angstroms (Å) to several dozen Å. Single molecules cannot usually be observed by light (as noted above), but small molecules and even the outlines of individual atoms may be traced in some circumstances by use of an atomic force microscope. Some of the largest molecules are macromolecules or supermolecules.

83.3.1 Smallest molecule diameter

The smallest molecule is the diatomic hydrogen (H_2), with a bond length of 0.74 Å.[10]

83.3.2 Largest molecule diameter

Mesoporous silica have been produced with a diameter of 1000 Å (100 nm)[11]

83.3.3 Radius

Effective molecular radius is the size a molecule displays in solution.[12][13] The table of permselectivity for different substances contains examples.

John Dalton

83.4 Formulas for molecules

83.4.1 Chemical formula types

Main article: Chemical formula

The chemical formula for a molecule uses a single line of chemical element symbols, numbers, and sometimes also other symbols, such as parentheses, dashes, brackets, and *plus* (+) and *minus* (−) signs. These are limited to a single typographic line of symbols, which may include subscripts and superscripts.

A compound's empirical formula is a very simple type of chemical formula. It is the simplest integer ratio of the chemical elements that constitute it. For example, water is always composed of a 2:1 ratio of hydrogen to oxygen atoms, and ethyl alcohol or ethanol is always composed of carbon, hydrogen, and oxygen in a 2:6:1 ratio. However, this does not determine the kind of molecule uniquely – dimethyl ether has the same ratios as ethanol, for instance. Molecules with the same atoms in different arrangements are called isomers. Also carbohydrates, for example, have the same ratio (carbon:hydrogen:oxygen = 1:2:1) (and thus the same empirical formula) but different total numbers of atoms in the molecule.

The molecular formula reflects the exact number of atoms that compose the molecule and so characterizes different molecules. However different isomers can have the same atomic composition while being different molecules.

The empirical formula is often the same as the molecular formula but not always. For example, the molecule acetylene has molecular formula C_2H_2, but the simplest integer ratio of elements is CH.

The molecular mass can be calculated from the chemical formula and is expressed in conventional atomic mass units equal to 1/12 of the mass of a neutral carbon-12 (^{12}C isotope) atom. For network solids, the term formula unit is used in stoichiometric calculations.

83.4.2 Structural formula

Main article: Structural formula

For molecules with a complicated 3-dimensional structure, especially involving atoms bonded to four different substituents, a simple molecular formula or even semi-structural chemical formula may not be enough to completely specify the molecule. In this case, a graphical type of formula called a structural formula may be needed. Structural formulas may in turn be represented with a one-dimensional chemical name, but such chemical nomenclature requires many words and terms which are not part of chemical formulas.

83.5 Molecular geometry

Main article: Molecular geometry

Molecules have fixed equilibrium geometries—bond lengths and angles— about which they continuously oscillate through vibrational and rotational motions. A pure substance is composed of molecules with the same average geometrical structure. The chemical formula and the structure of a molecule are the two important factors that determine its properties, particularly its reactivity. Isomers share a chemical formula but normally have very different properties because of their different structures. Stereoisomers, a particular type of isomers, may have very similar physico-chemical properties and at the same time different biochemical activities.

83.6 Molecular spectroscopy

Main article: Spectroscopy

Molecular spectroscopy deals with the response (spectrum) of molecules interacting with probing signals of known energy (or frequency, according to Planck's formula). Molecules have quantized energy levels that can be analyzed by detecting the molecule's energy exchange through absorbance or emission.[14] Spectroscopy does not generally refer to diffraction studies where particles such as neutrons, electrons, or high energy X-rays interact with a regular arrangement of molecules (as in a crystal).

83.7 Theoretical aspects

The study of molecules by molecular physics and theoretical chemistry is largely based on quantum mechanics and is essential for the understanding of the chemical bond. The simplest of molecules is the hydrogen molecule-ion, H_2^+, and the simplest of all the chemical bonds is the one-electron bond. H_2^+ is composed of two positively charged protons and one negatively charged electron, which means that the Schrödinger equation for the system can be solved more easily due to the lack of electron–electron repulsion. With the development of fast digital computers, approximate solutions for more complicated molecules became possible and are one of the main aspects of computational chemistry.

When trying to define rigorously whether an arrangement of atoms is "sufficiently stable" to be considered a molecule, IUPAC suggests that it "must correspond to a depression on the potential energy surface that is deep enough to confine at least one vibrational state".[2] This definition does not depend on the nature of the interaction between the atoms, but only on the strength of the interaction. In fact, it includes weakly bound species that would not traditionally be considered molecules, such as the helium dimer, He_2, which has one vibrational bound state[15] and is so loosely bound that it is only likely to be observed at very low temperatures.

Whether or not an arrangement of atoms is "sufficiently stable" to be considered a molecule is inherently an operational definition. Philosophically, therefore, a molecule is not a fundamental entity (in contrast, for instance, to an elementary particle); rather, the concept of a molecule is the chemist's way of making a useful statement about the strengths of atomic-scale interactions in the world that we observe.

83.8 See also

- Atom
- Van der Waals molecule
- Diatomic molecule
- Small molecule
- Chemical polarity
- Molecular geometry
- Covalent bond
- Noncovalent bonding
- list of compounds for a list of chemical compounds
- List of molecules in interstellar space
- Software for molecular mechanics modeling
- Molecular Hamiltonian

- Molecular ion
- Molecular orbital
- Molecular modelling
- Molecular design software
- WorldWide Molecular Matrix
- Periodic Systems of Small Molecules

83.9 References

[1] http://www.etymonline.com/index.php?term=molecule

[2] Template:GoldBookRefwhen they pook

[3] Ebbin, Darrell D. (1990). *General Chemistry* (3rd ed.). Boston: Houghton Mifflin Co. ISBN 0-395-43302-9.

[4] Brown, T.L.; Kenneth C. Kemp; Theodore L. Brown; Harold Eugene LeMay et al. (2003). *Chemistry – the Central Science* (9th ed.). New Jersey: Prentice Hall. ISBN 0-13-066997-0.

[5] Chang, Raymond (1998). *Chemistry* (6th ed.). New York: McGraw Hill. ISBN 0-07-115221-0.

[6] Zumdahl, Steven S. (1997). *Chemistry* (4th ed.). Boston: Houghton Mifflin. ISBN 0-669-41794-7.

[7] Chandra, Sulekh (2005). *Comprehensive Inorganic Chemistry*. New Age Publishers. ISBN 81-224-1512-1.

[8] Molecule, *Encyclopædia Britannica* on-line

[9] Molecule Definition (Frostburg State University)

[10] Roger L. DeKock; Harry B. Gray; Harry B. Gray (1989). *Chemical structure and bonding*. University Science Books. p. 199. ISBN 0-935702-61-X.

[11] http://pubs.acs.org/doi/abs/10.1021/ac303274w

[12] Chang RL; Deen WM; Robertson CR; Brenner BM. (1975). "Permselectivity of the glomerular capillary wall: III. Restricted transport of polyanions". *Kidney Int.* **8** (4): 212–218. doi:10.1038/ki.1975.104. PMID 1202253.

[13] Chang RL; Ueki IF; Troy JL; Deen WM et al. (1975). "Permselectivity of the glomerular capillary wall to macromolecules. II. Experimental studies in rats using neutral dextran". *Biophys J.* **15** (9): 887–906. Bibcode:1975BpJ....15..887C. doi:10.1016/6-3495(75)85863-2. PMC 1334749. PMID1182263.

[14] IUPAC, *Compendium of Chemical Terminology*, 2nd ed. (the "Gold Book") (1997). Online corrected version: (1997,2006) "spectroscopy".

[15] Anderson JB (May 2004). "Comment on "An exact quantum Monte Carlo calculation of the helium-helium intermolecular potential" [J. Chem. Phys. 115, 4546 (2001)]". *J Chem Phys* **120** (20): 9886–7. Bibcode:2004JChPh.120.9886A. doi:10.1063/1.1704638. PMID 15268005.

83.10 External links

- Molecule of the Month – School of Chemistry, University of Bristol

Chapter 84

Exotic baryon

Exotic baryons are composite particles that are bound states of four or more quarks and additional elementary particles, which may include antiquarks or gluons. An example would be pentaquarks, consisting of four quarks and one antiquark. This is to be contrasted with ordinary baryons, which are bound states of just three quarks.

So far, the only observed exotic baryons are the pentaquarks P+
c(4380) and P+
c(4450), discovered in 2015 by the LHCb collaboration.[1]

Several types of exotic baryons that require physics beyond the Standard Model have been conjectured in order to explain specific experimental anomalies. There is no independent experimental evidence for any of these particles. One example is supersymmetric R-baryons,[2] which are bound states of 3 quarks and a gluino. The lightest R-baryon is denoted as S^0 and consists of an up quark, a down quark, a strange quark and a gluino. This particle is expected to be long lived or stable and has been invoked to explain ultra-high-energy cosmic rays.[3][4] Stable exotic baryons are also candidates for strongly interacting dark matter.

It has been speculated by futurologist Ray Kurzweil that by the end of the 21st century it might be possible by using femtotechnology to create new chemical elements composed of exotic baryons that would eventually constitute a new periodic table of elements in which the elements would have completely different properties than the regular chemical elements.[5]

84.1 References

[1] R. Aaij et al. (LHCb collaboration) (2015). "Observation of J/ψp resonances consistent with pentaquark states in Λ0
b→J/ψK−
p decays". *Physical Review Letters* **115** (7). doi:10.1103/PhysRevLett.115.072001.

[2] G.R. Farrar (1996). "Detecting Gluino-Containing Hadrons". *Physical Review Letters* **76** (22): 4111–4114. arXiv:hep-ph/9603271. Bibcode:1996PhRvL..76.4111F. doi:10.1103/PhysRevLett.76.4111. PMID 10061204.

[3] D. Chung, G.R. Farrar, E.W. Kolb (1998). "Are ultra-high-energy cosmic rays signals of supersymmetry?". *Physical Review D* **57** (8): 4606. arXiv:astro-ph/9707036. Bibcode:1998PhRvD..57.4606C. doi:10.1103/PhysRevD.57.4606.

[4] I.F.M. Albuquerque, G. Farrar, E.W. Kolb (1999). "Exotic massive hadrons and ultra-high-energy cosmic rays". *Physical Review D* **59**: 015021. arXiv:hep-ph/9805288. Bibcode:1999PhRvD..59a5021A. doi:10.1103/PhysRevD.59.015021.

[5] Kurzweil, Ray *The Age of Spiritual Machines* 1999

Chapter 85

Exotic hadron

Exotic hadrons are subatomic particles composed of quarks and gluons, but which do not fit into the usual scheme of hadrons. While bound by the strong interaction they are not predicted by the simple quark model. That is, exotic hadrons do not have the same quark content as ordinary hadrons: **exotic baryons** have more than just the three quarks of ordinary baryons and **exotic mesons** do not have one quark and one antiquark like ordinary mesons. Exotic hadrons can be searched for by looking for S-matrix poles with quantum numbers forbidden to ordinary hadrons. Experimental signatures for such exotic hadrons have been seen recently[1] but remain a topic of controversy in particle physics.

Jaffe and Low [2] suggested that the exotic hadrons manifest themselves as poles of the P matrix, and not of the S matrix. Experimental P-matrix poles are determined reliably in both the meson-meson channels and nucleon-nucleon channels.

85.1 History

When the quark model was first postulated by Murray Gell-Mann and others in the 1960s, it was to organize the states known then to be in existence in a meaningful way. As Quantum Chromodynamics (QCD) developed over the next decade, it became apparent that there was no reason why only 3-quark and quark-antiquark combinations could exist. In addition, it seemed that gluons, the mediator particles of the strong interaction, could also form bound states by themselves (glueballs) and with quarks (hybrid hadrons). Several decades have passed without conclusive evidence of an exotic hadron that could be associated with the S-matrix pole.

In April 2014, The LHCb collaboration confirmed the existence of the $Z(4430)^-$. Examinations of the character of the particle suggest that it may be exotic.[3]

85.2 Candidates

There are several exotic hadron candidates:

- X(3872) – Discovered by the Belle detector at KEK in Japan, this particle has been variously hypothesized to be diquark or a mesonic molecule.
- Y(3940) – This particle fails to fit into the Charmonium spectrum predicted by theorists.
- Y(4140) – Discovered at Fermilab in March 2009 .
- Y(4260) – Discovered by the BaBar detector at SLAC in Menlo Park, California this particle is hypothesized to be made up of a gluon bound to a quark and antiquark.
- Zc(3900) – Discovered by Belle and BES III
- Z(4430) – Discovered by Belle and later confirmed by LHCb with 13.9σ significance

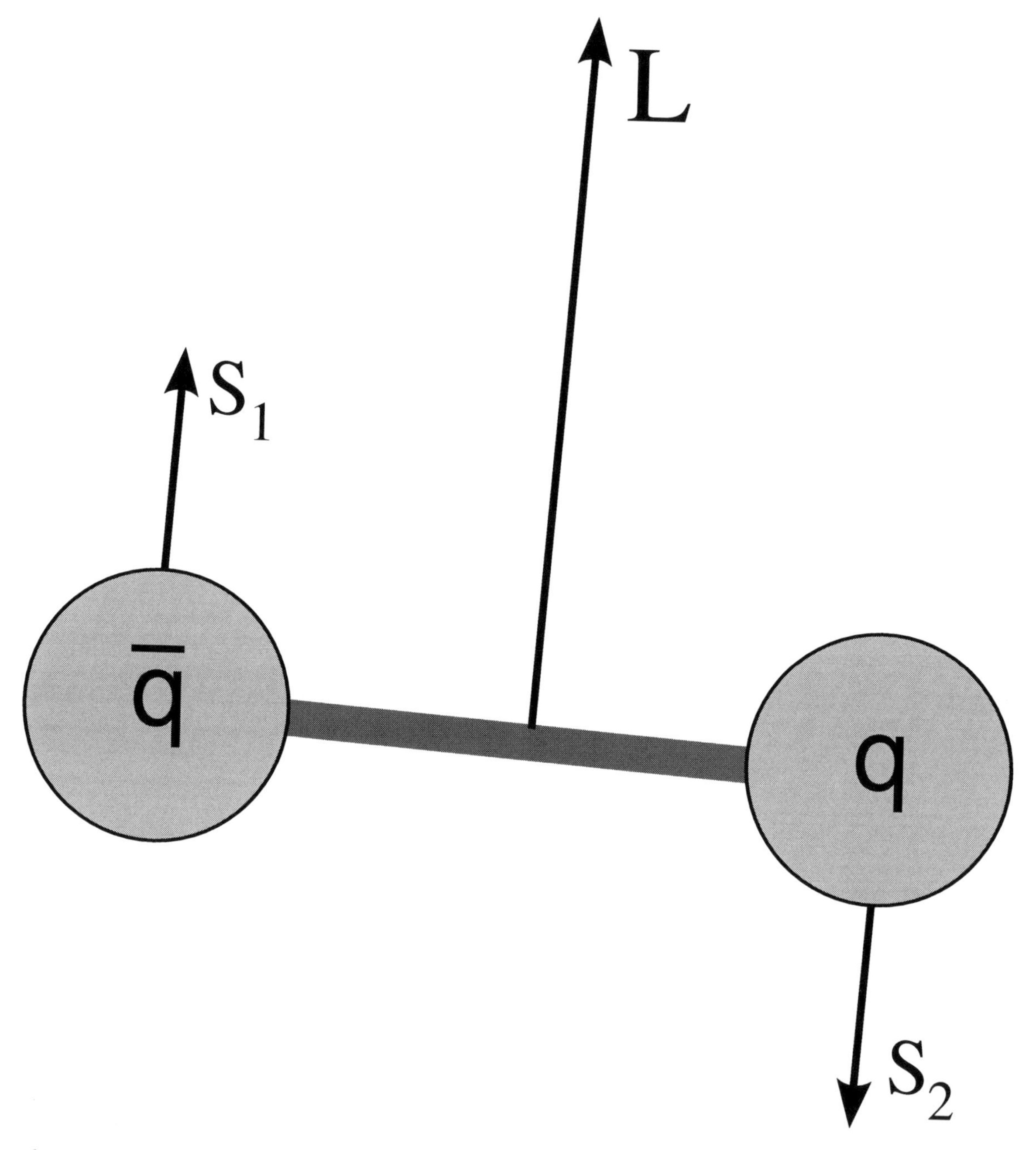

A regular meson made from a quark (q) and an antiquark (q) with spins s_2 and s_1 respectively and having an overall angular momentum L

85.3 See also

- Pentaquark
- Tetraquark

85.4 Notes

[1] See "note on non-q qbar mesons" in PDG 2006, Journal of Physics, G 33 (2006) 1.

[2] R. L. Jaffe and F. E. Low, Phys. Rev. D 19, 2105 (1979). doi:10.1103/PhysRevD.19.2105

[3] LHCb collaboration (7 April 2014). "Observation of the resonant character of the $Z(4430)^-$ state". arXiv:1404.1903.

Chapter 86

Exotic meson

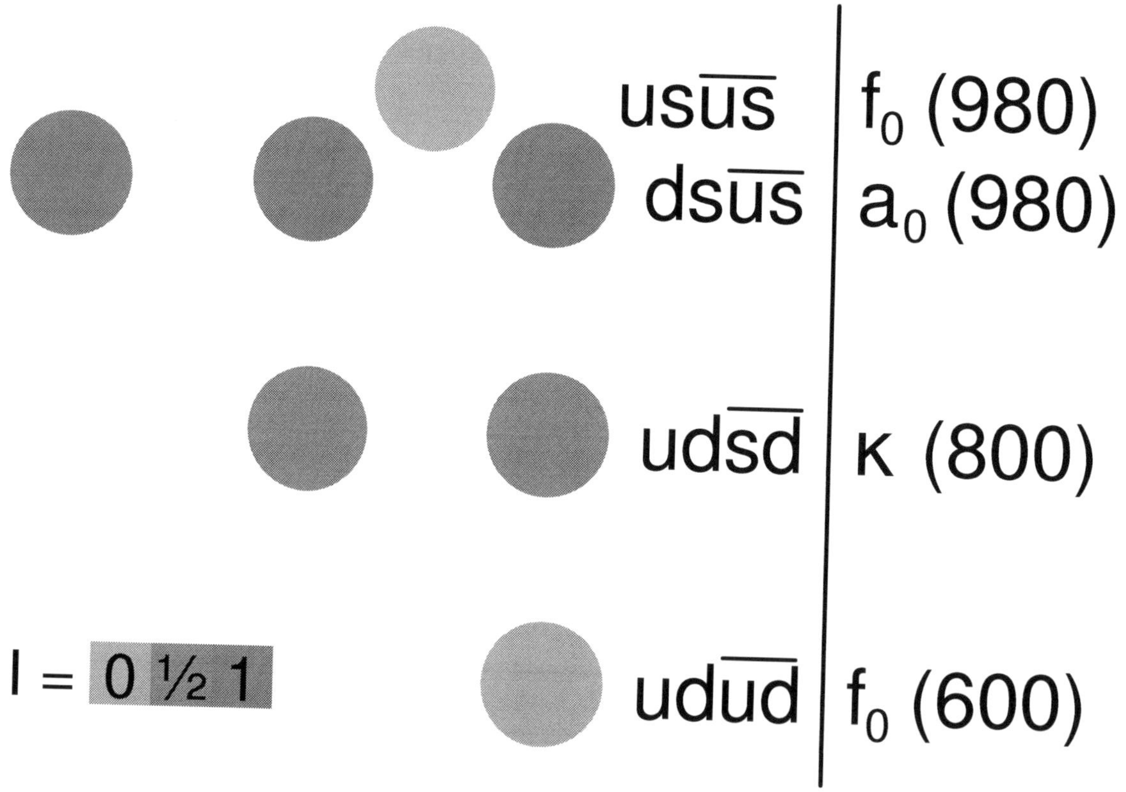

$us\overline{us}$ | f_0 (980)
$ds\overline{us}$ | a_0 (980)
$ud\overline{sd}$ | κ (800)
I = 0 ½ 1
$ud\overline{ud}$ | f_0 (600)

Identities and classification of possible tetraquark mesons. Green denotes I = *0 states, blue,* I = *1/2 and red,* I = *1. The vertical axis is the mass.*

Non-quark model mesons include

1. **exotic mesons**, which have quantum numbers not possible for mesons in the quark model;
2. **glueballs** or **gluonium**, which have no valence quarks at all;
3. **tetraquarks**, which have two valence quark-antiquark pairs; and
4. **hybrid mesons**, which contain a valence quark-antiquark pair and one or more gluons.

All of these can be classed as mesons, because they are hadrons and carry zero baryon number. Of these, glueballs must be flavor singlets; that is, have zero isospin, strangeness, charm, bottomness, and topness. Like all particle states, they are specified by the quantum numbers which label representations of the Poincaré symmetry, q.e., J^{PC} (where J is the angular momentum, P is the intrinsic parity, and C is the charge conjugation parity) and by the mass. One also specifies the isospin I of the meson.

Typically, every quark model meson comes in SU(3) flavor nonet: an octet and a flavor singlet. A glueball shows up as an extra (*supernumerary*) particle outside the nonet. In spite of such seemingly simple counting, the assignment of any given state as a glueball, tetraquark, or hybrid remains tentative even today. Even when there is agreement that one of several states is one of these non-quark model mesons, the degree of mixing, and the precise assignment is fraught with uncertainties. There is also the considerable experimental labor of assigning quantum numbers to each state and cross-checking them in other experiments. As a result, all assignments outside the quark model are tentative. The remainder of this article outlines the situation as it stood at the end of 2004.

86.1 Lattice predictions

Lattice QCD predictions for glueballs are now fairly stable, at least when virtual quarks are neglected. The two lowest states are

0^{++} with mass of 1611±163 MeV/c^2 and

2^{++} with mass of 2232±310 MeV/c^2

The 0^{-+} and exotic glueballs such as 0^{--} are all expected to lie above 2 GeV/c^2. Glueballs are necessarily isoscalar, with isospin $I = 0$.

The ground state *hybrid mesons* 0^{-+}, 1^{-+}, 1^{--}, and 2^{-+} all lie a little below 2 GeV/c^2. The hybrid with exotic quantum numbers 1^{-+} is at 1.9±0.2 GeV/c^2. The best lattice computations to date are made in the quenched approximation, which neglects virtual quarks loops. As a result, these computations miss mixing with meson states.

86.2 The 0^{++} states

The data show five isoscalar resonances: $f_0(500)$, $f_0(980)$, $f_0(1370)$, $f_0(1500)$, and $f_0(1710)$. Of these the $f_0(500)$ is usually identified with the σ of chiral models. The decays and production of $f_0(1710)$ give strong evidence that it is also a meson.

86.2.1 Glueball candidate

The $f_0(1370)$ and $f_0(1500)$ cannot both be a quark model meson, because one is supernumerary. The production of the higher mass state in two photon reactions such as $2\gamma \rightarrow 2\pi$ or $2\gamma \rightarrow 2K$ reactions is highly suppressed. The decays also give some evidence that one of these could be a glueball.

86.2.2 Tetraquark candidate

The $f_0(980)$ has been identified by some authors as a tetraquark meson, along with the $I = 1$ states $a_0(980)$ and $\kappa_0(800)$. Two long-lived (*narrow* in the jargon of particle spectroscopy) states: the scalar (0^{++}) state D*±
sJ(2317) and the vector (1^{+}) meson D*±
sJ(2460), observed at CLEO and BaBar, have also been tentatively identified as tetraquark states. However, for these, other explanations are possible.

86.3 The 2^{++} states

Two isoscalar states are definitely identified—$f_2(1270)$ and the $f'_2(1525)$. No other states have been consistently identified by all experiments. Hence it is difficult to say more about these states.

86.4 The 1^{-+} exotics and other states

The two isovector exotics $\pi_1(1400)$ and $\pi_1(1600)$ seem to be well established experimentally. They are clearly not glueballs, but could be either a tetraquark or a hybrid. The evidence for such assignments is weak.

The $\pi(1800)$ (0^{-+}), $\rho(1900)$ (1^{--}) and the $\eta_2(1870)$ (2^{-+}) are fairly well identified states, which have been tentatively identified as hybrids by some authors. If this identification is correct, then it is a remarkable agreement with lattice computations, which place several hybrids in this range of masses.

86.5 See also

- Quark model, mesons, baryons, quarks, and gluons
- Exotic hadrons and exotic baryons
- Quantum chromodynamics, flavor, and the QCD vacuum
- GlueX, an experiment which will explore the spectrum of glueballs and exotic mesons

86.6 References and external links

- W.-M. Yao *et al.* (Particle Data Group) (2006). "Review of Particle Physics: Non-qq mesons" (PDF). *Journal of Physics G* **33**: 1. arXiv:astro-ph/0601168. Bibcode:2006JPhG...33....1Y. doi:10.1088/0954-3899/33/1/001.

Chapter 87

Glueball

In particle physics, a **glueball** is a hypothetical composite particle.[1] It consists solely of gluon particles, without valence quarks. Such a state is possible because gluons carry color charge and experience the strong interaction. Glueballs are extremely difficult to identify in particle accelerators, because they mix with ordinary meson states.[2]

Theoretical calculations show that glueballs should exist at energy ranges accessible with current collider technology. However, due to the aforementioned difficulty (among others), they have (as of 2013) so far not been observed and identified with certainty.[3] The prediction that glueballs exist is one of the most important predictions of the Standard Model of particle physics that has not yet been confirmed experimentally.[4]

87.1 Properties of glueballs

In principle, it is theoretically possible for all properties of glueballs to be calculated exactly and derived directly from the equations and fundamental physical constants of quantum chromodynamics (QCD) without further experimental input. So, the predicted properties of these hypothetical particles can be described in exquisite detail using only Standard Model physics which have wide acceptance in the theoretical physics literature. But, the fact that QCD calculations are so difficult that solutions to these equations are almost always numerical approximations (reached by several very different methodologies) and the considerable uncertainty in the measurement of some of the relevant key physical constants can lead to variation in theoretical predictions of glueball properties like mass and branching ratios in glueball decays.

87.1.1 Constituent particles and color charge

Theoretical studies of glueballs have focused on glueballs consisting of either two gluons or three gluons, by analogy to mesons and baryons that have two and three quarks respectively. As in the case of mesons and baryons, glueballs would be QCD color charge neutral (aka isospin = 0). The baryon number of a glueball is zero.

87.1.2 Total angular momentum

Two gluon glueballs can have total angular momentum (J) of 0 (which are scalar or pseudo-scalar) or 2 (tensor). Three gluon glueballs can have total angular momentum (J) of 1 (vector boson) or 3. All glueballs have integer total angular momentum which implies that they are bosons rather than fermions.

Glueballs are the only particles predicted by the Standard Model with total angular momentum (J) (sometimes called "intrinsic spin") that could be either 2 or 3 in their ground states, although mesons made of two quarks with J=0 and J=1 with similar masses have been observed and excited states of other mesons can have these values of total angular momentum.

Fundamental particles with ground states having J=0 or J=2 are easily distinguished from glueballs. The hypothetical graviton, while having a total angular momentum J=2 would be massless and lack color charge, and so would be easily distinguished from glueballs. The Standard Model Higgs boson for which an experimentally measured mass of about 125-126 GeV/c^2 has been determined (although the status of the measured particle as a true Standard Model Higgs boson has not been definitively established), is the only fundamental particle with J=0 in the Standard Model, also lacks color charge and hence does not engage in strong force interactions. The Higgs boson is about 25-80 times as heavy as the mass of the various glueball states predicted by the Standard Model.

87.1.3 Electric charge

All glueballs would have electric charge, Q(e), of zero as gluons themselves do not have an electric charge.

87.1.4 Mass and parity

Glueballs are predicted by quantum chromodynamics to be massive, notwithstanding the fact that gluons themselves have zero rest mass in the Standard Model. Glueballs with all four possible combinations of quantum numbers P (parity) and C (c-parity) for every possible total angular momentum have been considered, producing at least fifteen possible glueball states including excited glueball states that share the same quantum numbers but have differing masses with the lightest states having masses as low as 1.4 GeV/c^2 (for a glueball with quantum numbers J=0, P=+, C=+), and the heaviest states having masses as great as almost 5 GeV/c^2 (for a glueball with quantum numbers J=0, P=+, C=-).[5]

These masses are on the same order of magnitude as the masses of many experimentally observed mesons and baryons, as well as to the masses of the tau lepton, charm quark, bottom quark, some hydrogen isotopes, and some helium isotopes.

87.1.5 Stability and decay channels

Just as all Standard Model mesons and baryons, except the proton, are unstable in isolation, all glueballs are predicted by the Standard Model to be unstable in isolation, with various QCD calculations predicting the total decay width (which is functionally related to half-life) for various glueball states. QCD calculations also make predictions regarding the expected decay patterns of glueballs.[6][7] For example, glueballs would not have radiative or two photon decays, but would have decays into pairs of pions, pairs of kaons, or pairs of eta mesons.[6]

87.2 Practical impact on macroscopic low energy physics

Because Standard Model glueballs are so ephemeral (decaying almost immediately into more stable decay products) and are only generated in high energy physics, glueballs only arise synthetically in the natural conditions found on Earth that humans can easily observe. They are scientifically notable mostly because they are a testable prediction of the Standard Model, and not because of phenomenological impact on macroscopic processes, or their engineering applications.

87.3 Lattice QCD simulations

Lattice field theory provides a way to study the glueball spectrum theoretically and from first principles. Some of the first quantities calculated using lattice QCD methods (in 1980) were glueball mass estimates.[9] Morningstar and Peardon[10] computed in 1999 the masses of the lightest glueballs in QCD without dynamical quarks. The three lowest states are tabulated below. The presence of dynamical quarks would slightly alter these data, but also makes the computations more difficult. Since that time calculations within QCD (lattice and sum rules) find the lightest glueball to be a scalar with mass in the range of about 1000–1700 MeV.[11]

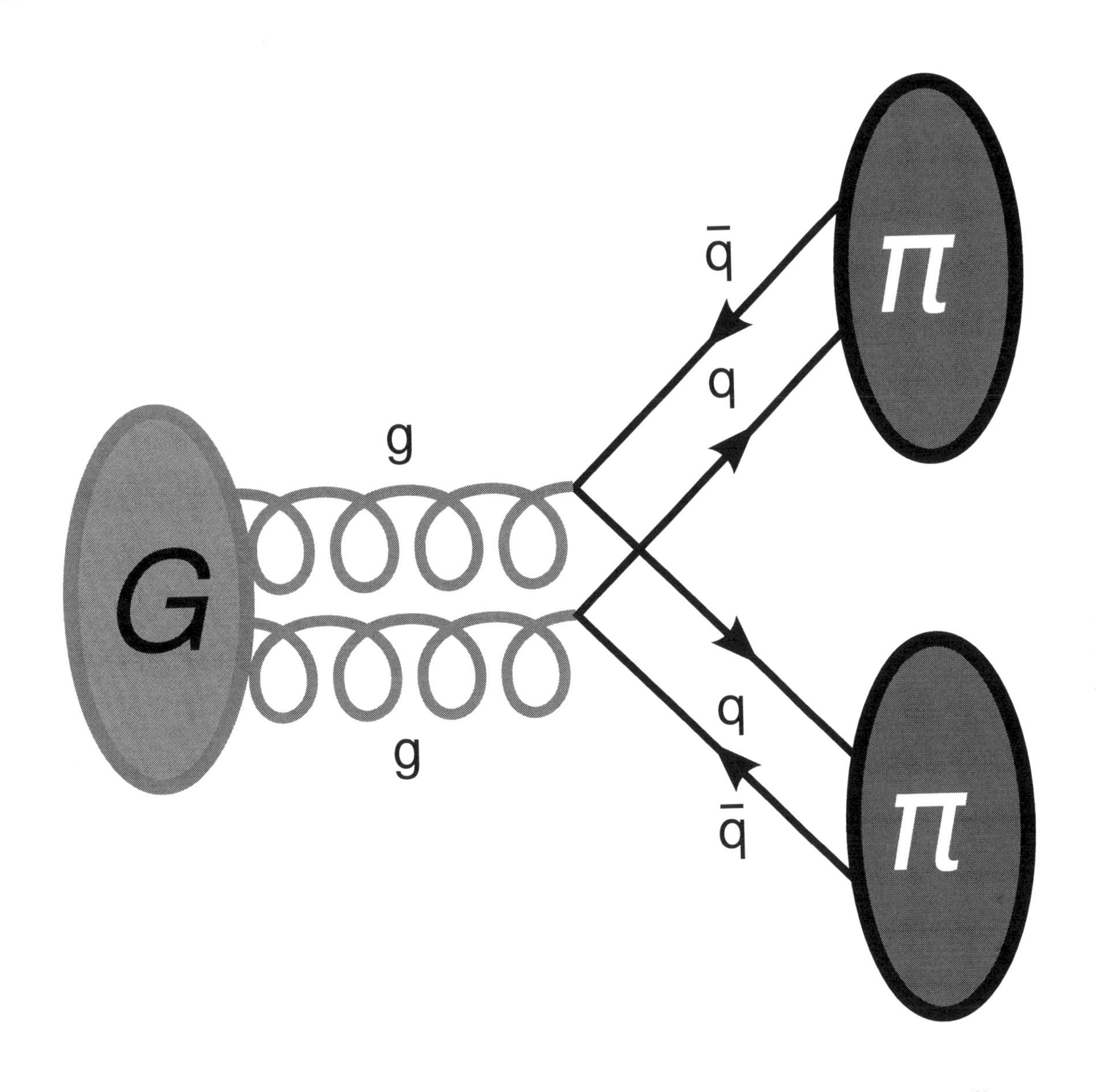

Feynman diagram of a glueball (G) decaying to two pions (π). Such decays help the study of and search for glueballs.[8]

87.4 Experimental candidates

Particle accelerator experiments are often able to identify unstable composite particles and assign masses to those particles to a precision of approximately 10 MeV/c^2, without being able to immediately assign to the particle resonance that is observed all of the properties of that particle. Scores of such particles have been detected, although particles detected in some experiments but not others can be viewed as doubtful. Some of the candidate particle resonances that could be glueballs, although the evidence is not definitive, include the following:

Vector, Pseudo-Vector, or Tensor Glueball Candidates:

- X(3020) observed by the BaBar collaboration is a candidate for an excited state of the 2-+, 1+- or 1-- glueball states with a mass of about 3.02 GeV/c^2.[4]

Scalar Glueball Candidates:

- $f_0(500)$ also known as σ -- the properties of this particle are possibly consistent with a 1000 MeV or 1500 MeV mass glueball.[12]
- $f_0(980)$ -- the structure of this composite particle is consistent with the existence of a light glueball.[12]
- $f_0(1370)$ -- existence of this resonance is disputed but is a candidate for a glueball-meson mixing state[12]
- $f_0(1500)$ -- existence of this resonance is undisputed but its status as a glueball-meson mixing state or pure glueball is not well established.[12]
- $f_0(1710)$ -- existence of this resonance is undisputed but its status as a glueball-meson mixing state or pure glueball is not well established.[12]

Other Glueball Candidates:

- Gluon jets at the LEP experiment show a 40% excess over theoretical expectations of electromagnetically neutral clusters which suggests that electromagnetically neutral particles expected in gluon rich environments such as glueballs are likely to be present.[12]

Many of these candidates have been the subject of active investigation for at least eighteen years.[6] The GlueX experiment, scheduled to begin in 2014, has been specifically designed to produce more definitive experimental evidence glueballs.[13]

87.5 See also

- Exotic meson
- GlueX
- Gluon
- Yang–Mills theory

87.6 References

[1] • Frank Close and Phillip R. Page, "Glueballs", *Scientific American*, vol. 279 no. 5 (November 1998) pp. 80–85

[2] Vincent Mathieu; Nikolai Kochelev; Vicente Vento (2009). "The Physics of Glueballs". *International Journal of Modern Physics E* **18**: 1–49. arXiv:0810.4453. Bibcode:2009IJMPE..18....1M. doi:10.1142/S0218301309012124. Glueball on arxiv.org

[3] Wolfgang Ochs, "The Status of Glueballs"J.Phys.G:Nuclear and Particle Physics40,67(2013)DOI:10.1088/0954- http://arxiv.org/pdf/1301.5183v3.pdf

[4] Y.K. Hsiao, C.Q. Geng, "Identifying Glueball at 3.02 GeV in Baryonic B Decays" (Version 2: October 9, 2013) http://arxiv.org/abs/1302.3331

[5] Wolfgang Ochs, "The Status of Glueballs"J.Phys.G:Nuclear and Particle Physics40,6(2013)DOI:10.1088/0954-3899 http://arxiv.org/pdf/1301.5183v3.pdf

[6] Walter Taki, "Search for Glueballs" (1996) http://www.slac.stanford.edu/cgi-wrap/getdoc/ssi96-006.pdf

[7] See, e.g., Walaa I. Eshraim, Stanislaus Janowski, "Branching ratios of the pseudoscalar glueball with a mass of 2.6 GeV", prepared for Proceedings of Confinement X - Conference on Quark Confinement and the Hadron Spectrum (Munich/Germany, 8–12 October 2012) (pre-print published January 15, 2013) http://arxiv.org/abs/1301.3345

[8] T. Cohen, F. J. Llanes-Estrada, J. R. Pelaez, J. Ruiz de Elvira (2014). "Non-ordinary light meson couplings and the 1/Nc expansion". arXiv:1405.4831 [hep-ph].

[9] B. Berg. Plaquette-plaquette correlations in the su(2) lattice gauge theory. Phys. Lett., B97:401, 1980.

[10] Colin J. Morningstar; Mike Peardon (1999). "Glueball spectrum from an anisotropic lattice study". *Physical Review D* **60** (3): 034509. arXiv:hep-lat/9901004. Bibcode:1999PhRvD..60c4509M. doi:10.1103/PhysRevD.60.034509.

[11] Wolfgang Ochs, "The status of glueballs" Source: JOURNAL OF PHYSICS G-NUCLEAR AND PARTICLE PHYSICS Volume: 40 Issue: 4 Article Number: 043001 DOI: 10.1088/0954-3899/40/4/043001 Published: APR 2013

[12] Wolfgang Ochs(2013). "The status of glueballs".*Journal of Physics G***40**(4): 043001. arXiv:1301.5183. Bibcode:2013JPhG. doi:10.1088/0954-3899/40/4/043001.

[13] "The Physics of GlueX".

Chapter 88

Hexaquark

In particle physics **hexaquarks** are a large family of hypothetical particles that would consist of six quarks or antiquarks of any flavours. With six constituent particles, there are several ways to combine quarks so that their colour charge is zero: a hexaquark can either contain six quarks, resembling two baryons bound together (a **dibaryon**), or three quarks and three antiquarks.[1] Dibaryons are predicted to be fairly stable once formed. Robert Jaffe proposed the existence of a possibly stable H dibaryon (with the quark composition udsuds), made by combining two uds hyperons, in 1977.[2]

A number of experiments have been suggested to detect dibaryon decays and interactions. Several candidate dibaryon decays were observed but not confirmed in the 1990s.[3][4][5]

There is a theory that strange particles such as hyperons [6] and dibaryons[7] could form in the interior of a neutron star, changing its mass–radius ratio in ways that might be detectable. Conversely, measurements of neutron stars set constraints on possible dibaryon properties.[8] A large fraction of the neutrons in a neutron star could turn into hyperons and merge into dibaryons during the early part of its collapse into a black hole . These dibaryons would very quickly dissolve into quark–gluon plasma during the collapse, or go into some currently unknown state of matter.

In 2014 a potential dibaryon was detected at the Jülich Research Center at about 2380 MeV. The particle existed for 10^{-23} seconds and was named d*(2380).[9]

88.1 See also

- Deuteron, the only known stable composite particle that consist of six quarks.
- Diproton, an extremely unstable dibaryon.
- Dineutron, another extremely unstable dibaryon.
- Pentaquark

88.2 References

[1] Vijande, J.; Valcarce, A; Richard, J.-M. (25 November 2011). "Stability of hexaquarks in the string limit of confinement". arXiv:1111.5921v1.

[2] R. L. Jaffe (1977). "Perhaps a Stable Dihyperon?". *Physical Review Letters* **38** (5): 195. Bibcode:1977PhRvL..38..195J. doi:10.1103/PhysRevLett.38.195.

[3] J. Belz et al. (BNL-E888 Collaboration) (1996). "Search for the weak decay of an H dibaryon". *Physical Review Letters* **76**: 3277–3280. arXiv:hep-ex/9603002. Bibcode:1996PhRvL..76.3277B. doi:10.1103/PhysRevLett.76.3277.

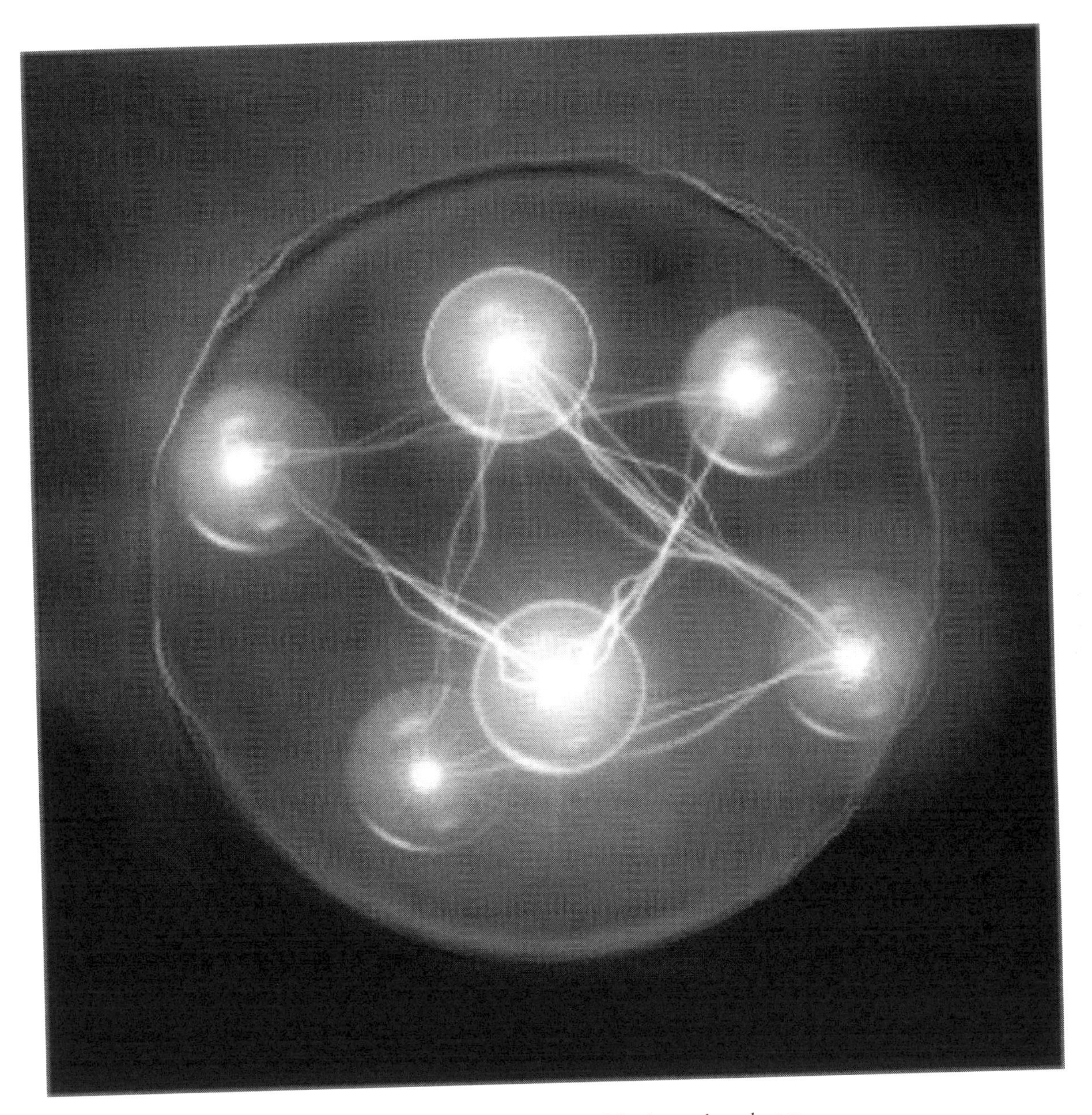

A dibaryon-type hexaquark. There are two constituent quarks for each of the three colour charges.

[4] R. W. Stotzer et al. (BNL-E836 Collaboration) (1997). "Search for H dibaryon in He-3 (K-, k+) Hn". *Physical Review Letters* **78**: 3646–36490. Bibcode:1997PhRvL..78.3646S. doi:10.1103/PhysRevLett.78.3646.

[5] A. Alavi-Harati et al. (KTeV Collaboration) (2000). "Search for the weak decay of a lightly bound H0 dibaryon". *Physical Review Letters* **84**: 2593–2597. arXiv:hep-ex/9910030. Bibcode:2000PhRvL..84.2593A. doi:10.1103/PhysRevLett.84.2593.

[6] V. A. Ambartsumyan; G. S. Saakyan (1960). "The Degenerate Superdense Gas of Elementary Particles". *Soviet Astronomy* **37**: 193. Bibcode:1960SvA.....4..187A.

[7] S. Kagiyama; A. Nakamura; T. Omodaka (1992). "Compressible bag model and dibaryon stars". *Zeitschrift für Physik C* **56** (4): 557–560. Bibcode:1992ZPhyC..56..557K. doi:10.1007/BF01474728.

[8] A. Faessler; A. J. Buchmann; M. I. Krivoruchenko (1997). "Constraints to coupling constants of the ω- and σ-mesons with dibaryons".*Physical Review C***56**: 1576. arXiv:nucl-th/9706080. Bibcode:1997PhRvC..56.1576F.doi:10.1103/PhysRevC.5.

[9] P. Adlarson et al. (2014). "Evidence for a New Resonance from Polarized Neutron-Proton Scattering". *Physical Review Letters* **112** (2): 202301. arXiv:1402.6844. Bibcode:2014PhRvL.112t2301A. doi:10.1103/PhysRevLett.112.202301.

Chapter 89

Mesonic molecule

A **mesonic molecule** is a set of two or more mesons bound together by the strong force. Unlike baryonic molecules, which form the nuclei of all elements in nature save hydrogen-1, a mesonic molecule has yet to be definitively observed. The X(3872) discovered in 2003 and the Z(4430) discovered in 2007 by the Belle experiment are the best candidates for such an observation.

89.1 References

89.2 See also

- Meson
- Tetraquark
- Pionium

Chapter 90

Pentaquark

Two models of a generic pentaquark

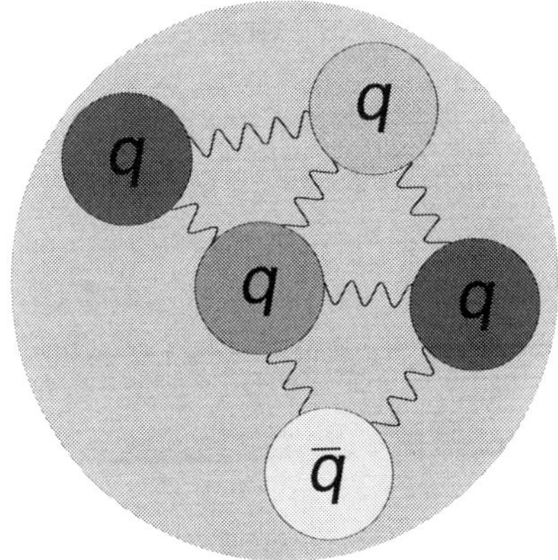

A five-quark "bag"

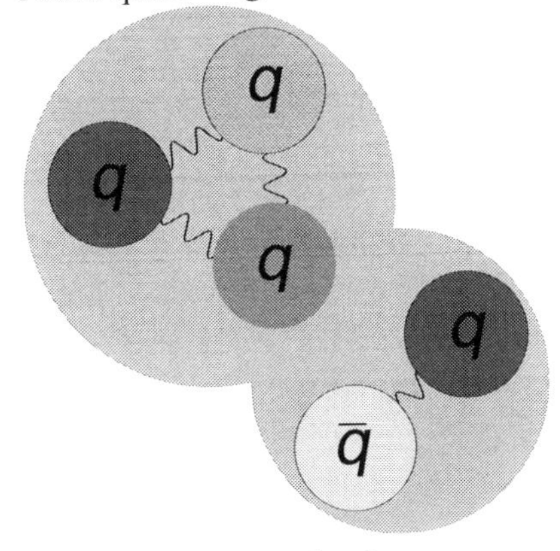

A meson-baryon molecule

q indicates a quark, whereas q indicates an antiquark. The wavy lines are gluons, which mediate the strong interaction between the quarks. The colours correspond to the various colour charges of quarks. The colours red, green, and blue must each be present and the remaining quark and antiquark must share corresponding colour and anticolour, here chosen to be blue and antiblue (shown as yellow).

A **pentaquark** is a subatomic particle consisting of four quarks and one antiquark bound together.

As quarks have a baryon number of $+1/3$, and antiquarks of $-1/3$, the pentaquark would have a total baryon number of 1, and thus would be a baryon. Further, because it has five quarks instead of the usual three found in regular baryons (aka 'triquarks'), it would be classified as an exotic baryon. The name pentaquark was coined by Harry J. Lipkin in 1987,[1] however, the possibility of five-quark particles was identified as early as 1964 when Murray Gell-Mann first postulated the existence of quarks.[2] Although predicted for decades, pentaquarks have proved surprisingly difficult to discover and some physicists were beginning to suspect that an unknown law of nature prevented their production.[3]

The first claim of pentaquark discovery was recorded at LEPS in Japan in 2003, and several experiments in the mid-2000s also reported discoveries of other pentaquark states.[4] Others were not able to replicate the LEPS results, however, and the other pentaquark discoveries were not accepted because of poor data and statistical analysis.[5] On 13 July 2015, the LHCb collaboration at CERN reported results consistent with pentaquark states in the decay of bottom Lambda baryons (Λ0
b).[6]

Outside of particle physics laboratories pentaquarks also could be produced naturally by supernovae as part of the process of forming a neutron star.[7] The scientific study of pentaquarks might offer insights into how these stars form, as well as, allowing more thorough study of particle interactions and the strong force.

90.1 Background

Main article: Quark

A quark is a type of elementary particle that has mass, electric charge, and colour charge, as well as an additional property called flavour, which describes what type of quark it is (up, down, strange, charm, top, or bottom). Due to an effect known as colour confinement, quarks are never seen on their own. Instead, they form composite particles known as hadrons so that their colour charges cancel out. Hadrons made of one quark and one antiquark are known as mesons, while those made of three quarks are known as baryons. These 'regular' hadrons are well documented and characterized, however, there is nothing in theory to prevent quarks from forming 'exotic' hadrons such as tetraquarks with two quarks and two antiquarks, or pentaquarks with four quarks and one antiquark.[3]

90.2 Structure

A wide variety of pentaquarks are possible, with different quark combinations producing different particles. To identify which quarks compose a given pentaquark, physicists use the notation *qqqqq*, where *q* and *q* respectively refer to any of the six flavours of quarks and antiquarks. The symbols u, d, s, c, b, and t stand for the up, down, strange, charm, bottom, and top quarks respectively, with the symbols of u, d, s, c, b, t corresponding to the respective antiquarks. For instance a pentaquark made of two up quarks, one down quark, one charm quark, and one charm antiquark would be denoted uudcc.

The quarks are bound together by the strong force, which acts in such a way as to cancel the colour charges within the particle. In a meson, this means a quark is partnered with an antiquark with an opposite colour charge – blue and antiblue, for example – while in a baryon, the three quarks have between them all three colour charges – red, blue, and green.[nb 1] In a pentaquark, the colours also need to cancel out, and the only feasible combination is to have one quark with one colour (e.g. red), one quark with a second colour (e.g. green), two quarks with the third colour (e.g. blue), and one antiquark to counteract the surplus colour (e.g. antiblue).[8]

The binding mechanism for pentaquarks is not yet clear. They may consist of five quarks tightly bound together, but it is also possible that they are more loosely bound and consist of a three-quark baryon and a two-quark meson interacting relatively weakly with each other via pion exchange (the same force that binds atomic nuclei) in a “meson-baryon molecule”.[9][2][10]

90.3 History

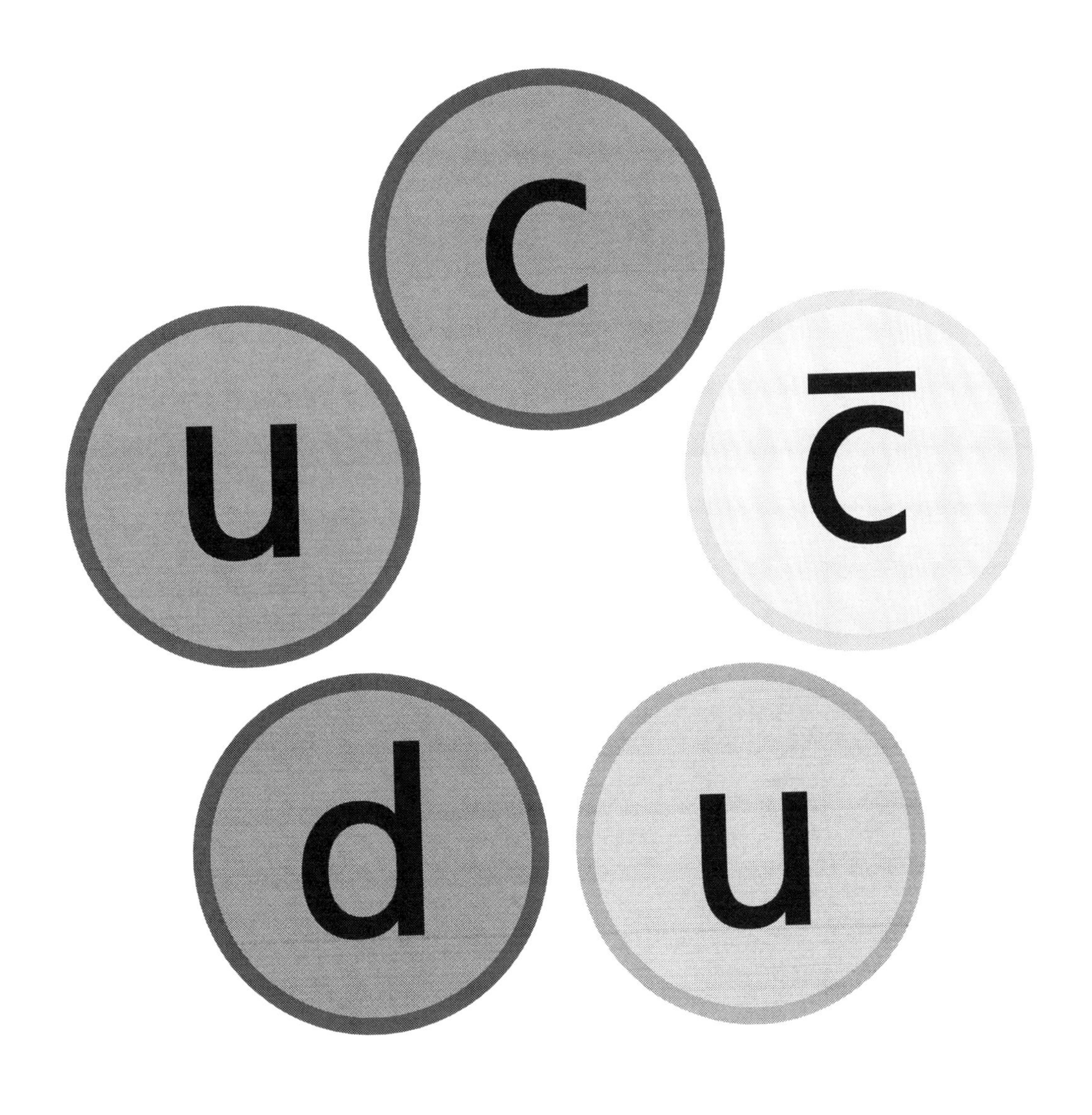

A diagram of the P+
c type pentaquark possibly discovered in July 2015, showing the flavours of each quark and one possible colour configuration.

90.3.1 Mid-2000s

The requirement to include an antiquark means that many classes of pentaquark are hard to identify experimentally – if the flavour of the antiquark matches the flavour of any other quark in the quintuplet, it will cancel out and the particle will resemble its three-quark hadron cousin. For this reason, early pentaquark searches looked for particles where the antiquark did not cancel.[8] In the mid-2000s, several experiments claimed to reveal pentaquark states. In particular, a resonance with a mass of 1540 MeV/c^2 (4.6 σ) was reported by LEPS in 2003, the Θ+.[11] This coincided with a pentaquark state with a mass of 1530 MeV/c^2 predicted in 1997.[12]

The proposed state was composed of two up quarks, two down quarks, and one strange antiquark (uudds). Following this announcement, nine other independent experiments reported seeing narrow peaks from nK+ and pK0, with masses between 1522 MeV/c^2 and 1555 MeV/c^2, all above 4 σ.[11] While concerns existed about the validity of these states, the

Particle Data Group gave the Θ+ a 3-star rating (out of 4) in the 2004 *Review of Particle Physics*.[11] Two other pentaquark states were reported albeit with low statistical significance—the Φ−− (ddssu), with a mass of 1860 MeV/c^2 and the Θ0 c (uuddc), with a mass of 3099 MeV/c^2. Both were later found to be statistical effects rather than true resonances.[11]

Ten experiments then looked for the Θ+, but came out empty-handed.[11] Two in particular (one at BELLE, and the other at CLAS) had nearly the same conditions as other experiments which claimed to have detected the Θ+ (DIANA and SAPHIR respectively).[11] The 2006 *Review of Particle Physics* concluded:[11]

> [T]here has not been a high-statistics confirmation of any of the original experiments that claimed to see the Θ+; there have been two high-statistics repeats from Jefferson Lab that have clearly shown the original positive claims in those two cases to be wrong; there have been a number of other high-statistics experiments, none of which have found any evidence for the Θ+; and all attempts to confirm the two other claimed pentaquark states have led to negative results. The conclusion that pentaquarks in general, and the Θ+, in particular, do not exist, appears compelling.

The 2008 *Review of Particle Physics* went even further:[5]

> There are two or three recent experiments that find weak evidence for signals near the nominal masses, but there is simply no point in tabulating them in view of the overwhelming evidence that the claimed pentaquarks do not exist... The whole story—the discoveries themselves, the tidal wave of papers by theorists and phenomenologists that followed, and the eventual "undiscovery"—is a curious episode in the history of science.

Despite these null results, LEPS results as of 2009 continue to show the existence of a narrow state with a mass of 1524±4 MeV/c^2, with a statistical significance of 5.1 σ.[13] Experiments continue to study this controversy.

90.3.2 2015 LHCb results

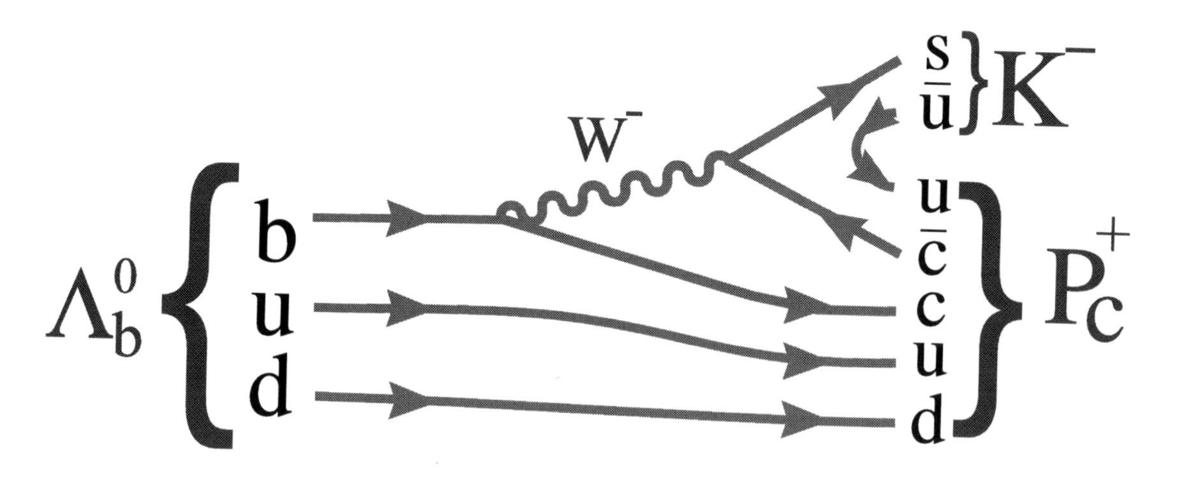

Feynman diagram representing the decay of a lambda baryon Λ0
b into a kaon K−
and a pentaquark P+
c.

In July 2015, the LHCb collaboration identified pentaquarks in the Λ0
b→J/ψK−
p channel, which represents the decay of the bottom lambda baryon (Λ0

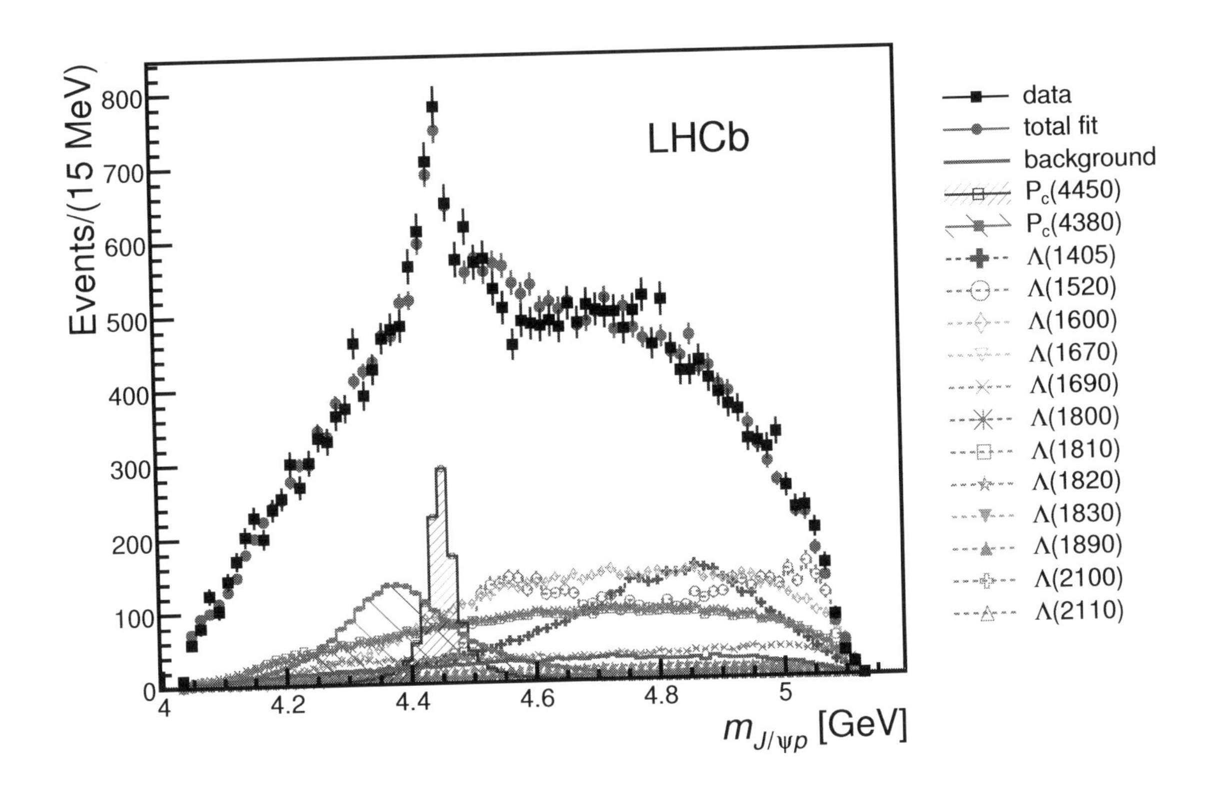

A fit to the J/ψp invariant mass spectrum for the Λ0
b→J/ψK−
p decay, with each fit component shown individually. The contribution of the pentaquarks are shown by hatched histograms.

b) into a J/ψ meson (J/ψ), a kaon (K−
) and a proton (p). The results showed that sometimes, instead of decaying directly into mesons and baryons, the Λ0
b decayed via intermediate pentaquark states. The two states, named P+
c(4380) and P+
c(4450), had individual statistical significances of 9 σ and 12 σ, respectively, and a combined significance of 15 σ — enough to claim a formal discovery. The analysis ruled out the possibility that the effect was caused by conventional particles.[2] The two pentaquark states were both observed decaying strongly to J/ψp, hence must have a valence quark content of two up quarks, a down quark, a charm quark, and an anti-charm quark (uudcc), making them charmonium-pentaquarks.[6][7][14]

The search for pentaquarks was not an objective of the LHCb experiment (which is primarily designed to investigate matter-antimatter asymmetry)[15] and the apparent discovery of pentaquarks was described as an “accident” and “something we’ve stumbled across” by a CERN spokesperson.[9]

90.4 Applications

The discovery of pentaquarks will allow physicists to study the strong force in greater detail and aid understanding of quantum chromodynamics. In addition, current theories suggest that some very large stars produce pentaquarks as they collapse. The study of pentaquarks might help shed light on the physics of neutron stars.[7]

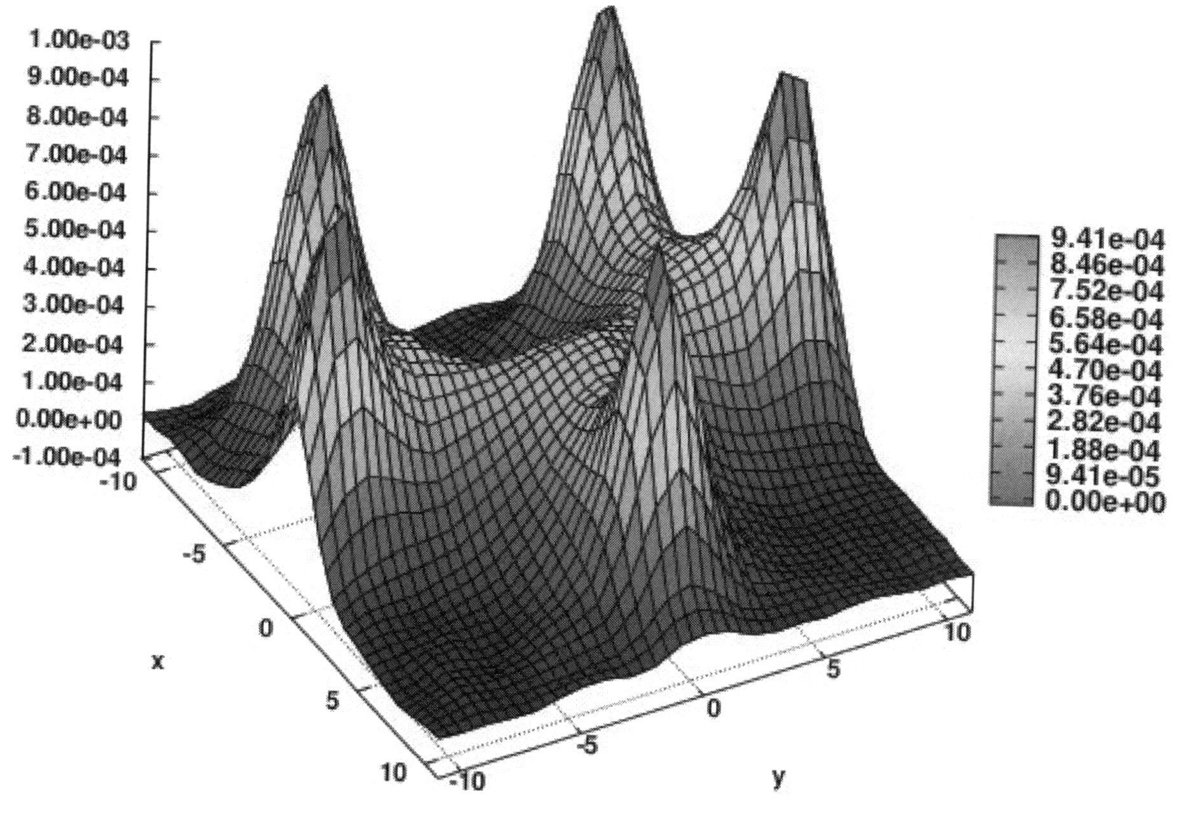

Colour flux tubes produced by five static quark and antiquark charges, computed in lattice QCD.[16] Confinement in Quantum Chromo Dynamics leads to the production of flux tubes connecting colour charges. The flux tubes act as attractive QCD string-like potentials.

90.5 See also

- Exotic matter
- List of particles
- Quark model
- Tetraquark
- Triquark

90.6 Footnotes

[1] The colour charges do not correspond to physical visible colours. They are arbitrary labels used to help scientists describe and visualise the charges of quarks.

90.7 References

[1] H. J. Lipkin (1987). "New possibilities for exotic hadrons — anticharmed strange baryons". *Physics Letters B* **195** (3): 484–488. Bibcode:1987PhLB..195..484L. doi:10.1016/0370-2693(87)90055-4.

[2] "Observation of particles composed of five quarks, pentaquark-charmonium states, seen in Λ0 b→J/ψpK⁻ decays". CERN/LHCb. 14 July 2015. Retrieved 2015-07-14.

[3] H. Muir (2 July 2003). "Pentaquark discovery confounds sceptics". *New Scientist*. Retrieved 2010-01-08.

[4] K. Hicks (23 July 2003). "Physicists find evidence for an exotic baryon". Ohio University. Retrieved 2010-01-08.

[5] See p. 1124 in C. Amsler et al. (Particle Data Group) (2008). "Review of particle physics" (PDF). *Physics Letters B* **667** (1-5): 1. Bibcode:2008PhLB..667....1A. doi:10.1016/j.physletb.2008.07.018.

[6] R. Aaij et al. (LHCb collaboration) (2015). "Observation of J/ψp resonances consistent with pentaquark states in Λ0 b→J/ψK−
p decays". *Physical Review Letters* **115** (7). doi:10.1103/PhysRevLett.115.072001.

[7] I. Sample (14 July 2015). "Large Hadron Collider scientists discover new particles: pentaquarks". *The Guardian*. Retrieved 2015-07-14.

[8] J. Pochodzalla (2005). "Duets of strange quarks". *Hadron Physics*. p. 268. ISBN 161499014X.

[9] G. Amit (14 July 2015). "Pentaquark discovery at LHC shows long-sought new form of matter". *New Scientist*. Retrieved 2015-07-14.

[10] T. D. Cohen, P. M. Hohler, R. F. Lebed (2005). "On the Existence of Heavy Pentaquarks: The large N_c and Heavy Quark Limits and Beyond".*Physical Review D***72**(7): 074010. arXiv:hep-ph/0508199. Bibcode:2005PhRvD..72g4010C.doi:10.1103/.

[11] W.-M. Yao et al. (Particle Data Group) (2006). "Review of particle physics: Θ+" (PDF). *Journal of Physics G* **33**: 1. arXiv:astro-ph/0601168. Bibcode:2006JPhG...33....1Y. doi:10.1088/0954-3899/33/1/001.

[12] D. Diakonov, V. Petrov, and M. Polyakov (1997). "Exotic anti-decuplet of baryons: prediction from chiral solitons". *Zeitschrift für Physik A* **359** (3): 305. arXiv:hep-ph/9703373. Bibcode:1997ZPhyA.359..305D. doi:10.1007/s002180050406.

[13] T. Nakano et al. (LEPS Collaboration) (2009). "Evidence of the Θ^+ in the $\gamma d \to K^+K^-pn$ reaction". *Physical Review C* **79** (2): 025210. arXiv:0812.1035. Bibcode:2009PhRvC..79b5210N. doi:10.1103/PhysRevC.79.025210.

[14] P. Rincon (14 July 2015). "Large Hadron Collider discovers new pentaquark particle". *BBC News*. Retrieved 2015-07-14.

[15] "Where has all the antimatter gone?". CERN/LHCb. 2008. Retrieved 2015-07-15.

[16] N. Cardoso, M. Cardoso, and P. Bicudo (2013). "Color fields of the static pentaquark system computed in SU(3) lattice QCD". *Physical Review D* **87** (3): 034504. arXiv:1209.1532. doi:10.1103/PhysRevD.87.034504.

90.8 Further reading

- David Whitehouse (1 July 2003). "Behold the Pentaquark (BBC News)". BBC News. Retrieved 2010-01-08.
- Thomas E. Browder, Igor R. Klebanov, Daniel R. Marlow (2004). "Prospects for Pentaquark Production at Meson Factories".*Physics Letters B***587**: 62. arXiv:hep-ph/0401115. Bibcode:2004PhLB..587...62B.doi:10.1016/j.
- Akio Sugamoto (2004). "An Attempt to Study Pentaquark Baryons in String Theory". arXiv:hep-ph/0404019 [hep-ph].
- Kenneth Hicks (2005). "An Experimental Review of the Θ+ Pentaquark". *Journal of Physics: Conference Series* **9**: 183. arXiv:hep-ex/0412048. Bibcode:2005JPhCS...9..183H. doi:10.1088/1742-6596/9/1/035.
- Mark Peplow (18 April 2005). "Doubt is Cast on Pentaquarks". *Nature*. doi:10.1038/news050418-1.
- Maggie McKie (20 April 2005). "Pentaquark hunt draws blanks". *New Scientist*. Retrieved 2010-01-08.
- Thomas Jefferson National Accelerator Facility (21 April 2005). "Is It Or Isn't It? Pentaquark Debate Heats Up". *Space Daily*. Retrieved 2010-01-08.
- Dmitri Diakonov (2005). "Relativistic Mean Field Approximation to Baryons". *European Physical Journal A* **24**: 3. Bibcode:2005EPJAS..24a...3D. doi:10.1140/epjad/s2005-05-001-3.
- Schumacher, R. A. (2006). "The Rise and Fall of Pentaquarks in Experiments". *AIP Conference Proceedings* **842**: 409. arXiv:nucl-ex/0512042. doi:10.1063/1.2220285.
- Kandice Carter (2006). "The Rise and Fall of the Pentaquark". *Symmetry Magazine* **3** (7): 16.

90.9 External links

- "Pentaquark on arxiv.org".

Chapter 91

Tetraquark

A **tetraquark**, in particle physics, is an exotic meson composed of four valence quarks. In principle, a tetraquark state may be allowed in quantum chromodynamics, the modern theory of strong interactions. Any established tetraquark state would be an example of an exotic hadron which lies outside the quark model classification.

91.1 History

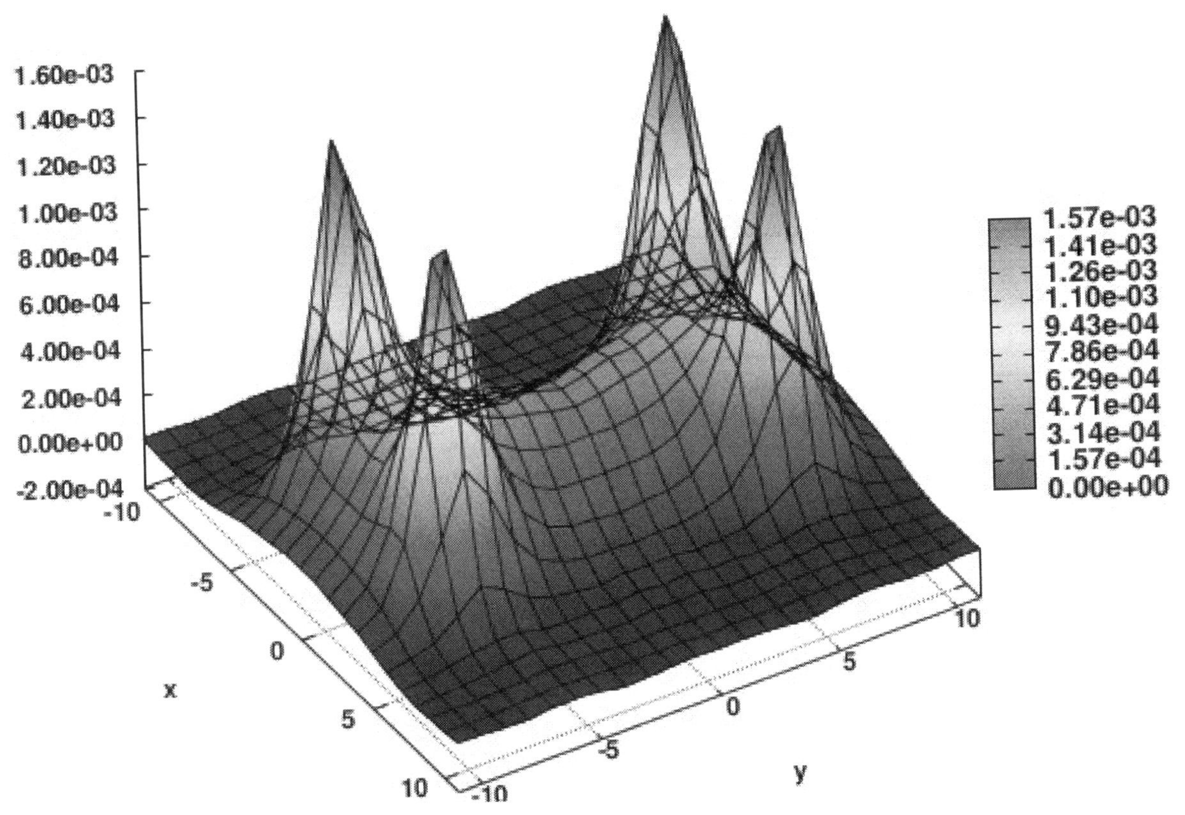

Colour flux tubes produced by four static quark and antiquark charges, computed in lattice QCD.[1] Confinement in Quantum Chromo Dynamics leads to the production of flux tubes connecting colour charges. The flux tubes act as attractive QCD string-like potentials.

In 2003 a particle temporarily called X(3872), by the Belle experiment in Japan, was proposed to be a tetraquark candidate,[2] as originally theorized.[3] The name X is a temporary name, indicating that there are still some questions about its properties to be tested. The number following is the mass of the particle in 10^0 MeV/c^2.

In 2004, the $D_sJ(2632)$ state seen in Fermilab's SELEX was suggested as a possible tetraquark candidate.

In 2007, Belle announced the observation of the Z(4430) state, a ccdu tetraquark candidate. In 2014, the Large Hadron Collider experiment LHCb confirmed this resonance with a significance of over 13.9σ.[4][5] There are also indications that the Y(4660), also discovered by Belle in 2007, could be a tetraquark state.[6]

In 2009, Fermilab announced that they have discovered a particle temporarily called Y(4140), which may also be a tetraquark.[7]

In 2010, two physicists from DESY and a physicist from Quaid-i-Azam University re-analyzed former experimental data and announced that, in connection with the ϒ(5S) meson (a form of bottomonium), a well-defined tetraquark resonance exists.[8][9]

In June 2013, two independent groups reported on $Z_c(3900)$.[10] [11]

91.2 See also

- Color confinement
- Hadron
- Pentaquark

91.3 References

[1] N. Cardoso, M. Cardoso, and P. Bicudo (2011). "Colour Fields Computed in SU(3) Lattice QCD for the Static Tetraquark System". *Physical Review D* **84** (5): 054508. arXiv:1107.1355. doi:10.1103/PhysRevD.84.054508.

[2] D. Harris (13 April 2008). "The charming case of X(3872)". *Symmetry Magazine*. Retrieved 2009-12-17.

[3] L. Maiani, F. Piccinini, V. Riquer and A.D. Polosa (2005). "Diquark-antidiquarks with hidden or open charm and the nature of X(3872)".*Physical Review D***71**: 014028. arXiv:hep-ph/0412098. Bibcode:2005PhRvD..71a4028M.doi:10.1103/PhysRev.

[4] "LHCb confirms existence of exotic hadrons".

[5] LHCb collaboration, LHCb; Aaij, R.; Adeva, B.; Adinolfi, M.; Affolder, A.; Ajaltouni, Z.; Albrecht, J.; Alessio, F.; Alexander, M.; Ali, S.; Alkhazov, G.; Alvarez Cartelle, P.; Alves Jr, A. A.; Amato, S.; Amerio, S.; Amhis, Y.; An, L.; Anderlini, L.; Anderson, J.; Andreassen, R.; Andreotti, M.; Andrews, J. E.; Appleby, R. B.; Aquines Gutierrez, O.; Archilli, F.; Artamonov, A.; Artuso, M.; Aslanides, E.; Auriemma, G. et al. (2014). "Observation of the resonant character of the Z(4430)– state". arXiv:1404.1903v1 [hep-ex].

[6] G. Cotugno, R. Faccini, A.D. Polosa and C. Sabelli (2010). "Charmed Baryonium". *Physical Review Letters* **104** (13): 132005. arXiv:0911.2178. Bibcode:2010PhRvL.104m2005C. doi:10.1103/PhysRevLett.104.132005.

[7] Anne Minard (2009-03-18). "New Particle Throws Monkeywrench in Particle Physics". Universetoday.com. Retrieved 2014-04-12.

[8] "Evidence grows for tetraquarks". physicsworld.com. Retrieved 2014-04-12.

[9] A. Ali, C. Hambrock, M.J. Aslam; Hambrock; Aslam (2010). "Tetraquark Interpretation of the BELLE Data on the Anomalous ϒ(1S)π+π- and ϒ(2S)π+π- Production near the ϒ(5S) Resonance". *Physical Review Letters* **104** (16): 162001. arXiv:0912.5016. Bibcode:2010PhRvL.104p2001A. doi:10.1103/PhysRevLett.104.162001.

[10] "Physics - New Particle Hints at Four-Quark Matter". Physics.aps.org. 2013-06-17. Retrieved 2014-04-12.

[11] Eric Swanson (2013). "Viewpoint: New Particle Hints at Four-Quark Matter". *Physics* **69** (6). Bibcode:2013PhyOJ...6...69S. doi:10.1103/Physics.6.69.

91.4 External links

- The Belle experiment (press release)
- O'Luanaigh, Cian. "LHCb confirms existence of exotic hadrons". *cern.ch*. Geneva, Switzerland: CERN. Retrieved 2014-04-12.

Chapter 92

Skyrmion

In particle theory, the **skyrmion** (/ˈskɜrmi.ɒn/) is a hypothetical particle related originally[1] to baryons . It was described by Tony Skyrme and consists of a quantum superposition of baryons and resonance states.[2] It could be predicted some of the nuclear matter properties, as well.[3]

Skyrmions as topological objects are also important in solid state physics, especially in the emerging technology of spintronics. A two-dimensional magnetic skyrmion, as a topological object, is formed, e.g., from a 3D effective-spin "hedgehog" (in the field of micromagnetics: out of a so-called "Bloch point" singularity of homotopy degree +1) by a stereographic projection, whereby the positive north-pole spin is mapped onto a far-off edge circle of a 2D-disk, while the negative south-pole spin is mapped onto the center of the disk.

92.1 Mathematical definition

In field theory, skyrmions are homotopically non-trivial classical solutions of a nonlinear sigma model with a non-trivial target manifold topology – hence, they are topological solitons. An example occurs in chiral models[4] of mesons, where the target manifold is a homogeneous space of the structure group

$$\left(\frac{SU(N)_L \times SU(N)_R}{SU(N)_{\text{diag}}}\right)$$

where $SU(N)L$ and $SU(N)R$ are the left and right parts of the $SU(N)$ matrix, and $SU(N)_{\text{diag}}$ is the diagonal subgroup.

If spacetime has the topology $S^3 \times \mathbf{R}$, then classical configurations can be classified by an integral winding number[5] because the third homotopy group

$$\pi_3\left(\frac{SU(N)_L \times SU(N)_R}{SU(N)_{\text{diag}}} \cong SU(N)\right)$$

is equivalent to the ring of integers, with the congruence sign referring to homeomorphism.

A topological term can be added to the chiral Lagrangian, whose integral depends only upon the homotopy class; this results in superselection sectors in the quantised model. A skyrmion can be approximated by a soliton of the Sine-Gordon equation; after quantisation by the Bethe ansatz or otherwise, it turns into a fermion interacting according to the massive Thirring model.

Skyrmions have been reported, but not conclusively proven, to be in Bose-Einstein condensates,[6] superconductors,[7] thin magnetic films[8] and in chiral nematic liquid crystals.[9]

92.2 Skyrmions in an emerging technology

One particular form of the skyrmions is found in magnetic materials that break the inversion symmetry and where the Dzyaloshinskii-Moriya interaction plays an important role. They form "domains" as small as a 1 nm (e.g. in Fe on

Ir(111)[10]). The small size of magnetic skyrmions makes them a good candidate for future data storage solutions. Physicists at the University of Hamburg have managed to read and write skyrmions using scanning tunneling microscopy.[11] The topological charge, representing the existence and non-existence of skyrmions, can represent the bit states "1" and "0". Room temperature skyrmions have been reported.[12]

92.3 References

[1] At later stages the model was also related to mesons.

[2] Wong, Stephen (2002). "What exactly is a Skyrmion?". arXiv:hep-ph/0202250 [hep/ph].

[3] M.R.Khoshbin-e-Khoshnazar,"Correlated Quasiskyrmions as Alpha Particles",*Eur.Phys.J.A* **14**,207-209 (2002).

[4] Chiral models stress the difference between "left-handedness" and "right-handedness".

[5] The same classification applies to the mentioned effective-spin "hedgehog" singularity": spin upwards at the northpole, but downward at the southpole.
See also Döring,W. (1968). "Point Singularities in Micromagnetism".*Journal of Applied Physic*.doi:10.1063/1.1656144.

[6] Al Khawaja, Usama; Stoof, Henk (2001). "Skyrmions in a ferromagnetic Bose–Einstein condensate". *Nature* **411** (6840): 918–20. Bibcode:2001Natur.411..918A. doi:10.1038/35082010. PMID 11418849.

[7] Baskaran,G. (2011). "Possibility of Skyrmion Superconductivity in Doped Antiferromagnet K$_2$Fe$_4$Se$_2 [cond-mat.supr-con].

[8] Kiselev, N. S.; Bogdanov, A. N.; Schäfer, R.; Rößler, U. K. (2011). "Chiral skyrmions in thin magnetic films: New objects for magnetic storage technologies?". *Journal of Physics D: Applied Physics* **44** (39): 392001. arXiv:1102.2726. Bibcode:20 .doi:10.1088/0022-3727/44/39/392001.

[9] Fukuda, J.-I.; Žumer, S. (2011). "Quasi-two-dimensional Skyrmion lattices in a chiral nematic liquid crystal". *Nature Communications* **2**: 246. Bibcode:2011NatCo...2E.246F. doi:10.1038/ncomms1250. PMID 21427717.

[10] Heinze, Stefan; Von Bergmann, Kirsten; Menzel, Matthias; Brede, Jens; Kubetzka, André; Wiesendanger, Roland; Bihlmayer, Gustav; Blügel, Stefan (2011). "Spontaneous atomic-scale magnetic skyrmion lattice in two dimensions". *Nature Physics* **7** (9): 713–718. Bibcode:2011NatPh...7..713H. doi:10.1038/NPHYS2045. Lay summary (Jul 31, 2011).

[11] Romming, N.; Hanneken, C.; Menzel, M.; Bickel, J. E.; Wolter, B.; Von Bergmann, K.; Kubetzka, A.; Wiesendanger, R. (2013). "Writing and Deleting Single Magnetic Skyrmions". *Science* **341** (6146): 636–9. Bibcode:2013Sci...341..636R. doi:10.1126/science.1240573. PMID 23929977. Lay summary – *phys.org* (Aug 8, 2013).

[12] "Blowing magnetic skyrmion bubbles". doi:10.1126/science.aaa1442. Lay summary.

Chapter 93

Pomeron

In physics, the **pomeron** is a Regge trajectory, a family of particles with increasing spin, postulated in 1961 to explain the slowly rising cross section of hadronic collisions at high energies. It's named after Isaak Pomeranchuk.

93.1 Overview

While other trajectories lead to falling cross sections, the pomeron can lead to logarithmically rising cross sections which experimentally are approximately constant ones. The identification of the pomeron and the prediction of its properties was a major success of the Regge theory of strong interaction phenomenology. In later years, a BFKL pomeron was derived in other kinematic regimes from perturbative calculations in QCD, but its relationship to the pomeron seen in soft high energy scattering is still not completely understood.

One consequence of the pomeron hypothesis is that the cross sections of proton–proton and proton–antiproton scattering should be equal at high enough energies. This was demonstrated by the Soviet physicist Isaak Pomeranchuk by analytic continuation assuming only that the cross sections do not fall. The pomeron itself was introduced by Vladimir Gribov, and it incorporated this theorem into Regge theory. Geoffrey Chew and Steven Frautschi introduced the pomeron in the west. The modern interpretation of Pomeranchuk's theorem is that the pomeron has no conserved charges—the particles on this trajectory have the quantum numbers of the vacuum.

The pomeron was well accepted in the 1960s despite the fact that the measured cross sections of proton–proton and proton–antiproton scattering at the energies then available were unequal. By the 1990s, the existence of the pomeron as well as some of its properties were experimentally well established, notably at Fermilab and DESY.

The pomeron carries no charges. The absence of electric charge implies that pomeron exchange does not lead to the usual shower of Cherenkov radiation, while the absence of color charge implies that such events do not radiate pions.

This is in accord with experimental observation. In high energy proton–proton and proton–antiproton collisions in which it is believed that pomerons have been exchanged, a rapidity gap is often observed. This is a large angular region in which no outgoing particles are detected.

93.2 Odderon

The odderon is the hypothetical counterpart of the pomeron that carries odd charge parity.

93.3 String theory

In early particle physics, the 'pomeron sector' was what is now called the 'closed string sector' while what was called the 'reggeon sector' is now the 'open string theory'.

93.4 See also

- Giuseppe Cocconi
- Alan Wetherell
- Bert Diddens

93.5 Further reading

- Otto Nachtmann (2003). “Pomeron Physics and QCD”. arXiv:hep-ph/0312279 [hep-ph].

93.6 External links

- Pomerons at Fermilab

Chapter 94

Quasiparticle

In physics, **quasiparticles** and **collective excitations** (which are closely related) are emergent phenomena that occur when a microscopically complicated system such as a solid behaves *as if* it contained different weakly interacting particles in free space. For example, as an electron travels through a semiconductor, its motion is disturbed in a complex way by its interactions with all of the other electrons and nuclei; however it *approximately* behaves like an electron with a *different mass* traveling unperturbed through free space. This "electron" with a different mass is called an "electron quasiparticle".[1] In another example, the aggregate motion of electrons in the valence band of a semiconductor is the same as if the semiconductor contained instead positively charged quasiparticles called holes. Other quasiparticles or collective excitations include phonons (particles derived from the vibrations of atoms in a solid), plasmons (particles derived from plasma oscillations), and many others.

These particles are typically called "quasiparticles" if they are related to fermions (like electrons and holes), and called "collective excitations" if they are related to bosons (like phonons and plasmons),[1] although the precise distinction is not universally agreed upon.[2]

The quasiparticle concept is most important in condensed matter physics, since it is one of the few known ways of simplifying the quantum mechanical many-body problem.

94.1 Overview

94.1.1 General introduction

Solids are made of only three kinds of particles: Electrons, protons, and neutrons. Quasiparticles are none of these; instead they are an *emergent phenomenon* that occurs inside the solid. Therefore, while it is quite possible to have a single particle (electron or proton or neutron) floating in space, a *quasi*particle can instead only exist inside the solid.

Motion in a solid is extremely complicated: Each electron and proton gets pushed and pulled (by Coulomb's law) by all the other electrons and protons in the solid (which may themselves be in motion). It is these strong interactions that make it very difficult to predict and understand the behavior of solids (see many-body problem). On the other hand, the motion of a *non-interacting* particle is quite simple: In classical mechanics, it would move in a straight line, and in quantum mechanics, it would move in a superposition of plane waves. This is the motivation for the concept of quasiparticles: The complicated motion of the *actual* particles in a solid can be mathematically transformed into the much simpler motion of imagined *quasi*particles, which behave more like non-interacting particles.

In summary, quasiparticles are a mathematical tool for simplifying the description of solids. They are not "real" particles inside the solid. Instead, saying "A quasiparticle is present" or "A quasiparticle is moving" is shorthand for saying "A large number of electrons and nuclei are moving in a specific coordinated way."

94.1.2 Relation to many-body quantum mechanics

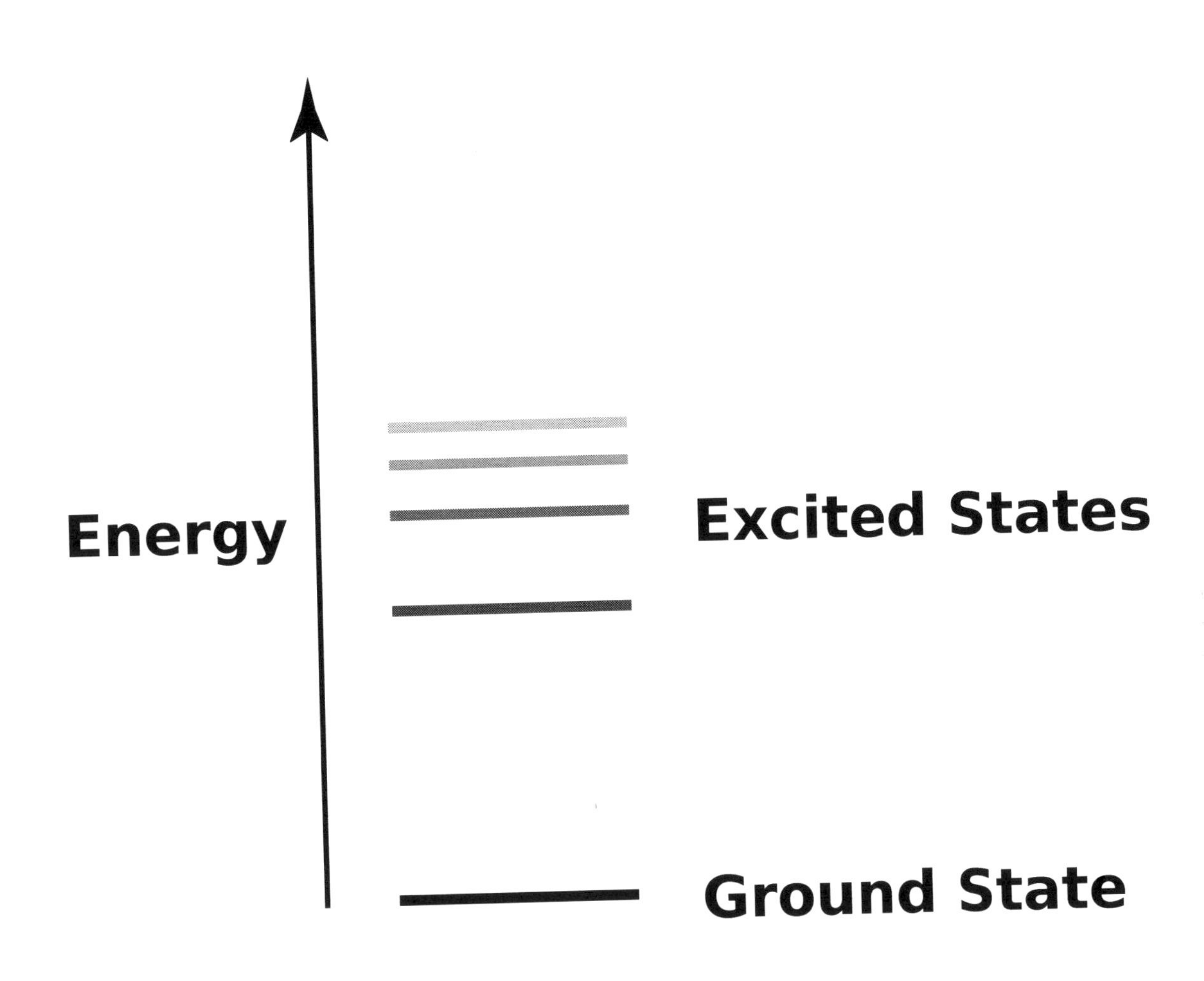

Any system, no matter how complicated, has a ground state along with an infinite series of higher-energy excited states.

The principal motivation for quasiparticles is that it is almost impossible to *directly* describe every particle in a macroscopic system. For example, a barely-visible (0.1mm) grain of sand contains around 10^{17} atoms and 10^{18} electrons. Each of these attracts or repels every other by Coulomb's law. In quantum mechanics, a system is described by a wavefunction, which, if the particles are interacting (as they are in our case), depends on the position of every particle in the system. So, each particle adds three independent variables to the wavefunction, one for each coordinate needed to describe the position of that particle. Because of this, directly approaching the many-body problem of 10^{18} interacting electrons by straightforwardly trying to solve the appropriate Schrödinger equation is impossible in practice, since it amounts to solving a partial differential equation not just in three dimensions, but in 3×10^{18} dimensions – one for each component of the position of each particle.

One simplifying factor is that the system as a whole, like any quantum system, has a ground state and various excited states with higher and higher energy above the ground state. In many contexts, only the "low-lying" excited states, with energy reasonably close to the ground state, are relevant. This occurs because of the Boltzmann distribution, which implies that very-high-energy thermal fluctuations are unlikely to occur at any given temperature.

Quasiparticles and collective excitations are a type of low-lying excited state. For example, a crystal at absolute zero is in the ground state, but if one phonon is added to the crystal (in other words, if the crystal is made to vibrate slightly at a particular frequency) then the crystal is now in a low-lying excited state. The single phonon is called an *elementary excitation*. More generally, low-lying excited states may contain any number of elementary excitations (for example, many phonons, along with other quasiparticles and collective excitations).[3]

When the material is characterized as having "several elementary excitations", this statement presupposes that the different excitations can be combined together. In other words, it presupposes that the excitations can coexist simultaneously and independently. This is never *exactly* true. For example, a solid with two identical phonons does not have exactly twice the excitation energy of a solid with just one phonon, because the crystal vibration is slightly anharmonic. However, in many materials, the elementary excitations are very *close* to being independent. Therefore, as a *starting point*, they are treated as free, independent entities, and then corrections are included via interactions between the elementary excitations, such as "phonon-phonon scattering".

Therefore, using quasiparticles / collective excitations, instead of analyzing 10^{18} particles, one needs only to deal with only a handful of somewhat-independent elementary excitations. It is therefore a very effective approach to simplify the many-body problem in quantum mechanics. This approach is not useful for *all* systems however: In strongly correlated materials, the elementary excitations are so far from being independent that it is not even useful as a starting point to treat them as independent.

94.1.3 Distinction between quasiparticles and collective excitations

Usually, an elementary excitation is called a "quasiparticle" if it is a fermion and a "collective excitation" if it is a boson.[1] However, the precise distinction is not universally agreed upon.[2]

There is a difference in the way that quasiparticles and collective excitations are intuitively envisioned.[2] A quasiparticle is usually thought of as being like a dressed particle: It is built around a real particle at its "core", but the behavior of the particle is affected by the environment. A standard example is the "electron quasiparticle": A real electron particle, in a crystal, behaves as if it had a different mass. On the other hand, a collective excitation is usually imagined to be a reflection of the aggregate behavior of the system, with no single real particle at its "core". A standard example is the phonon, which characterizes the vibrational motion of every atom in the crystal.

However, these two visualizations leave some ambiguity. For example, a magnon in a ferromagnet can be considered in one of two perfectly equivalent ways: (a) as a mobile defect (a misdirected spin) in a perfect alignment of magnetic moments or (b) as a quantum of a collective spin wave that involves the precession of many spins. In the first case, the magnon is envisioned as a quasiparticle, in the second case, as a collective excitation. However, both (a) and (b) are equivalent and correct descriptions. As this example shows, the intuitive distinction between a quasiparticle and a collective excitation is not particularly important or fundamental.

The problems arising from the collective nature of quasiparticles have also been discussed within the philosophy of science, notably in relation to the identity conditions of quasiparticles and whether they should be considered "real" by the standards of, for example, entity realism.[4][5]

94.1.4 Effect on bulk properties

By investigating the properties of individual quasiparticles, it is possible to obtain a great deal of information about low-energy systems, including the flow properties and heat capacity.

In the heat capacity example, a crystal can store energy by forming phonons, and/or forming excitons, and/or forming plasmons, etc. Each of these is a separate contribution to the overall heat capacity.

94.1.5 History

The idea of quasiparticles originated in Lev Landau's theory of Fermi liquids, which was originally invented for studying liquid helium-3. For these systems a strong similarity exists between the notion of quasi-particle and dressed particles in quantum field theory. The dynamics of Landau's theory is defined by a kinetic equation of the mean-field type. A similar equation, the Vlasov equation, is valid for a plasma in the so-called plasma approximation. In the plasma approximation, charged particles are considered to be moving in the electromagnetic field collectively generated by all other particles, and hard collisions between the charged particles are neglected. When a kinetic equation of the mean-field type is a valid first-order description of a system, second-order corrections determine the entropy production, and generally take the form of a Boltzmann-type collision term, in which figure only "far collisions" between virtual particles. In other words, every type of mean-field kinetic equation, and in fact every mean-field theory, involves a quasi-particle concept.

94.2 Examples of quasiparticles and collective excitations

This section contains examples of quasiparticles and collective excitations. The first subsection below contains common ones that occur in a wide variety of materials under ordinary conditions; the second subsection contains examples that arise in particular, special contexts.

94.2.1 More common examples

See also: List of quasiparticles

- In solids, an **electron quasiparticle** is an electron as affected by the other forces and interactions in the solid. The electron quasiparticle has the same charge and spin as a "normal" (elementary particle) electron, and like a normal electron, it is a fermion. However, its mass can differ substantially from that of a normal electron; see the article effective mass.[1] Its electric field is also modified, as a result of electric field screening. In many other respects, especially in metals under ordinary conditions, these so-called Landau quasiparticles closely resemble familiar electrons; as Crommie's "quantum corral" showed, an STM can clearly image their interference upon scattering.
- A **hole** is a quasiparticle consisting of the lack of an electron in a state; it is most commonly used in the context of empty states in the valence band of a semiconductor.[1] A hole has the opposite charge of an electron.
- A **phonon** is a collective excitation associated with the vibration of atoms in a rigid crystal structure. It is a quantum of a sound wave.
- A **magnon** is a collective excitation[1] associated with the electrons' spin structure in a crystal lattice. It is a quantum of a spin wave.
- A **roton** is a collective excitation associated with the rotation of a fluid (often a superfluid). It is a quantum of a vortex.
- In materials, a **photon** quasiparticle is a photon as affected by its interactions with the material. In particular, the photon quasiparticle has a modified relation between wavelength and energy (dispersion relation), as described by the material's index of refraction. It may also be termed a **polariton**, especially near a resonance of the material. For example, an **exciton-polariton** is a superposition of an exciton and a photon; a **phonon-polariton** is a superposition of a phonon and a photon.
- A **plasmon** is a collective excitation, which is the quantum of plasma oscillations (wherein all the electrons simultaneously oscillate with respect to all the ions).
- A **polaron** is a quasiparticle which comes about when an electron interacts with the polarization of its surrounding ions.
- An **exciton** is an electron and hole bound together.
- A **plasmariton** is a coupled optical phonon and dressed photon consisting of a plasmon and photon.

94.2.2 More specialized examples

- Composite fermions arise in a two-dimensional system subject to a large magnetic field, most famously those systems that exhibit the fractional quantum Hall effect.[6] These quasiparticles are quite unlike normal particles in two ways. First, their charge can be less than the electron charge *e*. In fact, they have been observed with charges of e/3, e/4, e/5, and e/7.[7] Second, they can be anyons, an exotic type of particle that is neither a fermion nor boson.[8]

- Stoner excitations in ferromagnetic metals
- Bogoliubov quasiparticles in superconductors. Superconductivity is carried by Cooper pairs—usually described as pairs of electrons—that move through the crystal lattice without resistance. A broken Cooper pair is called a Bogoliubov quasiparticle.[9] It differs from the conventional quasiparticle in metal because it combines the properties of a negatively charged electron and a positively charged hole (an electron void). Physical objects like impurity atoms, from which quasiparticles scatter in an ordinary metal, only weakly affect the energy of a Cooper pair in a conventional superconductor. In conventional superconductors, interference between Bogoliubov quasiparticles is tough for an STM to see. Because of their complex global electronic structures, however, high-Tc cuprate superconductors are another matter. Thus Davis and his colleagues were able to resolve distinctive patterns of quasiparticle interference in Bi-2212.[10]
- A Majorana fermion is a particle which equals its own antiparticle, and can emerge as a quasiparticle in certain superconductors.
- Magnetic monopoles arise in condensed matter systems such as spin ice and carry an effective magnetic charge as well as being endowed with other typical quasiparticle properties such as an effective mass. They may be formed through spin flips in frustrated pyrochlore ferromagnets and interact through a Coulomb potential.
- Skyrmions
- Spinon is represented by quasiparticle produced as a result of electron spin-charge separation, and can form both quantum spin liquid and strongly correlated quantum spin liquid in some minerals like Herbertsmithite. [11]

94.3 See also

- Fractionalization
- List of quasiparticles
- Mean field theory
- Pseudoparticle

94.4 References

[1] E. Kaxiras, *Atomic and Electronic Structure of Solids*, ISBN 0-521-52339-7, pages 65–69.

[2] *A guide to Feynman diagrams in the many-body problem*, by Richard D. Mattuck, p10. “As we have seen, the quasi particle consists of the original real, individual particle, plus a cloud of disturbed neighbors. It behaves very much like an individual particle, except that it has an effective mass and a lifetime. But there also exist other kinds of fictitious particles in many-body systems, i.e. 'collective excitations'. These do not center around individual particles, but instead involve collective, wavelike motion of *all* the particles in the system simultaneously.”

[3] *Principles of Nanophotonics* by Motoichi Ohtsu, p205 google books link

[4] A. Gelfert, 'Manipulative Success and the Unreal', *International Studies in the Philosophy of Science* Vol. 17, 2003, 245–263

[5] B. Falkenburg, *Particle Metaphysics* (The Frontiers Collection), Berlin: Springer 2007, esp. pp. 243–46

[6] Physics Today Article

[7] Cosmos magazine June 2008

[8] Nature article

[9] “Josephson Junctions”. *Science and Technology Review*. Lawrence Livermore National Laboratory.

[10] J. E. Hoffman; McElroy, K; Lee, DH; Lang, KM; Eisaki, H; Uchida, S; Davis, JC et al. (2002). "Imaging Quasiparticle Interference in Bi2Sr2CaCu2O8+". *Science* **297** (5584): 1148–51. arXiv:cond-mat/0209276. Bibcode:2002Sci...297.1148H. doi:10.1126/science.1072640. PMID 12142440.

[11] Shaginyan, V. R. et al. (2012). "Identification of Strongly Correlated Spin Liquid in Herbertsmithite". *EPL* **97** (5): 56001. arXiv:1111.0179. Bibcode:2012EL.....9756001S. doi:10.1209/0295-5075/97/56001.

94.5 Further reading

- L. D. Landau, *Soviet Phys. JETP.* 3:920 (1957)
- L. D. Landau, *Soviet Phys. JETP.* 5:101 (1957)
- A. A. Abrikosov, L. P. Gor'kov, and I. E. Dzyaloshinski, *Methods of Quantum Field Theory in Statistical Physics* (1963, 1975). Prentice-Hall, New Jersey; Dover Publications, New York.
- D. Pines, and P. Nozières, *The Theory of Quantum Liquids* (1966). W.A. Benjamin, New York. *Volume I: Normal Fermi Liquids* (1999). Westview Press, Boulder.
- J. W. Negele, and H. Orland, *Quantum Many-Particle Systems* (1998). Westview Press, Boulder
- Amusia, M., Popov, K., Shaginyan, V., Stephanovich, V. (2014). *Theory of Heavy-Fermion Compounds - Theory of Strongly Correlated Fermi-Systems*. Springer. ISBN 978-3-319-10825-4.

94.6 External links

- PhysOrg.com – Scientists find new 'quasiparticles'
- Curious 'quasiparticles' baffle physicists by Jacqui Hayes, Cosmos 6 June 2008. Accessed June 2008

Chapter 95

Davydov soliton

Davydov soliton is a quantum quasiparticle representing an excitation propagating along the protein α-helix self-trapped amide I. It is a solution of the Davydov Hamiltonian. It is named for the Soviet and Ukrainian physicist Alexander Davydov. The Davydov model describes the interaction of the amide I vibrations with the hydrogen bonds that stabilize the α-helix of proteins. The elementary excitations within the α-helix are given by the phonons which correspond to the deformational oscillations of the lattice, and the excitons which describe the internal amide I excitations of the peptide groups. Referring to the atomic structure of an α-helix region of protein the mechanism that creates the Davydov soliton (polaron, exciton) can be described as follows: vibrational energy of the C=O stretching (or amide I) oscillators that is localized on the α-helix acts through a phonon coupling effect to distort the structure of the α-helix, while the helical distortion reacts again through phonon coupling to trap the amide I oscillation energy and prevent its dispersion. This effect is called *self-localization* or *self-trapping*.[1][2][3] Solitons in which the energy is distributed in a fashion preserving the helical symmetry are dynamically unstable, and such symmetrical solitons once formed decay rapidly when they propagate. On the other hand, an asymmetric soliton which spontaneously breaks the local translational and helical symmetries possesses the lowest energy and is a robust localized entity.[4]

Davydov's Hamiltonian is formally similar to the Fröhlich-Holstein Hamiltonian for the interaction of electrons with a polarizable lattice. Thus the Hamiltonian of the energy operator $\hat{H}$ is

$$\hat{H} = \hat{H}_{\text{qp}} + \hat{H}_{\text{ph}} + \hat{H}_{\text{int}}$$

where $\hat{H}_{\text{qp}}$ is the quasiparticle (exciton) Hamiltonian, which describes the motion of the amide I excitations between adjacent sites; $\hat{H}_{\text{ph}}$ is the phonon Hamiltonian, which describes the vibrations of the lattice; and $\hat{H}_{\text{int}}$ is the interaction Hamiltonian, which describes the interaction of the amide I excitation with the lattice.[1][2][3]

The quasiparticle (exciton) Hamiltonian $\hat{H}_{\text{qp}}$ is:

$$\hat{H}_{\text{qp}} = \sum_{n,\alpha} E_0 \hat{A}^\dagger_{n,\alpha}\hat{A}_{n,\alpha} - J\sum_{n,\alpha}\left(\hat{A}^\dagger_{n,\alpha}\hat{A}_{n+1,\alpha} + \hat{A}^\dagger_{n,\alpha}\hat{A}_{n-1,\alpha}\right) + L\sum_{n,\alpha}\left(\hat{A}^\dagger_{n,\alpha}\hat{A}_{n,\alpha+1} + \hat{A}^\dagger_{n,\alpha}\hat{A}_{n,\alpha-1}\right)$$

where the index $n = 1, 2, \cdots, N$ counts the peptide groups along the α-helix spine, the index $\alpha = 1, 2, 3$ counts each α-helix spine, $E_0 = 3.28 \times 10^{-20}$ J is the energy of the amide I vibration (CO stretching), $J = 2.46 \times 10^{-22}$ J is the dipole-dipole coupling energy between a particular amide I bond and those ahead and behind along the same spine, $L = 1.55 \times 10^{-22}$ J is the dipole-dipole coupling energy between a particular amide I bond and those on adjacent spines in the same unit cell of the protein α-helix, $\hat{A}^\dagger_{n,\alpha}$ and $\hat{A}_{n,\alpha}$ are respectively the boson creation and annihilation operator for a quasiparticle at the peptide group. n, α [5][6]

The phonon Hamiltonian $\hat{H}_{\text{ph}}$ is

$$\hat{H}_{\text{ph}} = \frac{1}{2}\sum_{n,\alpha}\left[w(\hat{u}_{n+1,\alpha} - \hat{u}_{n,\alpha})^2 + \frac{\hat{p}^2_{n,\alpha}}{M}\right]$$

where $\hat{u}_{n,\alpha}$ is the displacement operator from the equilibrium position of the peptide group n, α , $\hat{p}_{n,\alpha}$ is the momentum operator of the peptide group n, α , M is the mass of each peptide group, and $w = 19.5$ N m $^{-1}$ is an effective elasticity coefficient of the lattice (the spring constant of a hydrogen bond).

Finally, the interaction Hamiltonian $\hat{H}_{\text{int}}$ is

$$\hat{H}_{\text{int}} = \chi \sum_{n,\alpha} \left[(\hat{u}_{n+1,\alpha} - \hat{u}_{n,\alpha}) \hat{A}^{\dagger}_{n,\alpha} \hat{A}_{n,\alpha} \right]$$

where $\chi = -30$ pN is an anharmonic parameter arising from the coupling between the quasiparticle (exciton) and the lattice displacements (phonon) and parameterizes the strength of the exciton-phonon interaction. The value of this parameter for α-helix has been determined via comparison of the theoretically calculated absorption line shapes with the experimentally measured ones.[7]

The mathematical techniques that are used to analyze Davydov's soliton are similar to some that have been developed in polaron theory. In this context the Davydov's soliton corresponds to a polaron that is (i) *large* so the continuum limit approximation is justified, (ii) *acoustic* because the self-localization arises from interactions with acoustic modes of the lattice, and (iii) *weakly coupled* because the anharmonic energy is small compared with the phonon bandwidth.[5]

The Davydov soliton is a *quantum quasiparticle* and it obeys Heisenberg's uncertainty principle. Thus any model that does not impose translational invariance is flawed by construction.[5] Supposing that the Davydov soliton is localized to 5 turns of the α-helix results in significant uncertainty in the velocity of the soliton $\Delta v = 133$ m/s, a fact that is obscured if one models the Davydov soliton as a classical object.

There are three possible fundamental approaches towards Davydov model:[6][8] (i) the quantum theory, in which both the amide I vibration (excitons) and the lattice site motion (phonons) are treated quantum mechanically; (ii) the mixed quantum-classical theory, in which the amide I vibration is treated quantum mechanically but the lattice is classical; and (iii) the classical theory, in which both the amide I and the lattice motions are treated classically.

95.1 References

[1] Davydov AS (1973). "The theory of contraction of proteins under their excitation". *Journal of Theoretical Biology* **38** (3): 559–569. doi:10.1016/0022-5193(73)90256-7. PMID 4266326.

[2] Davydov AS (1974). "Quantum theory of muscular contraction". *Biophysics* **19**: 684–691.

[3] Davydov AS (1977). "Solitons and energy transfer along protein molecules". *Journal of Theoretical Biology* **66** (2): 379–387. doi:10.1016/0022-5193(77)90178-3. PMID 886872.

[4] Brizhik L, Eremko A, Piette B, Zakrzewski W (2004). "Solitons in α-helical proteins". *Physical Review E* **70**: 031914, 1–16. arXiv:cond-mat/0402644. Bibcode:2004PhRvE..70a1914K. doi:10.1103/PhysRevE.70.011914.

[5] Scott AS(1992). "Davydov's soliton".*Physics Reports***217**: 1–67. Bibcode:1992PhR...217....1S.doi:10.1016/0370-1573(
F.

[6] Cruzeiro-Hansson L, Takeno S. (1997). "Davydov model: the quantum, mixed quantum-classical, and full classical systems". *Physical Review E* **56** (1): 894–906. Bibcode:1997PhRvE..56..894C. doi:10.1103/PhysRevE.56.894.

[7] Cruzeiro-Hansson L (2005). "Influence of the nonlinearity and dipole strength on the amide I band of protein α-helices". *The Journal of Chemical Physics* **123** (23): 234909, 1–7. Bibcode:2005JChPh.123w4909C. doi:10.1063/1.2138705. PMID 16392951.

[8] Cruzeiro-Hansson L (1997). "Short timescale energy transfer in proteins". *Solphys '97 Proceedings*.

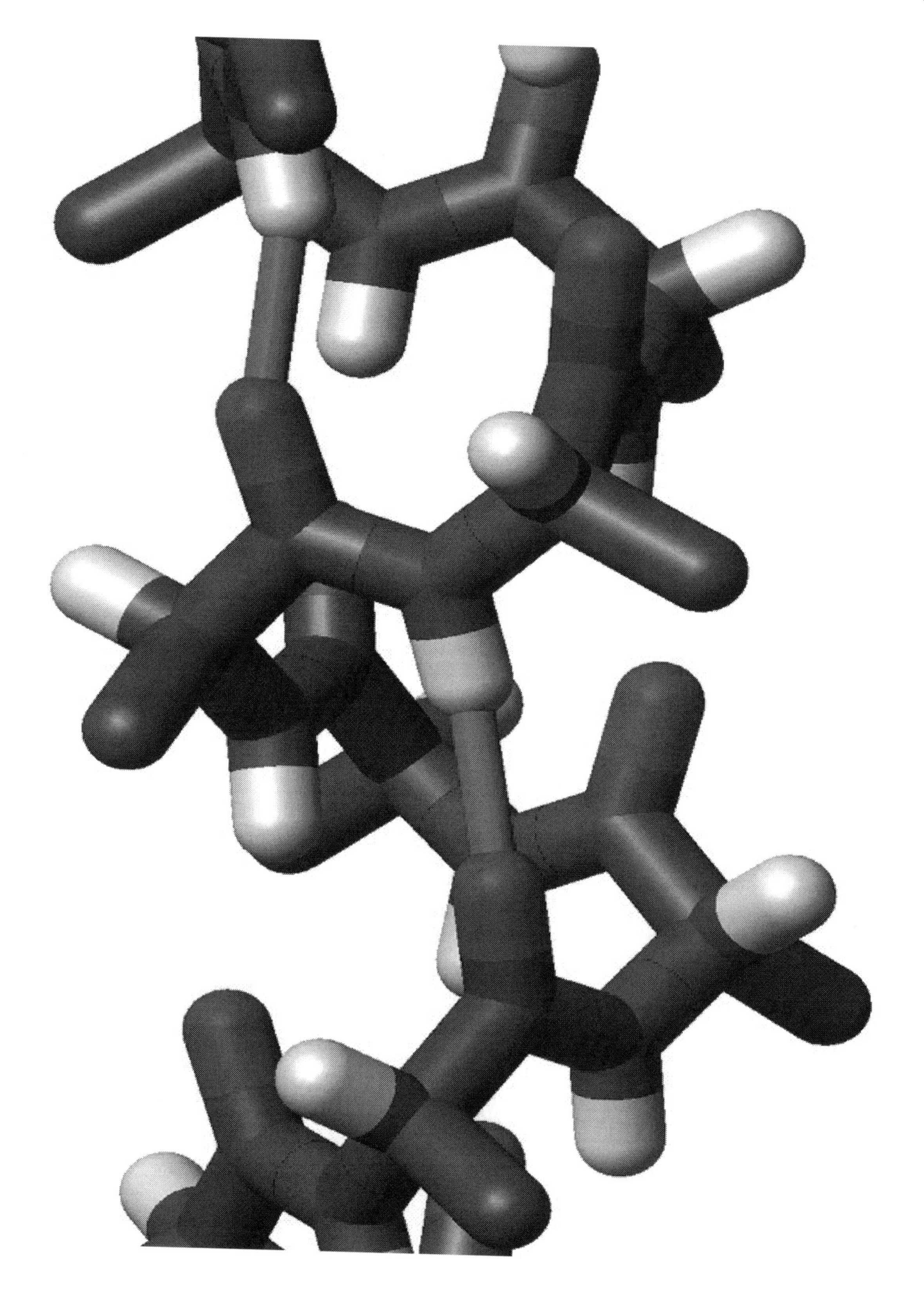

Side view of an α-helix of alanine residues in atomic detail. Protein α-helices provide substrate for Davydov soliton creation and propagation.

Chapter 96

Dropleton

A **dropleton** or **quantum droplet** is an artificial quasiparticle, constituting a collection of electrons and places without them inside a semiconductor. Dropleton is the first known quasiparticle that behaves like a liquid.[1] The creation of dropletons was announced on 26 February 2014 in a *Nature* article, that presented evidence for the creation of dropletons in an electron–hole plasma inside a gallium arsenide quantum well by ultrashort laser pulses.[2] The existence of dropletons was not predicted before the experiment.

Despite the relatively short lifetime of about 25 picoseconds, the dropletons are stable enough to be studied.[1] Dropletons possess favorable properties for studying quantum mechanics. Dropletons are approximately 200 nanometers wide, the size of the smallest bacteria. This fact offers the hope to the discoverers that they might one day actually see quantum droplets.[1]

96.1 References

[1] Clara Moskowitz (26 February 2014). "Meet the Dropleton—a "Quantum Droplet" That Acts Like a Liquid". *Scientific American*. Retrieved 26 February 2014.

[2] A. E. Almand-Hunter, H. Li, S. T. Cundiff, M. Mootz, M. Kira & S. W. Koch (26 February 2014). "Quantum droplets of electrons and holes". *Nature*. Bibcode:2014Natur.506..471A. doi:10.1038/nature12994. Retrieved 26 February 2014.

Chapter 97

Exciton

An **exciton** is a bound state of an electron and an electron hole which are attracted to each other by the electrostatic Coulomb force. It is an electrically neutral quasiparticle that exists in insulators, semiconductors and in some liquids. The exciton is regarded as an elementary excitation of condensed matter that can transport energy without transporting net electric charge.[1][2]

An exciton can form when a photon is absorbed by a semiconductor.[3] This excites an electron from the valence band into the conduction band. In turn, this leaves behind a positively charged electron hole (an abstraction for the location from which an electron was moved). The electron in the conduction band is then effectively attracted to this localized hole by the repulsive Coulomb forces from large numbers of electrons surrounding the hole and excited electron. This attraction provides a stabilizing energy balance. Consequently, the exciton has slightly less energy than the unbound electron and hole. The wavefunction of the bound state is said to be *hydrogenic*, an exotic atom state akin to that of a hydrogen atom. However, the binding energy is much smaller and the particle's size much larger than a hydrogen atom. This is because of both the screening of the Coulomb force by other electrons in the semiconductor (i.e., its dielectric constant), and the small effective masses of the excited electron and hole. The recombination of the electron and hole, i.e. the decay of the exciton, is limited by resonance stabilization due to the overlap of the electron and hole wave functions, resulting in an extended lifetime for the exciton.

The electron and hole may have either parallel or anti-parallel spins. The spins are coupled by the exchange interaction, giving rise to exciton fine structure. In periodic lattices, the properties of an exciton show momentum (k-vector) dependence.

The concept of excitons was first proposed by Yakov Frenkel in 1931,[4] when he described the excitation of atoms in a lattice of insulators. He proposed that this excited state would be able to travel in a particle-like fashion through the lattice without the net transfer of charge.

97.1 Classification

Excitons may be treated in two limiting cases, depending on the properties of the material in question.

97.1.1 Frenkel excitons

In materials with a small dielectric constant, the Coulomb interaction between an electron and a hole may be strong and the excitons thus tend to be small, of the same order as the size of the unit cell. Molecular excitons may even be entirely located on the same molecule, as in fullerenes. This *Frenkel exciton*, named after Yakov Frenkel, has a typical binding energy on the order of 0.1 to 1 eV. Frenkel excitons are typically found in alkali halide crystals and in organic molecular crystals composed of aromatic molecules, such as anthracene and tetracene.

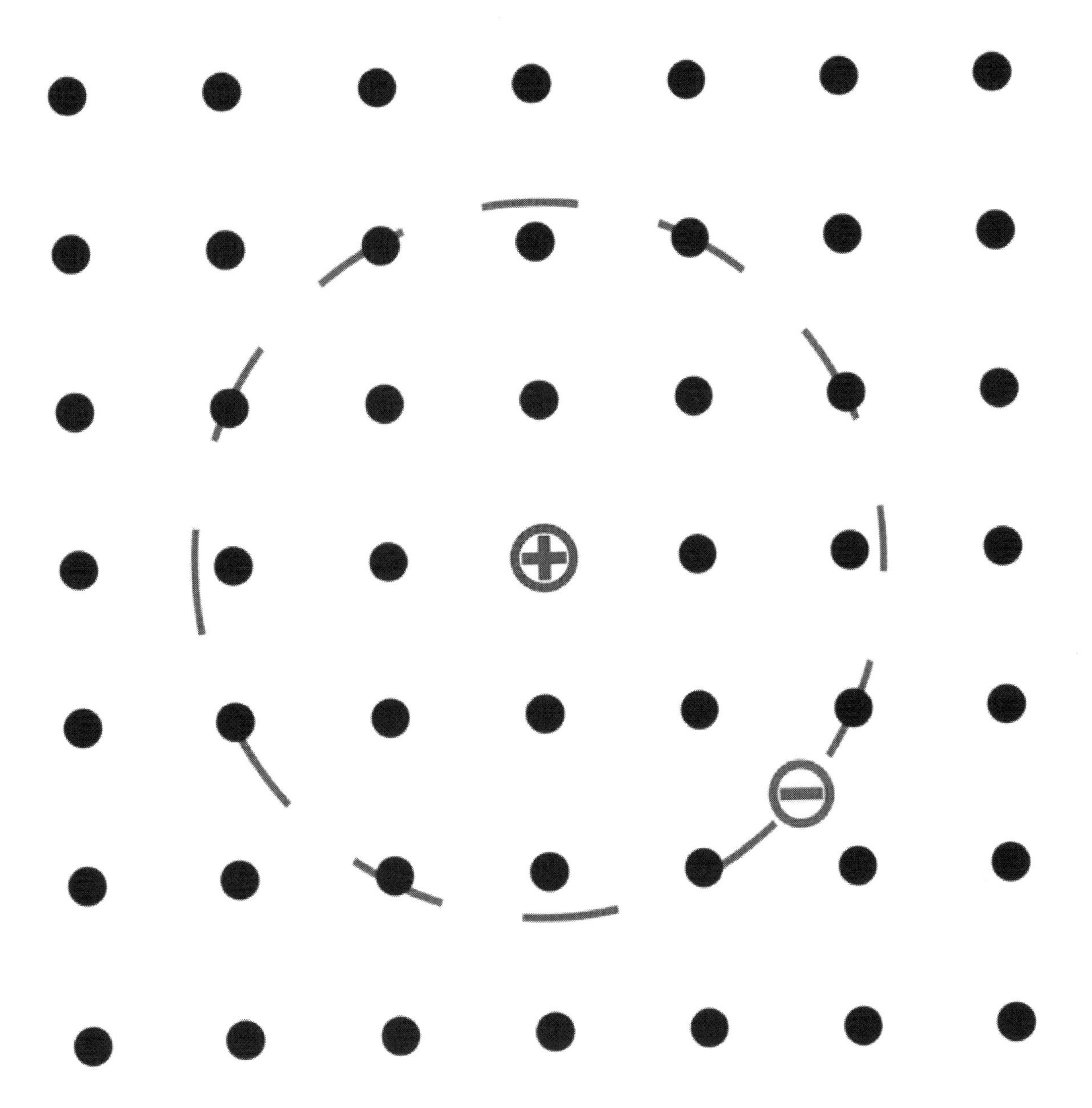

Frenkel exciton, bound electron-hole pair where the hole is localized at a position in the crystal represented by black dots

97.1.2 Wannier-Mott excitons

In semiconductors, the dielectric constant is generally large. Consequently, electric field screening tends to reduce the Coulomb interaction between electrons and holes. The result is a *Wannier exciton*,[5] which has a radius larger than the lattice spacing. Small effective mass of electrons that is typical of semiconductors also favors large exciton radii. As a result, the effect of the lattice potential can be incorporated into the effective masses of the electron and hole. Likewise, because of the lower masses and the screened Coulomb interaction, the binding energy is usually much less than that of a hydrogen atom, typically on the order of 0.01eV. This type of exciton was named for Gregory Wannier and Nevill Francis Mott. Wannier-Mott excitons are typically found in semiconductor crystals with small energy gaps and high dielectric constants, but have also been identified in liquids, such as liquid xenon. They are also known as *large excitons*.

In single-wall carbon nanotubes, excitons have both Wannier-Mott and Frenkel character. This is due to the nature of the Coulomb interaction between electrons and holes in one-dimension. The dielectric function of the nanotube itself is large enough to allow for the spatial extent of the wave function to extend over a few to several nanometers along the tube

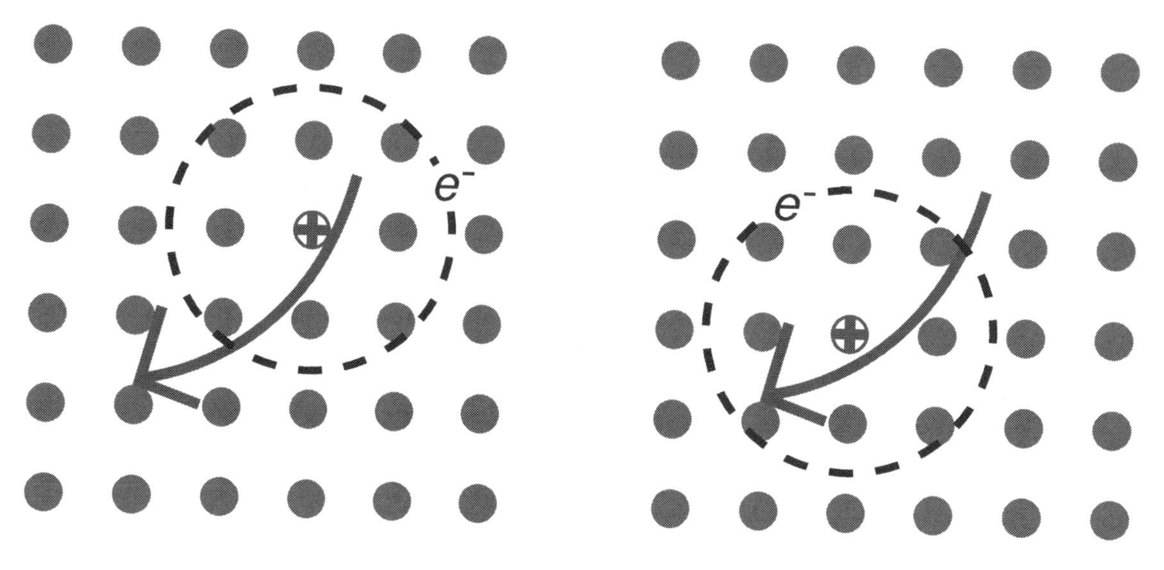

Wannier-Mott exciton, bound electron-hole pair that is not localized at a crystal position. This figure schematically shows diffusion of the exciton across the lattice.

axis, while poor screening in the vacuum or dielectric environment outside of the nanotube allows for large (0.4 to 1.0eV) binding energies.

Often there is more than one band to choose from for the electron and the hole leading to different types of excitons in the same material. Even high-lying bands can be effective as femtosecond two-photon experiments have shown. At cryogenic temperatures, many higher excitonic levels can be observed approaching the edge of the band,[6] forming a series of spectral absorption lines that are in principle similar to hydrogen spectral series.

97.1.3 Charge-transfer excitons

An intermediate case between Frenkel and Wannier excitons, *charge-transfer excitons* (sometimes called simply *CT excitons*) occur when the electron and the hole occupy adjacent molecules.[7] They occur primarily in ionic crystals.[8] Unlike Frenkel and Wannier excitons they display a static electric dipole moment.[9]

97.1.4 Surface excitons

At surfaces it is possible for so called *image states* to occur, where the hole is inside the solid and the electron is in the vacuum. These electron-hole pairs can only move along the surface.

97.1.5 Atomic and molecular excitons

Alternatively, an exciton may be an excited state of an atom, ion, or molecule, the excitation wandering from one cell of the lattice to another.

When a molecule absorbs a quantum of energy that corresponds to a transition from one molecular orbital to another molecular orbital, the resulting electronic excited state is also properly described as an exciton. An electron is said to be found in the lowest unoccupied orbital and an electron hole in the highest occupied molecular orbital, and since they are found within the same molecular orbital manifold, the electron-hole state is said to be bound. Molecular excitons typically have characteristic lifetimes on the order of nanoseconds, after which the ground electronic state is restored and the molecule undergoes photon or phonon emission. Molecular excitons have several interesting properties, one of which is energy transfer (see Förster resonance energy transfer) whereby if a molecular exciton has proper energetic matching to a second molecule's spectral absorbance, then an exciton may transfer (*hop*) from one molecule to another. The process is strongly dependent on intermolecular distance between the species in solution, and so the process has found application in sensing and *molecular rulers*.

The hallmark of molecular excitons in organic molecular crystals are doublets and/or triplets of exciton absorption bands strongly polarized along crystallographic axes. In these crystals an elementary cell includes several molecules sitting in symmetrically identical positions, which results in the level degeneracy that is lifted by intermolecular interaction. As a result, absorption bands are polarized along the symmetry axes of the crystal. Such multiplets were discovered by Antonina Prikhot'ko[10][11] and their genesis was proposed by Alexander Davydov. It is known as 'Davydov splitting'.[12][13]

97.1.6 Giant oscillator strength of bound excitons

Excitons are lowest excited states of the electronic subsystem of pure crystals. Impurities can bind excitons, and when the bound state is shallow, the oscillator strength for producing bound excitons is so high that impurity absorption can compete with intrinsic exciton absorption even at rather low impurity concentrations. This phenomenon is generic and applicable both to the large radius (Wannier-Mott) excitons and molecular (Frenkel) excitons. Hence, excitons bound to impurities and defects possess giant oscillator strength.[14]

97.1.7 Self-trapping of excitons

In crystals excitons interact with phonons, the lattice vibrations. If this coupling is weak as in typical semiconductors such as GaAs or Si, excitons are scattered by phonons. However, when the coupling is strong, excitons can be self-trapped.[15][16] **Self-trapping** results in dressing excitons with a dense cloud of virtual phonons which strongly suppresses the ability of excitons to move across the crystal. In simpler terms, this means a local deformation of the crystal lattice around the exciton. Self-trapping can be achieved only if the energy of this deformation can compete with the width of the exciton band. Hence, it should be of atomic scale, of about an electron-volt.

Self-trapping of excitons is similar to forming strong-coupling polarons but with three essential differences. First, self-trapped exciton states are always of a small radius, of the order of lattice constant, due to their electric neutrality. Second, there exists a **self-trapping barrier** separating free and self-trapped states, hence, free excitons are metastable. Third, this barrier enables **coexistence of free and self-trapped states** of excitons.[17][18] This means that spectral lines of free excitons and wide bands of self-trapped excitons can be seen simultaneously in absorption and luminescence spectra. It is interesting that while the self-trapped states are of lattice-spacing scale, the barrier has typically much larger scale. Indeed, its spacial scale is about $r_b \sim m\gamma^2/\omega^2$ where m is effective mass of the exciton, γ is the exciton-phonon coupling constant, and ω is the characteristic frequency of optical phonons. Excitons are self-trapped when m and γ are large, and then the spacial size of the barrier is large compared with the lattice spacing. Transforming a free exciton state into a self-trapped one proceeds as a collective tunneling of coupled exciton-lattice system (an instanton). Because r_b is large, tunneling can be described by a continuum theory.[19] The height of the barrier $W \sim \omega^4/m^3\gamma^4$. Because both m and γ appear in the denominator of W , the barriers are basically low. Therefore, free excitons can be seen in crystals with strong exciton-phonon coupling only in pure samples and at low temperatures. Coexistence of free and self-trapped excitons was observed in rare-gas solids,[20][21] alkali-halides,[22] and in molecular crystal of pyrene.[23]

97.1.8 Interaction

Excitons are the main mechanism for light emission in semiconductors at low temperature (when the characteristic thermal energy kT is less than the exciton binding energy), replacing the free electron-hole recombination at higher temperatures.

The existence of exciton states may be inferred from the absorption of light associated with their excitation. Typically, excitons are observed just below the band gap.

When excitons interact with photons a so-called polariton (also exciton-polariton) is formed. These excitons are sometimes referred to as *dressed excitons*.

Provided the interaction is attractive, an exciton can bind with other excitons to form a biexciton, analogous to a dihydrogen molecule. If a large density of excitons is created in a material, they can interact with one another to form an electron-hole liquid, a state observed in k-space indirect semiconductors.

Additionally, excitons are integer-spin particles obeying Bose statistics in the low-density limit. In some systems, where the interactions are repulsive, a Bose–Einstein condensed state is predicted to be the ground state. Exciton condensates have been seen in a double quantum well systems.[24]

97.1.9 Spatially Direct and Indirect Excitons

Normally, excitons in a semiconductor have a very short lifetime due to the close proximity of the electron and hole. However, by placing the electron and hole in spatially separated quantum wells with an insulating barrier layer in between so called 'spatially indirect' excitons can be created. In contrast to ordinary (spatially direct), these spatially indirect excitons can have large spatial separation between the electron and hole, and thus possess a much longer lifetime. This is often used to cool excitons to very low temperatures in order to study Bose Einstein condensation (or rather it's 2 dimensional analog).[25]

97.2 See also

- Polaron
- Phonon
- Polariton superfluid
- Oscillator strength
- Giant oscillator strength

97.3 References

[1] R. S. Knox, Theory of excitons, Solid state physics (Ed. by Seitz and Turnbul, Academic, NY), v. 5, 1963.

[2] Liang, W Y (1970). "Excitons". *Physics Education* **5** (125301). Bibcode:1970PhyEd...5..226L. doi:10.1088/0031-9120/5/4/003.

[3] Couto, ODD; Puebla, J (2011). "Charge control in InP/(Ga,In)P single quantum dots embedded in Schottky diodes". *Physical Rev. B* **84** (4): 226. Bibcode:1970PhyEd...5..226L. doi:10.1103/PhysRevB.84.125301.

[4] Frenkel, J. (1931). "On the Transformation of light into Heat in Solids. I". *Physical Review* **37**: 17. Bibcode:1931PhRv...37...17F. doi:10.1103/PhysRev.37.17.

[5] Wannier, Gregory (1937). "The Structure of Electronic Excitation Levels in Insulating Crystals". *Physical Review* **52** (3): 191. Bibcode:1937PhRv...52..191W. doi:10.1103/PhysRev.52.191.

[6] http://www.nature.com/nature/journal/v514/n7522/full/nature13832.html

[7] J. D. Wright (1995) [First published 1987]. *Molecular Crystals* (2nd ed.). Cambridge University Press. p. 108. ISBN 0-521-47730-1.

[8] Ivan Pelant, Jan Valenta (2012). *Luminescence Spectroscopy of Superconductors*. Oxford University Press. p. 161. ISBN 978-0-19-958833-6.

[9] Guglielmo Lanzani (2012). *The Photophysics Behind Photovoltaics and Photonics*. Wiley-VCH Verlag. p. 82.

[10] A. Prikhotjko, Absorption Spectra of Crystals at Low Temperatures, J. Physics USSR **8**, 257 (1944)

[11] A. F. Prikhot'ko, Izv, AN SSSR Ser. Fiz. **7**, 499 (1948) http://ujp.bitp.kiev.ua/files/journals/53/si/53SI18p.pdf

[12] A.S Davydov, Theory of Molecular Excitons (Plenum, NY) 1971

[13] V. L. Broude, E. I. Rashba, and E. F. Sheka, Spectroscopy of molecular excitons (Springer, NY) 1985

[14] E. I. Rashba, Giant Oscillator Strengths Associated with Exciton Complexes, Sov. Phys. Semicond. **8**, 807-816 (1975)

[15] N. Schwentner, E.-E. Koch, and J. Jortner, Electronic excitations in condensed rare gases, Springer tracts in modern physics, **107**, 1 (1985).

[16] M. Ueta, H. Kanzaki, K. Kobayashi, Y. Toyozawa, and E. Hanamura. Excitonic Processes in Solids, Springer Series in Solid State Sciences, Vol. **60** (1986).

[17] E. I. Rashba, "Theory of Strong Interaction of Electron Excitations with Lattice Vibrations in Molecular Crystals, Optika i Spektroskopiya **2**, 75, 88 (1957).

[18] E. I. Rashba, Self-trapping of excitons, in: Excitons (North-Holland, Amsterdam, 1982), p. 547.

[19] A. S. Ioselevich and E. I. Rashba, Theory of Nonradiative Trapping in Crystals, in: "Quantum tunneling in condensed media." Eds. Yu. Kagan and A. J. Leggett. (North-Holland, Amsterdam, 1992), p. 347-425.

[20] G. Zimmerer, "Excited-State Spectroscopy in Solids", in: Proceedings of the International School of Physics,`Enrico Fermi, Course XCVI, Varenna, Italy, 1985." (1987).

[21] I. Ya. Fugol', "Free and self-trapped excitons in cryocrystals: kinetics and relaxation processes." Advances in Physics **37**, 1-35 (1988).

[22] Ch. B. Lushchik, in "Excitons," edited by E. I. Rashba, and M. D. Sturge, (North Holland, Amsterdam, 1982), p. 505.

[23] M. Furukawa, Ken-ichi Mizuno, A. Matsui, N. Tamai and I. Yamazaiu, Branching of Exciton Relaxation to the Free and Self-Trapped Exciton States, Chemical Physics **138**, 423 (1989).

[24] "Exciton Condensation in Bilayer Quantum Hall Systems". *Annual Review of Condensed Matter Physics.* January 10, 2014. doi:10.1146/annurev-conmatphys-031113-133832.

[25] A. A. High (2012) "Spontaneous coherence in a cold exciton gas" Nature

Chapter 98

Electron hole

In physics, chemistry, and electronic engineering, an **electron hole** is the lack of an electron at a position where one could exist in an atom or atomic lattice. It is different from the positron, which is an actual particle of antimatter.

If an electron is excited into a higher state it leaves a hole in its old state. This meaning is used in Auger electron spectroscopy (and other x-ray techniques), in computational chemistry, and to explain the low electron-electron scattering-rate in crystals (metals, semiconductors).

In crystals, electronic band structure calculations lead to an effective mass for the electrons, which typically is negative at the top of a band. The negative mass is an unintuitive concept,[1] and in these situations a more familiar picture is found by considering a positive charge with a positive mass.

98.1 Solid-state physics

In solid-state physics, an **electron hole** (usually referred to simply as a **hole**) is the absence of an electron from an otherwise full valence band. A hole is essentially a way to conceptualize the interactions of the electrons within a nearly *full* system, which is *missing* just a few electrons. In some ways, the behavior of a hole within a semiconductor crystal lattice is comparable to that of the bubble in an otherwise full bottle of water.[2]

98.1.1 Simplified analogy: Empty seat in an auditorium

Hole conduction in a valence band can be explained by the following analogy. Imagine a row of people seated in an auditorium, where there are no spare chairs. Someone in the middle of the row wants to leave, so he jumps over the back of the seat into an empty row, and walks out. The empty row is analogous to the conduction band, and the person walking out is analogous to a free electron.

Now imagine someone else comes along and wants to sit down. The empty row has a poor view; so he does not want to sit there. Instead, a person in the crowded row moves into the empty seat the first person left behind. The empty seat moves one spot closer to the edge and the person waiting to sit down. The next person follows, and the next, et cetera. One could say that the empty seat moves towards the edge of the row. Once the empty seat reaches the edge, the new person can sit down.

In the process everyone in the row has moved along. If those people were negatively charged (like electrons), this movement would constitute conduction. If the seats themselves were positively charged, then only the vacant seat would be positive. This is a very simple model of how hole conduction works.

In reality, due to the crystal structure properties, the hole is not localized to a single position as described in the previous example. Rather, the hole spans an area in the crystal lattice covering many hundreds of unit cells. This is equivalent to being unable to tell which broken bond corresponds to the "missing" electron.

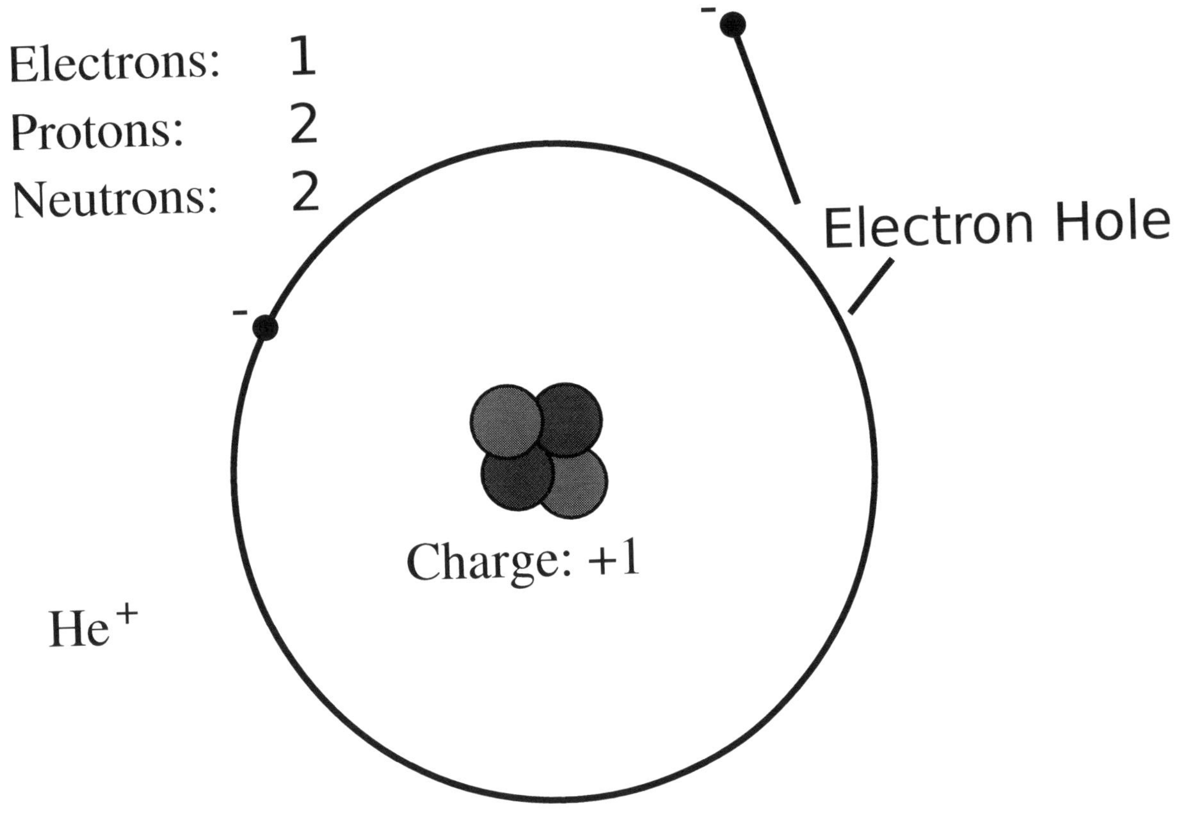

When an electron leaves a helium atom, it leaves an electron hole in its place. This causes the helium atom to become positively charged.

Instead of analyzing the movement of an empty state in the valence band as the movement of many separate electrons, a single equivalent imaginary particle called a "hole" is considered. In an applied electric field, the electrons move in one direction, corresponding to the hole moving in the other. If a hole associates itself with a neutral atom, that atom loses an electron and becomes positive. Therefore, the hole is taken to have positive charge of +e, precisely the opposite of the electron charge.

98.1.2 Detailed picture: A hole is the absence of a negative-mass electron

The analogy above is quite simplified, and cannot explain why holes create an opposite effect to electrons in the Hall effect and Seebeck effect. A more precise and detailed explanation follows.[3]

- *The dispersion relation determines how electrons respond to forces (via the concept of effective mass).*[3]

A dispersion relation is the relationship between wavevector (k-vector) and energy in a band, part of the electronic band structure. In quantum mechanics, the electrons are waves, and energy is the wave frequency. A localized electron is a wavepacket, and the motion of an electron is given by the formula for the group velocity of a wave. An electric field affects an electron by gradually shifting all the wavevectors in the wavepacket, and the electron moves because its wave group velocity changes. Therefore, again, the way an electron responds to forces is entirely determined by its dispersion relation. An electron floating in space has the dispersion relation $E=\hbar^2k^2/(2m)$, where m is the (real) electron mass and $\hbar$ is reduced Planck constant. In the conduction band of a semiconductor, the dispersion relation is instead $E=\hbar^2k^2/(2m^*)$ (m^* is the *effective mass*), so a conduction-band electron responds to forces *as if* it had the mass m^*.

- *Electrons near the top of the valence band behave as if they have negative mass.*[3]

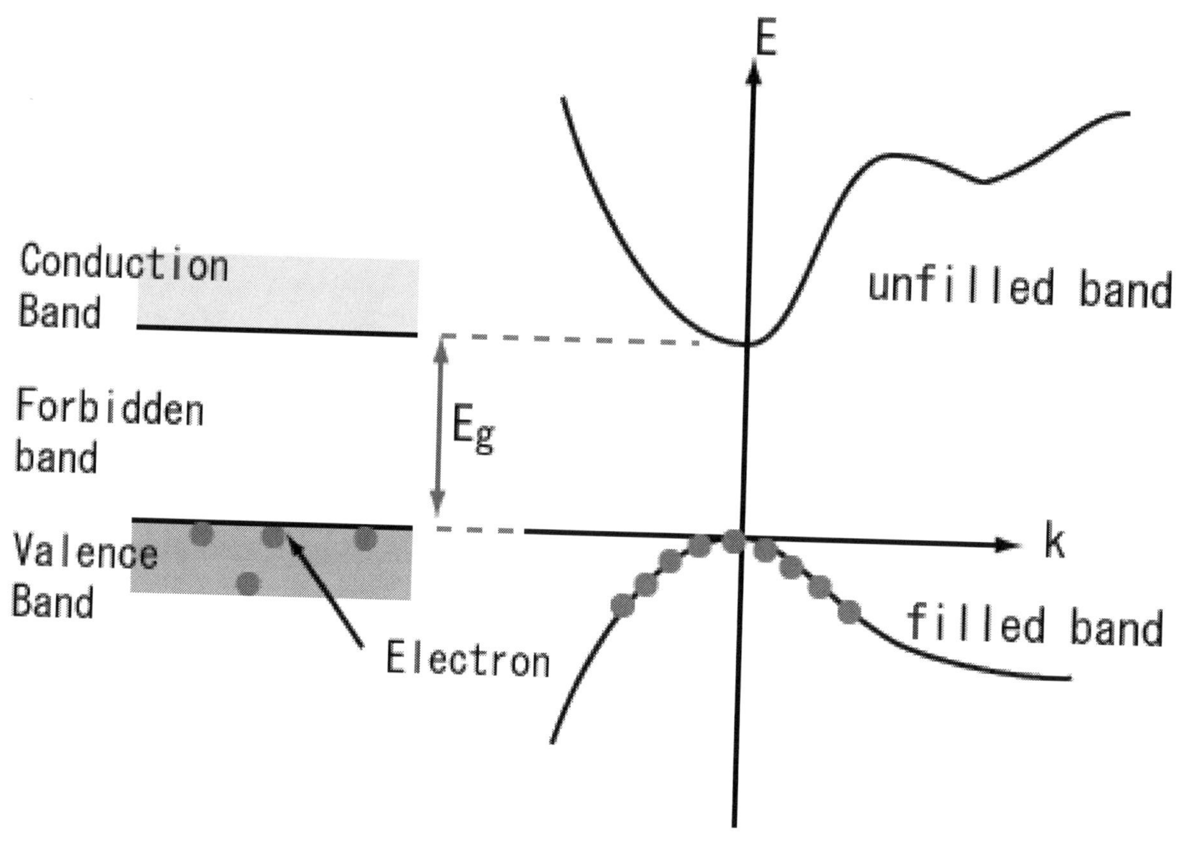

A semiconductor electronic band structure (right) includes the dispersion relation of each band, i.e. the energy of an electron E *as a function of the electron's wavevector* k. *The "unfilled band" is the semiconductor's conduction band; it curves upward indicating positive effective mass. The "filled band" is the semiconductor's valence band; it curves downward indicating negative effective mass.*

The dispersion relation near the top of the valence band is $E=\hbar^2 k^2/(2m^*)$ with *negative* effective mass. So electrons near the top of the valence band behave like they have negative mass. When a force pulls the electrons to the right, these electrons actually move left. This is solely due to the shape of the valence band, and is unrelated to whether the band is full or empty. If you could somehow empty out the valence band and just put one electron near the valence band maximum (an unstable situation), this electron would move the "wrong way" in response to forces.

- *Positively-charged holes as a shortcut for calculating the total current of an almost-full band.*[3]

A perfectly full band always has zero current. One way to think about this fact is that the electron states near the top of the band have negative effective mass, and those near the bottom of the band have positive effective mass, so the net motion is exactly zero. If an otherwise-almost-full valence band has a state *without* an electron in it, we say that this state is occupied by a hole. There is a mathematical shortcut for calculating the current due to every electron in the whole valence band: Start with zero current (the total if the band were full), and *subtract* the current due to the electrons that *would* be in each hole state if it wasn't a hole. Since *subtracting* the current caused by a *negative* charge in motion is the same as *adding* the current caused by a *positive* charge moving on the same path, the mathematical shortcut is to pretend that each hole state is carrying a positive charge, while ignoring every other electron state in the valence band.

- *A hole near the top of the valence band moves the same way as an electron near the top of the valence band* ***would*** *move*[3] (which is in the opposite direction compared to conduction-band electrons experiencing the same force.)

This fact follows from the discussion and definition above. This is an example where the auditorium analogy above is misleading. When a person moves left in a full auditorium, an empty seat moves right. But in this section we are imagining how electrons move through k-space, not real space, and the effect of a force is to move all the electrons through k-space in the same direction at the same time. So a better analogy is a bubble underwater in a river: The bubble moves the same direction as the water, not opposite. Since force = mass × acceleration, a negative-effective-mass electron near the top of the valence band would move the opposite direction as a positive-effective-mass electron near the bottom of the conduction band, in response to a given electric or magnetic force.

- *Conclusion: Hole as a positive-charge, positive-mass quasiparticle.*

From the above, a hole (A) carries a positive charge, and (B) responds to electric and magnetic fields as if it had a positive charge and positive mass. (The latter is because a particle with positive charge and positive mass responds to electric and magnetic fields in the same way as a particle with negative charge and negative mass.) That explains why holes can be treated in all situations as ordinary positively charged quasiparticles.

98.1.3 Role in semiconductor technology

In some semiconductors, such as silicon, the hole's effective mass is dependent on direction (anisotropic), however a value averaged over all directions can be used for some macroscopic calculations.

In most semiconductors, the effective mass of a hole is much larger than that of an electron. This results in lower mobility for holes under the influence of an electric field and this may slow down the speed of the electronic device made of that semiconductor. This is one major reason for adopting electrons as the primary charge carriers, whenever possible in semiconductor devices, rather than holes.

However, in many semiconductor devices, both electrons *and* holes play an essential role. Examples include p–n diodes and bipolar transistors.

98.2 Holes in quantum chemistry

An alternate meaning for the term **electron hole** is used in computational chemistry. In coupled cluster methods, the ground (or lowest energy) state of a molecule is interpreted as the "vacuum state"—conceptually, in this state there are no electrons. In this scheme, the absence of an electron from a normally filled state is called a "hole" and is treated as a particle, and the presence of an electron in a normally empty state is simply called an "electron". This terminology is almost identical to that used in solid-state physics.

98.3 Holes in superconductivity

Superconductors with highest transition temperatures (high T_c cuprates) have hole carriers in the normal state. Among the elements, the vast majority of superconductors have positive Hall coefficient in the normal state indicating dominant hole carrier transport (e.g. Pb, In, Nb), and elements with negative Hall coefficient (electron carriers) are almost never superconducting (e.g. Na, Cu, Au). It has been proposed that holes are essential to superconductivity in all materials.[4]

98.4 See also

- Band gap
- Carrier generation and recombination
- Effective mass
- Electrical resistivity and conductivity
- Semiconductor

98.5 References

[1] For these negative mass electrons, momentum is opposite to velocity, so forces acting on these electrons cause their velocity to change in the 'wrong' direction. As these electrons gain energy (moving towards the top of the band), they slow down.

[2] Weller,Paul F. (1967). "An analogy for elementary band theory concepts in solids".*J.Chem.Educ***44**(7): 391. Bibcode. doi:10.1021/ed044p391.

[3] Kittel, *Introduction to Solid State Physics* 8th edition, page 194-196

[4] http://physics.ucsd.edu/~{}jorge/hole.html

Chapter 99

Magnon

For other uses, see Magnon (disambiguation).

A **magnon** is a quasiparticle, a collective excitation of the electrons' spin structure in a crystal lattice. In the equivalent wave picture of quantum mechanics, a magnon can be viewed as a quantized spin wave. Magnons carry a fixed amount of energy and lattice momentum, and are spin-1, indicating they obey boson behavior.

99.1 Brief history

The concept of a magnon was introduced in 1930 by Felix Bloch in order to explain the reduction of the spontaneous magnetization in a ferromagnet. At absolute zero temperature, a Heisenberg ferromagnet reaches the state of lowest energy, in which all of the atomic spins (and hence magnetic moments) point in the same direction. As the temperature increases, more and more spins deviate randomly from the alignment, increasing the internal energy and reducing the net magnetization. If one views the perfectly magnetized state at zero temperature as the vacuum state of the ferromagnet, the low-temperature state with a few misaligned spins can be viewed as a gas of quasiparticles, in this case magnons. Each magnon reduces the total spin along the direction of magnetization by one unit of $\hbar$ and the magnetization by $\gamma\hbar$, where γ is the gyromagnetic ratio. This leads to Bloch's law for the temperature dependence of spontaneous magnetization:

$$M(T) = M_0(1 - (T/T_C)^{3/2})$$

The quantitative theory of magnons, quantized spin waves, was developed further by Theodore Holstein, Henry Primakoff (1940), and Freeman Dyson (1956). Using the second quantization formalism they showed that magnons behave as weakly interacting quasiparticles obeying Bose–Einstein statistics (bosons). A comprehensive treatment can be found in the solid state textbook by Charles Kittel or the early review article by Van Kranendonk and Van Vleck.

Direct experimental detection of magnons by inelastic neutron scattering in ferrite was achieved in 1957 by Bertram Brockhouse. Since then magnons have been detected in ferromagnets, ferrimagnets, and antiferromagnets.

The fact that magnons obey the Bose–Einstein statistics was confirmed by the light scattering experiments done during the 1960s through the 1980s. Classical theory predicts equal intensity of Stokes and anti-Stokes lines. However, the scattering showed that if the magnon energy is comparable to or smaller than the thermal energy, or $\hbar\omega < k_B T$, then the Stokes line becomes more intense, as follows from Bose–Einstein statistics. Bose–Einstein condensation of magnons was proven in an antiferromagnet at low temperatures by Nikuni *et al.* and in a ferrimagnet by Demokritov *et al.* at room temperature.[1] Recently Uchida *et al.* reported the generation of spin currents by surface plasmon resonance.[2]

99.2 Properties

Magnon behavior can be studied with a variety of scattering techniques. Magnons behave as a Bose gas with no chemical potential. Microwave pumping can be used to excite spin waves and create additional non-equilibrium magnons which thermalize into phonons. At a critical density, a condensate is formed, which appears as the emission of monochromatic microwaves. This microwave source can be tuned with an applied magnetic field.

99.3 See also

- Magnonics
- Spin wave

99.4 References

[1] "Bose–Einstein condensation of quasi-equilibrium magnons at room temperature under pumping". *Nature 443, 430-433.* 28 September 2006. Bibcode:2006Natur.443..430D. doi:10.1038/nature05117.

[2] Uchida, K.; Adachi, H.; Kikuchi, D.; Ito, S.; Qiu, Z.; Maekawa, S.; Saitoh, E. (January 8, 2015). "Generation of spin currents by surface plasmon resonance". *Nature Communications* **6**. arXiv:1308.3532. Bibcode:2015NatCo...6E5910U.do 0.PMC4354158. PMID 25569821.

- C. Kittel, *Introduction to Solid State Physics*, 7th edition (Wiley, 1995). ISBN 0-471-11181-3.
- F. Bloch, Z. Physik **61**, 206 (1930).
- T. Holstein and H. Primakoff, Phys. Rev. **58**, 1098 (1940). online
- F. J. Dyson, Phys. Rev. **102**, 1217 (1956). online
- B. N. Brockhouse, Phys. Rev. **106**, 859 (1957). online
- J. Van Kranendonk and J. H. Van Vleck, Rev. Mod. Phys. **30**, 1 (1958). online
- T. Nikuni, M. Oshikawa, A. Oosawa, and H. Tanaka, Phys. Rev. Lett. **84**, 5868 (1999). online
- S. O. Demokritov, V. E. Demidov, O. Dzyapko, G. A. Melkov, A. A. Serga, B. Hillebrands, and A. N. Slavin, Nature **443**, 430 (2006).online
- P. Schewe and B. Stein, Physics News Update **746**, 2 (2005). online
- A.V. Kimel, A. Kirilyuk and T.H. Rasing, Laser & Photon Rev. 1, No. 3, 275–287 (2007). online

Chapter 100

Phonon

For KDE Software Compilation 4's and KDE Frameworks 5's multimedia framework, see Phonon (software).
For Phonon Communications or Phonon.in company, see Phonon Communications.

In physics, a **phonon** is a collective excitation in a periodic, elastic arrangement of atoms or molecules in condensed matter, like solids and some liquids. Often designated a quasiparticle,[1] it represents an excited state in the quantum mechanical quantization of the modes of vibrations of elastic structures of interacting particles.

Phonons play a major role in many of the physical properties of condensed matter, like thermal conductivity and electrical conductivity. The study of phonons is an important part of condensed matter physics.

The concept of phonons was introduced in 1932 by Soviet physicist Igor Tamm. The name *phonon* comes from the Greek word *φωνή* (phonē), which translates to *sound* or *voice* because long-wavelength phonons give rise to sound. Shorter-wavelength higher-frequency phonons give rise to heat.

100.1 Definition

A phonon is a quantum mechanical description of an elementary vibrational motion in which a lattice of atoms or molecules uniformly oscillates at a single frequency. In classical mechanics this is designated a normal mode. Normal modes are important because any arbitrary lattice vibration can be considered to be a superposition of these *elementary* vibrations (cf. Fourier analysis). While normal modes are wave-like phenomena in classical mechanics, phonons have particle-like properties too, in a way related to the wave–particle duality of quantum mechanics.

100.2 Lattice dynamics

The equations in this section either do not use axioms of quantum mechanics or use relations for which there exists a direct correspondence in classical mechanics.

For example, a rigid regular, crystalline, i.e. not amorphous, lattice is composed of *N* particles. These particles may be atoms, but they may be molecules as well. *N* is a large number, say ~10^{23}, and on the order of Avogadro's number, for a typical sample of solid. If the lattice is rigid, the atoms must be exerting forces on one another to keep each atom near its equilibrium position. These forces may be Van der Waals forces, covalent bonds, electrostatic attractions, and others, all of which are ultimately due to the electric force. Magnetic and gravitational forces are generally negligible. The forces between each pair of atoms may be characterized by a potential energy function V that depends on the distance of separation of the atoms. The potential energy of the entire lattice is the sum of all pairwise potential energies:[2]

$$\sum_{i\neq j} V(r_i - r_j)$$

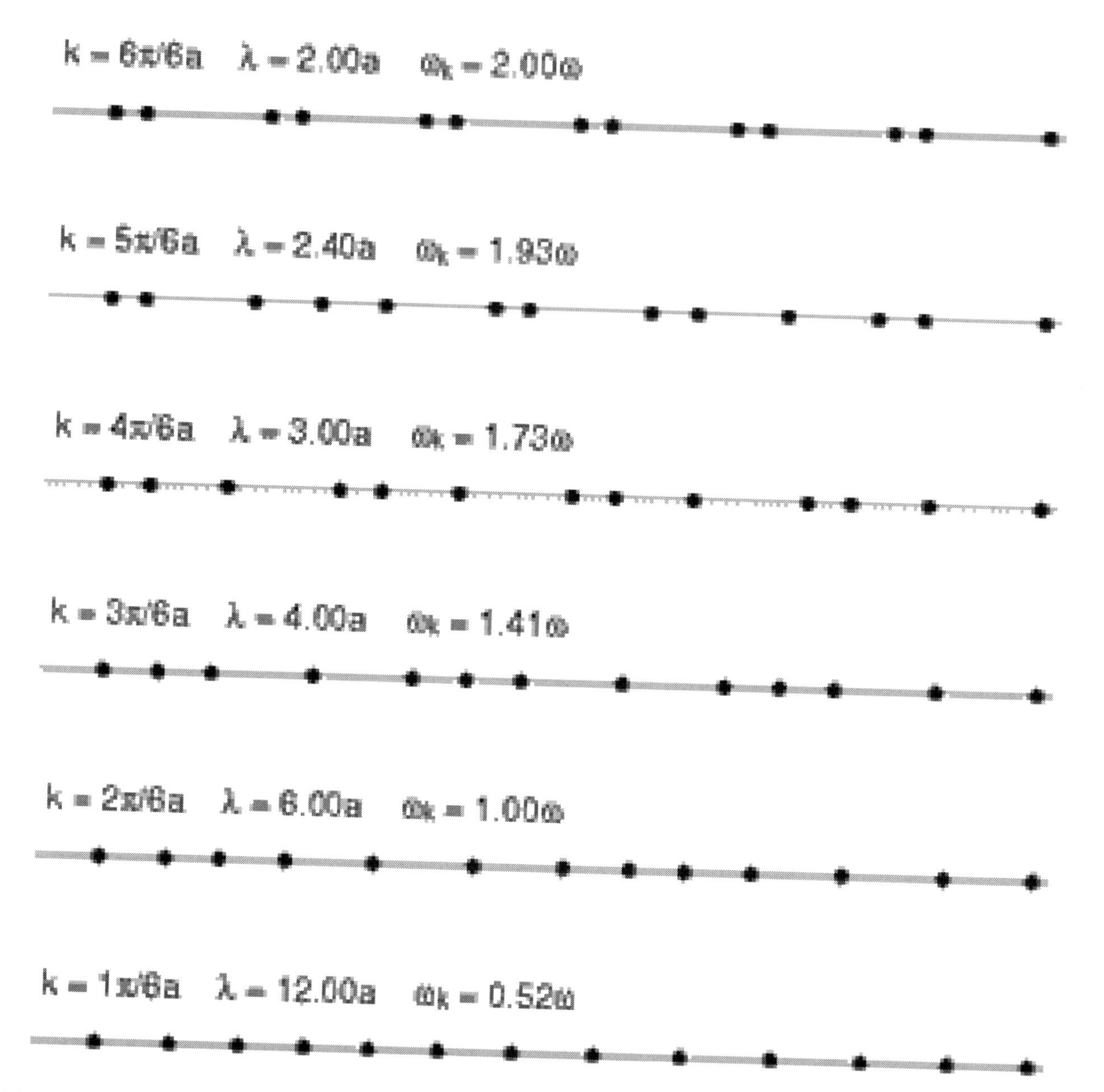

Normal modes of vibration progression through a crystal. The amplitude of the motion has been exaggerated for ease of viewing; in an actual crystal, it is typically much smaller than the lattice spacing.

where r_i is the position of the i th atom, and V is the potential energy between two atoms.

It is difficult to solve this many-body problem in full generality, in either classical or quantum mechanics. In order to simplify the task, two important approximations are usually imposed. First, the sum is only performed over neighboring atoms. Although the electric forces in real solids extend to infinity, this approximation is nevertheless valid because the fields produced by distant atoms are effectively screened. Secondly, the potentials V are treated as harmonic potentials. This is permissible as long as the atoms remain close to their equilibrium positions. Formally, this is accomplished by Taylor expanding V about its equilibrium value to quadratic order, giving V proportional to the displacement x^2 and the elastic force simply proportional to x . The error in ignoring higher order terms remains small if x remains close to the equilibrium position.

The resulting lattice may be visualized as a system of balls connected by springs. The following figure shows a cubic lattice, which is a good model for many types of crystalline solid. Other lattices include a linear chain, which is a very simple lattice which we will shortly use for modeling phonons. Other common lattices may be found under "crystal structure".

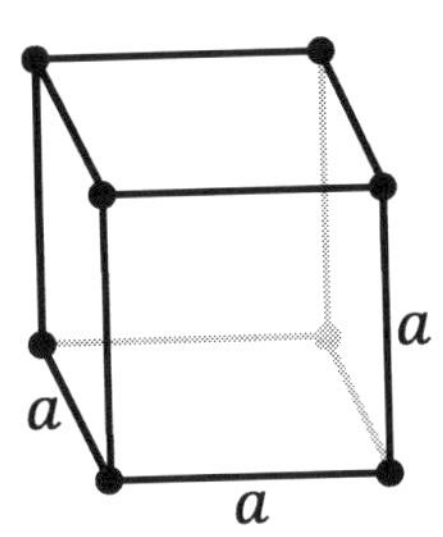

The potential energy of the lattice may now be written as

$$\sum_{\{ij\}(nn)} \frac{1}{2} m\omega^2 (R_i - R_j)^2.$$

Here, ω is the natural frequency of the harmonic potentials, which are assumed to be the same since the lattice is regular. R_i is the position coordinate of the i th atom, which we now measure from its equilibrium position. The sum over nearest neighbors is denoted as *(nn)*.

100.2.1 Lattice waves

Due to the connections between atoms, the displacement of one or more atoms from their equilibrium positions give rise to a set of vibration waves propagating through the lattice. One such wave is shown in the figure to the right. The amplitude of the wave is given by the displacements of the atoms from their equilibrium positions. The wavelength λ is marked.

There is a minimum possible wavelength, given by twice the equilibrium separation *a* between atoms. Any wavelength shorter than this can be mapped onto a wavelength longer than $2a$, due to the periodicity of the lattice.

Not every possible lattice vibration has a well-defined wavelength and frequency. However, the normal modes do possess well-defined wavelengths and frequencies.

100.2.2 One-dimensional lattice

In order to simplify the analysis needed for a 3-dimensional lattice of atoms it is convenient to model a 1-dimensional lattice or linear chain. This model is complex enough to display the salient features of phonons.

Classical treatment

The forces between the atoms are assumed to be linear and nearest-neighbour, and they are represented by an elastic spring. Each atom is assumed to be a point particle and the nucleus and electrons move in step (adiabatic approximation).

n−1 n n+1 ← d →

$\cdots$ o++++++o++++++o++++++o++++++o++++++o++++++o++++++o++++++o++++++o $\cdots$

$\to\to$ $\to$ $\to\to\to$

u_{n-1} u_n u_{n+1}

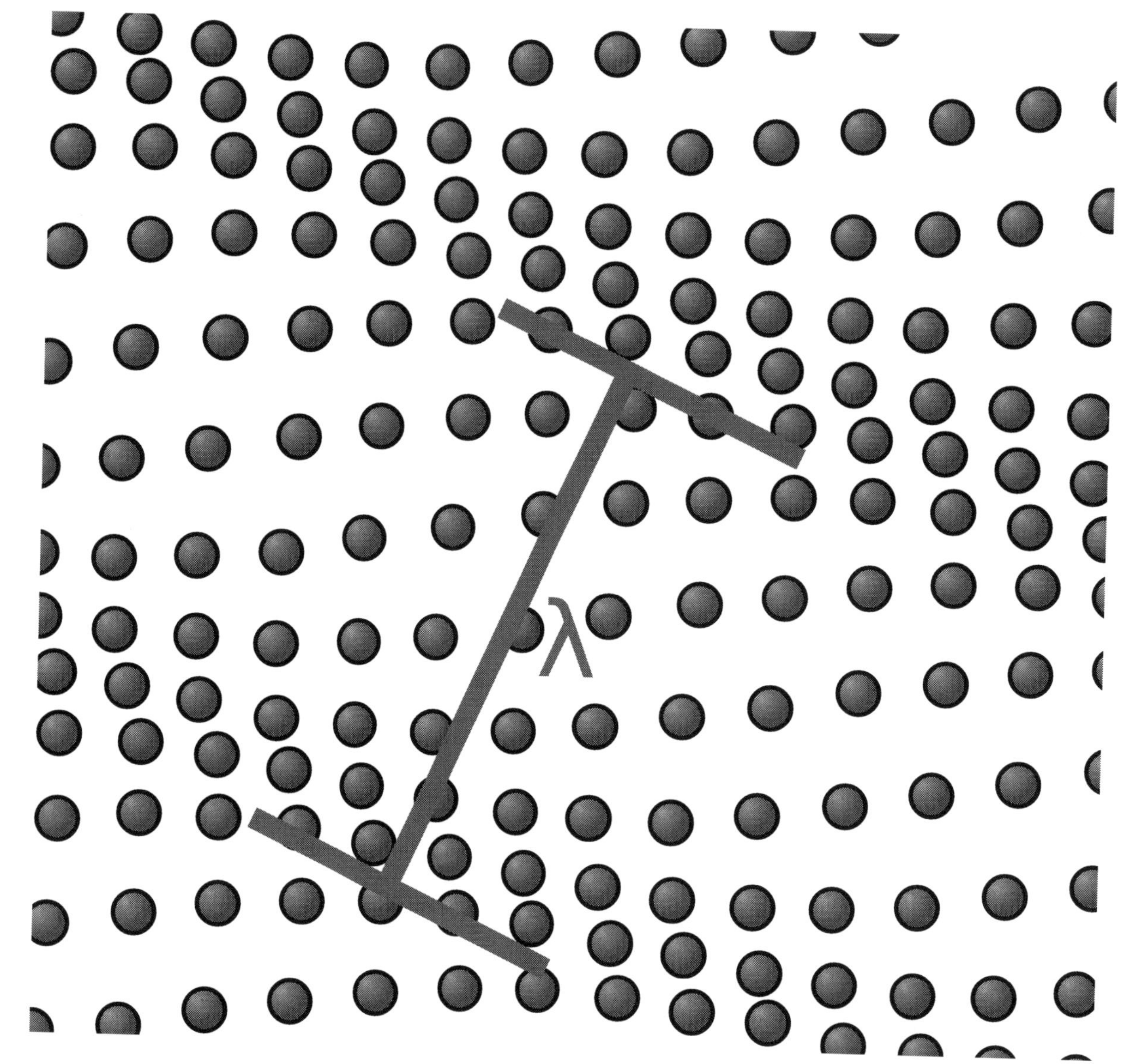

Phonon propagating through a square lattice (atom displacements greatly exaggerated)

Where n labels the n -th atom, d is the distance between atoms when the chain is in equilibrium and u_n the displacement of the n -th atom from its equilibrium position.
If C is the elastic constant of the spring and m the mass of the atom then the equation of motion of the n -th atom is :

$$-2Cu_n + C(u_{n+1} + u_{n-1}) = m\frac{\mathrm{d}^2 u_n}{\mathrm{d}t^2}$$

This is a set of coupled equations and since the solutions are expected to be oscillatory, new coordinates can be defined by a discrete Fourier transform, in order to de-couple them.[3]

Put

$$u_n = \sum_{k=1}^{N} U_k e^{iknd}$$

Here nd replaces the usual continuous variable x . The U_k are known as the normal coordinates. Substitution into the equation of motion produces the following decoupled equations.(This requires a significant manipulation using the orthonormality and completeness relations of the discrete Fourier transform [4])

$$2C(\cos kd - 1)U_k = m\frac{\mathrm{d}^2 U_k}{\mathrm{d}t^2}$$

These are the equations for harmonic oscillators which have the solution:

$$U_k = A_k e^{i\omega_k t}; \qquad \omega_k = \sqrt{\frac{2C}{m}(1 - \cos kd)}$$

Each normal coordinate U_k represents an independent vibrational mode of the lattice with wavenumber k which is known as a normal mode. The second equation for ω_k is known as the dispersion relation between the angular frequency and the wavenumber.[5]

Quantum treatment

A one-dimensional quantum mechanical harmonic chain consists of N identical atoms. This is the simplest quantum mechanical model of a lattice that allows phonons to arise from it. The formalism for this model is readily generalizable to two and three dimensions.

As in the previous section, the positions of the masses are denoted by $x_1, x_2, \ldots$, as measured from their equilibrium positions (i.e. $x_i = 0$ if particle i is at its equilibrium position.) In two or more dimensions, the x_i are vector quantities. The Hamiltonian for this system is

$$\mathbf{H} = \sum_{i=1}^{N} \frac{p_i^2}{2m} + \frac{1}{2}m\omega^2 \sum_{\{ij\}(nn)} (x_i - x_j)^2$$

where m is the mass of each atom (assuming is equal for all), and x_i and p_i are the position and momentum operators for the i th atom and the sum is made over the nearest neighbors (nn). However one expects that in a lattice there could also appear waves that behave like particles. It is customary to deal with waves in Fourier space which uses normal modes of the wavevector as variables instead coordinates of particles. The number of normal modes is same as the number of particles. However, the Fourier space is very useful given the periodicity of the system.

A set of N "normal coordinates" Q_k may be introduced, defined as the discrete Fourier transforms of the x 's and N "conjugate momenta" Π defined as the Fourier transforms of the p 's:

$$Q_k = \frac{1}{\sqrt{N}} \sum_l e^{ikal} x_l$$

$$\Pi_k = \frac{1}{\sqrt{N}} \sum_l e^{-ikal} p_l.$$

The quantity k_n turns out to be the wave number of the phonon, i.e. 2π divided by the wavelength.

This choice retains the desired commutation relations in either real space or wave vector space

$$[x_l, p_m] = i\hbar\delta_{l,m}$$

$$[Q_k, \Pi_{k'}] = \frac{1}{N}\sum_{l,m} e^{ikal}e^{-ik'am}[x_l, p_m]$$

$$= \frac{i\hbar}{N}\sum_{l} e^{ial(k-k')} = i\hbar\delta_{k,k'}$$

$$[Q_k, Q_{k'}] = [\Pi_k, \Pi_{k'}] = 0$$

From the general result

$$\sum_l x_l x_{l+m} = \frac{1}{N}\sum_{kk'} Q_k Q_k' \sum_l e^{ial(k+k')}e^{iamk'} = \sum_k Q_k Q_{-k} e^{iamk}$$

$$\sum_l {p_l}^2 = \sum_k \Pi_k \Pi_{-k}$$

The potential energy term is

$$\frac{1}{2}m\omega^2\sum_j (x_j - x_{j+1})^2 = \frac{1}{2}m\omega^2\sum_k Q_k Q_{-k}(2 - e^{ika} - e^{-ika}) = \frac{1}{2}\sum_k m{\omega_k}^2 Q_k Q_{-k}$$

where

$$\omega_k = \sqrt{2\omega^2\left[1 - \cos(ka)\right]} = 2\omega\left|\sin\left(\frac{ka}{2}\right)\right|$$

The Hamiltonian may be written in wave vector space as

$$\mathbf{H} = \frac{1}{2m}\sum_k \left(\Pi_k\Pi_{-k} + m^2\omega_k^2 Q_k Q_{-k}\right)$$

The couplings between the position variables have been transformed away; if the Q 's and Π 's were hermitian (which they are not), the transformed Hamiltonian would describe N *uncoupled* harmonic oscillators.

The form of the quantization depends on the choice of boundary conditions; for simplicity, *periodic* boundary conditions are imposed, defining the $(N+1)$ th atom as equivalent to the first atom. Physically, this corresponds to joining the chain at its ends. The resulting quantization is

$$k = k_n = \frac{2\pi n}{Na} \quad \text{for } n = 0, \pm 1, \pm 2, ..., \pm\frac{N}{2}.$$

The upper bound to n comes from the minimum wavelength, which is twice the lattice spacing a , as discussed above.

The harmonic oscillator eigenvalues or energy levels for the mode ω_k are :

$$E_n = \left(\frac{1}{2} + n\right)\hbar\omega_k \qquad n = 0, 1, 2, 3......$$

The levels are evenly spaced at:

$$\frac{1}{2}\hbar\omega, \quad \frac{3}{2}\hbar\omega, \quad \frac{5}{2}\hbar\omega \quad \ldots\ldots$$

Where $\frac{1}{2}\hbar\omega$ is the zero-point energy of a quantum harmonic oscillator.

An **exact** amount of energy $\hbar\omega$ must be supplied to the harmonic oscillator lattice to push it to the next energy level. In comparison to the photon case when the electromagnetic field is quantized, the quantum of vibrational energy is called a phonon.

All quantum systems show wave-like and particle-like properties simultaneously. The particle-like properties of the phonon are best understood using the methods of second quantization and operator techniques described later.[6]

100.2.3 Three-dimensional lattice

This may be generalized to a three-dimensional lattice. The wave number k is replaced by a three-dimensional wave vector **k**. Furthermore, each **k** is now associated with three normal coordinates.

The new indices s = *1, 2, 3* label the polarization of the phonons. In the one-dimensional model, the atoms were restricted to moving along the line, so the phonons corresponded to longitudinal waves. In three dimensions, vibration is not restricted to the direction of propagation, and can also occur in the perpendicular planes, like transverse waves. This gives rise to the additional normal coordinates, which, as the form of the Hamiltonian indicates, we may view as independent species of phonons.

100.2.4 Dispersion relation

For a one-dimensional alternating array of two types of ion or atom of mass m_1, m_2 repeated periodically at a distance a, connected by springs of spring constant K, two modes of vibration result:[8]

$$\omega_{\pm}^2 = K\left(\frac{1}{m_1}+\frac{1}{m_2}\right) \pm K\sqrt{\left(\frac{1}{m_1}+\frac{1}{m_2}\right)^2 - \frac{4\sin^2(ka/2)}{m_1 m_2}},$$

where k is the wave-vector of the vibration related to its wavelength by $k=2\pi/\lambda$. The connection between frequency and wave-vector, $\omega=\omega(k)$, is known as a dispersion relation. The plus sign results in the so-called *optical* mode, and the minus sign to the *acoustic* mode. In the optical mode two adjacent different atoms move against each other, while in the acoustic mode they move together.

The speed of propagation of an acoustic phonon, which is also the speed of sound in the lattice, is given by the slope of the acoustic dispersion relation, $\frac{\partial \omega_k}{\partial k}$ (see group velocity.) At low values of k (i.e. long wavelengths), the dispersion relation is almost linear, and the speed of sound is approximately ωa, independent of the phonon frequency. As a result, packets of phonons with different (but long) wavelengths can propagate for large distances across the lattice without breaking apart. This is the reason that sound propagates through solids without significant distortion. This behavior fails at large values of k, i.e. short wavelengths, due to the microscopic details of the lattice.

For a crystal that has at least two atoms in its primitive cell (which may or may not be different), the dispersion relations exhibit two types of phonons, namely, optical and acoustic modes corresponding to the upper blue and lower red of curve in the diagram, respectively. The vertical axis is the energy or frequency of phonon, while the horizontal axis is the wave-vector. The boundaries at $-\pi/a$ and π/a are those of the first Brillouin zone.[8] It is also interesting that for a crystal with N (> 2) different atoms in a primitive cell, there are always three acoustic modes: one longitudinal acoustic mode and two transverse acoustic modes. The number of optical modes is $3N - 3$. The lower figure shows the dispersion relations for several phonon modes in GaAs as a function of wavevector ***k*** in the principal directions of its Brillouin zone.[7]

Many phonon dispersion curves have been measured by neutron scattering.

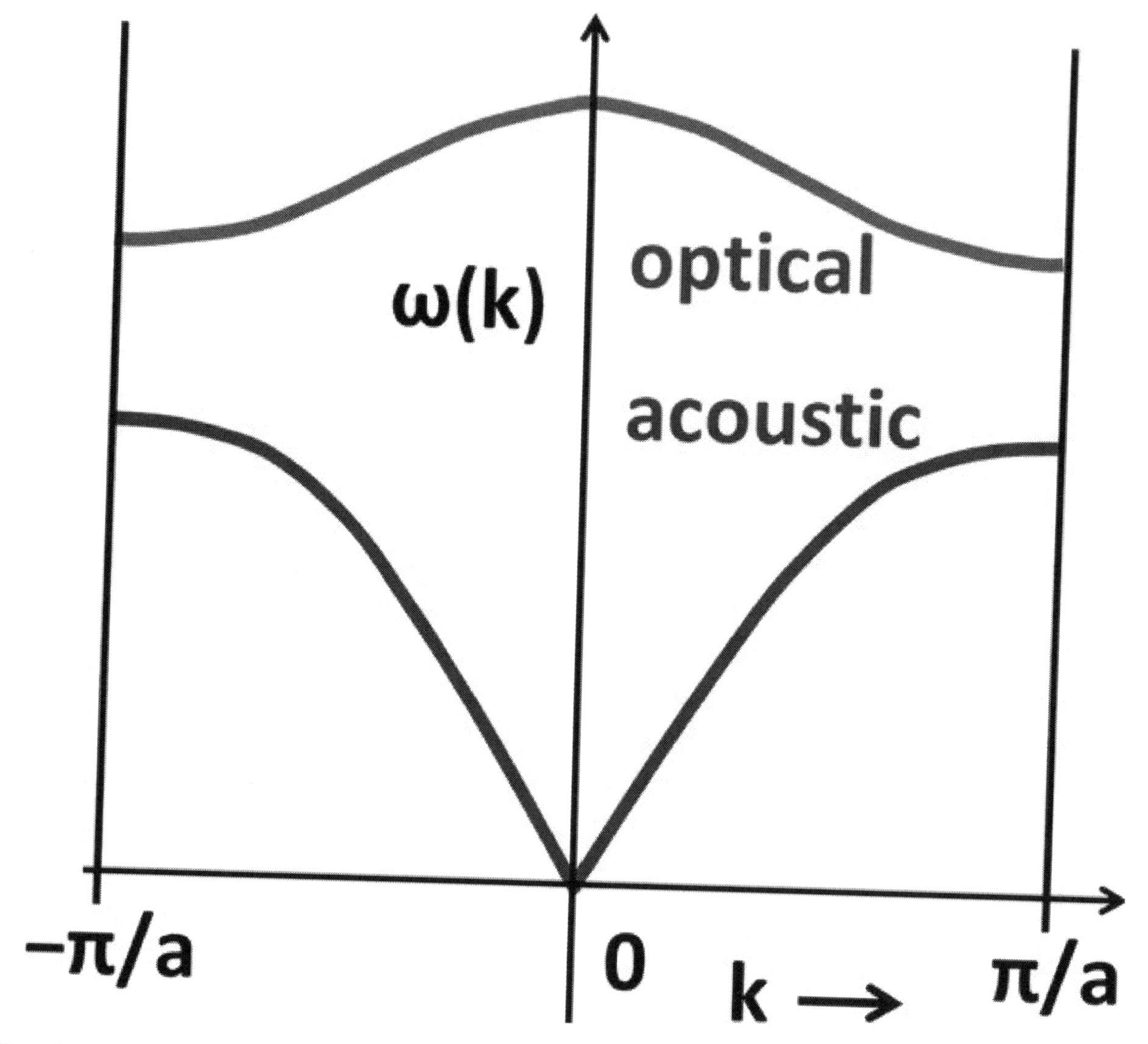

Dispersion curves in linear diatomic chain

The physics of sound in fluids differs from the physics of sound in solids, although both are density waves: sound waves in fluids only have longitudinal components, whereas sound waves in solids have longitudinal and transverse components. This is because fluids can't support shear stresses (but see viscoelastic fluids, which only apply to high frequencies, though).

100.2.5 Interpretation of phonons using second quantization techniques

In fact, the above-derived Hamiltonian looks like the classical Hamiltonian function, but if it is interpreted as an operator, then it describes a quantum field theory of non-interacting bosons.

The energy spectrum of this Hamiltonian is easily obtained by the method of ladder operators, similar to the quantum harmonic oscillator problem. We introduce a set of ladder operators defined by:

$$b_k = \frac{1}{\sqrt{2}}\left(\frac{Q_k}{l_k} + i\frac{\Pi_{-k}}{\hbar/l_k}\right) \quad , \quad Q_k = l_k\frac{1}{\sqrt{2}}({b_k}^\dagger + b_{-k})$$

$${b_k}^\dagger = \frac{1}{\sqrt{2}}\left(\frac{Q_{-k}}{l_k} - i\frac{\Pi_k}{\hbar/l_k}\right) \quad , \quad \Pi_k = \frac{\hbar}{l_k}\frac{i}{\sqrt{2}}({b_k}^\dagger - b_{-k})$$

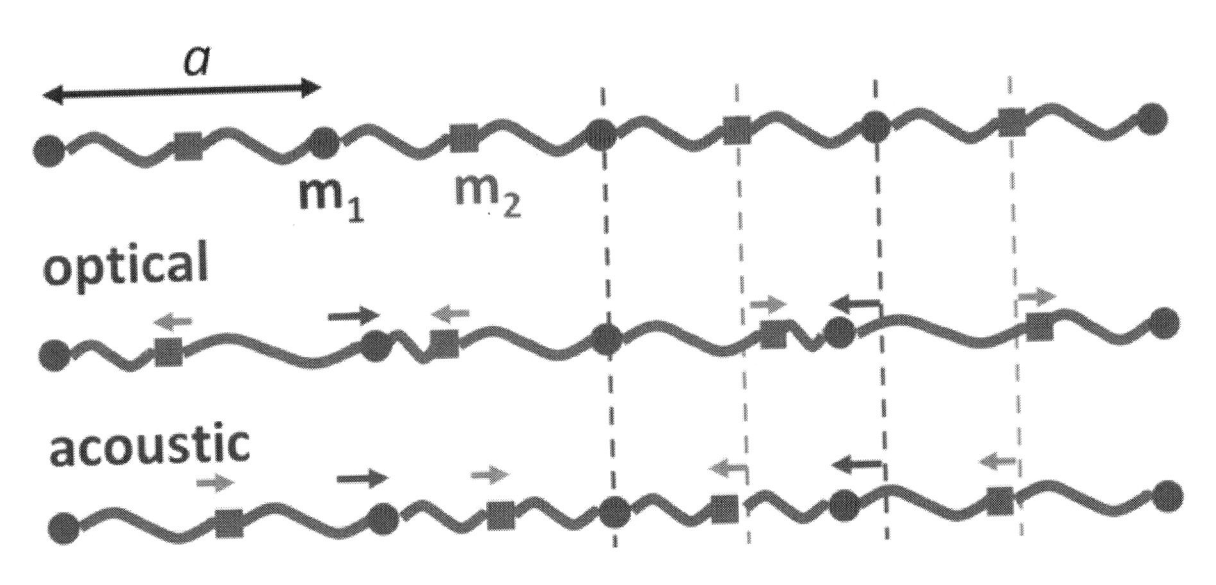

Optical and acoustic vibrations in linear diatomic chain.

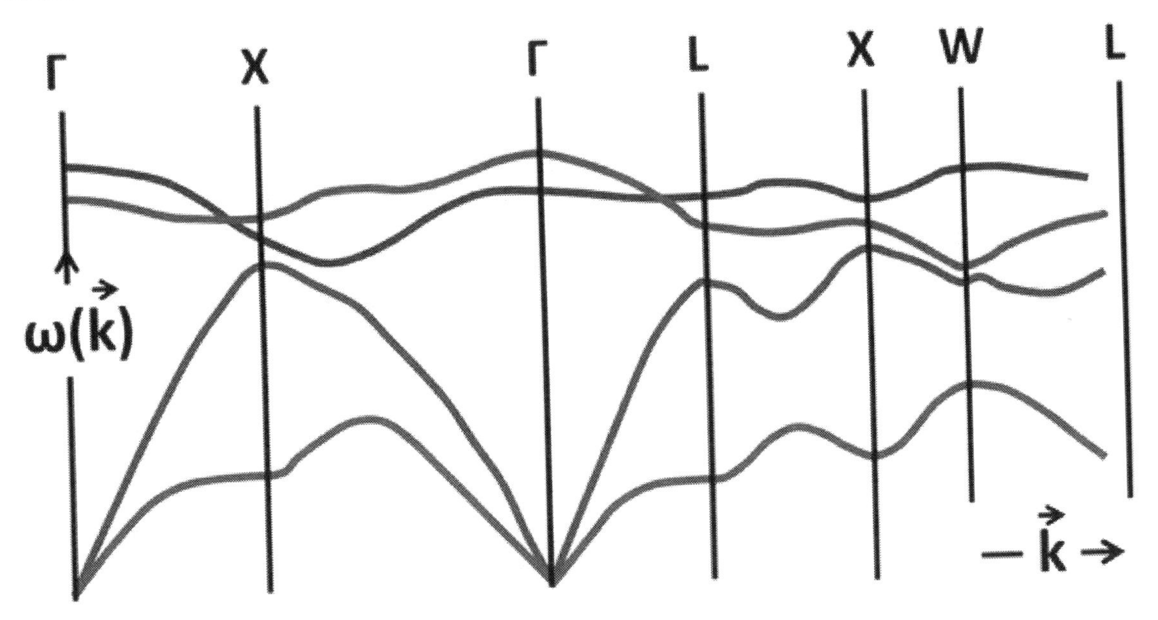

*Dispersion relation ω=ω(**k**) for some waves corresponding to lattice vibrations in GaAs.*[7]

$$l_k = \sqrt{\frac{\hbar}{m\omega_k}}$$

By direct insertion on the Hamiltonian, it is readily verified that

$$H_{ph} = \sum_k \hbar\omega_k \left({b_k}^\dagger b_k + \frac{1}{2} \right)$$

$[b_k, b_{k'}^\dagger] = \delta_{k,k'}, [b_k, b_{k'}] = [b_k^\dagger, b_{k'}^\dagger] = 0.$

As with the quantum harmonic oscillator, one can show that $b_k^\dagger$ and b_k respectively create and destroy one excitation of energy $\hbar\omega_k$. These excitations are phonons.

Two important properties of phonons may be deduced. Firstly, phonons are bosons, since any number of identical excitations can be created by repeated application of the creation operator $b_k^\dagger$. Secondly, each phonon is a "collective mode" caused by the motion of every atom in the lattice. This may be seen from the fact that the ladder operators contain sums over the position and momentum operators of every atom.

It is not *a priori* obvious that these excitations generated by the b operators are literally waves of lattice displacement, but one may convince oneself of this by calculating the *position-position correlation function*. Let $|k\rangle$ denote a state with a single quantum of mode k excited, i.e.

$$|k\rangle = b_k^\dagger|0\rangle.$$

One can show that, for any two atoms j and ℓ ,

$$\langle k|x_j(t)x_\ell(0)|k\rangle = \frac{\hbar}{Nm\omega_k}\cos\left[k(j-\ell)a - \omega_k t\right] + \langle 0|x_j(t)x_\ell(0)|0\rangle$$

which is exactly what we would expect for a lattice wave with frequency ω_k and wave number k .

In three dimensions the Hamiltonian has the form

$$\mathbf{H} = \sum_k \sum_{s=1}^3 \hbar\omega_{k,s}\left(b_{k,s}^\dagger b_{k,s} + 1/2\right).$$

100.3 Acoustic and optical phonons

Solids with more than one atom in the smallest unit cell, exhibit two types of phonons: acoustic phonons and optical phonons.

Acoustic phonons are coherent movements of atoms of the lattice out of their equilibrium positions. If the displacement is in the direction of propagation, then in some areas the atoms will be closer, in others farther apart, as in a sound wave in air (hence the name acoustic). Displacement perpendicular to the propagation direction is comparable to waves in water. If the wavelength of acoustic phonons goes to infinity, this corresponds to a simple displacement of the whole crystal, and this costs zero energy. Acoustic phonons exhibit a linear relationship between frequency and phonon wavevector for long wavelengths. The frequencies of acoustic phonons tend to zero with longer wavelength. Longitudinal and transverse acoustic phonons are often abbreviated as LA and TA phonons, respectively.

Optical phonons are out-of-phase movement of the atoms in the lattice, one atom moving to the left, and its neighbour to the right. This occurs if the lattice basis consists of two or more atoms. They are called *optical* because in ionic crystals, like sodium chloride, they are excited by infrared radiation. The electric field of the light will move every positive sodium ion in the direction of the field, and every negative chloride ion in the other direction, sending the crystal vibrating. Optical phonons have a non-zero frequency at the Brillouin zone center and show no dispersion near that long wavelength limit. This is because they correspond to a mode of vibration where positive and negative ions at adjacent lattice sites swing against each other, creating a time-varying electrical dipole moment. Optical phonons that interact in this way with light are called *infrared active*. Optical phonons that are *Raman active* can also interact indirectly with light, through Raman scattering. Optical phonons are often abbreviated as LO and TO phonons, for the longitudinal and transverse modes respectively; the splitting between LO and TO frequencies is often described accurately by the Lyddane–Sachs–Teller relation.

When measuring optical phonon energy by experiment, optical phonon frequencies are sometimes given in spectroscopic wavenumber notation, where the symbol ω represents ordinary frequency (not angular frequency), and is expressed in units of cm^{-1}. The value is obtained by dividing the frequency by the speed of light in vacuum. In other words, the frequency in cm^{-1} units corresponds to the inverse of the wavelength of a photon in vacuum, that has the same frequency as the measured phonon.[9] The cm^{-1} is a unit of energy used frequently in the dispersion relations of both acoustic and optical phonons, see units of energy for more details and uses.

100.4 Crystal momentum

Main article: Crystal momentum

By analogy to photons and matter waves, phonons have been treated with wave vector k as though it has a momentum $\hbar k$

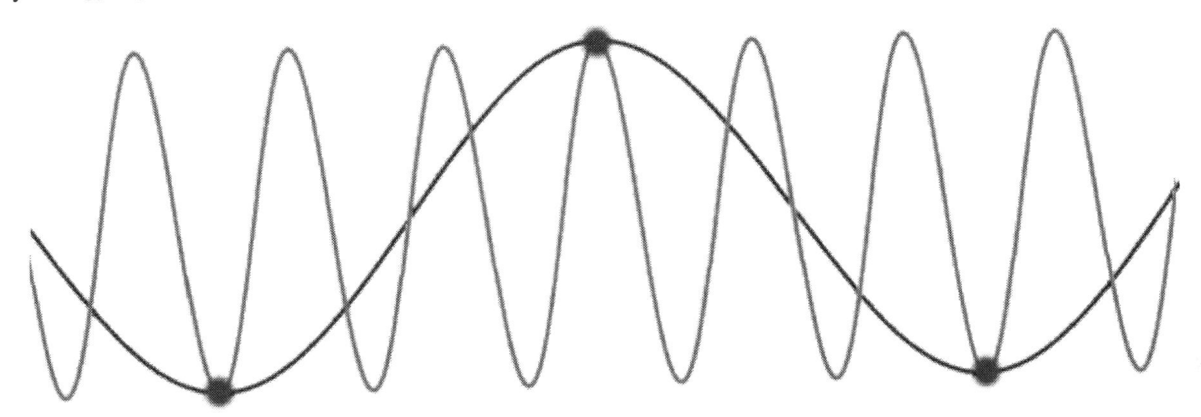

k-vectors exceeding the first Brillouin zone (red) do not carry any more information than their counterparts (black) in the first Brillouin zone.

, however, this is not strictly correct, because $\hbar k$ is not actually a physical momentum; it is called the *crystal momentum* or *pseudomomentum*. This is because k is only determined up to addition of constant vectors (the reciprocal lattice vectors and integer multiples thereof). For example, in the one-dimensional model, the normal coordinates Q and Π are defined so that

$$Q_k \stackrel{\text{def}}{=} Q_{k+K} \quad ; \quad \Pi_k \stackrel{\text{def}}{=} \Pi_{k+K}$$

where

$$K = 2n\pi/a$$

for any integer n. A phonon with wave number k is thus equivalent to an infinite "family" of phonons with wave numbers $k \pm \frac{2\pi}{a}$, $k \pm \frac{4\pi}{a}$, and so forth. Physically, the reciprocal lattice vectors act as additional "chunks" of momentum which the lattice can impart to the phonon. Bloch electrons obey a similar set of restrictions.

It is usually convenient to consider phonon wave vectors k which have the smallest magnitude ($|k|$) in their "family". The set of all such wave vectors defines the *first Brillouin zone*. Additional Brillouin zones may be defined as copies of the first zone, shifted by some reciprocal lattice vector.

100.5 Thermodynamics

The thermodynamic properties of a solid are directly related to its phonon structure. The entire set of all possible phonons that are described by the above phonon dispersion relations combine in what is known as the phonon density of states which determines the heat capacity of a crystal.

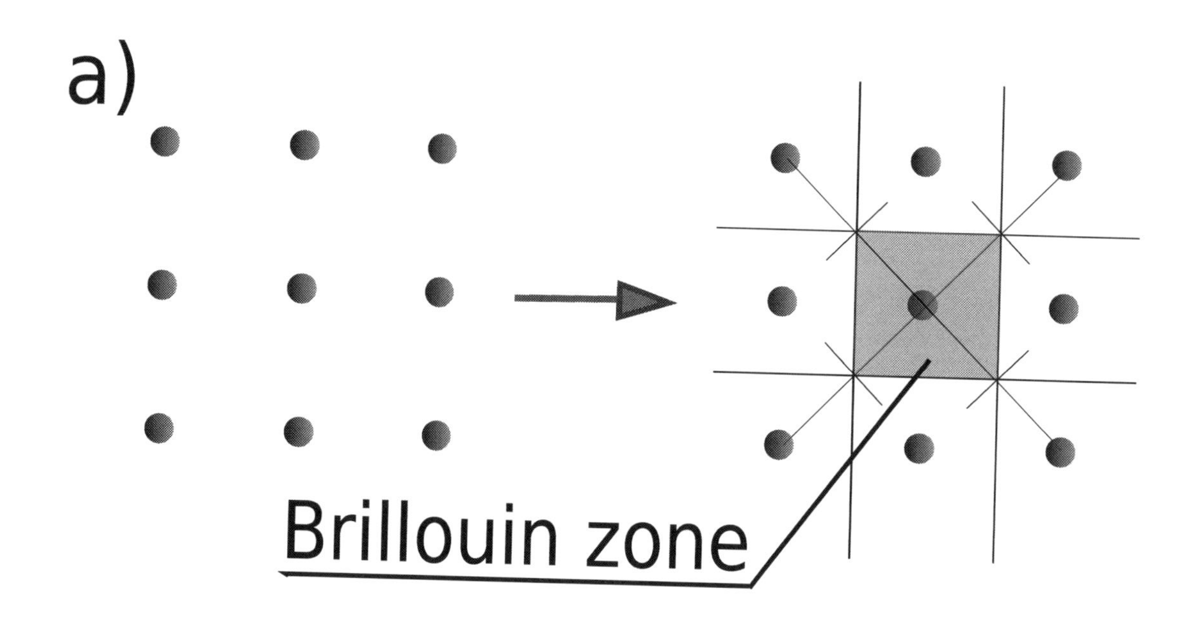

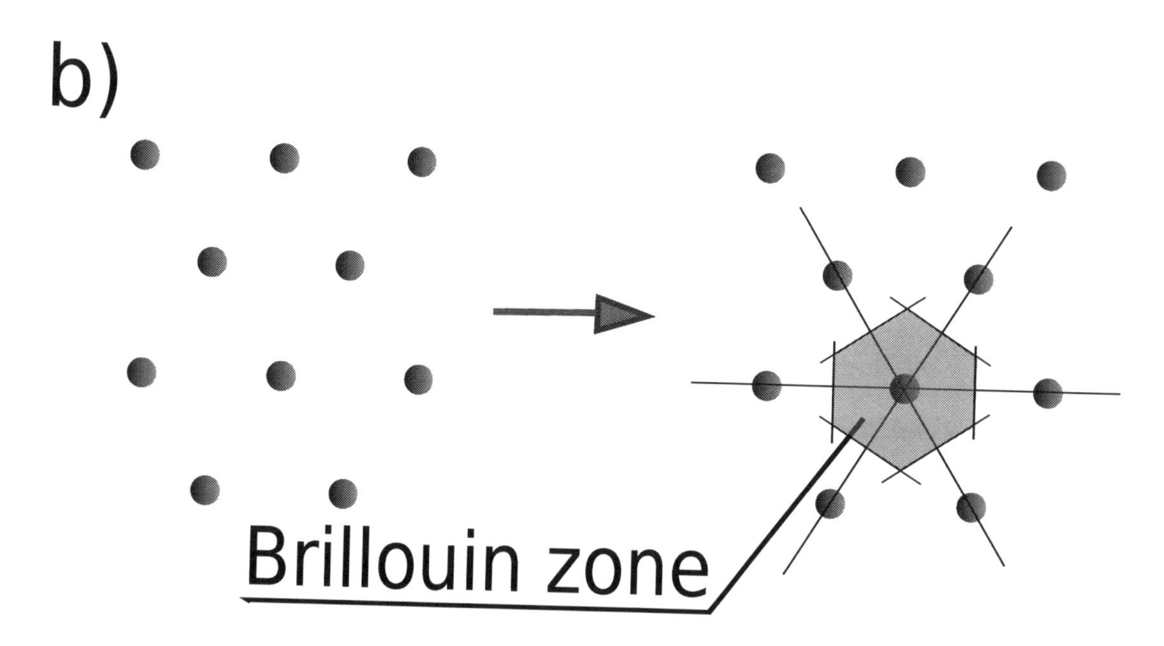

Brillouin zones, a) in a square lattice, and b) in a hexagonal lattice

At absolute zero temperature, a crystal lattice lies in its ground state, and contains no phonons. A lattice at a non-zero temperature has an energy that is not constant, but fluctuates randomly about some mean value. These energy fluctuations are caused by random lattice vibrations, which can be viewed as a gas of phonons. (The random motion of the atoms in the lattice is what we usually think of as heat.) Because these phonons are generated by the temperature of the lattice, they are sometimes designated thermal phonons.

Unlike the atoms which make up an ordinary gas, thermal phonons can be created and destroyed by random energy fluctuations. In the language of statistical mechanics this means that the chemical potential for adding a phonon is zero. This behavior is an extension of the harmonic potential, mentioned earlier, into the anharmonic regime. The behavior of thermal phonons is similar to the photon gas produced by an electromagnetic cavity, wherein photons may be emitted or absorbed by the cavity walls. This similarity is not coincidental, for it turns out that the electromagnetic field behaves like a set of harmonic oscillators; see Black-body radiation. Both gases obey the Bose–Einstein statistics: in thermal equilibrium and within the harmonic regime, the probability of finding phonons (or photons) in a given state with a given angular frequency is:

$$n(\omega_{k,s}) = \frac{1}{\exp(\hbar\omega_{k,s}/k_B T) - 1}$$

where $\omega_{k,s}$ is the frequency of the phonons (or photons) in the state, k_B is Boltzmann's constant, and T is the temperature.

100.6 Operator formalism

The phonon Hamiltonian is given by

$$\mathbf{H} = \frac{1}{2}\sum_{\alpha}(p_\alpha^2 + \omega_\alpha^2 q_\alpha^2 - \frac{1}{2}\hbar\omega_\alpha)$$

In terms of the operators, these are given by

$$\mathbf{H} = \sum_{\alpha} \hbar\omega_\alpha a_\alpha^\dagger a_\alpha$$

Here, in expressing the Hamiltonian in operator formalism, we have not taken into account the $\frac{1}{2}\hbar\omega_q$ term, since if we take an infinite lattice or, for that matter a continuum, the $\frac{1}{2}\hbar\omega_q$ terms will add up giving an infinity. Hence, it is "renormalized" by putting the factor of $\frac{1}{2}\hbar\omega_q$ to 0 arguing that the difference in energy is what we measure and not the absolute value of it. Hence, the $\frac{1}{2}\hbar\omega_q$ factor is absent in the operator formalised expression for the Hamiltonian.
The ground state also called the "vacuum state" is the state composed of no phonons. Hence, the energy of the ground state is 0. When, a system is in state $|n_1 n_2 n_3 ...\rangle$, we say there are n_α phonons of type α . The n_α are called the occupation number of the phonons. Energy of a single phonon of type α being $\hbar\omega_q$, the total energy of a general phonon system is given by $n_1\hbar\omega_1 + n_2\hbar\omega_2 + ...$. In other words, the phonons are non-interacting. The action of creation and annihilation operators are given by

$$a_\alpha^\dagger |n_1 ... n_{\alpha-1} n_\alpha n_{\alpha+1} ...\rangle = \sqrt{n_\alpha + 1}|n_1 ..., n_{\alpha-1}, n_\alpha + 1, n_{\alpha+1} ...\rangle$$

and,

$$a_\alpha |n_1 ... n_{\alpha-1} n_\alpha n_{\alpha+1} ...\rangle = \sqrt{n_\alpha}|n_1 ..., n_{\alpha-1}, (n_\alpha - 1), n_{\alpha+1}, ...\rangle$$

i.e. $a_\alpha^\dagger$ creates a phonon of type α while a_α annihilates. Hence, they are respectively the creation and annihilation operator for phonons. Analogous to the Quantum harmonic oscillator case, we can define particle number operator as $N = \sum_\alpha a_\alpha^\dagger a_\alpha$. The number operator commutes with a string of products of the creation and annihilation operators if, the number of a 's are equal to number of $a^\dagger$'s.
Phonons are bosons since, $|\alpha, \beta\rangle = |\beta, \alpha\rangle$ i.e. they are symmetric under exchange.[10]

100.7 Nonlinearity

As well as photons, phonons can interact via parametric down conversion[11] and form squeezed coherent states.[12]

100.8 Phononic computing

As phonons carry information, it is theoretically possible to build a quantum computer using phonons.[13][14][15]

100.9 See also

- Boson
- Brillouin scattering
- Fracton
- Linear elasticity
- Phonon scattering
- Phononic crystal
- Rayleigh wave
- Relativistic heat conduction
- Rigid unit modes
- SASER
- Second sound
- Surface acoustic wave
- Surface phonon
- Thermal conductivity

100.10 References

[1] F. Schwabl, *Advanced Quantum Mechanics*, 4th Ed., Springer (2008), p. 253

[2] Krauth, Werner (April 2006). *Statistical mechanics: algorithms and computations*. International publishing locations: Oxford University Press. pp. 231–232. ISBN 978-0-19-851536-4.

[3] Mattuck R. A guide to Feynman Diagrams in the many-body problem

[4] Greiner & Reinhardt. Field Quantisation

[5] Donovan B. & Angress J.; Lattice Vibrations

[6] Mahan, GD (1981). *many particle physics*. New York: springer. ISBN 0306463385.

[7] Peter Y. Yu, Manuel Cardona (2010). "Fig. 3.2: Phonon dispersion curves in GaAs along high-symmetry axes". *Fundamentals of Semiconductors: Physics and Materials Properties* (4th ed.). Springer. p. 111. ISBN 3-642-00709-0.

[8] For a discussion see Prasanta Kumar Misra (2010). "§2.1.3 Normal modes of a one-dimensional chain with a basis". *Physics of Condensed Matter*. Academic Press. pp. 44 *ff*. ISBN 0-12-384954-3.

[9] Mahan, Gerald (2010). *Condensed Matter in a Nutshell*. Princeton: Princeton University Press. ISBN 0-691-14016-2.

[10] Feynman, Richard P. (1982). *Statistical Mechanics, A Set of Lectures*. Reading, Massachusetts: The Benjamin/Cummings Publishing Company, Inc. p. 159. ISBN 0-8053-2508-5.

[11] Phonon-phonon interactions due to non-linear effects in a linear ion trap

[12] Reiter, D E et al. (2009). "Generation of squeezed phonon states by optical excitation of a quantum dot" (PDF). Institute of Physics. Retrieved February 28, 2014.

[13] "How To Build A Phononic Computer". *www.technologyreview.com* (MIT). August 13, 2012. Retrieved 23 June 2014.

[14] "Thermal Memory: A Storage of Phononic Information."

[15] "'Phononic Computer' Could Process Information with Heat."

100.11 External links

- Explained: Phonons, MIT News, 2010.
- Optical and acoustic modes
- Phonons in a One Dimensional Microfluidic Crystal and with movies in .

Chapter 101

Plasmaron

In physics, a **plasmaron** is a quasiparticle arising in a system that has strong plasmon-electron interactions. It is a quasiparticle formed by quasiparticle-quasiparticle interactions, since both plasmons and electron holes are collective modes of different kinds. It has recently been observed in graphene and earlier in elemental bismuth.[1][2]

101.1 References

[1] Bostwick; Speck, F.; Seyller, T.; Horn, K.; Polini, M.; Asgari, R.; MacDonald, A. H.; Rotenberg, E. et al. (2010-05-21). "Observation of Plasmarons in Quasi-Freestanding Doped Graphene". *Science* **328** (5981): 999–1002. Bibcode:2010Sci...32 .doi:10.1126/science.1186489. PMID 20489018. Retrieved2010-05-25.

[2] Riccardo Tediosi, N. P. Armitage, E. Giannini, and D. van der Marel (1971-09-13). "Charge Carrier Interaction with a Purely Electronic Collective Mode: Plasmarons and the Infrared Response of Elemental Bismuth". *Phys. Rev. Lett.* **27** (11): 711–714. arXiv:cond-mat/0701447. Bibcode:2007PhRvL..99a6406T. doi:10.1103/PhysRevLett.99.016406. Retrieved 2010-06-17.

Chapter 102

Plasmon

This article is about the physics particle. For the brand of dried milk biscuit, see Plasmon biscuit.

In physics, a **plasmon** is a quantum of plasma oscillation. As the light consists of photons, the plasma oscillation consists of plasmons. The plasmon can be considered a quasiparticle since it arises from the quantization of plasma oscillations, just like phonons are quantizations of mechanical vibrations. Thus, plasmons are collective (a discrete number) oscillations of the free electron gas density, for example, at optical frequencies. Plasmons can couple with a photon to create another quasiparticle called a plasma polariton.

102.1 Derivation

The plasmon was initially proposed in 1952 by David Pines and David Bohm[1] and was shown to arise from a Hamiltonian for the long-range electron-electron correlations.[2]

Since plasmons are the quantization of classical plasma oscillations, most of their properties can be derived directly from Maxwell's equations.

102.2 Explanation

Plasmons can be described in the classical picture as an oscillation of free electron density with respect to the fixed positive ions in a metal. To visualize a plasma oscillation, imagine a cube of metal placed in an external electric field pointing to the right. Electrons will move to the left side (uncovering positive ions on the right side) until they cancel the field inside the metal. If the electric field is removed, the electrons move to the right, repelled by each other and attracted to the positive ions left bare on the right side. They oscillate back and forth at the plasma frequency until the energy is lost in some kind of resistance or damping. Plasmons are a quantization of this kind of oscillation.

102.2.1 Role of plasmons

Plasmons play a large role in the optical properties of metals. Light of frequencies below the plasma frequency is reflected, because the electrons in the metal screen the electric field of the light. Light of frequencies above the plasma frequency is transmitted, because the electrons cannot respond fast enough to screen it. In most metals, the plasma frequency is in the ultraviolet, making them shiny (reflective) in the visible range. Some metals, such as copper[3] and gold,[4] have electronic interband transitions in the visible range, whereby specific light energies (colors) are absorbed, yielding their distinct color. In semiconductors, the valence electron plasma frequency is usually in the deep ultraviolet,[5][6] which is why they are reflective.

The plasmon energy can often be estimated in the free electron model as

$$E_p = \hbar \sqrt{\frac{ne^2}{m\epsilon_0}} = \hbar\, \omega_p,$$

where n is the conduction electron density, e is the elementary charge, m is the electron mass, ϵ_0 the permittivity of free space, $\hbar$ the reduced Planck constant and ω_p the plasmon frequency.

102.3 Surface plasmons

Main article: Surface plasmon

Surface plasmons are those plasmons that are confined to surfaces and that interact strongly with light resulting in a polariton.[7] They occur at the interface of a vacuum and material with a small positive imaginary and large negative real dielectric constant (usually a metal or doped dielectric). They play a role in surface-enhanced Raman spectroscopy and in explaining anomalies in diffraction from metal gratings (Wood's anomaly), among other things. Surface plasmon resonance is used by biochemists to study the mechanisms and kinetics of ligands binding to receptors (i.e. a substrate binding to an enzyme).

Surface plasmon may also be observed in the X-ray emission spectra of metals. A dispersion relation of surface plasmon in the X-ray emission spectra of metals has been derived (Harsh and Agarwal).[8]

More recently surface plasmons have been used to control colors of materials.[9] This is possible since controlling the particle's shape and size determines the types of surface plasmons that can couple to it and propagate across it. This in turn controls the interaction of light with the surface. These effects are illustrated by the historic stained glass which adorn medieval cathedrals. In this case, the color is given by metal nanoparticles of a fixed size which interact with the optical field to give the glass its vibrant color. In modern science, these effects have been engineered for both visible light and microwave radiation. Much research goes on first in the microwave range because at this wavelength material surfaces can be produced mechanically as the patterns tend to be of the order a few centimeters. To produce optical range surface plasmon effects involves producing surfaces which have features <400 nm. This is much more difficult and has only recently become possible to do in any reliable or available way.

Recently, graphene has also shown to accommodate surface plasmons, observed via near field infrared optical microscopy techniques[10][11] and infrared spectroscopy.[12] Potential applications of graphene plasmonics mainly addressed the terahertz to midinfrared frequencies, such as optical modulators, photodetectors, biosensors.[13]

102.4 Possible applications

Position and intensity of plasmon absorption and emission peaks are affected by molecular adsorption, which can be used in molecular sensors. For example, a fully operational prototype device detecting casein in milk has been fabricated. The device is based on detecting a change in absorption of a gold layer.[14] Localized surface plasmons of metal nanoparticles can be used for sensing different types molecules, proteins, etc.

Plasmons are being considered as a means of transmitting information on computer chips, since plasmons can support much higher frequencies (into the 100 THz range, while conventional wires become very lossy in the tens of GHz). However, for plasmon-based electronics to be useful, the analog to the transistor, called a plasmonster, first needs to be created.[15]

Plasmons have also been proposed as a means of high-resolution lithography and microscopy due to their extremely small wavelengths. Both of these applications have seen successful demonstrations in the lab environment. Finally, surface plasmons have the unique capacity to confine light to very small dimensions which could enable many new applications.

Surface plasmons are very sensitive to the properties of the materials on which they propagate. This has led to their use to measure the thickness of monolayers on colloid films, such as screening and quantifying protein binding events.

Gothic stained glass rose window of Notre-Dame de Paris. The colors were achieved by colloids of gold nano-particles.

Companies such as Biacore have commercialized instruments which operate on these principles. Optical surface plasmons are being investigated with a view to improve makeup by L'Oréal among others.[16]

In 2009, a Korean research team found a way to greatly improve organic light-emitting diode efficiency with the use of plasmons.[17]

A group of European researchers led by IMEC has begun work to improve solar cell efficiencies and costs through incorporation of metallic nanostructures (using plasmonic effects) that can enhance absorption of light into different types of solar cells: crystalline silicon (c-Si), high-performance III-V, organic, and dye-sensitized solar cells. [18] However, in order for plasmonic solar photovoltaic devices to function optimally ultra-thin transparent conducting oxides are necessary.[19] Full color holograms using *plasmonics*[20] have been demonstrated.

102.5 See also

- Surface plasmon resonance
- Waves in plasmas
- Plasma oscillation
- Spinplasmonics
- Transformation optics
- Extraordinary optical transmission
- Phonon
- List of plasma (physics) articles

102.6 References

[1] David Pines, David Bohm: *A Collective Description of Electron Interactions: II. Collective vs Individual Particle Aspects of the Interactions*, Phys. Rev. 85, 338, 15 January 1952. Cited after: Dror Sarid; William Challener (6 May 2010). *Modern Introduction to Surface Plasmons: Theory, Mathematica Modeling, and Applications*. Cambridge University Press. p. 1. ISBN 978-0-521-76717-0.

[2] David Bohm, David Pines: *A Collective Description of Electron Interactions: III. Coulomb Interactions in a Degenerate Electron Gas*, Phys. Rev. 92, 609, 1 November 1953. Cited after: N. J. Shevchik, *Alternative derivation of the Bohm-Pines theory of electron-electron interactions*, 1974, J. Phys. C: Solid State Phys., vol. 7, pp. 3930 ff., DOI 10.1088/0022-3719/7/21/013 (abstract)

[3] Burdick, Glenn (1963). "Energy Band Structure of Copper". *Physical Review* **129**: 138–150. Bibcode:1963PhRv..129..138B. doi:10.1103/PhysRev.129.138.

[4] S.Zeng et al. (2011). "A review on functionalized gold nanoparticles for biosensing applications". *Plasmonics* **6** (3): 491–506. doi:10.1007/s11468-011-9228-1.

[5] , have electronic interband transitions in the visible range, whereby specific light energies (colors) are absorbed, yielding their distinct color Kittel, C. (2005). *Introduction to Solid State Physics* (8th ed.). John Wiley & Sons. p. 403, table 2.

[6] Böer, K. W. (2002). *Survey of Semiconductor Physics* **1** (2nd ed.). John Wiley & Sons. p. 525.

[7] Zeng, Shuwen; Yu, Xia; Law, Wing-Cheung; Zhang, Yating et al. (2013). "Size dependence of Au NP-enhanced surface plasmon resonance based on differential phase measurement". *Sensors and Actuators B: Chemical* **176**: 1128–1133. doi:10.1016/j.snb.2012.09.073.

[8] "Surface plasmon dispersion relation in the X-ray emission spectra of a semi-infinite rectangular metal bounded by a plane". *Physica B+C* **150**: 378–384. doi:10.1016/0378-4363(88)90078-2.

[9] "LEDs work like butterflies' wings". *BBC News*. November 18, 2005. Retrieved May 22, 2010.

[10] http://www.nature.com/nature/journal/v487/n7405/abs/nature11254.html

[11] http://www.nature.com/nature/journal/v487/n7405/full/nature11253.html

[12] http://www.nature.com/nphoton/journal/v7/n5/abs/nphoton.2013.57.html

[13] T. Low and P. Avouris, ACS Nano 8, p1086 (2014) http://pubs.acs.org/doi/abs/10.1021/nn406627u

[14] Heip, H. M. et al. (2007). "A localized surface plasmon resonance based immunosensor for the detection of casein in milk". *Science and Technology of Advanced Materials* **8** (4): 331–338. Bibcode:2007STAdM...8..331M. doi:10.1016/j.stam.2006.12.010.

[15] Lewotsky, Kristin (2007). "The Promise of Plasmonics". *SPIE Professional*. doi:10.1117/2.4200707.07.

[16] "The L'Oréal Art & Science of Color Prize – 7th Prize Winners".

[17] "Prof. Choi Unveils Method to Improve Emission Efficiency of OLED". KAIST. 9 July 2009.

[18] "EU partners eye metallic nanostructures for solar cells". ElectroIQ. 30 March 2010.

[19] Gwamuri, et al. Limitations of ultra-thin transparent conducting oxides for integration into plasmonic-enhanced thin-film solar photovoltaic devices. *Materials for Renewable and Sustainable Energy* 4:12 (2015). DOI: 10.1007/s40243-015-0055-8

[20] Kawata, Satoshi. "New technique lights up the creation of holograms". Phys.org. Retrieved 24 September 2013.

- Stefan Maier (2007). *Plasmonics: Fundamentals and Applications*. Springer. ISBN 978-0-387-33150-8.
- Michael G. Cottam & David R. Tilley (1989). *Introduction to Surface and Superlattice Excitations*. Cambridge University Press. ISBN 0-521-32154-9.
- Heinz Raether (1980). *Excitation of plasmons and interband transitions by electrons*. Springer-Verlag. ISBN 0-387-09677-9.
- Barnes, W. L.; Dereux, A.; Ebbesen T.W. (2003). "Surface plasmon subwavelength optics". *Nature* **424** (6950): 824–830. Bibcode:2003Natur.424..824B. doi:10.1038/nature01937. PMID 12917696.
- Zayats, A. V.; Smolyaninov, I. I.; Maradudin, A. A. (2005). "Nano-optics of surface plasmon polaritons". *Physics Reports* **408** (3–4): 131–314. Bibcode:2005PhR...408..131Z. doi:10.1016/j.physrep.2004.11.001.
- Atwater,Harry A. (2007). "The Promise of Plasmonics".*Scientific American***296**(4): 56–63. doi:10.1038/scien-56. PMID 17479631.
- Ozbay, Ekmel (2006). "Plasmonics: Merging Photonics and Electronics at Nanoscale Dimensions". *Science* **311** (5758): 189–193. Bibcode:2006Sci...311..189O. doi:10.1126/science.1114849. PMID 16410515.
- Schuller, Jon; Barnard, Edward; Cai, Wenshan; Jun, Young Chul et al. (2010). "Plasmonics for Extreme Light Concentration and Manipulation". *Nature Materials* **9** (3): 193–204. Bibcode:2010NatMa...9..193S. doi:10.0.PMID 20168343.
- Brongersma,Mark;Shalaev,Vladimir(2010). "The case for plasmonics".*Science***328**: 440–441. Bibcode:2010Sci.B. doi:10.1126/science.1186905.

102.7 External links

- A selection of free-download papers on Plasmonics in New Journal of Physics
- http://www.plasmonicfocus.com
- http://www.sprpages.nl
- http://www.qub.ac.uk/mp/con/plasmon/sp1.html
- http://www.nano-optics.org.uk
- Plasmonic computer chips move closer
- Progress at Stanford for use in computers
- Slashdot: A Plasmonic Revolution for Computer Chips?
- A Microscope from Flatland *Physical Review Focus*, January 24, 2005
- Wikinews:Invisibility shield gets blueprint
- http://www.plasmonanodevices.org

- http://www.eu-pleas.org
- http://www.plasmocom.org
- Test the limits of plasmonic technology
- http://www.activeplasmonics.org
- http://www.plaisir-project.eu

Chapter 103

Polariton

In physics, **polaritons** /pəˈlærɪtɒnz/ are quasiparticles resulting from strong coupling of electromagnetic waves with an electric or magnetic dipole-carrying excitation. They are an expression of the common quantum phenomenon known as level repulsion, also known as the avoided crossing principle. Polaritons describe the crossing of the dispersion of light with any interacting resonance.

103.1 History

Oscillations in ionized gases were observed by Tonks and Langmuir in 1929. Collective interactions were published by David Pines and David Bohm in 1952 and plasmons were described in silver by Fröhlich and Pelzer in 1955. Ritchie predicted surface plasmons in 1957, then Ritchie and Eldridge published experiments and predictions of emitted photons from irradiated metal foils in 1962. Andreas Otto first published on surface plasmon-polaritons in Zeitschrift fur Physik in 1968.

103.2 Types

A **polariton** is the result of the mixing of a photon with an excitation of a material. The following are types of polaritons:

- Phonon polaritons result from coupling of an infrared photon with an optic phonon;
- Exciton polaritons result from coupling of visible light with an exciton
- Intersubband polaritons result from coupling of an infrared or terahertz photon with an intersubband excitation.
- Surface plasmon polaritons result from coupling of surface plasmons with light (the wavelength depends on the substance and its geometry).
- Bragg polaritons ("Braggoritons") result from coupling of Bragg photon modes with bulk excitons.[1]

103.3 Principles

Whenever the polariton picture is valid, the model of photons propagating freely in crystals is insufficient. A major feature of polaritons is a strong dependency of the propagation speed of light through the crystal on the frequency. For exciton-polaritons, rich experimental results on various aspects have been gained in copper (I) oxide.

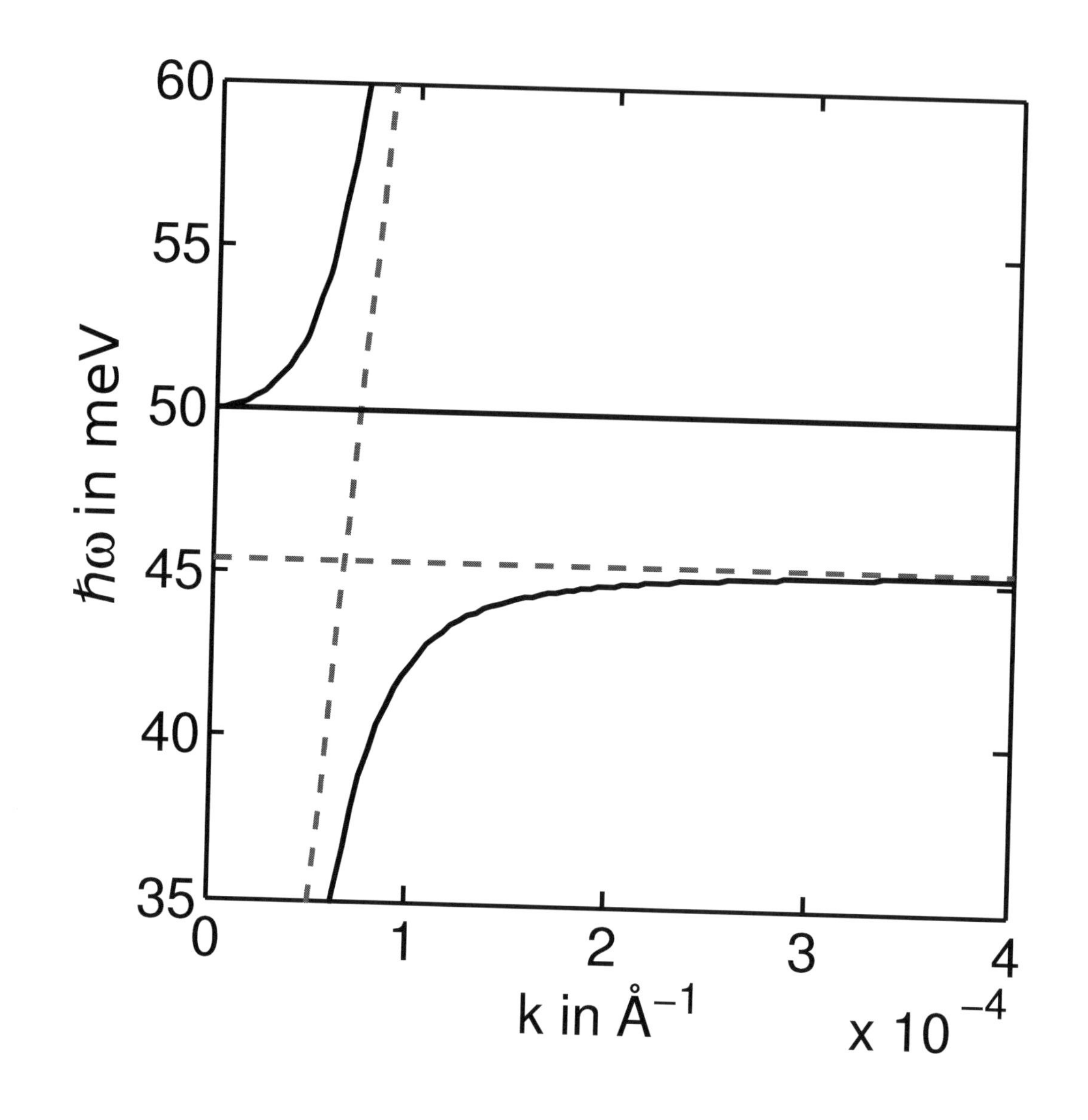

Dispersion relation of polaritons in GaP. Red curves are the uncoupled phonon and photon dispersion relations, black curves are the result of coupling (from top to bottom: upper polariton, LO phonon, lower polariton).

The polariton is a bosonic quasiparticle, and should not be confused with the polaron (a fermionic one), which is an electron plus an attached phonon cloud. Polaritons were first considered theoretically by Kirill Borisovich Tolpygo,[2][3] a Ukrainian physicist, and were initially termed light-excitons in Ukrainian and Russian scientific literature.

103.4 See also

- Atomic coherence
- Polariton laser
- Polariton superfluid

- Polaritonics

103.5 References

[1] N. Eradat "etal", *Evidence for braggoriton excitations in opal photonic crystals infiltrated with highly polarizable dyes,* Appl. Phys. Lett. **80**, 3491 (2002).

[2] Tolpygo, Kirill Borisovich

[3] K.B. Tolpygo, "Physical properties of a rock salt lattice made up of deformable ions," Zh.Eks.Teor.Fiz. v.20, No 6, pp.497–509 (1950), in Russian. English translation: Ukrainian Journal of Physics, v.53, special issue (2008); http://www.ujp.bitp.kiev.ua/files/file/papers/53/special_issue/53SI21p.pdf

- Fano, U. (1956). "Atomic Theory of Electromagnetic Interactions in Dense Materials". *Physical Review* **103** (5): 1202–1218. Bibcode:1956PhRv..103.1202F. doi:10.1103/PhysRev.103.1202.
- Hopfield, J. J. (1958). "Theory of the Contribution of Excitons to the Complex Dielectric Constant of Crystals". *Physical Review* **112** (5): 1555–1567. Bibcode:1958PhRv..112.1555H. doi:10.1103/PhysRev.112.1555.
- Otto, A. Excitation of nonradiative surface plasma waves in silver by the method of frustrated total reflection. Z. Phys. 216, 398–410 (1968)

103.6 Further reading

- Baker-Jarvis, J. (2012). "The Interaction of Radio-Frequency Fields With Dielectric Materials at Macroscopic to Mesoscopic Scales" (PDF). *Journal of Research of the National Institute of Standards and Technology* (National Institute of Science and Technology) **117**: 1. doi:10.6028/jres.117.001.

103.7 External links

- YouTube animation explaining what is polariton in a semiconductor micro-resonator.

Chapter 104

Polaron

Energy spectrum of an electron moving in a periodical potential of rigid crystal lattice consists of allowed and forbidden bands and is known as the Bloch spectrum. An electron with energy inside an allowed band moves as a free electron but with effective mass (solid-state physics) that differs from the electron mass in vacuum. However, crystal lattice is deformable and displacements of atoms (ions) from their equilibrium positions are described in terms of phonons. Electrons interact with these displacements, and this interaction is known as electron-phonon coupling. One of possible scenarios was proposed in the seminal 1933 paper by Lev Landau, it includes production of a lattice defect such as an F-center and trapping the electron by this defect. A different scenario was proposed by Solomon Pekar that envisions dressing the electron with lattice deformation (a cloud of virtual phonons). Such an electron with the accompanying deformation moves freely across the crystal, but with increased effective mass.[1] Pekar coined for this charge carrier the term **polaron**.

The general concept of a polaron has been extended to describe other interactions between the electrons and ions in metals that result in a bound state, or a lowering of energy compared to the non-interacting system. Major theoretical work has focused on solving Fröhlich and Holstein Hamiltonians. This is still an active field of research to find exact numerical solutions to the case of one or two electrons in a large crystal lattice, and to study the case of many interacting electrons.

Experimentally, polarons are important to the understanding of a wide variety of materials. The electron mobility in semiconductors can be greatly decreased by the formation of polarons. Organic semiconductors are also sensitive to polaronic effects, and is particularly relevant in the design of organic solar cells that effectively transport charge. The electron phonon interactions that form cooper pairs in type-I superconductors can also be modelled as a polaron, and two opposite spin electrons may form a bipolaron sharing a phonon cloud. This has been suggested as a mechanism for cooper pair formation in type-II superconductors. Polarons are also important for interpreting the optical conductivity of these types of materials.

The polaron, a fermionic quasiparticle, should not be confused with the polariton, a bosonic quasiparticle analogous to a hybridized state between a photon and an optical phonon.

104.1 Polaron theory

L. D. Landau [2] and S. I. Pekar [3] formed the basis of polaron theory. A charge placed in a polarizable medium will be screened. Dielectric theory describes the phenomenon by the induction of a polarization around the charge carrier. The induced polarization will follow the charge carrier when it is moving through the medium. The carrier together with the induced polarization is considered as one entity, which is called a polaron (see Fig. 1).

A conduction electron in an ionic crystal or a polar semiconductor is the prototype of a polaron. Herbert Fröhlich proposed a model Hamiltonian for this polaron through which its dynamics are treated quantum mechanically (Fröhlich Hamiltonian).[6][7] This model assumes that electron wavefunction is spread out over many ions which are all somewhat displaced from their equilibrium positions, or the continuum approximation. The strength of the electron-phonon interaction is expressed by a dimensionless coupling constant α introduced by Fröhlich.[7] In Table 1 the Fröhlich coupling

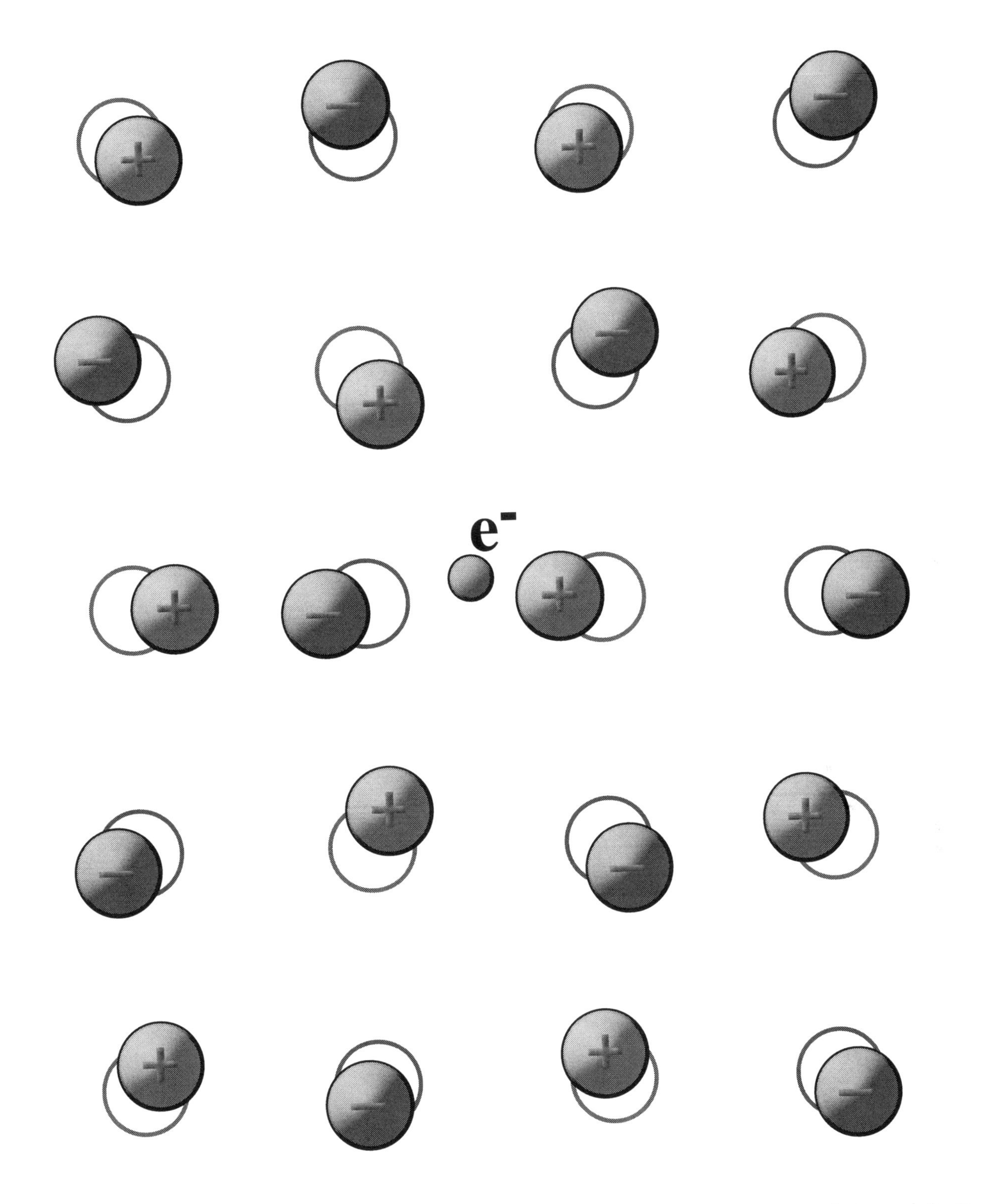

Fig. 1: Artist view of a polaron.[4] *A conduction electron in an ionic crystal or a polar semiconductor repels the negative ions and attracts the positive ions. A self-induced potential arises, which acts back on the electron and modifies its physical properties.*

constant is given for a few solids. The Fröhlich Hamiltonian for a single electron in a crystal using second quantization notation is: $H = H_e + H_{ph} + H_{e-ph}$

$H_e = \sum_{k,s} \xi(k,s) c^{\dagger}_{k,s} c_{k,s}$

$H_{ph} = \sum_{q,v} \omega_{q,v} a^{\dagger}_{q,v} a_{q,v}$

$H_{e-ph} = \frac{1}{\sqrt{2N}} \sum_{k,s,q,v} \gamma(\alpha, q, k, v) \omega_{qv} (c^{\dagger}_{k,s} c_{k-q,s} a_{q,v} + c^{\dagger}_{k-q,s} c_{k,s} a^{\dagger}_{q,v})$

The exact form of gamma depends on the material and the type of phonon being used in the model. A detailed advanced discussion of the variations of the Fröhlich Hamiltonian can be found in J. T. Devreese and A. S. Alexandrov [8] The terms Fröhlich polaron and large polaron are sometimes used synonymously, since the Fröhlich Hamiltonian includes the continuum approximation and long range forces. There is no known exact solution for the Fröhlich Hamiltonian with longitudinal optical (LO) phonons and linear γ (the most commonly considered variant of the Fröhlich polaron) despite extensive investigations.[3][5][6][7][9][10][11][12][13][14]

Despite the lack of an exact solution, some approximations of the polaron properties are known.

The physical properties of a polaron differ from those of a band-carrier. A polaron is characterized by its *self-energy* ΔE , an *effective mass* $m*$ and by its characteristic *response* to external electric and magnetic fields (e. g. dc mobility and optical absorption coefficient).

When the coupling is weak (α small), the self-energy of the polaron can be approximated as:[15]

and the polaron mass $m*$, which can be measured by cyclotron resonance experiments, is larger than the band mass m of the charge carrier without self-induced polarization:[16]

When the coupling is strong (α large), a variational approach due to Landau and Pekar indicates that the self-energy is proportional to α^2 and the polaron mass scales as α^4. The Landau-Pekar variational calculation [3] yields an upper bound to the polaron self-energy $E < -C_{PL}\alpha^2$, valid for *all* α, where C_{PL} is a constant determined by solving an integro-differential equation. It was an open question for many years whether this expression was asymptotically exact as α tends to infinity. Finally, Donsker and Varadhan,[17] applying large deviation theory to Feynman's path integral formulation for the self-energy, showed the large α exactitude of this Landau-Pekar formula. Later, Lieb and Thomas [18] gave a shorter proof using more conventional methods, and with explicit bounds on the lower order corrections to the Landau-Pekar formula.

Feynman [19] introduced a variational principle for path integrals to study the polaron. He simulated the interaction between the electron and the polarization modes by a harmonic interaction between a hypothetical particle and the electron. The analysis of an exactly solvable ("symmetrical") 1D-polaron model,[20][21] Monte Carlo schemes [22][23] and other numerical schemes [24] demonstrate the remarkable accuracy of Feynman's path-integral approach to the polaron ground-state energy. Experimentally more directly accessible properties of the polaron, such as its mobility and optical absorption, have been investigated subsequently.

In the strong coupling limit, $\alpha >> 1$, the spectrum of excited states of a polaron begins with polaron-phonon bound states with energies less than $\hbar\omega_0$, where ω_0 is the frequency of optical phonons.[25]

104.2 Polaron optical absorption

The expression for the magnetooptical absorption of a polaron is:[26]

Here, ω_c is the cyclotron frequency for a rigid-band electron. The magnetooptical absorption Γ(Ω) at the frequency Ω takes the form Σ(Ω) is the so-called "memory function", which describes the dynamics of the polaron. Σ(Ω) depends also on α, $\beta_{\text{what is beta}}$? and ω_c .

In the absence of an external magnetic field ($\omega_c = 0$) the optical absorption spectrum (3) of the polaron at weak coupling is determined by the absorption of radiation energy, which is reemitted in the form of LO phonons. At larger coupling, $\alpha \geq 5.9$, the polaron can undergo transitions toward a relatively stable internal excited state called the "relaxed excited state" (RES) (see Fig. 2). The RES peak in the spectrum also has a phonon sideband, which is related to a Franck-Condon-type transition.

A comparison of the DSG results [27] with the optical conductivity spectra given by approximation-free numerical [28] and approximate analytical approaches is given in ref.[29]

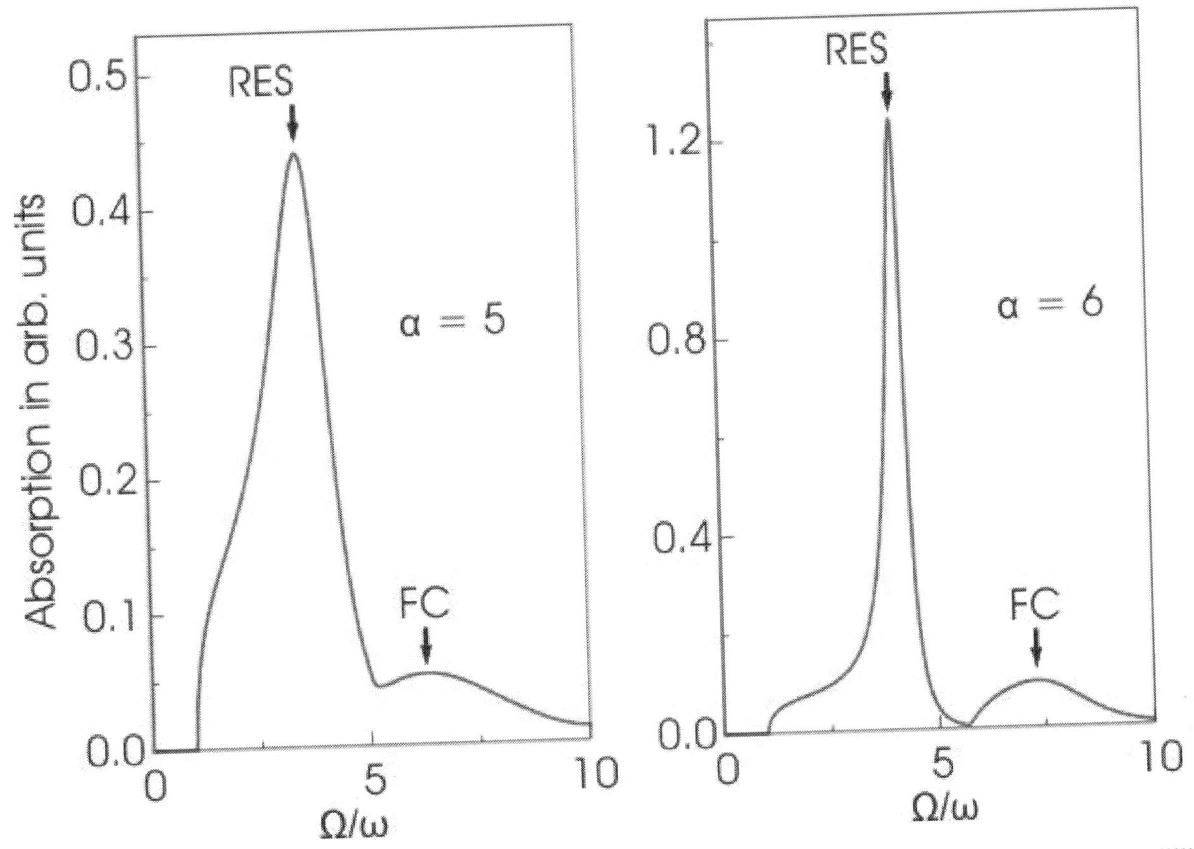

Fig.2. Optical absorption of a polaron at $\alpha = 5$ and 6. The RES peak is very intense compared with the Franck-Condon (FC) peak.[11][27]

Calculations of the optical conductivity for the Fröhlich polaron performed within the Diagrammatic Quantum Monte Carlo method,[28] see Fig. 3, fully confirm the results of the path-integral variational approach [27] at $\alpha \lesssim 3$. In the intermediate coupling regime $3 < \alpha < 6$, the low-energy behavior and the position of the maximum of the optical conductivity spectrum of ref.[28] follow well the prediction of ref.[27] There are the following qualitative differences between the two approaches in the intermediate and strong coupling regime: in ref.,[28] the dominant peak broadens and the second peak does not develop, giving instead rise to a flat shoulder in the optical conductivity spectrum at $\alpha = 6$. This behavior can be attributed to the optical processes with participation of two [30] or more phonons. The nature of the excited states of a polaron needs further study.

The application of a sufficiently strong external magnetic field allows one to satisfy the resonance condition $\Omega = \omega_c + \mathrm{Re}\Sigma(\Omega)$, which {(for $\omega_c < \omega$)} determines the polaron cyclotron resonance frequency. From this condition also the polaron cyclotron mass can be derived. Using the most accurate theoretical polaron models to evaluate $\Sigma(\Omega)$, the experimental cyclotron data can be well accounted for.

Evidence for the polaron character of charge carriers in AgBr and AgCl was obtained through high-precision cyclotron resonance experiments in external magnetic fields up to 16 T.[31] The all-coupling magneto-absorption calculated in ref.,[26] leads to the best quantitative agreement between theory and experiment for AgBr and AgCl. This quantitative interpretation of the cyclotron resonance experiment in AgBr and AgCl [31] by the theory of ref.[26] provided one of the most convincing and clearest demonstrations of Fröhlich polaron features in solids.

Experimental data on the magnetopolaron effect, obtained using far-infrared photoconductivity techniques, have been applied to study the energy spectrum of shallow donors in polar semiconductor layers of CdTe.[32]

The polaron effect well above the LO phonon energy was studied through cyclotron resonance measurements, e. g., in II-VI semiconductors, observed in ultra-high magnetic fields.[33] The resonant polaron effect manifests itself when the cyclotron frequency approaches the LO phonon energy in sufficiently high magnetic fields.

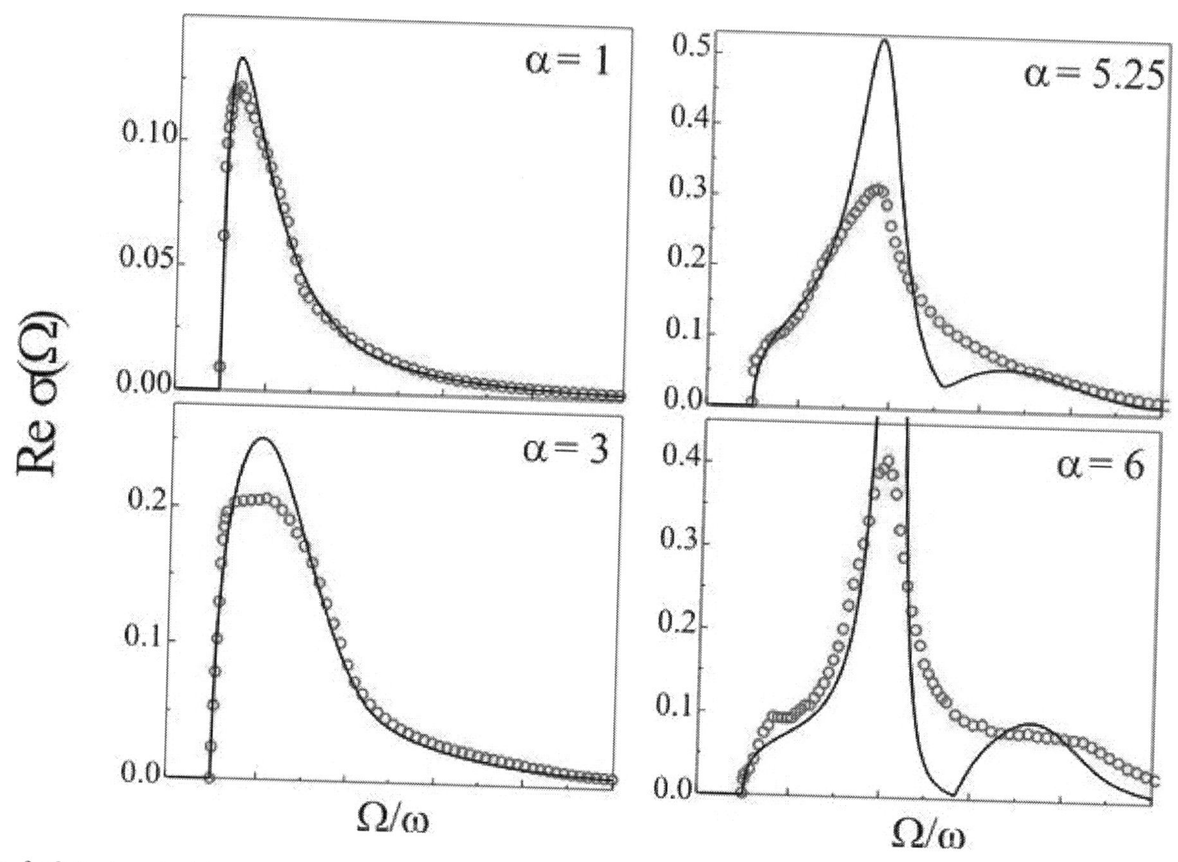

Fig. 3: Optical conductivity spectra calculated within the Diagrammatic Quantum Monte Carlo method (open circles) compared to the DSG calculations (solid lines).[27][28]

104.3 Polarons in two dimensions and in quasi-2D structures

The great interest in the study of the two-dimensional electron gas (2DEG) has also resulted in many investigations on the properties of polarons in two dimensions.[34][35][36] A simple model for the 2D polaron system consists of an electron confined to a plane, interacting via the Fröhlich interaction with the LO phonons of a 3D surrounding medium. The self-energy and the mass of such a 2D polaron are no longer described by the expressions valid in 3D; for weak coupling they can be approximated as:[37][38]

It has been shown that simple scaling relations exist, connecting the physical properties of polarons in 2D with those in 3D. An example of such a scaling relation is:[36]

where m^*_{2D} (m^*_{3D}) and m_{2D} (m_{3D}) are, respectively, the polaron and the electron-band masses in 2D (3D).

The effect of the confinement of a Fröhlich polaron is to enhance the *effective* polaron coupling. However, many-particle effects tend to counterbalance this effect because of screening.[34][39]

Also in 2D systems cyclotron resonance is a convenient tool to study polaron effects. Although several other effects have to be taken into account (nonparabolicity of the electron bands, many-body effects, the nature of the confining potential, etc.), the polaron effect is clearly revealed in the cyclotron mass. An interesting 2D system consists of electrons on films of liquid He.[40][41] In this system the electrons couple to the ripplons of the liquid He, forming "ripplopolarons". The effective coupling can be relatively large and, for some values of the parameters, self-trapping can result. The acoustic nature of the ripplon dispersion at long wavelengths is a key aspect of the trapping.

For GaAs/$Al_xGa_{1-x}As$ quantum wells and superlattices, the polaron effect is found to decrease the energy of the shallow donor states at low magnetic fields and leads to a resonant splitting of the energies at high magnetic fields. The energy spectra of such polaronic systems as shallow donors ("bound polarons"), e. g., the D_0 and D^- centres, constitute the most complete and detailed polaron spectroscopy realised in the literature.[42]

In GaAs/AlAs quantum wells with sufficiently high electron density, anticrossing of the cyclotron-resonance spectra has been observed near the GaAs transverse optical (TO) phonon frequency rather than near the GaAs LO-phonon frequency.[43] This anticrossing near the TO-phonon frequency was explained in the framework of the polaron theory.[43]

Besides optical properties,[5][13][44] many other physical properties of polarons have been studied, including the possibility of self-trapping, polaron transport,[45] magnetophonon resonance, etc.

104.4 Extensions of the polaron concept

Significant are also the extensions of the polaron concept: acoustic polaron, piezoelectric polaron, electronic polaron, bound polaron, trapped polaron, spin polaron, molecular polaron, solvated polarons, polaronic exciton, Jahn-Teller polaron, small polaron, bipolarons and many-polaron systems.[5] These extensions of the concept are invoked, e. g., to study the properties of conjugated polymers, colossal magnetoresistance perovskites, high- T_c superconductors, layered MgB_2 superconductors, fullerenes, quasi-1D conductors, semiconductor nanostructures.

The possibility that polarons and bipolarons play a role in high- T_c superconductors has renewed interest in the physical properties of many-polaron systems and, in particular, in their optical properties. Theoretical treatments have been extended from one-polaron to many-polaron systems.[5][46][47]

A new aspect of the polaron concept has been investigated for semiconductor nanostructures: the exciton-phonon states are not factorizable into an adiabatic product Ansatz, so that a *non-adiabatic* treatment is needed.[48] The *non-adiabaticity* of the exciton-phonon systems leads to a strong enhancement of the phonon-assisted transition probabilities (as compared to those treated adiabatically) and to multiphonon optical spectra that are considerably different from the Franck-Condon progression even for small values of the electron-phonon coupling constant as is the case for typical semiconductor nanostructures.[48]

In biophysics Davydov soliton is a propagating along the protein α-helix self-trapped amide I excitation that is a solution of the Davydov Hamiltonian. The mathematical techniques that are used to analyze Davydov's soliton are similar to some that have been developed in polaron theory. In this context the Davydov soliton corresponds to a *polaron* that is (i) *large* so the continuum limit approximation in justified, (ii) *acoustic* because the self-localization arises from interactions with acoustic modes of the lattice, and (iii) *weakly coupled* because the anharmonic energy is small compared with the phonon bandwidth.[49]

More recently it was shown that the system of an impurity in a Bose–Einstein condensate is also a member of the polaron family.[50] This is very promising for experimentally probing the hitherto inaccessible strong coupling regime since in this case interaction strengths can be externally tuned through the use of a Feshbach resonance.

104.5 See also

- Sigurd Zienau
- exciton

104.6 References

[1] L. D. Landau and S. I. Pekar, Effective mass of a polaron, Zh. Eksp. Teor. Fiz. **18**, 419–423 (1948) [in Russian], English translation: Ukr. J. Phys., Special Issue, **53**, p.71-74 (2008), http://ujp.bitp.kiev.ua/files/journals/53/si/53SI15p.pdf

[2] Landau LD (1933). "Über die Bewegung der Elektronen in Kristalgitter". *Phys. Z. Sowjetunion* **3**: 644–645.

[3] Pekar SI (1951). "Issledovanija po Elektronnoj Teorii Kristallov". *Gostekhizdat, Moskva*.. English translation: Research in Electron Theory of Crystals, AEC-tr-555, US Atomic Energy Commission (1963)

[4] Devreese JTL (1979). “Moles agitat mentem. Ontwikkelingen in de fysica van de vaste stof”. *Rede uitgesproken bij de aanvaarding van het ambt van buitengewoon hoogleraar in de fysica van de vaste stof, in het bijzonder de theorie van de vaste stof, bij de afdeling der technische natuurkunde aan de Technische Hogeschool Eindhoven.*

[5] Devreese, Jozef T. (2005). “Polarons”. In Lerner, R.G.; Trigg, G.L. *Encyclopedia of Physics* **2** (Third ed.). Weinheim: Wiley-VCH. pp. 2004–2027. OCLC 475139057.

[6] Fröhlich H; Pelzer H; Zienau S (1950). “Properties of slow electrons in polar materials”. *Phil. Mag.* **41**: 221.

[7] Fröhlich H(1954). “Electrons in latticefields”.*Adv.Phys.***3**(11): 325. Bibcode:1954AdPhy...3..325F.doi:10.1080/00018735.

[8] J. T. Devreese & A. S. Alexandrov (2009). “Fröhlich polaron and bipolaron: recent developments”. *Rep. Prog. Phys.* **72** (6): 066501. arXiv:0904.3682. Bibcode:2009RPPh...72f6501D. doi:10.1088/0034-4885/72/6/066501.

[9] Kuper GC; Whitfield GD, eds. (1963). “Polarons and Excitons”. *Oliver and Boyd, Edinburgh.*

[10] Appel J (1968). “Polarons”. *In: Solid State Physics, F. Seitz, D. Turnbull, and H. Ehrenreich (eds.), Academic Press, New York* **21**: pp.193–391.

[11] Devreese JTL, ed. (1972). “Polarons in Ionic Crystals and Polar Semiconductors”. *North-Holland, Amsterdam.*

[12] Mitra TK; Chatterjee A; Mukhopadhyay S (1987). “Polarons”. *Phys. Rep.* **153** (2–3): 91. Bibcode:1987PhR...153...91M. doi:10.1016/0370-1573(87)90087-1.

[13] Devreese JTL (1996). *In" Encyclopedia of Applied Physics, G. L. Trigg (ed.), VCH, Weinheim* **14**: pp.383–413. Missing or empty |title= (help)

[14] Alexandrov AS; Mott N (1996). “Polarons and Bipolarons”. *World Scientific, Singapore.*

[15] Smondyrev MA (1986). “Diagrams in the polaron model”. *Theor. Math. Phys.* **68** (1): 653. Bibcode:1986TMP....68..653S. doi:10.1007/BF01017794.

[16] Röseler J(1968). “A new variational ansatz in the polaron theory”.*Physica Status Solidi(b)***25**(1): 311. Bibcode:1968PSSBR. doi:10.1002/pssb.19680250129.

[17] M. Donsker and R.Varadhan(1983). “Asymptotics for the Polaron”, *Commun. Pure Appl. Math.* **36**, 505–528.

[18] Lieb E.H.; Thomas L.E. (1997). “Exact Ground State Energy of the Strong Coupling Polaron”. *Commun. Math. Physics* **183** (3): 511–519. arXiv:cond-mat/9512112. Bibcode:1997CMaPh.183..511L. doi:10.1007/s002200050040.

[19] Feynman RP(1955). “Slow Electrons in a Polar Crystal”.*Phys.Rev.***97**(3): 660. Bibcode:1955PhRv...97..660F.doi:10.11030.

[20] Devreese JTL;Evrard R(1964). “On the excited states of a symmetrical polaron model”.*Phys.Lett.***11**(4): 278. BibcodeD. doi:10.1016/0031-9163(64)90324-5.

[21] Devreese JTL; Evrard R (1968). *Proceedings of the British Ceramic Society* **10**: 151. Missing or empty |title= (help)

[22] Mishchenko AS; Prokof'ev NV; Sakamoto A; Svistunov BV (2000). “Diagrammatic quantum Monte Carlo study of the Fröhlich polaron”. *Phys. Rev. B* **62** (10): 6317. Bibcode:2000PhRvB..62.6317M. doi:10.1103/PhysRevB.62.6317.

[23] Titantah JT; Pierleoni C; Ciuchi S (2001). “Free Energy of the Fröhlich Polaron in Two and Three Dimensions”. *Phys. Rev. Lett.* **87** (20): 206406. arXiv:cond-mat/0010386. Bibcode:2001PhRvL..87t6406T. doi:10.1103/PhysRevLett.87.206406. PMID 11690499.

[24] De Filippis G; Cataudella V; Marigliano Ramaglia V; Perroni CA et al. (2003). “Ground state features of the Fröhlich model”. *Eur. Phys. J. B* **36** (1): 65. arXiv:cond-mat/0309309. Bibcode:2003EPJB...36...65D. doi:10.1140/epjb/e2003-00317-x.

[25] V.I.Mel'nikov and E.I.Rashba. ZhETF Pis Red.,**10**1969,95,359(1959),JETP Lett**10**,60(1969).. ac.ru/ps/1687/article_25692.pdf

[26] Peeters FM;Devreese JTL(1986). “Magneto-optical absorption of polarons”.*Phys.Rev.B***34**(10): 7246. Bibcode. doi:10.1103/PhysRevB.34.7246.

[27] Devreese JTL; De Sitter J; Goovaerts M (1972). “Optical Absorption of Polarons in the Feynman-Hellwarth-Iddings-Platzman Approximation”. *Phys. Rev. B* **5** (6): 2367. Bibcode:1972PhRvB...5.2367D. doi:10.1103/PhysRevB.5.2367.

[28] Mishchenko AS; Nagaosa N; Prokof'ev NV; Sakamoto A et al. (2003). "Optical Conductivity of the Fröhlich Polaron". *Phys. Rev. Lett.* **91** (23): 236401. arXiv:cond-mat/0312111. Bibcode:2003RvL..91w6401M.doi:10.1103/PhysRevLett.91.PMID 14683203.

[29] De Filippis G; Cataudella V; Mishchenko AS; Perroni CA et al. (2006). "Validity of the Franck-Condon Principle in the Optical Spectroscopy: Optical Conductivity of the Fröhlich Polaron". *Phys. Rev. Lett.* **96** (13): 136405. arXiv:cond-mat/0603219. Bibcode:2006PhRvL..96m6405D. doi:10.1103/PhysRevLett.96.136405. PMID 16712012.

[30] Goovaerts MJ; De Sitter J; Devreese JTL (1973). "Numerical Study of Two-Phonon Sidebands in the Optical Absorption of Free Polarons in the Strong-Coupling Limit".*Phys.Rev.*7(6): 2639. Bibcode:1973PhRvB...7.2639G.doi:10.1103/PhysRevB.7..

[31] Hodby JW; Russell GP; Peeters F; Devreese JTL et al. (1987). "Cyclotron resonance of polarons in the silver halides: AgBr and AgCl". *Phys. Rev. Lett.* **58** (14): 1471–1474. Bibcode:1987PhRvL..58.1471H. doi:10.1103/PhysRevLett.58.1471. PMID 10034445.

[32] Grynberg M; Huant S; Martinez G; Kossut J et al. (15 July 1996). "Magnetopolaron effect on shallow indium donors in CdTe". *Physical Review B-Condensed Matter* (APS) **54** (3): 1467–70.

[33] Miura N; Imanaka Y (2003). "Polaron cyclotron resonance in II–VI compounds at high magnetic fields". *Physica Status Solidi (b)* **237** (1): 237. Bibcode:2003PSSBR.237..237M. doi:10.1002/pssb.200301781.

[34] Devreese JTL; Peeters FM, eds. (1987). "The Physics of the Two-Dimensional Electron Gas". *ASI Series, Plenum, New York* **B157**.

[35] Wu XG; Peeters FM; Devreese JTL (1986). "Effect of screening on the optical absorption of a two-dimensional electron gas in GaAs-Al$_x$Ga$_{1-x}$As heterostructures".*Phys.Rev.B*34(4): 2621. Bibcode:1986PhRvB..34.2621W.doi:10.1103/PhysRevB.341.

[36] Peeters FM; Devreese JTL (1987). "Scaling relations between the two- and three-dimensional polarons for static and dynamical properties". *Phys. Rev. B* **36** (8): 4442. Bibcode:1987PhRvB..36.4442P. doi:10.1103/PhysRevB.36.4442.

[37] Sak J(1972). "Theory of Surface Polarons".*Phys.Rev.B*6(10): 3981. Bibcode:1972PhRvB...6.3981S.doi:10.1103/PhysRevB81.

[38] Peeters FM; Wu XG; Devreese JTL (1988). "Exact and approximate results for the mass of a two-dimensional polaron". *Phys. Rev. B* **37** (2): 933. Bibcode:1988PhRvB..37..933P. doi:10.1103/PhysRevB.37.933.

[39] Das Sarma S; Mason BA (1985). "Optical phonon interaction effects in layered semiconductor structures". *Ann. Phys. (New York)* **163** (1): 78. Bibcode:1985AnPhy.163...78S. doi:10.1016/0003-4916(85)90351-3.

[40] Shikin VB; Monarkha YP (1973). *Sov. Phys. – JETP* **38**: 373. Missing or empty |title= (help)

[41] Jackson SA; Platzman PM (1981). "Polaronic aspects of two-dimensional electrons on films of liquid He". *Phys. Rev. B* **24** (1): 499. Bibcode:1981PhRvB..24..499J. doi:10.1103/PhysRevB.24.499.

[42] Shi JM; Peeters FM; Devreese JTL (1993). "Magnetopolaron effect on shallow donor states in GaAs". *Phys. Rev. B* **48** (8): 5202. Bibcode:1993PhRvB..48.5202S. doi:10.1103/PhysRevB.48.5202.

[43] Poulter AJL; Zeman J; Maude DK; Potemski M et al. (2001). "Magneto Infrared Absorption in High Electron Density GaAs Quantum Wells".*Phys.Rev.Lett.*86(2): 336–9. arXiv:cond-mat/0012008. Bibcode:2001PhRvL..86..33.PMID11177825.

[44] Calvani P (2001). "Optical Properties of Polarons". *Editrice Compositori, Bologna*.

[45] Feynman RP; Hellwarth RW; Iddings CK; Platzman PM (1962). "Mobility of Slow Electrons in a Polar Crystal". *Phys. Rev.* **127** (4): 1004. Bibcode:1962PhRv..127.1004F. doi:10.1103/PhysRev.127.1004.

[46] Bassani FG; Cataudella V; Chiofalo ML; De Filippis G et al. (2003). "Electron gas with polaronic effects: beyond the mean-field theory". *Physica Status Solidi (b)* **237** (1): 173. Bibcode:2003PSSBR.237..173B. doi:10.1002/pssb.200301763.

[47] Hohenadler M; Hager G; Wellein G; Fehske H (2007). "Carrier-density effects in many-polaron systems". *J. Phys.: Condens. Matter* **19** (25): 255210. arXiv:cond-mat/0611586. Bibcode:2007JPCM...19y5210H. doi:10.1088/0953-8984/19/25/255210.

[48] Fomin VM; Gladilin VN; Devreese JTL; Pokatilov EP et al. (1998). "Photoluminescence of spherical quantum dots". *Phys. Rev. B* **57** (4): 2415. Bibcode:1998PhRvB..57.2415F. doi:10.1103/PhysRevB.57.2415.

[49] Scott AS (1992). "Davydov's soliton". *Physics Reports* **217** (1): 1–67. Bibcode:1992PhR...217....1S. doi:10.1016/0370-1573(92)90093-F.

[50] Tempere J; Casteels W; Oberthaler M; Knoop S et al. (2009). "Feynman path-integral treatment of the BEC-impurity polaron". *Phys. Rev. B* **80** (18): 184504. arXiv:0906.4455. Bibcode:2009PhRvB..80r4504T. doi:10.1103/PhysRevB.80.184504.

104.7 External links

Chapter 105

Roton

For other uses, see Roton (disambiguation).

In theoretical physics, a **roton** is an elementary excitation, or quasiparticle, in superfluid helium-4. The dispersion relation of elementary excitations in this superfluid shows a linear increase from the origin, but exhibits first a maximum and then a minimum in energy as the momentum increases. Excitations with momenta in the linear region are called phonons; those with momenta close to the minimum are called rotons. Excitations with momenta near the maximum are sometimes called maxons.

Originally, the roton spectrum was phenomenologically introduced by Lev Landau. Currently there exist different models which try to explain the roton spectrum, with a different degree of success and fundamentality.[1][2] The requirement for any model of such kind is that it must explain not only the shape of the spectrum itself but also other related observables, such as the speed of sound and structure factor of superfluid helium-4. Microwave and Bragg spectroscopy has been conducted on helium to study roton spectrum.[3]

Bose-Einstein condensation of rotons has been proposed and studied, but not yet detected.[4]

The term "roton" is also used for the quantised eigenmode of a freely rotating molecule.

105.1 Bibliography

- Feynman, RP, *Superfluidity and Superconductivity*, Rev. Mod. Phys. 29, 205 (1957)

105.2 See also

- Superfluid helium-4
- Superfluid
- Macroscopic quantum phenomena
- Bose–Einstein condensate

105.3 References

[1] "Fingerprinting Rotons in a Dipolar Condensate: Super-Poissonian Peak in the Atom-Number Fluctuations". *Phys. Rev. Lett. 110, 265302*. 26 June 2013. arXiv:1304.3605. Bibcode:2013PhRvL.110z5302B. doi:10.1103/PhysRevLett.110.265302.

[2] "Roton spectroscopy in a harmonically trapped dipolar Bose-Einstein condensate". *Phys. Rev. A 86, 021604(R)*. Aug 15, 2012. arXiv:1206.2770. Bibcode:2012PhRvA..86b1604B. doi:10.1103/PhysRevA.86.021604.

[3] "Microwave Spectroscopy of Condensed Helium at the Roton Frequency". *Journal of Low Temperature Physics*. 4 Nov 2009. Bibcode:2010JLTP..158..244R. doi:10.1007/s10909-009-0025-6.

[4] "The role of the condensate in the existence of phonons and rotons". *Journal of Low Temperature Physics*. December 1993. Bibcode:1993JLTP...93..861G. doi:10.1007/BF00692035.

•

Chapter 106

Trion (physics)

A **trion** is a localized excitation which consists of three charged quasiparticles. A negative trion consists of two electrons and one hole and a positive trion consists of two holes and one electron. The trion itself is a quasiparticle and is somewhat similar to an exciton, which is a complex of one electron and one hole. The trion has a ground singlet state (spin s = 1/2) and an excited triplet state (s = 3/2). Here singlet and triplet degeneracies originate not from the whole system but from the two identical particles in it. The half-integer spin value distinguishes trions from excitons in many phenomena; for example, energy states of trions, but not excitons, are split in an applied magnetic field. Trion states were predicted theoretically and then observed experimentally in various optically excited semiconductors, especially in quantum dots and quantum well structures.[1][2] There is evidence of their existence in nanotubes.[3] At least in zigzag-type carbon nanotubes trions are an excimer compounds.[4]

Despite numerous reports of experimental trion observations in different semiconductor heterostructures, there are serious concerns on the exact physical nature of the detected complexes. The originally foreseen 'true' trion particle has a delocalized wavefunction (at least at the scales of several Bohr radii) while recent studies reveal significant binding from charged impurities in real semiconductor quantum wells [5]

106.1 References

[1] S. A. Moskalenko et al. (2000). *Bose-Einstein condensation of excitons and biexcitons: and coherent nonlinear optics with excitons*. Cambridge University Press. p. 140. ISBN 0-521-58099-4.

[2] Dieter Bimberg (2008). *Semiconductor Nanostructures*. Springer. pp. 243–245. ISBN 3-540-77898-5.

[3] Matsunaga, R.; Matsuda, K.; Kanemitsu, Y. (2011). "Observation of Charged Excitons in Hole-doped Carbon Nanotubes Using Photoluminescence and Absorption Spectroscopy". *Phys. Rev. Lett.* **106** (037404): 1. arXiv:1009.2297. Bibcode:2011Ph .doi:10.1103/PhysRevLett.106.037404.

[4] Marchenko,Sergey(2012). "Stability of rionic States in Zigzag Carbon Nanotubes". *Ukr.J.Phys.* **57**(10): 1055. arXiv:1211.. Bibcode:2012arXiv1211.5754M.

[5] Solovyev, V.V.; Kukushkin, I.V. (2009). "Measurement of binding energy of negatively charged excitons in GaAs/Al0.3Ga0.7As quantum wells". *Phys.Rev.B.* **79**: 233306. arXiv:0906.5612. Bibcode:2009PhRvB..79w3306S.doi:10.1103/PhysRevB.79.23.

Chapter 107

List of baryons

Baryons are composite particles made of three quarks, as opposed to mesons, which are composite particles made of one quark and one antiquark. Baryons and mesons are both hadrons, which are particles composed solely of quarks or both quarks and antiquarks. The term *baryon* is derived from the Greek *"βαρύς"* (*barys*), meaning "heavy", because, at the time of their naming, it was believed that baryons were characterized by having greater masses than other particles that were classed as matter.

Until a few years ago, it was believed that some experiments showed the existence of pentaquarks – baryons made of four quarks and one antiquark.[1][2] The particle physics community as a whole did not view their existence as likely by 2006.[3] On 13 July 2015, the LHCb collaboration at CERN reported results consistent with pentaquark states in the decay of bottom Lambda baryons (Λ0
b).[4]

Since baryons are composed of quarks, they participate in the strong interaction. Leptons, on the other hand, are not composed of quarks and as such do not participate in the strong interaction. The most famous baryons are the protons and neutrons that make up most of the mass of the visible matter in the universe, whereas electrons, the other major component of atoms, are leptons. Each baryon has a corresponding antiparticle known as an antibaryon in which quarks are replaced by their corresponding antiquarks. For example, a proton is made of two up quarks and one down quark, while its corresponding antiparticle, the antiproton, is made of two up antiquarks and one down antiquark.

107.1 Lists of baryons

These lists detail all known and predicted baryons in total angular momentum $J = 1/2$ and $J = 3/2$ configurations with positive parity.

- Baryons composed of one type of quark (uuu, ddd, ...) can exist in $J = 3/2$ configuration, but $J = 1/2$ is forbidden by the Pauli exclusion principle.
- Baryons composed of two types of quarks (uud, uus, ...) can exist in both $J = 1/2$ and $J = 3/2$ configurations
- Baryons composed of three types of quarks (uds, udc, ...) can exist in both $J = 1/2$ and $J = 3/2$ configurations. Two $J = 1/2$ configurations are possible for these baryons.

The symbols encountered in these lists are: *I* (isospin), *J* (total angular momentum), *P* (parity), u (up quark), d (down quark), s (strange quark), c (charm quark), b (bottom quark), *Q* (charge), *B* (baryon number), *S* (strangeness), *C* (charm), *B′* (bottomness), as well as a wide array of subatomic particles (hover for name). (See the *baryon* article for a detailed explanation of these symbols.)

Antibaryons are not listed in the tables; however, they simply would have all quarks changed to antiquarks, and *Q*, *B*, *S*, *C*, *B′*, would be of opposite signs. Particles with † next to their names have been predicted by the Standard Model but not

A diagram of a proton, one of the most famous baryons, containing two up quarks and one down quark

yet observed. Values in red have not been firmly established by experiments, but are predicted by the quark model and are consistent with the measurements.[5][6]

107.1.1 $J^P = \frac{1}{2}^+$ baryons

† ^ Particle has not yet been observed.

[a] ^ The masses of the proton and neutron are known with much better precision in atomic mass units (u) than in MeV/c^2, due to the relatively poorly known value of the elementary charge. In atomic mass unit, the mass of the proton is 1.007276466812(90) u whereas that of the neutron is 1.00866491600(43) u.

[b] ^ At least 10^{35} years. See proton decay.

[c] ^ For free neutrons; in most common nuclei, neutrons are stable.

[d] ^ PDG reports the resonance width (Γ). Here the conversion $\tau = {}^{\hbar}/_{\Gamma}$ is given instead.

[e] ^ Some controversy exists about this data.[23]

107.1.2 $J^P = 3/2^+$ baryons

† ^ Particle has not yet been observed.
[h] ^ PDG reports the resonance width (Γ). Here the conversion $\tau = \hbar/\Gamma$ is given instead.

107.1.3 Baryon resonance particles

This table gives the name, quantum numbers (where known), and experimental status of baryons resonances confirmed by the PDG.[36] Baryon resonance particles are excited baryon states with short half lives and higher masses. Despite significant research, the fundamental degrees of freedom behind baryon excitation spectra are still poorly understood.[37] The spin-parity J^P (when known) is given with each particle. For the strongly decaying particles, the J^P values are considered to be part of the names, as is the mass for all resonances.

107.2 See also

- Eightfold way (physics)
- List of mesons
- List of particles
- Timeline of particle discoveries

107.3 References

[1] H. Muir (2003)

[2] K. Carter (2003)

[3] W.-M. Yao *et al.* (2006): Particle listings – Positive Theta

[4] R. Aaij et al. (LHCb collaboration) (2015). "Observation of J/ψp resonances consistent with pentaquark states in Λ0
b→J/ψK–
p decays". *Physical Review Letters* **115** (7). doi:10.1103/PhysRevLett.115.072001.

[5] J. Beringer *et al.* (2012) and 2013 partial update for the 2014 edition: Particle summary tables – Baryons

[6] J.G. Körner *et al.* (1994)

[7] J. Beringer *et al.* (2012): Particle listings – p+

[8] J. Beringer *et al.* (2012): Particle listings – n0

[9] J. Beringer *et al.* (2012): Particle listings – Λ

[10] J. Beringer *et al.* (2012): Particle listings – Λ
c

[11] J. Beringer *et al.* (2012): Particle listings – Λ
b

[12] J. Beringer *et al.* (2012): Particle listings – Σ+

[13] J. Beringer *et al.* (2012): Particle listings – Σ0

[14] J. Beringer *et al.* (2012): Particle listings – Σ–

[15] J. Beringer *et al.* (2012): Particle listings – Σ
c

[16] J. Beringer *et al.* (2012): Particle listings – Σ
b

[17] J. Beringer *et al.* (2012): Particle listings – Ξ0

[18] J. Beringer *et al.* (2012): Particle listings – Ξ–

[19] J. Beringer *et al.* (2012): Particle listings – Ξ+
c

[20] J. Beringer *et al.* (2012): Particle listings – Ξ0
c

[21] J. Beringer *et al.* (2012): Particle listings – Ξ′+
c

[22] J. Beringer *et al.* (2012): Particle listings – Ξ′0
c

[23] J. Beringer *et al.* (2012): Particle listings – Ξ+
cc

[24] J. Beringer *et al.* (2012): Particle listings – Ξ
b

[25] J. Beringer *et al.* (2012): Particle listings – Ω0
c

[26] J. Beringer *et al.* (2012): Particle listings – Ω–
b

[27] J. Beringer *et al.* (2012): Particle listings – Δ(1232)

[28] J. Beringer *et al.* (2012): Particle listings – Σ(1385)

[29] J. Beringer *et al.* (2012): Particle listings – Σ
c(2520)

[30] J. Beringer *et al.* (2012): Particle listings – Σ*
b

[31] J. Beringer *et al.* (2012): Particle listings – Ξ(1530)

[32] J. Beringer *et al.* (2012): Particle listings – Ξ
c(2645)

[33] J. Beringer *et al.* (2012): Particle listings – Ξ0
b(5945)

[34] J. Beringer *et al.* (2012): Particle listings – Ω–

[35] J. Beringer *et al.* (2012): Particle listings – Ω0
c(2770)

[36] http://pdg.lbl.gov/2014/tables/rpp2014-qtab-baryons.pdf

[37] Crede, Volker; Roberts, Winston (2013). "Progress Toward Understanding Baryon Resonances". *Rep. Prog. Phys.* **76**. doi:10.1088/0034-4885/76/7/076301. Retrieved 23 July 2015.

107.3.1 Bibliography

- R. Aaij et al. (LHCb collaboration) (2015). "Observation of J/ψp resonances consistent with pentaquark states in Λ0
b→J/ψK−
p decays". arXiv:1507.03414 [hep-ex].
- J. Beringer *et al.* (Particle Data Group) (2012). "Review of Particle Physics". *Physical Review D* **86** (01): 010001. Bibcode:2012PhRvD...86a0001B. doi:10.1103/PhysRevD.86.010001.
- K. Nakamura *et al.* (Particle Data Group) (2010). "Review of Particle Physics". *Journal of Physics G* **37** (7A): 075021. Bibcode:2010JPhG...37g5021N. doi:10.1088/0954-3899/37/7A/075021.
- C. Amsler *et al.* (Particle Data Group) (2008). "Review of Particle Physics". *Physics Letters B* **667** (1): 1–1340. Bibcode:2008PhLB..667....1P. doi:10.1016/j.physletb.2008.07.018.
- V.M. Abazov (DØ Collaboration) (2008). "Observation of the doubly strange b baryon Ω−
b" (PDF). Fermilab-Pub-08/335-E.
- K. Carter (2006). "The rise and fall of the pentaquark". *Symmetry Magazine*. Fermilab/SLAC. Retrieved 2008-05-27.
- W.-M. Yao *et al.* (Particle Data Group) (2006). "Review of Particle Physics". *Journal of Physics G* **33**: 1–1232. arXiv:astro-ph/0601168. Bibcode:2006JPhG...33....1Y. doi:10.1088/0954-3899/33/1/001.
- H. Muir (2003). "Pentaquark discovery confounds sceptics". *New Scientist*. Retrieved 2008-05-27.
- J.G. Körner, M. Krämer, and D. Pirjol (1994). "Heavy Baryons". *Progress in Particle and Nuclear Physics* **33**: 787–868. arXiv:hep-ph/9406359. Bibcode:1994PrPNP..33..787K. doi:10.1016/0146-6410(94)90053-1.

107.4 Further reading

- H. Garcilazo, J. Vijande, and A. Valcarce (2007). "Faddeev study of heavy-baryon spectroscopy". *Journal of Physics G* **34** (5): 961–976. doi:10.1088/0954-3899/34/5/014.
- S. Robbins (2006). "Physics Particle Overview – Baryons". *Journey Through the Galaxy*. Retrieved 2008-04-20.
- D.M. Manley (2005). "Status of baryon spectroscopy". *Journal of Physics: Conference Series* **5**: 230–237. Bibcode:2005JPhCS...9..230M. doi:10.1088/1742-6596/9/1/043.
- S.S.M. Wong (1998). *Introductory Nuclear Physics* (2nd ed.). New York (NY): John Wiley & Sons. ISBN 0-471-23973-9.
- R. Shankar (1994). *Principles of Quantum Mechanics* (2nd ed.). New York (NY): Plenum Press. ISBN 0-306-44790-8.
- E. Wigner (1937). "On the Consequences of the Symmetry of the Nuclear Hamiltonian on the Spectroscopy of Nuclei". *Physical Review* **51** (2): 106–119. Bibcode:1937PhRv...51..106W. doi:10.1103/PhysRev.51.106.
- M.Gell-Mann(1964). "A Schematic of Baryons and Mesons".*Physics Letters***8**(3): 214–215. Bibcode:1964PhL..G. doi:10.1016/S0031-9163(64)92001-3.
- W.Heisenberg(1932). "Über den Bau der Atomkerne I".*Zeitschrift für Physik*(in German)**77**: 1–11. BibcodH. doi:10.1007/BF01342433.
- W. Heisenberg (1932). "Über den Bau der Atomkerne II". *Zeitschrift für Physik* (in German) **78** (3–4): 156–164. Bibcode:1932ZPhy...78..156H. doi:10.1007/BF01337585.
- W. Heisenberg (1932). "Über den Bau der Atomkerne III". *Zeitschrift für Physik* (in German) **80** (9–10): 587–596. Bibcode:1933ZPhy...80..587H. doi:10.1007/BF01335696.

107.5 External links

- Particle Data Group – Review of Particle Physics (2008).
- Georgia State University – HyperPhysics
- Baryons made thinkable, an interactive visualisation allowing physical properties to be compared

Chapter 108

List of quasiparticles

This is a **list of quasiparticles**.

108.1 References

[1] Angell, C.A.; Rao, K.J. Configurational excitations in condensed matter, and "bond lattice" model for the liquid-glass transition. J. Chem. Phys. 1972, 57, 470-481

[2] Clara Moskowitz (26 February 2014). "Meet the Dropleton—a "Quantum Droplet" That Acts Like a Liquid". *Scientific American*. Retrieved 26 February 2014.

[3] J. Schlappa, K. Wohlfeld, K. J. Zhou, M. Mourigal, M. W. Haverkort, V. N. Strocov, L. Hozoi, C. Monney, S. Nishimoto, S. Singh, A. Revcolevschi, J.-S. Caux, L. Patthey, H. M. Rønnow, J. van den Brink, and T. Schmitt; (2012-04-18). "Spin–orbital separation in the quasi-one-dimensional Mott insulator Sr2CuO3". *Nature,* Advance Online Publication **485** (7396): 82–5. arXiv:1205.1954. Bibcode:2012Natur.485...82S. doi:10.1038/nature10974. PMID 22522933.

[4] "Introducing the Phoniton: a tool for controlling sound at the quantum level". University of Maryland Department of Physics. Retrieved 26 Feb 2014.

[5] Amusia, M., Popov, K., Shaginyan, V., Stephanovich, V. (2014). "Theory of Heavy-Fermion Compounds - Theory of Strongly Correlated Fermi-Systems". Springer. ISBN 978-3-319-10825-4.

[6] Johnson, Hamish. "Introducing the 'wrinklon'". *Physics World*. Retrieved 26 Feb 2014.

[7] Meng, Lan; Su, Ying; Geng, Dechao; Yu, Gui; Liu, Yunqi; Dou, Rui-Fen; Nie, Jia-Cai; He, Lin (2013). "Hierarchy of graphene wrinkles induced by thermal strain engineering". *Applied Physics Letters* **103** (25): 251610. arXiv:1306.0171. Bibcode:2013ApPhL.103y1610M. doi:10.1063/1.4857115. Retrieved 22 March 2014.

108.2 Text and image sources, contributors, and licenses

108.2.1 Text

- **Elementary particle** *Source:* https://en.wikipedia.org/wiki/Elementary_particle?oldid=679470800 *Contributors:* CYD, Mav, Bryan Derksen, XJaM, Heron, Stevertigo, Patrick, Fbjon, Looxix~enwiki, Александър, Julesd, Glenn, AugPi, Mxn, Timwi, Reddi, Tpbradbury, Furrykef, Bevo, Donarreiskoffer, Robbot, Craig Stuntz, Nurg, Papadopc, Wikibot, Jimduck, Anthony, Ancheta Wis, Giftlite, DavidCary, Mikez, Haselhurst, Monedula, Xerxes314, Alison, Guanaco, Greydream, Anythingyouwant, Bodnotbod, Kate, Brianjd, Mormegil, Urvabara, Rich Farmbrough, Guanabot, Qutezuce, Hidaspal, Dmr2, Goplat, RJHall, RoyBoy, Robotje, Neonumbers, ליאור, Dirac1933, DV8 2XL, Azmaverick623, Blaxthos, Kay Dekker, Joriki, Simetrical, TomTheHand, Mpatel, Isnow, Ggonnell, Palica, Strait, Miserlou, Ligulem, Naraht, DannyWilde, Lmatt, Srleffler, Chobot, Cactus.man, Roboto de Ajvol, YurikBot, Hairy Dude, NTBot~enwiki, Ohwilleke, Albert Einsteins pipe, Stephenb, Chaos, Vibritannia, SCZenz, Edwardlalone, Larsobrien, Bota47, BraneJ, Dna-webmaster, Arthur Rubin, Oyvind, GrinBot~enwiki, SmackBot, Mrcoolbp, Bomac, GrGBL~enwiki, Chris the speller, MalafayaBot, George Rodney Maruri Game, Silly rabbit, Complexica, MovGP0, Fmalan, Scwlong, Amazins490, Cybercobra, EPM, Garry Denke, Drphilharmonic, Sadi Carnot, ArglebargleIV, Tktktk, NongBot~enwiki, WhiteHatLurker, Jonhall, Dekaels~enwiki, Jynus, Newone, Courcelles, Laplace's Demon, SchmittM, J Milburn, Fordmadoxfraud, Cydebot, Bvcrist, Kozuch, Thijs!bot, Lord Hawk, Headbomb, MichaelMaggs, Escarbot, Ssr, JAnDbot, Eurobas, Acroterion, VoABot II, Appraiser, BatteryIncluded, R'n'B, Sgreddin, MikeBaharmast, Lk69, Acalamari, DraakUSA, TomasBat, Joshua Issac, Kenneth M Burke, Ken g6, Idioma-bot, VolkovBot, SarahLawrence Scott, Nxavar, JhsBot, Abdullais4u, Lejarrag, Antixt, PGWG, SieBot, Timb66, Sonicology, PlanetStar, Bamkin, Dhatfield, Byrialbot, Svick, Perfectapproach, Thorncrag, Big55e, ClueBot, Jmorris84, Maxtitan, Alexbot, Dekisugi, Paradoxalterist, Saintlucifer2008, Cockshut12345, Rreagan007, RP459, Truthnlove, Addbot, Yakiv Gluck, Draco 2k, Mac Dreamstate, Funky Fantom, CarsracBot, HerculeBot, Legobot, Blah28948, Luckas-bot, Zhitelew, KamikazeBot, Kulmalukko, Orion11M87, AnomieBOT, Girl Scout cookie, Templatehater, Icalanise, Citation bot, Onesius, Vuerqex, Bci2, ArthurBot, Rightly, Xqbot, Phazvmk, Kirin13, FrescoBot, Delphinus1997, Steve Quinn, Robo37, SuperJew, HRoestBot, Sthyne, Hellknowz, Yahia.barie, Skyerise, Tobi - Tobsen, FoxBot, Physics therapist, Think!97, Bjnorge, RjwilmsiBot, Beyond My Ken, EmausBot, John of Reading, Mnkyman, GoingBatty, Mthorndill, ZéroBot, Bollyjeff, StringTheory11, Markinvancouver, Quantumor, Maschen, RolteVolte, Negovori, NTox, I hate whitespace, ClueBot NG, CocuBot, Widr, Micah.yannatos1, Helpful Pixie Bot, Guzman.c, Bibcode Bot, BG19bot, Spaceawesome, Rainbot, Leaverward, Let'sBuildTheFuture, Eduardofeld, Sha-256, Dr.RobertTweed, ZX95, Joeinwiki, Mark viking, Cephas Atheos, Yo butt, Snakeboy666, Psyruby42, Haminoon, Sardeth42, TaiSakuma, LadyCailin, Morph dtlr, Delbert7, Karam adel, Liance, KasparBot, Kurousagi and Anonymous: 186
- **Timeline of particle discoveries** *Source:* https://en.wikipedia.org/wiki/Timeline_of_particle_discoveries?oldid=679101637 *Contributors:* Rmhermen, Tempshill, Harp, Xerxes314, Bodhitha, Perey, Discospinster, Cmdrjameson, JohnAlbertRigali, Crosbiesmith, Rjwilmsi, Strait, Bubba73, Goudzovski, David H Braun (1964), Yamara, SCZenz, Tony1, GrinBot~enwiki, Attilios, SmackBot, Onionmon, Dl2000, JeffW, Lottamiata, Newone, Jhlawr, Headbomb, D.H, Pkoppenb, JNW, Joshmt, Chronitis, SieBot, A. Carty, BartekChom, Muhends, Wprlh, Addbot, DOI bot, MizzoulaB, Luckas-bot, Icalanise, Pepo13, Citation bot, Gogiva, Xqbot, Omnipaedista, Ulm, Carlog3, Citation bot 1, Wbm1058, Bibcode Bot, Dexbot, Trinitresque, Makecat-bot, JanJaeken, ElŞahin, Revolution1221, Monkbot and Anonymous: 24
- **List of particles** *Source:* https://en.wikipedia.org/wiki/List_of_particles?oldid=682746251 *Contributors:* AxelBoldt, Danny, Rmhermen, Stevertigo, Bdesham, Ahoerstemeier, Stan Shebs, Docu, Salsa Shark, Nikai, Evercat, Schneelocke, Charles Matthews, Jitse Niesen, CBDunkerson, Bevo, Raul654, Donarreiskoffer, Robbot, Sanders muc, Merovingian, Pengo, Giftlite, Herbee, Xerxes314, Dratman, Jeremy Henty, Alensha, Bodhitha, Physicist, Hayne, Quadell, RetiredUser2, Mysidia, Icairns, Asbestos, D6, Urvabara, Discospinster, Rich Farmbrough, FT2, Qutezuce, ArnoldReinhold, Neko-chan, El C, Laurascudder, Susvolans, EmilJ, Physicistjedi, Minghong, Gbrandt, Eddideigel, Axl, Mac Davis, David Ko, Radical Mallard, RJFJR, Count Iblis, Dirac1933, TenOfAllTrades, LFaraone, Oleg Alexandrov, Linas, JarlaxleArtemis, Duncan.france, GregorB, Cedrus-Libani, Karam.Anthony.K, Palica, Rjwilmsi, Zbxgscqf, JLM~enwiki, Strait, Ems57fcva, Krash, Dan Guan, DannyWilde, Lmatt, Goudzovski, Chobot, YurikBot, Bambaiah, Vuvar1, Madkayaker, Hydrargyrum, Presscorr, Chaos, Salsb, Tavilis, SCZenz, Lexicon, TUSHANT JHA, Dna-webmaster, Tomvds, Poulpy, Cstmoore, TLSuda, NeilN, MacsBug, Tom Lougheed, McGeddon, Bazza 7, WookieInHeat, Derdeib, Yamaguchi??, Betacommand, Bluebot, Master of Puppets, DHN-bot~enwiki, Raistuumum, Juancnuno, Kittybrewster, Acepectif, Ligulembot, TriTertButoxy, ArglebargleIV, Khazar, John, FrozenMan, JorisvS, 041744, Dr Greg, Slakr, Mets501, Scorpion0422, Cbuckley, Iridescent, TwistOfCain, Happy-melon, JRSpriggs, Flickboy, Van helsing, Lithium6, Neelix, Rotiro, Cydebot, Quibik, Christian75, Omicronpersei8, Thijs!bot, Qwyrxian, TauLibrus, Headbomb, Inner Earth, 49, Guptasuneet, Scottmsg, WinBot, Elmoosecapitan, Tyco.skinner, AubreyEllenShomo, Arch dude, Johnman239, Mwarren us, TheEditrix2, CalamusFortis, MartinBot, Sadisticsuburbanite, Bissinger, Anaxial, CommonsDelinker, Maurice Carbonaro, Zojj, OliverHarris, Joshmt, Adanadhel, Lseixas, Graphite Elbow, VolkovBot, Jmrowland, Quilbert, Anonymous Dissident, Dstary, Escalona, JPMasseo, Figureskatingfan, Inx272, Meters, Antixt, Hamish a e fowler, GodersUK, Bluetryst, SieBot, Ishvara7, WereSpielChequers, Audrius u, VovanA, Paolo.dL, RSStockdale, Anchor Link Bot, StewartMH, Explicit, ClueBot, Unbuttered Parsnip, Nolimitownass, DragonBot, Atomic7732, TimothyRias, SkyLined, Addbot, DOI bot, Jojhutton, Favonian, LinkFA-Bot, OlEnglish, Teles, Legobot, Luckas-bot, Yobot, Dov Henis, Azcolvin429, AnomieBOT, Götz, Icalanise, Flewis, Materialscientist, OllieFury, Vuerqex, ArthurBot, Vulcan Hephaestus, Blennow, Reality006, Coretheapple, Jcimorra, RibotBOT, Ernsts, A. di M., Axelfoley12, Zosterops, FrescoBot, Paine Ellsworth, Citation bot 1, JIK1975, Tom.Reding, Diffequa, WikitanvirBot, Racerx11, 112358sam, Aegnor.erar, Hops Splurt, HESUPERMAN, Hhhippo, AvicBot, JSquish, StringTheory11, Waperkins, Bamyers99, Suslindisambiguator, L Kensington, DennisIsMe, RockMagnetist, ClueBot NG, Snotbot, Primergrey, Vio45lin, Widr, MsFionnuala, Oklahoma3477, Bibcode Bot, CityOfSilver, Cap'n G, BML0309, Dan653, Twocount, Penguinstorm300, Dexbot, LightandDark2000, Ohiggy, TwoTwoHello, Andyhowlett, Printersmoke, Orion 2013, ARUNEEK, Seino van Breugel, AspaasBekkelund, TheMagikCow, Vyom27, ParkersComments, Selva Ganapathy and Anonymous: 290
- **Fermion** *Source:* https://en.wikipedia.org/wiki/Fermion?oldid=682673767 *Contributors:* AxelBoldt, Chenyu, Derek Ross, CYD, Mav, Bryan Derksen, The Anome, Ben-Zin~enwiki, Alan Peakall, Dominus, Dcljr, Looxix~enwiki, Glenn, Nikai, Andres, Wikiborg, David Latapie, Phys, Bevo, Stormie, Olathe, Donarreiskoffer, Robbot, Merovingian, Rorro, Wikibot, HaeB, Giftlite, Fropuff, Xerxes314, Vivektewary, JoJan, Karol Langner, Tothebarricades.tk, Icairns, Hidaspal, Vsmith, Laurascudder, Lysdexia, Ashlux, Graham87, Magister Mathematicae, Kbdank71, Syndicate, Strait, Protez, Drrngrvy, FlaBot, Srleffler, Chobot, YurikBot, RobotE, Jimp, Bhny, Captaindan, SpuriousQ, Salsb, Lomn, Enormousdude, CharlesHBennett, Federalist51, Tom Lougheed, Unyoyega, Jrockley, MK8, BabuBhatt, Complexica, Zachorious, Shalom Yechiel, QFT, Garry Denke, Daniel.Cardenas, SashatoBot, Flipperinu, Dan Gluck, LearningKnight, Happy-melon, Paulfriedman7, Cydebot, Meno25,

Zalgo, Thijs!bot, Mbell, Headbomb, Nick Number, Orionus, Shlomi Hillel, CosineKitty, NE2, Mwarren us, ZPM, Vanished user ty12kl89jq10, Joshua Davis, R'n'B, Tensegrity, Rod57, Dgiraffes, Alpvax, VolkovBot, TXiKiBoT, Red Act, Anonymous Dissident, Abdullais4u, כל יכול, Tanhueiming, Antixt, Haiviet~enwiki, EmxBot, Kbrose, SieBot, Likebox, Jojalozzo, Dhatfield, Oxymoron83, TubularWorld, ClueBot, Seervoitek, Rodhullandemu, Jorisverbiest, Feebas factor, ChandlerMapBot, Nilradical, Wikeepedian, Stephen Poppitt, Addbot, Vectorboson, Luckas-bot, Yobot, Planlips, Dickdock, AnomieBOT, Icalanise, Materialscientist, Xqbot, Br77rino, Balaonair, ??, Paine Ellsworth, Blackoutjack, Kikeku, Rameshngbot, Tom.Reding, RedBot, Alarichus, Michael9422, Silicon-28, TjBot, EmausBot, WikitanvirBot, Quazar121, Solomonfromfinland, JSquish, Fimin, Quondum, AManWithNoPlan, EdoBot, ClueBot NG, PBot1, EthanChant, Bibcode Bot, BG19bot, Petermahlzahn, KingKhan85, ChrisGualtieri, BoethiusUK, DerekWinters, Tentinator, JNrgbKLM, Mohit rajpal, KasparBot, Jiswin1992 and Anonymous: 120

- **Quark** *Source:* https://en.wikipedia.org/wiki/Quark?oldid=681925889 *Contributors:* AxelBoldt, Derek Ross, Vicki Rosenzweig, Mav, Bryan Derksen, The Anome, Gareth Owen, Andre Engels, PierreAbbat, Peterlin~enwiki, Ben-Zin~enwiki, Zoe, Heron, Montrealais, Hfastedge, Edward, Dante Alighieri, Ixfd64, CesarB, Card~enwiki, NuclearWinner, Looxix~enwiki, Ahoerstemeier, Elliot100, Docu, J-Wiki, Nanobug, Aarchiba, Julesd, Glenn, Schneelocke, Jengod, A5, Timwi, Dysprosia, DJ Clayworth, Phys, Ed g2s, Bevo, Olathe, MD87, Jni, Phil Boswell, Sjorford, Donarreiskoffer, Robbot, Sanders muc, Moncrief, Merovingian, PxT, Texture, Bkell, UtherSRG, Widsith, Ancheta Wis, Giftlite, ShaunMacPherson, Harp, Nunh-huh, Lupin, Herbee, Leflyman, Monedula, 0x6D667061, Xerxes314, Anville, Hoho~enwiki, Alison, Beardo, Moogle10000, Wronkiew, Jackol, Bobblewik, Bodhitha, Piotrus, Kaldari, Elroch, Icairns, Zfr, TonyW, Ukexpat, BrianWilloughby, Grunt, O'Dea, Jiy, Discospinster, Rich Farmbrough, Guanabot, T Long, Vsmith, Saintswithin, SocratesJedi, Mani1, Bender235, Lancer, RJHall, Mr. Billion, El C, Kwamikagami, Laurascudder, Susvolans, Triona, Axezz, Bobo192, Army1987, C S, Ziggurat, Rangelov, Matt McIrvin, Jojit fb, Nk, Pentalis, Obradovic Goran, Fwb22, Lysdexia, Benjonson, Alansohn, Gary, Gintautasm, Guy Harris, Keenan Pepper, MonkeyFoo, Lectonar, Mac Davis, Wdfarmer, Snowolf, Schapel, Knowledge Seeker, Evil Monkey, VivaEmilyDavies, CloudNine, Kusma, Kazvorpal, Kay Dekker, Crosbiesmith, Mogigoma, Linas, Mindmatrix, JarlaxleArtemis, ScottDavis, LOL, Wdyoung, Before My Ken, Tylerni7, Jwanders, Dataphiliac, AndriyK, Noetica, Wayward, Wisq, Palica, Marudubshinki, Calréfa Wéná, GSlicer, Graham87, Deltabeignet, Kbdank71, Yurik, Crzrussian, Rjwilmsi, Bremen, Marasama, SpNeo, Mike Peel, Bubba73, DoubleBlue, Matt Deres, Yamamoto Ichiro, Algebra, Dsnow75, RobertG, Nihiltres, Jeff02, RexNL, TeaDrinker, Chobot, DVdm, Jpacold, Gwernol, Elfguy, Roboto de Ajvol, YurikBot, Wavelength, Bambaiah, Sceptre, Hairy Dude, Jimp, Phantomsteve, TheDoober, Dobromila, JabberWok, CambridgeBayWeather, Chaos, Salsb, Wimt, Ugur Basak, NawlinWiki, Spike Wilbury, Bossrat, SCZenz, Randolf Richardson, Danlaycock, Tony1, DRosenbach, Robertbyrne, Dna-webmaster, WAS 4.250, Closedmouth, Pietdesomere, Heathhunnicutt, Kevin, Banus, RG2, Kamickalo, That Guy, From That Show!, Veinor, MacsBug, Smack-Bot, Aigarius, BBandHB, Incnis Mrsi, InverseHypercube, C.Fred, Bazza 7, Ikip, Anastrophe, Jrockley, Eskimbot, AnOddName, Jonathan Karlsson, Edgar181, Gilliam, Dauto, NickGarvey, Vvarkey, Bluebot, KaragouniS, Keegan, Dahn, Bigfun, Miquonranger03, OrangeDog, Silly rabbit, Metacomet, Tripledot, Nbarth, DHN-bot~enwiki, Sbharris, Colonies Chris, Hallenrm, Scwlong, Gsp8181, Can't sleep, clown will eat me, Mallorn, Jeff DLB, TKD, Addshore, Mqjjb30e, Cybercobra, Khukri, B jonas, Jdlambert, Lpgeffen, Nrcprm2026, Akriasas, Zadignose, Jóna Þórunn, Bdushaw, Beyazid, TriTertButoxy, SashatoBot, SciBrad, Doug Bell, Soap, Richard L. Peterson, John, Mgiganteus1, SpyMagician, Edconrad, Loadmaster, 2T, Waggers, SandyGeorgia, Ravi12346, Dbzfrk15146, Peyre, Newone, GDallimore, Happy-melon, Majora4, Chovain, Tawkerbot2, Cryptic C62, JForget, Vaughan Pratt, Hello789, ZICO, SUPRATIM DEY, Ruslik0, CuriousEric, Paulfriedman7, Logical2u, Myasuda, RoddyYoung, Typewritten, Cydebot, Abeg92, Mike Christie, Grahamec, Gogo Dodo, Jayen466, 879(CoDe), Michael C Price, Tawkerbot4, Ameliorate!, Akcarver, Gimmetrow, SallyScot, Casliber, Thijs!bot, Epbr123, NeoPhyteRep, LeBofSportif, Markus Pössel, Anupam, Sopranosmob781, Headbomb, Marek69, John254, KJBurns, MichaelMaggs, Escarbot, Eleuther, Ice Ardor, Aadal, AntiVandalBot, SmokeyTheCat, Tyco.skinner, Exteray, RobJ1981, Rsocol, Ke garne, Deflective, Husond, MER-C, CosineKitty, Andonic, East718, Pkoppenb, DanPMK, Magioladitis, WolfmanSF, Thasaidon, Bongwarrior, VoABot II, باسم, Inertiatic076, Kevinmon, Christoph Scholz~enwiki, Aka042, Giggy, Tanvirzaman, Johnbibby, Cyktsui, ArchStanton69, Ace42, Allstarecho, Shijualex, DerHexer, Elandra, Denis tarasov, MartinBot, Poeloq, Dorvaq, CommonsDelinker, HEL, J.delanoy, Nev1, Ops101ex, DrKiernan, Hgpot, Ferdyshenko, Jigesh, DJ1AM, Tarotcards, Coppertwig, TomasBat, Nikbuz, SJP, FJPB, Vainamainien, Tiggydong, Robprain, Sheliak, Cuzkatzimhut, Lights, X!, VolkovBot, CWii, ABF, John Darrow, Holme053, Nousernamesleft, Ryan032, GimmeBot, Davehi1, A4bot, Captain Courageous, Guillaume2303, Anonymous Dissident, Drestros power, Qxz, Anna Lincoln, Eldaran~enwiki, Leafyplant, Don4of4, PaulTanenbaum, Abdullais4u, Jbryancoop, Mbalelo, Gilisa, Eubulides, Chronitis, Seresin, Dustybunny, Insanity Incarnate, Upquark, Edge1212, Ollieho, AOEU Warrior, SieBot, Graham Beards, WereSpielChequers, Csmart287, Guguma5, Winchelsea, Jbmurray, Caltas, Vanished User 8a9b4725f8376, Keilana, Bentogoa, Aillema, RadicalOne, Arbor to SJ, Elcobbola, Physics one, Dhatfield, RSStockdale, Son of the right hand, Ngexpert5, Ngexpert6, Ngexpert7, Psycherevolt, Sean.hoyland, Mygerardromance, Dabomb87, Nergaal, Muhends, Romit3, SallyForth123, Atif.t2, ClueBot, The Thing That Should Not Be, Wwheaton, Xeno malleus, Harland1, Piledhigheranddeeper, Maxtitan, DragonBot, Glopso, Choonkiat.lee, Himynameisdumb, Worth my salt, Arthur Quark, Estirabot, Brews ohare, Jotterbot, PhySusie, Brianboulton, Dekisugi, ANOMALY-117, Sallicio, Yomangan, Jtle515, Katanada, DumZiBoT, TimothyRias, XLinkBot, Vayalir, Oldnoah, Saintlucifer2008, Nathanwesley3, Dragonfiremage, Devilist666, Mancune2001, Jbeans, WikiDao, SkyLined, Truthnlove, Airplaneman, Eklipse, Addbot, Eric Drexler, AVand, Some jerk on the Internet, Captain-tucker, Giants2008, Iceblock, Ronhjones, Quarksci, Mseanbrown, Looie496, LaaknorBot, Peti610botH, AgadaUrbanit, Tide rolls, Vicki breazeale, Gail, Extruder~enwiki, Abduallah mohammed, Dealer77, Luckas-bot, Yobot, Fraggle81, Cflm001, Legobot II, Amble, Mmxx, Superpenguin1984, Worm That Turned, The Vector Kid, Planlips, Fangfyre, TestEditBot, Azcolvin429, Vroo, Synchronism, Bility, Orion11M87, AnomieBOT, Xi rho, Rubinbot, Jim1138, Bookaneer, Yotcmdr, Crystal whacker, Sonic h, Materialscientist, Citation bot, Pitke, Vuerqex, Bci2, ArthurBot, LilHelpa, Xqbot, Jeffrey Mall, AbigailAbernathy, Srich32977, Alex2510, Almabot, Uscbino, Pmlineditor, RibotBOT, Shmomuffin, Gunjan verma81, Chotarocket, Ernsts, Renverse, A. di M., Weekendpartier, FrescoBot, Paine Ellsworth, DelphinidaeZeta, Steve Quinn, Citation bot 1, AstaBOTh15, Pinethicket, Jonesey95, Calmer Waters, Skyerise, Pmokeefe, Jschnur, Searsshoesales, Jrobbinz123, Lissajous, Turian, Lando Calrissian, Wotnow, Ansumang, Reaper Eternal, 564dude, Jackvancs, Bobotast, MINTOPOINT, TjBot, DexDor, Антон Гліністы, Daggersteel10, Chiechiecheist, EmausBot, John of Reading, WikitanvirBot, Duskbrood, FergalG, Slightsmile, Barak90, Wikipelli, TheLemon1234, Manofgrass, Brazmyth, H3llBot, Stoneymufc29, GeorgeBarnick, Brandmeister, Ego White Tray, TYelliot, ClueBot NG, Gilderien, A520, Cheeseequalsyum, Timothy jordan, 123Hedgehog456, Maplelanefarm, 336, Helpful Pixie Bot, Jeffreyts11, 123456789malm, Bibcode Bot, BG19bot, Hurricanefan25, MusikAnimal, Davidiad, MosquitoBird11, Mydogpwnsall, MrBill3, Njavallil, Glacialfox, Walterpfeifer, Thebannana, CE9958, Marioedesouza, Mediran, Dexbot, Rishab021, Cjean42, Sriharsh1234, Sam boron100, Wankybanky, Wikitroll12345, RojoEsLardo, Jwratner1, NottNott, Saebre, JNrgbKLM, KheltonHeadley, AspaasBekkelund, HectorCabreraJr, Hazinho93, Quadrupedi, QuantumMatt101, Philipphilip0001, Monkbot, RiderDB, Egfraley, Tetra quark, Weed305, KasparBot and Anonymous: 705

- **Up quark** *Source:* https://en.wikipedia.org/wiki/Up_quark?oldid=666803858 *Contributors:* Bryan Derksen, Alfio, Jni, Giftlite, Xerxes314, Kjoonlee, Bookandcoffee, CharlesC, Rjwilmsi, Mike Peel, Chobot, Hairy Dude, Rt66lt, Spike Wilbury, SCZenz, Poulpy, Eog1916, Bluebot, Tamfang, T-borg, Eric Saltsman, Hetar, Lottamiata, Laplace's Demon, Merryjman, CmdrObot, Myasuda, Raoul NK, Headbomb, JAnDbot, Abyssoft, I310342~enwiki, Idioma-bot, Sheliak, Wilmot1, VolkovBot, TXiKiBoT, Anonymous Dissident, Gekritzl, AlleborgoBot, SieBot, Muhends, Bobathon71, DragonBot, DumZiBoT, TimothyRias, Addbot, Eivindbot, LaaknorBot, ChenzwBot, Naidevinci, Ehrenkater, Lightbot, Luckas-bot, Citation bot, ArthurBot, Xqbot, DSisyphBot, Ditimchanly, Almabot, A. di M., Paine Ellsworth, Citation bot 1, Trappist the monk, TjBot, Ripchip Bot, EmausBot, WikitanvirBot, StringTheory11, Quondum, Helpful Pixie Bot, Bibcode Bot, BG19bot, P76837, Oznitecki, Alexzhang2, The Great Leon, Monkbot and Anonymous: 38
- **Down quark** *Source:* https://en.wikipedia.org/wiki/Down_quark?oldid=663284022 *Contributors:* Bryan Derksen, Alfio, Timwi, Jni, Herbee, Xerxes314, Rich Farmbrough, Kjoonlee, Rjwilmsi, Mike Peel, Chobot, YurikBot, Rt66lt, Acidsaturation, Spike Wilbury, SCZenz, Poulpy, Otto ter Haar, Skizzik, Bluebot, Tamfang, Llwang, Eric Saltsman, MTSbot~enwiki, Hetar, Laplace's Demon, Myasuda, Raoul NK, Thijs!bot, Headbomb, Davidhorman, JAnDbot, Abyssoft, MartinBot, Jvineberg, I310342~enwiki, Sheliak, VolkovBot, TXiKiBoT, Anonymous Dissident, SieBot, Ngexpert6, Muhends, Bobathon71, Lawrence Cohen, Daigaku2051, Auntof6, NuclearWarfare, TimothyRias, Addbot, Lightbot, Luckas-bot, Yobot, Citation bot, ArthurBot, Xqbot, DSisyphBot, Paine Ellsworth, Citation bot 1, Tim1357, Trappist the monk, EmausBot, ZéroBot, Quondum, Rezabot, Helpful Pixie Bot, Bibcode Bot, TheMan4000, 786b6364, Monkbot and Anonymous: 22
- **Charm quark** *Source:* https://en.wikipedia.org/wiki/Charm_quark?oldid=663286006 *Contributors:* Bryan Derksen, Alfio, Bogdangiusca, Xerxes314, Bodhitha, Perey, Kjoonlee, Rjwilmsi, Mike Peel, Chobot, YurikBot, Bambaiah, Conscious, Salsb, SCZenz, Scottfisher, Poulpy, SmackBot, Delldot, Warhol13, Rezecib, Vina-iwbot~enwiki, Happy-melon, Laplace's Demon, CRGreathouse, Michael C Price, Thijs!bot, Headbomb, Nisselua, JAnDbot, Abyssoft, Uncle.wink, Bryanhiggs, HEL, I310342~enwiki, Qoou.Anonimu, Idioma-bot, Sheliak, Anonymous Dissident, Kumorifox, BelsKr, AlleborgoBot, SieBot, Muhends, TimothyRias, Addbot, Mjamja, Lightbot, Luckas-bot, Yobot, Nallimbot, Citation bot, ArthurBot, Quebec99, Xqbot, DSisyphBot, GrouchoBot, RibotBOT, SassoBot, A. di M., Paine Ellsworth, Dogposter, D'ohBot, Citation bot 1, Citation bot 4, RedBot, MastiBot, Trappist the monk, EarthCom1000, Alph Bot, EmausBot, ZéroBot, Quondum, Anita5192, CocuBot, Rezabot, Helpful Pixie Bot, Bibcode Bot, Penguinstorm300, Hoppeduppeanut, Leowestland and Anonymous: 35
- **Strange quark***Source:* https://en.wikipedia.org/wiki/Strange_quark?oldid=663285537*Contributors:* Bryan Derksen, Alfio, Jni, Owain, Xerxes314,Soman, Kjoonlee, Kwamikagami, Rsholmes, Esb82, Neonumbers, Rjwilmsi, Mike Peel, Gurch, Erik4, Chobot, YurikBot, Jimp, Salsb , SCZenz,Poulpy, SmackBot, Bluebot, NCurse, Vina-iwbot~enwiki, Yevgeny Kats, *Zzzzzzzzzz*, Laplace's Demon, MightyWarrior, Myasuda, Thijs!bot,Headbomb, Chillysnow, JAnDbot, Abyssoft, Bongwarrior, Albmont, McSly, I310342~enwiki, Pdcook, Sheliak, VolkovBot, SieBot, Muhends,Auntof6, Iohannes Animosus, TimothyRias, IngerAlHaosului, Addbot, ProbablyAmbiguous, Luckas-bot, Yobot, AnomieBOT , Citation bot,Sarah12sarah, Erik9bot, Thehelpfulbot, Paine Ellsworth, Rkr1991, Citation bot 1, Skyerise, Johann137, Trappist the monk, Puzl bustr, Agrasa,Wikiborg4711, EmausBot, Hhhippo, ZéroBot, Quondum, CocuBot, Helpful Pixie Bot, Bibcode Bot, Vkpd11, P76837, Matthew gib, Glaisher,RhinoMind and Anonymous: 38
- **Top quark** *Source:* https://en.wikipedia.org/wiki/Top_quark?oldid=679102575 *Contributors:* Damian Yerrick, Bryan Derksen, HPA, Haryo, Bkell, Giftlite, Xerxes314, Edcolins, Bodhitha, David Schaich, Kjoonlee, Axl, Woohookitty, Rjwilmsi, Strait, Mike Peel, Vegaswikian, Wikiliki, Goudzovski, Chobot, YurikBot, Bambaiah, JabberWok, Gaius Cornelius, Salsb, Howcheng, SCZenz, Emijrp, Physicsdavid, SmackBot, Incnis Mrsi, ZerodEgo, Mr.Z-man, Jgwacker, Pulu, Stikonas, Mets501, Peyre, RekishiEJ, Banedon, הסרפד, Headbomb, Davidhorman, Oreo Priest, AntiVandalBot, JAnDbot, Abyssoft, Maliz, HEL, Fatka, I310342~enwiki, Idioma-bot, Sheliak, Biggus Dictus, TXiKiBoT, Reibot, Kachuak, Ptrslv72, SieBot, Hatster301, Muhends, ClueBot, Niceguyedc, Noca2plus, Choonkiat.lee, Brews ohare, Kakofonous, Jtle515, TimothyRias, Prostarplayer321, SkyLined, Cockatoot, Addbot, Mr0t1633, Mjamja, ChenzwBot, Ginosbot, Zorrobot, Luckas-bot, Naudefjbot~enwiki, Dreamer08, AnomieBOT, Icalanise, Citation bot, ArthurBot, LilHelpa, DSisyphBot, Unready, GrouchoBot, RibotBOT, Soandos, Paine Ellsworth, Citation bot 1, Jonesey95, Thinking of England, Nomis2k, Higgshunter, RjwilmsiBot, Mophoplz, EmausBot, John of Reading, WikitanvirBot, Barak90, StringTheory11, Peter.poier, Quondum, Samlever, Whoop whoop pull up, Reify-tech, Helpful Pixie Bot, Bibcode Bot, Glevum, Kephir, Mmitchell10, Quadrupedi, Monkbot, BrunoUbaldo and Anonymous: 64
- **Bottom quark** *Source:* https://en.wikipedia.org/wiki/Bottom_quark?oldid=676070719 *Contributors:* Bryan Derksen, Xerxes314, Bodhitha, Icairns, Kjoonlee, Bobo192, Pinar, WadeSimMiser, Rjwilmsi, Mike Peel, Erkcan, FlaBot, Itinerant1, Chobot, YurikBot, Bambaiah, Jimp, Conscious, Ozabluda, SpuriousQ, Salsb, SCZenz, Lexicon, Poulpy, Physicsdavid, SmackBot, Hmains, Luís Felipe Braga, Laplace's Demon, CmdrObot, Outriggr, Niubrad, הסרפד, Thijs!bot, Headbomb, JAnDbot, Abyssoft, Pkoppenb, Dr. Morbius, I310342~enwiki, Joshmt, Idioma-bot, Sheliak, VolkovBot, Antixt, AlleborgoBot, BartekChom, Muhends, Auntof6, TimothyRias, Lockalbot, Addbot, Mr Sme, Luckas-bot, THEN WHO WAS PHONE?, Citation bot, ArthurBot, Xqbot, GrouchoBot, StevenVerstoep, Thehelpfulbot, Paine Ellsworth, Citation bot 1, Jonesey95, Double sharp, TjBot, EmausBot, Barak90, TuHan-Bot, ZéroBot, StringTheory11, Quondum, Chris857, ChuispastonBot, Widr, Helpful Pixie Bot, Bibcode Bot, P76837, ChrisGualtieri, Ajd268, Mfb, Monkbot, Axel Azzopardi, Kenijr and Anonymous: 33
- **Lepton***Source:* https://en.wikipedia.org/wiki/Lepton?oldid=679886360*Contributors:* Bryan Derksen, Andre Engels, PierreAbbat, Ben-Zin~enwiki,Heron, Xavic69, Fruge~enwiki, Fwappler, Ahoerstemeier, Julesd, Glenn, Mxn, A5, Wikiborg, Dysprosia, Radiojon, Imc, Morwen, Fibonacci,Bcorr, Phil Boswell, Donarreiskoffer, Robbot, Merovingian, Wikibot, Giftlite, Smjg, DocWatson42, Harp, Herbee, Xerxes314, Sysin , Knutux,LiDaobing, LucasVB, ClockworkLunch, RetiredUser2, Icairns, Mike Rosoft, Chris j wood, Martinl~enwiki, Smalljim, Giraffedata, Jumbuck,RobPlatt, Neonumbers, Ahruman, Computerjoe, Simon M, Woohookitty, Mindmatrix, Rjwilmsi, Strait, Erkcan, FlaBot, DannyWilde, Mas-torrent, Celebere, Peterl, YurikBot, Bambaiah, Jimp, Salsb, Spike Wilbury, Jaxl, SCZenz, DeadEyeArrow, Tetracube, Smoggyrob, Dmuth,Jaysbro, Sbyrnes321, That Guy, From That Show!, SmackBot, Bazza 7, KocjoBot~enwiki, Jrockley, Mom2jandk, Cool3, Hmains, Complex-ica, DHN-bot~enwiki, Mesons, Yevgeny Kats, TriTertButoxy, SashatoBot, Ouzo~enwiki, Happy-melon, Kurtan~enwiki, Myasuda, Cydebot,Meno25, Photocopier, Michael C Price, Casliber, Thijs!bot, Headbomb, Newton2, Mentifisto, Autotheist, Steveprutz, NeverWorker, NicoSan,MartinBot, Arjun01, HEL, J.delanoy, Numbo3, Gombang, Num1dgen, Ceoyoyo, VolkovBot, Macedonian, Mocirne, TXiKiBoT, Anony-mous Dissident, Abdullais4u, Antixt, Jhb110, Thanatos666, AlleborgoBot, SieBot, ToePeu.bot, RadicalOne, Ngexpert7, Jacob.jose, Hamilton-daniel, TubularWorld, Muhends, ClueBot, ICAPTCHA, UniQue tree, Snigbrook, Fyyer, IceUnshattered, Cmj91uk, LieAfterLie, Manu-ve ProSki, TimothyRias, Addbot, Betterusername, AgadaUrbanit, Ehrenkater, OlEnglish, Zorrobot, Andy2308, Legobot, Luckas-bot, Ptbotgourou,Maxim Sabalyauskas, Planlips, JackieBot, Icalanise, Citation bot, MastiBot, ArthurBot, Almabot, Omnipaedista, Alexeymorgunov, ??,Tormine, MathFacts, Citation bot 1, MastiBot, .يدماغ.دمحأ24, ArthurBot, Almabot, Omnipaedista, Alexeymorgunov, ??,Tormine, MathFacts, Citation bot 1, MastiBot, Earthandmoon, EmausBot, John of Reading, Az29, Galaktiker, StringTheory11, Quondum,Surajt88, I hate whitespace, ClueBot NG, Scimath Genius, Braincricket, Widr, Helpful Pixie Bot, Bibcode Bot, Tyler6360534, Katagun5,Melenc, DerekWinters, Prasanna4s, Machosquirrel, Devinhorn, KasparBot and Anonymous: 149

- **Positron** *Source:* https://en.wikipedia.org/wiki/Positron?oldid=679431995 *Contributors:* Bryan Derksen, Andre Engels, Hhanke, Peterlin~enwiki,Patrick, Looxix~enwiki, Hashar, Stismail, Pstudier, BenRG, Donarreiskoffer, Robbot, Merovingian, LGagnon, David Gerard, Decumanus,Giftlite, Jmnbpt, Xerxes314, Utcursch, Xmnemonic, Knutux, Icairns, Tumbarumba, Mike Rosoft, Vsmith, Jpk, Murtasa, Bender235, El C,Asierra~enwiki, La goutte de pluie, Tra, Atlant, Spangineer, Wtmitchell, SidP, RogerBarnett, Gene Nygaard, WojciechSwiderski~enwiki,UTSRelativity, Richard Arthur Norton (1958-), MartinSpacek, Jannex, Ma Baker, Robert K S, Palica, MassGalactusUniversum, V8rik,BD2412, Canderson7, Seraphimblade, Mike Peel, FlaBot, DannyWilde, Kolbasz, Fresheneesz, Tardis, Wrightbus, Chobot, YurikBot, Bam-baiah, Jimp, Arado, Chuck Carroll, Bergsten, Damato, Hellbus, Astriks, Salsb, MidnightWolf, Vanished user kjdioejh329io3rksdkj, Com-plainer, Robertvan1, Howcheng, SCZenz, Zwobot, Scottfisher, Lt-wiki-bot, Arthur Rubin, Terbospeed, Vicarious, GrinBot~enwiki, Smack-Bot, Unyoyega, Edgar181, KingRaptor, Skizzik, Kmarinas86, Complexica, DHN-bot~enwiki, Sbharris, Rodri316, JorisvS, Mr. Vernon,Rock4arolla, MTSbot~enwiki, Iridescent, Michaelbusch, Darth Sader, CapitalR, CmdrObot, Eric, Cofax48, Yzphub, HalJor, Cydebot, NickY., Bvcrist, Yolocavo, HPaul, Quibik, Thijs!bot, Gamer007, Headbomb, Rlupsa, Davidhorman, Shadow Blaziken, Griba2010, Escarbot,Poshzombie, Harrylentil, JAnDbot, CosineKitty, Rob Mahurin, MegX, Recurring dreams, Mother.earth, Jetterman, Maliz, Gwern, Geboy,Boddey, Rustyfence, CliffC, Rigmahroll, CommonsDelinker, HEL, Solarswordsman, AntiSpamBot, Y2H, Sheliak, TreasuryTag, Philip True-man, TXiKiBoT, DavidRThomas, Corticopia, Agradada, Corvus cornix, Lerdthenerd, Gabriel Vidal, Seraphita~enwiki, AlleborgoBot, SieBot,Winchelsea, Triwbe, RadicalOne, Flyer22, WingkeeLEE, BenoniBot~enwiki, Drgarden, ClueBot, Fyyer, Gabriel Vidal Álvarez, Piledhigh-eranddeeper, Maxtitan, Djr32, Tyler, EncyclopediaUpdaticus, DumZiBoT, AgnosticPreachersKid, TheRealVolucrix, SkyLined, Addbot, DOlbot, AkhtaBot, CanadianLinuxUser, Jim10701, AgadaUrbanit, Positroni, Tide rolls, Luckas-bot, Yobot, Allowgolf~enwiki, Götz, Jim1138, Taupositron, Materialscientist, Citation bot, Yathimc, Jakouso, Stephen.G.McAteer, FrescoBot, Paine Ellsworth, Dscraggs, Citation bot 1, Idream of horses, Tom.Reding, Wdanbae, RjwilmsiBot, Samdacruel, EmausBot, WikitanvirBot, Harddk, H3llBot, Quondum, I kabir, MG-Weatherman08, Carmichael, Teapeat, ClueBot NG, Snotbot, Helpful Pixie Bot, Ieditpagesincorrectly, Bibcode Bot, Slaughter182, SergeantCribb, Drizzt182, Dexbot, Crocgandhi, Bigdumpy, Planetguy2345, KasparBot, Kobiej100, People73, Person420 and Anonymous: 148
- **Muon** *Source:* https://en.wikipedia.org/wiki/Muon?oldid=680138424 *Contributors:* AxelBoldt, The Epopt, CYD, Mav, Bryan Derksen, Zundark, Roadrunner, Bkellihan, Youandme, Tim Starling, EddEdmondson, Looxix~enwiki, Ahoerstemeier, Angela, Rob Hooft, Kbk, Donarreiskoffer, AlexPlank, Robbot, Merovingian, Rholton, Ojigiri~enwiki, Auric, Roscoe x, Bkell, Millosh, Wikibot, Ruakh, Diberri, Giftlite, Wizzy, Herbee, Xerxes314, Bodhitha, LiDaobing, Pcarbonn, DragonflySixtyseven, Deglr6328, Eb.hoop, Rich Farmbrough, Pjacobi, Vsmith, Mani1,STGM,Kjoonlee,RJHall,Army1987,Danski14,Anthony Appleyard,RobPlatt,Keenan Pepper,RJFJR,Ceyockey,Falcorian,Woohookitty, Xinghuei,Bennetto,Graham87,Vanderdecken,Rjwilmsi,Strait,Mike Peel,Bubba73,DoubleBlue,Dougluce,FlaBot,Fivemack,DannyWilde,Sp00n, Lmatt, Goudzovski, Srleffler, Chobot, YurikBot, Wavelength, Bambaiah, Limulus, JabberWok, Hellbus, Salsb, SCZenz, Ravedave,Scottfisher, Tetracube, Lt-wiki-bot, E Wing, Roberto DR, CrniBombarder!!!, Sbyrnes321, Eog1916, SmackBot, Incnis Mrsi, Melchoir, Sti-fle, Gilliam, Dauto, Bluebot, Tigerhawkvok, Sbharris, Colonies Chris, Can't sleep, clown will eat me, Yevgeny Kats, Dane Sorensen, JorisvS,Mets501, JoeBot, CapitalR, SchmittM, Ruslik0, Ken Gallager, Rotiro, A876, Corpx, Thijs!bot, Headbomb, D.H, Bm gub, Andrew Carlssin,Spencer, Kariteh, Deflective, Belg4mit, Swpb, Mother.earth, Nono64, HEL, Hans Dunkelberg, 5Q5, Tarotcards, Coppertwig, Bermy88,Jarry1250, Thecinimod,Sheliak,Cuzkatzimhut,VolkovBot,Larryisgood,VasilievVV,TXiKiBoT,Anonymous Dissident,Mihaip,Graymorn-ings, SalomonCeb , SieBot, Csmart287, Gerakibot, Statue2, Mhouston, StewartMH, ClueBot, Polyamorph, DnetSvg, Esbboston, Saritepe,Stefan Ritt, BarretB, Kajabla, Addbot, Roentgenium111, Toyokuni3, Download, Ehrenkater, Lightbot, Luckas-bot, Yobot, Evaders99, Kul-malukko, AnomieBOT, Icalanise, Kingpin13, Materialscientist, Citation bot, Kotika98, ArthurBot, Xqbot, Cjxc92, Gilo1969, Srich32977,Misterigloo, Kyng, A. di M., Paine Ellsworth, Citation bot 1, Citation bot 4, Rameshngbot, Isofox, Jetstoknowhere, TobeBot, Trappist themonk, Puzl bustr, RjwilmsiBot, EmausBot, John of Reading, WikitanvirBot, Dewritech, GoingBatty, Milledit, Naviguessor, StringTheory11,Medeis,Suslindisambiguator, Quondum,Timetraveler3.14,Layona1,Aerthis,Mikhail Ryazanov,Frietjes,CaroleHenson,Kebil,Bibcode Bot,Jesusmonkey,NotWith,BattyBot, Kisokj,Liam135,MuonRay,Tony Mach,Telfordbuck,Krotera,Ajdigregorio,Seoman2snowlock,Monkbot,Jromerofontalvo,KasparBot, Corrupt Titan,QzPhysics and Anonymous: 191
- **Tau (particle)** *Source:* https://en.wikipedia.org/wiki/Tau_(particle)?oldid=680139275 *Contributors:* Bryan Derksen, Iluvcapra, Ahoerstemeier, Bueller 007, Schneelocke, Dysprosia, Donarreiskoffer, Merovingian, Rorro, Davidl9999, Millosh, Harp, Herbee, Codepoet, Xerxes314, Bodhitha, CryptoDerk, Icairns, Rich Farmbrough, Pjacobi, Martpol, Sunborn, Kjoonlee, El C, Reuben, JellyWorld, RobPlatt, RJFJR, Falcorian, Dmitry Brant, Christopher Thomas, Palica, Rjwilmsi, Strait, Mike Peel, FlaBot, DannyWilde, Goudzovski, Chobot, RobotE, Bambaiah, AcidHelmNun, JabberWok, Eleassar, Salsb, SCZenz, Zwobot, Ospalh, PS2pcGAMER, Bota47, Someones life, Poulpy, Physicsdavid, Incnis Mrsi, Dauto, Pieter Kuiper, Loodog, JorisvS, MTSbot~enwiki, WISo, Q43, Thijs!bot, Headbomb, Davidhorman, Hcobb, Escarbot, RogueNinja, Yill577, Soulbot, Kostisl, STBotD, Sheliak, Joyko~enwiki, VolkovBot, Fences and windows, TXiKiBoT, Awl, Jba138, SieBot, OKBot, ImageRemovalBot, Plastikspork, Djr32, Alexbot, TimothyRias, Assosiation, BodhisattvaBot, SkyLined, J Hazard, Addbot, Eric Drexler, Ronhjones, ChenzwBot, Jklukas, Theozzfancometh, Skippy le Grand Gourou, Luckas-bot, Yobot, Grebaldar, AnomieBOT, Icalanise, Citation bot, Xqbot, Blennow, Franco3450, ??, Paine Ellsworth, Jonesey95, Three887, Plasticspork, 3ph, Miracle Pen, RjwilmsiBot, TjBot, Ripchip Bot, EmausBot, Dcirovic, Suslindisambiguator, Quondum, Rezabot, Helpful Pixie Bot, Bibcode Bot, BG19bot, Sudsguest, YFdyh-bot, Redcliffe maven, TwoTwoHello, Akro7, KasparBot, JPPepper, QzPhysics and Anonymous: 48
- **Neutrino** *Source:* https://en.wikipedia.org/wiki/Neutrino?oldid=681496713 *Contributors:* AxelBoldt, Chenyu, Bryan Derksen, Zundark, The Anome, Tarquin, Andre Engels, Xaonon, XJaM, William Avery, Roadrunner, DrBob, Heron, Cwitty, MimirZero, Spiff~enwiki, Edward, Patrick, Ken Arromdee, EddEdmondson, Ezra Wax, Gdarin, Meekohi, Bcrowell, Cyde, Arpingstone, Alfio, Looxix~enwiki, Strebe, JWSchmidt, Julesd, Glenn, Nikai, Andres, Evercat, Rob Hooft, TheSeez, Crissov, Wikiborg, Reddi, Lfh, Cos111, Tpbradbury, Fibonacci, Warofdreams, Twang, Donarreiskoffer, Drxenocide, Robbot, Findel, Zandperl, Nurg, Masao, Merovingian, Bobunf, Rursus, Meelar, Matty j, Intangir, Wikibot, Wereon, Duien, Jimduck, Bbx, David Gerard, Giftlite, Graeme Bartlett, DocWatson42, Laudaka, Mikez, Harp, Lethe, HangingCurve, Xerxes314, Anville, Dratman, Curps, Jorge Stolfi, Eequor, Mdob, Espetkov, LiDaobing, Elroch, Icairns, Doug Danner, Nickptar, Fg2, Lrenh, Deglr6328, Hmmm~enwiki, Mattman723, Helohe, Rich Farmbrough, Hydrox, Cacycle, Pjacobi, Vsmith, Dbachmann, Mani1, Pavel Vozenilek, Ralfoide, Sunborn, Neko-chan, Kharhaz, RJHall, Charm, Haxwell, RoyBoy, Smalljim, Cje~enwiki, Viriditas, Cwolfsheep, Foobaz, I9Q79oL78KiL0QTFHgyc, La goutte de pluie, Thewayforward, Thuktun, Fleurot~enwiki, Quaoar, Alansohn, Anthony Appleyard, ChristopherWillis, Calton, Axl, Mac Davis, Hdeasy, RJFJR, Dirac1933, TenOfAllTrades, Vuo, Cmprince, Pauli133, Gene Nygaard, Lyuokdea, Flying fish, Richard Arthur Norton (1958-), Woohookitty, Swamp Ig, Insaneinside, Benhocking, Nakos2208~enwiki, GregorB, SDC, Joke137, Fxer, Palica, RedBLACKandBURN, Ashmoo, Graham87, Qwertyus, Raymond Hill, Drbogdan, Rjwilmsi, Coemgenus, Strait, John187, Staecker, Jmcc150, Salix alba, Mike Peel, Vegaswikian, Oblivious, Ligulem, R.e.b., Jehochman, The wub, FlaBot, Ian Pitchford, DannyWilde, Itiner-

ant1, RexNL, Gurch, Kolbasz, Goudzovski, Sperxios, Scythe33, Smithbrenon, Chobot, Nagytibi, DVdm, YurikBot, Bambaiah, Vuvar1, Phmer, RussBot, Ohwilleke, Witan, Xihr, Bhny, Chris Capoccia, JabberWok, Gaius Cornelius, Salsb, Grafen, Długosz, Gillis, SCZenz, Ravedave, Abb3w, CecilWard, Santaduck, Bota47, Maunus, Dna-webmaster, Ms2ger, Rhynchosaur, DrWorm, Alias Flood, Ilmari Karonen, Nimbex, Phr en, Otto ter Haar, Fragman, That Guy, From That Show!, AndrewWTaylor, Palapa, Morgan wascko, SmackBot, Trainbrain27, Reedy, Tom Lougheed, Melchoir, The Monster, Arbe, Mscuthbert, BiT, Ohnoitsjamie, Dauto, Kmarinas86, GregRM, Decowski, Yurigerhard, DHN-bot~enwiki, Sbharris, Colonies Chris, Hengsheng120, Nap~enwiki, Sergio.ballestrero, Милан Јелисавчић, Cophus, Mayrel, Wen D House, Engwar, Jdlambert, Webmaster Pete, TheMaster42, Pwjb, Akriasas, DenisRS, Rjn~enwiki, Kukini, Yevgeny Kats, Ged UK, Jjpcondor, Gobonobo, ThorAvaTahr, JorisvS, Makyen, Libera~enwiki, Aeluwas, Dicklyon, Mets501, MTSbot~enwiki, Galactor213, Fredil Yupigo, Masoninman, Newone, Richard75, Chalnoth, Mssgill, Valoem, Abeneal, Rszasz, CmdrObot, Calmargulis, Olaf Davis, Vyznev Xnebara, MrFizyx, Thubsch, Myasuda, Alton, Astralusenet, Icek~enwiki, Szdori~enwiki, Hyperdeath, Gogo Dodo, HPaul, RC Master, Q43, Michael C Price, Quibik, Christian75, DumbBOT, Joe Chick, Thijs!bot, Martin Hogbin, Naucer, Headbomb, Dtgriscom, WVhybrid, Esemono, James086, Second Quantization, Davidhorman, Weasel5i2, Jonny-mt, D.H, Greg L, Mentifisto, Luna Santin, Guy Macon, Seaphoto, Alphachimpbot, Astavats, Parande, DagosNavy, JAnDbot, Deflective, MER-C, CosineKitty, Savant13, Magioladitis, WolfmanSF, VoABot II, Nyq, Websterwebfoot, Bakken, DMcanada, Christoph Scholz~enwiki, Seleucus, Dirac66, Adrian J. Hunter, LorenzoB, NJR ZA, Khalid Mahmood, Squidonius, Pavel Jelínek, Gwern, Denis tarasov, Glrx, R'n'B, Fatka, Maurice Carbonaro, MrBell, Aqwis, Salih, Nalumc, Plasticup, Warut, Nwbeeson, Rosenknospe, Juliancolton, Mike Clough, Bonadea, Lseixas, Sheliak, Cuzkatzimhut, Jharris1993, VolkovBot, Camrn86, AlnoktaBOT, DrJohnPCostella, TXiKiBoT, The Original Wildbear, Nxavar, MinotAuruS, Cgr1123, Awl, Michael H 34, HannesHultgren~enwiki, SuperLonghorn, BotKung, SwordSmurf, Norbu19, Richwil, Tomaxer, The assassin 47, Morangm, AlleborgoBot, Thunderbird2, Angelastic, SieBot, Fredelige, PlanetStar, Zelab, Laoris, Ergateesuk, Stratman07, Jerryobject, RadicalOne, Aaarnooo, ScAvenger lv, John fromer, Thehotelambush, ShadowPhox, Ergo4sum, AWeishaupt, Jahilia, Extensive~enwiki, Sagredo, Martarius, ClueBot, Justin W Smith, Plastikspork, Apparentslug, Der Golem, Mild Bill Hiccup, Msgarrett, Hyh1048576, Mostargue, Paulcmnt, Ajoykt, Rhmtsang~enwiki, Xmantis, JayVora, Excirial, Kain Nihil, Wacko375, TonyBermanseder, Cenarium, Alastair301, Jwfvalle, Billrob458, Askahrc, Bouhadef, Termatt56, Johnuniq, PSimeon, Ost316, Nepenthes, Jprw, PL290, Jht4060, Buchler, MystBot, SkyLined, NCDane, Stephen Poppitt, Rical, Addbot, DOI bot, Download, Tide rolls, Lightbot, Taketa, SPat, Zorrobot, GDK, Gameseeker, Olsen-Fan, Legobot, Luckas-bot, Yobot, Bunnyhop11, Zagothal, Amble, Wikipedian Penguin, Azcolvin429, LibrarianofBabel, Dickdock, Robert Treat, AnomieBOT, Fatal!ty, Wrongfilter, Jim1138, Icalanise, Dakarateka, Mahmudmasri, Citation bot, Tano-kun, Vuerqex, GB fan, Xqbot, Blennow, JimVC3, Nickkid5, Cydelin, Srich32977, GrouchoBot, Abce2, Baba476, Backpackadam, QMarion II, 78.26, Aashaa, Ernsts, A. di M., Eldudarino, FrescoBot, LucienBOT, Paine Ellsworth, Tobby72, Ajgw56, WurzelT, Citation bot 1, Merongb10, RandomDSdevel, Pinethicket, HRoestBot, Rameshngbot, Tom.Reding, Swamper777, Phil John Hawkins, PRONIZ, Rknop, Jkforde, Nieuwenh, Puzl bustr, Tawe, Higgshunter, Beladee, Lotje, Persian knight shiraz, DrSinn, Miracle Pen, EngineerFromVega, RjwilmsiBot, MalapropX14, DexDor, Ripchip Bot, Phlegat, John of Reading, MindBlender, Architeuthidae, Racerx11, RA0808, Theonhwiki, 8digits, Themorrissey, NorthernRaven, Hhhippo, ZéroBot, همنشین بهار, StringTheory11, Waperkins, Herp Derp, H3llBot, Quondum, UniversumExNihilo, Almatinez, Aschwole, Mayur, Kranix, Maschen, Eg-T2g, LarsJanZeeuwRules, LikeLakers2, Rocketrod1960, H1tchh1ker4, Mikhail Ryazanov, ClueBot NG, PeterKirk69, 4Jmaster, Gilderien, Navasj, Law of Entropy, Confuddledone, 336, Helpful Pixie Bot, Mightyname, Electriccatfish2, Kronn8, Curb Chain, Bibcode Bot, Tirebiter78, Rm1271, Neutral current, Stehgdop, Watson system, Koska One, Cs1791, Chewkaflax, Ddanndt, Johan.lundberg, Dragonami, Ownedroad9, Quickcrazy78, Rajibganguly01, Uioplk, Acalloni, Zedshort, TF SHaDowMAn, Yaroslav Nikitenko, Neutrinoread, Jakekong, Achowat, Pritombose, Ysawires, GodsAccident, Blakee911, Pieceofchit1, BattyBot, Layth888, Friedncrispy, Dja1979, Adyyy, Dexbot, Jdjwright, Mogism, Jaxcp3, Reatlas, Provacitu74, Coladar, CensoredScribe, Juan tanesia12, Rmohapat, Rmohapatra, Christophe1946, Klingerdinger, EWPage, Monkbot, Richard Henry Eckert, Jazzwhiz101, Crystallizedcarbon, Fimatic, PerpetuaLux, Isambard Kingdom, Lxplot, TheHecster, DN-boards1, Matan Kovac, KasparBot, Alarana, Saturn comes back around, CumbleSpuzz and Anonymous: 576

- **Boson** *Source:* https://en.wikipedia.org/wiki/Boson?oldid=682396358 *Contributors:* CYD, The Anome, Xaonon, Aldie, Enchanter, Roadrunner, Ben-Zin~enwiki, Lisiate, Michael Hardy, Tim Starling, Kroose, Looxix~enwiki, Andrewa, Glenn, Andres, Kaihsu, Samw, Panoramix, Schneelocke, CAkira, Wikiborg, The Anomebot, Saltine, Phys, Drxenocide, Robbot, Altenmann, Bkalafut, Merovingian, Rorro, Hadal, Robinh, VanishedUser kfljdfjsg33k, Giftlite, Fropuff, Pharotic, Isidore, Alexf, Beland, Jossi, Icairns, Zfr, Cructacean, Ornil, Mormegil, DanielCD, Noisy, Discospinster, Rich Farmbrough, Guanabot, Hidaspal, Sunborn, Kbh3rd, Jensbn, Alxndr, La goutte de pluie, Anthony Appleyard, Jlandahl, Leoadec, H2g2bob, Rocastelo, Benbest, Mpatel, Nakos2208~enwiki, GregorB, SDC, Palica, Ashmoo, Graham87, Kbdank71, Zzedar, Drbogdan, Strait, Master Justin, Wragge, FlaBot, Srleffler, Chobot, YurikBot, Bhny, The1physicist, Salsb, NawlinWiki, Welsh, Pyg, Dna-webmaster, Enormousdude, NeilN, Finell, Hal peridol, SmackBot, Incnis Mrsi, Ashley thomas80, Melchoir, Gilliam, MK8, MalafayaBot, Complexica, Epastore, DHN-bot~enwiki, Scienz Guy, Sbharris, QFT, Voyajer, Grover cleveland, Philvarner, Bradenripple, SashatoBot, Lambiam, Turbothy, T-dot, MagnaMopus, Candamir, WhiteHatLurker, Dicklyon, Treyp, Focomoso, Dan Gluck, UltraHighVacuum, Iridescent, Mathninja, Buckyboy314, Ianji, Cydebot, Stebbins, W.F.Galway, VashiDonsk, Tenbergen, Ward3001, Abtvctkto61, Thijs!bot, Barticus88, Mbell, Frozenport, Headbomb, MichaelMaggs, Escarbot, Orionus, Shan23, Alomas, JAnDbot, Deflective, CosineKitty, Pkoppenb, TheEditrix2, Magioladitis, VoABot II, Inertiatic076, Vanished user ty12kl89jq10, CodeCat, MartinBot, STBot, R'n'B, Tarotcards, Uberdude85, RuneSylvester, The Wild Falcon, Asnr 6, TXiKiBoT, Hqb, Anonymous Dissident, Abdullais4u, Bertrem, Moutane, Dirkbb, Antixt, Jeraaldo, BriEnBest, SieBot, Jim E. Black, Gerakibot, RadicalOne, Flyer22, Radon210, Sunayanaa, Jojalozzo, Tpvibes, Nsajjansajja, Owhanow~enwiki, Mike2vil, Mgurgan, VanishedUser sdu9aya9fs787sads, Danthewhale, PipepBot, Rodhullandemu, ChandlerMapBot, Excirial, PixelBot, Nilradical, Cenarium, Wikeepedian, Ouchitburns, Addbot, Bwr6, Minami Kana, Aboctok, Numbo3-bot, OlEnglish, David0811, WikiDreamer Bot, Jack who built the house, Luckas-bot, Yobot, Ptbotgourou, Senator Palpatine, AnomieBOT, Jim1138, JackieBot, Kingpin13, Materialscientist, Xqbot, Gravitivistically, Daners, Tomwsulcer, GrouchoBot, MeDrewNotYou, ⁇, Ace of Spades, Alarics, Paul Laroque, Rameshngbot, RedBot, FoxBot, DixonDBot, Michael9422, Weedwhacker128, Tbhotch, TjBot, Ripchip Bot, EmausBot, JSquish, Kkm010, HiW-Bot, ZéroBot, StringTheory11, Lagomen, Robhenry9, Tls60, RockMagnetist, ClueBot NG, Raghavankl, GioGziro95, HBook, Helpful Pixie Bot, Bibcode Bot, 2001:db8, AvocatoBot, Nickni28, Minsbot, Blogger 20, Protomaestro, Abitoby, Darryl from Mars, NoRwEgIaNbAcTeRiUm, Jason7898, Valluvan888, Ov.kulkarni, Crpandya, Enamex, Lugia2453, Graphium, Federicoaolivieri, 77Mike77, Rltb, 314Username, Dllaughingwang, Codeusirae, Sometree, DR ROBERT HALT, KasparBot, Jiswin1992 and Anonymous: 223
- **Gauge boson** *Source:* https://en.wikipedia.org/wiki/Gauge_boson?oldid=662605501 *Contributors:* Bryan Derksen, Andre Engels, Michael Hardy, Ahoerstemeier, Bueller 007, LouI, Phys, Robbot, Gwrede, Rholton, Rursus, Davidl9999, Giftlite, Xerxes314, Alison, JeffBobFrank, Chinasaur, Andris, Garth 187, Beland, Setokaiba, Icairns, AmarChandra, Lumidek, Vsmith, Roybb95~enwiki, Mal~enwiki, La goutte de

pluie, Nk, Kusma, Ringbang, Mpatel, Nakos2208~enwiki, Tevatron~enwiki, Kbdank71, Chobot, Roboto de Ajvol, Hairy Dude, Salsb, StuRat, ArielGold, RG2, InverseHypercube, Niels Olson, Sadi Carnot, TriTertButoxy, Ekjon Lok, Bjankuloski06en~enwiki, Phatom87, Headbomb, Tyco.skinner, Knotwork, Swpb, Maurice Carbonaro, Gombang, TXiKiBoT, Odellus, Antixt, AlleborgoBot, SieBot, Jim E. Black, Homonihilis, BOTarate, DumZiBoT, SilvonenBot, Addbot, Bertman600, NjardarBot, Numbo3-bot, Lightbot, Zorrobot, Luckas-bot, Yobot, Citation bot, ArthurBot, A. di M., Rameshngbot, RedBot, RobinK, Mary at CERN, TjBot, EmausBot, ZéroBot, StringTheory11, Mentibot, Dsperlich, CeraBot, Galactic Messiah, DerekWinters, Fisherv, KasparBot and Anonymous: 41

- **Photon** *Source:* https://en.wikipedia.org/wiki/Photon?oldid=682075782 *Contributors:* AxelBoldt, WojPob, Mav, Bryan Derksen, The Anome, Tarquin, Koyaanis Qatsi, Ap, Josh Grosse, Ben-Zin~enwiki, Heron, Youandme, Spiff~enwiki, Bdesham, Michael Hardy, Ixfd64, TakuyaMurata, NuclearWinner, Looxix~enwiki, Snarfies, Ahoerstemeier, Stevenj, Julesd, Glenn, AugPi, Mxn, Smack, Pizza Puzzle, Wikiborg, Reddi, Lfh, Jitse Niesen, Kbk, Laussy, Bevo, Shizhao, Raul654, Jusjih, Donarreiskoffer, Robbot, Hankwang, Fredrik, Eman, Sanders muc, Altenmann, Bkalafut, Merovingian, Gnomon Kelemen, Hadal, Wereon, Anthony, Wjbeaty, Giftlite, Art Carlson, Herbee, Xerxes314, Everyking, Dratman, Michael Devore, Bensaccount, Foobar, Jaan513, DÅ,ugosz, Zeimusu, LucasVB, Beland, Setokaiba, Kaldari, Vina, RetiredUser2, Icairns, Lumidek, Zondor, Randwicked, Eep², Chris Howard, Zowie, Naryathegreat, Discospinster, Rich Farmbrough, Yuval madar, Pjacobi, Vsmith, Ivan Bajlo, Dbachmann, Mani1, SpookyMulder, Kbh3rd, RJHall, Ben Webber, El C, Edwinstearns, Laurascudder, RoyBoy, Spoon!, Dalf, Drhex, Bobo192, Foobaz, I9Q79oL78KiL0QTFHgyc, La goutte de pluie, Zr40, Apostrophe, Minghong, Rport, Alansohn, Gary, Sade, Corwin8, PAR, UnHoly, Hu, Caesura, Wtmitchell, Bucephalus, Max rspct, BanyanTree, Cal 1234, Count Iblis, Egg, Dominic, Gene Nygaard, Ghirlandajo, Kazvorpal, UTSRelativity, Falcorian, Drag09, Boothy443, Richard Arthur Norton (1958-), Woohookitty, Linas, Gerd Breitenbach, StradivariusTV, Oliphaunt, Cleonis, Pol098, Ruud Koot, Mpatel, Nakos2208~enwiki, Dbl2010, Ch'marr, SDC, CharlesC, Alan Canon, Reddwarf2956, Mandarax, BD2412, Kbdank71, Zalasur, Sjakkalle, Rjwilmsi, Саша Стефановић, Strait, MarSch, Dennis Estenson II, Trlovejoy, Mike Peel, HappyCamper, Bubba73, Brighterorange, Cantorman, Egopaint, Noon, Godzatswing, FlaBot, RobertG, Arnero, Mathbot, Nihiltres, Fresheneesz, TeaDrinker, Srleffler, BradBeattie, Chobot, Jaraalbe, DVdm, Elfguy, EamonnPKeane, YurikBot, Bambaiah, Splintercellguy, Jimp, RussBot, Supasheep, JabberWok, Wavesmikey, KevinCuddeback, Stephenb, Gaius Cornelius, Salsb, Trovatore, Długosz, Tailpig, Joelr31, SCZenz, Randolf Richardson, Ravedave, Tony1, Roy Brumback, Gadget850, Dna-webmaster, Enormousdude, Lt-wiki-bot, Oysteinp, JoanneB, Ligart, John Broughton, GrinBot~enwiki, Sbyrnes321, Itub, SmackBot, Moeron, Incnis Mrsi, KnowledgeOfSelf, CelticJobber, Melchoir, Rokfaith, WilyD, Jagged 85, Jab843, Cessator, AnOddName, Skizzik, Dauto, JSpudeman, Robin Whittle, Ati3414, Persian Poet Gal, MK8, Jprg1966, Complexica, Sbharris, Colonies Chris, Ebertek, WordLife565, V1adis1av, RWincek, Aces lead, Stangbat, Cybercobra, Valenciano, EVula, A.R., Mini-Geek, AEM, DMacks, N Shar, Sadi Carnot, FlyHigh, The Fwanksta, Drunken Pirate, Yevgeny Kats, Lambiam, Harryboyles, IronGargoyle, Ben Moore, A. Parrot, Mr Stephen, Fbartolom, Dicklyon, SandyGeorgia, Mets501, Ceeded, Ambuj.Saxena, Ryulong, Vincecate, Astrobayes, Newone, J Di, Lifeverywhere, Tawkerbot2, JRSpriggs, Chetvorno, Luis A. Veguilla-Berdecia, CalebNoble, Xod, Gregory9, CmdrObot, Wafulz, Van helsing, John Riemann Soong, Rwflammang, Banedon, Wquester, Outriggr, Logical2u, Myasuda, Howardsr, Cydebot, Krauss, Kanags, A876, WillowW, Bvcrist, Hyperdeath, Hkyriazi, Rracecarr, Difluoroethene, Edgerck, Michael C Price, Tawkerbot4, Christian75, Ldussan, RelHistBuff, Waxigloo, Kozuch, Thijs!bot, Epbr123, Opabinia regalis, Markus Pössel, Mglg, 24fan24, Headbomb, Newton2, John254, J.christianson, Escarbot, Stannered, AntiVandalBot, Luna Santin, Jtrain4469, Normanmargolus, Tyco.skinner, TimVickers, NSH001, Dodecahedron~enwiki, Tim Shuba, Gdo01, Sluzzelin, Abyssoft, CosineKitty, AndyBloch, Bryanv, ScottStearns, Hroðulf, Bongwarrior, VoABot II, B&W Anime Fan, SHCarter, Lgoger, I JethroBT, Dirac66, Hveziris, Maliz, Lord GaleVII, TRWBW, Shijualex, Glen, DerHexer, Patstuart, Gwern, Taborgate, MartinBot, MNAdam, Jay Litman, HEL, Ralf 58, J.delanoy, DrKiernan, Trusilver, C. Trifle, AstroHurricane001, Numbo3, Pursey, CMDadabo, Kevin aylward, UchihaFury, Pirate452, H4xx0r, Iamthewalrus35, Iamthewalrus36, Gee Eff, Chimpy07, Dirkdiggler69, Lk69, Hallamfm, Annoying editter, Yehoodig, Acalamari, Foreigner1, McSly, Samtheboy, Tarotcards, Rominandreu, ARTE, Tanaats, Potatoswatter, Y2H, Divad89, Scott Illini, Stack27, THEblindwarrior, VolkovBot, AlnoktaBOT, Hyperlinker, DoorsAjar, TXiKiBoT, Oshwah, Cosmic Latte, The Original Wildbear, Davehi1, Chiefwaterfall, Vipinhari, Hqb, Anonymous Dissident, HansMair, Predator24, BotKung, Luuva, Calvin4986, Improve~enwiki, Kmhkmh, Richwil, Antixt, Gorank4, Falcon8765, GlassFET, Cryptophile, MattiasAndersson, AlleborgoBot, Carlodn6, NHRHS2010, Relilles~enwiki, Tpb, SieBot, Timb66, Graham Beards, WereSpielChequers, ToePeu.bot, JerrySteal, Android Mouse, Likebox, RadicalOne, Paolo.dL, Lightmouse, PbBot, Spartan-James, Duae Quartunciae, Hamiltondaniel, StewartMH, Dstebbins, ClueBot, Bobathon71, The Thing That Should Not Be, Mwengler, EoGuy, Jagun, RODERICKMOLASAR, Wwheaton, Dmlcyal8er, Razimantv, Mild Bill Hiccup, Feebas factor, J8079s, Rotational, MaxwellsLight, Awickert, Excirial, PixelBot, Sun Creator, NuclearWarfare, PhySusie, El bot de la dieta, DerBorg, Shamanchill, PoofyPeter99, J1.grammar natz, Laserheinz, TimothyRias, XLinkBot, Jovianeye, Petedskier, Hess88, Addbot, Mathieu Perrin, DOI bot, DougsTech, Download, James thirteen, AndersBot, LinkFA-Bot, Barak Sh, AgadaUrbanit, Тиверополник, Dayewalker, Quantumobserver, Kein Einstein, Legobot, Luckas-bot, Yobot, Kilom691, Allowgolf~enwiki, AnomieBOT, Ratul2000, Kingpin13, Materialscientist, Citation bot, Xqbot, Ambujarind69, Mananay, Emezei, Sharhalakis, Shirik, RibotBOT, Rickproser, SongRenKai, Max derner, Merrrr, A. di M., ⁇, CES1596, Paine Ellsworth, Gsthae with tempo!, Nageh, TimonyCrickets, WurzelT, Steve Quinn, Spacekid99, Radeksonic, Citation bot 1, Pinethicket, I dream of horses, HRoestBot, Tanweer Morshed, Eno crux, Tom.Reding, Jschnur, RedBot, IVAN3MAN, Gamewizard71, FoxBot, TobeBot, Earthandmoon, PleaseStand, Marie Poise, RjwilmsiBot, Антон Гліністы, Ripchip Bot, Ofercomay, Chemyanda, EmausBot, Bookalign, WikitanvirBot, Roxbreak, Word2need, Gcastellanos, Tommy2010, Dcirovic, K6ka, Hhhippo, Cogiati, 1howardsr1, StringTheory11, Waperkins, Jojojlj, Access Denied, Quondum, AManWithNoPlan, Raynor42, L Kensington, Maschen, HCPotter, Haiti333, RockMagnetist, Rocketrod1960, ClueBot NG, JASMEET SINGH HAFIST, Schicagos, Snotbot, Vinícius Machado Vogt, Helpful Pixie Bot, SzMithrandir, Bibcode Bot, BG19bot, Roberticus, Paolo Lipparini, Wzrd1, Rifath119, Davidiad, Mark Arsten, Peter.sujak, Wikarchitect, Hamish59, Caypartisbot, Penguinstorm300, KSI ROX, Bhargavuk1997, Chromastone1998, TheJJJunk, Nimmo1859, EagerToddler39, Dexbot, EZas3pt14, Webclient101, Chrisanion, Vanquisher.UA, Tony Mach, PREMDASKANNAN, Meghas, Reatlas, Profb39, Zerberos, Thesuperseo, The User 111, Eyesnore, Ybidzian, Tentinator, Illusterati, Celso ad, Quenhitran, Manul, DrMattV, Anrnusna, Wyn.junior, K0RTD, Monkbot, Vieque, BethNaught, Markmizzi, Garfield Garfield, Smokey2022, Zargol Rejerfree, RAL2014, Shahriar Kabir Pavel, Sdjncskdjnfskje, Anshul1908, Professor Flornoy, Thatguytestw, Tetra quark, Harshit100, KasparBot, Chinta 01, Geek3, TheKingOfPhysics and Anonymous: 496

- **Gluon** *Source:* https://en.wikipedia.org/wiki/Gluon?oldid=681546689 *Contributors:* AxelBoldt, CYD, Bryan Derksen, Gdarin, TakuyaMurata, Card~enwiki, Looxix~enwiki, Ellywa, Ahoerstemeier, Med, Schneelocke, Phys, Phil Boswell, Donarreiskoffer, Fredrik, Merovingian, Hadal, Giftlite, Herbee, Xerxes314, Eequor, Darrien, Keith Edkins, RetiredUser2, Icairns, Mike Rosoft, AlexChurchill, HedgeHog, Kenny TM~~enwiki, David Schaich, Ioliver, Mashford, El C, Kwamikagami, Ardric47, Obradovic Goran, Alansohn, Guy Harris, Dachannien, Ricky81682, Batmanand, Velella, Kazvorpal, April Arcus, Forteblast, Mpatel, Palica, BD2412, Kbdank71, Rjwilmsi, Macumba, Strait, Mike

Peel, Bubba73, Klortho, FlaBot, Srleffler, Chobot, YurikBot, Wavelength, Bambaiah, Hairy Dude, Jimp, JabberWok, Zelmerszoetrop, Salsb, SCZenz, Randolf Richardson, Ravedave, Danlaycock, Bota47, LeonardoRob0t, Anclation~enwiki, Physicsdavid, Erudy, GrinBot~enwiki, Kgf0, SmackBot, Melchoir, Cessator, Benjaminevans82, Abtal, MK8, Colonies Chris, Can't sleep, clown will eat me, Decltype, Qcdmaestro, Edconrad, Darkpoison99, FredrickS, Omsharan, Pegasusbot, Gregbard, ProfessorPaul, Thijs!bot, Headbomb, Rriegs, Oreo Priest, AntiVandal-Bot, Shambolic Entity, Deflective, Mujokan, Yill577, Happycool, Mother.earth, Martynas Patasius, WiiWillieWiki, HEL, Hans Dunkelberg, Gombang, Inwind, Sheliak, Jonthaler, VolkovBot, TXiKiBoT, Davehi1, Kriak, Anonymous Dissident, Imasleepviking, AlleborgoBot, EJF, SieBot, Steven Crossin, OKBot, ClueBot, Wwheaton, Qsaw, Nucularphysicist, Ottava Rima, Gordon Ecker, Rhododendrites, Brews ohare, Ca-cadril, RexxS, JKeck, Against the current, SkyLined, Addbot, DOI bot, Lightbot, Skippy le Grand Gourou, Luckas-bot, Planlips, AnomieBOT, Jim1138, JackieBot, Citation bot, Bci2, ArthurBot, Xqbot, Neil95, Triclops200, Omnipaedista, TorKr, ??, Paine Ellsworth, Ivoras, Citation bot 1, Pekayer11, Rameshngbot, PNG, RjwilmsiBot, TjBot, Lilcal89012, EmausBot, Socob, JSquish, StringTheory11, Quondum, TyA, Maschen, RolteVolte, ClueBot NG, Timothy jordan, Maplelanefarm, Bibcode Bot, BG19bot, Gravitoweak, Cadiomals, Tropcho, Fraulein451, DrHjmHam, Rhlozier, D.shinkaruk, Yaara dildaara, BronzeRatio, Monkbot, Yikkayaya, KasparBot and Anonymous: 142

- **W and Z bosons** *Source:* https://en.wikipedia.org/wiki/W_and_Z_bosons?oldid=676803444 *Contributors:* AxelBoldt, Sodium, Mav, Bryan Derksen, The Anome, Ap, Andre Engels, Danny, Roadrunner, DrBob, Michael Hardy, Tim Starling, Karada, Egil, Ahoerstemeier, Ryan Cable, Julesd, Mxn, Charles Matthews, Ike9898, Saltine, Phys, Topbanana, BenRG, Finlay McWalter, Twang, Phil Boswell, Donarreiskoffer, Robbot, Pigsonthewing, Nurg, DHN, Xanzzibar, M-Falcon, Giftlite, Tremolo, Harp, Herbee, Xerxes314, Jeremy Henty, Bodhitha, LiDaobing, RetiredUser2, Icairns, Mike Rosoft, Vsmith, Gianluigi, Kjoonlee, Drhex, Obradovic Goran, Jérôme, Fkbreitl, Cameron.simpson, Gene Nygaard, Linas, LoopZilla, Graham87, Kbdank71, Rjwilmsi, Strait, Mike Peel, Lmatt, Goudzovski, Chobot, FrankTobia, Roboto de Ajvol, Ugha, Mushin, Bambaiah, Wester, Hairy Dude, Hellbus, Salsb, Seb35, Długosz, Turbolinux999, Ravedave, Scottfisher, Dna-webmaster, Modify, Argo Navis, Teply, Sbyrnes321, SmackBot, Tom Lougheed, Jagged 85, ZerodEgo, Dauto, Bluebot, Shaggorama, Sbharris, Niels Olson, Radagast83, Acdx, John, Lottamiata, Happy-melon, Tubezone, MightyWarrior, Joelholdsworth, Tangobot, Michael C Price, Quibik, Dchristle, Realjanuary, Headbomb, Davidhorman, Nosirrom, Certain, Gökhan, JAnDbot, Tigga, Omeganian, Brimofinsanity, TheEditrix2, Trapezoidal, Magioladitis, ThoHug, Leyo, Lilac Soul, HEL, Rod57, Y2H, HiEv, Adam Zivner, Madblueplanet, Sheliak, Dextrose, Anonymous Dissident, Synthebot, Antixt, Coronellian~enwiki, SieBot, STANMAR725, Jim E. Black, Gerakibot, Martin Kealey, CutOffTies, Fratrep, ClueBot, Mild Bill Hiccup, Alexbot, Carsrac, SkyLined, Dieppu, Stephen Poppitt, Addbot, Eric Drexler, Toyokuni3, Mjamja, Ronkonkaman, Download, CarsracBot, ChenzwBot, Lightbot, M sotirov, Luckas-bot, Yobot, Jim1138, MehrdadAfshari, ArthurBot, Ernsts, A. di M., Howard McCay, FrescoBot, Paine Ellsworth, D'ohBot, Citation bot 1, Gil987, Tom.Reding, Swallerick, FoxBot, Earthandmoon, Tm1729, TjBot, Антон Гліністы, Newty23125, EmausBot, Mnkyman, StringTheory11, Quondum, MisterDub, WaterCrane, Whoop whoop pull up, ClueBot NG, Helpful Pixie Bot, Bibcode Bot, BG19bot, Bakkedal, JYBot, Mamaphyskerin, Anrnusna, MartinNicklin, Boidal-Quantized and Anonymous: 137
- **Scalar boson** *Source:* https://en.wikipedia.org/wiki/Scalar_boson?oldid=653639224 *Contributors:* AugPi, Phys, Giftlite, Linas, Mpatel, Russ-Bot, Bhny, SmackBot, Sergio.ballestrero, QFT, WLevine, Magioladitis, Tdadamemd, Sigmundur, Antixt, Jim E. Black, JL-Bot, ClueBot, Naradawickramage, Addbot, Dr. Universe, Anypodetos, J04n, Erik9bot, Fortdj33, Tomville219, RedBot, GoingBatty, Carbosi, ZéroBot, RolteVolte, Darine Of Manor, Parcly Taxel, Beaumont877, Mogism and Anonymous: 15
- **Higgs boson** *Source:* https://en.wikipedia.org/wiki/Higgs_boson?oldid=680354810 *Contributors:* AxelBoldt, CYD, ClaudeMuncey, Bryan Derksen, Manning Bartlett, Roadrunner, David spector, Heron, Ewen, Stevertigo, Edward, Boud, TeunSpaans, Dante Alighieri, Ixfd64, Gaurav, TakuyaMurata, CesarB, Anders Feder, Mgimpel~enwiki, Bueller 007, Mark Foskey, Kaihsu, Samw, Cherkash, Lee M, Mxn, Ehn, Timwi, Dcoetzee, Wikiborg, Kbk, Tpbradbury, Phys, Bevo, Topbanana, JonathanDP81, AnonMoos, Bcorr, Jerzy, BenRG, Slawojarek, Phil Boswell, Donarreiskoffer, Robbot, Josh Cherry, ChrisO~enwiki, Owain, Iwpg, Goethean, Altenmann, Nurg, Lowellian, Merovingian, Rursus, Caknuck, Hadal, Alba, Mattflaschen, David Gerard, M-Falcon, Giftlite, Graeme Bartlett, Harp, ShaneCavanaugh, Lethe, Herbee, Jrquinlisk, Xerxes314, Ds13, Fleminra, Dratman, Muzzle, Varlaam, Jason Quinn, Foobar, DÅ,ugosz, Golbez, Bodhitha, Mmm~enwiki, Aughtandzero, Quadell, Selva, Kaldari, Fred Stober, Johnflux, RetiredUser2, Thincat, Elektron, Bbbl67, Icairns, J0m1eisler, Cructacean, Tdent, TJSwoboda, John-Armagh, Safety Cap, ProjeX, Njh@bandsman.co.uk, Mike Rosoft, Chris Howard, Jkl, Discospinster, Rich Farmbrough, FT2, Qutezuce, Vsmith, Pie4all88, Kooo, David Schaich, Xgenei, Mal~enwiki, Dbachmann, Mani1, Bender235, ESkog, RJHall, Ylee, Pt, El C, Lycurgus, Lars~enwiki, Laurascudder, Art LaPella, Bookofjude, Brians, TheMile, Dragon76, Smalljim, C S, Reuben, La goutte de pluie, Rangelov, Sasquatch, Bawolff, Tritium6, Eritain, HasharBot~enwiki, Jumbuck, Yoweigh, Alansohn, Andrew Gray, JohnAlbertRigali, Axl, Sligocki, Kocio, Mlm42, Tom12519, Chuckupd, Atomicthumbs, Wtmitchell, KapilTagore, Endersdouble, Dirac1933, DrGaellon, Falcorian, Itinerant, DarTar, Joriki, Reinoutr, Linas, Mindmatrix, Jamsta, Sburke, Benbest, Jonburchel, Thruston, TotoBaggins, GregorB, J M Rice, CharlesC, Waldir, Christopher Thomas, Karam.Anthony.K, Tevatron~enwiki, RichardWeiss, Ashmoo, Fleisher, Kbdank71, GrundyCamellia, Drbogdan, Rjwilmsi, Nightscream, Koavf, Strait, XP1, Martaf, BlueMoonlet, MZMcBride, Mike Peel, NeonMerlin, R.e.b., Jehochman, Bubba73, Afterwriting, A Man In Black, Splarka, RobertG, Nihiltres, Norvy, Itinerant1, Gurch, Mark J, Nimur, Shawn@garbett.org, ElfQrin, Danny-DaWriter, Goudzovski, Diza, Consumed Crustacean, Srleffler, Sbove, Chobot, DVdm, Bgwhite, Zentropa, Bambaiah, Wester, Hairy Dude, Huw Powell, Wikky Horse, Pip2andahalf, RussBot, Jacques Antoine, Bhny, JabberWok, Hellbus, Archelon, Eleassar, Rsrikanth05, Salsb, Big Brother 1984, NawlinWiki, Folletto, Buster79, Trovatore, Neutron, SCZenz, Daniel Mietchen, Gadget850, Bota47, Karl Andrews, Dna-webmaster, Jezzabr, Thor Waldsen, Crisco 1492, Deeday-UK, Daniel C, WAS 4.250, Paul Magnussen, Closedmouth, D'Agosta, Bondegezou, Netrapt, Egumtow, LeonardoRob0t, Ilmari Karonen, NeilN, Kgf0, Maryhit, Dragon of the Pants, SmackBot, Nahald, Moeron, Ashley thomas80, Slashme, InverseHypercube, Melchoir, Cinkcool, Baad, Jagged 85, Nickst, Frymaster, AnOddName, ZerodEgo, Giandrea, Gilliam, Ohnoitsjamie, Skizzik, Carl.bunderson, Aurimas, Dauto, JCSantos, TimBentley, RevenDS, Jprg1966, Rick7425, Cadmasteradam, Roscelese, Epastore, DHN-bot~enwiki, Sbharris, Eusebeus, Scwlong, Modest Genius, Famspear, V1adis1av, Rhodesh, Fiziker, Lantrix, Grover cleveland, Jmnbatista, Wen D House, Flyguy649, Jgwacker, Daqu, Mesons, Rezecib, Martijn Hoekstra, Pulu, BullRangifer, Andrew c, Gildir, Kendrick7, Marcus Brute, Vina-iwbot~enwiki, Yevgeny Kats, Frglee, TriTertButoxy, CIS, SashatoBot, Lambiam, Mukadderat, Hi2lok, Kuru, Khazar, Shirifan, Eikern, Tktktk, JorisvS, DMurphy, Mgiganteus1, Bjankuloski06en~enwiki, IronGargoyle, Aardvark23, Loadmaster, Smith609, Deceglie, Hvn0413, Xiphoris, Norm mit, Keith-264, Kencf0618, Britannica~enwiki, Paul venter, Newone, Twas Now, GDallimore, Benplowman, Airstrike~enwiki, Chetvorno, DKqwerty, Harold f, JForget, Laplacian, Er ouz, Jtuggle, Banedon, Ruslik0, Krioni, McVities, Keithh, Rotiro, Yaris678, Slazenger, Cydebot, Martinthoegersen, Gogo Dodo, Anonymi, Lewisxxxusa, Mat456, Jlmorgan, Hippypink, Michael C Price, Quibik, AndersFeder, Raoul NK, PKT, Thijs!bot, Keraunos, Anupam, Headbomb, Kathovo, James086, Hcobb, D.H, Logicat, Jo-moal99, Northumbrian, Oreo Priest, JitendraS, -dennis-~enwiki, Widefox, Seaphoto, Orionus, QuiteUnusual, Readro, Hsstr8, Tlabshier, Tim

Shuba, Yellowdesk, TuvicBot, JAnDbot, Asmeurer, Tigga, Jde123, Roman à clef, Zekemurdock, Mcorazao, Mozart998, Kborland, Bongwarrior, NeverWorker, Ronstew, Marcel Kosko, Jpod2, J mcandrews, Walter Wpg, Trugster, JMBryant, Vanished user ty12kl89jq10, CodeCat, Allstarecho, Brian Fenton, JaGa, GermanX, Alangarr, WLU, TimidGuy, Mr Shark, Pagw, Andre.holzner, Sigmundg, Ben MacDui, David Nicoson, Anaxial, JTiago, CommonsDelinker, Leyo, Gah4, Fconaway, Oddz, Tgeairn, J.delanoy, Fatka, Pharaoh of the Wizards, Maurice Carbonaro, Stephanwehner, Foober, Aveh8, McSly, Memory palace, NewEnglandYankee, Policron, 83d40m, Usp, Lamp90, Austinian, Izno, SoCalSuperEagle, Robprain, Cuzkatzimhut, Deor, Schucker, VolkovBot, Off-shell, ABF, Eliga~enwiki, JohnBlackburne, AlnoktaBOT, Tburket, Davidwr, Philip Trueman, Spemble, TXiKiBoT, Quatschman, The Original Wildbear, Gwib, Fatram, Kipb9, Andrius.v, Matan568, Nxavar, Nafhan, Photonh2o, Impunv, Peterbullockismyname, Cerebellum, Martin451, Praveen pillay, LoverOfArt, Abdullais4u, Justinrossetti, Cgwaldman, Bcody80, BotKung, Tennisnutt92, Dirkbb, Antixt, Francis Flinch, Moose-32, Ptrslv72, TheBendster, Masterofpsi, Jonbutterworth, Adrideba, SieBot, StAnselm, Manyugarg, PlanetStar, Jor63, Meldor, OlliffeObscurity, Jdcanfield, Yintan, Abhishikt, Flyer22, Graycrow, Infestor, Hrishirise, Cablehorn, Arthur Smart, Aperseghin, Mattmeskill, Gobbledygeek, Cthomas3, Steven Crossin, Nskillen, Sunrise, Afernand74, Jimtpat, Iknowyourider, StaticGull, Jfromcanada, MvL1234, Sphilbrick, Nergaal, Denisarona, Escape Orbit, Quinling, Martarius, PhysicsGrad2013, ClueBot, Victor Chmara, The Thing That Should Not Be, TomRed, Alyjack, Infrasonik, Mx3, Master1228, Drmies, Frmorrison, Loves martyr, Polyamorph, Nobaddude, Sjdunn9, Kitsunegami, Ktr101, Excirial, Joeyfjj, Wmlschlotterer, Pawan ctn, Lartoven, Artur80, Sun Creator, BobertWABC, PeterTheWall, Nondisclosure, M.O.X, SchreiberBike, JasonAQuest, Another Believer, Scf1984, 1ForTheMoney, Anoopan, Wnt, Darkicebot, TimothyRias, XLinkBot, Rreagan007, Resonance cascade, JinJian, Jabberwoch, Hess88, Hybirdd, Tayste, Addbot, Proofreader77, Mortense, Jacopo Werther, DBGustavson, DOI bot, Betterusername, Ocdnctx, OttRider, Cgd8d, Leszek Jańczuk, WikiUserPedia, NjardarBot, Download, LaaknorBot, AndersBot, Favonian, AgadaUrbanit, HandThatFeeds, Tide rolls, Lightbot, ScAvenger, SPat, Zorrobot, Jarble, ScienceApe, Dnamanish, Luckas-bot, Yobot, Chreod, EchetusXe, Nsbinsnj, Evans1982, Amble, Now dance, fu.cker, dance!, Anypodetos, Nallimbot, Trinitrix, SkepticalPoet, Pulickkal, Fernandosmission, Apollo reactor, Csmallw, AnomieBOT, November-rain94, 1exec1, Jim1138, JackieBot, Gc9580, LlywelynII, Materialscientist, Citation bot, Brightgalrs, Onesius, ArthurBot, Northryde, LilHelpa, Xqbot, Konor org, Noonehasthisnameithink, Engineering Guy, Yutenite, Newzebras, Universalsuffrage, DeadlyMETAL, Tomdo08, Professor J Lawrence, Br77rino, Srich32977, Arni.leibovits, StevenVerstoep, ProtectionTaggingBot, Vdkdaan, Omnipaedista, RibotBOT, Kyng, Waleswatcher, WissensDürster, Ace111, Kristjan.Jonasson, MerlLinkBot, Ernsts, Chaheel Riens, A. di M., A.amitkumar, Markdavid2000, ??, Dave3457, FrescoBot, Weyesr1, Paine Ellsworth, Kenneth Dawson, Cdw1952, CamB424, CamB4242, Steve Quinn, N4tur4le, Jc odcsmf, Cannolis, Dolyn, Citation bot 1, Openmouth, Gil987, OriumX, Biker Biker, Gautier lebon, Pinethicket, Edderso, Boson15, Jonesey95, Three887, CarsonsDad, Calmer Waters, Jusses2, RedBot, BiObserver, Aknochel, Meier99, Trappist the monk, Puzl bustr, Proffsl, Higgshunter, Mary at CERN, Periglas, Zanhe, Lotje, Callanecc, Comet Tuttle, Jdigitalbath, Vrenator, SeoMac, ErikvanB, Tbhotch, Minimac, Coolpranjal, Mean as custard, RjwilmsiBot, TjBot, Olegrog, 123Mike456Winston789, Weaselpit, Newty23125, Techhead7890, Tesseract2, Skamecrazy123, Northern Arrow, Mukogodo, J36miles, EmausBot, John of Reading, WikitanvirBot, Stryn, Dadaist6174, Nuujinn, Montgolfière, Fotoni, GoingBatty, RA0808, Bengt Nyman, Bt8257, Gimmetoo, KHamsun, LHC Tommy, Slightsmile, NikiAnna, TeeTylerToe, Dekker451, Hhhippo, Evanh2008, JSquish, Kkm010, ZéroBot, John Cline, Liquidmetalrob, Fæ, Bollyjeff, Érico Júnior Wouters, StringTheory11, Stevengoldfarb, Sgerbic, Opkdx, Quondum, AndrewN, Tbushman, Makecat, Timetraveler3.14, Foonle77, Tolly4bolly, Wiggles007, Irenan, Nobleacuff, Brandmeister, Baseballrocks538, Chris81w, Inswoon, Maschen, Donner60, Ontyx, Angelo souti, ChuispastonBot, ChiZeroOne, Ninjalectual, Exsmokey, Herk1955, I hate whitespace, Rocketrod1960, Whoop whoop pull up, Ajuvr, Petrb, Grapple X, ClueBot NG, Perfectlight, Aaron Booth, Gareth Griffith-Jones, Siswick, MelbourneStar, Gilderien, PhysicsAboveAll, Manu.ajm, Muon, Parcly Taxel, O.Koslowski, Widr, Mohd. Toukir Hamid, Diyar se, Helpful Pixie Bot, Popcornduff, Aesir.le, Bibcode Bot, 2001:db8, Lowercase sigmabot, BG19bot, Scottaleger, Mcarmier, Jibu8, Loupatriz67, Dave4478, Frze, Ervin Goldfain, Reader505, Mark Arsten, Lovetrivedi, BarbaraMervin, Silvrous, Drcooljoe, Cadiomals, Joydeep, Altaïr, Piet De Pauw, Jeancey, Sovereign8, Visuall, Ownedroad9, Brainssturm, Jw2036, Writ Keeper, DPL bot, Nickni28, Philpill691, Lee.boston, Scientist999, Benjiboy187, Duxwing, Cengime, Skiret girdet njozet, GRighta, Downtownclay-tonbrown, Diasjordan, Ghsetht, Marioedesouza, BattyBot, 1narendran, LORDCOTTINGHAM2, NO SOPA, Tchaliburton, Wijnburger, StarryGrandma, Mdann52, Dilaton, Magikal Samson, Samuelled, Dja1979, Georgegroom, BecurSansnow, EuroCarGT, MSUGRA, Rhlozier, Pscott558, Turullulla, Blueprinteditor, Misterharris~enwiki, AstroDoc, Bigbear213, Dexbot, Randomizer3, Daggerbot, DoctorLazarusLong, Caroline1981, Nitpicking polisher, SoledadKabocha, Gsmanu007, Windows.dll, Mogism, Prabal123koirala, Abitoby, Clidog, Rongended, Darryl from Mars, Cerabot~enwiki, MuonRay, TheTruth72, Capt. Mohan Kuruvilla, Gatheringstorm2, Jason7898, Mumbai999999, SkepticalKid, Cjean42, Nmrzuk, Lugia2453, Mafuee, Frosty, SFK2, Thegodparticlebook, Rijensky, Mishra866868, Rockstar999999999, Toddbeck911, Nilaykumar07, Thepalerider2012, WikiPhysTech, The Anonymouse, Ahmar Saeed, Pincrete, Apidium23, Prahas.wiki, Exenola, Pdotpwns, Epicgenius, Fireballninja, Greengreengreenred, ??, Technogeek101, NicoPosner, Apurva Godghase, Durfyy, Soumya Mittal, American In Brazil, SaifAli13, Qwerkysteve, Spatiandas, Retroherb, Tango303, Hoppeduppeanut, Redplain, Shaelote, Quadrum, AntiguanAcademic, Simpsonojsim, Agyeyaankur, DavidLeighEllis, Ethanthevelociraptor, Qfang12, Comp.arch, Eletro1903, E8xE8, HeineBOB, Kahtar, Depthdiver, JAaron95, Mfb, Anrnusna, Stamptrader, Man of Steel 85, Cteirmn, AiraCobra, MyNameIsn'tElvis, Meganlock8, Sxxximf, Drsoumyadeepb, 22merlin, Ndidi Okonkwo Nwuneli, Monkbot, Dialga5555, Fred1810, Akro7, Pewpewpewpapapa, BradNorton1979, 21bhar-gav, Whistlemethis, Thinking Skeptically, Amk365, Gagnonlg, Knowledgebattle, L21234, TheNextMessiah, Naterealm224, Joey van Helsing, Adrian Lamplighter, Arnab santra, Gemadi, BATMAN1021, Isambard Kingdom, Mercedes321, DrKitts, KasparBot, JJMC89, GBjun3, TheRoamer64, Firstcause, Seventhorbitday, RobeDM and Anonymous: 962

- **Faddeev–Popov ghost** *Source:* https://en.wikipedia.org/wiki/Faddeev%E2%80%93Popov_ghost?oldid=666463549 *Contributors:* Michael Hardy, 6birc, Charles Matthews, Phys, Giftlite, Lumidek, Fwb22, TheParanoidOne, Linas, Strait, Godzatswing, YurikBot, Conscious, Archelon, Anomalocaris, Caco de vidro, Chris the speller, PDD, MalafayaBot, Sadads, Sbharris, Yevgeny Kats, Mgiganteus1, Newone, Michael C Price, AntiVandalBot, Colin Kiegel, Jpod2, Tgeairn, Александр Сигачёв, Igob8a2, Tbone762, Cuzkatzimhut, Antixt, Mild Bill Hiccup, Addbot, Ross Rhodes, Malmax, Luckas-bot, Yobot, Amirobot, AnomieBOT, Srich32977, Carlog3, FrescoBot, Ysyoon, RedBot, Trappist the monk, EmausBot, ZéroBot, Isocliff, Luizpuodzius, QuantumMasterF, MittensR and Anonymous: 16
- **Superpartner** *Source:* https://en.wikipedia.org/wiki/Superpartner?oldid=677538417 *Contributors:* Roadrunner, SimonP, Phys, Donarreiskoffer, Giftlite, Kocio, Alai, Duncan.france, Mpatel, Rjwilmsi, R.e.b., Drrngrvy, FlaBot, KFP, Conscious, SCZenz, SmackBot, Reedy, Dauto, Jgwacker, Thijs!bot, Headbomb, Maliz, Hans Dunkelberg, LovroZitnik, Agharo, Antixt, AlleborgoBot, Madacs, Bobathon71, Niceguyedc, Alexbot, SilvonenBot, SkyLined, Addbot, Barak Sh, Luckas-bot, ArthurBot, Xqbot, Erik9bot, Carlog3, Paine Ellsworth, Haeinous, Cracrunch, RedBot, EmausBot, Hydroxonium, Flloater, ClueBot NG, Bibcode Bot, Hrttu523, Rolf h nelson, Akro7 and Anonymous: 14
- **Gaugino** *Source:* https://en.wikipedia.org/wiki/Gaugino?oldid=600368183 *Contributors:* Bryan Derksen, Timwi, BenRG, Northgrove, Lu-

midek, Pjacobi, Jag123, Plumbago, Tabletop, Ligulem, Lmatt, YurikBot, Bambaiah, Conscious, Stephan Schneider, Bluebot, V1adis1av, Pulu, Thijs!bot, Headbomb, Pamputt, Antixt, Yartsa, BartekChom, Armin Rigo, MystBot, SkyLined, Addbot, Chahoo, LaaknorBot, Luckas-bot, Omnipaedista, Emaus, Gil987, EmausBot, JSquish, Flloater, ChuispastonBot, Bibcode Bot, QuarkyPi and Anonymous: 17

- **Gluino***Source:* https://en.wikipedia.org/wiki/Gluino?oldid=659020757*Contributors:* Schneelocke, David Latapie, Icairns, YUL89YYZ, Army 1987,Jag123, Vslashg, Theofilatos, Kbdank71, Strait, YurikBot, Conscious, SCZenz, V1adis1av, Jgwacker, Bjankuloski06en~enwiki , Quaeler,Thijs!bot, Dbeattie, Dgiraffes, Antixt, Djr32, Alexbot, NuclearWarfare, SkyLined, Addbot, Luckas-bot, ArthurBot, Ernsts, Gil987, EmausBot,Flloater, Comicboy1996, MSUGRA, Akro7, Kifmaster11235 and Anonymous: 4
- **Gravitino** *Source:* https://en.wikipedia.org/wiki/Gravitino?oldid=658304515 *Contributors:* EddEdmondson, Schneelocke, Maximus Rex, BenRG, Donarreiskoffer, Icairns, Lumidek, Urvabara, Army1987, Jag123, Physicistjedi, Pauli133, Jef-Infojef, Kbdank71, Rjwilmsi, FlaBot, Ewlyahoocom, Conscious, Pigman, DavidConrad, Martinwilke1980, Vald, Bluebot, V1adis1av, T-borg, Yevgeny Kats, Dan Gluck, Mssgill, DJBullfish, Thijs!bot, Headbomb, Spartaz, Shambolic Entity, Hair Commodore, Tarotcards, Telecomtom, Cuzkatzimhut, Antixt, AnonyScientist, SkyLined, Addbot, F Notebook, Luckas-bot, ArthurBot, Acky69, FrescoBot, Three887, StringTheory11, Ebrambot, Suslindisambiguator and Anonymous: 17
- **Photino** *Source:* https://en.wikipedia.org/wiki/Photino?oldid=682538898 *Contributors:* Bryan Derksen, Alfio, Jni, Donarreiskoffer, Moink, Pharotic, Roybb95~enwiki, Jag123, Bambaiah, Necessary Evil, Michael C Price, Lamro, Keynell, Addbot, Americanplus and Anonymous: 3
- **Higgsino** *Source:* https://en.wikipedia.org/wiki/Higgsino?oldid=676111955 *Contributors:* Schneelocke, Jni, Robbot, Wurblzap~enwiki, Lumidek, Jag123, MarSch, Bambaiah, Conscious, Bluebot, V1adis1av, QFT, Jgwacker, Mgiganteus1, Dan Gluck, Thijs!bot, Dan Pelleg, HEL, VolkovBot, Antixt, Yartsa, SkyLined, Addbot, Luckas-bot, Sailorleo, AnomieBOT, Ernsts, JanderClamber, Thehelpfulbot, FrescoBot, EmausBot, JSquish, Ebrambot, BattyBot, YFdyh-bot, Spudeino, Akro7, Lxplot and Anonymous: 12
- **Neutralino** *Source:* https://en.wikipedia.org/wiki/Neutralino?oldid=671536470 *Contributors:* Angela, Julesd, Schneelocke, Charles Matthews, Saltine, Donarreiskoffer, Robbot, Rursus, Moink, Awolf002, Herbee, Xerxes314, Waltpohl, Pharotic, Eequor, Icairns, Urvabara, Pjacobi, Slicky, Physicistjedi, JPFlip, SDC, Theofilatos, Kbdank71, FlaBot, Roboto de Ajvol, YurikBot, Conscious, SCZenz, AndrewWTaylor, SmackBot, Stepa, Nickst, V1adis1av, Pulu, Lester, Newone, Wadoli Itse, Thijs!bot, Barticus88, Headbomb, Stannered, Squantmuts, NicZ~enwiki, Choihei, Antixt, ClueBot, SkyLined, Addbot, Prim Ethics, Yobot, AnomieBOT, Citation bot, ArthurBot, Br77rino, Ernsts, A. di M., Erik9bot, Tom.Reding, ZéroBot, David C Bailey, Helpful Pixie Bot, Bibcode Bot, Halfb1t, Manar al Zraiy, Makecat-bot, Phseek and Anonymous: 24
- **Chargino***Source:* https://en.wikipedia.org/wiki/Chargino?oldid=666472766*Contributors:* Schneelocke, Donarreiskoffer, David Gerard, Icairns,Lumidek, Pjacobi, Jag123, Keenan Pepper, FlaBot, Bambaiah, Conscious, SCZenz, Zwobot, SmackBot, Dauto, V1adis1av, BWDuncan , Mat-tisse, Thijs!bot, Yonidebot, Antixt, Yartsa, SkyLined, Addbot, Luckas-bot, Yobot, ArthurBot, Omnipaedista, Erik9bot, HighFlyingFish, Car-log3, Sławomir Biały, Gil987, JSquish, ZéroBot and Anonymous: 5
- **Axino** *Source:* https://en.wikipedia.org/wiki/Axino?oldid=636844075 *Contributors:* BenRG, Lumidek, Dirac1933, Lmatt, Nickst, V1adis1av, Meno25, Thijs!bot, Headbomb, Serpent's Choice, Maliz, Freeboson, VolkovBot, Antixt, Alexbot, Addbot, Luckas-bot, ArthurBot, Carlog3, Citation bot 1, Three887, EmausBot, Optiguy54, Comicboy1996, Bibcode Bot, Joeinwiki and Anonymous: 1
- **Sfermion** *Source:* https://en.wikipedia.org/wiki/Sfermion?oldid=580181478 *Contributors:* BenRG, Xerxes314, Blaxthos, TheMatt, SmackBot, V1adis1av, Soap, Michael C Price, Christian75, Thijs!bot, Headbomb, TomS TDotO, Dgiraffes, VolkovBot, Anonymous Dissident, Antixt, Niceguyedc, WikHead, Addbot, Luckas-bot, Anypodetos, Xqbot, Susiemorgan, Ernsts, Erik9bot, Carlog3, Gil987, Calmer Waters, JSquish, StringTheory11, Manar al Zraiy, Spinfermion and Anonymous: 9
- **Stop squark** *Source:* https://en.wikipedia.org/wiki/Stop_squark?oldid=666472875 *Contributors:* Lumidek, Malcolma, SmackBot, Pulu, MER-C, Beagel, Jakebathman, Dgiraffes, Logan, Addbot, Omnipaedista, Anti-Chronon, Gejyspa, Arbnos, Teaktl17, Andyhowlett, Joeinwiki, Akro7 and Anonymous: 3
- **Planck particle** *Source:* https://en.wikipedia.org/wiki/Planck_particle?oldid=646843749 *Contributors:* Davidl9999, Bradeos Graphon, Hidaspal, Wtmitchell, Jheald, Kazvorpal, Linas, Rjwilmsi, Jaraalbe, Hairy Dude, SmackBot, Headbomb, I do not exist, Silver seren, QuantumEngineer, Talon Artaine, Leyo, Thurth, Jim E. Black, Yerpo, CharlesGillingham, Cirt, Alexbot, MetaBohemian~enwiki, Addbot, Download, Ersik, AnomieBOT, Δ, Preon, Latifahphysics, Helpful Pixie Bot, Gigarose, Bm gub2, Gamesgamesgames, Joeinwiki, Yardimsever and Anonymous: 12
- **Axion***Source:* https://en.wikipedia.org/wiki/Axion?oldid=680013713*Contributors:* Bryan Derksen, Roadrunner, Maury Markowitz, Heron,Stevertigo, Edward, Ahoerstemeier, J'raxis, Jengod, Dragons flight, Saltine, Phys, Thue, BenRG, Seglea, Pmineault, Rursus, Herbee, Xerxes314,JeffBobFrank, Eequor, Pjacobi, David Schaich, Bender235, JustinWick, Themusicgod1, John Vandenberg, .:Ajvol:., Reuben, I9Q79oL78KiL0QTFHgyc,Physicistjedi, Fwb22, Guy Harris, Axl, Mac Davis, Wiccan Quagga, RJFJR, Count Iblis, Poseidon^3, ????, Gene Nygaard, Betsythedevine,Kbdank71, Rjwilmsi, Captmondo, Mike Peel, AndyKali, Erkcan, Itinerant1, Lmatt, Peri~enwiki, Ggb667, Chobot, Peter Grey, Uriah923,YurikBot, Spacepotato, Bambaiah, RussBot, Salsb, Oni Lukos, Buster79, Długosz, Dogcow, Ravedave, 2over0, A13ean, SmackBot, TomLougheed, Nickst, Colonies Chris, Martin Blank, V1adis1av, OrphanBot, Pwjb, A5b, Lambiam, JorisvS, Anescient, Ossipewsk, Dsspiegel,Danlev, Friendly Neighbour, Foice, Runningonbrains, Phatom87, Thijs!bot, Epbr123, Wikid77, Headbomb, Davidhorman, Escarbot, OreoPriest, Orionus, Billdad, Bubsir, Dricherby, Mwzappe, Qev , Swpb, Homunq, Otivaeey, Maliz, Alsee, HEL, William H. Kinney, 0-Jenny-0, Adamwang, Markhealey, Voorlandt, Antixt, SieBot, Senor Cuete, Martarius, ArdClose, Uxorion, FOARP, Kcren, SkyLined, Addbot,Uruk2008, DOI bot, Howard Landman, Zorrobot, KusterM, Luckas-bot, Yobot, Ptbotgourou, AnomieBOT, RoundPanda, JackieBot, Cita-tion bot, Xqbot, Reality006, Omnipaedista, FrescoBot, Chuli2802, Citation bot 1, Citation bot 4, Three887, Governus, Trappist the monk,Dinamik-bot, Earthandmoon, Sideways713, Marie Poise, StringTheory11, Suslindisambiguator, Zhskyy, Timetraveler3.14, ClueBot NG, Alessandro.de.angelis, Koornti, Asstrak, Bibcode Bot, BG19bot, Trevayne08, Cerabot~enwiki, Wipark, Andyhowlett, Ajbilan, Monkbot, TerryAlex, PS83, BradNorton1979 and Anonymous: 86
- **Dilaton** *Source:* https://en.wikipedia.org/wiki/Dilaton?oldid=647593052 *Contributors:* Michael Hardy, Charles Matthews, Mporter, Lupin, Fropuff, Lumidek, Pjacobi, Bender235, Jérôme, RJFJR, Jeff3000, Mpatel, Kyleca, GregorB, Ems57fcva, FlaBot, Roboto de Ajvol, Wavelength, SmackBot, Pdaoust, Colonies Chris, V1adis1av, QFT, CRGreathouse, Thijs!bot, Keraunos, Headbomb, Jpod2, Anonymous Dissident, Antixt, Paradoctor, Alexbot, Yosefverbin, SkyLined, Addbot, Luckas-bot, AnomieBOT, Hep thinker, MastiBot, Jakeukalane, ZéroBot, Cogiati, Throwmeaway, Ebrambot, TonyMath, Fluctuating metric, Raidr, Helpful Pixie Bot, Bibcode Bot, Jeremy112233 and Anonymous: 12

- **Graviton** *Source:* https://en.wikipedia.org/wiki/Graviton?oldid=677652863 *Contributors:* CYD, Bryan Derksen, Timo Honkasalo, XJaM, Fubar Obfusco, Maury Markowitz, Kaczor~enwiki, Jketola, TakuyaMurata, Eric119, Looxix~enwiki, Glenn, Cyan, Wooster, Charles Matthews, Timwi, Wik, BenRG, Donarreiskoffer, Scott McNay, Stephan Schulz, Arkuat, Chris Roy, Merovingian, Davidl9999, Giftlite, Xerxes314, Jason Quinn, Matt Crypto, CryptoDerk, RetiredUser2, Icairns, Zfr, Lumidek, Ukexpat, Urvabara, Discospinster, Pjacobi, Vapour, Brian0918, El C, Joanjoc~enwiki, Dalf, Army1987, Mpvdm, La goutte de pluie, Physicistjedi, Daniel Arteaga~enwiki, Zenosparadox, Dethtron5000, Keenan Pepper, Viridian, SidP, Falcorian, Skeejay, Simetrical, Dr Archeville, Mpatel, Kyleca, Tmassey, Christopher Thomas, Tevatron~enwiki, Kbdank71, Nightscream, Koavf, Mike Peel, Ems57fcva, FlaBot, RexNL, Chobot, DVdm, Roboto de Ajvol, Spacepotato, Anonymous editor, SnoopY~enwiki, Salsb, Bachrach44, Hyperbrand, NickBush24, Pnrj, RL0919, EEMIV, IslandGyrl, Bota47, C h fleming, Petri Krohn, Mario23, Alias Flood, Tim314, Teply, GrinBot~enwiki, SmackBot, Amcbride, Melchoir, Eskimbot, Gilliam, Skizzik, Timneu22, Complexica, Villarinho, Colonies Chris, V1adis1av, Chlewbot, Xyzzyplugh, Jmnbatista, Fuhghettaboutit, Sadi Carnot, Yevgeny Kats, TenPoundHammer, Lambiam, Zaphraud, JorisvS, Mr Stephen, Ramuman, Quasar Jarosz, Lottamiata, Firewall62, Kurtan~enwiki, CmdrObot, BeenAroundAWhile, WeggeBot, Shultz IV, UncleBubba, Michael C Price, Anthmoo, Thijs!bot, Epbr123, Headbomb, KevinS06, Opelio, Spartaz, JAnDbot, Xoneca, SHCarter, Pikazilla, Robin S, STBot, Kostisl, J.delanoy, Tarotcards, Coppertwig, Wesino, Sava ankit2006, Tygrrr, Idioma-bot, Sheliak, JoAnneThrax, TXiKiBoT, WilliamSommerwerck, Hqb, Anonymous Dissident, Antixt, SieBot, Flyer22, Henry Delforn (old), ClueBot, Ergn, Darkicebot, DenverRedhead, Addbot, Eric Drexler, Uruk2008, DOI bot, BrianBop, PJonDevelopment, F Notebook, Legobot, Picturesofnothing, Dov Henis, Alfredschrader, Eric-Wester, AnomieBOT, VanishedUser sdu9aya9fasdsopa, Jim1138, Materialscientist, Citation bot, Tomflaherty, ProtectionTaggingBot, Waleswatcher, FrescoBot, Juto20, LucienBOT, Paine Ellsworth, I dream of horses, Tom.Reding, RedBot, Omar.tigereyes, IVAN3MAN, Ashish.kotwal, Michael9422, D0wnfalle, EmausBot, Octaazacubane, 8digits, Slightsmile, K6ka, Thecheesykid, User10 5, Rcsprinter123, Orbjeeples, Puffin, Herk1955, ClueBot NG, Raidr, Masssly, Helpful Pixie Bot, Bibcode Bot, BG19bot, Shapoopy178, ServiceAT, PhnomPencil, Trevayne08, Brainssturm, Tjamcclain2, ChrisGualtieri, Ariscod, TheUyulala, LightandDark2000, Jessybun, Makecatbot, Kryomaxim, JRYon, Andyhowlett, Mark viking, Yorsh07, CensoredScribe, WPratiwi, Monkbot, Bryan Paul Senior, Dr.Begich, Nompynuthead, Jacobflarsen and Anonymous: 196

- **Majoron** *Source:* https://en.wikipedia.org/wiki/Majoron?oldid=680768709 *Contributors:* Bryan Derksen, Maury Markowitz, Maximus Rex, Fibonacci, Pharotic, Icairns, Pjacobi, Bender235, Count Iblis, Allen3, Kbdank71, Conscious, CWenger, V1adis1av, Happy-melon, Headbomb, Kevinwiatrowski, Maliz, STBot, R'n'B, HEL, Phoenix1177, Lseixas, Antixt, Ishvara7, Ngebendi, SkyLined, Addbot, Omnipedian, AnomieBOT, ArthurBot, Paine Ellsworth, ZéroBot, Timetraveler3.14, Bibcode Bot, Dja1979, SteenthIWbot, Monkbot, Zchacko, Lxplot and Anonymous: 7

- **Majorana fermion***Source:* https://en.wikipedia.org/wiki/Majorana_fermion?oldid=680786745*Contributors:* Pablo Mayrgundter, Jerzy, BenRG,Finlay McWalter, Lumos3, Phil Boswell, Giftlite, Dmmaus, Chris Howard, ArnoldReinhold, Bender235, Sburke, Rjwilmsi, Bubba73, HairyDude, Chris Capoccia, Buster79, SCZenz, Larsobrien, Thnidu, Teply, Incnis Mrsi, Modest Genius, JorisvS, Brienanni, Tmangray, Vttoth,Cydebot, Ntsimp, Difluoroethene, Quibik, Whatever1111, Headbomb, Davidhorman, Tjmayerinsf, Bpmullins, R sirahata, HEL, MistyMorn,Ljgua124, Satani, Pamputt, Afernand74, Coinmanj, Jwfvalle, Another Believer, Dthomsen8, Addbot, Roentgenium111, GDK, Luckas-bot,Yobot, Ptbotgourou, Amirobot, AnomieBOT, Citation bot, Obersachsebot, MIRROR, Omnipaedista, The Interior, Astiburg, Nicolas PerraultIII, Paine Ellsworth, Abductive, Tom.Reding, Amonet, Trappist the monk, RRBiswas, RobinPolt, JSquish, Quondum, AlbertusmagnusOP,Brandmeister, KarlsenBot, Claradea, ClaudeDes, Editør, Tyzoid, Bibcode Bot, BG19bot, Ymblanter, Moguns, ????, YFdyh-bot, M Krikke,Neutrinomajorana, MrCondense, Anton.akhmerov, Kowtje , Kolen Cheung, Anrnusna, Alien Putsch resistant, IRW0, ScrapIronIV, DrKittsand Anonymous: 55

- **Magnetic monopole** *Source:* https://en.wikipedia.org/wiki/Magnetic_monopole?oldid=682682748 *Contributors:* Bryan Derksen, The Anome, Ap, Andre Engels, Roadrunner, Maury Markowitz, Heron, Camembert, Patrick, Michael Hardy, Tim Starling, EddEdmondson, Dominus, Ixfd64, Skysmith, Looxix~enwiki, Mkweise, Ahoerstemeier, Stevenj, Aarchiba, Cyan, Hollgor, Charles Matthews, Timwi, Phys, Jerzy, BenRG, Jeffq, Henrygb, Rasmus Faber, Pengo, Cutler, Enochlau, Giftlite, Mintleaf~enwiki, Xerxes314, Rapjo, Waltpohl, Pharotic, Peter Ellis, Nova77, ConradPino, Gzuckier, Beland, MFNickster, Anythingyouwant, Elektron, Icairns, Lumidek, Karl Dickman, Mike Rosoft, Urvabara, Jkl, Rich Farmbrough, TedPavlic, Pjacobi, ArnoldReinhold, MuDavid, Bender235, ESkog, Kjoonlee, El C, Sasquatch, Thuktun, Congruence, Alansohn, Anthony Appleyard, Cmprince, Pauli133, Nick Mks, Falcorian, Linas, JarlaxleArtemis, Ruud Koot, Mpatel, Tabletop, GregorB, CharlesC, TheAlphaWolf, Emerson7, Mandarax, Aarghdvaark, BD2412, Rjwilmsi, HonoluluMan, MarSch, Eyu100, Seraphimblade, DonSiano, Gareth McCaughan, R.e.b., Erkcan, Drrngrvy, Mathbot, Nihiltres, Tardis, Adarsh116098, Chobot, DVdm, Amaurea, YurikBot, Phmer, RussBot, Xihr, JabberWok, Gaius Cornelius, PoorLeno, DragonHawk, Wiki alf, Welsh, Długosz, Gillis, Dchoulette, Jstrater, Crasshopper, Tony1, Crumley, SamuelRiv, 2over0, Reyk, KingCarrot, Sbyrnes321, Mhardcastle, SmackBot, Michaelliv, Melchoir, Jonathan Karlsson, Octahedron80, Skatche, V1adis1av, QFT, Alex Fix, Ianmacm, Khukri, Mohseng, Skiminki, Yevgeny Kats, Nat2, JorisvS, Loadmaster, Rock4arolla, Stephen B Streater, Norm mit, Gorog, Courcelles, Piccor, Achoo5000, Chetvorno, Disambiguator, Randall Nortman, GRB, Capefeather, Moyerjax, Michael C Price, Quibik, Doug Weller, DumbBOT, Karl-H, Difty, Wikid77, Headbomb, Luna Santin, Thranduil, Fru1tbat, Spartaz, JAnDbot, Igodard, Catslash, Bakken, Sjanusz, David Eppstein, Stevvers, WLU, 2bithacker, C.R.Selvakumar, Abecedare, Nsande01, Nlalic, NerdyNSK, JA.Davidson, Rod57, Dawright12, Aoosten, Tarotcards, Plasticup, Loohcsnuf, Barraki, Ross Fraser, Ratfox, Dorftrottel, Trmatthe, VolkovBot, FDominec, Rei-bot, Lixo2, Mathfreak11235, Wingedsubmariner, Antixt, RaseaC, Stigin, SieBot, CatherS, Likebox, Pit-trout, Henke37, Lisatwo, Dickontoo, Skeptical scientist, Maxime.Debosschere, Martarius, Balashpersia~enwiki, Unbuttered Parsnip, Razimantv, Mild Bill Hiccup, LonelyBeacon, Wrsh11, Sfitzsi, SchreiberBike, DumZiBoT, XLinkBot, Oldnoah, Avoided, Addbot, Jacopo Werther, DOI bot, Мыша, Barak Sh, 84user, Qaswqaswgd, Skippy le Grand Gourou, Luckas-bot, Munkel Davidson, Yobot, KamikazeBot, AnomieBOT, Floquenbeam, Materialscientist, Citation bot, Flying hazard, Xqbot, Renaissancee, Kbodouhi, Charvest, Nagualdesign, FrescoBot, Goodbye Galaxy, Citation bot 1, Relke, DrilBot, Cwedhrin, Q0k, MarcelB612, RedBot, Trappist the monk, YURI-21century, Morphotomy, Splartmaggot, FKLS, Deanmullen09, Wrotesolid, Waylah, Giscard2, John of Reading, WikitanvirBot, Wikipelli, ZéroBot, Prayerfortheworld, Cogiati, N0RND123, StringTheory11, Quondum, Maschen, Particle hep, Zooooooooaa, Isocliff, David Thorne, ClueBot NG, Andrija radovic, MerlIwBot, Bibcode Bot, BG19bot, PearlSt82, Gorthian, F=q(E+v^B), Niqomi, BattyBot, ChrisGualtieri, Khazar2, MaxwellDecoherence, Enyokoyama, JRYon, Jaxcp3, Andyhowlett, GabeIglesia, Razibot, Consecutor, Monkbot, BethNaught, BoltNinja, Cc universe and Anonymous: 237

- **Tachyon** *Source:* https://en.wikipedia.org/wiki/Tachyon?oldid=678997047 *Contributors:* CYD, Bryan Derksen, Jeronimo, Wayne Hardman, Youssefsan, XJaM, Rgamble, Ben-Zin~enwiki, B4hand, Olivier, Michael Hardy, Skysmith, William M. Connolley, Smack, Schneelocke, Timwi, David Latapie, VeryVerily, Phys, Dysfunktion, Nosebud, MD87, BenRG, Carbuncle, Chuunen Baka, AlexPlank, Robbot, MrJones,

Fredrik, Seano1, Xanzzibar, GreatWhiteNortherner, Giftlite, DocWatson42, Inter, Xerxes314, Anville, Varlaam, Neilc, Farside~enwiki, Armaced, M.e, Sam Hocevar, Tdent, Urhixidur, Joyous!, Neale Monks, Flex, Mike Rosoft, Cypa, Urvabara, Discospinster, Rich Farmbrough, Gussi, FT2, Pjacobi, Eric Shalov, Dbachmann, Pavel Vozenilek, Bender235, Mr. Billion, El C, Wolfy, Kwamikagami, RoyBoy, Gershwinrb, Muskrat Collective, Enric Naval, Cwolfsheep, Ghoseb, Alansohn, Keenan Pepper, Jeltz, Ducttapeavenger, Iris lorain, Xaphan9966, Cal 1234, Count Iblis, Scottishmatt, TeamUnderhill, Adrian.benko, WilliamKF, Mindmatrix, Miss Madeline, Kyleca, GregorB, Tslocum, Graham87, Sjö, Rjwilmsi, Tim!, Arabani, Arunkumar, Mike Peel, Hsriniva, Sumanch, Musical Linguist, Nihiltres, Enon, Jeremygbyrne, Diza, Turbinator, Snailwalker, Chobot, EamonnPKeane, YurikBot, Laurentius, RobotE, Xoet, Jimp, Wikky Horse, Phantomsteve, Bhny, Stalmannen, Ksyrie, RandallJones, Salsb, Big Brother 1984, Alextangent, Długosz, Justin Biggs, BlackAndy, RiesstiuIV, Froth, E2mb0t~enwiki, JPMcGrath, DeadEyeArrow, Dna-webmaster, Virogthecong, Richardcavell, Deville, Tim314, FyzixFighter, SmackBot, Reedy, Jagged 85, Clpo13, Zyxw, Timeshifter, BiT, Apers0n, Septegram, Yamaguchi??, Cuddlyopedia, Schmiteye, Teemu Ruskeepää, Bluebot, E afshin, SynergyBlades, The Benefactor, Kungming2, Colonies Chris, V1adis1av, Sloverlord, Cybercobra, EPM, Candorwien, Jheriko, JorisvS, Mgiganteus1, The Man in Question, Helloelan, Starfyredragon, Dr Greg, Hypnosifl, Silverthorn, Doczilla, AdultSwim, Grapplequip, Quaeler, Dan Gluck, Iridescent, Michaelbusch, Newone, Katharsis~enwiki, RekishiEJ, JRSpriggs, Jonathan W, Petr Matas, Grungekid262, Chronodin, MORBis, Rwflammang, Rotiro, Difluoroethene, Michael C Price, Tawkerbot4, Doc W, Unknown entity, Wikid77, Nitchell, Pstanton, Martin Hogbin, Headbomb, D.H, Eddysoto, SandChigger, Spartaz, Storkk, Deflective, Kaobear, Smiddle, BaileyZRose, JediRRT, TassadarAlpha, Bongwarrior, VoABot II, Jpod2, TxAlien, Nyttend, Hveziris, Heliac, War wizard90, Rickard Vogelberg, Joshua Davis, Gwern, Imonawave, Gadzirai, Anaxial, R'n'B, Ash, AstroHurricane001, Maurice Carbonaro, Don Cuan, Sigmundur, Barraki, Adam Zivner, Cuzkatzimhut, VolkovBot, Preston47, John-Blackburne, Maghnus, JL12~enwiki, Barneca, Tkloes, SeanNovack, A4bot, Anonymous Dissident, John Carter, Lamro, Antixt, Monty845, EverGreg, AmigoNico, SieBot, Rangutan, Gilsinan, Op56589988745, Thefreemarket, RadicalOne, Flyer22, Sunayanaa, Sfreedkin, OKBot, S2000magician, PerryTachett, Skald the Rhymer, Jimqbob, LoveMed, Ernstwall, Martarius, Ackshatt, ClueBot, The Thing That Should Not Be, EoGuy, Drmies, Thegeneralguy, Maanush, Ajoykt, BHARATH PRINCE, 12 Noon, Rollinsk, Msinkm, Rostheskunk, Thingg, Dhroova, BgJff, Pzoxicuvybtnrm, JederCoulious, Ano-User, TimothyRias, KingRuchaka, Gchackerguy, XLinkBot, EastTN, Mifter, MaizeAndBlue86, CalumH93, Addbot, DOI bot, Melab-1, Mabdul, Ronhjones, MagnusA.Bot, Instant Ramen, LaaknorBot, Troyguerrero, PFSLAKES1, Chzz, Doniago, Numbo3-bot, Tide rolls, MartinBennett Aus, FakenMC, Pianoman06, Dc4life78, Luckas-bot, Yobot, StarTroll, DallozDoppler, Gelbukh, Kulmalukko, AnomieBOT, Esmeribetheda, Sz-iwbot, Citation bot, Xqbot, Ekwos, PrometheusDesmotes, 4twenty42o, Andersæøå, Vzty, Omnipaedista, RibotBOT, SassoBot, Gbruin, Waleswatcher, Saalstin, Mtk180, Look Busy, FreeKnowledgeCreator, FrescoBot, Pepper, Andrewhayes, Brentfuller, Citation bot 1, Merongb10, Gil987, Jonesey95, Tom.Reding, ErnieD24, TedderBot, Rausch, Double sharp, Trappist the monk, Strojar 88, RjwilmsiBot, Noommos, EmausBot, Timatnet, Dancedanceyo, 8digits, Slightsmile, K6ka, StringTheory11, DacodaNelson, H3llBot, TonyMath, Iiar, L Kensington, Ico555, Holbenilord, Long.wang.2000, ClueBot NG, Cwmhiraeth, Drpilotti, Gilderien, Dfarrell07, MZ1191, Widr, Bopomofo, Dylanhansch1, Writerchic99, Helpful Pixie Bot, Oklahoma3477, Curb Chain, VatiBear, Bibcode Bot, BG19bot, Genesisrmint, Raybob95, Copernicus01, Zgstehdyp, Hstdgrypk, Wang1979, Mr.viktor.stepanov, Dyetsu, Bobo123456, Cdeatly, Ownedroad9, Brainssturm, PsiEpsilon, Guanghuilin, Unicorn234, BattyBot, ChrisGualtieri, Enyokoyama, Makecat-bot, Abits52, DmVdx, Reatlas, Timspraguetls, Penitence, Wikiuser13, Aorshahar, Luxure, Arlene47, Dbtech, Reykjavik aattrreeyyee, MOFO69696, Kaneq, FivePillarPurist and Anonymous: 404

- **Leptoquark** *Source:* https://en.wikipedia.org/wiki/Leptoquark?oldid=674819966 *Contributors:* Bcorr, Herbee, Xerxes314, Pjacobi, David Schaich, Bender235, Erkcan, Timboe, Hairy Dude, Conscious, PJTraill, Colonies Chris, Mesons, Linus M., Headbomb, Sanitycult, Calwiki, Waltoncats, Addbot, Mjamja, Wireader, Citation bot, Ernsts, Citation bot 1, Slightsmile, ResidentAnthropologist, Bibcode Bot and Anonymous: 9
- **X and Y bosons** *Source:* https://en.wikipedia.org/wiki/X_and_Y_bosons?oldid=621650282 *Contributors:* XJaM, Timwi, Rich Farmbrough, Laurascudder, Anthony Appleyard, Bubba73, V1adis1av, Quaeler, Michael C Price, Thijs!bot, Headbomb, JAnDbot, Choihei, VolkovBot, Antixt, Dawn Bard, BartekChom, Yaybob, DragonBot, SkyLined, Addbot, Yobot, Ernsts, A. di M., Griffinofwales, Paine Ellsworth, Gil987, Double sharp, ZéroBot, Primergrey, Latifahphysics, Helpful Pixie Bot and Anonymous: 9
- **W′ and Z′ bosons** *Source:* https://en.wikipedia.org/wiki/W%E2%80%B2_and_Z%E2%80%B2_bosons?oldid=677151189 *Contributors:* Rjwilmsi, R.e.b., Smurrayinchester, SmackBot, Incnis Mrsi, Valoem, Michael C Price, Headbomb, Magioladitis, Antixt, Ptrslv72, BartekChom, Myst-Bot, SkyLined, Addbot, Scientryst, Yobot, AnakngAraw, Citation bot, ArthurBot, Carlog3, Paine Ellsworth, Citation bot 1, Trappist the monk, Heurisko, RjwilmsiBot, StringTheory11, Ethaniel, Bibcode Bot, Monkbot and Anonymous: 9
- **Sterile neutrino** *Source:* https://en.wikipedia.org/wiki/Sterile_neutrino?oldid=679151361 *Contributors:* Rursus, Giftlite, Xerxes314, Robert Brockway, Jkl, Rich Farmbrough, Pjacobi, I9Q79oL78KiL0QTFHgyc, Cgmusselman, Count Iblis, Ceyockey, GregorB, Rjwilmsi, Strait, Goudzovski, Tinlad, SCZenz, Reyk, Nickst, Fredvanner, Colonies Chris, V1adis1av, QFT, Mayrel, JorisvS, Dan Gluck, Foice, Michael C Price, Thijs!bot, Headbomb, Qwerty Binary, Wdtaylor1066, DanPMK, .anacondabot, Jayanthtn, Ccrummer, R'n'B, Mikek999, Fordi, Rod57, Sheliak, VolkovBot, Paradoctor, Wing gundam, BartekChom, Wwheaton, Alexbot, Jwfvalle, DumZiBoT, BodhisattvaBot, MystBot, SkyLined, Addbot, Eric Drexler, Yobot, Romul~enwiki, Stuffed cat, Citation bot, Omnipaedista, Ernsts, Paine Ellsworth, DrilBot, Jonesey95, Tom.Reding, Michael9422, JLincoln, Bj norge, Dskrvk, EmausBot, ZéroBot, StringTheory11, Timetraveler3.14, Maschen, Bibcode Bot, BG19bot, ChrisGualtieri, Egofofo, AloisKabelschacht, Ajbilan, Monkbot, ??, Lxplot and Anonymous: 34
- **Preon** *Source:* https://en.wikipedia.org/wiki/Preon?oldid=646794027 *Contributors:* Maury Markowitz, Heron, Ewen, Edward, Kickaha~enwiki, Dcljr, Timwi, David Latapie, Chrisjj, BenRG, Altenmann, Merovingian, Xanzzibar, David Gerard, Giftlite, Graeme Bartlett, Herbee, Monedula, Semorrison, Mboverload, Eequor, Icairns, Urhixidur, Rich Farmbrough, Pjacobi, Drhex, John Vandenberg, Jag123, Calton, Alai, Uncle G, GregorB, BD2412, Ketiltrout, Rjwilmsi, Fragglet, Mathrick, Smithbrenon, CJLL Wright, Chobot, ScottAlanHill, Jpkotta, YurikBot, Ugha, Bambaiah, Phmer, Ohwilleke, Merick, Gcapp1959, Dialectric, Buster79, Trovatore, Długosz, Closedmouth, Iellwood, Paul D. Anderson, Lserni, SmackBot, RockMaestro, Bayardo, Stepa, GwydionM, Kmarinas86, DocKrin, JesseStone, Trekphiler, V1adis1av, Fatla00, SilverStar, Jaganath, Md2perpe, Will314159, Friendly Neighbour, CmdrObot, Doc W, AlphaNumeric, LactoseTI, Mglg, Keraunos, Headbomb, Thadius856, Joe Schmedley, Ph.eyes, GurchBot, Yill577, Randyfurlong, Dr. Morbius, Lexivore, Experiential, ChauriCh, Tanaats, OliverHarris, VolkovBot, Fences and windows, Calwiki, Anonymous Dissident, Thrawn562, Dirkbb, Synthebot, Antixt, Pegasus1965, AHMartin, PlanetStar, Work permit, WereSpielChequers, 1ForTheMoney, SkyLined, Addbot, Lightbot, OlEnglish, Zorrobot, The Bushranger, Luckas-bot, Yobot, EnochBethany, AnomieBOT, Icalanise, Materialscientist, Citation bot, Jsharpminor, GrouchoBot, FrescoBot, Goodbye Galaxy, Citation bot 1, MastiBot, Bj norge, RjwilmsiBot, Ofercomay, Detogain, WikitanvirBot, Slightsmile, The Mysterious El Willstro, Hhhippo, Wyvern

Rex., Suslindisambiguator, Gilderien, Preon, Helpful Pixie Bot, Bibcode Bot, BG19bot, Marioedesouza, Andyhowlett, I am One of Many, FrigidNinja, Draconnis caput, Delbert7 and Anonymous: 101

- **Bound state** *Source:* https://en.wikipedia.org/wiki/Bound_state?oldid=651409696 *Contributors:* Smack, Loren Rosen, Phys, Dratman, Karol Langner, H Padleckas, Lumidek, Dmr2, Jag123, Guy Harris, Keenan Pepper, Lectonar, Tony Sidaway, SeventyThree, Rjwilmsi, Mushin, Amyst, Conscious, SmackBot, Incnis Mrsi, Tom Lougheed, A876, Davius, Thijs!bot, Mbell, Headbomb, Second Quantization, Grimlock, Hbent, TXiKiBoT, Antixt, Neparis, SieBot, Reuqr, Prince Max (scientist), DumZiBoT, Ejakku, Addbot, Mjamja, Yobot, Inalokasimera, AnomieBOT, RibotBOT, Erik9bot, StringTheory11, RolteVolte, Bibcode Bot, Orentago, Pelotin~enwiki, Wdlang, BattyBot, ChrisGualtieri, Melonkelon, Monkbot and Anonymous: 19
- **Hadron***Source:* https://en.wikipedia.org/wiki/Hadron?oldid=681027720*Contributors:* Bryan Derksen, Manning Bartlett, Peterlin~enwiki,Edward, Erik Zachte, ESnyder2, Fruge~enwiki, TakuyaMurata, Darkwind, Glenn, Nikai, Ehn, Olya, Phys, Bevo, Topbanana, BenRG, Twang,Donarreiskoffer, Korath, Wjhonson, Merovingian, Ojigiri~enwiki, Sunray, JesseW, Xanzzibar, Giftlite, Xerxes314, Dratman, Physicist, Mikro2nd,LiDaobing, Pthompson, Icairns, Jimaginator, Mike Rosoft, Vsmith, Goochelaar, Sunborn, Livajo, El C, Kwamikagami , Shanes, Fwb22,Jumbuck, Cookiemobsta, Velella, Rebroad, Vuo, Kusma, DV8 2XL, Linas, GrouchyDan, Palica, Marudubshinki, Kbdank71, Mana Excal-ibur, Kinu, Strait, FlaBot, RexNL, Goudzovski, FrankTobia, YurikBot, Radishes, Bambaiah, Hydrargyrum, Salsb, NawlinWiki, Wiki alf,SCZenz, Davemck, Bota47, Scriber~enwiki, Modify, Katieh5584, Eog1916, SmackBot, McGeddon, Gilliam, Benjaminevans82, Dingar, Per-sian Poet Gal, Telempe, DHN-bot~enwiki, Audriusa, Acepectif, Kokot.kokotisko, JorisvS, JarahE, BranStark, SJCrew, Eratticus, Chrumps,Jtuggle, Q43, Epbr123, Wikid77, Headbomb, Escarbot, Deflective, Gcm, NE2, Trapezoidal, Naval Scene, ΚΕΚΡΩΨ, NeverWorker, Wwmbes,Alexllew, Lvwarren, Jebus0, DariusU, Khalid Mahmood, Adriaan, Rustyfence, Ron2, Leyo, J.delanoy, Maurice Carbonaro, JVersteeg, Rod57,Way2Smart22, Hugh Hudson, Y2H, Ansans, Bobxii, Chris Longley, Useight, Dylan bossart, VolkovBot, TXiKiBoT, Kinkydarkbird, Anony-mous Dissident, Don4of4, Wordsmith, LeaveSleaves, Antixt, Enviroboy, Insanity Incarnate, Nibios, AlleborgoBot, SieBot, Yintan, Lead-SongDog, RadicalOne, Paolo.dL, OKBot, JohnSawyer, Lazarus1907, Pinkadelica, Danthewhale, Martarius, ClueBot, Amaamaddq, Authori-tative Physicist, Wwheaton, Rotational, DragonBot, Sciencedude9998, Tuchomator, El planeto, Kaiba, Thingg, Koshoid, Aitias, Apparition11,Rishi.bedi, TimothyRias, InternetMeme, Jbeans, MystBot, Sgpsaros, Tayste, Addbot, Pkkphysicist, Ehrenkater, Lightbot, Luckas-bot, Yobot,Nallimbot, Dagus2000, Fangfyre, LOLx 9000, Thisaccountwillbebanned, Citation bot, Xqbot, Drilnoth, Br77rino, Wikiedit33, Ajahnjohn,Omnipaedista, RibotBOT, Mashmeister, Tjbright2, My cat's breath smells like catfood, Haeinous, Citation bot 1, Javert, Gil987, I dreamof horses, Jonesey95, Rameshngbot, Thinking of England, Alarichus, SkyMachine, FoxBot, Johnshnappay, Антон Гліністы, Teravolt, Rac-erx 11, Naznin farhah, Tommy2010, Rafabaez, Wikipelli, ZéroBot, StringTheory11, Hadron12, Donner60, Bobogoobo, Petrb, ClueBot NG,Gareth Griffith-Jones, Bibcode Bot, BG19bot, Dwightboone, Njavallil, Walterpfeifer, Pfeiferwalter, ChrisGualtieri, Ugog Nizdast, Lithelimbs,RoKo89, Michikohundred, KasparBot, Wwilliam726 and Anonymous: 169
- **Baryon** *Source:* https://en.wikipedia.org/wiki/Baryon?oldid=681412960 *Contributors:* AxelBoldt, Tobias Hoevekamp, Bryan Derksen, Ben-Zin~enwiki, Heron, Tim Starling, Alan Peakall, Paul A, Salsa Shark, Glenn, Mxn, Charles Matthews, The Anomebot, ElusiveByte, Phys, Bevo, Traroth, Donarreiskoffer, Robbot, Korath, Kristof vt, Merovingian, Ojigiri~enwiki, Sunray, Wikibot, Giftlite, DocWatson42, Shaun-MacPherson, Herbee, Xerxes314, Dratman, DÅ,ugosz, Kaldari, OwenBlacker, Icairns, JohnArmagh, Rich Farmbrough, Guanabot, Mani1, E2m, Tompw, El C, Bobo192, I9Q79oL78KiL0QTFHgyc, Giraffedata, Physicistjedi, Jumbuck, Gary, ABCD, Oleg Alexandrov, Woohookitty, Tevatron~enwiki, BD2412, Kbdank71, Nightscream, Ae77, MZMcBride, Chekaz, R.e.b., Erkcan, Maxim Razin, Oo64eva, Chobot, Roboto de Ajvol, YurikBot, Bambaiah, Jimp, Salsb, Ergzay, DragonHawk, SCZenz, E2mb0t~enwiki, Bota47, Simen, Sbyrnes321, Lainagier, Timotheus Canens, Bluebot, Colonies Chris, Kingdon, Shadow1, Bigmantonyd, Drphilharmonic, Kseferovic, Wierdw123, Physicsdog, Torrazzo, Verdy p, Michael C Price, Thijs!bot, Headbomb, Hcobb, Orionus, QuiteUnusual, Spartaz, Plantsurfer, Amateria1121, Diamond2, Swpb, BatteryIn-cluded, Hveziris, Saxophlute, Gwern, Ben MacDui, R'n'B, Ash, Tgeairn, Maurice Carbonaro, STBotD, VolkovBot, GimmeBot, NoiseEHC, Tearmeapart, BotKung, BrianADesmond, Antixt, AlleborgoBot, Lou427, SieBot, VVVBot, Gerakibot, LeadSongDog, Keilana, Paolo.dL, Doctorfluffy, TrufflesTheLamb, OKBot, Hamiltondaniel, TubularWorld, ClueBot, Artichoker, ChandlerMapBot, CalumH93, Addbot, Laa-knorBot, CarsracBot, Jonhstone12, Legobot, Luckas-bot, Bugbrain 04, AnomieBOT, JackieBot, Materialscientist, Citation bot, ArthurBot, Xqbot, Omnipaedista, SassoBot, Spellage, WaysToEscape, FrescoBot, Citation bot 1, FoxBot, Noommos, EmausBot, John of Reading, JSquish, ZéroBot, StringTheory11, Stibu, Ethaniel, Markinvancouver, ClueBot NG, Koornti, Kasirbot, Rezabot, Bibcode Bot, Atomician, Zedshort, Marioedesouza, ChrisGualtieri, WorldWideJuan, CoolHandLouis, Monkbot, KasparBot and Anonymous: 106
- **Hyperon** *Source:* https://en.wikipedia.org/wiki/Hyperon?oldid=661765936 *Contributors:* Dominus, CesarB, Donarreiskoffer, Robbot, Her-bee, Xerxes314, Srbauer, Chbarts, Cherlin, HenkvD, Kusma, Woohookitty, Kbdank71, Ketiltrout, Rjwilmsi, Koavf, Erkcan, Baryonic Being, Goudzovski, Roboto de Ajvol, YurikBot, Bambaiah, X42bn6, Phmer, Salsb, David R. Ingham, Tetracube, SmackBot, Melchoir, Gribeco, Skrewtape, Kmarinas86, Colonies Chris, Pwjb, Physicsdog, Thijs!bot, Headbomb, Sobreira, JAnDbot, Conundrumer, I310342~enwiki, Lit-tlealien182, Venny85, Senemmar, Antixt, SieBot, Ideal gas equation, SkyLined, Addbot, Yobot, AnomieBOT, JackieBot, Citation bot, Arthur-Bot, Ernsts, RedBot, Heavyion, PNG, EmausBot, AvicBot, JSquish, ZéroBot, StringTheory11, Quondum, Gilderien, YFdyh-bot and Anony-mous: 15
- **Nucleon** *Source:* https://en.wikipedia.org/wiki/Nucleon?oldid=681131517 *Contributors:* Kpjas, Andre Engels, Peterlin~enwiki, Pichai Asokan, Twilsonb, Fruge~enwiki, Ellywa, Evercat, Jusjih, BenRG, Gromlakh, Donarreiskoffer, Robbot, Waelder, Merovingian, Ojigiri~enwiki, Kagre-don, Herbee, Xerxes314, Karol Langner, Icairns, D6, Rich Farmbrough, Igorivanov~enwiki, Vsmith, El C, Whosyourjudas, La goutte de pluie, Jumbuck, Msh210, Keenan Pepper, Kwikwag, Kusma, Gene Nygaard, Linas, Palica, Marudubshinki, Emerson7, Kbdank71, JIP, Miq, Rjwilmsi, Strait, Ttwaring, FlaBot, Eubot, OSt~enwiki, Srleffler, Wrightbus, Guliolopez, YurikBot, Dirigible, Bambaiah, Chuck Carroll, Salsb, Welsh, Długosz, Daniel Mietchen, CPColin, Bota47, Dna-webmaster, Katieh5584, GrinBot~enwiki, Sbyrnes321, SmackBot, Melchoir, Sasha-toBot, MTSbot~enwiki, Peyre, Heartofgoldfish, WeggeBot, Headbomb, John254, Wiki fanatic, Aadal, KrakatoaKatie, JAnDbot, Gcm, BenB4, DanPMK, Nyttend, Mother.earth, TheEgyptian, Leyo, Mike.lifeguard, TraceyR, VolkovBot, FantasticAsh, JhsBot, Don4of4, Antixt, Allebor-goBot, SieBot, Citizen, OKBot, Anchor Link Bot, Li4kata, Florentyna, Djr32, Keithbowden, SoxBot, Qwfp, SkyLined, Addbot, Hakan Kayı, Chamal N, Lightbot, Verazzano, Luckas-bot, Yobot, Tannkrem, The High Fin Sperm Whale, Citation bot, Bci2, ArthurBot, Xqbot, Melmann, DJWolfy, ??, FrescoBot, LucienBOT, Citation bot 1, Merongb10, Thinking of England, Trappist the monk, PNG, Jonkerz, Begoon, Rjwilm-siBot, EmausBot, WikitanvirBot, Dcirovic, StringTheory11, Ethaniel, Quondum, RolteVolte, Efiiamagus, ClueBot NG, Gilderien, Satellizer, Bibcode Bot, 4Jays1034, ChrisGualtieri, TwoTwoHello, Marekich, Trompedo, Rayhartung, Monkbot, Internucleon, KasparBot, Snackbag and Anonymous: 69

- **Proton** *Source:* https://en.wikipedia.org/wiki/Proton?oldid=682675320 *Contributors:* Mav, Bryan Derksen, Zundark, Manning Bartlett, Ap, Andre Engels, Josh Grosse, Danny, XJaM, Ellmist, Heron, Jaknouse, Stevertigo, Dwmyers, Bdesham, Patrick, Kku, Stewacide, TakuyaMurata, Egil, NuclearWinner, Looxix~enwiki, Mkweise, Ahoerstemeier, Sobekhotep, Salsa Shark, Glenn, Kaihsu, Jordi Burguet Castell, Mxn, Denny, Timwi, Paul-L~enwiki, Omegatron, Geraki, David.Monniaux, Donarreiskoffer, Robbot, Josh Cherry, Jakohn, RedWolf, Altenmann, Yelyos, Merovingian, Flauto Dolce, Hadal, Giftlite, Mikez, Tom harrison, Lupin, Herbee, Xerxes314, Fleminra, Bensaccount, Foobar, PlatinumX, Utcursch, Knutux, Antandrus, Beland, Eroica, Melikamp, Rdsmith4, DragonflySixtyseven, Icairns, Cglassey, Deglr6328, Grunt, Thorwald, Jenlight, Mike Rosoft, Diagonalfish, Discospinster, Cacycle, Vsmith, Jpk, Wikiacc, Mani1, Bender235, Kjoonlee, RJHall, El C, Shrike, Femto, Bobo192, O18, Army1987, Smalljim, GTubio, Vortexrealm, Elipongo, Foobaz, Kjkolb, Obradovic Goran, Nsaa, Eddideigel, Anthony Appleyard, Mattpickman, Apoc2400, Carmelbuck, Spangineer, Wtmitchell, Saga City, Uucp, Crobzub, Vcelloho, RainbowOfLight, TenOfAllTrades, Computerjoe, Kusma, Itsmine, Falcorian, Richard Arthur Norton (1958-), Firien, GregorB, Macaddct1984, Mayz, Karam.Anthony.K, Marudubshinki, Bebenko, Rtcpenguin, Graham87, Kbdank71, Ketiltrout, Drbogdan, Rjwilmsi, Strait, NeonMerlin, RadicalJester, Bubba73, Yamamoto Ichiro, Rangek, FlaBot, Nivix, RexNL, Fresheneesz, Srleffler, Imnotminkus, King of Hearts, CiaPan, Chobot, Deyyaz, Roboto de Ajvol, The Rambling Man, YurikBot, Bambaiah, JWB, Anuran, Pip2andahalf, RussBot, Wigie, Jumbo Snails, Raquel Baranow, Hellbus, Salsb, Oni Lukos, Anomalocaris, NawlinWiki, Injinera, Welsh, Długosz, Martin Ulfvik, Moe Epsilon, BOT-Superzerocool, DeadEyeArrow, Bota47, D-Day, Mtu, Pooryorick~enwiki, J S Ayer, Theodolite, Bayerischermann, Closedmouth, Reyk, Petri Krohn, DGaw, Paul D. Anderson, Katieh5584, RG2, SDS, GrinBot~enwiki, Nekura, Orii, Luk, Sycthos, Itub, Attilios, SmackBot, Moeron, Incnis Mrsi, Melchoir, CyclePat, Edgar181, HalfShadow, Dhochron, Munky2, Gilliam, Chris the speller, Rajeevmass~enwiki, Rkitko, AndrewBuck, Bethling, SchfiftyThree, Complexica, Sbharris, Rogermw, Can't sleep, clown will eat me, Shalom Yechiel, PeteShanosky, Writtenright, Homestarmy, Wikiwikiwiki3~enwiki, SundarBot, COMPFUNK2, Dreadstar, Orczar, Drphilharmonic, Dvorak729, DMacks, Mion, Vina-iwbot~enwiki, Bdushaw, SashatoBot, ArglebargleIV, Khazar, Cholerashot, Rijkbenik, Spacecadethailey, Herr apa, Deathcakes, Noah Salzman, Aeluwas, Waggers, Mozzura, Mattabat, Elb2000, Newone, MOBle, Igoldste, Rhetth, Frank Lofaro Jr., Tawkerbot2, CmdrObot, Ale jrb, Sir Vicious, KyraVixen, Ruslik0, McVities, TheTito, Cydebot, Nick Y., Gogo Dodo, Red Director, Umdunno, Difluoroethene, Odie5533, Q43, Tawkerbot4, Dwool99f, Narayanese, Rasheedy, Zalgo, Lo2u, Jenswort, Thijs!bot, Epbr123, Tsogo3, Headbomb, Marek69, Electron9, Mnemeson, Dfrg.msc, Philippe, Aadal, AntiVandalBot, Seaphoto, HairyDan, Shirt58, EarthPerson, Gregnx, Jj137, Naturalnumber, Myanw, Ellissound, Leuko, MER-C, CosineKitty, Fetchcomms, Andonic, Kerotan, .anacondabot, Acroterion, Plynn9, Casmith 789, Bongwarrior, VoABot II, Astrangequark, Swpb, WODUP, Recurring dreams, Avicennasis, Catgut, Dirac66, 28421u2232nfenfcenc, Hveziris, Fang 23, The Real Marauder, Oddworth, JaGa, MartinBot, Rettetast, Pbroks13, Artaxiad, J.delanoy, WeglarczykJ, Silverxxx, C.A.T.S. CEO, Maurice Carbonaro, 12dstring, WarthogDemon, Acalamari, Exdejesus, TomasBat, Antony-22, Potatoswatter, KylieTastic, Vanished user 39948282, Treisijs, S, SoCalSuperEagle, Xiahou, Specter01010, Idioma-bot, Ciju, 28bytes, VolkovBot, Doc7777777777, Jeff G., Soliloquial, Tuffcarrot, Philip Trueman, TXiKiBoT, The Original Wildbear, Vipinhari, Bjman, Bigyaks, Alexalexalex123~enwiki, Meters, Antixt, Spinningspark, Insanity Incarnate, Upquark, AlleborgoBot, Vitalikk, B41988, Petergans, Demmy100, SieBot, Accounting4Taste, Jauerback, Studnic12, Xe1881, Yintan, GlassCobra, Keilana, RadicalOne, Flyer22, Sbowers3, Prestonmag, Oxymoron83, BenoniBot~enwiki, Jacob.jose, Mygerardromance, Rajbboy69, ClueBot, The Thing That Should Not Be, RODERICKMOLASAR, Tigerboy1966, Regibox, ChandlerMapBot, Mr blabla, Excirial, Alexbot, Robbie098, Poopmister91191, Ploft, NuclearWarfare, Lunchscale, PhySusie, SoxBot, El pobre Pedro, Thehelpfulone, La Pianista, Thingg, Kanxkawii, Aitias, Subash.chandran007, Johnuniq, XLinkBot, Avoided, WikHead, SilvonenBot, SkyLined, Mls1492, Weletahoozyzog, Addbot, Zrules, Arcturus87, Ronhjones, Cst17, Glane23, AndersBot, Wandering Traveler, Omnipedian, LinkFA-Bot, Numbo3-bot, Tide rolls, Thermalimage, Luckas-bot, Yobot, Велетень, Wickedwizardofoz, Newportm, Kilom691, Heart of a Lion, Eric-Wester, Jay0205, AnomieBOT, LeftyAce, Götz, Jim1138, Sp eloc, Judoc, Materialscientist, The High Fin Sperm Whale, Citation bot, OllieFury, Apollo, Xqbot, Phazvmk, Blennow, Cureden, Wyklety, Aa77zz, Squishywushy123, Srich32977, Rueyfgugdtj, RibotBOT, PM800, A. di M., ??, Rain bowell, CES1596, FrescoBot, Paine Ellsworth, Tobby72, Citation bot 1, Pinethicket, HRoestBot, Calmer Waters, Bejinhan, Impala2009, Nicklcms, Blckmgc, SkyMachine, Gryllida, Double sharp, TobeBot, Ilovefatchicks, Keegscee, DARTH SIDIOUS 2, TjBot, StudentDoc73, Nachos0123, EmausBot, WikitanvirBot, ANDREVV, Bencbartlett, Zues zeus kratos, Pcorty, Sterrettc, K6ka, JSquish, Stuffness12, John Cline, Harddk, Fæ, StringTheory11, Brazmyth, Suslindisambiguator, Wayne Slam, Zach444, Rcsprinter123, Wiggles007, Brandmeister, Donner60, Sarthak 94, Xonqnopp, ClueBot NG, Timelord360, This lousy T-shirt, IHopeThisNameIsntTaken, Corusant, Cntras, Dictabeard, Rezabot, Helpful Pixie Bot, Wbm1058, Bibcode Bot, BG19bot, ArthropodOfDoom, RadioActiveKitKat, AvocatoBot, Metricopolus, JacobTrue, Mlkamitso, Toccata quarta, Blaspie55, Mhutchison43, 220 of Borg, Anbu121, Hitheresir, RudolfRed, Knodir, BattyBot, ChrisGualtieri, Hower64, Ducknish, Stephen Glass, BrightStarSky, Dexbot, Mogism, Sheehan Cein14, Frosty, Pidotclan, Marcoapc.84, Delnium strex, Faizan, Huddydakota, Prof.Professer, Dustin V. S., Borreswafflertron, Ugog Nizdast, Jwratner1, Javierha, JaconaFrere, AspaasBekkelund, Bballbro62, Melcous, Monkbot, Jayakumar RG, Scorpion1045, Krebs49, Yollowswagger19, Maddie005, Junchuann, Chrisbrownthathoe, Orgasam069, Tktobykerby, Interpuncts, Tetra quark, Cjohnson2020, KasparBot, Muzammil Alam Baig, Rambunctious Racoon, Teo boruch and Anonymous: 638

- **Neutron** *Source:* https://en.wikipedia.org/wiki/Neutron?oldid=682674206 *Contributors:* AxelBoldt, Tobias Hoevekamp, Chenyu, Trelvis, Calypso, Mav, Bryan Derksen, The Anome, AstroNomer~enwiki, Malcolm Farmer, Andre Engels, Xaonon, Danny, XJaM, Roadrunner, Jaknouse, Olivier, Patrick, Michael Hardy, Valery Beaud, Ixfd64, TakuyaMurata, NuclearWinner, Looxix~enwiki, ArnoLagrange, Mkweise, Ellywa, Ahoerstemeier, Cyp, Andrewa, Aarchiba, Julesd, Glenn, Nikai, Andres, Stone, Denni, Kbk, Tarosan~enwiki, Maximus Rex, Donarreiskoffer, Gentgeen, Robbot, Fredrik, Romanm, Merovingian, Rursus, Wikibot, Alan Liefting, Dave6, Giftlite, Mikez, Art Carlson, Herbee, Xerxes314, Everyking, Dratman, NeoJustin, Bensaccount, Poupoune5, Jorge Stolfi, Christofurio, Knutux, Karol Langner, Aecarol, Icairns, Zfr, Cglassey, Peter bertok, Frau Holle, M1ss1ontomars2k4, Sparky2002b, Mike Rosoft, Guanabot, Vsmith, Dbachmann, Bender235, Kjoonlee, AlDragon, Geoking66, Neko-chan, RJHall, CanisRufus, El C, Susvolans, Femto, CDN99, Bobo192, O18, Smalljim, SpeedyGonsales, Kjkolb, Obradovic Goran, Sam Korn, Nsaa, Jakew, Eddideigel, Jumbuck, Patsw, Alansohn, Interiot, Riana, Wtmitchell, BRW, NickMartin, Vuo, DV8 2XL, HenryLi, Tchaika, Forteblast, Falcorian, Richard Arthur Norton (1958-), JarlaxleArtemis, WadeSimMiser, Sega381, SDC, Jon Harald Søby, Prashanthns, Abd, LexCorp, Graham87, Magister Mathematicae, Doughboy, Ketiltrout, Rjwilmsi, Nightscream, Zbxgscqf, Strait, AySz88, Oo64eva, Rangek, FlaBot, Nihiltres, Goudzovski, Srleffler, Ronebofh, King of Hearts, Chobot, DVdm, YurikBot, RobotE, Bambaiah, JWB, TSO1D, Jimp, Phantomsteve, KyleDantarin, Stephenb, Gaius Cornelius, Yyy, Salsb, NawlinWiki, Tupungato, Wiki alf, Complainer, Grafen, Długosz, Voidxor, Scottfisher, Kkmurray, Spute, Dna-webmaster, Wknight94, Stefan Udrea, Mike Serfas, Closedmouth, Reyk, Modify, Alchie1, CWenger, RG2, Paul Erik, Triple333, Attilios, SmackBot, Caiyern, Melchoir, Wiki Tiki God, Unyoyega, Jrockley, Dr.Science, Edgar181, Yamaguchi??, Kdliss, Wigren, Chris the speller, Rajeevmass~enwiki, Persian Poet Gal, SchfiftyThree, Complexica, DHN-bot~enwiki, Sbharris, Colonies Chris, Brainblaster52, Can't sleep, clown will eat me, DéRahier, Juancnuno, SundarBot, DFriend, Aldaron, KunalKathuria, Nakon, Mwtoews, DMacks, Soarhead77, Bdushaw, Pilotguy, Renafaye77, SashatoBot, Demicx, Tim bates, Mgiganteus1, Slakr, Citicat, Asyndeton, BranStark, Shoeofdeath, Newone, Tawkerbot2, Atomobot, Mosaffa, CmdrObot, Wafulz, Dycedarg, Rwflammang, Joelholdsworth, Lokal Profil, Karenjc, Myasuda, Safalra, Icek~enwiki, Badseed, Nick Y., Gogo Dodo, Chasingsol, Phydend, Gimmetrow, Thijs!bot, Epbr123, Montazmeahii, Goods21, Tsogo3, N5iln, Oerjan, Headbomb, Marek69, SouthernMan, RoboServien, Escarbot, Aadal, WikiSlasher, AntiVandalBot, Seaphoto, Naturalnumber, Spencer, Astavats, Husond, CosineKitty, Medconn, TheEditrix2, Bongwarrior, VoABot II, Kuyabribri, JamesBWatson, WODUP, Mother.earth, Animum, BatteryIncluded, Dirac66, LorenzoB, DerHexer, JaGa, Hans Moravec, Hyray, Patstuart, MartinBot, Church of emacs, Gnuarm, Mennoblaauw, Andre.holzner, Rettetast, J.delanoy, Dbiel, Extransit, Acalamari, Ncmvocalist, TomasBat, MetsFan76, Joshmt, Heavens is the world, Scott Illini, TraceyR, Idioma-bot, Mviduka4197, VolkovBot, Tourbillon, Thedjatclubrock, Jeff G., Mocirne, Seattle Skier, TXiKiBoT, DoctorPiouk, Dev 176, Martin451, ABigGreenHippo, Abdullais4u, FreeFull, Wikiisawesome, Scarymaryfwfc, RadiantRay, Roomyt, W1k13rh3nry, Antixt, Deanlsinclair, Enviroboy, Burntsauce, Brianga, AllEborgoBot, EmxBot, Neparis, D. Recorder, Ponyo, YohanN7, SieBot, Cwkmail, Yintan, Agesworth, JerrySteal, Keilana, RadicalOne, Toddst1, Tiptoety, JetLover, Arjen Dijksman, Sbowers3, Aruton, Oxymoron83, AnonGuy, Beej175560, Techman224, Anyeverybody, Nergaal, Denisarona, Lord Shivan, Naturespace, ClueBot, RudolfSchmidt, PipepBot, Fasettle, Fyyer, The Thing That Should Not Be, Starkiller88, Industrieman, Mild Bill Hiccup, Polyamorph, Shjacks45, ChandlerMapBot, DragonBot, Gnome de plume, Jusdafax, Ju7kik8ol568r, Cenarium, Jotterbot, Vboo-belarus, Subash.chandran007, Plasmic Physics, Versus22, XLinkBot, Dark Mage, PL290, SkyLined, Addbot, Taschna, DOI bot, Ronhjones, Mr. Wheely Guy, LaaknorBot, CarsracBot, JBukon, Favonian, LinkFA-Bot, 5 albert square, AgadaUrbanit, Morgrimm, Numbo3-bot, Ehrenkater, LarryFrank, Tide rolls, Lightbot, Teles, Legobot, Luckas-bot, Yobot, Велетень, 2D, Tohd8BohaithuGh1, Cabb99, AnakngAraw, AnomieBOT, Bsimmons666, Jim1138, AdjustShift, Bluerasberry, Materialscientist, Hdehuer, The High Fin Sperm Whale, Citation bot, Satan's Kitchen, Maxis ftw, Raven1977, ArthurBot, Marshallsumter, Xqbot, Gopal81, Capricorn42, Drilnoth, DSisyphBot, Gilo1969, Paula Pilcher, Faatoafe90, Goostyyy, WaveEtherSniffer, GrouchoBot, Abce2, Amaury, Doulos Christos, Gordonrox24, Shadowjams, A. di M., Samwb123, R8R Gtrs, FrescoBot, LucienBOT, Paine Ellsworth, Cannolis, Citation bot 1, Ecko15, Biker Biker, Pinethicket, HRoestBot, Jonesey95, Nicklcms, Seattle Jörg, Abhinav paulite, Double sharp, Darrell cosare, عقیل کاشف, Mr.98, Diannaa, Ironnickel, Andrea105, Onel5969, TjBot, MagnInd, Jackehammond, Jimmy be, Robert Johnson 10, EmausBot, Green Day143, WikitanvirBot, Unkenruf, GoingBatty, Illdz, Psturm~enwiki, Pcorty, Wikipelli, Hhhippo, ZéroBot, John Cline, Brazmyth, Quondum, GianniG46, Copper.nanotube, Brandmeister, L Kensington, Epicstonemason, Sjkimminau, Chris857, VictorianMutant, DASHBotAV, Whoop whoop pull up, ClueBot NG, Nebulosus, CocuBot, Satellizer, Letoya123, TruPepitoM, OverQuantum, Heyheyheyhohoho, Rezabot, Android1188, Widr, Diyar se, Ieditpagesincorrectly, Bibcode Bot, Neutronscattering, Wiki13, Metricopolus, Contact '97, Universuminkeisari, Nathanrohler, Zedshort, Hamish59, Nitrobutane, Hobos-r-us12, Oznitecki, BattyBot, MeowMeowArf, Dansalmo, ChrisGualtieri, GeorgEhlers, Ducknish, Gladiator222, Dexbot, Mogism, 331dot, TwoTwoHello, Lugia2453, Graphium, FaerieChilde, Fossilsnout, Morg00, Cldorian, Xuanmingzi, DihllonJessie, Jesse.johns, The Herald, Zenibus, Darkch2, Jwratner1, Javierha, My name is not dave, Cytokinetics, EtymAesthete, DudeWithAFeud, Abitslow, Aspaas-Bekkelund, Bballbro62, Mahusha, Light on the wall, Monkbot, Profesionalpretzels, Jayakumar RG, Haftswinch532, Selmatoed50, Istillcant, HMSLavender, Petahr, Orduin, Kethrus, Pulkit 4325, DiscantX, TSchonfeldt, Matan Kovac, KasparBot, Kafishabbir, Lord Wingus The Third and Anonymous: 554

- **Delta baryon** *Source:* https://en.wikipedia.org/wiki/Delta_baryon?oldid=680402583 *Contributors:* Donarreiskoffer, Robbot, Rasmus Faber, Woohookitty, Rjwilmsi, MZMcBride, Jimp, Hellbus, Buster79, Retired username, Simen, Poulpy, Sbyrnes321, Incnis Mrsi, Tom Lougheed, Hmains, Farry, Can't sleep, clown will eat me, Muriel R, Happy-melon, Rwflammang, Skittleys, Headbomb, Tim Shuba, JAnDbot, Reedy Bot, Adam Zivner, Sheliak, TXiKiBoT, Antixt, BartekChom, MystBot, SkyLined, Addbot, Zorrobot, Jack who built the house, Luckas-bot, Yobot, AnomieBOT, Citation bot, ArthurBot, Xqbot, A. di M., D'ohBot, John85, Citation bot 1, PNG, EmausBot, Tommy2010, Hhhippo, StringTheory11, Quondum, Bibcode Bot and Anonymous: 18
- **Lambda baryon***Source:* https://en.wikipedia.org/wiki/Lambda_baryon?oldid=668924014*Contributors:* Donarreiskoffer, Voyager640, RScheiber,Gene Nygaard, Falcorian, Woohookitty, Rjwilmsi, Kaluza81, Teleolurian, Hydrargyrum, Hawkeye7, SmackBot, Stepa, JorisvS, Happy-melon,Headbomb, Bigbill2303, JAnDbot, Gwern, STBot, Squids and Chips, Sheliak, VolkovBot, TXiKiBoT, Pamputt, Antixt, Demize, Wing gun-dam, OKBot, Muro Bot, MystBot, SkyLined, Addbot, Substar, LaaknorBot, Luckas-bot, Ptbotgourou, Anypodetos, Joule36e5, Citation bot,ArthurBot, TorKr, Citation bot 1, PNG, EmausBot, Ida Shaw, StringTheory11, Quondum, Alcazar84, Rgelpke, Bibcode Bot, Reculet, Mfb,Anrnusna and Anonymous: 13
- **Sigma baryon** *Source:* https://en.wikipedia.org/wiki/Sigma_baryon?oldid=668924086 *Contributors:* Donarreiskoffer, Nickptar, Woohookitty, Rjwilmsi, Mongreilf, Jimp, Ilmari Karonen, SmackBot, Kingdon, Happy-melon, Ruslik0, Alaibot, Thijs!bot, Headbomb, Gwern, Jondaman21, Sheliak, CSumit, TXiKiBoT, Pamputt, PipepBot, MystBot, SkyLined, Addbot, Quick Fists, Luckas-bot, Yobot, Citation bot, GrouchoBot, Rasmus.mackeprang, Citation bot 1, Thinking of England, PNG, Tollerach, Akesich, EmausBot, WikitanvirBot, Jirka62, AvicBot, ZéroBot, StringTheory11, Quondum, Teaktl17, Bibcode Bot, EdwardH and Anonymous: 8
- **Xi baryon** *Source:* https://en.wikipedia.org/wiki/Xi_baryon?oldid=679118830 *Contributors:* Donarreiskoffer, Academic Challenger, Asparagus, Rpyle731, Beland, Physicistjedi, Falcorian, Woohookitty, Rjwilmsi, Chekaz, Bubba73, Erkcan, Bgwhite, Xihr, Simen, F15mos, Ilmari Karonen, Poulpy, SmackBot, Bluebot, HLwiKi, Happy-melon, Ruslik0, Phatom87, Headbomb, Liquid-aim-bot, T e r o, Balcerzak, Kostisl, Fleebo, CommonHartpoole, Sheliak, TXiKiBoT, SieBot, DumZiBoT, SkyLined, Addbot, DOI bot, Ehrenkater, Luckas-bot, Yobot, Citation bot, ArthurBot, Xqbot, FrescoBot, Citation bot 1, BasvanPelt, Heavyion, Double sharp, PNG, Egemont, EmausBot, WikitanvirBot, Gfoley4, Av11235, ZéroBot, StringTheory11, AManWithNoPlan, Timetraveler3.14, Brandmeister, Whoop whoop pull up, Bibcode Bot, Corwil, The helicity, SteenthIWbot, Monkbot and Anonymous: 16
- **Omega baryon** *Source:* https://en.wikipedia.org/wiki/Omega_baryon?oldid=680023977 *Contributors:* The Anome, Icarus~enwiki, Michael Hardy, Poor Yorick, Mxn, Jeoth, Pakaran, BenRG, Donarreiskoffer, Merovingian, Daibhid C, Erik Garrison, Sam Hocevar, MakeRocketGoNow, Jkl, Khaldei, Kjoonlee, Bletch, Cburnett, Stephen, Woohookitty, Astrowob, Kbdank71, Rjwilmsi, Strait, Philten, E Pluribus Anthony, YurikBot, Hellbus, SCZenz, Netrapt, F15mos, Poulpy, SmackBot, Stepa, Chris the speller, Beetstra, Happy-melon, Ruslik0, Lokal Profil, Thijs!bot, Headbomb, Cerrigno, Magioladitis, Vanished user ty12kl89jq10, T e r o, Balcerzak, Matgoth, Sheliak, Wi-kiry-lan, DumZiBoT, SkyLined, Addbot, DOI bot, Mjamja, Eivindbot, Zorrobot, Luckas-bot, Yobot, AnomieBOT, Bowser1226, ArthurBot, Xqbot, Citation bot 2, Citation bot 1, Gil987, Double sharp, PNG, EmausBot, John of Reading, ZéroBot, StringTheory11, Quondum, Tolly4bolly, HupHollandHup, Helpful Pixie Bot, Bibcode Bot, Glevum, Monkbot and Anonymous: 32
- **Meson** *Source:* https://en.wikipedia.org/wiki/Meson?oldid=672777660 *Contributors:* AxelBoldt, Bryan Derksen, Josh Grosse, PierreAbbat, Ben-Zin~enwiki, Xavic69, TakuyaMurata, Fwappler, Ahoerstemeier, Ping, Phys, Bcorr, Jeffq, Donarreiskoffer, Robbot, Fredrik, Sanders muc,

Merovingian, Rursus, Ojigiri~enwiki, Davidl9999, DocWatson42, Harp, Marcika, Xerxes314, Niteowlneils, Eequor, Physicist, Eroica, Icairns, Sam Hocevar, Lehi, Rich Farmbrough, Pjacobi, Tjic, Robotje, Nicke Lilltroll~enwiki, Pearle, Jumbuck, Jérôme, Bucephalus, Falcorian, Palica, Tevatron~enwiki, Mandarax, Kbdank71, Strait, Titoxd, FlaBot, Jeremygbyrne, Chobot, YurikBot, Wavelength, Bambaiah, Phmer, Jimp, Ozabluda, JabberWok, Salsb, Leutha, Długosz, SCZenz, Ravedave, Gadget850, Antiduh, Tetracube, SmackBot, Melchoir, Eskimbot, Chris the speller, DHN-bot~enwiki, Sbharris, Kevinpurcell, Mesons, DMacks, Jashank, JorisvS, Mgiganteus1, Geologyguy, Ryulong, JarahE, Myasuda, ChrisKennedy, Michael C Price, Thijs!bot, Headbomb, Escarbot, Orionus, Spartaz, Gökhan, Deflective, Magioladitis, Swpb, Khalid Mahmood, Tercer, Kostisl, Hans Dunkelberg, Tarotcards, Xiahou, JeffreyRMiles, VolkovBot, Prizrak, TXiKiBoT, Muro de Aguas, Martin451, LeaveSleaves, Antixt, SieBot, Majeston, Gerakibot, Graf Von Crayola, Humanityisthedisease, Mimihitam, Fratrep, OKBot, ClueBot, Terrorist96, Diagramma Della Verita, Brews ohare, Neville35, RMFan1, WikHead, Stephen Poppitt, Addbot, Gtakanis, Chzz, Debresser, CosmiCarl, AgadaUrbanit, Dickdock, Magog the Ogre, AnomieBOT, StratoWiki, Altruism2010, Citation bot, ArthurBot, Xqbot, Omnipaedista, WaysToEscape, FrescoBot, Paine Ellsworth, Ironboy11, Steve Quinn, 000ojjo000, Yehoshua2, Citation bot 1, Wdcf, Thinking of England, Puzl bustr, Ale And Quail, Discovery4, Mean as custard, Dkzico007, John of Reading, WikitanvirBot, GoingBatty, Hanretty, ZéroBot, StringTheory11, Markinvancouver, ClueBot NG, Christian.kolen, Wallace Kneeland, Helpful Pixie Bot, Bibcode Bot, Glevum, DerekWinters, Mark viking, Justin567Hicks, Prokaryotes, SJ Defender, Monkbot, KasparBot and Anonymous: 87

- **Quarkonium** *Source:* https://en.wikipedia.org/wiki/Quarkonium?oldid=677791042 *Contributors:* AxelBoldt, Xavic69, Rursus, Mako098765, Icairns, Mike Rosoft, Anthony Appleyard, Kbdank71, Rjwilmsi, MZMcBride, MatthewMastracci, FlaBot, Goudzovski, YurikBot, Bambaiah, Jim-Hairy Dude, Salsb, Długosz, SCZenz, Physicsdavid, Reedy, Colonies Chris, Teki D, Akriasas, Kurtan~enwiki, Thijs!bot, Headbomb, Jimbobl, Magioladitis, Choihei, Reedy Bot, Tarotcards, Joshmt, Muro de Aguas, Antixt, Neparis, SieBot, BartekChom, Bobathon71, MystBot, Addbot, Amirobot, Citation bot, MIRROR, Metrictensor, Ricolai, Citation bot 1, Jonesey95, Johann137, 564dude, RjwilmsiBot, EmausBot, SanchoOoPansa, ZéroBot, Chris857, Bibcode Bot, Khazar2, Illia Connell, Giu8888, JJ1976, Dbjergaard and Anonymous: 19
- **Pion** *Source:* https://en.wikipedia.org/wiki/Pion?oldid=679013958 *Contributors:* AxelBoldt, Vicki Rosenzweig, Bryan Derksen, Dragon Dave, Josh Grosse, Marcustacitus, Rsabbatini, Michael Hardy, TakuyaMurata, Glenn, Magnus.de, Donarreiskoffer, Robbot, Merovingian, Davidl9999, Mattflaschen, GreatWhiteNortherner, M-Falcon, Giftlite, Harp, Herbee, Xerxes314, Everyking, Alison, Bobblewik, Icairns, Rich Farmbrough, Kjoonlee, CDN99, Nk, LostLeviathan, Alai, Y0u, Linas, Kbdank71, Zbxgscqf, Strait, Cakedamber, FlaBot, Goudzovski, Krishnavedala, Bgwhite, Pezlogd, Algebraist, YurikBot, Bambaiah, RussBot, JabberWok, Salsb, Brandon, LifeStar, Tetracube, Simen, Poppy, Besselfunctions, Infinity0, Timothyarnold85, SmackBot, Tom Lougheed, Eskimbot, RDBrown, Complexica, Sbharris, Fiziker, Writtenright, Brennan Milligan, Newone, Trialsanderrors, JRSpriggs, Usgnus, HPaul, Pedro Fonini, Thijs!bot, Headbomb, Davidhorman, MichaelMaggs, Escarbot, Orionus, Shambolic Entity, Magioladitis, Michael Goodyear, Tarotcards, Sheliak, Cuzkatzimhut, TXiKiBoT, Shureg, Anonymous Dissident, Go2slash, TobiasS, Mclee2007, SieBot, Hatster301, ClueBot, Pi zero, Alexbot, Robertpdot, SchreiberBike, SilvonenBot, SkyLined, Truthnlove, Addbot, Chzz, SamatBot, Numbo3-bot, Tide rolls, ScAvenger, Luckas-bot, Yobot, AnomieBOT, JackieBot, Citation bot, ArthurBot, Cydelin, GrouchoBot, GVilKa, A. di M., Paine Ellsworth, Ironboy11, Merongb10, Xiglofre, Py4nf, Double sharp, Juhko, Miracle Pen, RjwilmsiBot, Ripchip Bot, WikitanvirBot, Faustina.minor, Splibubay, Maschen, QuantumSquirrel, Teaktl17, Jcn94, Bibcode Bot, Hrttu523, Dexbot, Andyhowlett, G7fernandes, Abitslow, Monkbot, Kenijr and Anonymous: 104
- **Rho meson** *Source:* https://en.wikipedia.org/wiki/Rho_meson?oldid=653181212 *Contributors:* Michael Hardy, Phys, Donarreiskoffer, Filemon, Estel~enwiki, Lumidek, Rich Farmbrough, Fwb22, Marudubshinki, Kbdank71, Strobilomyces, Wavelength, Gaius Cornelius, Salsb, Poulpy, Sbyrnes321, Jgwacker, Happy-melon, Thijs!bot, Headbomb, Madmarigold, Magioladitis, Choihei, Sheliak, Lozuk, Senderovich, RafaAzevedo, Speshuldusty, SkyLined, Addbot, Mjamja, Luckas-bot, Yobot, AnomieBOT, GrouchoBot, HRoestBot, David.c.stone, EmausBot, WikitanvirBot, ZéroBot, Ebrambot, QuantumSquirrel, Isocliff and Anonymous: 13
- **Eta meson** *Source:* https://en.wikipedia.org/wiki/Eta_meson?oldid=665372791 *Contributors:* Palfrey, Auric, Mike Rosoft, Rich Farmbrough, David Schaich, Strobilomyces, Goudzovski, Chobot, Headbomb, VolkovBot, CristianCantoro, Callmederek, Jcline1, SkyLined, Addbot, Luckasbot, Yobot, Citation bot, ArthurBot, Xqbot, Carlog3, Paine Ellsworth, Thinking of England, Trappist the monk, RjwilmsiBot, TjBot, EmausBot, ZéroBot, Bibcode Bot, Lbrlieuo, Andyhowlett, Der Schmitzi, Stamptrader, Samproctor125, Corrupt Titan and Anonymous: 4
- **Phi meson** *Source:* https://en.wikipedia.org/wiki/Phi_meson?oldid=665815403 *Contributors:* Edcolins, Rich Farmbrough, Nihiltres, Casliber, Headbomb, Magioladitis, MystBot, Addbot, Luckas-bot, Yobot, Ptbotgourou, ArthurBot, Carlog3, Paine Ellsworth, Dinamik-bot, EmausBot, ZéroBot and Anonymous: 1
- **List of mesons** *Source:* https://en.wikipedia.org/wiki/List_of_mesons?oldid=679943368 *Contributors:* Cherkash, Donarreiskoffer, Giftlite, Xerxes314, Michael Devore, Eequor, Rich Farmbrough, ZeroOne, Tompw, Physicistjedi, Pearle, Keenan Pepper, Zyqqh, TenOfAllTrades, Woohookitty, Ch'marr, Kbdank71, JVz, Strait, Nihiltres, Agerom, RussBot, David McCormick, SCZenz, Gadget850, Sbyrnes321, That Guy, From That Show!, SmackBot, JorisvS, Happy-melon, Charles Baynham, Chrumps, Usgnus, Cydebot, Christian75, Coccoinomane, Headbomb, Stannered, JAnDbot, Magioladitis, Mollwollfumble, Gwern, Leyo, Potatoswatter, VolkovBot, Antixt, Ocsenave, Muhends, Mikaey, SkyLined, Addbot, Yobot, Kan8eDie, 4th-otaku, Rubinbot, Citation bot, ArthurBot, Xqbot, Ulm, Carlog3, W-C, Yehoshua2, Citation bot 1, Thinking of England, John of Reading, StringTheory11, Markinvancouver, Helpful Pixie Bot, Bibcode Bot, Maysens, Srebre, YiFeiBot, Monkbot and Anonymous: 16
- **J/psi meson** *Source:* https://en.wikipedia.org/wiki/J/psi_meson?oldid=678932131 *Contributors:* Kowloonese, Cyp, Andrewa, Schneelocke, Doradus, Thue, Jeffq, Donarreiskoffer, Sanders muc, Sverdrup, JB82, Giftlite, Eequor, Plutor, DragonflySixtyseven, Sam Hocevar, Eb.hoop, Grutter, Kjoonlee, Kwamikagami, Menscher, Tripodics, Palica, Kbdank71, Rjwilmsi, Strait, Erkcan, Bgwhite, YurikBot, Bambaiah, Jimp, Salsb, SCZenz, Ravedave, Poulpy, Physicsdavid, Wizofaus, SmackBot, Vald, Hmains, Colonies Chris, OrphanBot, JorisvS, Dicklyon, Adriferr, Norm mit, Kurtan~enwiki, Headbomb, WinBot, Certain, JAnDbot, Matthew Fennell, Igodard, Jondaman21, HEL, MITBeaverRocks, STBotD, Sheliak, VolkovBot, TimAdye, Senemmar, Tomaxer, SieBot, Nihil novi, TubularWorld, Gabor Denes, SkyLined, Addbot, LaaknorBot, Lightbot, Luckas-bot, Yobot, Citation bot, ArthurBot, TheAMmollusc, GrouchoBot, Carlog3, Paine Ellsworth, Xiglofre, Johann137, Trappist the monk, EmausBot, StringTheory11, SporkBot, Timetraveler3.14, Bibcode Bot, Hmainsbot1, Bill2239, Trwood, Monkbot, Grand'mere Eugene and Anonymous: 37
- **Upsilon meson** *Source:* https://en.wikipedia.org/wiki/Upsilon_meson?oldid=679109367 *Contributors:* Donarreiskoffer, Rich Farmbrough, Kjoonlee, GregorB, Kbdank71, Rjwilmsi, Erkcan, Goudzovski, Agerom, SCZenz, Zwobot, JLaTondre, Poulpy, Sbharris, Ligulembot, Mgiganteus1, Beetstra, Happy-melon, CRGreathouse, Headbomb, Joshmt, Sheliak, Alexbot, SkyLined, Addbot, DOI bot, Mjamja, Zorrobot,

Luckas-bot, ArthurBot, Obersachsebot, GrouchoBot, Carlog3, Paine Ellsworth, HRoestBot, RedBot, ZéroBot, EWikist, Makecat, Bibcode Bot, Monkbot and Anonymous: 9

- **Theta meson** *Source:* https://en.wikipedia.org/wiki/Theta_meson?oldid=607156016 *Contributors:* Rich Farmbrough, Goudzovski, Headbomb, Addbot, Skippy le Grand Gourou, Yobot, ArthurBot, GrouchoBot, Carlog3, ZéroBot and Anonymous: 1
- **Kaon** *Source:* https://en.wikipedia.org/wiki/Kaon?oldid=680631374 *Contributors:* Bryan Derksen, Stevertigo, Ahoerstemeier, Salsa Shark, Denny, Traroth, Donarreiskoffer, Merovingian, Davidl9999, Matt Gies, Giftlite, Xerxes314, Jason Quinn, Icairns, Kate, Perey, YUL89YYZ, Kwamikagami, Longhair, Jérôme, Andrew Gray, Count Iblis, Reaverdrop, SeventyThree, BD2412, Kbdank71, FlaBot, Goudzovski, Chobot, YurikBot, Bambaiah, Gaius Cornelius, Salsb, David R. Ingham, Gillis, SCZenz, Larsobrien, SmackBot, Tom Lougheed, MattOates, Tydus Arandor, Clarityfiend, Elfrah, Michael C Price, Thijs!bot, Headbomb, Dawnseeker2000, Luna Santin, Tim Shuba, Storkk, Deflective, Thasaidon, Septuagent, Dr. Morbius, NikNaks, Tarotcards, Adam Zivner, Sheliak, VolkovBot, TXiKiBoT, Anonymous Dissident, SieBot, TJRC, Lazarus1907, ClueBot, Binksternet, Wwheaton, Mild Bill Hiccup, Auntof6, Virginia-American, SkyLined, Addbot, Luckas-bot, Yobot, AnomieBOT, JackieBot, Citation bot, Quebec99, Shmomuffin, A. di M., Omar35880, Paine Ellsworth, Grinevitski, Citation bot 1, Citation bot 4, Plucas58, Phasespace, Thinking of England, Eekerz, ZéroBot, ClueBot NG, Widr, Bibcode Bot, JPaestpreornJeolhlna, Danny Quark, Mowsseep, Monkbot, Ishwadut2, Garfield Garfield, Soham92, Markjamescapella and Anonymous: 55
- **B meson** *Source:* https://en.wikipedia.org/wiki/B_meson?oldid=668085904 *Contributors:* TakuyaMurata, Rich Farmbrough, Rjwilmsi, Goudzovski, Jimp, F15mos, Hmains, Happy-melon, Christian75, Headbomb, Hcobb, Magioladitis, DWIII, Pbroks13, Antixt, ToePeu.bot, BartekChom, SkyLined, Addbot, Roentgenium111, Mjamja, Zorrobot, Yobot, AnomieBOT, Ipatrol, GrouchoBot, Carlog3, D'ohBot, Citation bot 1, DASHBot, EmausBot, Shipbegan, ZéroBot, ChuispastonBot, Ebehn, Law of Entropy, Bibcode Bot, IDenni, KasparBot and Anonymous: 11
- **D meson** *Source:* https://en.wikipedia.org/wiki/D_meson?oldid=667958261 *Contributors:* Rich Farmbrough, Goudzovski, Eleassar, SmackBot, Owlbuster, Happy-melon, Headbomb, Alexllew, Wing gundam, BartekChom, DumZiBoT, SkyLined, Addbot, Zorrobot, Luckas-bot, Yobot, Ptbotgourou, Tonyrex, JackieBot, ArthurBot, Obersachsebot, Carlog3, Newty23125, EmausBot, ZéroBot, Theopolisme, Colbert Nation 111, KasparBot and Anonymous: 4
- **T meson** *Source:* https://en.wikipedia.org/wiki/T_meson?oldid=679091348 *Contributors:* Schewek, Rich Farmbrough, Rjwilmsi, Headbomb, Addbot, Yobot, Carlog3, D'ohBot, TjBot, Marek Koudelka, ZéroBot, Monkbot and Anonymous: 2
- **Atomic nucleus** *Source:* https://en.wikipedia.org/wiki/Atomic_nucleus?oldid=682627820 *Contributors:* Kpjas, Andre Engels, XJaM, Merphant, Graft, Stevertigo, Patrick, JohnOwens, Michael Hardy, Tim Starling, Nixdorf, Bcrowell, Mcarling, Looxix~enwiki, Ellywa, Ahoerstemeier, Александър, Andres, Smack, Rednblu, The Anomebot, Topbanana, Cvaneg, Palefire, Gentgeen, Robbot, Sander123, Arkuat, Merovingian, Meelar, Wikibot, GarnetRChaney, Anthony, Giftlite, Graeme Bartlett, Christopher Parham, Awolf002, Mikez, Fastfission, Xerxes314, Bensaccount, Yath, Antandrus, Beland, OverlordQ, Icairns, Joyous!, JohnArmagh, Deglr6328, Trevor MacInnis, Mike Rosoft, Chris Howard, Perey, Guanabot, Igorivanov~enwiki, Vsmith, Gianluigi, Paul August, El C, Koenige, Madhu p, Shanes, Bookofjude, Mickeymousechen~enwiki, Foobaz, Obradovic Goran, Nsaa, Jumbuck, Disneyfreak96, Alansohn, Gintautasm, Neonumbers, Riana, Kfitzgib, Bart133, Velella, EvenT, H2g2bob, DV8 2XL, Mattbrundage, Kay Dekker, Flying fish, Stemonitis, Firsfron, AndriyK, Mandarax, Graham87, FreplySpang, Martinevos~enwiki, Rjwilmsi, Syndicate, Strait, Tangotango, Ttwaring, Lcolson, Nivix, OSt~enwiki, Maustrauser, Fresheneesz, Srleffler, Chobot, DVdm, Ahpook, Gwernol, Roboto de Ajvol, YurikBot, RobotE, Bambaiah, JWB, RussBot, Sillybilly, SpuriousQ, Hellbus, Okedem, Gaius Cornelius, CambridgeBayWeather, Wiki alf, FFLaguna, Dbfirs, Bota47, Supspirit, Tachyon01, Jess Riedel, Zzuuzz, Tsunaminoai, CWenger, Anclation~enwiki, Curpsbot-unicodify, Kungfuadam, RG2, AssistantX, SmackBot, FocalPoint, Incnis Mrsi, Bomac, Wogsland, Jrockley, Edgar181, Gilliam, Betacommand, Schmiteye, Chris the speller, Keegan, SchfiftyThree, DHN-bot~enwiki, Sbharris, Klacquement, Blake-, Itchjones, Dreadstar, DMacks, Daniel.Cardenas, Mion, Sadi Carnot, Bdushaw, Pilotguy, Kukini, Clicketyclack, Serein (renamed because of SUL), Kuru, Olin, Zarniwoot, NongBot~enwiki, Ekrub-ntyh, Funnybunny, MTSbot~enwiki, Cbuckley, Caiaffa, Dan Gluck, Kelvinaom, Joseph Solis in Australia, Lottamiata, Tubezone, Tawkerbot2, Flubeca, Sxim, Ale jrb, Scohoust, Shernren, Rowellcf, Engelmann15~enwiki, Kanags, MC10, Gogo Dodo, My Flatley, Christian75, DumbBOT, Bieeanda, Thijs!bot, Headbomb, Marek69, Tellyaddict, Wildthing61476, CTZMSC3, AntiVandalBot, Widefox, Quintote, TimVickers, Ilovescience, Gdo01, Gmarsden, JAnDbot, D99figge, Leuko, MER-C, Gfsheppard, .anacondabot, Bongwarrior, VoABot II, Rajb245, JamesBWatson, CalamusFortis, Mother.earth, Dirac66, Kopovoi, Dravick, Vssun, JaGa, Philg88, Hbent, Goodynotion, Akhil999in, MartinBot, Schmloof, Tiger-Smith, Roastytoast, J.delanoy, Classicalclarinet, Trusilver, Fnordius, Maurice Carbonaro, WarthogDemon, Lol nubs, Fylwind, Eshywiki, Tygrrr, TraceyR, Dylan bossart, VolkovBot, Indubitably, Stefan Kruithof, The Original Wildbear, Rei-bot, Anonymous Dissident, Piperh, Anna Lincoln, Shonenknifefan1, Venny85, Synthebot, Antixt, Enviroboy, PGWG, EmxBot, NEIL4737, SieBot, Sonicology, Scarian, Gerakibot, Caltas, Tiptoety, Onesspite, BenoniBot~enwiki, Afernand74, Cyfal, ClueBot, The Thing That Should Not Be, Jan1nad, Jekatz, Drmies, Boing! said Zebedee, Lainy8, CounterVandalismBot, Excirial, Alexbot, LordFoppington, SpikeToronto, Rhododendrites, Brews ohare, PhySusie, Tinymonty, Kakofonous, Versus22, InternetMeme, XLinkBot, Rangel lucy, Mr beeg lol, Addbot, Peyton.gaumer, Praseprase, Some jerk on the Internet, Leszek Jańczuk, Cst17, LinkFA-Bot, 5 albert square, CuteHappyBrute, Numbo3-bot, Tide rolls, Jan eissfeldt, Luckas-bot, Yobot, WikiDan61, 2D, TaBOT-zerem, Anypodetos, IW.HG, محبوب عالم, Podlif, Synchronism, AnomieBOT, Kingpin13, Flewis, Bluerasberry, Materialscientist, The High Fin Sperm Whale, Citation bot, Carlsotr, Quebec99, Xqbot, SouthH, Sionus, Capricorn42, Jeffrey Mall, RibotBOT, BSTemple, Shrikeangel, A. di M., Ironboy11, Kobewetnaps, Citation bot 1, Rylee118, Pinethicket, Tinton5, Games 101 wiki, RedBot, Minivip, White Shadows, FoxBot, Lionslayer, TheBFG, Dinamik-bot, Reaper Eternal, Specs112, Onel5969, RjwilmsiBot, Killaoftoast, DASHBot, EmausBot, Ornithikos, Mariov0288, The Pineapple, Cedar T., Thecheesykid, Hhhippo, JSquish, StringTheory11, Arbnos, Wayne Slam, Ocaasi, Kim cupcake, Sunshine4921, Mjbmrbot, Petrb, ClueBot NG, Magic Wizard, MelbourneStar, Fukushimayoshiho, Schunck, IOPhysics, Widr, Reify-tech, Helpful Pixie Bot, Downtowntrollin, Bibcode Bot, Lowercase sigmabot, AvocatoBot, Flying hippo705, Jordanf7, Gimp 11, Zedshort, Uopchem2511, Jburk711, DarafshBot, Akbask, JYBot, BrightStarSky, Dexbot, TwoTwoHello, Lugia2453, Bulba2036, Marekich, SassyLilNugget, Wlad2000, Mark viking, Siddhantsingh123, DennouNeko, DavidLeighEllis, Sladeb, Spyglasses, Y-S.Ko, KasparBot, Aless Val M and Anonymous: 441
- **Atom** *Source:* https://en.wikipedia.org/wiki/Atom?oldid=679943114 *Contributors:* AxelBoldt, Trelvis, Lee Daniel Crocker, Mav, Zundark, The Anome, Tarquin, AstroNomer~enwiki, Stokerm, Andre Engels, Youssefsan, Branden, Ben-Zin~enwiki, Drbug, Heron, Comte0, PeterBohne, Stevertigo, Hfastedge, Lir, Patrick, Infrogmation, D, Michael Hardy, Tim Starling, FrankH, Pit~enwiki, Wapcaplet, Ixfd64, Bcrowell, Dcljr, Shoaler, Delirium, PingPongBoy, Card~enwiki, NuclearWinner, Looxix~enwiki, Mdebets, Ahoerstemeier, LoonBB, Suisui, Angela, Jebba, Darkwind, Александър, Julesd, Glenn, Poor Yorick, Andres, Kaihsu, Evercat, Samw, Rob Hooft, Denny, Schneelocke, Ddoherty,

Frieda, Timwi, Wikiborg, Stone, Dysprosia, The Anomebot, Doradus, Hr oskar, Big Bob the Finder, Morwen, VeryVerily, SEWilco, Flockmeal, Ldo, BenRG, Gromlakh, Donarreiskoffer, Gentgeen, Robbot, ChrisO~enwiki, Arkuat, Merovingian, Sverdrup, Der Eberswalder, DHN, Caknuck, Rebrane, Paul G, Hadal, UtherSRG, Wikibot, Roozbeh, Mandel, Anthony, Diberri, Cutler, Dina, Timemutt, Tobias Bergemann, David Gerard, Ancheta Wis, Vonkwink, Giftlite, Christopher Parham, Awolf002, Mikez, Palapala, Yuri koval, Inter, BenFrantzDale, Tom harrison, MSGJ, Xerxes314, Everyking, Curps, Michael Devore, Bensaccount, Frencheigh, Guanaco, Andrea Parri, Jorge Stolfi, SWAdair, Jurema Oliveira, Kandar, Stevietheman, Gadfium, Utcursch, Bact, Pcarbonn, Antandrus, OverlordQ, MisfitToys, Piotrus, Kaldari, Jossi, Karol Langner, JimWae, DragonflySixtyseven, Icairns, Tail, Sam Hocevar, Darksun, Engleman, Deglr6328, Achven, M1ss1ontomars2k4, Adashiel, Trevor MacInnis, Grunt, Bluemask, Freakofnurture, Amxitsa, DanielCD, Discospinster, Rich Farmbrough, Guanabot, Huffers, Hidaspal, Rama, Vsmith, Guanabot2, Wadewitz, MarkS, Jlcooke, ESkog, Kbh3rd, Pmetzger, RJHall, El C, Edward Z. Yang, Shanes, Art LaPella, RoyBoy, Bookofjude, CDN99, Afed, Bobo192, Army1987, Harley peters, Whosyourjudas, Shenme, Viriditas, Brim, Maurreen, Richi, Boxed, Timl, Joe Jarvis, Nk, Deryck Chan, Thewayforward, PiccoloNamek, Obradovic Goran, MPerel, Sam Korn, Pearle, Benbread, Nsaa, Ogress, HasharBot~enwiki, Jumbuck, Storm Rider, JYolkowski, Mennato, Richard Harvey, Thebeginning, Slugmaster, Iothiania, Riana, InShaneee, Spangineer, Malo, Snowolf, Saga City, Dirac1933, Geraldshields11, Henry W. Schmitt, Bsadowski1, BlastOButter42, Computerjoe, Gene Nygaard, Drbreznjev, Iustinus, Bookandcoffee, Dan100, Ceyockey, Tm1000, Rusticnature, Flying fish, Feezo, Blaze Labs Research, Thryduulf, Velho, Simetrical, Karnesky, Cimex, LOL, Kurzon, WadeSimMiser, Jwanders, Mpatel, Miss Madeline, Eleassar777, GregorB, Wayward, ??????, Gimboid13, Palica, Dysepsion, MrSomeone, Paxsimius, RichardWeiss, Graham87, BD2412, Zeroparallax, Deadcorpse, FreplySpang, Cmsg, Jclemens, Edison, Canderson7, Ketiltrout, Sjakkalle, Rjwilmsi, Quiddity, Sdornan, Gudeldar, Nneonneo, Bhadani, Sarg, Ucucha, Sango123, Matjlav, Lionelbrits, Titoxd, FlaBot, Pediadeep, RobertG, Nihiltres, Nivix, Chanting Fox, RexNL, Ewlyahoocom, Gurch, DannyDaWriter, TeaDrinker, Terrx, Fervidfrogger, Bmicomp, SteveBaker, Srleffler, BradBeattie, Le Anh-Huy, Physchim62, Mallocks, Nicholasink, Chobot, DVdm, Gwernol, YurikBot, Wavelength, Spacepotato, JWB, Sceptre, Hairy Dude, Jimp, Phantomsteve, Russ-Bot, DMahalko, EDM, Ericorbit, SpuriousQ, Lucinos~enwiki, Kirill Lokshin, Stephenb, Gaius Cornelius, Ihope127, Pseudomonas, Wimt, RadioKirk, NawlinWiki, TEB728, SEWilcoBot, Trovatore, Muppetmaster, Tailpig, Ashwinr, Dureo, Seegoon, Cleared as filed, Irishguy, Deodar~enwiki, Nucleusboy, Brandon, Matticus78, Vb, Bobak, Zwobot, BOT-Superzerocool, DeadEyeArrow, Derek.cashman, Elkman, Kkmurray, Dna-webmaster, Wknight94, Zero1328, 21655, Zzuuzz, Closedmouth, Kriscotta, Jody Burns, BorgQueen, Dr U, JoanneB, Vicarious, Fram, Kevin, Spliffy, Enkauston, Katieh5584, Kungfuadam, Junglecat, TLSuda, RG2, GrinBot~enwiki, SkerHawx, Jade Knight, DVD R W, Finell, Msyjsm, Luk, ChemGardener, Sycthos, Itub, Iorek85, SmackBot, Meshach, Rex the first, Incnis Mrsi, Herostratus, Prodego, KnowledgeOfSelf, David.Mestel, Unyoyega, C.Fred, Jacek Kendysz, Yuyudevil, Jagged 85, Thunderboltz, Stepa, Pandion auk, Jrockley, Delldot, Jihiro, Bgrech, BiT, Edgar181, Alsandro, Mak17f, Moralis, Gilliam, Betacommand, Skizzik, Richfife, JAn Dudík, Constan69, Chris the speller, Bluebot, SlimJim, Quinsareth, BabuBhatt, Miquonranger03, SchfiftyThree, Complexica, Alink, Kevin Ryde, Wykis, Jfsamper, Robth, Sbharris, Hallenrm, Newmanbe, Can't sleep, clown will eat me, Scott3, Mitsuhirato, OrphanBot, Sephiroth BCR, Voyajer, Caleb Murdock, Rrburke, VMS Mosaic, HBow3, Addshore, SundarBot, Phaedriel, Monty2, Flyguy649, Fuhghettaboutit, Iapetus, Nakon, Jdlambert, Jiddisch~enwiki, Hoof Hearted, Dream out loud, TrogdorPolitiks, Hgilbert, Jameshales, DMacks, Wizardman, Evlekis, Sadi Carnot, Vinaiwbot~enwiki, Pilotguy, Kukini, Ceoil, Will Beback, Kuzaar, DJIndica, The undertow, Lambiam, Vanished user 9i39j3, Kuru, MarcMacé, MagnaMopus, Dracion, Scientizzle, Vgy7ujm, Buchanan-Hermit, Ascend, Heimstern, Calum MacÙisdean, Shadowlynk, Hemmingsen, Ospinad, Accurizer, CaptainVindaloo, KronosI, Aleenf1, IronGargoyle, Beefball, Ckatz, Sunni Jeskablo, Hohomanofsteel, Slakr, Kfsung, Beetstra, Noah Salzman, Martinp23, Alethiophile, Mr Stephen, GilbertoSilvaFan, Waggers, Eridani, Mets501, Funnybunny, LaMenta3, Darry2385, KJS77, LenW, BranStark, Iridescent, Theone00, Igge, Mark Oakley, The editor1, Igoldste, RekishiEJ, Saku kodo, Mr Chuckles, Willy Skillets, Tubezone, Tawkerbot2, Dlohcierekim, Bubbha, MightyWarrior, MarkAlldridge, Switchercat, Scienceloser09, JForget, Stifynsemons, CmdrObot, Tanthalas39, Wafulz, David s graff, Dycedarg, The ed17, Megaboz, Ruslik0, Benwildeboer, Jsd, Dgw, Yarnalgo, McVities, Outriggr, MarsRover, No1lakersfan, TheAdventMaster, Chris83, Nauticashades, Cydebot, Nick Y., MC10, Mziebell, SyntaxError55, Astrochemist, Rifleman 82, Meno25, Gogo Dodo, Zginder, KnightMove, Rracecarr, Skittleys, Strom, Jack Phoenix, DumbBOT, Ameliorate!, Narayanese, Optimist on the run, SteveMcCluskey, Vanished User jdksfajlasd, Daniel Olsen, Gimmetrow, Click23, FrancoGG, Thijs!bot, Epbr123, Daa89563, Qwyrxian, Goods21, Ramananrv123, Keraunos, PerfectStorm, Mojo Hand, Headbomb, Jojan, Yzmo, Marek69, John254, Tapir Terrific, A3RO, Ujm, Doyley, CharlotteWebb, Lithpiperpilot, Hempfel, Mactin, I already forgot, Hmrox, Cyclonenim, DrowningInRoyalty, AntiVandalBot, Majorly, Luna Santin, XxX-Teddy-XxX, Blue Tie, Bigtimepeace, Quintote, Doc Tropics, Kbthompson, Jj137, Danger, MECU, Istartfires, Indian Chronicles, Robert A. Mitchell, Gökhan, JAnDbot, Narssarssuaq, Leuko, Husond, DuncanHill, MER-C, Instinct, Hello32020, Andonic, $?, Pkoppenb, Savant13, Kirrages, Xact, Steveprutz, Acroterion, Freshacconci, Penubag, Magioladitis, Pedro, Bongwarrior, VoABot II, InvertedCommas, Yandman, CattleGirl, Think outside the box, Sikory, ThoHug, WODUP, Brusegadi, Midgrid, Catgut, Indon, LeeBuescher, Kevinwiatrowski, Orangemonster2k1, Joe hill, Alexllew, Schumi555, Cpl Syx, Lord mrazon, Gimlei, Vssun, Just James, Glen, DerHexer, JaGa, Andythelovell, Markco1, Wikinger, DGG, 0612, PsyMar, MartinBot, Schmloof, BetBot~enwiki, ChemNerd, APT, Naohiro19, Nanotrix, Rettetast, TechnoFaye, Mschel, R'n'B, Dr. Hfuhruhurr, AlexiusHoratius, Mikejones11693, Dragonwings158, Cyrus Andiron, Worldedixor, Zarathura, Manticore, J.delanoy, Captain panda, Sasajid, DrKiernan, Rgoodermote, EscapingLife, Bogey97, Abby, Maurice Carbonaro, All Is One, Irfanhasmit, Awertx, 5Q5, Eliz81, Extransit, Aryoc, Bluesquareapple, Eskimospy, Gzkn, Acalamari, Katalaveno, Mousehunter, DarkFalls, Slippered sleep, Hawkmaster9, Battyboy69, Notapotato, Drumbeatz, JayJasper, Pyrospirit, AntiSpamBot, Xmonicleman, Floaterfluss, Lastchance4onelastdances, Pcfjr9, Volk282, Cheyeric, TomasBat, NewEnglandYankee, Antony-22, Healy6991, Hennessey, Patrick, Fieryiceissweet, Shoessss, Zaswert, KylieTastic, Adamd1008, Cometstyles, DorganBot, Cite needed, Sjwk, Mike V, Sarregouset, Bonadea, WinterSpw, Arock09300, Divad89, Ja 62, Jwmmorley, Inwind, Guess001, Useight, Vinsfan368, Xiahou, Nyquist562, CardinalDan, Pqwo, Idioma-bot, The Simonator, Wikieditor06, Lights, Termasueder, VolkovBot, ABHISHEKARORA, Gofannon, Macedonian, Koombayah, Ndsg, Jeff G., JohnBlackburne, VasilievVV, CGupte, Wolfnix, Philip Trueman, Wildshack, TXiKiBoT, Tbonepower07, GimmeBot, Katoa, Maximillion Pegasus, Nxavar, Ann Stouter, Anonymous Dissident, Wildshack300, BlueLint, Arnon Chaffin, TheGuyInTheIronMask, Pinkalfonzochief19, Voorlandt, Mr. Hallman, Littlealien182, Corvus cornix, Gharbad, JhsBot, LeaveSleaves, Ripepette, Psyche825, Freak104, Katimawan2005, Shanata, V81, Mwilso24, Greswik, Brian Huffman, Enigmaman, Synthebot, Antixt, Falcon8765, Hellsangel 5000, Burntsauce, Insanity Incarnate, Agüeybaná, Brianga, Yamster, Skarz, Gunnville, Mary quite contrary, Sue Rangell, Aaronownesfrod, Onceonthisisland, Nathan M. Swan, AlleborgoBot, Louis.pure, Ekmr, Zhangyijiang, EmxBot, Poneman, D. Recorder, Tylerofmaine, Psymun747, The Random Editor, Vex879, Zdavid408, Demmy100, EthanCoryHarris, SieBot, Coffee, Delorian8, Augustus Rookwood, Tresiden, PlanetStar, JamesA, Tiddly Tom, Scarian, Keymolen, Jauerback, Evud, Studnic12, Winchelsea, Exxy, Viskonsas, Caltas, RJaguar3, Triwbe, Brandon1919, Amill, Clew25, Wolfstevej, Mothmolevna, Xelgen, Keilana, Maddiekate, Anglicanus, Aillema, RadicalOne, Radon210, Exert, Lavers, NOK4987, Oda Mari, Arbor to SJ, Momo san, Elnutter, Argentix, Mimihitam, Oxymoron83, Nuttycoconut, Demosis, Designed for me and me, Steven Crossin, Lightmouse, Freirec, Buttpants, KathrynLybarger, Hobartimus, Jakeng, This-namestaken, Experiment6, Etx123456789, Pappapippa, Emesee, Traec1000, Mattyhoops, Spartan-James, Wingtale, StaticGull, Mike2vil, Some1random, Bogwhistle, Randomblue, Kunkyis, Maralia, Hello evry1, Nn123645, Pinkadelica, DRTllbrg, Efe, JL-Bot, Escape Orbit, Blakedriver123, Hridi339, Hadseys, Church, Loren.wilton, Separa, Amdragman, ClueBot, Deviator13, Binksternet, GorillaWarfare, Pyromaniac122, PipepBot, The Thing That Should Not Be, Endoxa, Ndenison, Drphallus, Wangaroo jamportha, Meekywiki, Botodo, Arroni, Mild Bill Hiccup, Uncle Milty, J8079s, JTBX, CounterVandalismBot, Niceguyedc, Harland1, Dylan620, Piledhigheranddeeper, Ottava Rima, Puchiko, Say2anniyan, Inter132, Manishearth, MindstormsKid, DragonBot, Martine23, Justaccount1, Excirial, Alexbot, Jusdafax, Coadgeorge, Kurt-cobain911, Pspmichel2, Neuenglander, Slavko19, Eeekster, Taxa, Andriolo, Feline Hymnic, Winston365, Abrech, California847, Lartoven, Bairdy619, Tyler, NuclearWarfare, 10"weiner, Cenarium, Nmoo, Arjayay, Jotterbot, PhySusie, Soccerchicslhok, Razorflame, Firestarterrulz, Thingg, Aitias, Jesse.Senior, Katanada, SoxBot III, CTSW, Party, Alpixpakjian, Devaes, Darkicebot, Crazy Boris with a red beard, Willi-pod, Hornthecheck, XLinkBot, Kevind08, Pobob, Zacefron223962239622396, BodhisattvaBot, Jovianeye, Ozwego, NellieBly, Unclefrizzle, Mifter, Manfi, Walter.doug, Alyajohnson, NHJG, Vianello, Lemmey, Ab811996, Miagirljmw14, Freestyle-69, Eklipse, Siripswich, Willi-sis2, Addbot, Jennyvu96, Professor Calculus, Lewjam18, Joee8001, AVand, DOI bot, Sedsa1, Landon1980, Cmk1994, Dmjkmb, 15lsoucy, Ultraorange260, Vishnava, Mac Dreamstate, WFPM, Apemaster3000, Cst17, Download, LaaknorBot, Chamal N, CarsracBot, Glane23, Tech30, Akazme93, Debresser, AnnaFrance, Favonian, Chateau Brillant, AtheWeatherman, LinkFA-Bot, Sin Aura, Omg123123, IOLJeff, Eas4200c.team0, Numbo3-bot, Psproots, Koliri, Tide rolls, Bfigura's puppy, Luckas Blade, Jarble, Emperor Genius, GiantPea, KarenEdda, Georgieboi123, Angrysockhop, Legobot, Northexit182, Drpickem, XxGOWxx, Luckas-bot, Yobot, Googins, Senator Palpatine, Julia W, Yiplop stick stop, Cepheiden, Wikipedian Penguin, Nerdguy, Milesarise, The Bacon Machine, Jmproductionss, Cowdudemanxboyguymale-senorninoxxxxxxx, Ayrton Prost, Blk48, Yami89~enwiki, Chadisasexybeast, Skin and Bones, Kingpin13, Abshirdheere, Pokehen, Csigabi, Materialscientist, The High Fin Sperm Whale, Citation bot, E2eamon, Xqbot, TinucherianBot II, Intelati, GeometryGirl, Jeffrey Mall, Tol-manator, Tad Lincoln, Grim23, P99am, Almabot, Quixotex, GrouchoBot, Harley2121, Capecodsamm, Omnipaedista, Bahahs, Brutaldeluxe, GhalyBot, MerlLinkBot, N419BH, SchnitzelMannGreek, Sesu Prime, Pauswa, Legobot III, Tobby72, Dogbert66, Astronomyinertia, Cargok-ing, Machine Elf 1735, Finalius, Saiarcot895, Diremarc, Citation bot 1, Kobrabones, Clevercheetah123, Biker Biker, Pinethicket, MBirkholz, HRoestBot, Tom.Reding, RedBot, Bigad1, Footwarrior, White Shadows, Bsece010, FoxBot, Trappist the monk, Sweet xx, Sheogorath, Mrrstatham, Vrenator, Extra999, EzraNemo123, Burnthefairy56, Theproatlearningstuff, Loki2022, Reach Out to the Truth, Minimac, Marie Poise, DARTH SIDIOUS 2, RjwilmsiBot, TjBot, Ling cao, Edouard.darchimbaud, DASHBot, B20180, EmausBot, Thecreator09, Orphan Wiki, Immunize, Gfoley4, Pete Hobbs, 271196samantha, Smallchief, Hhhippo, Werieth, JSquish, CanonLawJunkie, Arik Islam, Mubasher55, Kbop451, Chaitana, Sambam340, H3llBot, Quondum, DanDao, Ahmetkilit, Lizzy chic38, Kobinks, Wayne Slam, Acdcfan1223, Bender176, Coasterlover1994, Pussvock, Thomasroper, Maschen, Dcgunasekar, Epicstonemason, Lilgas52, Puffin, Happiestmuackz11, ChuispastonBot, RockMagnetist, Oliozzicle, Lo79y123, Benjthedivine, DASHBotAV, ResearchRave, Petrb, ClueBot NG, Ulflund, Hiperfelix, Mitch09876, Sagenfowler, Big-Jason99, Hazhk, Moneya, O.Koslowski, Rezabot, Danim, Helpful Pixie Bot, Art and Muscle, Jubobroff, Bibcode Bot, Branthecan, Leonxlin, Malvel, Mysterytrey, Goodfluff, Xyzmathrules, Begman5, Fuppamaster, Jrobbinz1, Rgbc2000, Typoltion, Dr.Nadon, Connorbishop, IMn00b, Arghya33, DGK1318, Shawn Worthington Laser Plasma, Alarbus, Life421, BattyBot, Jimw338, Hebert Peró, Chris-Gualtieri, GoShow, 786b6364, JYBot, Dexbot, Mogism, Falktan, GeeBIGS, MeteMetheus, Jamesx12345, RandomLittleHelper, Reatlas, Anas-tronomer, Raghav Bhalerao, Cowfdashkli, Pirtert, Homaa, AmericanLemming, Praemonitus, Tedsanders, PianoEngineer18, Mandruss, Any-thingcouldhappen, Susan.grayeff, Anrnusna, Stamptrader, Joshdude356, Jazzmusician94, IStoleThePies, Mahusha, Monkbot, IiKkEe, Suntrax south, Mario Castelán Castro, Narky Blert, Y-S.Ko, Tetra quark, Isambard Kingdom, Forscienceonly, CV9933, Nøkkenbuer, KasparBot, White909090lightning, Myaccount7 and Anonymous: 1481

- **Diquark** *Source:* https://en.wikipedia.org/wiki/Diquark?oldid=604233153 *Contributors:* Rjwilmsi, Phmer, Conscious, Salsb, Incnis Mrsi, Wogsland, Hmains, Mbell, Headbomb, Oreo Priest, Signalhead, Sue H. Ping, Addbot, AnnaFrance, Stylus881, Citation bot, W-C, RjwilmsiBot, WikitanvirBot, Bibcode Bot and Anonymous: 6
- **Exotic atom** *Source:* https://en.wikipedia.org/wiki/Exotic_atom?oldid=659999329 *Contributors:* AxelBoldt, Bryan Derksen, The Anome, Gareth Owen, Andre Engels, Marcustacitus, RTC, Tim Starling, David Latapie, Robbot, Rursus, Ojigiri~enwiki, Xerxes314, Everyking, Eequor, Pgan002, Geni, Karol Langner, DragonflySixtyseven, Urhixidur, Vicsun, Grunt, Mike Rosoft, Rich Farmbrough, Cacycle, Arnol-dReinhold, CDN99, Bobo192, Anthony Appleyard, CCK, Radical Mallard, Dirac1933, DV8 2XL, Marasama, Goudzovski, Erik4, YurikBot, Spacepotato, Bambaiah, Hairy Dude, Hellbus, Gaius Cornelius, Quadraxis, Shanel, Długosz, Miraculouschaos, Ninly, Geoffrey.landis, Ilmari Karonen, CassiusBilbao, SmackBot, Sbharris, Lchiarav, Chlewbot, BadgerBadger, Pwjb, SpiderJon, Vina-iwbot~enwiki, Robofish, Physis, JoeBot, Happy-melon, CmdrObot, Chrumps, A876, Headbomb, Mentifisto, Ninjajake, JAnDbot, Albmont, Thibbs, VolkovBot, Chronitis, SkyLined, Addbot, Zorrobot, Luckas-bot, Yobot, Amirobot, AnomieBOT, ArthurBot, Theirs15elements, Carlog3, Bob Saint Clar, RedBot, ZéroBot, StringTheory11, Helpful Pixie Bot, Penguinstorm300, Reatlas and Anonymous: 32
- **Positronium** *Source:* https://en.wikipedia.org/wiki/Positronium?oldid=672601923 *Contributors:* Bryan Derksen, Xavic69, Michael Hardy, Julesd, Omegatron, Intangir, Giftlite, Graeme Bartlett, Curps, Eequor, Pgan002, Urhixidur, Mike Rosoft, Pjacobi, ArnoldReinhold, Tri-tium6, Helix84, Arthena, Keenan Pepper, !!, ABCD, Radical Mallard, Tylerni7, Mpatel, IIBewegung, Rjwilmsi, Erkcan, Ewlyahoocom, Choess, Goudzovski, Chobot, YurikBot, Spacepotato, Bambaiah, JabberWok, Salsb, Thatoneguy, Trovatore, Chewyrunt, SmackBot, Autarch, MalafayaBot, Tamfang, V1adis1av, Chlewbot, Eyeball kid, Manticorp, Bob the Hamster, Rglovejoy, Tmangray, Chetvorno, Ralph Purtcher, Koeplinger, Headbomb, Ioeth, Madmarigold, Cherrywood, Hillgentleman, Maliz, HEL, Craigheinke, CaptinJohn, Broadbot, Antixt, PhysPhD, SieBot, Haavikon, Nergaal, Brhestir, Michał Sobkowski, WestwoodMatt, Stepheng3, Ladsgroup, MystBot, SkyLined, Addbot, DOI bot, חצרוני, Download, Xario, Luckas-bot, Quasar1826, AnomieBOT, Citation bot, ArthurBot, LilHelpa, Margenau, Citation bot 1, 10metreh, Tom.Reding, EmausBot, John of Reading, WikitanvirBot, ZéroBot, Quondum, TonyMath, Timetraveler3.14, Yclept:Berr, Rafael.ferragut, FeatherPluma, Mikhail Ryazanov, Bibcode Bot, BattyBot, Timothy Gu, ChrisGualtieri, 786b6364, JYBot, Hcrater, Makecat-bot, Mfb, Uclmaps, Monkbot, Yikkayaya, Prisencolin and Anonymous: 68
- **Muonium** *Source:* https://en.wikipedia.org/wiki/Muonium?oldid=678976362 *Contributors:* Tim Starling, Egil, Julesd, Emperorbma, Reddi, Stone, Lumos3, RedWolf, Csibert, Karol Langner, Rich Farmbrough, Pavel Vozenilek, ZeroOne, Jag123, Walkerma, Caesura, Rjwilmsi, Strait, JHBrewer, Chobot, YurikBot, Spacepotato, Bambaiah, Hairy Dude, Salsb, Groyolo, SmackBot, OrangeDog, Can't sleep, clown will eat me, Vina-iwbot~enwiki, Kirit.makwana, Calvero JP, Thijs!bot, Headbomb, Ninjajake, Tarotcards, Steel1943, McM.bot, AlleborgoBot, Nergaal, Mimosveta, SkyLined, TStein, 84user, Legobot, Luckas-bot, AnomieBOT, Tuvalkin, Citation bot, Gilo1969, Gap9551, Bob Saint Clar, Citation

bot 1, RjwilmsiBot, Fcy, EmausBot, ZéroBot, AbigwikiFan, Helpful Pixie Bot, Bibcode Bot, BG19bot, Monkbot, Skirtland and Anonymous: 21

- **Onium** *Source:* https://en.wikipedia.org/wiki/Onium?oldid=673474690 *Contributors:* Mike Rosoft, Anthony Appleyard, Kazvorpal, Strait, Spacepotato, Malcolma, Geoffrey.landis, SmackBot, JorisvS, A876, Headbomb, SGGH, Belg4mit, Antixt, Jeremiah Mountain, Addbot, Luckas-bot, AnomieBOT, Erik9bot, Carlog3, Paine Ellsworth and Anonymous: 7

- **Superatom** *Source:* https://en.wikipedia.org/wiki/Superatom?oldid=648443121 *Contributors:* Bryan Derksen, Michael Hardy, Tim Starling, Rursus, Gtrmp, Nickptar, Peter bertok, Rich Farmbrough, Pavel Vozenilek, NeilTarrant, Blotwell, Keenan Pepper, Kzollman, V8rik, Rjwilmsi, Marasama, Mattman00000, Rogertudor, Tone, Gaius Cornelius, Pegship, Tetracube, New guy, SmackBot, Incnis Mrsi, EncycloPetey, Pwjb, Serein (renamed because of SUL), John, JorisvS, Tawkerbot2, Headbomb, RogueNinja, Steveprutz, Rhadamante, Coppertwig, Minesweeper.007, WingkeeLEE, Cyfal, Addbot, Alfie66, Fraggle81, Hyju, Fentlehan, Dude1818, Double sharp, TobeBot, Elium2, EmausBot, Kpufferfish, ZéroBot, StringTheory11, Jureveles, Whoop whoop pull up, Poeta3d, Pclayborne, Jeremy112233, ChrisGualtieri, Sharma yamijala, M.ashwin111, IA-chemist, Décio pinto de jesus and Anonymous: 23

- **Molecule** *Source:* https://en.wikipedia.org/wiki/Molecule?oldid=682123432 *Contributors:* Sodium, Tarquin, WillWare, Andre Engels, Christian List, Ben-Zin~enwiki, Lir, Patrick, Michael Hardy, Fred Bauder, Liftarn, Looxix~enwiki, Ahoerstemeier, Theresa knott, Nanobug, Jebba, Александър, Julesd, Nikai, Andres, Samuel~enwiki, Ddoherty, Coren, 4lex, Piolinfax, Saltine, Shizhao, Gakrivas, Phil Boswell, Gentgeen, Robbot, Sander123, Fredrik, Zandperl, Arkuat, Postdlf, Merovingian, Lsy098~enwiki, Academic Challenger, Caknuck, Kagredon, Hadal, Guy Peters, Marc Venot, Centrx, Giftlite, Dbenbenn, Graeme Bartlett, DocWatson42, Jmnbpt, Ævar Arnfjörð Bjarmason, Everyking, Fleminra, Curps, Bensaccount, Unconcerned, Jorge Stolfi, Bobblewik, SonicAD, Karol Langner, Icairns, Golnazfotohabadi, JohnArmagh, Deglr6328, Bluemask, Mike Rosoft, Discospinster, Hydrox, Cacycle, Vsmith, Wk muriithi, ESkog, Eric Forste, RJHall, Livajo, Art LaPella, Bobo192, Smalljim, Viriditas, Elipongo, Giraffedata, SpeedyGonsales, Kjkolb, Deryck Chan, Obradovic Goran, Jumbuck, Alansohn, Mykej, BodyTag, Walkerma, Bart133, Wtmitchell, Cburnett, John W. Kennedy, Ron Ritzman, Linas, TigerShark, Scjessey, Tygar, CiTrusD, TotoBaggins, SDC, Graham87, V8rik, Jclemens, Sjö, Drbogdan, HappyCamper, Ligulem, FlaBot, Naraht, Ian Pitchford, Admp~enwiki, RexNL, Gurch, OSt~enwiki, McDogm, Srleffler, Chobot, DVdm, Antiuser, The Rambling Man, Andel, YurikBot, Wavelength, RobotE, Brandmeister (old), Anonymous editor, Lucinos~enwiki, Ironist, Yyy, Mclparker, Haranoh, THB, Vb, Azazell0, Ospalh, Kyle Barbour, DeadEyeArrow, Zzuuzz, Lt-wiki-bot, Ketsuekigata, Pietdesomere, Jolt76, Livitup, JoanneB, Katieh5584, Ásgeir IV.~enwiki, GrinBot~enwiki, Jade Knight, DVD R W, ChemGardener, Itub, SmackBot, Unschool, CarbonCopy, Prodego, Hydrogen Iodide, David Shear, Phaldo, Bomac, Davewild, BMunage, Lowzeewee, Edgar181, M stone, Yamaguchi??, Gilliam, Ohnoitsjamie, Skizzik, JAn Dudík, Bduke, Miquonranger03, MalafayaBot, Droll, Moshe Constantine Hassan Al-Silverburg, Ctbolt, DHN-bot~enwiki, Sbharris, Colonies Chris, Hallenrm, Can't sleep, clown will eat me, Rrburke, Addshore, SundarBot, EdGl, DMacks, SteveLower, Sadi Carnot, Akendall, John, Scientizzle, Heimstern, Disavian, Accurizer, IronGargoyle, A. Parrot, MarkSutton, Munita Prasad, Hetar, Iridescent, Shoeofdeath, Llydawr, Sahuagin, Mulder416sBot, MightyWarrior, JForget, JohnCD, Nunquam Dormio, Exzakin, KnightLago, NickW557, FlyingToaster, Moreschi, Road Wizard, Rifleman 82, Michaelas10, Gogo Dodo, Flowerpotman, Dcandy, Tawkerbot4, Christian75, Pokeman, Omicronpersei8, Epbr123, Pajz, N5iln, Callmarcus, Headbomb, Marek69, Kendal Ozzel, GMiranda, D.H, Greg L, CTZMSC3, Escarbot, AntiVandalBot, Skynet1216, Gökhan, Res2216firestar, JAnDbot, Deflective, Husond, Captain538, MER-C, Avaya1, Andonic, PhilKnight, Magioladitis, Pedro, Bongwarrior, VoABot II, CattleGirl, CTF83!, Avicennasis, Animum, Johnbibby, Eldumpo, DerHexer, JaGa, Otvaltak, Yobol, MartinBot, Mythealias, Mermaid from the Baltic Sea, NAHID, ChemNerd, Halsall, CommonsDelinker, AlexiusHoratius, Fusion7, Nono64, PrestonH, KBlott, J.delanoy, Erajda, Abecedare, Rgoodermote, Numbo3, Hans Dunkelberg, Eliz81, Icseaturtles, James A. Stewart, Ncmvocalist, Ignatzmice, Mikael Häggström, SasukeX, Jcwf, TomasBat, Bobianite, GDW13, KylieTastic, Joshua Issac, Linshukun, Useight, CardinalDan, Idioma-bot, Deor, VolkovBot, AlnoktaBOT, Atrip, Cursayer, Philip Trueman, TXiKiBoT, Cosmic Latte, Crohnie, Qxz, Someguy1221, Littlealien182, Anna Lincoln, Lradrama, KyleRGiggs, Jackfork, LeaveSleaves, Mannafredo, Katimawan2005, Thebigbendizzle, Vladsinger, Synthebot, Antixt, Jason Leach, Enviroboy, Burntsauce, Insanity Incarnate, Bobo The Ninja, AlleborgoBot, Daveberos, SieBot, Whiskey in the Jar, Restre419, Scarian, Happysailor, Flyer22, Antonio Lopez, KPH2293, Steven Crossin, Lightmouse, OKBot, Troy 07, Explicit, ClueBot, LAX, PipepBot, The Thing That Should Not Be, Starkiller88, Gawaxay, Wysprgr2005, Russellboyd, Drmies, Mild Bill Hiccup, SuperHamster, CounterVandalismBot, LizardJr8, DragonBot, Excirial, -Midorihana-, Jusdafax, Abrech, Lartoven, Rhododendrites, Tyler, Footballfan190, Versus22, SoxBot III, SteelMariner, Strangey0, Alchemist Jack, Aj00200, XLinkBot, Ivan Akira, Gnowor, Rror, Fishhead15, Locumele, Noctibus, TravisAF, ZooFari, HexaChord, Addbot, American Eagle, Some jerk on the Internet, Guoguo12, Landon1980, Otisjimmy1, Binary TSO, Looie496, Favonian, Funnybunny123456789, LinkFA-Bot, Numbo3-bot, Ehrenkater, Bluestar232, Tide rolls, Lolimgay, Legobot, Middayexpress, Luckas-bot, ZX81, 2D, Fraggle81, THEN WHO WAS PHONE?, KamikazeBot, AnakngAraw, Suntag, Eric-Wester, Jaytoneypwnz, AnomieBOT, AUG, Jim1138, IRP, Galoubet, NickK, Materialscientist, Edguy99, The High Fin Sperm Whale, Citation bot, Mcpazzo, Roboticoman, Range125, GB fan, LilHelpa, Gsmgm, Vulcan Hephaestus, Xqbot, TinucherianBot II, Sionus, JimVC3, Capricorn42, Bihco, Renaissancee, DR.smallprick, P99am, Mileycyrushater9, BoxingWear3, Jayg101, GrouchoBot, Sophus Bie, Eugene-elgato, Erik9, Magic.Wiki, LucienBOT, Tobby72, TimonyCrickets, Recognizance, Alarics, DivineAlpha, Wireless Keyboard, Shortydude3630, Citation bot 1, Rmiller8, Amplitude101, Pinethicket, Bernarddb, Tom.Reding, Akmkgp, Jschnur, RedBot, Chembrain~enwiki, SpaceFlight89, Barras, Jauhienij, Juliobryant24, Jonkerz, MrX, Izzymello, David Hedlund, Tbhotch, Sideways713, DARTH SIDIOUS 2, RjwilmsiBot, TjBot, Agent Smith (The Matrix), Justinashley, Muttenjeff, DASHBot, EmausBot, John of Reading, ScottyBerg, MarioFanNo1, GoingBatty, Onegumas, Winner 42, Wikipelli, K6ka, Thecheesykid, Werieth, 15turnsm, ZéroBot, CanonLawJunkie, Everard Proudfoot, Kingkamata, L Kensington, Donner60, Negovori, DASHBotAV, Mattdj2, Jimmywolfe100, E. Fokker, ClueBot NG, CocuBot, Corusant, O.Koslowski, Widr, Spannerjam, Antiqueight, Diyar se, Helpful Pixie Bot, Art and Muscle, Gob Lofa, Bibcode Bot, BG19bot, JacobTrue, Mark Arsten, Joydeep, 15chongw1, Youfailalot, NotWith, YVSREDDY, MuAlphaTheta, Rob Hurt, Oogashocka, Ducknish, JYBot, Smartypants436, Lugia2453, James Ayling, Frosty, Scooters are gay, Sammyreedy, Kevin12xd, 069952497a, Reatlas, Darth Sitges, Elie.nasrallah, Maonato, CopyEditor998, Ophara~enwiki, Briitters, Triolysat, Jarrod127, Vesperthevoid, Joystickpenguin97, Monkbot, Aamupala86, Paparoach1301, Rgesses, Hailey Girges, BetaFrisco, TROLLCOPTER, KasparBot, Mayah011213, Machkata, SSTflyer and Anonymous: 731

- **Exotic baryon** *Source:* https://en.wikipedia.org/wiki/Exotic_baryon?oldid=675789200 *Contributors:* Bryan Derksen, Glenn, Ehn, Phys, Codepoet, Xerxes314, Siroxo, Gzornenplatz, Physicist, Icairns, D6, Cherlin, Lectonar, Count Iblis, Kyleca, Kbdank71, Rjwilmsi, Amaurea, YurikBot, Bambaiah, Salsb, Poulpy, Lainagier, Hmains, V1adis1av, Lambiam, JorisvS, Thijs!bot, Keraunos, Headbomb, Antixt, MystBot, Addbot, Citation bot, ArthurBot, TheOmnipotentLemur, Carlog3, Citation bot 1, Carbosi, JSquish, Bibcode Bot and Anonymous: 14

- **Exotic hadron** *Source:* https://en.wikipedia.org/wiki/Exotic_hadron?oldid=675789447 *Contributors:* AxelBoldt, Glenn, Xerxes314, Rich Farmbrough, YUL89YYZ, Keenan Pepper, Count Iblis, April Arcus, Linas, Bambaiah, Wogsland, V1adis1av, Acjohnson55, Happy-melon, JRSpriggs, Postmodern Beatnik, ZICO, Thijs!bot, Whatever1111, Headbomb, Stannered, Leyo, VolkovBot, Antixt, Muhends, Curtis95112, MystBot, SkyLined, Addbot, Tide rolls, Luckas-bot, Yobot, Carlog3, Tarsilia, Acather96, Carbosi, ZéroBot, StringTheory11, Ethaniel, ChuispastonBot, Comicboy1996, Trompedo and Anonymous: 16
- **Exotic meson** *Source:* https://en.wikipedia.org/wiki/Exotic_meson?oldid=671307183 *Contributors:* Xavic69, Michael Hardy, Glenn, Phys, Xerxes314, Gzornenplatz, Physicist, Icairns, D6, Bender235, Gauge, Apyule, Pearle, Fwb22, Kbdank71, Strait, Erkcan, YurikBot, Bambaiah, Salsb, Garion96, Banus, SmackBot, Bluebot, V1adis1av, Ziusudra, ChrisCork, Physic sox, Difluoroethene, Thijs!bot, Headbomb, Fallschirmjäger, VolkovBot, RedAndr, Antixt, Addbot, TaBOT-zerem, FrescoBot, Jennyxie, ClueBot NG, Parcly Taxel, Rezabot, Bibcode Bot and Anonymous: 14
- **Glueball** *Source:* https://en.wikipedia.org/wiki/Glueball?oldid=671829421 *Contributors:* Paul A, Loren Rosen, Phys, Sanders muc, Xerxes314, Mennonot, MuDavid, Jeodesic, Bambaiah, Hairy Dude, Ohwilleke, Xaxafrad, Smurrayinchester, Triple333, Saravask, Kmarinas86, V1adis1av, Sasata, Zaphody3k, Thijs!bot, Headbomb, Magioladitis, Nyq, Idioma-bot, Anonymous Dissident, Antixt, YonaBot, Avidallred, Boemmels, Alexbot, SchreiberBike, Addbot, Mpfiz, Lightbot, Luckas-bot, Dreamer08, AnomieBOT, Pra1998, Tom.Reding, Loqueelvientoajuarez, RjwilmsiBot, Carbosi, Drummermean, JSquish, ZéroBot, Suslindisambiguator, Whoop whoop pull up, Bibcode Bot, Vkpd11, Retnuh66, ChrisGualtieri, Richardbernstein and Anonymous: 10
- **Hexaquark** *Source:* https://en.wikipedia.org/wiki/Hexaquark?oldid=682556611 *Contributors:* Giftlite, Cherlin, Pol098, Miss Madeline, Rjwilmsi,Goudzovski, Smurrayinchester, SmackBot, Dark Formal, JorisvS, Runningonbrains, ShelfSkewed, Headbomb, Leyo, Lamro, Addbot, Luckas-bot, Citation bot, ArthurBot, Carlog3, Citation bot 1, HRoestBot, Jakeukalane, Cutelyaware, StringTheory11, Gerasime, Bibcode Bot, RodzynKa, Mi Tatara Buela, Monkbot, Prioritat im Vordergrund, Bodhisattwa and Anonymous: 8
- **Mesonic molecule** *Source:* https://en.wikipedia.org/wiki/Mesonic_molecule?oldid=603995173 *Contributors:* Rich Farmbrough, Miss Madeline, Goudzovski, Leptictidium, SmackBot, Wogsland, Wdspann, Antixt, BartekChom, MystBot, Addbot, Erik9bot, Carlog3, ZéroBot and Anonymous: 6
- **Pentaquark** *Source:* https://en.wikipedia.org/wiki/Pentaquark?oldid=682556411 *Contributors:* Bryan Derksen, The Anome, Bdesham, EddEdmondson, J'raxis, Glenn, Wfeidt, Schneelocke, Joquarky, Tpbradbury, Phys, Bevo, Phil Boswell, Robbot, Sanders muc, Merovingian, Herbee, Codepoet, Xerxes314, DÅ,ugosz, Elektron, Icairns, Metahacker, Kenny TM~~enwiki, Pjacobi, Bender235, Kaganer, Fwb22, Mac Davis, Lokedhs, Dismas, Mpatel, GregorB, Mekong Bluesman, Ashmoo, Kbdank71, Marasama, Bubba73, Bambaiah, Wester, Phmer, 4C~enwiki, Salsb, Welsh, Simen, Arthur Rubin, Smurrayinchester, Physicsdavid, Bluebot, Hgrosser, George Ho, V1adis1av, Doug Bell, JorisvS, Thijs!bot, Headbomb, Sobreira, Davidhorman, Oreo Priest, Widefox, Ericoides, BatteryIncluded, Hweimer, Chiswick Chap, 83d40m, VolkovBot, FourteenDays, Lamro, Antixt, Spinningspark, Nubiatech, JackSchmidt, Denisarona, Muhends, Curtis95112, BobKawanaka, Muro Bot, Facts707, Addbot, Masur, LaaknorBot, Lightbot, PV=nRT, Yobot, Dreamer08, AnomieBOT, Materialscientist, Citation bot, ArthurBot, Ace111, Citation bot 2, Citation bot 1, Redrose64, Thinking of England, Winner 42, ZéroBot, Mandula, Brandmeister, Chris857, Ad Orientem, ClueBot NG, Ben morphett, Bibcode Bot, Pedro.bicudo, BG19bot, Enervation, Khazar2, AutisticCatnip, Jamietwells, KevinLiu, Qwerty123uiop, 22merlin, Belle, Oiyarbepsy, DN-boards1, GeneralizationsAreBad, Lollipop, Pentaquarksuperfan, Aashish Khadka, Vcydx and Anonymous: 67
- **Tetraquark** *Source:* https://en.wikipedia.org/wiki/Tetraquark?oldid=676856701 *Contributors:* Bryan Derksen, Phys, Phil Boswell, Merovingian, Herbee, Varlaam, Physicist, Setokaiba, Icairns, SeaDour, Drbogdan, Rjwilmsi, Bubba73, Nhussein, Bambaiah, Todd Vierling, Antiduh, 2over0, Teply, SmackBot, V1adis1av, Wiki me, Yevgeny Kats, Newone, Headbomb, Sobreira, Hcobb, Ron2, VolkovBot, Antixt, Muhends, Psyden, Addbot, Luckas-bot, AnomieBOT, Citation bot, ArthurBot, Omnipaedista, Citation bot 1, Jonesey95, Jakeukalane, Meier99, WikitanvirBot, Peaceray, JSquish, ZéroBot, ChuispastonBot, Polosa, Bibcode Bot, Pedro.bicudo, Mesonic Interference, Tony Mach, Faizan, Bicyclegeek, Monkbot and Anonymous: 29
- **Skyrmion** *Source:* https://en.wikipedia.org/wiki/Skyrmion?oldid=681892425 *Contributors:* Michael Hardy, Charles Matthews, Phys, Icairns, Lumidek, Brianhe, Pjacobi, Jag123, Fwb22, Rjwilmsi, Conscious, Wikid77, Headbomb, Widefox, Lincoln F. Stern, Tarotcards, KylieTastic, PixelBot, Doprendek, Addbot, Luckas-bot, Yobot, Citation bot, Obersachsebot, Omnipaedista, FrescoBot, Citation bot 1, Merongb10, Meier99, Korepin, EmausBot, JSquish, ZéroBot, StringTheory11, AManWithNoPlan, Isocliff, Parcly Taxel, Bibcode Bot, BG19bot, BattyBot, ChrisGualtieri, Andyhowlett, 1andreasse, Zimboras, Nicohoho, NorskMaelstrom, Noah Van Horne, Farank olamaeian and Anonymous: 8
- **Pomeron** *Source:* https://en.wikipedia.org/wiki/Pomeron?oldid=681458027 *Contributors:* Phys, Ccady, Icairns, Rich Farmbrough, Igorivanov~enwiki, Pjacobi, .:Ajvol:., Jag123, Kay Dekker, Erkcan, Goudzovski, Conscious, Salsb, ILDuceMas, Closedmouth, Chymicus, JarahE, Luis Sanchez, Headbomb, Sobreira, WinBot, DanPMK, MarkJefferys, Likebox, ClueBot, Bobathon71, Alexbot, MagnusPI, Addbot, Lmbduk, Mjamja, Lightbot, Omnipaedista, Carlog3, Anterior1, Citation bot 2, Dskrvk, HFEO and Anonymous: 6
- **Quasiparticle** *Source:* https://en.wikipedia.org/wiki/Quasiparticle?oldid=678833031 *Contributors:* CYD, Glenn, Charles Matthews, Phys, Topbanana, Lumos3, Donarreiskoffer, Giftlite, Dratman, Jason Quinn, Sysin, Lumidek, Vsmith, RoyBoy, Lysdexia, Gene Nygaard, Wafry, Rjwilmsi, Arnero, Rune.welsh, Srleffler, Chobot, YurikBot, Wavelength, Shaddack, Sbyrnes321, SmackBot, Stepa, Chris the speller, Complexica, UNV, Vanished user 9i39j3, Euchiasmus, JorisvS, Brienanni, Iridescent, WilliamDParker, CmdrObot, Cydebot, Mbell, Hazmat2, Headbomb, BehnamFarid, James Slezak, Yellowdesk, Michael.j.sykora, David Eppstein, Stevvers, Connor Behan, Gernewvic, Venny85, Antixt, AlleborgoBot, Kbrose, Goeie, ClueBot, Mr Accountable, Alexbot, Addbot, Mathieu Perrin, SPat, Luckas-bot, Yobot, Genius.scholar, LilHelpa, Doraemonpaul, Freddy78, RedAcer, Citation bot 1, Logiolgeirss, ZéroBot, StringTheory11, Diego Grez Bot, Tls60, RockMagnetist, ClueBot NG, PBot1, Asi013, Helpful Pixie Bot, Bibcode Bot, Rolancito, F=q(E+v^B), Shaginyan, Mark viking, LeoKadanoff, Susan.grayeff and Anonymous: 48
- **Davydov soliton** *Source:* https://en.wikipedia.org/wiki/Davydov_soliton?oldid=544626124 *Contributors:* Danko Georgiev, ^demon, Rjwilmsi, Betacommand, Chris the speller, Alaibot, Headbomb, Top.Squark, Wasell, Antixt, Lisatwo, Addbot, DOI bot, Tassedethe, Yobot, Carlog3, Citation bot 1, DrilBot, NULL, Bibcode Bot and Anonymous: 6
- **Dropleton** *Source:* https://en.wikipedia.org/wiki/Dropleton?oldid=665319792 *Contributors:* Kjkolb, Peyre, Headbomb, M.O.X, Snaily, Brandmeister, Bibcode Bot, Everymorning and Buckrogers24

- **Exciton** *Source:* https://en.wikipedia.org/wiki/Exciton?oldid=682198255 *Contributors:* Bryan Derksen, Arvindn, PierreAbbat, Michael Hardy, Tim Starling, Ellywa, Cyan, Tantalate, Donarreiskoffer, Robbot, Ojigiri~enwiki, Giftlite, Darrien, Roo72, Gianluigi, La goutte de pluie, Gbrandt, Gene Nygaard, Snafflekid, Ecalman, R.e.b., Arnero, Chobot, Jaraalbe, YurikBot, Dobromila, Shaddack, Salsb, Długosz, Vatassery, Petri Krohn, Sam Diener, The Photon, Eskimbot, Kmarinas86, Chris the speller, Bluebot, MK8, HRyanjones, Colonies Chris, Jaganath, Mgiganteus1, Tmangray, CmdrObot, Jetblack101, Thijs!bot, Barticus88, N5iln, Headbomb, Ferritecore, Avjoska, LorenzoB, Tgeairn, P4k, Casmaia, VolkovBot, Larryisgood, MenasimBot, TXiKiBoT, Antoni Barau, Matthias Buchmeier, Lamro, Antixt, Kbrose, Wing gundam, Henry Delforn (old), Daniel.Schmidt~enwiki, Alkamid, Coinmanj, Hess88, Addbot, Eric Drexler, Мыша, LaaknorBot, WikiDreamer Bot, Luckas-bot, Yobot, Jcrochet, Materialscientist, Oloinsigh, Citation bot, ArthurBot, Ibid.Ibought, Tripodian, Kstueve, Jbae, Steve Quinn, Citation bot 1, Jordgette, John of Reading, StringTheory11, Timetraveler3.14, Superdelocalizable, ClueBot NG, Bibcode Bot, BG19bot, Vahidvms, Wikih101, Toni 001, Mikaylashley, PootisHeavy, Cerabot~enwiki, JudgeDeadd, TooOldMan, Kalpathy, Temahukn and Anonymous: 78
- **Electron hole** *Source:* https://en.wikipedia.org/wiki/Electron_hole?oldid=682317931 *Contributors:* Maury Markowitz, Tim Starling, Lkesteloot,Omegatron, Bevo, Robbot, Ojigiri~enwiki, Wjbeaty, LiDaobing, Karol Langner, DragonflySixtyseven, Discospinster, Roo 72, Tirthajyoti,PhilHibbs, Nigelj, Robotje, Ranveig, Capi crimm, Palica, Mandarax, Salleman, FlaBot, Arnero, Ironside@elec.gla. ac.uk, Chobot, Yurik-Bot, Wavelength, RobotE, Bambaiah, Bhny, Archelon, Shaddack, Spike Wilbury, Ninly, Sbyrnes321, The Photon, Shai-kun, Betacommand,Jcarroll, Chris the speller, Bluebot, Jprg1966, Lagrangian, DMacks, Jaganath, Robofish, JorisvS, Mgiganteus1, Noah Salzman, Iridescent,Tawkerbot2, Jh12, Chetvorno, Gogo Dodo, Delta Spartan, Envy0, Thijs!bot, Barticus88, Salgueiro~enwiki, JAnDbot, Britcom, Savant13,Sir Link, Sqush101, Aboutmovies, Andejons, Kurosa~enwiki, Captainlavender, JhsBot, דוד שי, Jack Naven Rulez, Lamro, SieBot, Extremecircuitz,No such user, Addbot, Out of Phase User, Bob K31416, OlEnglish, Uroboros, Meisam, Legobot, Luckas-bot, Materialscientist, Lil-Helpa, Xqbot, Leonardo Da Vinci, Erik9bot, Citation bot 1, Lissajous, EmausBot, K6ka, ClueBot NG, Crazymonkey1123, Bibcode Bot,Williammathew30, Iplaycards, Vanquisher.UA, YimmyYohnson, Qqwweerrttyy22, KasparBot and Anonymous: 63
- **Magnon** *Source:* https://en.wikipedia.org/wiki/Magnon?oldid=682003415 *Contributors:* Pumpie, Donarreiskoffer, Herbee, Laurascudder, Muntfish, LCD~enwiki, Saga City, FlaBot, Chobot, Roboto de Ajvol, Hellbus, Shaddack, Neil.steiner, Z-Scorpion, Chaiken, SmackBot, Olegt1, Kmarinas86, Jjalexand, Colonies Chris, BWDuncan, 'Ff'lo, Headbomb, Osquar F, Speaker to wolves, Matthias Buchmeier, A4bot, Anonymous Dissident, Antixt, AlleborgoBot, MystBot, IsmaelLuceno, Addbot, Nratter711, Luckas-bot, FrescoBot, MsPorterAtFHS, Miracle Pen, Demokritov, GoingBatty, JSquish, StringTheory11, Timetraveler3.14, RockMagnetist, Bibcode Bot, BG19bot, ChrisGualtieri, Crispulop and Anonymous: 18
- **Phonon** *Source:* https://en.wikipedia.org/wiki/Phonon?oldid=680926420 *Contributors:* CYD, Heron, FlorianMarquardt, Camembert, Michael Hardy, Breakpoint, TakuyaMurata, SebastianHelm, Looxix~enwiki, Stevenj, Fuck You, Glenn, AugPi, Schneelocke, Hollgor, Charles Matthews, Tantalate, Saltine, Phys, Omegatron, Phil Boswell, Rogper~enwiki, Donarreiskoffer, Robbot, Giftlite, Donvinzk, CyborgTosser, Pcarbonn, Mikko Paananen, MuDavid, El C, Kjkolb, Atlant, Dschwen, H2g2bob, Philthecow, Firsfron, LOL, Isnow, SeventyThree, Pharmacomancer, Magister Mathematicae, IIBewegung, Nanite, Koavf, Eubot, Arnero, RexNL, Ewlyahoocom, Lmatt, Srleffler, Kri, Physchim62, Chobot, Pelleapa~enwiki, Roboto de Ajvol, Wavelength, Bhny, Okedem, Archelon, Shaddack, Deskana, Długosz, Davemck, Mysid, Dna-webmaster, Guillom, SmackBot, Kopaka649, Moocowpong1, Aram.harrow, Bluebot, Jjalexand, MK8, AndrewBuck, Fredvanner, Sbharris, Colonies Chris, OrphanBot, MattOates, Wiki me, Bigmantonyd, LoveEncounterFlow, Chymicus, AThing, Karakal, JorisvS, Stephen B Streater, Marcusl, Gregstortz, Shaind, WilliamDParker, N2e, Myasuda, CumbiaDude, Iliank, Shepplestone, Quantyz, JamesAM, Btball, Mbell, Headbomb, I do not exist, Hcobb, Greg L, Nick Number, Austin Maxwell, Serenity id, Tyco.skinner, JAnDbot, 100110100, Wasell, Gumby600, Kostisl, CommonsDelinker, Maurice Carbonaro, Loohcsnuf, Sigmundur, DavidCBryant, Gwen Gale, Telecomtom, Fimbulfamb, VolkovBot, AlnoktaBOT, HiraV, Matthias Buchmeier, FDominec, Judge Nutmeg, Antixt, Recot, Why Not A Duck, Kbrose, SieBot, Wing gundam, Trumpsternator, WingkeeLEE, OKBot, CultureDrone, Chocobo93, Martarius, Sun Creator, Brews ohare, DumZiBoT, Nettings, Nathan Johnson, Minimag23, Stephen Poppitt, Addbot, Download, Tide rolls, Zorrobot, Quantumobserver, Wiso, Luckas-bot, AdamSiska, AnomieBOT, Materialscientist, ArthurBot, Tripodian, Nanog, Logger9, Echisolm, CES1596, Johannes919, Steve Quinn, RedAcer, Andrestand, IVAN3MAN, Vrenator, Tranh Nguyen, Marie Poise, RjwilmsiBot, TjBot, Alisha.4m, JSquish, Splibubay, Debangshu0, Borders999, GianniG46, Fizicist, Sonygal, Tls60, ClueBot NG, Starshipenterprise, Kgordiz, Zak.estrada, Helpful Pixie Bot, Skarmenadius, Berkecelik, Boston1034, BattyBot, ChrisGualtieri, Mwchalmers, Lugia2453, Datta research, Cesaranieto~enwiki, Mark viking, DungeonSiegeAddict510, ScotXW, Monkbot, KasparBot, Macky0209, Sushins94 and Anonymous: 160
- **Plasmaron** *Source:* https://en.wikipedia.org/wiki/Plasmaron?oldid=662184864 *Contributors:* Rjwilmsi, SmackBot, DanPMK, Addbot, Yobot, Citation bot, Sopher99, ZéroBot, QuantumSquirrel, Bibcode Bot and Anonymous: 4
- **Plasmon** *Source:* https://en.wikipedia.org/wiki/Plasmon?oldid=676152160 *Contributors:* Bryan Derksen, The Anome, AdamRetchless, Karada,Julesd, Glenn, Tantalate, Reddi, LMB, Donarreiskoffer, Alexwcovington, Giftlite, Dratman, Niteowlneils, St3vo, Chris Howard, D6, Zombieje-sus, Bender235, Violetriga, Wareh, Bobo192, Donny11, Neologism, Melaen, Brookie, Elvenlord Elrond, Pfalstad, Marudubshinki, Rjwilmsi,Vegaswikian, Zizzybaluba, Oo64eva, FlaBot, Arnero, Rune.welsh, Srleffler, Chobot, YurikBot, Shaddack, Cryptic, Joel7687, Light current,Aremisasling, Oysteinp, Closedmouth, Arthur Rubin, Reyk, Eno-ja, That Guy, From That Show !, SmackBot, Kmarinas86, Papa Novem-ber, Droll, Colonies Chris, Kukini, Vincenzo.romano, Porterjoh, JesseChisholm, Thijs!bot, Tomio-, Headbomb, Sobreira, Klyk, Beathovn,Michael.j.sykora, Sangak, Ajfeist, Laserboy1969, Aoosten, Lyctc, Pdcook, Idiomabot, TXiKiBoT, FDominec, Liquidcentre, AnonymousDissident, CaptinJohn, Tysanner, Felipebm, Antixt, Ptrslv72, The Mad Genius , Sanjaya mala, Ceanothus, ClueBot, ScottTParker, XLinkBot,Hess88, Addbot, DOI bot, MrOllie, Lightbot, Luckas-bot, Yobot, Amirobot, 4th-otaku, Materialscientist, Citation bot, ArthurBot, Grou-choBot, FrescoBot, Steve Quinn, Matthieu.berthome, Tom. Reding, RobinK, RjwilmsiBot, EmausBot, JSquish, Eg-T2g, Mikhail Ryazanov,HBook, Plasmon man, Axel.hallinder, Paulzubrinich, CasualVisitor, Helpful Pixie Bot, Bibcode Bot, GKFX, Stenlly69, Shawn Worthing-ton Laser Plasma, Elixirbouncybounce, SantoshBot, M0532062613, Mark viking, 20M030810, Monkbot, Jeancartier1982, Oharsh, Wiki-bot123456789, Yannick Zondag and Anonymous: 126
- **Polariton** *Source:* https://en.wikipedia.org/wiki/Polariton?oldid=679378282 *Contributors:* Michael Hardy, Tantalate, Dysprosia, Laussy, Jeffq, Donarreiskoffer, Dratman, Rchandra, Darrien, TheBlueWizard, Bender235, Pearle, Cruccone, Dwward, Nezumidumousseau, Kerowyn, Srleffler, Jaraalbe, YurikBot, Shaddack, Kmarinas86, Amalas, Headbomb, Dougher, Deflective, Magioladitis, Glrx, Hans Dunkelberg, VolkovBot, Antixt, M.qrius, XLinkBot, Asynchro, WikiInnocent~enwiki, Addbot, Mathieu Perrin, DOI bot, LaaknorBot, JJJinks, Luckas-bot, Yobot, Ventsolaire, Jo3sampl, FrescoBot, StringTheory11, Timetraveler3.14, Bibcode Bot, Srodrig, Sabrinaspringer, Sergey Tolpygo, Mark viking, Yikkayaya and Anonymous: 22

- **Polaron** *Source:* https://en.wikipedia.org/wiki/Polaron?oldid=682301180 *Contributors:* Nolambar, Bronger, Charles Matthews, Greenrd, Chuunen Baka, Donarreiskoffer, R3m0t, Noplasma, Jrdioko, Danko Georgiev, Rich Farmbrough, RJHall, Laurascudder, Keenan Pepper, Saga City, Woohookitty, V8rik, Rjwilmsi, FlaBot, Chobot, Shaddack, 2over0, Reyk, Sbyrnes321, SmackBot, Kmarinas86, HRyanjones, Thumperward, Colonies Chris, JorisvS, Wafulz, Barticus88, Headbomb, Easchiff, Nyttend, Laikh, Leyo, VolkovBot, TXiKiBoT, Elieb, Steven J. Anderson, Venny85, Antixt, AlleborgoBot, Mr Accountable, Alexbot, Addbot, Mathieu Perrin, DOI bot, Luckas-bot, Yobot, Steamturn, ArthurBot, J04n, Olivier d'ALLIVY KELLY, Citation bot 1, Tom.Reding, EmausBot, GoingBatty, StringTheory11, AManWithNoPlan, Widr, Bibcode Bot, BG19bot, ChrisGualtieri, Mogism, Alerebola, TooOldMan, Rydbergite, Anrnusna, MSEgirl and Anonymous: 28
- **Roton** *Source:* https://en.wikipedia.org/wiki/Roton?oldid=667334493 *Contributors:* Salsa Shark, Cryoboy, WolfgangRieger, Secretlondon, Steve Leach, Giftlite, Neutrality, Gurkha, UnHoly, Lockley, Shaddack, Chaos, SmackBot, Bluebot, Colonies Chris, Eynar, Torrazzo, Alaibot, Headbomb, Escarbot, Goldenrowley, Melkor23, VolkovBot, Yomach, McM.bot, Venny85, SieBot, DragonBot, Jovianeye, MystBot, Addbot, Luckas-bot, Yobot, Tinton5, MastiBot, EmausBot, Timetraveler3.14, Bibcode Bot, BattyBot, Agonbroke and Anonymous: 4
- **Trion (physics)** *Source:* https://en.wikipedia.org/wiki/Trion_(physics)?oldid=679104668 *Contributors:* Kkmurray, Addbot, Eutactic, Yobot, Crissxx2, AnomieBOT, Materialscientist, OneOfTwo, Jonesey95, RedBot, QuantumSquirrel, Helpful Pixie Bot, Bibcode Bot, Monkbot and Anonymous: 2
- **List of baryons** *Source:* https://en.wikipedia.org/wiki/List_of_baryons?oldid=675785999 *Contributors:* Cherkash, GPHemsley, Donarreiskoffer, Giftlite, Rich Farmbrough, ZeroOne, Tompw, Keenan Pepper, Oleg Alexandrov, Woohookitty, Astrowob, Kbdank71, Strait, Mike Peel, R.e.b., Arctic.gnome, YurikBot, Jonrock, Jimp, Cryptic, SCZenz, Gadget850, F15mos, Banus, GrinBot~enwiki, Sbyrnes321, EvanJPW, SmackBot, Tom Lougheed, Oceanh, Jiminy pop, JorisvS, UncleDouggie, Happy-melon, JRSpriggs, Neelix, Cydebot, WillowW, Mike Christie, Abtract, Wikid77, Headbomb, Magioladitis, Mollwollfumble, Randyfurlong, Leyo, Gogobera, Beatnik Party, MichaelSchoenitzer, GimmeBot, Anonymous Dissident, Sascha.baumeister~enwiki, Antixt, PaddyLeahy, Richard Ye, Wing gundam, Thisisnotatest, Lightmouse, Dabomb87, Muhends, NuclearWarfare, SkyLined, Addbot, DOI bot, Mjamja, Debresser, Vectorboson, SamatBot, Luckas-bot, Yobot, Materialscientist, Citation bot, Carlog3, Citation bot 1, Thinking of England, EmausBot, Markinvancouver, Rmashhadi, Bibcode Bot, BG19bot, P76837, Tony Mach, YiFeiBot, Monkbot and Anonymous: 37
- **List of quasiparticles** *Source:* https://en.wikipedia.org/wiki/List_of_quasiparticles?oldid=678938935 *Contributors:* Daniel.inform, Falcorian, Rjwilmsi, Malcolma, Brienanni, Xxanthippe, Thijs!bot, Headbomb, Mintz l, Choihei, Synthebot, ClueBot, Chhe, Addbot, Mathieu Perrin, Ka Faraq Gatri, SPat, Luckas-bot, Materialscientist, Pepo13, Marshallsumter, Xqbot, CES1596, RedAcer, RjwilmsiBot, EmausBot, Brandmeister, Helpful Pixie Bot, Bibcode Bot, Xamnidar, Atomician, BattyBot, Shaginyan, Toni 001, Steven James Sutcliffe, Logical1004, Monkbot, ?? and Anonymous: 7

108.2.2 Images

- **File:1D_normal_modes_(280_kB).gif** *Source:* https://upload.wikimedia.org/wikipedia/commons/9/9b/1D_normal_modes_%28280_kB%29.gif *License:* CC-BY-SA-3.0 *Contributors:* This is a compressed version of the Image:1D normal modes.gif phonon animation on Wikipedia Commons that was originally created by Régis Lachaume and freely licensed. The original was 6,039,343 bytes and required long-duration downloads for any article which included it. This version is 4.7% the size of the original and loads *much* faster. This version also has an interframe delay of 40 ms (v.s. the original's 100 ms). Including processing time for each frame, this version runs at a frame rate of about 20–22.5 Hz on a typical computer, which yields a more fluid motion. Greg L 00:41, 4 October 2006 (UTC). (from http://en.wikipedia.org/wiki/Image:1D_normal_modes_%28280_kB%29.gif) *Original artist:* Original Uploader was Greg L (talk) at 00:41, 4 October 2006.
- **File:1e0657_scale.jpg** *Source:* https://upload.wikimedia.org/wikipedia/commons/a/a8/1e0657_scale.jpg *License:* Public domain *Contributors:* Chandra X-Ray Observatory: 1E 0657-56 *Original artist:* NASA/CXC/M. Weiss
- **File:2-photon_Higgs_decay.svg** *Source:* https://upload.wikimedia.org/wikipedia/commons/3/32/2-photon_Higgs_decay.svg *License:* CC BY-SA 3.0 *Contributors:* Own work *Original artist:* Parcly Taxel
- **File:4-lepton_Higgs_decay.svg** *Source:* https://upload.wikimedia.org/wikipedia/commons/b/b2/4-lepton_Higgs_decay.svg *License:* CC BY-SA 3.0 *Contributors:* Own work *Original artist:* Parcly Taxel
- **File:AIP-Sakurai-best.JPG** *Source:* https://upload.wikimedia.org/wikipedia/commons/2/2b/AIP-Sakurai-best.JPG *License:* Public domain *Contributors:* Own work *Original artist:* self
- **File:Alpha_helix_neg60_neg45_sideview.png** *Source:* https://upload.wikimedia.org/wikipedia/commons/e/ec/Alpha_helix_neg60_neg45_sideview.png *License:* CC-BY-SA-3.0 *Contributors:* No machine readable source provided. Own work assumed (based on copyright claims). *Original artist:* No machine readable author provided. WillowW assumed (based on copyright claims).
- **File:Ambox_important.svg** *Source:* https://upload.wikimedia.org/wikipedia/commons/b/b4/Ambox_important.svg *License:* Public domain *Contributors:* Own work, based off of Image:Ambox scales.svg *Original artist:* Dsmurat (talk · contribs)
- **File:Asymmetricwave2.png** *Source:* https://upload.wikimedia.org/wikipedia/commons/0/0d/Asymmetricwave2.png *License:* CC BY 3.0 *Contributors:* Own work *Original artist:* TimothyRias
- **File:Atisane3.png** *Source:* https://upload.wikimedia.org/wikipedia/commons/a/a4/Atisane3.png *License:* CC-BY-SA-3.0 *Contributors:* en:Image:Atisane3.png, en:Image:Atisane.png (molecule on the left) *Original artist:* en:User:Unconcerned, en:User:Ddoherty
- **File:Atomic_orbital_energy_levels.svg** *Source:* https://upload.wikimedia.org/wikipedia/commons/9/98/Atomic_orbital_energy_levels.svg *License:* CC BY-SA 3.0 *Contributors:* File:High School Chemistry.pdf, page 301 *Original artist:* Richard Parsons (raster), Adrignola (vector)
- **File:Atomic_orbitals_and_periodic_table_construction.ogv** *Source:* https://upload.wikimedia.org/wikipedia/commons/6/61/Atomic_and_periodic_table_construction.ogv *License:* CC BY-SA 3.0 *Contributors:* Own work *Original artist:* Jubobroff
- **File:Atomic_resolution_Au100.JPG** *Source:* https://upload.wikimedia.org/wikipedia/commons/e/ec/Atomic_resolution_Au100.JPG *License:* Public domain *Contributors:* ? *Original artist:* ?

- **File:BandDiagram-Semiconductors-E.PNG** *Source:* https://upload.wikimedia.org/wikipedia/commons/8/85/BandDiagram-Semiconductors-E.PNG *License:* CC BY-SA 2.5 *Contributors:* No machine readable source provided. Own work assumed (based on copyright claims). *Original artist:* No machine readable author provided. S-kei assumed (based on copyright claims).
- **File:Baryon-decuplet-small.svg** *Source:* https://upload.wikimedia.org/wikipedia/commons/7/78/Baryon-decuplet-small.svg *License:* Public domain *Contributors:* Own work *Original artist:* Trassiorf
- **File:Baryon-octet-small.svg** *Source:* https://upload.wikimedia.org/wikipedia/commons/b/b5/Baryon-octet-small.svg *License:* Public domain *Contributors:* Own work *Original artist:* Trassiorf
- **File:Baryon_decuplet.svg** *Source:* https://upload.wikimedia.org/wikipedia/commons/f/f6/Baryon_decuplet.svg *License:* Public domain *Contributors:* Own work (Original text: *self-made*) *Original artist:* Wierdw123 at English Wikipedia
- **File:Beta-minus_Decay.svg** *Source:* https://upload.wikimedia.org/wikipedia/commons/a/aa/Beta-minus_Decay.svg *License:* Public domain *Contributors:* This vector image was created with Inkscape. *Original artist:* Inductiveload
- **File:Beta_Negative_Decay.svg** *Source:* https://upload.wikimedia.org/wikipedia/commons/8/89/Beta_Negative_Decay.svg *License:* Public domain *Contributors:* This vector image was created with Inkscape. *Original artist:* Joel Holdsworth (Joelholdsworth)
- **File:Binding_energy_curve_-_common_isotopes.svg** *Source:* https://upload.wikimedia.org/wikipedia/commons/5/53/Binding_energy_curve_-_common_isotopes.svg *License:* Public domain *Contributors:* ? *Original artist:* ?
- **File:Blausen_0342_ElectronEnergyLevels.png** *Source:* https://upload.wikimedia.org/wikipedia/commons/2/2c/Blausen_.png *License:* CC BY 3.0 *Contributors:* Own work *Original artist:* BruceBlaus
- **File:Bohr-atom-PAR.svg** *Source:* https://upload.wikimedia.org/wikipedia/commons/5/55/Bohr-atom-PAR.svg *License:* CC-BY-SA-3.0 *Contributors:* Transferred from en.wikipedia to Commons. *Original artist:* Original uplo:JabberWok]] at en.wikipedia
- **File:Bohr_atom_animation_2.gif** *Source:* https://upload.wikimedia.org/wikipedia/commons/1/17/Bohr_atom_animation_2.gif *License:* CC BY-SA 3.0 *Contributors:* Own work *Original artist:* Kurzon
- **File:Bose_Einstein_condensate.png** *Source:* https://upload.wikimedia.org/wikipedia/commons/a/af/Bose_Einstein_condensate.png *License:* Public domain *Contributors:* NIST Image *Original artist:* NIST/JILA/CU-Boulder
- **File:Brillouin_zone.svg** *Source:* https://upload.wikimedia.org/wikipedia/commons/2/22/Brillouin_zone.svg *License:* CC-BY-SA-3.0 *Contributors:* Wikipedia en *Original artist:* Gang65
- **File:Buckminsterfullerene-perspective-3D-balls.png** *Source:* https://upload.wikimedia.org/wikipedia/commons/0/0f/Buckminsterballs.png *License:* Public domain *Contributors:* Own work *Original artist:* Benjah-bmm27
- **File:CERN-20060225-24.jpg** *Source:* https://upload.wikimedia.org/wikipedia/commons/6/6f/CERN-20060225-24.jpg *License:* CC-BY-SA-3.0 *Contributors:* ? *Original artist:* ?
- **File:CERN_LHC_Tunnel1.jpg** *Source:* https://upload.wikimedia.org/wikipedia/commons/f/fc/CERN_LHC_Tunnel1.jpg *License:* CC BY-SA 3.0 *Contributors:* Own work *Original artist:* Julian Herzog (website)
- **File:Charmed-dia-w.png** *Source:* https://upload.wikimedia.org/wikipedia/en/6/6d/Charmed-dia-w.png *License:* Fair use *Contributors:* http://www.bnl.gov/bnlweb/history/charmed.asp *Original artist:* ?
- **File:Clyde_Cowan.jpg** *Source:* https://upload.wikimedia.org/wikipedia/commons/5/53/Clyde_Cowan.jpg *License:* Public domain *Contributors:* ? *Original artist:* ?
- **File:Cold_Neutron_Source.jpg** *Source:* https://upload.wikimedia.org/wikipedia/commons/0/0d/Cold_Neutron_Source.jpg *License:* Public domain *Contributors:* http://www.ncnr.nist.gov/coldgains/ *Original artist:* Bill Kamitakahara
- **File:Commons-logo.svg** *Source:* https://upload.wikimedia.org/wikipedia/en/4/4a/Commons-logo.svg *License:* ? *Contributors:* ? *Original artist:* ?
- **File:Cubic.svg** *Source:* https://upload.wikimedia.org/wikipedia/commons/5/55/Cubic.svg *License:* CC-BY-SA-3.0 *Contributors:* donated work *Original artist:* Original PNGs by Daniel Mayer, traced in Inkscape by User:Stannered
- **File:CuttingABarMagnet.svg** *Source:* https://upload.wikimedia.org/wikipedia/commons/4/43/CuttingABarMagnet.svg *License:* CC0 *Contributors:* Own work *Original artist:* Sbyrnes321
- **File:Dalton_John_desk.jpg** *Source:* https://upload.wikimedia.org/wikipedia/commons/3/3f/Dalton_John_desk.jpg *License:* Public domain *Contributors:* Frontispiece of *John Dalton and the Rise of Modern Chemistry* by Henry Roscoe *Original artist:* Henry Roscoe (author), William Henry Worthington (engraver), and Joseph Allen (painter)
- **File:Daltons_symbols.gif** *Source:* https://upload.wikimedia.org/wikipedia/commons/3/39/Daltons_symbols.gif *License:* Public domain *Contributors:* ? *Original artist:* ?
- **File:Diatomic_phonons.png** *Source:* https://upload.wikimedia.org/wikipedia/commons/0/04/Diatomic_phonons.png *License:* CC BY-SA 3.0 *Contributors:* Own work *Original artist:* Brews ohare
- **File:Edit-clear.svg** *Source:* https://upload.wikimedia.org/wikipedia/en/f/f2/Edit-clear.svg *License:* Public domain *Contributors:* The *Tango! Desktop Project*. *Original artist:*
  The people from the Tango! project. And according to the meta-data in the file, specifically: "Andreas Nilsson, and Jakub Steiner (although minimally)."
- **File:Electron-hole.svg** *Source:* https://upload.wikimedia.org/wikipedia/commons/d/d6/Electron-hole.svg *License:* Public domain *Contributors:* Own derivative work of wikipedia:Image:ElectronHole.JPG *Original artist:* User:Sir_Link

- **File:Elementary_particle_interactions.svg** *Source:* https://upload.wikimedia.org/wikipedia/commons/4/4c/Elementary_particle.svg *License:* Public domain *Contributors:* en:Image:Interactions.png *Original artist:* en:User:TriTertButoxy, User:Stannered
- **File:Em_dipoles.svg** *Source:* https://upload.wikimedia.org/wikipedia/commons/f/f0/Em_dipoles.svg *License:* CC0 *Contributors:* Own work *Original artist:* Maschen
- **File:Em_monopoles.svg** *Source:* https://upload.wikimedia.org/wikipedia/commons/2/2f/Em_monopoles.svg *License:* CC0 *Contributors:* Own work *Original artist:* Maschen
- **File:Energy_levels.svg** *Source:* https://upload.wikimedia.org/wikipedia/commons/a/a8/Energy_levels.svg *License:* CC BY-SA 3.0 *Contributors:* This file was derived from: Energylevels.png
Original artist: SVG: Hazmat2 Original: Rozzychan
- **File:Enrico_Fermi_1943-49.jpg** *Source:* https://upload.wikimedia.org/wikipedia/commons/d/d4/Enrico_Fermi_1943-49.jpg *License:* Public domain *Contributors:* This media is available in the holdings of the National Archives and Records Administration, cataloged under the ARC Identifier (National Archives Identifier) **558578**. *Original artist:* Department of Energy. Office of Public Affairs
- **File:Ethanol-3D-balls.png** *Source:* https://upload.wikimedia.org/wikipedia/commons/b/b0/Ethanol-3D-balls.png *License:* Public domain *Contributors:* ? *Original artist:* ?
- **File:Ettore_Majorana.jpg** *Source:* https://upload.wikimedia.org/wikipedia/commons/5/59/Ettore_Majorana.jpg *License:* Public domain *Contributors:* [1] [2] *Original artist:* Unknown (Mondadori Publishers)
- **File:Exciton.png** *Source:* https://upload.wikimedia.org/wikipedia/commons/a/aa/Exciton.png *License:* Public domain *Contributors:* Own work (Original text: *selbst erstellt*) *Original artist:* Axelfoley12
- **File:Exotic_mesons.svg** *Source:* https://upload.wikimedia.org/wikipedia/commons/c/c0/Exotic_mesons.svg *License:* CC BY-SA 3.0 *Contributors:* Own work *Original artist:* Parcly Taxel
- **File:Feynman_Diagram_Y-3g.PNG** *Source:* https://upload.wikimedia.org/wikipedia/commons/1/1c/Feynman_Diagram_Y-3g.PNG *License:* CC BY-SA 3.0 *Contributors:* Own work *Original artist:* DrHjmHam
- **File:Feynman_diagram_of_decay_of_tau_lepton.svg** *Source:* https://upload.wikimedia.org/wikipedia/commons/1/11/Feynman_diagram_of_decay_of_tau_lepton.svg *License:* CC-BY-SA-3.0 *Contributors:* Modified and corrected from en:Image:TauDecays.svg by en:User:JabberW *Original artist:* en:User:JabberWok andTime3000
- **File:Feynmann_Diagram_Glueball-to-Pion.svg** *Source:* https://upload.wikimedia.org/wikipedia/commons/a/af/Feynmann_Diagram.svg *License:* CC BY-SA 4.0 *Contributors:* Own work *Original artist:* Smurrayinchester
- **File:Feynmann_Diagram_Gluon_Radiation.svg** *Source:* https://upload.wikimedia.org/wikipedia/commons/1/1f/Feynmann_Diagram_Gluon_Radiation.svg *License:* CC BY 2.5 *Contributors:* Non-Derived SVG of Radiate_gluon.png, originally the work of SilverStar at Feynmann-diagram-gluon-radiation.svg, updated byjoelholdsworth. *Original artist:* Joel Holdsworth (Joelholdsworth)
- **File:FigP24-2.jpg** *Source:* https://upload.wikimedia.org/wikipedia/en/a/ab/FigP24-2.jpg *License:* Cc-by-sa-3.0 *Contributors:* ? *Original artist:* ?
- **File:FirstNeutrinoEventAnnotated.jpg** *Source:* https://upload.wikimedia.org/wikipedia/commons/5/57/FirstNeutrinoEventAnnotated.jpg *License:* Public domain *Contributors:* Image courtesy of Argonne National Laboratory *Original artist:* Argonne National Laboratory
- **File:Folder_Hexagonal_Icon.svg** *Source:* https://upload.wikimedia.org/wikipedia/en/4/48/Folder_Hexagonal_Icon.svg *License:* Cc-by-sa-3.0 *Contributors:* ? *Original artist:* ?
- **File:Fraunhofer_lines.svg** *Source:* https://upload.wikimedia.org/wikipedia/commons/2/2f/Fraunhofer_lines.svg *License:* Public domain *Contributors:*
- Fraunhofer_lines.jpg *Original artist:* Fraunhofer_lines.jpg: nl:Gebruiker:MaureenV
- **File:Fusion_rxnrate.svg** *Source:* https://upload.wikimedia.org/wikipedia/commons/d/d0/Fusion_rxnrate.svg *License:* CC BY 2.5 *Contributors:* Own work *Original artist:* Dstrozzi
- **File:Gamma-ray-microscope.svg** *Source:* https://upload.wikimedia.org/wikipedia/commons/6/63/Gamma-ray-microscope.svg *License:* CC-BY-SA-3.0 *Contributors:* Own work *Original artist:* Radeksonic
- **File:Gnome-preferences-desktop-accessibility2.svg** *Source:* https://upload.wikimedia.org/wikipedia/commons/4/4c/Gnome-preferey2.svg *License:* CC BY-SA 3.0 *Contributors:* HTTP / FTP *Original artist:* GNOME icon artists
- **File:Gnome-searchtool.svg** *Source:* https://upload.wikimedia.org/wikipedia/commons/1/1e/Gnome-searchtool.svg *License:* LGPL *Contributors:* http://ftp.gnome.org/pub/GNOME/sources/gnome-themes-extras/0.9/gnome-themes-extras-0.9.0.tar.gz *Original artist:* David Vignoni
- **File:Gold_foil_experiment_conclusions.svg** *Source:* https://upload.wikimedia.org/wikipedia/commons/9/9b/Gold_foil_experiment.svg *License:* Public domain *Contributors:* Own work *Original artist:* Kurzon
- **File:GothicRayonnantRose003.jpg** *Source:* https://upload.wikimedia.org/wikipedia/commons/4/49/GothicRayonnantRose003.jpg *License:* GFDL *Contributors:* Based on File:Rozeta Paryż notre-dame chalger.jpg *Original artist:* Krzysztof Mizera, changed by Chagler and **<a href='//commons.wikimedia.org/wiki/User:MathKnight' title='User:MathKnight'>MathKnight</a>**
- **File:H_dibaryon.jpg** *Source:* https://upload.wikimedia.org/wikipedia/commons/9/97/H_dibaryon.jpg *License:* Public domain *Contributors:* Foxman *Original artist:* Linfoxman
- **File:Hadron_colors.svg** *Source:* https://upload.wikimedia.org/wikipedia/commons/e/e4/Hadron_colors.svg *License:* CC BY-SA 3.0 *Contributors:*
- Hadron_colors.png *Original artist:* Hadron_colors.png: Army1987

- **File:Helium_atom_QM.svg** *Source:* https://upload.wikimedia.org/wikipedia/commons/2/23/Helium_atom_QM.svg *License:* CC-BY-SA-3.0 *Contributors:* Own work *Original artist:* User:Yzmo
- **File:Higgs,_Peter_(1929)_cropped.jpg** *Source:* https://upload.wikimedia.org/wikipedia/commons/2/21/Higgs%2C_Peter_%281929%29_cropped.jpg *License:* CC BY-SA 2.0 de *Contributors:* Mathematisches Institut Oberwolfach (MFO), http://owpdb.mfo.de/detail?photo_id=12812 *Original artist:* Gert-Martin Greuel
- **File:Higgs-Mass-MetaStability.svg** *Source:* https://upload.wikimedia.org/wikipedia/commons/7/70/Higgs-Mass-MetaStability.svg *License:* CC BY-SA 3.0 *Contributors:* Transferred from en.wikipedia to Commons. *Original artist:* Folletto at English Wikipedia
- **File:HiggsBR.svg** *Source:* https://upload.wikimedia.org/wikipedia/commons/0/07/HiggsBR.svg *License:* CC BY-SA 3.0 *Contributors:* Own work *Original artist:* TimothyRias
- **File:Higgsdecaywidth.svg** *Source:* https://upload.wikimedia.org/wikipedia/commons/3/36/Higgsdecaywidth.svg *License:* CC BY-SA 3.0 *Contributors:* Own work *Original artist:* TimothyRias
- **File:Hydrogen.svg** *Source:* https://upload.wikimedia.org/wikipedia/commons/3/3f/Hydrogen.svg *License:* CC-BY-SA-3.0 *Contributors:* Own work *Original artist:* Mets501
- **File:Hydrogen300.png** *Source:* https://upload.wikimedia.org/wikipedia/commons/a/ad/Hydrogen300.png *License:* Public domain *Contributors:* Transferred from en.wikipedia; transferred to Commons by User:OverlordQ using CommonsHelper. *Original artist:* PoorLeno (talk) Original uploader was PoorLeno at en.wikipedia
- **File:Institut_Laue–Langevin_(ILL)_in_Grenoble,_France.jpg** *Source:* https://upload.wikimedia.org/wikipedia/commons/6/68/Institut_Laue%E2%80%93Langevin_%28ILL%29_in_Grenoble%2C_France.jpg *License:* CC BY 3.0 *Contributors:* Own work *Original artist:* Marek Ślusarczyk (Tupungato) Photo gallery
- **File:Isotopes_and_half-life.svg** *Source:* https://upload.wikimedia.org/wikipedia/commons/8/80/Isotopes_and_half-life.svg *License:* Public domain *Contributors:* Own work *Original artist:* BenRG
- **File:J-psi_p_pentaquark_mass_spectrum.svg** *Source:* https://upload.wikimedia.org/wikipedia/commons/4/44/J-psi_p_pentaquark_mass_spectrum.svg *License:* CC BY 4.0 *Contributors:* Figure 3b in <a data-x-rel='nofollow' class='external text' href='http://arxiv.org/pdf/1507.03414v1.pdf'>"Observation of J/ψp resonances consistent with pentaquark states in $\Lambda^0{}_b \to J/\psi K^- p$ decays"</a> (arXiv:1507.03414, LHCb collaboration *Original artist:* CERN on behalf of the LHCb collaboration,
- **File:Jpsi-fit-mass.gif** *Source:* https://upload.wikimedia.org/wikipedia/commons/4/4d/Jpsi-fit-mass.gif *License:* Public domain *Contributors:* Fermilab *Original artist:* CDF Collaboration
- **File:Kaon-box-diagram-with-bar.svg** *Source:* https://upload.wikimedia.org/wikipedia/commons/8/82/Kaon-box-diagram-with-bar.svg *License:* CC BY-SA 3.0 *Contributors:*
- Kkbar.png *Original artist:* Kkbar.png: Skaller
- **File:Kaon-box-diagram.svg** *Source:* https://upload.wikimedia.org/wikipedia/commons/8/8e/Kaon-box-diagram.svg *License:* CC-BY-SA-3.0 *Contributors:* ? *Original artist:* ?
- **File:Kaon-decay.png** *Source:* https://upload.wikimedia.org/wikipedia/commons/7/75/Kaon-decay.png *License:* CC-BY-SA-3.0 *Contributors:* Own work *Original artist:* User JabberWok on en.wikipedia
- **File:Lattice_wave.svg** *Source:* https://upload.wikimedia.org/wikipedia/commons/2/27/Lattice_wave.svg *License:* CC-BY-SA-3.0 *Contributors:* Created by en:User:FlorianMarquardt (en:Image:Phonon.png), vectorized in Inkscape by Mysid. *Original artist:* en:User:FlorianMarquardt, Mysid
- **File:Lepton-interaction-vertex-eeg.svg** *Source:* https://upload.wikimedia.org/wikipedia/commons/b/b5/Lepton-interaction-vertex-eeg.svg *License:* CC BY 3.0 *Contributors:* Own work *Original artist:* TimothyRias
- **File:Lepton_isodoublets_fixed.png** *Source:* https://upload.wikimedia.org/wikipedia/en/9/93/Lepton_isodoublets_fixed.png *License:* CC-BY-SA-3.0 *Contributors:*
  I (HEL (talk)) created this work entirely by myself. *Original artist:*
  HEL (talk)
- **File:Light-wave.svg** *Source:* https://upload.wikimedia.org/wikipedia/commons/a/a1/Light-wave.svg *License:* CC-BY-SA-3.0 *Contributors:* No machine readable source provided. Own work assumed (based on copyright claims). *Original artist:* No machine readable author provided. Gpvos assumed (based on copyright claims).
- **File:Light_cone_colour.svg** *Source:* https://upload.wikimedia.org/wikipedia/commons/5/56/Light_cone_colour.svg *License:* Public domain *Contributors:* Own work *Original artist:* Incnis Mrsi 10:15, 4 June 2008 (UTC)
- **File:Light_dispersion_of_a_mercury-vapor_lamp_with_a_flint_glass_prism_IPNr°0125.jpg** *Source:* https://upload.wikimedia.org commons/1/1f/Light_dispersion_of_a_mercury-vapor_lamp_with_a_flint_glass_prism_IPNr%C2%B00125.jpg *License:* CC BY-SA 3.0 at *Contributors:* Own work *Original artist:* D-Kuru
- **File:Liquid_drop_model.svg** *Source:* https://upload.wikimedia.org/wikipedia/commons/5/5b/Liquid_drop_model.svg *License:* CC BY-SA 3.0 *Contributors:* http://de.wikipedia.org/wiki/Datei:Tröpfchenmodell.svg *Original artist:* Daniel FR
- **File:Liquid_helium_Rollin_film.jpg** *Source:* https://upload.wikimedia.org/wikipedia/commons/f/f8/Liquid_helium_Rollin_film.jpg *License:* Public domain *Contributors:* Own work *Original artist:* I, AlfredLeitner, took this photograph as part of my movie "Liquid Helium,Superfluid"
- **File:Mach-Zehnder_photons_animation.gif** *Source:* https://upload.wikimedia.org/wikipedia/commons/a/a0/Mach-Zehnder_photons.gif *License:* CC BY 3.0 *Contributors:* Own work *Original artist:* user:Geek3
- **File:Mecanismo_de_Higgs_PH.png** *Source:* https://upload.wikimedia.org/wikipedia/commons/4/44/Mecanismo_de_Higgs_PH.png *License:* CC-BY-SA-3.0 *Contributors:* ? *Original artist:* ?

- **File:Mergefrom.svg** *Source:* https://upload.wikimedia.org/wikipedia/commons/0/0f/Mergefrom.svg *License:* Public domain *Contributors:* ? *Original artist:* ?
- **File:Meson-Baryon-molecule-generic.svg***Source:* https://upload.wikimedia.org/wikipedia/commons/0/0d/Meson-Baryon-molecule.svg *License:* CC BY-SA 4.0 *Contributors:* Own work *Original artist:* Smurrayinchester
- **File:Meson.svg** *Source:* https://upload.wikimedia.org/wikipedia/commons/f/f9/Meson.svg *License:* Public domain *Contributors:* en:Image:Meson.gif *Original artist:* en:User:Wogsland, traced by User:Stannered
- **File:Meson_nonet_-_spin_0.svg** *Source:* https://upload.wikimedia.org/wikipedia/commons/c/c0/Meson_nonet_-_spin_0.svg *License:* Public domain *Contributors:* Image:Noneto mesônico de spin 0.png *Original artist:* User:E2m, User:Stannered
- **File:Meson_nonet_-_spin_1.svg** *Source:* https://upload.wikimedia.org/wikipedia/commons/1/13/Meson_nonet_-_spin_1.svg *License:* Public domain *Contributors:* Image:Noneto mesônico de spin 1.png *Original artist:* User:E2m, User:Stannered
- **File:Moon'{}s_shadow_in_muons.gif** *Source:* https://upload.wikimedia.org/wikipedia/en/6/6f/Moon%27s_shadow_in_muons.gif *License:* Fair use *Contributors:*

  http://hepweb.rl.ac.uk/ppUKpics/POW/pr_990602.html *Original artist:* ?
- **File:Moving_Wannier_exciton.svg** *Source:* https://upload.wikimedia.org/wikipedia/commons/4/43/Moving_Wannier_exciton.svg *License:* CC BY-SA 3.0 *Contributors:* Drawn by Inkscape *Original artist:* ?? ?
- **File:Muon-Electron-Decay.svg** *Source:* https://upload.wikimedia.org/wikipedia/en/6/6f/Muon-Electron-Decay.svg *License:* Cc-by-sa-3.0 *Contributors:* ? *Original artist:* ?
- **File:Muon_Decay.svg** *Source:* https://upload.wikimedia.org/wikipedia/commons/1/13/Muon_Decay.svg *License:* Public domain *Contributors:*
- Muon_Decay.png *Original artist:* Muon_Decay.png: Thymo
- **File:Murray_Gell-Mann.jpg** *Source:* https://upload.wikimedia.org/wikipedia/commons/8/87/Murray_Gell-Mann.jpg *License:* CC BY 2.0 *Contributors:* http://flickr.com/photos/jurvetson/414368314/ *Original artist:* jurvetson of flickr.com
- **File:Neutron.svg** *Source:* https://upload.wikimedia.org/wikipedia/commons/d/dd/Neutron.svg *License:* Public domain *Contributors:* ? *Original artist:* ?
- **File:Nitrous-oxide-3D-balls.png***Source:* https://upload.wikimedia.org/wikipedia/commons/9/93/Nitrous-oxide-3D-balls.png *License:* Public domain *Contributors:* Own work *Original artist:* Ben Mills
- **File:Nobel_Prize_24_2013.jpg** *Source:* https://upload.wikimedia.org/wikipedia/commons/5/5f/Nobel_Prize_24_2013.jpg *License:* CC BY 2.0 *Contributors:* Flickr: IMG_7469 *Original artist:* Bengt Nyman
- **File:Noneto_mesônico_de_spin_0.png***Source:* https://upload.wikimedia.org/wikipedia/commons/c/cd/Noneto_mes%C3%B4nico_de_0.png *License:* Public domain *Contributors:* ? *Original artist:* ?
- **File:NuclearReaction.png** *Source:* https://upload.wikimedia.org/wikipedia/commons/7/7d/NuclearReaction.png *License:* CC BY-SA 3.0 *Contributors:* Own work *Original artist:* Michalsmid
- **File:Nuclear_Force_anim_smaller.gif** *Source:* https://upload.wikimedia.org/wikipedia/commons/3/35/Nuclear_Force_anim_smaller.gif *License:* CC BY-SA 3.0 *Contributors:* Own work *Original artist:* Manishearth
- **File:Nuclear_fission.svg** *Source:* https://upload.wikimedia.org/wikipedia/commons/1/15/Nuclear_fission.svg *License:* Public domain *Contributors:* ? *Original artist:* ?
- **File:Nucleus_drawing.svg** *Source:* https://upload.wikimedia.org/wikipedia/commons/f/f0/Nucleus_drawing.svg *License:* CC BY-SA 3.0 *Contributors:* Own work (vector version of PNG image) *Original artist:* Marekich
- **File:Nuvola_apps_edu_science.svg** *Source:* https://upload.wikimedia.org/wikipedia/commons/5/59/Nuvola_apps_edu_science.svg *License:* LGPL *Contributors:* http://ftp.gnome.org/pub/GNOME/sources/gnome-themes-extras/0.9/gnome-themes-extras-0.9.0.tar.gz *Original artist:* David Vignoni / ICON KING
- **File:Nuvola_apps_katomic.png** *Source:* https://upload.wikimedia.org/wikipedia/commons/7/73/Nuvola_apps_katomic.png *License:* LGPL *Contributors:* http://icon-king.com *Original artist:* David Vignoni / ICON KING
- **File:Office-book.svg** *Source:* https://upload.wikimedia.org/wikipedia/commons/a/a8/Office-book.svg *License:* Public domain *Contributors:* This and myself. *Original artist:* Chris Down/Tango project
- **File:Omega_Baryon.svg** *Source:* https://upload.wikimedia.org/wikipedia/commons/d/d6/Omega_Baryon.svg *License:* Public domain *Contributors:* ? *Original artist:* ?
- **File:One-loop-diagram.svg** *Source:* https://upload.wikimedia.org/wikipedia/commons/5/57/One-loop-diagram.svg *License:* CC-BY-SA-3.0*Contributors:* Self-made(Originally uploaded on en.wikipedia)*Original artist:* Originally uploaded byJabberWok(Transferred bygronde)
- **File:Optical_&_acoustic_vibrations.png***Source:* https://upload.wikimedia.org/wikipedia/commons/b/b2/Optical_%26_acoustic_.png *License:* CC BY-SA 3.0 *Contributors:* Own work *Original artist:* Brews ohare
- **File:PQ_EB_ape_hyp_geom5_B_3D.jpg** *Source:* https://upload.wikimedia.org/wikipedia/commons/8/85/PQ_EB_ape_hyp_geom5_B_3D.jpg *License:* CC BY-SA 4.0 *Contributors:* Own work *Original artist:* Pedro.bicudo
- **File:Particle_overview.svg** *Source:* https://upload.wikimedia.org/wikipedia/commons/7/7f/Particle_overview.svg *License:* CC BY-SA 3.0 *Contributors:* Own work *Original artist:* Headbomb
- **File:Pentaquark-Feynman.svg** *Source:* https://upload.wikimedia.org/wikipedia/commons/3/31/Pentaquark-Feynman.svg *License:* CC BY 4.0 *Contributors:* Figure 1b in <a data-x-rel='nofollow' class='external text' href='http://arxiv.org/pdf/1507.03414v1.pdf'>"Observation of J/ψp resonances consistent with pentaquark states in $\Lambda^0_b \to J/\psi K^- p$ decays"</a> (arXiv:1507.03414, LHCb collaboration *Original artist:* CERN on behalf of the LHCb collaboration,

- **File:Pentaquark-generic.svg** *Source:* https://upload.wikimedia.org/wikipedia/commons/5/5b/Pentaquark-generic.svg *License:* CC BY-SA 4.0 *Contributors:* Own work *Original artist:* Headbomb
- **File:Pentaquark.svg** *Source:* https://upload.wikimedia.org/wikipedia/commons/c/cd/Pentaquark.svg *License:* CC0 *Contributors:* Own work *Original artist:* Smurrayinchester
- **File:Phonon_dispersion_relations_in_GaAs.png***Source:* https://upload.wikimedia.org/wikipedia/commons/4/45/Phonon_dispersion in_GaAs.png *License:* CC BY-SA 3.0 *Contributors:* Own work *Original artist:* Brews ohare
- **File:Phonon_k_3k.gif** *Source:* https://upload.wikimedia.org/wikipedia/commons/7/7b/Phonon_k_3k.gif *License:* Public domain *Contributors:* Transferred from en.wikipedia to Commons. *Original artist:* Shaind at English Wikipedia
- **File:Phonon_polaritons.svg** *Source:* https://upload.wikimedia.org/wikipedia/commons/e/e7/Phonon_polaritons.svg *License:* CC BY-SA 3.0 *Contributors:* Own work *Original artist:* Mathieu Perrin
- **File:PiPlus_muon_decay.svg** *Source:* https://upload.wikimedia.org/wikipedia/commons/6/69/PiPlus_muon_decay.svg *License:* CC0 *Contributors:* Own work *Original artist:* Krishnavedala
- **File:Pn_Scatter_Quarks.svg** *Source:* https://upload.wikimedia.org/wikipedia/commons/2/2e/Pn_Scatter_Quarks.svg *License:* CC BY-SA 4.0*Contributors:* <a href='//commons.wikimedia.org/wiki/File:Pn_scatter_quarks.png' class='image'><img alt='Pn scatter quarks.png' src=' https://upload.wikimedia.org/wikipedia/commons/thumb/3/31/Pn_scatter_quarks.png/100px-Pn_scatter_quarks.png' width='100' height=' 64' srcset='https://upload.wikimedia.org/wikipedia/commons/thumb/3/31/Pn_scatter_quarks.png/150px-Pn_scatter_quarks.png 1.5x, https ://upload.wikimedia.org/wikipedia/commons/thumb/3/31/Pn_scatter_quarks.png/200px-Pn_scatter_quarks.png 2x' data-file-width='810' data-file-height='516' /></a>*Original artist:* Fred the Oyster
- **File:PolaronOptCond.jpg** *Source:* https://upload.wikimedia.org/wikipedia/en/a/a9/PolaronOptCond.jpg *License:* Cc-by-sa-3.0 *Contributors:* ? *Original artist:* ?
- **File:Polaron_scheme1.svg** *Source:* https://upload.wikimedia.org/wikipedia/commons/e/ed/Polaron_scheme1.svg *License:* GFDL *Contributors:* en:File:Polaron_scheme1.jpg *Original artist:*
- en:User_talk:S_klimin
- **File:Portal-puzzle.svg** *Source:* https://upload.wikimedia.org/wikipedia/en/f/fd/Portal-puzzle.svg *License:* Public domain *Contributors:* ? *Original artist:* ?
- **File:PositronDiscovery.jpg** *Source:* https://upload.wikimedia.org/wikipedia/commons/6/69/PositronDiscovery.jpg *License:* Public domain *Contributors:* Anderson, Carl D. (1933). “The Positive Electron”. *Physical Review* **43** (6): 491–494. DOI:10.1103/PhysRev.43.491. *Original artist:* Carl D. Anderson (1905–1991)
- **File:Positronium.svg** *Source:* https://upload.wikimedia.org/wikipedia/commons/d/db/Positronium.svg *License:* CC BY-SA 3.0 *Contributors:* Own work *Original artist:* Original: Manticorp
- **File:Positronium_Beam.jpg** *Source:* https://upload.wikimedia.org/wikipedia/commons/3/37/Positronium_Beam.jpg *License:* CC BY 2.0 *Contributors:* https://www.flickr.com/photos/uclmaps/19436059256 *Original artist:* UCL Faculty of Mathematical & Physical Sciences
- **File:Potential_energy_well.svg** *Source:* https://upload.wikimedia.org/wikipedia/commons/c/c5/Potential_energy_well.svg *License:* Public domain *Contributors:* Based upon Image:Potential well.png, created by User:Koantum. This version created by bdesham in Inkscape. *Original artist:* Benjamin D. Esham (bdesham)
- **File:Proton_proton_cycle.svg** *Source:* https://upload.wikimedia.org/wikipedia/commons/a/ac/Proton_proton_cycle.svg *License:* CC BY 2.5 *Contributors:* file:Proton proton cycle.png *Original artist:* Dorottya Szam
- **File:QCDphasediagram.svg** *Source:* https://upload.wikimedia.org/wikipedia/commons/b/bc/QCDphasediagram.svg *License:* CC BY-SA 3.0 *Contributors:* Own work *Original artist:* TimothyRias
- **File:Quark_masses_as_balls.svg** *Source:* https://upload.wikimedia.org/wikipedia/commons/b/b5/Quark_masses_as_balls.svg *License:* CC BY-SA 3.0 *Contributors:* Own work *Original artist:* Incnis Mrsi
- **File:Quark_structure_antineutron.svg** *Source:* https://upload.wikimedia.org/wikipedia/en/4/47/Quark_structure_antineutron.svg *License:* CC-BY-SA-3.0 *Contributors:*
  Self created using Inkscape from File:Quark_structure_proton.svg as a template. *Original artist:*
  <a href='//en.wikipedia.org/wiki/User:Spinningspark' title='User:Spinningspark'>**Spinning**Spark</a> real life identity: SHA-1 commitment ba62ca25da3fee2f8f36c101994f571c151abee7
- **File:Quark_structure_antiproton.svg** *Source:* https://upload.wikimedia.org/wikipedia/en/2/24/Quark_structure_antiproton.svg *License:* CC-BY-SA-3.0 *Contributors:*
  Self created using Inkscape from File:Quark_structure_proton.svg as a template. *Original artist:*
  <a href='//en.wikipedia.org/wiki/User:Spinningspark' title='User:Spinningspark'>**Spinning**Spark</a> real life identity: SHA-1 commitment ba62ca25da3fee2f8f36c101994f571c151abee7
- **File:Quark_structure_neutron.svg** *Source:* https://upload.wikimedia.org/wikipedia/commons/8/81/Quark_structure_neutron.svg *License:* CC BY-SA 2.5 *Contributors:* ? *Original artist:* ?
- **File:Quark_structure_pion.svg** *Source:* https://upload.wikimedia.org/wikipedia/commons/6/62/Quark_structure_pion.svg *License:* CC BY-SA 2.5 *Contributors:* ? *Original artist:* ?
- **File:Quark_structure_proton.svg** *Source:* https://upload.wikimedia.org/wikipedia/commons/9/92/Quark_structure_proton.svg *License:* CC BY-SA 2.5 *Contributors:* Own work *Original artist:* Arpad Horvath

- **File:Quark_weak_interactions.svg** *Source:* https://upload.wikimedia.org/wikipedia/commons/6/66/Quark_weak_interactions.svg *License:* Public domain *Contributors:* Derivative work, from public down work uploaded to en.wikipedia. original *Original artist:*
- Original work: [1]
- **File:Question_book-new.svg** *Source:* https://upload.wikimedia.org/wikipedia/en/9/99/Question_book-new.svg *License:* Cc-by-sa-3.0 *Contributors:*

  Created from scratch in Adobe Illustrator. Based on Image:Question book.png created by User:Equazcion *Original artist:* Tkgd2007
- **File:Right_left_helicity.svg** *Source:* https://upload.wikimedia.org/wikipedia/commons/a/a9/Right_left_helicity.svg *License:* Public domain *Contributors:* en:Image:Right left helicity.jpg *Original artist:* en:User;HEL, User:Stannered
- **File:Rutherford_1911_Solvay.jpg** *Source:* https://upload.wikimedia.org/wikipedia/commons/3/3b/Rutherford_1911_Solvay.jpg *License:* Public domain *Contributors:* ? *Original artist:* ?
- **File:S-p-Orbitals.svg** *Source:* https://upload.wikimedia.org/wikipedia/commons/0/0f/S-p-Orbitals.svg *License:* CC-BY-SA-3.0 *Contributors:* selfmade from [1] *Original artist:* This file was made by **User:Sven**
- **File:Sasahara.svg** *Source:* https://upload.wikimedia.org/wikipedia/commons/c/cc/Sasahara.svg *License:* CC-BY-SA-3.0 *Contributors:* Transferred from en.wikipedia *Original artist:* Original uploader was JWB at en.wikipedia
- **File:SatyenBose1925.jpg** *Source:* https://upload.wikimedia.org/wikipedia/commons/f/fe/SatyenBose1925.jpg *License:* Public domain *Contributors:* Picture in Siliconeer *Original artist:* Unknown
- **File:Science.jpg** *Source:* https://upload.wikimedia.org/wikipedia/commons/5/54/Science.jpg *License:* Public domain *Contributors:* ? *Original artist:* ?
- **File:Speakerlink-new.svg** *Source:* https://upload.wikimedia.org/wikipedia/commons/3/3b/Speakerlink-new.svg *License:* CC0 *Contributors:* Own work *Original artist:* Kelvinsong
- **File:Spontaneous_symmetry_breaking_(explanatory_diagram).png** *Source:* https://upload.wikimedia.org/wikipedia/commons/a/a5/symmetry_breaking_%28explanatory_diagram%29.png *License:* CC BY-SA 3.0 *Contributors:* Own work *Original artist:* FT2
- **File:Standard_Model_of_Elementary_Particles.svg** *Source:* https://upload.wikimedia.org/wikipedia/commons/0/00/Standard_Model_of_Elementary_Particles.svg *License:* CC BY 3.0 *Contributors:* Own work by uploader, PBS NOVA [1], Fermilab, Office of Science, United States Department of Energy, Particle Data Group *Original artist:* MissMJ
- **File:Stimulatedemission.png** *Source:* https://upload.wikimedia.org/wikipedia/commons/8/8a/Stimulatedemission.png *License:* CC-BY-SA-3.0 *Contributors:* en:Image:Stimulatedemission.png *Original artist:* User:(Automated conversion),User:DrBob
- **File:Strong_force_charges.svg** *Source:* https://upload.wikimedia.org/wikipedia/commons/b/b6/Strong_force_charges.svg *License:* CC BY-SA 3.0 *Contributors:* Own work, Created from Garret Lisi's Elementary Particle Explorer *Original artist:* Cjean42
- **File:Stylised_Lithium_Atom.svg** *Source:* https://upload.wikimedia.org/wikipedia/commons/e/e1/Stylised_Lithium_Atom.svg *License:* CC-BY-SA-3.0 *Contributors:* based off of Image:Stylised Lithium Atom.png by Halfdan. *Original artist:* SVG by Indolences. Recoloring and ironing out some glitches done by Rainer Klute.
- **File:Supernova-1987a.jpg** *Source:* https://upload.wikimedia.org/wikipedia/commons/4/43/Supernova-1987a.jpg *License:* CC BY 3.0 *Contributors:* Supernova 1987A: Halo for a Vanished Star, Mosaic of Supernova 1987A. *Original artist:* First image: Dr. Christopher Burrows, ESA/STScI and NASA; Second image: Hubble Heritage team.
- **File:Symbol_book_class2.svg** *Source:* https://upload.wikimedia.org/wikipedia/commons/8/89/Symbol_book_class2.svg *License:* CC BY-SA 2.5 *Contributors:* Mad by Lokal_Profil by combining: *Original artist:* Lokal_Profil
- **File:Symmetricwave2.png** *Source:* https://upload.wikimedia.org/wikipedia/commons/1/1d/Symmetricwave2.png *License:* CC BY 3.0 *Contributors:* Own work *Original artist:* TimothyRias
- **File:TPI1_structure.png** *Source:* https://upload.wikimedia.org/wikipedia/commons/1/1c/TPI1_structure.png *License:* Public domain *Contributors:* based on 1wyi (http://www.pdb.org/pdb/explore/explore.do?structureId=1WYI), made in pymol *Original artist:* →<a href='//commons.wikimedia.org/wiki/User:AzaToth' title='User:AzaToth'>Aza</a><a href='//commons.wikimedia.org/wiki/User_talk:AzaToth' title='User talk:AzaToth'>Toth</a>
- **File:TQ_EB_ape_hyp_r1_8_r2_14_Act_3D_Sim.jpg** *Source:* https://upload.wikimedia.org/wikipedia/commons/9/93/TQ_EB_ape_hyp_r1_8_r2_14_Act_3D_Sim.jpg *License:* CC BY-SA 4.0 *Contributors:* Own work *Original artist:* Pedro.bicudo
- **File:Tachyon04s.gif** *Source:* https://upload.wikimedia.org/wikipedia/commons/6/64/Tachyon04s.gif *License:* CC-BY-SA-3.0 *Contributors:* Transferred from en.wikipedia to Commons by User:Sumanch. *Original artist:* TxAlien at en.wikipedia
- **File:Text_document_with_red_question_mark.svg** *Source:* https://upload.wikimedia.org/wikipedia/commons/a/a4/Text_document_with_red_question_mark.svg *License:* Public domain *Contributors:* Created by bdesham with Inkscape; based upon Text-x-generic.svg from the Tango project. *Original artist:* Benjamin D. Esham (bdesham)
- **File:The_incomplete_circle_of_everything.svg** *Source:* https://upload.wikimedia.org/wikipedia/commons/0/0d/The_incomplete_circle_of_everything.svg *License:* CC BY 3.0 *Contributors:* Own work *Original artist:* Zhitelew
- **File:Top_antitop_quark_event.svg** *Source:* https://upload.wikimedia.org/wikipedia/commons/3/35/Top_antitop_quark_event.svg *License:* Public domain *Contributors:* Own work *Original artist:* Raeky
- **File:Upsilon_peak.jpg** *Source:* https://upload.wikimedia.org/wikipedia/commons/0/03/Upsilon_peak.jpg *License:* Public domain *Contributors:* ? *Original artist:* ?
- **File:Vertex_correction.svg** *Source:* https://upload.wikimedia.org/wikipedia/commons/8/87/Vertex_correction.svg *License:* Public domain *Contributors:* ? *Original artist:* User:Harmaa

- **File:VisibleEmrWavelengths.svg** *Source:* https://upload.wikimedia.org/wikipedia/commons/e/e2/VisibleEmrWavelengths.svg *License:* Public domain *Contributors:* created by me *Original artist:* maxhurtz
- **File:Wiki_letter_w_cropped.svg** *Source:* https://upload.wikimedia.org/wikipedia/commons/1/1c/Wiki_letter_w_cropped.svg *License:* CC-BY-SA-3.0 *Contributors:*
- Wiki_letter_w.svg *Original artist:* Wiki_letter_w.svg: Jarkko Piiroinen
- **File:Wikibooks-logo.svg** *Source:* https://upload.wikimedia.org/wikipedia/commons/f/fa/Wikibooks-logo.svg *License:* CC BY-SA 3.0 *Contributors:* Own work *Original artist:* User:Bastique, User:Ramac et al.
- **File:Wikinews-logo.svg** *Source:* https://upload.wikimedia.org/wikipedia/commons/2/24/Wikinews-logo.svg *License:* CC BY-SA 3.0 *Contributors:* This is a cropped version of Image:Wikinews-logo-en.png. *Original artist:* Vectorized by Simon 01:05, 2 August 2006 (UTC) Updated by Time3000 17 April 2007 to use official Wikinews colours and appear correctly on dark backgrounds. Originally uploaded by Simon.
- **File:Wikiquote-logo.svg** *Source:* https://upload.wikimedia.org/wikipedia/commons/f/fa/Wikiquote-logo.svg *License:* Public domain *Contributors:* ? *Original artist:* ?
- **File:Wikisource-logo.svg** *Source:* https://upload.wikimedia.org/wikipedia/commons/4/4c/Wikisource-logo.svg *License:* CC BY-SA 3.0 *Contributors:* Rei-artur *Original artist:* Nicholas Moreau
- **File:Wikiversity-logo-Snorky.svg** *Source:* https://upload.wikimedia.org/wikipedia/commons/1/1b/Wikiversity-logo-en.svg *License:* CC BY-SA 3.0 *Contributors:* Own work *Original artist:* Snorky
- **File:Wiktionary-logo-en.svg** *Source:* https://upload.wikimedia.org/wikipedia/commons/f/f8/Wiktionary-logo-en.svg *License:* Public domain *Contributors:* Vector version of Image:Wiktionary-logo-en.png. *Original artist:* Vectorized by Fvasconcellos (talk · contribs), based on original logo tossed together by Brion Vibber
- **File:Wpdms_physics_proton_proton_chain_1.svg** *Source:* https://upload.wikimedia.org/wikipedia/commons/7/74/Wpdms_physics_proton_proton_chain_1.svg *License:* Public domain *Contributors:* Own work *Original artist:* see below
- **File:Young_Diffraction.png** *Source:* https://upload.wikimedia.org/wikipedia/commons/8/8a/Young_Diffraction.png *License:* Public domain *Contributors:* ? *Original artist:* ?

108.2.3 Content license

Made in the USA
San Bernardino, CA
21 December 2015